TROPICAL GRASSES

"I think it may be fairly claimed that the pivotal crop on which the world's agricultural prosperity depends is without doubt, the herbaceous sward, and at long last it is beginning to be realized that the character and money-earning ability of this sward are more amenable to control by scientific management than perhaps any other crop with which the farmer is concerned."

Professor Stapledon, M.B.E., M.A.
University College of Wales

FAO Plant Production and Protection Series   No. 23

Tropical grasses

by
P.J. Skerman
and
F. Riveros

**Food and
Agriculture
Organization
of the
United Nations**

Rome, 1990

The designations employed and the presentation of material in this publication do not imply the expression of any opinion whatsoever by the Food and Agriculture Organization of the United Nations concerning the legal status of any country, city or area or of its authorities, or concerning the delimitation of its frontiers or boundaries. The views expressed are those of the author.

The mention of specific companies or of their products or brand names does not imply any endorsement or recommendation on the part of the Food and Agriculture Organization of the United Nations.

P-10
ISBN 92-5-101128-1

David Lubin Memorial Library Cataloguing in Publication Data

Skerman, P.J.
   Tropical grasses.
   (FAO Plant Production and Protection Series, no. 23)

1. Grasses.   2. Gramineae.   3. Tropical zones.
I. Riveros, F.   II. Title.   III. Series.

FAO code: 11   AGRIS: F01   1989
ISBN 92-5-101128-1

The copyright in this book is vested in the Food and Agriculture Organization of the United Nations. The book may not be reproduced, in whole or in part, by any method or process, without written permission from the copyright holder. Applications for such permission, with a statement of the purpose and extent of the reproduction desired, should be addressed to the Director, Publications Division, Food and Agriculture Organization of the United Nations, Via delle Terme di Caracalla, 00100 Rome, Italy.

© FAO 1989

Printed in Italy

Foreword

This book, written at the request of the Food and Agriculture Organization of the United Nations, is a companion volume to *Tropical forage legumes,* written by the same author and published by FAO in 1977.

It is not a specialized book, apart from Chapter 14, *The chemical composition and nutritive value of tropical grasses,* by D.J. Minson of the Division of Tropical Crops and Pastures, Commonwealth Scientific and Industrial Research Organization (CSIRO), Australia. The book is essentially a source of ready information for all those interested in the role of grasses in agriculture and the grass species of economic importance in agronomy.

The world's major natural grasslands and their dominant species are described briefly and their annual pattern of production indicated. Grasses are inexpensive in terms of initial outlay and still provide most of the nutriment for the world's grazing animals, despite their obvious seasonal limitations. In the face of increasing urbanization and the expansion of cropping, the world's population cannot afford to allow land to remain at its present level of productivity if it is possible to make it more productive.

More than usual attention has been given to grasses for special purposes such as soil conservation, stabilizing and utilizing saline areas, grazing under plantation crops, and for soils of high aluminium content. There is also a chapter dealing with the handling of difficult grasses. In the case of *Imperata cylindrica* (blady grass, kunai, alang-alang, cogon, illuk), the final solution to its control depends above all on the economic benefit to be derived and the ability of the small farmer to handle subsistence crops, weed control, and pasture improvement with limited land, finances and labour. Farm crop residues will always play a part in the nutrition of animals.

Fodder conservation has received fairly detailed treatment as the protracted dry season poses special problems in animal nutrition in the subhumid and semi-arid tropics.

The establishment of improved pastures in suitable areas, reseeding the drier rangelands, and grass and legume species which might be used, are dealt with in Chapters 9 and 10. The breeding of improved grasses has not been covered as this is a specialized field. However, the known cultivars now available as a result of plant exploration, adaptation and breeding are listed and described.

Finally, the catalogue of grasses contains important forage, fodder and weedy species and has been drawn up to highlight individual characteristics which are mainly agronomic. Only brief botanical descriptions are given and emphasis is placed on the visual presentation of species. To this end many pen-and-ink sketches and photographs have been included.

Fernando Riveros **P.J. Skerman**

Acknowledgements

Many people and organizations are thanked for their assistance in compiling this book. Thanks are due to FAO and especially Mr J. Grossman, former chief of FAO's Editorial Branch, for his excellent presentation of the material in the publication and his cooperation with illustrations; Fiorella Marcon-D'Andrea and Sandro Cassola for producing the plant drawings and graphic material; and the librarian for full access to the library.

The director of the Royal Botanical Gardens, Kew, United Kingdom, the director-general, Queensland Department of Primary Industries, the chief, Division of Tropical Crops and Pastures, Commonwealth Scientific and Industrial Research Organization (CSIRO), Australia, and the librarian, University of Queensland, made their staff available. The director of the Botany Branch and the director of the Agrostology Branch of the Queensland Department of Primary Industries were extremely helpful. Mr B.L. Simon joined with Dr W.E. Clayton of Kew in assisting with taxonomy, and W.J. Bisset obtained many of the black-and-white photographs from his departmental colleagues. Dr J.G. McIvor of CSIRO and Mr R.E. Harrison supplied some of the colour slides, and Mr C.V. Malcolm of the Western Australia Department of Agriculture forwarded information and photographs dealing with salinity.

Finally, sincere thanks are due to Joan Skerman for her contributions and support; Mrs Joan Holmwood, for meticulous care in checking and typing the manuscript; and to her husband, who helped in numerous ways during its preparation.

Contents

Foreword v

1. **The importance of grasslands** 1

2. **The classification and distribution of grasses** 7
   Distribution of some of the tribes of the Gramineae 8

3. **The world's major tropical grasslands** 11
   Africa 11
   India 15
   Sri Lanka 19
   Southeast Asia 20
   Australia 21
   The Pacific Islands 23
   Southern United States of America 24
   Central America and the West Indies 24
   Latin America 25

4. **Performance and management of natural pasture** 31
   Live-weight progress on the natural range 31
   Management of indigenous grassland 34

5. **The case for improved pastures to replace indigenous species** . 45

6.	**Pasture improvement by introducing new species**	49
	Land clearing	49
	Establishment of grass in pure and mixed swards in areas of adequate rainfall	55
7.	**Selection of pasture grass species, seed purchase and storage, and fertilizer needs**	61
	Pure grass pastures or grass-legume mixtures	63
	Species recommended for sown grasses in humid and subhumid zones	65
	Seed purchase and storage	65
	Fertilizers for establishment and maintenance	68
	Fertilizers for the small farmer	71
	Nitrogen fixation by grasses	71
8.	**Pasture leys**	73
9.	**Management of improved grassland in semi-intensive and intensive production systems**	77
	Grazing methods	77
	Kind of grazing animal	83
	Night kraaling	86
10.	**Reseeding the arid and semi-arid range**	87
	Selection of species	87
	Seeding	89
	Machinery	89
	Grazing	91
11.	**Handling difficult grasses**	93
	Axonopus affinis (narrow-leaved carpet or mat grass)	93
	Cymbopogon nardus (false citronella)	96
	Digitaria abyssinica (African couch grass)	97
	Eleusine jaegeri (manyatta grass)	98

	Imperata cylindrica (blady grass or alang-alang)	100
	Methods of suppression or control	102
	Recommended measures for control of *Imperata*	106
	Pennisetum polystachyon (mission grass, khachornchob)	108
	Sorghum halepense (Johnson grass)	109
12.	**Grasses for special purposes**	113
	Tropical grasses exhibiting some tolerance to salinity or alkalinity	113
	Pastures under plantation crops	122
	Grasses suitable for erosion control and their rainfall ranges	126
	Grasses suitable to aquatic, semi-aquatic and moist areas (hydrophilous grasses)	129
	Grasses that tolerate high levels of soil aluminium	130
13.	**Utilization and conservation of forage**	133
	Grazing	133
	Green chop	134
	Conservation as standing hay or deferred feed	135
	Stored fodder	137
	Hay	138
	Pelleting	145
	Silage	146
	Choice of species for ensilage	154
	Silage for the small farmer	157
	Haylage	158
	The economics of fodder conservation	160
14.	**The chemical composition and nutritive value of tropical grasses** *by D.J. Minson*	163
	Voluntary intake	163
	Digestibility	167
	Protein	171
	Mineral composition	172
	Undesirable factors	180
	Conclusion	180

15. The tropical grasses catalogue 181

Acroceras sp.	181	*Imperata* sp.	468
Andropogon sp.	185	*Ischaemum* spp.	474
Anthephora sp.	191	*Iseilema* spp.	480
Aristida spp.	193	*Ixophorus* sp.	488
Astrebla spp.	199	*Lasiurus* sp.	491
Axonopus spp.	206	*Leersia* sp.	494
Bothriochloa spp.	217	*Leptochloa* sp.	497
Brachiaria spp.	234	*Leptothrium* sp.	499
Cenchrus spp.	263	*Loudetia* sp.	501
Chloris spp.	283	*Melinis* sp.	504
Chrysopogon spp.	296	*Oryza* sp.	508
Coix sp.	300	*Panicum* spp.	512
Cymbopogon sp.	303	*Paspalidium* sp.	555
Cynodon spp.	306	*Paspalum* spp.	557
Dactyloctenium spp.	322	*Pennisetum* spp.	596
Dichanthium spp.	328	*Phragmites* spp.	635
Digitaria spp.	345	*Saccharum* spp.	640
Diplachne sp.	368	*Sehima* sp.	652
Echinochloa spp.	370	*Setaria* spp.	655
Eleusine spp.	397	*Sorghum* spp.	670
Enteropogon spp.	407	*Spinifex* sp.	698
Entolasia sp.	410	*Sporobolus* spp.	701
Eragrostis spp.	411	*Stenotaphrum* sp.	711
Eriochloa spp.	422	*Themeda* spp.	714
Eustachys sp.	439	*Trachypogon* sp.	725
Exotheca sp.	441	*Triodia* sp.	727
Hemarthria sp.	443	*Tripsacum* spp.	729
Heteropogon sp.	446	*Urochloa* spp.	735
Hymenachne spp.	450	*Vetiveria* sp.	747
Hyparrhenia spp.	454	*Vossia* sp.	749
Hyperthelia sp.	465	*Zea* spp.	752

Bibliography 761

Common names of tropical grasses 811

Common names of other plants 819

Index 821

Illustrations

1.1.	Climatic zones of the tropics and subtropics	2
1.2	Major grazing areas of the world	4
1.3.	World distribution of the principal zonal soil groups	5
2.1.	Distribution of some major tribes	9
3.1.	Western and western equatorial regions, Africa	12
3.2.	Eastern and central regions, Africa	14
3.3.	Distribution of grass covers in India	16
3.4.	Pasture distribution in Australia	22
4.1.	Average live-weight changes of native cattle over three years, Mpwapwa, Tanzania	32
4.2.	Growth rates of cattle grazing *Heteropogon* pasture in northern Queensland, Australia	32
4.3.	Pasture quality and cattle performance, Katherine, Northern Territory, Australia	33
4.4.	Live-weight gains of zebu steers introduced to new season grazing	35
5.1.	Cumulative live-weight changes in grazing pastures with varying percentages of legumes and a pure grass sward fertilized with nitrogen, Beearwah, southeastern Queensland, Australia	46
6.1.	Wind-rowing without dead running	54
6.2.	Heavy duty stump-jump disc plough removing *Acacia harpophylla* (brigalow) regrowth before overseeding the pasture	57
6.3.	Agitators attached to the fertilizer plates in a grain/fertilizer drill sowing *Cenchrus ciliaris* (buffel grass)	58
9.1.	Cumulative live-weight gains from rotationally and continuously grazed pastures	78

9.2.	Effect of stocking rate on production per animal and per hectare	81
9.3.	Relationship of grazing pressure to gain per animal and gain per unit area of land	81
9.4.	Proposed relationship between stocking rate and live-weight gain per animal	82
9.5.	Hypothetical response of gain per animal to stocking rate for each month and for the entire season	83
9.6.	Relationship between gain per animal and gain per hectare in response to increasing stocking rate on a setaria/siratro pasture	84
10.1.	A grass seeder mounted on a tandem offset scalloped-disc cultivator for seeding rangeland	90
11.1	Effect of three years' nitrogen fertilization at different states on the botanical composition of a Kikuyu/paspalum/carpet grass sward	95
11.2.	World distribution of *Imperata cylindrica*	101
12.1.	The Mallen Niche Seeder	116
13.1.	A combined chopper/catcher forage harvester for green chop	135
13.2.	Growth cycle of crops	136
13.3.	Elevation of haystack, showing foundations, stack and weights	142
13.4.	A frame for joining the iron sheets	142
13.5.	Plan and elevation of a trench silo with a capacity of 125 tonnes	151
13.6.	A top-unloading device for tower silos	154
14.1.	Frequency distribution of voluntary intake observations of tropical and temperate grasses	164
14.2.	Relation of voluntary intake to digestibility for six varieties of *Panicum* species	166
14.3.	Frequency distribution of digestibilities for 543 cuts of tropical grasses and 592 cuts of temperate grasses	168
14.4.	Relation of mean temperature to dry-matter digestibility	169
14.5.	Frequency distribution of crude protein for tropical grasses, tropical legumes and temperate grasses	173

14.6.	Relation of the age of regrowth to the crude protein contents of stem and leaf fractions of tropical grasses	174
14.7.	Frequency distribution of six elements in tropical grasses	175
14.8.	Effect of age and plant part on mineral composition of five tropical grasses	177
15.1.	*Acroceras macrum*	182
15.2.	*Andropogon gayanus*	186
15.3.	*Aristida adscensionis*	194
15.4.	*Aristida latifolia*	197
15.5.	*Aristida latifolia* invading *Astrebla* spp. at Blackall, central Queensland, Australia	198
15.6.	Distribution of *Astrebla* spp. pastures in Australia	200
15.7.	*Astrebla* spp.	201
15.8.	*Axonopus affinis*	207
15.9.	*Axonopus compressus*	212
15.10.	*Bothriochloa bladhii*	218
15.11.	*Bothriochloa insculpta* (creeping blue grass)	223
15.12.	*Bothriochloa insculpta* growing with *Macroptilium atropurpureum* (siratro) at Rockhampton, Queensland, Australia	224
15.13.	*Bothriochloa ischaemum*	228
15.14.	*Bothriochloa pertusa*	231
15.15.	*Brachiaria brizantha*	235
15.16.	*Brachiaria decumbens*	239
15.17.	*Brachiaria humidicola*	246
15.18.	*Brachiaria mutica*	250
15.19.	*Brachiaria ruziziensis*	256
15.20.	*Brachiaria subquadripara*	261
15.21.	*Cenchrus biflorus*	264
15.22.	*Cenchrus ciliaris*	267
15.23.	Vertical section of the root system of *Cenchrus ciliaris* cv. Biloela, seven months after planting	268
15.24.	*Cenchrus pennisetiformis*	276

15.25.	*Cenchrus setigerus*	280
15.26.	*Chloris gayana*	284
15.27.	*Chloris mosambicensis*	290
15.28.	*Chloris roxburghiana*	292
15.29.	*Chloris virgata*	294
15.30.	*Chrysopogon aciculatus*	297
15.31.	*Chrysopogon aucheri*	299
15.32.	*Coix lacryma-jobi*	301
15.33.	*Cymbopogon nardus*	304
15.34.	Distribution of *Cynodon* spp. in Africa	304
15.35.	*Cynodon aethiopicus*	309
15.36.	*Cynodon dactylon*	311
15.37.	*Cynodon nlemfuensis*	317
15.38.	*Cynodon plectostachyus*	320
15.39.	*Dactyloctenium aegyptium*	323
15.40.	*Dactyloctenium giganteum*	326
15.41.	*Dichanthium annulatum*	329
15.42.	*Dichanthium aristatum*	334
15.43.	*Dichanthium caricosum*	337
15.44.	Effects of supplementary fertilized *Leucaena* spp. on cumulative annual live-weight gain of steers grazing *Dichanthium caricosum* in Fiji	340
15.45.	*Dichanthium sericeum*	342
15.46.	*Digitaria abyssinica*	346
15.47.	*Digitaria ciliaris*	349
15.48.	*Digitaria decumbens*	352
15.49.	*Digitaria didactyla*	358
15.50.	*Digitaria milanjiana*	361
15.51.	*Digitaria pentzii*	364
15.52.	*Diplachne fusca*	369
15.53.	*Echinochloa colona*	371

15.54.	*Echinochloa crus-galli*	374
15.55.	*Echinochloa frumentacea*	377
15.56.	*Echinochloa haploclada*	380
15.57.	*Echinochloa pyramidalis*	384
15.58.	*Echinochloa scabra*	388
15.59.	*Echinochloa turneriana*	392
15.60.	*Echinochloa utilis*	394
15.61.	*Eleusine coracana*	398
15.62.	*Eleusine indica*	402
15.63.	*Eleusine jaegeri*	406
15.64.	*Enteropogon macrostachyus*	408
15.65.	*Eragrostis caespitosa*	412
15.66.	*Eragrostis chloromelas*	414
15.67.	*Eragrostis cilianensis*	416
15.68.	*Eragrostis curvula*	418
15.69.	*Eragrostis lehmanniana*	423
15.70.	*Eragrostis superba*	426
15.71.	*Eragrostis tef*	429
15.72.	*Eragrostis tremula*	432
15.73.	*Eriochloa fatmensis*	435
15.74.	*Eriochloa punctata*	437
15.75.	*Eustachys paspaloides*	440
15.76.	*Exotheca abyssinica*	442
15.77.	*Hemarthria altissima*	444
15.78.	*Heteropogon contortus*	447
15.79.	*Hymenachne acutigluma*	451
15.80.	*Hymenachne amplexicaulis*	453
15.81.	*Hyparrhenia filipendula*	455
15.82.	*Hyparrhenia hirta*	458
15.83.	*Hyparrhenia rufa*	461
15.84.	*Hyperthelia dissoluta*	466

15.85.	*Imperata cylindrica*	469
15.86.	*Ischaemum indicum*	475
15.87.	*Iseilema laxum*	481
15.88.	*Iseilema membranaceum*	484
15.89.	*Iseilema vaginiflorum*	487
15.90.	*Ixophorus unisetus*	489
15.91.	*Lasiurus hirsutus*	492
15.92.	*Leersia hexandra*	495
15.93.	*Leptochloa obtusiflora*	498
15.94.	*Leptothrium senegalense*	500
15.95.	*Loudetia simplex*	502
15.96.	*Melinis minutiflora*	505
15.97.	*Oryza sativa*	509
15.98.	*Panicum coloratum* var. *makarikariense*	518
15.99.	*Panicum coloratum* var. *makarikariense* cv. Bambatsi showing good growth one month after a flood	519
15.100.	*Panicum maximum*	523
15.101.	Vertical section showing the root system of *Panicum maximum* seven months after planting	524
15.102.	A nil-phosphorus plot in a sown pasture mixture of *Panicum maximum* cv. Hamil and *Stylosanthes guianensis*, Belyana, northern Queensland, Australia	526
15.103.	*Panicum maximum* growing with *Pueraria phaseoloides* at Tully, Queensland, Australia	527
15.104.	Feed supply patterns of some useful pasture species on the Atherton Tableland, Queensland, Australia	528
15.105.	Harvesting *Panicum maximum* var. *trichoglume* on the Darling Downs, Queensland, Australia	535
15.106.	*Panicum miliaceum*	539
15.107.	*Panicum miliaceum*	540
15.108.	*Panicum repens*	544
15.109.	*Panicum trichocladum*	548
15.110.	Distribution of *Panicum turgidum*	550

15.111.	*Panicum turgidum*	551
15.112.	*Panicum whitei*	554
15.113.	*Paspalidium desertorum*	556
15.114.	*Paspalum dilatatum*	558
15.115.	Growth of *Lolium perenne* and *Paspalum dilatatum* over a range of constant temperatures	560
15.116.	*Paspalum distichum*	566
15.117.	High-quality Angus cattle grazing *Paspalum distichum* on salt marsh rangeland in Louisiana, United States	567
15.118.	*Paspalum nicorae*	570
15.119.	*Paspalum notatum*	572
15.120.	*Paspalum paspaloides*	577
15.121.	*Paspalum plicatulum*	580
15.122.	*Paspalum plicatulum* cv. Bryan growing with *Macroptilium atropurpureum* at Bundaberg, Queensland, Australia	582
15.123.	*Paspalum scrobiculatum*	586
15.124.	*Paspalum urvillei*	591
15.125.	*Pennisetum americanum*	597
15.126.	*Pennisetum clandestinum*	605
15.127.	Growth rhythm of *Pennisetum clandestinum* at Crow's Nest, Queensland, Australia and at Wollongbar, New South Wales, Australia	607
15.128.	*Pennisetum pedicellatum*	613
15.129.	*Pennisetum polystachyon*	617
15.130.	Seasonal dry-matter production of *Pennisetum polystachyon* in Fiji	619
15.131.	*Pennisetum purpureum*	622
15.132.	*Pennisetum hohenackeri*	630
15.133.	*Phragmites australis*	636
15.134.	*Phragmites karka*	639
15.135.	*Saccharum officinarum*	641
15.136.	*Saccharum spontaneum*	650
15.137.	*Sehima nervosum*	653

15.138.	*Setaria italica*	656
15.139.	*Setaria italica* cv. Panorama	657
15.140.	*Setaria sphacelata*	663
15.141.	*Setaria sphacelata* var. *sericea* cv. Kazungula	664
15.142.	*Sorghum almum*	671
15.143.	*Sorghum bicolor*	678
15.144.	Water consumption of grain sorghum during the crop cycle	680
15.145.	*Sorghum halepense*	687
15.146.	*Sorghum sudanense*	691
15.147.	*Spinifex hirsutus*	699
15.148.	*Sporobolus airoides*	702
15.149.	*Sporobolus helvolus*	704
15.150.	*Sporobolus marginatus*	707
15.151.	*Sporobolus virginicus*	710
15.152.	*Stenotaphrum secundatum*	712
15.153.	*Themeda australis*	715
15.154.	*Themeda quadrivalvis*	719
15.155.	*Themeda triandra*	722
15.156.	*Trachypogon spicatus*	726
15.157.	*Triodia pungens*	728
15.158.	*Tripsacum dactyloides*	730
15.159.	*Urochloa mosambicensis*	736
15.160.	*Urochloa mosambicensis*	737
15.161.	*Urochloa oligotricha*	741
15.162.	*Urochloa panicoides*	744
15.163.	*Vetiveria zizanioides*	748
15.164.	*Vossia cuspidata*	750
15.165.	*Zea mays*	753
15.166.	Uptake of nutrients in relation to dry weight of *Zea mays*	754
15.167.	*Zea mexicana*	759

Colour plates

1. A severely overgrazed area of thornbush, Riwa, Kenya
2. Effect of exclosure from grazing animals for 12 months, Riwa, Kenya
3. A scrub rake pushes fallen timber into wind-rows
4. The King Ranch root-plough
5. Seed-harvesting ants at work in the Sahel, Kordofan Province, the Sudan
6. The shallow trough or niche made by the Mallen Niche Seeder
7. Santa Gertrudis heifers grazing Para grass under coconuts, the Philippines
8. Mobile field hay-stacking frame, Nebraska, United States
9. Large bales of rolled hay left in the field for winter feeding, United States
10. Consolidating silage in a trench silo with a tractor
11. Silage trenches sealed with soil to provide a camber to shed rain
12. A grab attached to the end-loader of a tractor to excavate silage
13. An excavator-elevator unit for removing silage from a trench
14. A hay barn, tower silo and trench silo
15. Feeding silage on the ground to Merino sheep, Queensland, Australia
16. A crawler-tractor, forage harvester and trailer harvesting sweet sorghum
17. *Andropogon gayanus* (gamba grass)
18. *Anthephora pubescens*
19. A 10 000-bale stack of *Astrebla* spp. hay

20. Emus stroll across Mitchell grass, Queensland, Australia
21. Mitchell grass grassland in the early wet season, Queensland, Australia
22. Mitchell grass grassland in early autumn, Queensland, Australia
23. A mixture of *Heteropogon triticeus, H. contortus* and *Themeda australis,* with *Bothriochloa pertusa*
24. *Brachiaria decumbens* and *Centrosema pubescens*
25. *Brachiaria mutica* and *B. humidicola,* Fiji
26. Para grass (*Brachiaria mutica*), centro and calopo, the Philippines
27. *Cenchrus biflorus* in a gum arabic plantation, the Sudan
28. *Cenchrus ciliaris* established under a *Eucalyptus populnea* tree
29. *Cenchrus ciliaris* cv. Gayndah (Gayndah buffel grass) Charleville, Queensland, Australia
30. A home-made buffel seed harvester
31. Buffel grass stabilizing a clay soil around a watering point
32. Rhodes grass suppressing regrowth of brigalow
33. Boran steers grazing star grass (*Cynodon* sp.), Gilgil, Kenya
34. Shorthorn steers grazing irrigated pangola grass, Parada, northern Queensland
35. *Echinochloa pyramidalis* (antelope grass)
36. *Eleusine jaegeri* (manyatta grass), invading Kikuyu grass in the Kenya highlands
37. A "cut-and-carry" load of *Eragrostis tremula,* Kordofan Province, the Sudan
38. Seed clusters of bunch spear grass, *Heteropogon contortus*
39. *Heteropogon contortus* spreading into *Chloris-Paspalidium* after annual burning
40. *Imperata cylindrica* (lalang or blady grass), Sulawesi, Indonesia
41. *Melinis minutiflora* (molasses grass), stabilizing a steep slope, formerly rain forest, Queensland
42. *Panicum antidotale* (blue panic) in former gidgea scrub country
43. New pasture of Guinea grass and *Stylosanthes guianensis,* Tully, northern Queensland

Colour plates

44. A *Paspalum dilatatum/Trifolium repens* pasture, renovated and untreated
45. *Pennisetum americanum* (pearl or bulrush millet)
46. *Pennisetum americanum*, Kordofan Province, the Sudan
47. Breeding plots of *Pennisetum americanum*, Tifton, Georgia, United States
48. Kikuyu grass, *Desmodium intortum* and *D. uncinatum*
49. *Pennisetum purpureum* (elephant grass)
50. Mexican variety of *Pennisetum purpureum*, South Kalimantan, Indonesia
51. Sugar cane and Para grass, Mt Bartle Frere, northern Queensland
52. *Sorghum almum* (Columbus grass)
53. A crop of *Sorghum bicolor* and cowpea (*Vigna unguiculata*) for ensilage
54. The root parasite *Striga hermonthica* on sorghum, Kordofan Province, the Sudan
55. Johnson grass (*Sorghum halepense*) invading a peach orchard, Georgia, United States
56. *Triodia pungens*, Charleville, Queensland
57. Young regrowth of *Triodia pungens* after burning, northwest Queensland
58. Teosinte × maize hybrids, United States

1. The importance of grasslands

> The staple foods of the great majority of mankind come from grass. The grains of grasses such as maize in the Americas, rice in Asia, wheat, rye, barley and oats in Europe, and the sorghums in Africa (and India) provide the main basis of man's carbohydrate diet while the flesh of animals that graze on pastures provides the main source of his proteins and fats. Thus the nurture of grasses and grasslands is a matter of great importance to him and has always been so. Many of his migrations and invasions have resulted from his search for grasslands. Increasing skill in growing cereal crops and maintaining grazing animals has been a fundamental nature of his progress in civilization. Indeed, the future of his present-day civilization may depend largely on his ability to extend these basic resources to his nourishment. (Barnard & Frankel, 1964).

Although the great importance of grasslands lies in providing sustenance, grasses also serve humanity in other ways. Grass may be used for building homes and furniture (walls, thatch, matting, brooms), lawns, sportsfields, and as components of some cosmetics and medicines.

Shantz (1954) estimated that grasslands now extend over about 46 million km^2 and comprise 24 percent of the world's vegetation. They occur most typically in the interior of the great continental land masses, ranging northwards to about 55°N in Asia and North America and southwards to 40°S in South America. In the tropics and subtropics, species extend to about 28°N and 30°S of the equator (Thomas, 1978).

In the tropical regions of the world (those with a wet season of more than four and a half months) and subtropical regions (with year-round rainfall or hot, wet summers), 4.5×10^9 hectares, or 23 percent, is grazing land, according to 't Mannetje's 1978 estimates, based on a map prepared in 1966 by Troll (see Fig. 1.1). Thus, 't Mannetje concluded that Troll's tropical zones — with a wet season of more than four and a half months — humid subtropical zones and wet-summer-and-dry-winter zone would be able to sustain productive pastures. He excluded those zones with a wet season of less than four and a half months because they are semi-arid or arid.

Treeless prairies, steppes and pampas predominate in the temperate regions and are replaced by savannah, with varying proportions of trees and shrubs in the tropics and subtropics. All regions with fewer than 100 species of flowering plants have more than 14 percent grasses and a low percentage of grasses (below 7 percent) occur only in regions with a total flora of more

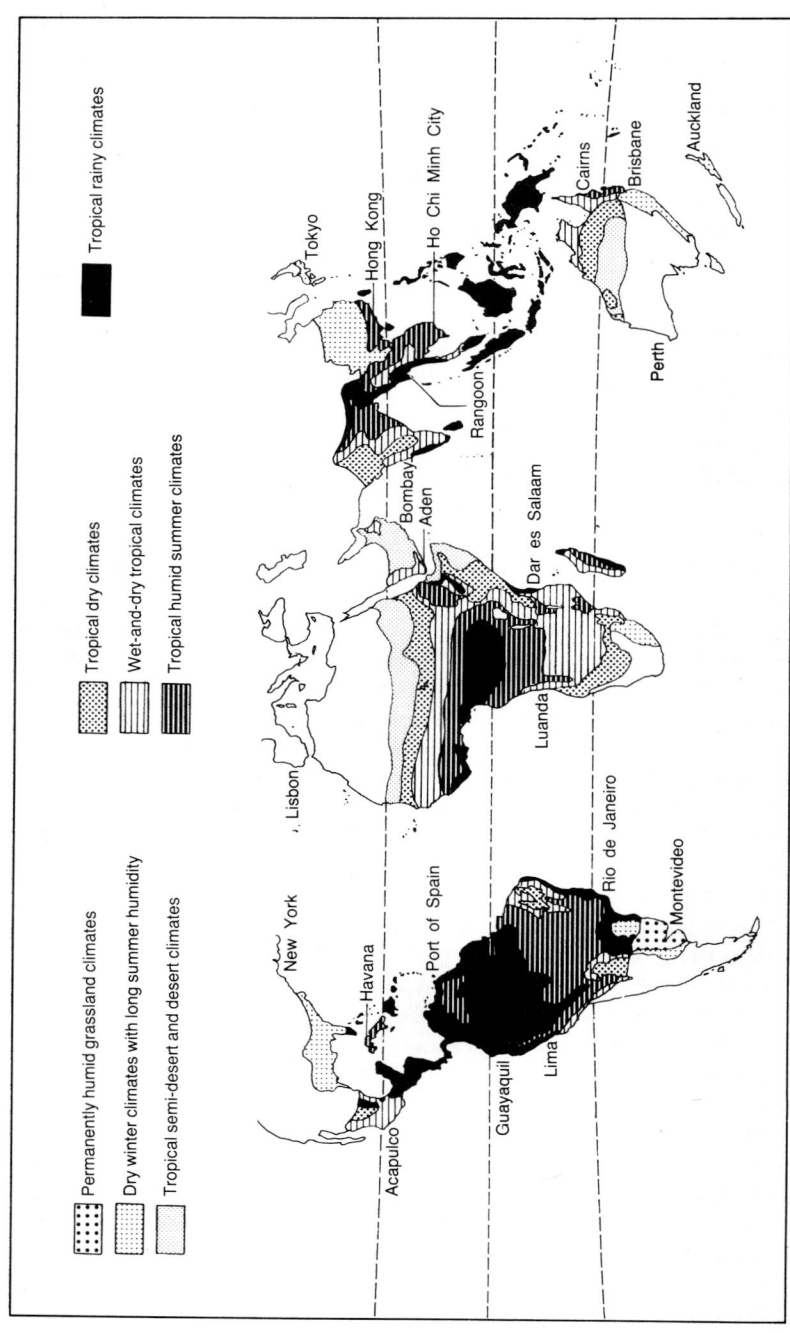

Figure 1.1. Climatic zones of the tropics and subtropics (**Source:** Troll, 1966)

than 4 000 species (Hartley, 1964). In the tropical Curaçao group (11-12°N), there is 12.4 percent grass. In a tropical rain forest, grass may constitute less than 0.5 percent of the ground cover, where light can penetrate the forest canopy; in grassland, grass may make up to 99.5 percent of the ground cover.

A natural grassland is a plant community in which the dominant species are perennial grasses, there are few or no shrubs and trees are absent. Usually associated with the dominant grasses are less abundant grass species and a variety of other herbaceous plants, both annual and perennial types which at certain times of the year give a characteristic aspect to the plant community (Moore, 1964).

Grassland is one of a number of seral phases of vegetation. The vegetation structure is dynamic, rather than static. One ecological association follows upon, and grows in consequence of, its predecessor in a well-marked and orderly sequence. One association therefore acts as a nursery to its immediate successor. This series of successional phases, from the first to the last, is referred to as the "sere", grassland forming one characteristic phase of that sere.

The development of the sere may be arrested at any given point if environmental conditions are such that further development is retarded. The sere may thus end at a subclimax rather than at its climax, e.g. in semi-arid areas the natural vegetation may be steppe or open grassland with no trees of any kind. In areas of higher rainfall, forest is the climax. Thus, while humanity is striving to maintain grassland, nature is striving toward the development of forest. In grassland studies the biotic factor is stressed in determining the character of vegetation, the process of grassland creation, grassland stabilization and the progression from grassland to forest, especially the important part played by domesticated grazing animals.

In regions of high rainfall, the tendency to revert toward forest is particularly marked and confronts the pioneer with difficult problems of stock and pasture management. Continued understocking will allow a normal reversion, first to weeds and then to shrubs and scrub; habitual overgrazing will tend to weaken the sward so that the establishment of weeds is made easier.

Many of the large grassland areas, such as the prairies and plains of North America, the pampas of South America, the steppes of Asia and the velds of Africa are believed to be of great antiquity and are climax formations determined by soil and climate. Other grasslands are of more recent origin and have replaced forests that have been destroyed mainly by cutting and fire; these have been maintained largely through grazing animals (Barnard & Frankel, 1964). In East Africa the destruction of trees and shrubs by elephants is a significant factor in this process, and Lamprey (1979) has stated that in the Serengeti game park, elephants have been converting woodland (large trees) to grassland at an annual rate of 6 percent since 1935.

The control and use of fire has been a very important feature associated with the development of grassland. The use of fire to flush out wild animals

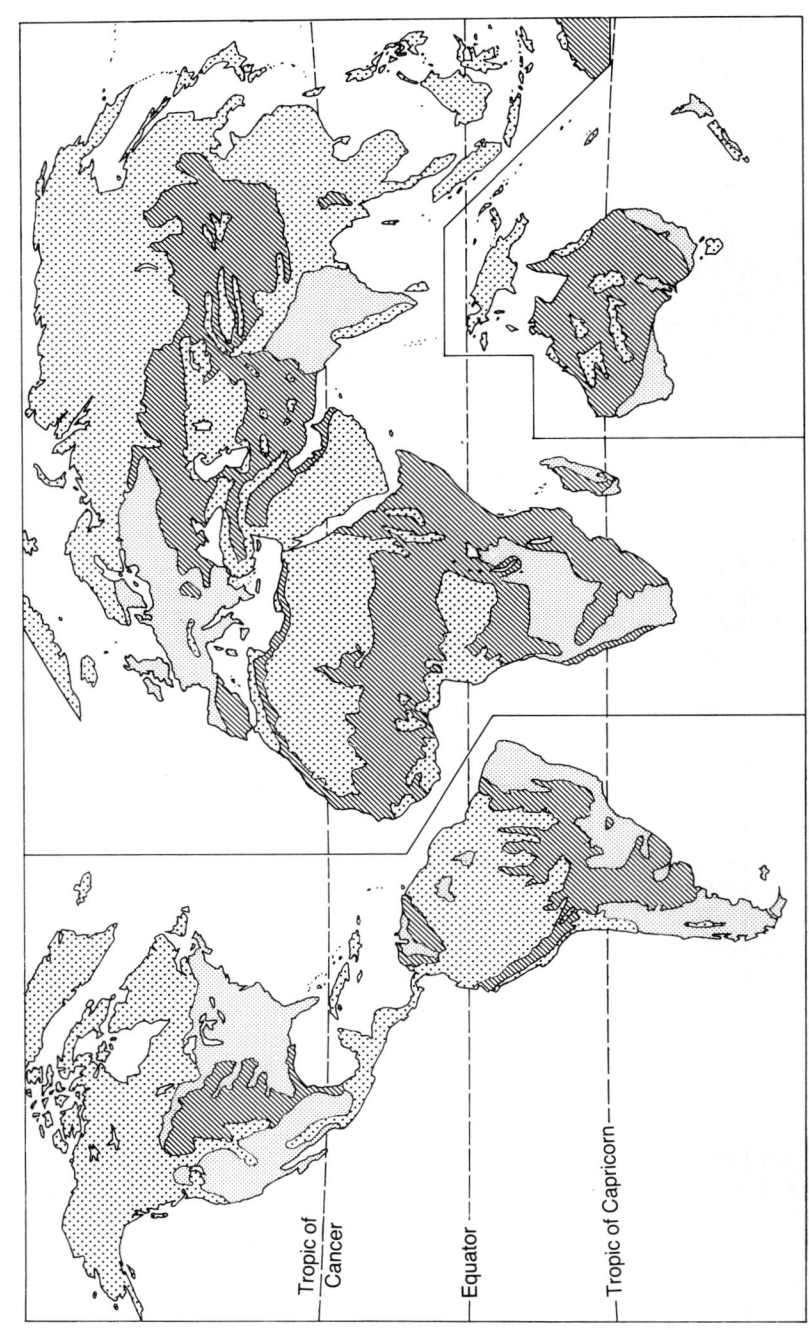

Figure 1.2. Major grazing areas of the world

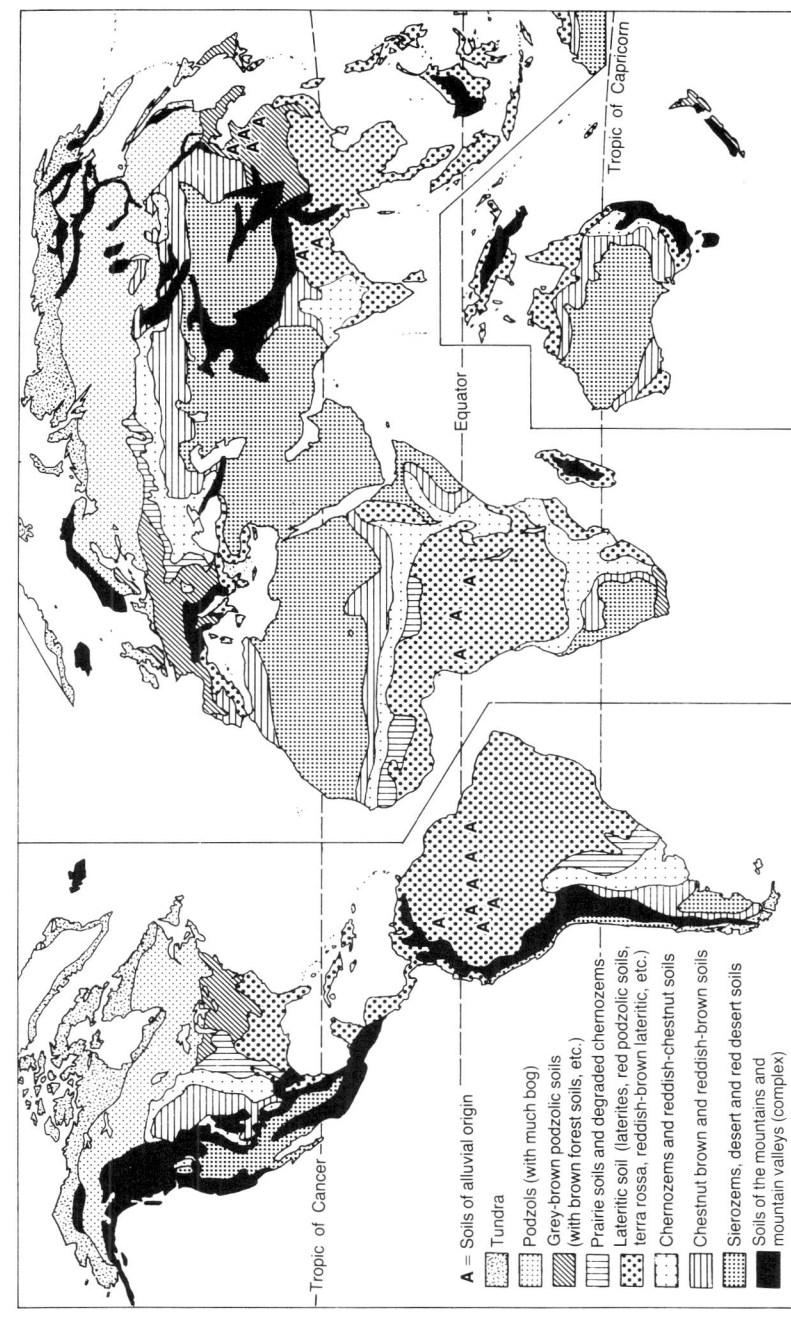

Figure 1.3. World distribution of the principal zonal soil groups

and, later, to provide young green regrowth for grazing animals, has developed large areas of grassland that constitute a fire "disclimax". Examples are the extensive *Imperata cylindrica* grasslands of Southeast Asia, the *Pennisetum polystachion* grasslands of Thailand and Fiji, and the *Trachypogon* savannahs of Venezuela and Colombia.

The major grazing areas of the world are shown in Figure 1.2. Vegetation type is greatly influenced by the soil type upon which it grows; Figure 1.3 indicates the world's principal zonal soil groups.

2. The classification and distribution of grasses

Since the publication of Linnaeus' sexual system of classification of plants in 1753 in his *Species Plantarum*, there has been great development in grass systematics. However, there is still some confusion. Most taxonomists include a tribal classification, and in this book ten tribal groupings are presented, which include the genera listed in Table 2.1.

TABLE 2.1 **Tribes and genera of the family Gramineae (grasses)**

Tribe	Genera
Andropogoneae	*Andropogon, Bothriochloa, Chrysopogon, Coix, Cymbopogon, Dichanthium, Hemarthria, Heteropogon, Hyparrhenia, Hyperthelia, Imperata, Ischaemum, Iseilema, Lasiurus, Saccharum, Sehima, Sorghum, Themeda, Trachypogon, Tripsacum, Vetiveria, Vossia, Zea*
Aristideae	*Aristida*
Arundineae	*Phragmites*
Arundinelleae	*Loudetia, Tristachya*
Chlorideae	*Astrebla, Chloris, Cynodon, Enteropogon*
Eragrostideae	*Dactyloctenium, Diplachne, Eleusine, Eragrostis, Triodia*
Oryzeae	*Leersia, Oryza*
Paniceae	*Acroceras, Anthephora, Axonopus, Brachiaria, Cenchrus, Digitaria, Echinochloa, Eriochloa, Hymenachne, Melinis, Panicum, Paspalidium, Paspalum, Pennisetum, Setaria, Spinifex, Stenotaphrum, Tricholaena, Urochloa*
Sporoboleae	*Sporobolus*
Zoysieae	*Leptothrium*

The normal spectrum for grass distribution throughout the world is:

Agrosteae	8.2%	Temperate zone grasses
Andropogoneae	11.9%	The sorghum tribe

Aveneae	6.3%	The oat tribe
Eragrostideae	8.1%	The love grass tribe
Festuceae	16.5%	The fescue tribe
Paniceae	24.7%	The millet tribe
Minor tribes	24.3%	

The apparent factors determining the relative abundance of the species of major grass tribes in the grass flora of different regions are set out in Table 2.2.

TABLE 2.2 Factors determining the relative abundance of major grass tribes

	Agrosteae	Andropogoneae	Aveneae	Eragrostideae	Festuceae	Paniceae
Climatic factors:						
Temperature	+++	++	++	++	+++	++
Rainfall	—	(+)	(+)	+	(+)	++
Historical factors	—	+	+	—	(+)	+
Taxonomic heterogeneity	—	—	++	—	—	—

NOTE: +++ = predominant factor; ++ = major factor; + = minor factor

Distribution of some of the tribes of the Gramineae

Agrosteae. Distribution seems to be determined by factors of latitude, including day length and temperature. The area of distribution is divided by the 10°C minimum temperature isotherm (mean temperature of the coldest month). Winter temperatures in particular have a dominating influence. Total rainfall and seasonal distribution seem to have little effect. The tribe is abundant in the temperate areas of the Northern Hemisphere, particularly in the main continental areas (Hartley, 1950).

Andropogoneae. Distribution is closely related to temperature. Rainfall is apparently unimportant, as members of the tribe are found in areas with rainfall varying from 125 to 2 250 mm annually. They are abundant in the tropical savannahs of India, Africa and South America (see Fig. 2.1).

Aveneae. Most abundant in temperate and cold regions, scarce in the Western Hemisphere. The three main centres are Europe and north and south

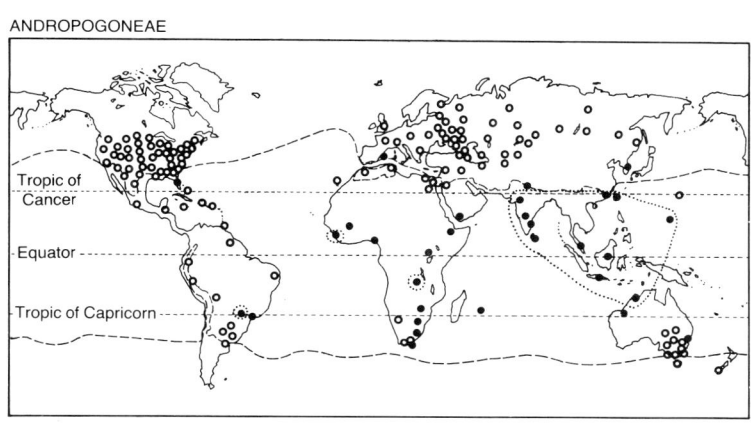

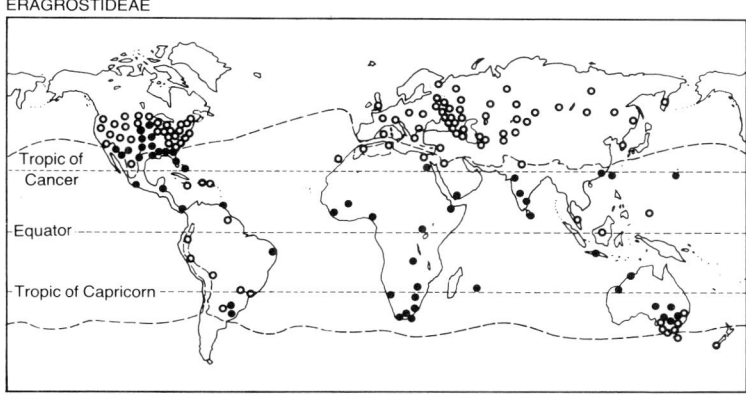

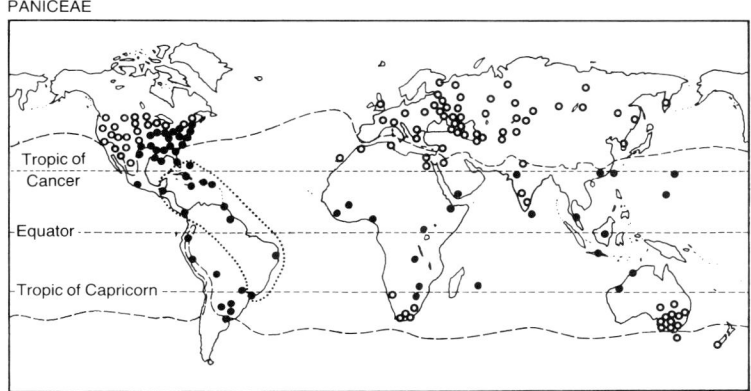

- ● Above-normal percentage of grass flora   ――― 12° C mean temperature of midwinter month
- ○ Below-normal percentage of grass flora

Figure 2.1. Distribution of some major tribes (**Source:** Hartley, 1950)

Asia; the Cape Province of South Africa; and southeast Australia and southern New Zealand.

Eragrostideae. Temperature has a marked effect on the distribution of this tribe (Fig. 2.1). The greatest concentration of species occurs where the mean winter temperature is above 10°C. This is correlated with a lower annual rainfall of 1 000 mm.

Festuceae. Distribution follows that of Agrosteae.

Paniceae. This tribe is abundant in the tropics and subtropics, and greatest where temperatures are high and rainfall low (see Fig. 2.1).

3. The world's major tropical grasslands

Numerous genera of grasses are spread throughout the tropics and subtropics. An attempt has been made in this chapter to indicate the major genera in various associations within these areas. Examples of the genera are listed with illustrations in the catalogue of this volume (see Chapter 15). Beginning with West Africa, the various countries are dealt with from west to east, although the list is not exhaustive.

Africa

Rattray (1960b) produced a book entitled *The grass cover of Africa,* including a relevant map. Using this survey and map, Whyte (1968) drew up a chart of

TABLE 3.1 **Dominant grass genera in the ecoclimatic zones of western and equatorial Africa**

Zone	Approx. annual rainfall (mm)	Dry season (months)	Dominant grass association	Country
Saharan	—	12	*Aristida, Panicum turgidum*	Chad, Mali, Mauritania, Niger, Sudan
Sub-Saharan	100-250	9-12	*Aristida, P. turgidum*	Chad, Mali, Mauritania, Niger, Sudan
Sahelian	250-600	9	*Cenchrus*	Burkina Faso, Chad, Mali, Mauritania, Niger, Senegal, Sudan
Sudanian	600-1 250	6-8	*Andropogon*	Burkina Faso, Chad, Mali, Niger, northern Nigeria, Senegal, Sudan
Guinean	>1 250	3-6	*Hyparrhenia*	Benin, Burkina Faso, Cameroon, Central African Republic, Chad, Côte d'Ivoire, Gambia, Ghana, Guinea, Mali, Nigeria, Senegal, Sudan, Togo, Zaire
Guinean equatorial	>1 800	Short	*Pennisetum*	Cameroon, Gabon, Ghana, Côte d'Ivoire, Liberia, Nigeria, Sierra Leone, Zaire

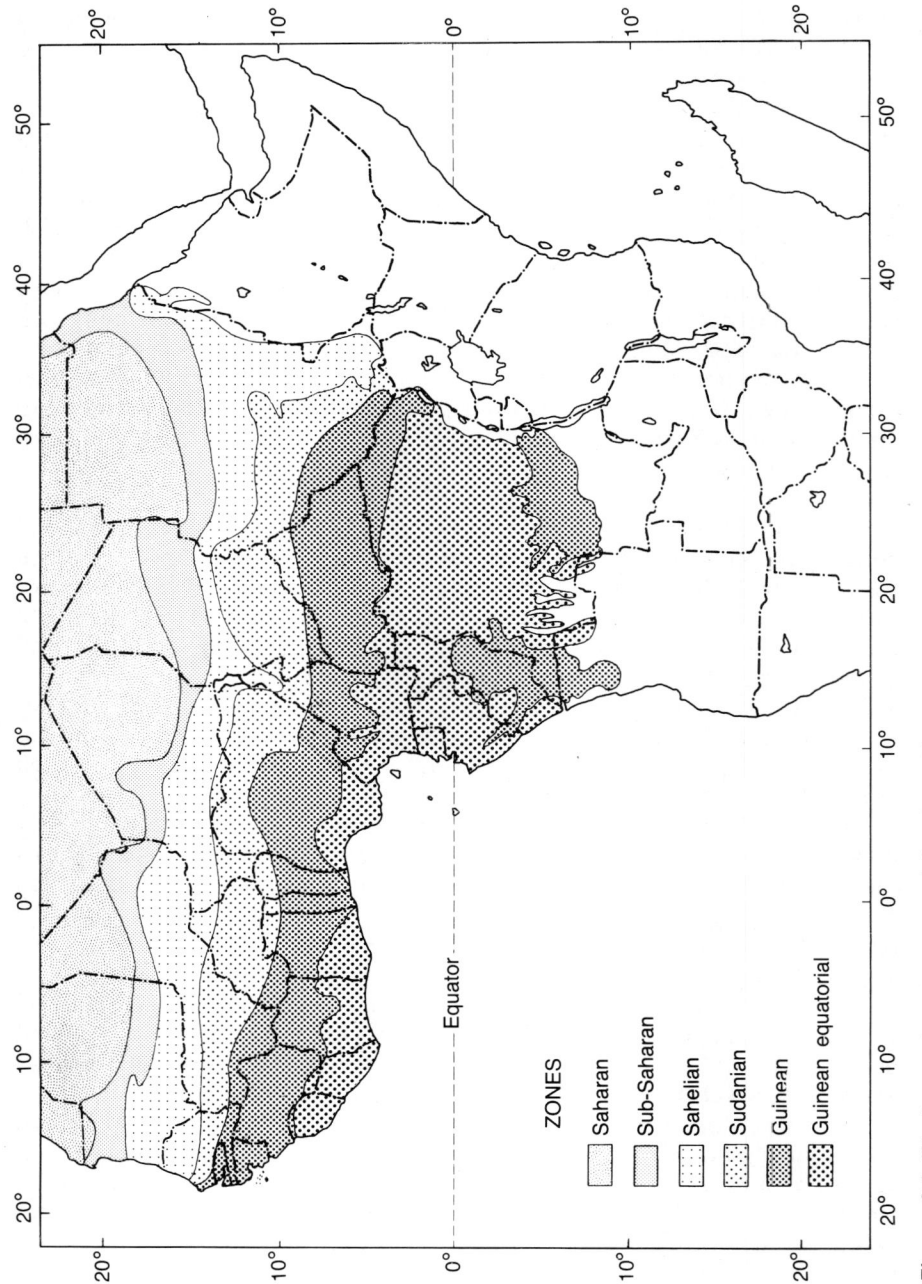

Figure 3.1. Western and western equatorial regions, Africa

the ecoclimatic gradient in western and equatorial regions in Africa in his book *Grasslands of the monsoon*. Table 3.1 gives a modified and simplified chart. The climatic zones of northwest Africa lie almost parallel from west to east, as is shown in Fig. 3.1.

Vesey-Fitzgerald (1963) published a fairly detailed paper on the grasslands of the Central African Plateau embracing the headwater catchment area of the Congo River and the interior drainage basins of the Rukwa Valley and the Mweru-wa-Ntipa Depression (see Fig. 3.1). The area is dominated by the Great Rift Valley and associated highlands and depressions. The rainfall pattern is a six-month dry season from May to October, with heavy downpours in the rainy season from November to April. Soils have a catenary sequence from well-drained red soils on the highlands to poorly drained black clays of the valley bottoms, where edaphic grasslands develop.

Dambo or vlei soils occur in the valleys of the upper reaches and support small-valley grassland, while the flood plains in the lower reaches give rise to flood-plain grasslands. The natural groups into which the valley grasslands fall are as follows.

- *Dambos,* or headwater valley grasslands, carry fine-leaved perennial bunch grasslands of *Andropogon, Hyparrhenia* and *Loudetia* with *Loudetia simplex* (Fig. 15.95) as the dominant grass over extensive areas, which flower early in the wet season, followed by *Hyparrhenia* spp. later in the season. At the end of the season all the grass flora dries off and is usually burnt. Several genera, such as *Acroceras* and *Leersia,* grow in the bogs at the lowest points;
- Riverine grasslands are zoned in relation to drainage. *Hyparrhenia, Loudetia* and *Themeda* occur in the better-drained areas and *Acroceras, Leersia, Oryza, Vossia* and *Echinochloa* on the wetter water meadows, which are seasonally flooded;
- Flood-plain grasslands, or *mbuga,* exhibit a sequence from the shallowly flooded edges to the almost permanently wet flood-plain bottoms. *Hyparrhenia rufa* (Fig. 15.83) is characteristic of the perimeter grasses, *Echinochloa pyramidalis* (Fig. 15.57) of the deeply flooded country and *Vossia cuspidata* (Fig. 15.164) of the sump areas where the water lies longest.

Swamps of reeds and *Oryza, Leersia, Vossia, Echinochloa scabra* (Fig. 15.58) and *Panicum repens* (see Fig. 15.108) are scattered across the flood plain, while on alkaline and saline pans, *Sporobolus spicatus* (and other *Sporobolus* species) and *Diplachne fusca* (Fig. 15.52) are found.

The East African highlands (see Fig. 3.2) of Ethiopia and central Kenya, and of the Ngorongoro Crater, carry a natural grassland of *Pennisetum clandestinum,* or Kikuyu grass (Fig. 15.126), while in Zaire and the southern highlands of the United Republic of Tanzania, the high-lying savannahs are dominated by *Loudetia* spp., *Themeda triandra* (Fig. 15.155), *Hyparrhenia, Cymbopogon nardus* (Fig. 15.33), *Setaria sphacelata* (Fig. 15.140), *Digitaria abys-*

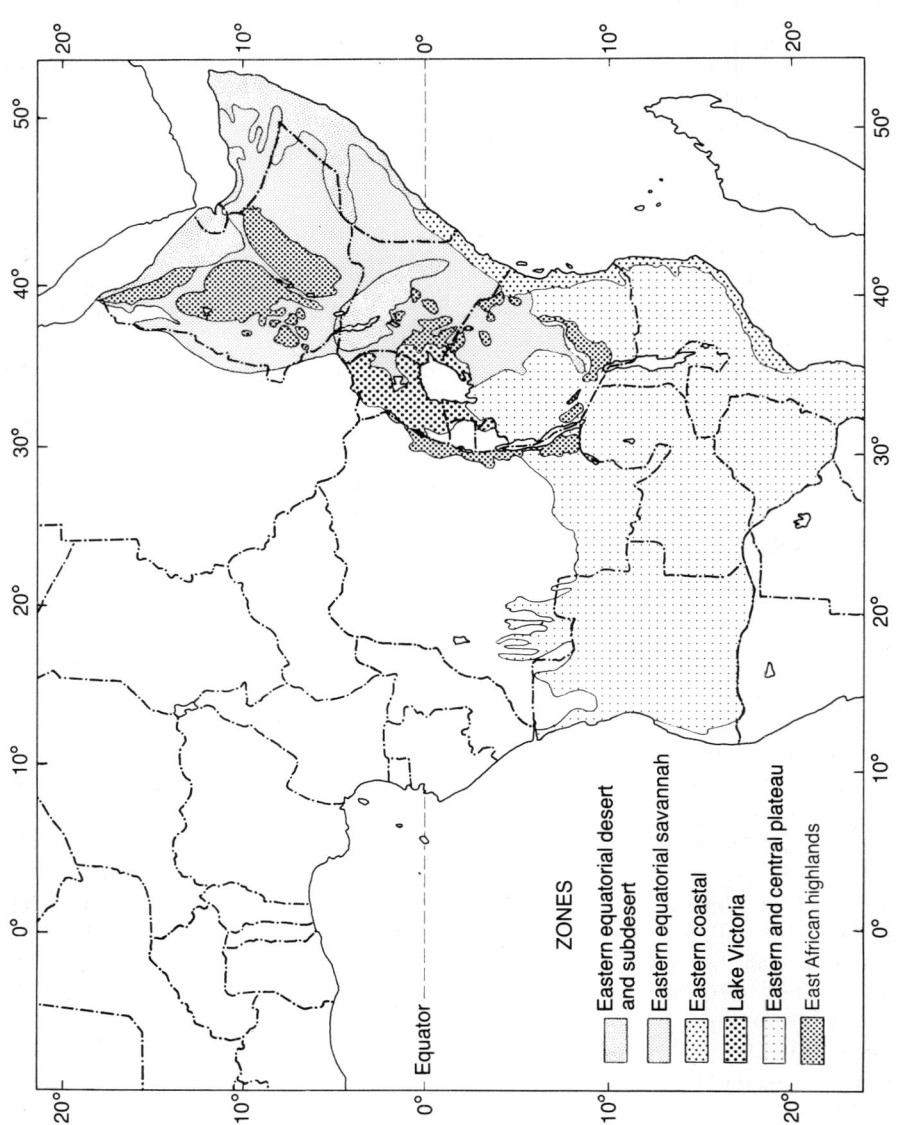

Figure 3.2.
Eastern and central regions, Africa

sinica (Fig. 15.46), *Pennisetum trisetum* and *Exotheca abyssinica* (Fig. 15.76). In the fertile crescent around Lake Victoria in Uganda, *Pennisetum purpureum* (Fig. 15.131) is indigenous in areas of higher fertility and rainfall.

In the arid Horn of Africa, the grasses on the Ethiopian Plain are mainly *Eremopogon foveolatus* with species of *Eragrostis, Panicum* and *Aristida*. On the arid Red Sea coast the main grass is *Panicum turgidum* (Fig. 15.111). *Chrysopogon aucheri* (Fig. 15.31) is the dominant grass in the arid areas of Somalia and northern Kenya.

The eastern equatorial savannah (see Fig. 3.2), stretching from the United Republic of Tanzania through Kenya to Ethiopia, just east of the extensive *miombo* of central Africa, carries *Themeda triandra* on the red latosolic soils, *Hyparrhenia* spp. on the sandy surfaced soils and *Cenchrus* spp. on the heavy black soils with *Cynodon* spp. in the Rift Valley.

In southern Africa the "sweet" and the "sour" veld are recognized. The sweet veld is made up of summer grasses that retain their palatability and thus provide useful grazing in the dry winter season. This grazing is made up of species of *Hyparrhenia, Eragrostis, Themeda triandra, Setaria* spp., *Bothriochloa insculpta* (Fig. 15.11), *Brachiaria* spp. and *Digitaria* spp. The soil is fertile. The sour veld grows on infertile, acid, sandy soil and at maturity the grasses provide poor winter grazing. The veld consists of species of *Hyparrhenia, Cymbopogon, Heteropogon, Andropogon* and *Loudetia* spp. The sweet veld grows in lower rainfall areas while the sour veld occurs in areas of high rainfall where the grasses mature readily but are only palatable as young plants. A mixed veld, containing elements of both sweet and sour veld, is of value during the growing season only.

India

Dabadghao and Shankarnarayan (1973) reduced the major grassland types in India to four groups (see Fig. 3.3). These are *Sehima/Dichanthium; Dichanthium/Cenchrus/Lasiurus; Phragmites/Saccharum/Imperata;* and *Themeda/Arundinella*. They include the tropical and subtropical types and the temperate Alpine type.

Sehima/Dichanthium, the largest of the five grass zones of India (Dabadghao, 1960), occurs south of the Northern Great Plains between longitudes 68 and 87°E and latitudes 8 and 24°N. The topography is undulating to hilly with intervening valleys; the elevation ranges between 300 and 1 200 m. The rainfall is monsoonal and varies from 300 mm to over 2 000 mm annually. Soils are gneissic or basaltic in origin and range from shallow coarse-textured yellow or red to deep, fine-textured black.

Twenty-four perennial grass communities with different ecological statuses occur in this zone. The principal ones are *Sehima, Dichanthium, Iseilema, Ischaemum, Chrysopogon, Bothriochloa, Heteropogon* and

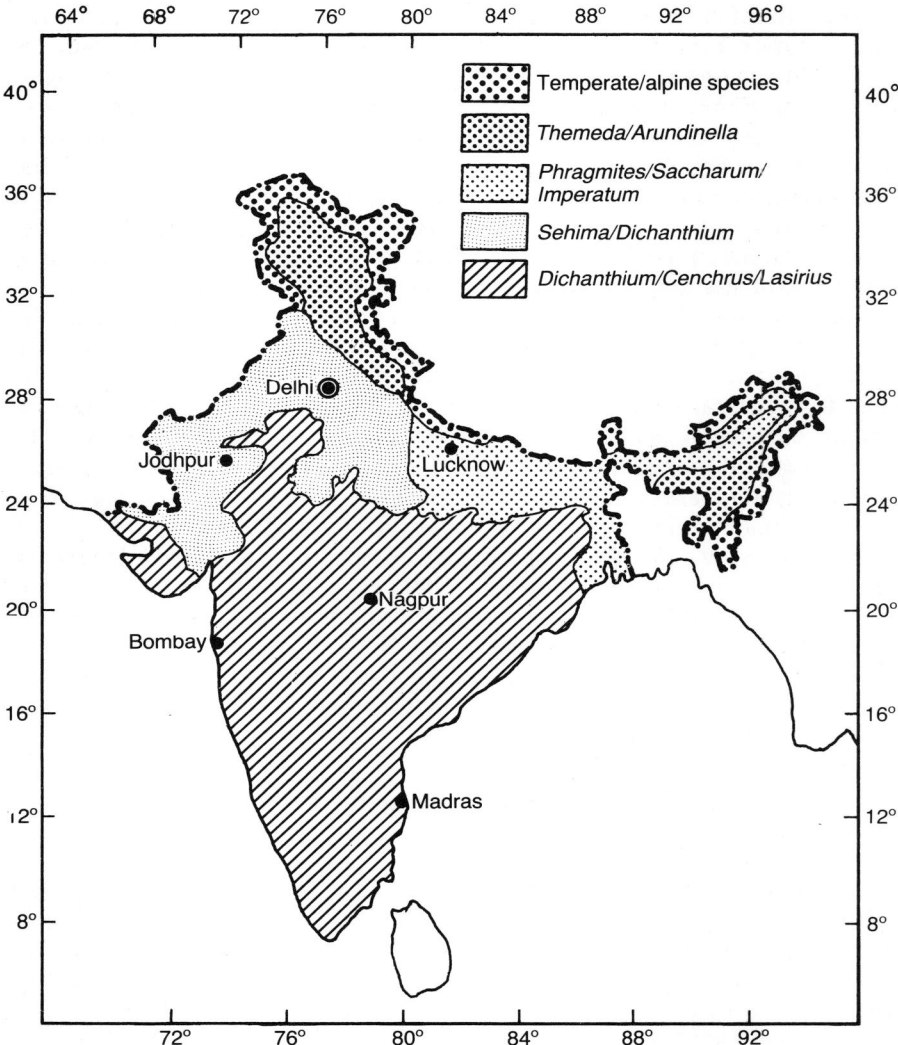

Figure 3.3. Distribution of grass covers in India (**Source:** Dabadghao & Shankarnarayan, 1973)

Themeda. These communities have a common growth pattern. Growth starts after the outbreak of the monsoon in June/July and attains a peak in September. The grasses mature at the end of October and remain dormant during the eight-month dry period. The grazing animals therefore have dry forage for most of the year. Pasture growth and production fluctuate as a result of erratic rainfall, causing feeding problems during periods of low production. Legumes are either absent, or so sparse that they do not contribute to soil fertility or forage quality. The high livestock population places tremendous pressure on pastures, leading to a depletion of vegetation and to erosion. The particular significance of cattle, and the almost negligible area under fodder production, create grassland problems peculiar to India (Dabadghao & Shankarnarayan, 1970).

The *Dichanthium/Cenchrus/Lasiurus* type of grassland occurs on the plains of Punjab, western Uttar Pradesh and northern Rajasthan, which comprise the arid and semi-arid region of western India. These areas have below 750 mm of rain annually, with high summer temperatures and relatively severe winters, frosts being frequent in December and January. Alluvial soils, sandy to sandy loam in texture, are drained imperfectly to moderately well. There are 40 grass species characterizing this type of grassland and 14 perennial species. The most characteristic in order of importance are *Dichanthium annulatum, Cenchrus ciliaris, C. setigerus, Lasiurus hirsutus, Eleusine flagellifera, Cynodon dactylon, Sporobolus marginatus, S. pallidus, Panicum turgidum, Heteropogon contortus* and *Dactyloctenium sindicum*.

The following changes of progression and regression occur within this type.

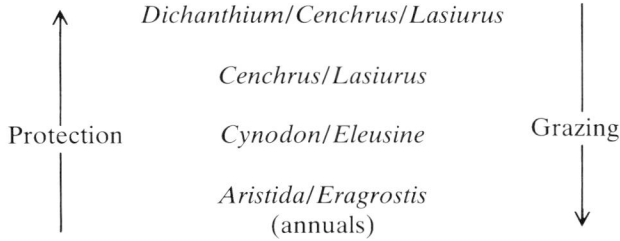

When highly deteriorated grassland of this type is protected, *Sporobolus pallidus* increases gradually during the first two years while the annual species of *Aristida* and *Eragrostis* slowly disappear. From the third year onward, *Cynodon dactylon* or *Eleusine flagellifera* begin to appear. The grasses which are highest in this progression, *Cenchrus ciliaris, C. setigerus* and *Dichanthium annulatum* generally start to appear after about five years of protection.

The management of this type of grass cover should aim at a good balance between *D. annulatum* and *Cenchrus* spp. where rainfall exceeds 250 mm. Below this level, the immediate objective would be a stand of *Lasiurus hirsutus,* since during prolonged droughts this species can always be relied upon to hold the soil and provide grazing for cattle. The *Cynodon* stage represents the last resistant perennial cover in the regression. When this stage is encountered the intensity of utilization must be reduced considerably and immediately to enable the grassland to recover. *Aristida/Eragrostis* grasslands are too poor to provide any grazing and should be completely closed for four or five years to allow the succession to progress toward the fair condition represented by *Cynodon dactylon* as a first step toward improvement.

The *Dichanthium/Cenchrus/Lasiurus* type occurs in subtropical arid and semi-arid regions constituting the northern portion of Gujarat, the whole of Rajasthan, western Uttar Pradesh, Delhi State and Punjab, with a coverage of 436 000 km². This region is situated between 23 and 32°N; the topography is more or less level, although broken by sand dunes in western Rajasthan, with a mean height above sea level of 150-300 m. The rainfall between July and September ranges from 100 mm in the west to 750 mm in the east. The mean June temperature is 32-33°C and the mean winter temperature in January between 10 and 18°C; frosts can occur in December and January. The soils are coarse-textured, with a pH range of 7.5 to 8.6. Important species are *Dichanthium annulatum, Cenchrus ciliaris, C. setigerus, Cynodon dactylon, Dactyloctenium sindicum, Lasiurus hirsutus* and *Sporobolus marginatus;* these are the preferred species for grazing animals. As the most palatable species, *D. annulatum* disappears first under grazing to be replaced by *Cenchrus* and *Lasiurus*.

The *Phragmites/Saccharum/Imperata* type covers the Gangetic Plain, the Brahmaputra Valley and westward to Punjab, covering 2 800 000 km². It is situated between latitudes 26 and 32°N, with elevations between 150 m in the east and 300 m in Punjab. It consists of low-lying, ill-drained lands with a high water-table. Rainfall is variable, up to 4 000 mm annually. The mean June temperature is 29-38°C and the winter temperature in January varies from 13 to 19°C. Soils include most textures, with pH values of 4.5-7.5. The dominant grass species are *Phragmites karka, Saccharum arundinaceum, S. spontaneum* and *Imperata cylindrica*. These tall grasses (particularly *Phragmites*) are cut for thatch; with cutting and burning, they are replaced by *Saccharum* and *Imperata*. Burning and heavy grazing induce changes to a *Sporobolus/Paspalum/Chrysopogon* cover.

The *Themeda/Arundinella* type of grassland occupies the entire northern and northwestern montane tract of 230 400 km² in Manipur, Assam, West Bengal, Uttar Pradesh, Punjab, Himachal Pradesh, Jammu and Kashmir. It is found from latitudes 29 to 37°N in the west and 22 to 28.5°N in the east, and occurs between 350 and 2 100 m above sea-level. The annual rainfall is from 1 000 mm in the west to 2 000 mm in the east. The mean June temperature is 27-32°C, and the winter temperature in January 13-17°C. Snow is common. The soils are shallow and gravelly with a pH of 6.2. The grass vegetation is dominated by *Arundinella bengalensis, Bothriochloa bladhii, B. pertusa, Cynodon dactylon, Heteropogon contortus, Ischaemum barbatum* and *Themeda anathera. T. anathera* is the highest community within the type. Degradation under grazing proceeds from *T. anathera* to *Arundinella* spp. to *Chrysopogon fulvus* to *Heteropogon contortus* and *Bothriochloa* spp. to *Cynodon dactylon*.

Sri Lanka

The montane grasslands. A considerable area of the jungle vegetation (rain forest) of the montane zone has been destroyed and is now either tea plantation or native patana grassland. The patanas occur above 1 000 m and extend to some 64 000 ha; the larger part of the patana grassland lies within the semi-arid zone.

Native patana grassland occurs with wide tongues protruding into the surrounding jungles. It appears to be a fire subclimax that has developed after the clearing of jungle vegetation. Dry patanas are much more extensive than wet patanas.

Generally, patanas are dominated by coarse, tussocky grasses of very low fodder value. Three main communities have been recognized.

- *Mana*, dominated by *Cymbopogon confertiflorus*, which grows to a height of about 1.25 m. As only the tender shoots are eaten by animals, this grass is commonly burnt to produce fresh growth. It occurs in almost pure stands at the forest margins and in depressions;
- *Gavara*, in which *Chrysopogon zeylanicus* is the dominant species, produces a tall coarse herbage of low quality, similar to *mana* grass;
- *Pini baru tana*, dominated by *Themeda tremula*, which does not grow as tall as the first two grasses, but has herbage of better quality. It is commonly found on exposed hillsides and represents an intermediate stage of succession to the taller grass communities described above.

A few legumes, such as *Parochetus communis*, *Smithia blanda* and *Desmodium microphyllum*, occur in the patanas (Andrew, 1971).

The hill country. This area consists of tea and some rubber estates and large areas of improved Kikuyu grass grassland which produces most of the milk supplies. Abandoned tea estates are being sown to improved pastures. Species include *Panicum maximum, Pennisetum purpureum, Brachiaria decumbens* and *B. brizantha* (Kannegieter, 1970); a good deal of Guatemala grass (*Tripsacum laxum*) is also grown to build up organic matter for replanting tea. Although cut in the young stage and fed to animals, this grass has very poor nutritional value and low digestibility (Andrew, 1971), but could be made into silage. *Paspalum urvillei* (vasey grass) is also grown to be cut for silage.

The *villus*. These marsh grasslands are submerged for long periods and thus count as seasonal grasslands. They provide cattle with succulent palatable feed for eight months of the year, following the receding waterline, and need no fertilizer. They are dominated by *Brachiaria mutica*, which competes with weeds and stands heavy grazing. The carrying capacity of the *villu* grazing during the dry season is 2.5 cows per hectare, but near villages it is often overgrazed. Associated grasses are *Cynodon dactylon* and *Digitaria longiflora* (Kannegieter, 1970).

The low country wet zone coconut area. There are an estimated 440 000 ha of land under coconuts, which also support some 500 000 head of cattle and buffaloes in the wet southwestern part of the island. The area receives 1 875-2 000 mm of rainfall annually, with only a short dry spell during February-March. Although the smallholder coconut farmer often keeps a few animals for draught power, meat and manure, husbandry is usually poor.

The grasses *Brachiaria subquadripara (miliiformis), B. brizantha, Pennisetum purpureum* and *Panicum maximum* grow well, but in areas receiving less than 1 625 mm of rain annually, they compete with the coconut (Santhirasegaram, 1966).

The dry zone. This area receives less than 1 875 mm of rain annually, but supports 60 percent of the country's cattle for draught power, manure, and prestige (a walking bank account). They graze on the edges of roads, around irrigation tanks and ditches, bunds, and around villages on scant pickings.

Southeast Asia

A good deal of Southeast Asia — an estimated area of 200 million hectares — is covered by *Imperata cylindrica* as a fire subclimax. It has invaded areas cleared of original rain forest and wet sclerophyll forests, and land abandoned after cultivation. Other secondary grasses include *Pennisetum polystachion, Chrysopogon aciculatus* and *Cynodon dactylon* in previously burnt or overgrazed areas. In Thailand, *Arundinaria pusilla* is common in the hill country. Tall grasses of the wetter areas include *Saccharum spontaneum, S. robustum, Themeda arundinacea* with *Phragmites* spp. in swamps. Areas with slightly less rainfall support *Sehima nervosum, Sorghum* spp., *Heteropogon contortus* and *Themeda australis*.

In Papua New Guinea, cattle are grazed mostly on unimproved *Imperata cylindrica* and *Themeda australis*. There are about 4.5-5.5 million hectares of natural grassland. The main types of grassland are mid-height, containing species of *Ophiurus, Imperata cylindrica* and *Themeda australis* in the littoral and fluvial plains; tall, *Saccharum spontaneum, S. robustum, Imperata cylindrica* and *Phragmites karka* on the plains and in swamps; and savannah, *Ophiurus* spp., *Imperata cylindrica* and *Themeda australis* in the coastal hills, foothills and uplands.

Australia

In the tropical and subtropical summer rainfall areas of northern Australia, the Andropogoneae (sorghums and related plants), Paniceae (panics), Eragrostideae (*Eragrostis* spp.) and Chlorideae (*Chloris* spp.) are dominant.

A marked seasonal incidence of growth is a universal feature of Australian pastures, and northern Australia has a rainfall with summer dominance. Fig. 3.4 shows a map of Australian pastures; the chief types north of the Tropic of Capricorn include the following major groups.

Tropical tall grass. The grasses that occur usually in open forest or woodlands are mainly perennial tussock grasses, 1-2.5 cm in height. They are of low nutritive value except for short periods after rain or fire. The major species are the annual *Sorghum australiense* and the perennials *Sehima nervosum, Chrysopogon fallax, Themeda australis, Sorghum plumosum, Heteropogon triticeus* and *H. contortus*. On many skeletal soils of dissected country, species of *Triodia* and *Plectrachne* are dominant. There are large areas of low quality *Aristida ingrata* and *A. hygrometrica* in the gulf country of Queensland. The estuarine plains carry saltwater couch (*Sporobolus virginicus*), and the heavy soils in the depressions and valleys carry *Dichanthium tenuiculum, Bothriochloa bladhii, Eulalia fulva* and *Chrysopogon fallax*. *Imperata cylindrica* (blady grass) is a weed in areas cleared of rain forest.

Bunch spear grass. This group is found in the open forest zone continuous with the tropical tall grass zone of the north. The dominant perennial grasses are *Heteropogon contortus* (black, or bunch, spear grass), *Themeda australis* and *Bothriochloa ewartiana* in the north, and *B. bladhii* in the south.

Brigalow. This is a zone of mixed country in which the major vegetation types are scrubs of brigalow (*Acacia harpophylla*), belah (*Casuarina cristata*) and wilga (*Geijera parviflora*) with a sparse ground flora of tussocks of *Paspalidium caespitosum*. After ring-barking or clearing the brigalow, native *Chloris, Eragrostis* and *Dichanthium* species take over.

Blue grass. Geographically associated with the brigalow zone, and at the higher rainfall limits of the *Astrebla* (Mitchell grass) zone, there are areas of open tufted grasslands on heavy cracking clay soils dominated by *Dichanthium sericeum, Eriochloa* spp., *Paspalidium globoideum* and *Panicum queenslandicum*.

Mitchell/Flinders grass. This is a tussock grassland association with the tussock of *Astrebla* spp., 0.5-1 m apart, the roots of which intermingle below ground level. During the rainy season on these heavy, cracking, dark grey and brown calcareous clays, annual species such as Flinders grasses (*Iseilema*

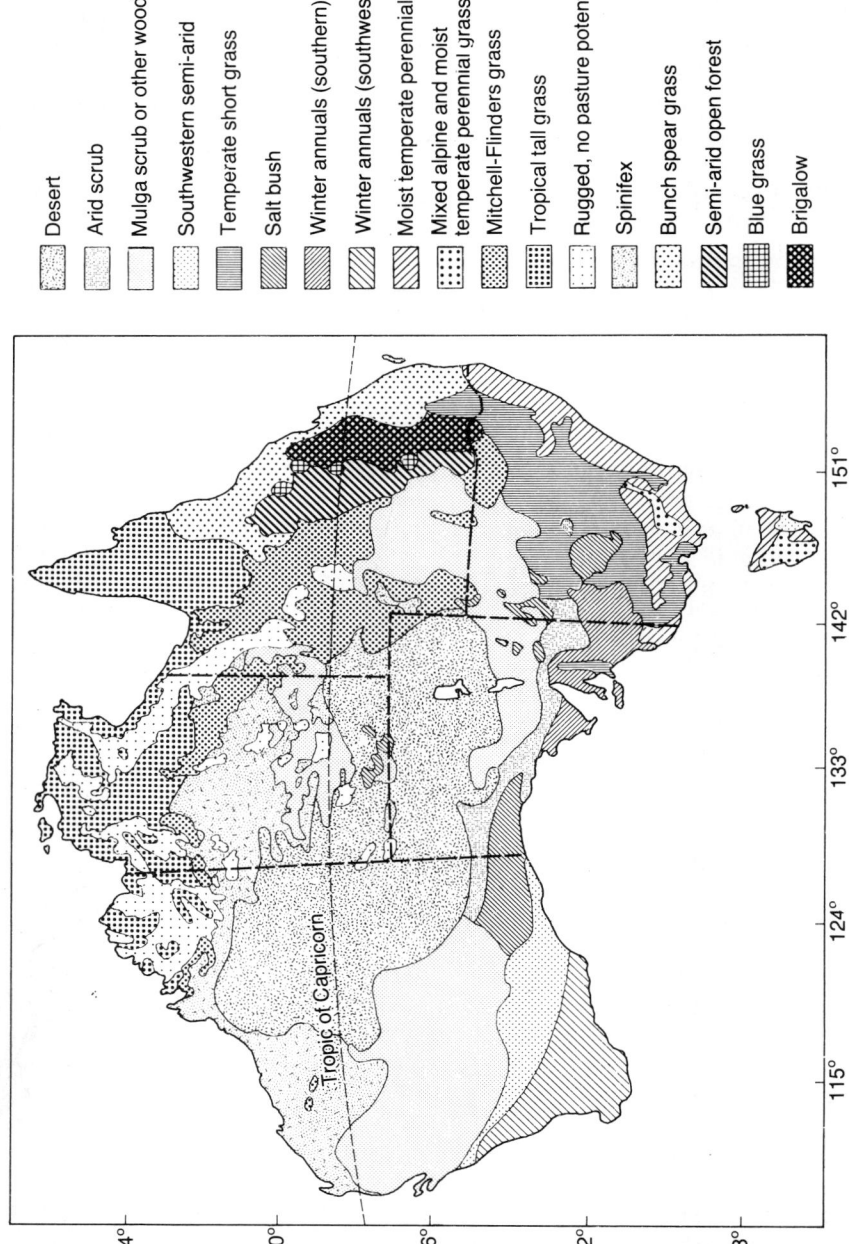

Figure 3.4. Pasture distribution in Australia

spp.), *Dactyloctenium radulans, Panicum whitei* and *Brachyachne convergens* appear. The perennial Mitchell grasses consist of four species, *Astrebla elymoides, A. lappacea, A. pectinata* and *A. squarrosa*. The grasslands are virtually treeless, but there may be some tree savannah with isolated trees of boree (*Acacia cana*), whitewood (*Atalaya hemiglauca*), vine tree (*Ventilago viminalis*), leopardwood (*Flindersia maculosa*), emu apple (*Owenia acidula*) and boonaree (*Heterodendron oleifolium*).

The channel country. During heavy rains the flood plain alluvia of the Georgina, Mulligan, Diamantina, and Bulloo rivers and Cooper's Creek are covered with varying depths of water. As the water recedes, ephemeral grasses and herbage spring up to provide valuable grazing. With flood waters from a summer flood, the temporary pasture is dominated by the annual *Echinochloa turnerana* and, on the shallow sites, *Panicum whitei*.

Semi-arid open forest (or desert country). This is a mixed zone made up mainly of open eucalypt forest. The soils are very poor and the indigenous grass cover is made up mainly of species of *Aristida* and *Plectrachne*.

Mulga scrub. A fairly dense zone of low woodland occupies lateritic red earth (ultisols) in an area of erratic rainfall, receiving in all 400-450 mm of summer and (some) winter rain. The zone supports *Acacia aneura* (mulga), and the sparse ground cover embraces a mixture of temperate and subtropical species of *Danthonia, Stipa, Aristida, Neurachne, Eragrostis, Eriachne, Digitaria, Enneapogon* and *Dactyloctenium*.

Spinifex. This is a steppe area dominated by spinifex (*Triodia* spp.) with occasional shrubs (*Acacia* spp.) and low trees (*Eucalyptus* spp.). The spinifex is burnt annually to encourage young shoots.

The Pacific Islands

Fiji. Dairying commenced in Fiji only in about 1920, and farmers depended on grazing two indigenous species, Batiki blue grass (*Ischaemum indicum*) on the wetter, windward sides of the main islands, and Nadi blue grass (*Dichanthium caricosum*) on the drier foothills of Vita Levu and smaller eastern islands. Para grass (*Brachiaria mutica*) was used on the wetter alluvial flats. *Pennisetum polystachyon* was introduced and has now invaded the mountainous areas since the clearing of the rain forest, and has become a fire climax. Koronivia grass (*Brachiaria humidicola*), also introduced, has become useful on hill country (Roberts, 1970b).

The Hawaiian islands. Above an elevation of 1 000 m, temperate species occur. Only shrubs and annual grasses can survive the semi-arid conditions of the lee side of the islands below an altitude of 180 m. *Setaria verticillata, Chloris inflata, C. virgata* (Fig. 15.29), *Panicum* spp. and *Heteropogon contortus* (acting as an annual) provide scant grazing. At an average altitude of 650 m, *H. contortus* is the dominant perennial, with shrubs and various annual species of *Digitaria, Cenchrus, Eragrostis* and *Panicum*. Above this zone, and reaching up to 1 350 m, *Cynodon dactylon* is closely grazed along with the inferior *Chrysopogon aciculatus, Setaria geniculata, Paspalum orbiculare* and *Sporobolus capensis*. In higher rainfall areas at an altitude of 1 200-2 000 m, the indigenous low-quality perennials *Paspalum conjugatum* and *Sacciolepis contracta* occur. In most of these areas the natural grasses have been replaced by crops or sown pasture species (Ripperton & Hosaka, 1942).

Southern United States of America

In the United States, tropical grasses occur only in the southern states. In the heavier soils of Texas various *Andropogon* and *Bothriochloa* species (bluestems) are common warm-season grasses, but these have been introduced. On the acidic, infertile soils of Florida, natural pastures include *Panicum hemitomon, Andropogon stolonifer* and *A. capillipes* with a carrying capacity of 1 animal unit (AU) to 7-15 ha. Farther south on the muck (peat) soils of Florida, St Augustine grass (*Stenotaphrum secundatum*) and carib grass (*Eriochloa punctata*) provide most of the grazing.

The desert grasslands of southeast Arizona, south-central and southwestern New Mexico and southwestern Texas, with a rainfall varying from 300 mm to 450 mm annually, support various native species of *Aristida* (three-awn), *Andropogon* (blue-stems) and *Eragrostis* (love grasses) (Humphrey, 1960a).

Central America and the West Indies

Mexico. In the arid areas of Mexico, shrubs and cacti predominate, with species of *Sporobolus, Andropogon* and annual *Cenchrus* in steppes and sandy areas, and *Paspalum paspaloides, Echinochloa crus-galli* and *Panicum dichotomiflorum* in swamps (Roseveare, 1948). In the savannah grasslands, *Imperata brasiliensis*, species of *Andropogon, Paspalum virgatum* and *Homolepis* provide extensive grazing. Flooded alluvial soils are commonly invaded by *Leersia hexandra*. Where the evergreen forests have been cleared in the high rainfall zones, indigenous *Paspalum notatum, P. conjugatum, P. minus, Axonopus affinis* and *A. compressus* develop.

Panama. In Panama, most of the indigenous grasses have been replaced by sown Jaragua grass (*Hyparrhenia rufa*) which is burnt annually and now constitutes a fire subclimax.

West Indies. The natural vegetation is broad-leaved tropical forest and the grasses are mainly introduced. In Jamaica, one-third of the land is under grass and two-thirds of this area is under low-yielding *Bothriochloa pertusa, Stenotaphrum secundatum, Axonopus compressus, Cynodon dactylon, Paspalum notatum, P. conjugatum* and other *Paspalum* spp. Indigenous pastures with *Stenotaphrum secundatum* and *Paspalum notatum* dominance have a carrying capacity range of from 0.8 to 2 steers per hectare, with an average daily live-weight gain of 0.6 kg (Motta, 1956).

Latin America

Grasslands in Latin America have been dealt with in some detail by Roseveare (1948), who divided them into four types. These are good natural grasslands; cool mountain grasslands; more or less arid grazings in both hot and cold climates; and savannahs of hot climates, alternately excessively dry and excessively wet.

In a book on tropical grasses, details of the cool mountain grasses can be omitted. However, they form a most important ecological and economic unit in many countries where dairying, sheep raising and llama, alpaca and vicuña grazing are undertaken. They include the Andean grazings of central Chile, Peru and Ecuador; the mountain grasslands of Colombia and Venezuela, western Argentina and southern Chile; the upland pastures of the central plateau of Costa Rica; and the *páramos*. These areas are mostly planted to improved temperate grass species, but Kikuyu grass (*Pennisetum clandestinum*) is a major dairying grass in the cool mountain pastures of Colombia (up to 3 500 m) and Costa Rica, and molasses grass (*Melinis minutiflora*) is planted in reforestation zones up to 2 000 m in the Andes.

Good natural grasslands. These areas are wholly or mostly tick-free, with good herbage and a more or less temperate climate without extremes or regular seasons of excessive rain or drought.

The humid pampa of Argentina. The Argentine pampas are the great plains lying south of the Chaco and east of the Andean foothills. Two divisions are usually recognized, the wetter eastern part (known as the humid pampa), and a drier, western part (the dry pampa). The humid pampa includes practically the whole province of Buenos Aires and parts of Sante Fe, Córdoba and La Pampa, comprising some 37 million hectares, mostly of loess.

The grasses are a mixture of temperate and subtropical species belonging to the genera *Briza, Bromus, Paspalum, Panicum, Stipa, Andropogon, Eragrostis, Eleusine, Hordeum, Lolium* and *Poa* with very few legumes, for example, *Rhynchosia, Desmanthus* and *Vicia*.

Uruguay, the purple land. This is how Hudson (1885) described the country. Here the grasses have a faint purplish tinge. The prevailing cover of Uruguay is grass; only 3 percent is woodland. Rainfall is evenly distributed with 950 mm at Montevideo and 1 250 mm in the north annually. Natural pastures occupy 82.5 percent of the country. Overgrazing has converted a large part of the native *Paspalum* spp. to poor quality *espartillo* grasses (*Stipa* spp.). Formerly cultivated arable land reverts to *Digitaria ciliaris, Echinochloa colona, Cynodon dactylon, Paspalum paspaloides* and *Setaria geniculata*.

The south of Brazil. This area comprises the states of Paraná, Santa Catarina and Rio Grande do Sul. At the junction of the states of São Paulo and Paraná, there is an important boundary between the rainfall systems common to the highlands of tropical Brazil — where a winter dry season occurs — and the southern rainfall system, which has rain all the year round. A fairly sharp line running a little south of the town of Sorocaba (23°31'S and 47°17'W) divides the two zones; it is also the northern limit of frosts. Frosts are common on the uplands of Paraná. The river Paraná on the west forms a sharp vegetation boundary to the southern Brazilian grasslands. To the west, scrub with open patches of savannah begins. To the east, the great forest-covered escarpment separates the upland prairies from the coastal lands. From Sorocaba south there is a gradual transition from tropical semi-deciduous forest to pure grasslands; the prairies, which occur in patches throughout São Paulo, Paraná and Santa Catarina, become more extensive than the forests in Rio Grande do Sul south of the Uruguay Valley. Paraná, Santa Catarina and the hill lands of Rio Grande do Sul have partly grazing on natural range and partly improved pastures on cultivated land. In the extreme south of Rio Grande do Sul are the flat herbaceous pastures that rear the best quality cattle in Brazil. The pastures are the *campo* or grasslands. Where the land is manured near the farms, *Paspalum notatum* dominates along with other *Paspalum* spp., *Eleusine indica,* and sedges and small legumes; farther from the towns a short, heath-like community of short legumes, *Zornia, Stylosanthes* and shrubs is common. A tall bunch-grass community of Andropogoneae occupies this area in large communities, including *Elyonurus latiflorus, Erianthus angustifolius, Andropogon incanus* and in poor areas, *Aristida*. Before the summer is over, such swards are burnt and in autumn they are useless for grazing, while the *Paspalum* swards are still fresh and green and *Erianthus* and *Andropogon* are still growing. Cattle raising is the main occupation in this region.

The semi-arid grasslands. *Northeast Brazil.* This area is hot and, for the most part, semi-arid. Thirty-five percent is equatorial and 65 percent subtropical with a winter rainy season and a hot dry summer; 8 percent receives not more than 250 mm of rain annually, 25 percent from 250 to 625 mm, and 65 percent more than 625 mm. Ninety percent of the rain falls between December and April, with poor distribution. Severe droughts occur. Along the coast are limited areas of good rainfall for dairying, but elsewhere cattle are raised. Flat to undulating land consisting of *caatinga* vegetation (spiny woody plants, trees and shrubs) carries a sparse natural grazing of *Aristida adscensionis, Digitaria ciliaris* and cacti, with limited browse from *Cassia* and *Bauhinia* spp. and a few legumes belonging to *Teramnus, Zornia, Centrosema* and *Stylosanthes* spp. The *caatingas* lose their leaf in the dry season but cattle harvest the litter.

Central Mexico. This is a transition zone with a dry and burning heat in summer, excessive cold in winter, high evaporation, violent winds and a shortage of water. *Zea mexicana* and *Tripsacum dactyloides* are used for fodder. Among the Andropogoneae in the dry areas are *Andropogon saccharoides* and *A. glomeratus* of the steppes; Paniceae provides *Cenchrus pauciflorus* on sands, *Paspalum paspaloides* and *Echinochloa crus-galli* in wet areas and *Panicum dichotomiflorum* in the swamps. *Prosopis glandulosa* (mesquite) provides browse.

The campos of central Brazil. These occupy an intermediate position between arid and semi-arid grasslands and are composed of two kinds, the treeless savannah known as *campo limpo* and the savannah with scattered trees, *campo cerrado.* Seasonal aridity in winter is a hazard and summers are rainy and hot.

Savannahs of the hot climates. *The Orinoco flood plains, or llanos of Venezuela and Colombia.* The Venezuelan llanos are vast plains that stretch from the Andes to the Orinoco River, 1 600 km east. The topography rises only 100-300 m above sea-level. Some areas vary from partly arid to almost sterile, while other parts are fertile and very rich. There are vast savannahs bare of trees, dense rain forests or groups of trees in vast meadow lands, and fringe forests. In summer they can be crossed, in winter they are obstructed by water channels and lakes.

The llanos north of the Orinoco occupy an area 960 km by 320 km, forming a rough semicircle from the river Guavire in Colombia to the Atlantic coast of Venezuela; they seldom exceed 200-230 m above sea-level. The rainy season lasts from May to October. The mean annual rainfall is 800-1 800 mm, with 75-150 rainy days. The rainy season is always accompanied by the inundation of vast tracts of land, especially in the vicinity of the Orinoco. The soils are all alluvials varying from gravels to silts.

In the central llanos near Caracas, the grasses consist of *Paspalum* (nine spp.), *Andropogon* (nine spp.), *Panicum* (six spp.), *Eragrostis* and *Sporobolus* (four spp. each), with some legumes species of *Meiboma* (11 spp.), *Cassia* (nine spp.), *Mimosa* and *Aeschynomene* (four spp. each). High grass savannahs are dominated by *Paspalum fasciculatum* with creeping interlacing stems, growing 1-2 m tall. The northern bunch-grass savannah consists of *Cymbopogon rufus* over very large areas, with *Andropogon condensatus* and *Sporobolus capensis* over smaller, but still extensive, areas. In the flatter seaside regions *Panicum elephantipes, Echinochloa polystachya* and *Eragrostis maypurensis* are common.

On the eastern llanos at Santami near the Tigre River are vast areas of *Trachypogon plumosus, T. montufari, Aristida riparia, A. cognata, Paspalum gardnerianum, Axonopus anceps* and *Arundinella hispida*, which are coarse forage and form a fire climax. Mixed with them is the good forage provided by *Andropogon hirtiflorus, A. selloanus, A. bicornis, Elyonurus adjustus, Sporobolus cubensis* and *Axonopus purpusii;* the latter endures extreme drought and spreads by vigorous stolons, but seeds little and is propagated by cuttings.

In the Mesa de Guanipa tableland, at 300 m, three main vegetation types exist. These are the arrow grass savannahs dominated by *Trachypogon* species — not eaten by stock as the sharp points damage the animals — which are burnt each year and constitute a fire climax; the periodically flooded savannahs on the left bank of the Cani River with good grasses dominated by *Paspalum* spp., which are fully submerged during the rainy season (but become arid when the ground dries in summer); and the palm savannahs, dominated by the palm *Mauritia minor,* associated with moist spongy lands which also support six species of *Paspalum* and six species of *Panicum*, as well as the legumes *Desmodium adscendens* and *D. barbatum*.

The Colombian llanos are similar to the Venezuelan llanos. Improved species introduced include *Brachiaria mutica, Panicum maximum* and *Hyparrhenia rufa.*

The Bolivar savannahs of Colombia and the lowland savannahs of Costa Rica. The Bolivar savannahs form a vast lowland plain less than 200 m in altitude, west of the river Magdalena. East of the river, the land is permanently flooded and carries *Leersia, Paspalum* and *Panicum* spp.; to the west, the flooded land dries out during the dry season from October to March and is grazed. The most important grasses are *Brachiaria mutica* and *Panicum maximum* on the non-flooded ground. *Hyparrhenia rufa* is also used on non-flooded ground. *Melinis minutiflora* is used to cover slopes and to suppress weeds. *Paspalum* spp. and *Axonopus scoparius* (a local grass) are also used.

The lowlands of Costa Rica, the Guanacaste grasslands, have an altitude of 0-750 m. The winter wet season lasts from May to November and the summer dry season is from December to April. Annual rainfall varies from 1 000 to 3 000 mm. Stock move from the lowlands to the uplands in the wet season. Natural grasses are of two types, tall *Panicum maximum, Hyparrhenia rufa* and the semi-tall *Brachiaria mutica;* and the short grass, *Paspalum notatum*.

Savannahs of the Amazon Basin and the Guyanas. These are generally dense rain forests, but grass fires in the clearings extend the savannah areas. The average temperature is 25.6°C and the daily temperature range is only 4°C; humidity is high. Rainfall is heavy but there is a so-called dry season from July to December; in the state of Pará there is a longer dry season. The most important Amazon savannahs are in the northeast on Rio Branco toward Brazil's northern frontier and in the lower Amazon reach between the Rio Negro and Rio Xie tributaries.

There are flooded and dry savannahs according to topography. The flooded ones carry mainly grasses, whereas the dry areas support grasses, legumes and other dicotyledons. *Brachiaria mutica* is planted widely.

The undulating upland savannahs of the Guyanas, with an altitude of 100-150 m, have an annual rainfall of 1 500 mm received over four months. They are mainly bunch grass savannahs of *Andropogon, Cymbopogon, Trachypogon, Elyonurus, Paspalum, Arundinella* and *Heteropogon* spp., which constitute a fire climax. There are severe mineral deficiencies in the livestock.

The savannahs of Bolivia, Mato Grosso and the Gran Chaco. In Bolivia the wet savannah llanos occur in Yacuma and Mojos in the northern riverine lands. The eastern llanos of the Gran Chaco are shared by Bolivia, Paraguay, Brazil and Argentina and are subject to ticks.

Three-quarters of the Mato Grosso are extensively and partially flooded natural grasslands; the climate is subtropical with a cold, dry, rainless winter and a hot, rainy summer. The altitude is 116 m in the lowlands and 906 m in the central elevated plain. The most important grasses are species of *Paspalum, Eragrostis, Andropogon, Chloris* and *Sporobolus*. In the uplands *Tristachya chrysothrix* and *T. leiostachya* are dominant. The Gran Chaco is a region of scrub forest interspersed with patches of savannah. It extends north from latitude 30°S into Paraguay, east Bolivia and north Brazil, and forms a lowland plain of alluvium from the Andes, subject to very high temperatures and heavy flooding.

The Cuban savannahs. The whole of the central part of Cuba is savannah alternating with pasture. It is usually flat or undulating, non-arable, grass-covered and often has underlying serpentine rock. The most important cattle areas are in the east-central part on poorer soils but with a mild, semi-tropical climate. *Brachiaria mutica* is planted on the lowlands and *Panicum maximum* cultivated. Other *Panicum, Paspalum (P. plicatulum* and *P. notatum)* and *Dichanthium* spp. are used.

4. Performance and management of natural pasture

Live-weight progress on the natural range

The indigenous tropical grasslands have one thing in common, that the live weight of cattle grazing the pastures increases with age in a step-like direction. Thus, the cattle lose weight in the wet season in the humid tropics, and in the dry season (winter and early spring) in the subhumid tropics. The extent of the loss is greatest during a long dry season.

There is usually a small weight loss immediately after the break of the season, when the cattle seek the young shoots of the rejuvenated grass on burnt country; intake is low and the cattle scour, while the old, fibrous, rain-soaked material is unattractive (French, 1932). Over the following few months there is a rapid weight increase corresponding to the seasonal growth of the dominant pastures, and finally a decline after the pasture begins to mature. This results in animal live weights which rise in steps, gradually increasing over several years (French, 1932; Mawson, 1956) (see Figs 4.1 and 4.2). The live-weight gains reflect not only the availability of pasture, but also grass quality.

Norman (1963b) has illustrated the various factors operating in northern Australia (Fig. 4.3). Cattle grazing native *Themeda australis/Sorghum plumosum/Chrysopogon fallax* pasture begin to gain weight at the start of the main flush of wet season pasture growth in late November and continue to do so until shortly after the wet season in late May. Maximum rates of gain occur in January when pastures are high in nitrogen and phosphorus, and the peak live weight occurs in March, coinciding with maximum dry matter. The live weight then declines, the maximum live-weight loss occurring in November at the end of the dry season. The average net annual live-weight gain is 120 kg/ha. Without supplementary sown pasture and fodder crops, cattle took five to six years to reach slaughter weight.

The actual live-weight gain can be quite high on *Heteropogon contortus* pastures during the growing season and Mawson (1956) recorded a daily rate of up to 1.3 kg per head by a British-type group of cattle in February 1954, and 1.17 kg a head in December 1954. But this level of gain does not last, and

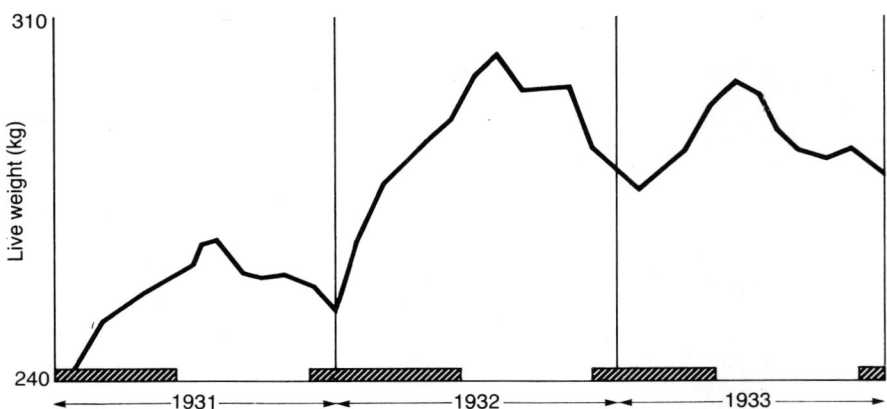

Figure 4.1. Average live-weight changes of native cattle over three years, Mpwapwa, Tanzania (**Source:** French, 1932)

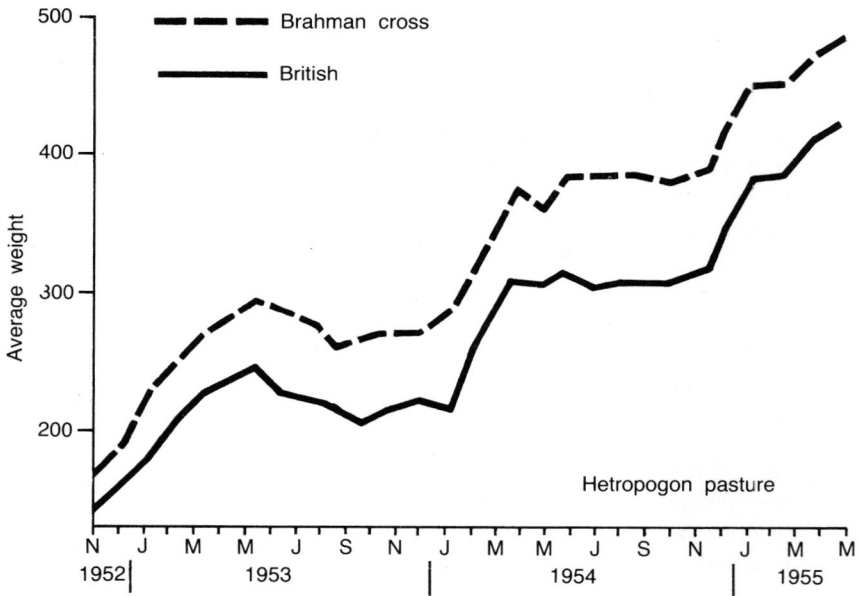

Figure 4.2. Growth rates of cattle grazing *Heteropogon* pasture in northern Queensland, Australia (**Source:** Mawson, 1956)

Figure 4.3. Pasture quality and cattle performance, Katherine, Northern Territory, Australia

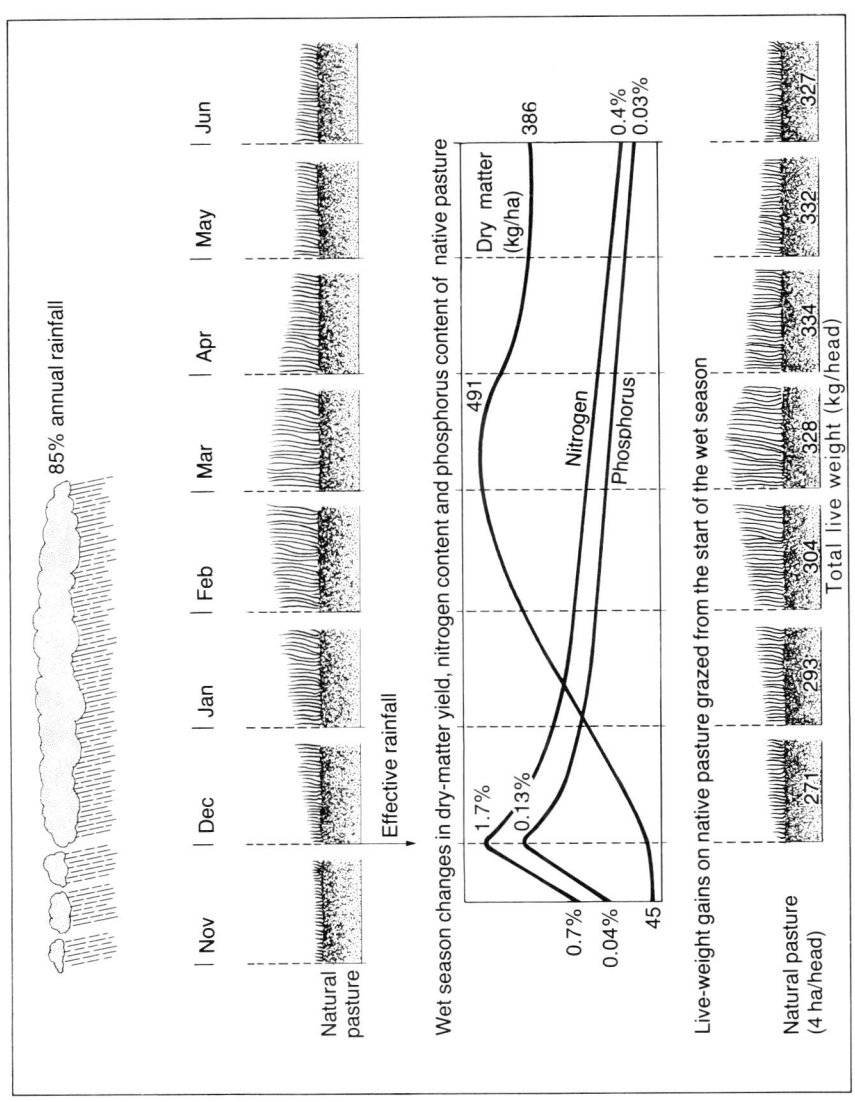

Heteropogon contortus produces 90 percent of its growth during the warmer months between mid-October and mid-April. Its winter contribution is small.

The rapid increase in weight at the beginning of the wet season is enhanced by a high compensatory gain (Fig. 4.4) by animals that are in poor condition prior to the wet season, compared with animals that were better nourished at the outset (French & Ledger, 1957).

Grazing cattle require a diet of 10 percent crude protein on a dry-matter basis and mature cattle 7 percent for maintenance. Early in the season many natural pastures contain this amount, but deterioration is rapid. In frequency-of-cutting trials by Miles (1949) at Fitzroyvale near Rockhampton (on the Tropic of Capricorn in Queensland), it took eight cuts a year to prevent the crude protein level of native pastures from falling to very low levels and even then it remained between 4 and 6 percent, which is insufficient for adequate nutrition.

Management of indigenous grassland

> The object of range management is to maintain the vegetation at the best stage in the succession for the grazing animal, not necessarily the climax, and to ensure the continued vigour of the palatable and nutritious species; the system of management must also safeguard against soil erosion at all times of the year. (Rains, 1963)

It can be seen, therefore, that range management is an integrated system within the soil-plant-animal complex, with emphasis on management of the plant and soil on the one hand, and herd management on the other. None of these elements can be considered in isolation. The basis of plant management — grass in this case — is the provision of rest or recovery periods between grazing cycles, during which the plant builds up reserves that will ensure continued vigour. In addition, the care of palatable species is necessary in the control of the growth of unpalatable species and scrub or bush which compete for light and nutrients.

On any grassland, some plants are more valuable than others; these more useful plants are called "key species". Range managers encourage their growth, and use the key species to measure pasture quality. Their prevalence ensures both the continuity of the sward and the good nutrition of the grazing animals. In the United States some range managers apply a "50-50" rule, that is, to allow no more than half the top growth of the key species to be grazed. This permits the key species to increase in the sward and to crowd out weedy plants. The "50-50" grazing rule does not apply to all the plants in a pasture, but only to the key species. If the pasture is grazed until half of all the forage is grazed, the more palatable species will have been seriously over-used. Periodic checks of the "increasers" and "decreasers" in a pasture species complex will indicate how the botanical composition is changing.

Grasses transfer photosynthates from leaves to roots at any time during

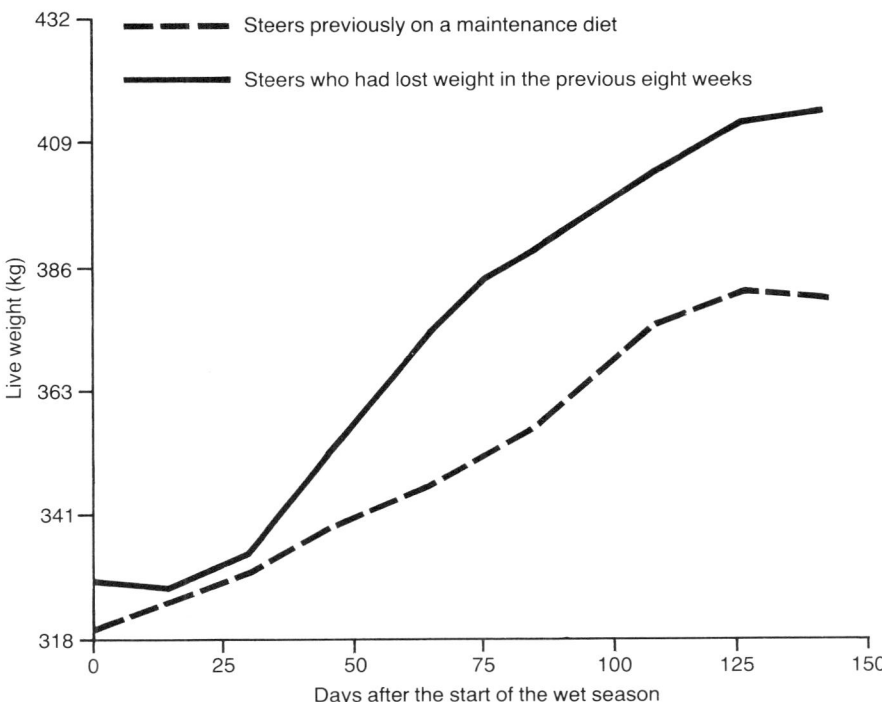

Figure 4.4. Live-weight gains of zebu steers introduced to new season grazing (**Source:** French & Ledger, 1957)

the growing season when the rate of carbohydrate production in the plant exceeds the rate of demand. The root reserves thus accumulated help to initiate new spring growth after the mature above-ground material has been grazed, burnt, or disintegrated. A plant may use up to three-quarters of its root reserves to produce the first few centimetres of new spring growth; it continues to reduce reserves until translocation of reserves back to the roots as the plant nears flowering. Heavy defoliation of preferred species during this active growing period weakens the plants' competitive ability in relation to less palatable and nutritious species, causing them to disappear gradually. This happens in overgrazed rangeland and results in nutritious perennial species being replaced, initially by inferior perennials, then by annuals and finally by deserts. Pasture ecologists, particularly in India, have recorded such sequences when climax grasslands are overgrazed (see Chapter 3).

After such deterioration, the speed of regeneration under protection measures depends upon the residual seed supply of the soil of the climax species, rainfall and the extent of topsoil erosion caused by overgrazing. To prevent deterioration, grazing systems have been devised that incorporate rest periods (called "deferred" grazing) on a rotational basis (see Plates 1 and 2). The aim is to preserve the key species, specifically by:

- Allowing for seed production by promoting the establishment of seedlings which develop from seeds produced;
- Promoting vegetative proliferation by encouraging tiller development and tiller survival;
- Promoting plant vigour either through the development of a large leaf system during periods of a potentially high photosynthetic rate, or by promoting the translocation of carbohydrates to the storage organs;
- Reducing differences in vigour between the preferred (and therefore heavily grazed) plants and the less preferred (and therefore lightly grazed) plants in the sward;
- Allowing for the accumulation of large quantities of herbage, either to provide direct competition for developing tree seedlings or to provide sufficient fuel for a hot burn to kill the developing tree seedlings (Booysen & Tainton, 1978).

Grazing methods. Two major grazing systems are in use for extensive and semi-intensive grassland utilization: continuous grazing and rotational grazing.

Continuous grazing. Early settlers in areas where extensive pastoralism in semi-arid areas was undertaken found the system of continuous grazing of large paddocks, with minimum fencing, to be the cheapest method of grassland utilization. Grazing could be controlled by water supplies and animals could select their diet. Fluctuating seasonal conditions did not permit long-term stocking programmes. Breeding was mainly uncontrolled. In a good season stock numbers would build up; in a bad season a part of surplus stock could be sold and the remainder allowed to spread out, scavenging for the sparse grass, browse and leaf fall. This system currently has many adherents. Numerous trials in several countries have shown that continuous grazing is as productive in terms of animal products as rotational grazing. A key strategy is to control stock numbers in relation to available feed. Roe and Allen (1945) reported the results of a three-year grazing experiment on Mitchell grass (*Astrebla* spp.) dominant grassland at Cunnamulla in southwestern Queensland (average rainfall 350 mm, summer dominance). Continuous year-round grazing with sheep was compared with a six-month rotation, the sheep being confined to one-half of the grazing area during winter and the other half during the summer. Each management treatment was compared at three stocking rates of Merino wethers: 0.33, 0.50 (district average) and 1.00 per hectare. In his summary Roe states that "the total forage available was slightly greater under continuous grazing throughout, and the advantage tended to be greater the heavier the rate of stocking. ... No significant differences in live weight or wool production were recorded for sheep on continuous and rotational grazing at comparable rates."

Similar experiences have been recorded in several other countries. Experience in South Africa has shown that when continuous grazing was combined with a heavy stocking rate, sward degeneration followed because of the over-use of preferred areas, species or individual plants. Thus, ungrazed or lightly grazed inferior species gradually became dominant. In central Australia, Hindmarsh (1980) quotes Lowe *et al.* (1977), who state that arid rangelands are not homogeneous areas that can all be handled within one fence. For example, in one paddock in Alice Springs, 12 different types of country occur, each requiring different management. Table 4.1 gives the major plant communities and their uses.

Hilder and Mottershead (1963) have shown that there is a significant transfer of nutrients from distant grazing areas to the environs of watering points and livestock resting (camping) areas. In a continuous grazing system in extensive areas this can lead to nitrophilous weeds being established in such fertile areas unless species such as buffel grass — which can take advantage of such situations — are specifically sown (Ebersohn & Lucas, 1965). Similar cases of nutrient transfer occur in Africa where the night kraaling of animals is common. The heavy trampling of such areas by concentrations of livestock may finally denude them of all vegetation. This is common around

TABLE 4.1 **Plant communities and grazing performance in one paddock, Alice Springs, Australia (average paddock grazing density – 3.5 beasts/km^2)**

Plant community	Description	Percentage of area	Percentage of cattle carried	Grazing density (beasts/km^2)
Hills	*Acacia* shrubs and short grasses	8	5	1.9
Foothill fans	Short grass under scattered trees	5	12	8.0
Flood plains	Transient short grasses	13	24	6.6
Gilgaied plains	Perennial grass	5	7	5.5
Woodland	Short grass under scattered trees	13	25	6.9
Mulga (annual grass)	Scattered mulga over short annual grasses	27	20	2.4
Mulga (perennial grass)	Mulga trees in groves over perennial grass and shrubs	26	3	0.4

permanent watering points in semi-arid and arid areas. For example, in the treeless central Queensland *Astrebla* grasslands, the sheep gradually congregate in the southeast corner of paddocks, as animals (and particularly sheep) prefer to graze upwind. Since the prevailing wind is from the southeast, these southeast corners are usually denuded by camping and trampling.

Thus under a continuous grazing system, while a variety of forage will give animals an opportunity to select a more nutritious diet, and while labour and fencing requirements are minimal, unless an intelligent policy of light stocking (or "put-and-take") is adopted in relation to animal numbers, the long-term results may be far from satisfactory.

In India, the continuous grazing of the arid *Lasiurus-Eleusine-Aristida* grassland in Rajasthan has given significantly higher live-weight gains than other treatments. The next-best production was from deferred rotational grazing which was considered more reliable as it gave steadier, more consistent and less fluctuating live-weight gains (Prajapita, 1970).

Rotational grazing. This system involves the successive grazing of a number of enclosures so that at any given time the animals are rotationally concentrated on a portion of the total grazing land, usually by fencing. The main objectives of the system are to maintain high stocking rates and as complete a utilization of the available material as possible; maintain sward composition through a reduction in the selective use of preferred areas, species and individual plants; maintain sward height and presentation of more nutritious feed; and improve animal management including classifying and segregating stock, breeding and parasite control (such as against tick infestation), and easier mustering and provision of specific animal needs.

In South Africa, ecological studies have shown that the rotational resting of paddocks for specific purposes — such as seeding, or the translocation of nutrients and the restoration of vigour and productivity — should be incorporated in rotational grazing systems. Rotational grazing involves a greater financial investment in fencing and watering facilities than does continuous grazing.

While pasture workers in South Africa have investigated "multi-camp" rotational grazing systems for many years, their productivity has not been compared with that of continuous grazing. Two systems have now been adopted (Booysen & Tainton, 1978): the HPG/CSG system (high production, high performance grazing system, incorporating controlled selective grazing), which emphasizes light use to retain sufficient actively photosynthesizing leaf on the pasture at all times to maximize regrowth, and the HUG/NSG system (high utilization, non-selective grazing), which emphasizes as complete a utilization of available herbage as possible before animals are moved to the next paddock in the rotation.

The authors' rationale for these two philosophies may be summarized as follows. There are two ways to stimulate increased production: *a*) by increasing the utilization of available herbage, which can be achieved by severe defoliation of all plants by severe grazing; and *b*) by increasing the growth rate of individual plants, achieved by incomplete defoliation of even the preferred plants of the sward by light grazing. It should be noted that short grass is good grass and that long grass is neither palatable nor nutritious (Hassall, 1976).

Similarly there are two ways in which sward composition can be maintained: by the forced utilization of even the less preferred plants, which can be achieved by intensive severe grazing; and by a deliberate under-utilization of less-preferred plants to induce them to become moribund. This can be achieved by the moderate defoliation of preferred plants and under-utilization of less-preferred plants.

Booysen and Tainton suggest that for climax grasslands of the more arid regions, HPG might be expected to perform better than HUG as

– these areas are subject to frequent periods of drought so that recovery from heavy grazing may often be slow and the grass may not recover by the time the paddock is again due to be grazed;
– the grasses of these areas are usually sweet, and thus still acceptable to animals when mature, so that the residual material from one grazing cycle may still be readily grazed in the subsequent cycle;
– these grassland types are inherently sensitive to misuse because their cover breaks down readily and soil loss may be excessive; and
– successional development is slow and does not lead to the replacement of the climax grassland community by less useful communities even under light use.

On the other hand, the HUG system seems to be generally more appropriate in the fire/grazing climax grasslands of the humid regions, where

– summer droughts are infrequent so that in most seasons recovery is assured between successive grazing periods;
– the grasses are predominantly sour (often unpalatable and of low nutritive value), so that the residual material from one grazing cycle is not readily grazed by animals during subsequent cycles;
– the grasslands are moderately to extremely resistant to heavy grazing pressure, provided that intermittent rests are incorporated in the system to allow for recovery; and
– the fire/grazing subclimax grasslands are readily replaced by less useful tall grass and scrub communities if the canopy is allowed to remain dense for any length of time.

The general approach to veld management should be dictated by *a*) the degree to which species selection occurs and *b*) the susceptibility of the veld to overgrazing. Together these indicate whether the emphasis should be placed on rotational grazing (where selection is serious), or on rotational resting (where denudation is a threat). It has been shown that over a long period rotational resting results in increased production for each animal and a higher carrying capacity in the sweet veld areas (Roberts & Opperman, 1966).

The use of fire in grazing management. Fire is a convenient and simple tool for controlling both the growth and botanical composition of pastures. In most tropical grazing land ecosystems, management is aimed at establishing, or maintaining, a permanent and productive grass cover in areas where the climax vegetation is some type of forest, woodland or shrub community. Grass fires play a large part in maintaining the open nature of these pastures; a reduction in fire frequency or intensity resulting from using the grass as forage rather than as fuel normally leads to an increase in woody species (Coaldrake, 1961). The problem of bush or brush encroachment is widespread in the tropics whenever an increase in the utilization of rangeland resources is attempted. Africans used fire more than 50 000 years ago (Oakley, 1956; 1962). Primitive peoples used fire to flush out game and smaller animals for food or protection; to destroy old, dried vegetation in order to encourage more nutritious growth for livestock and to improve visibility, collect seeds and clear debris from cropping land, especially in shifting cultivation; and for warmth and cooking. Modern societies have attempted to harness fire as a tool in range management, as it is inexpensive and, under suitable climatic conditions, effective.

Fire can be caused by lightning, a common occurrence in semi-arid grasslands where "dry" storms occur. Lightning-induced fires occur frequently in the *Astrebla* (Mitchell grass) grassland of central western Queensland. However, most fires are started by people, either on purpose or accidentally.

Accidental fires are few in the evergreen forest zone and rarely get out of control. In shifting cultivation these forests are cleared and fire used to consume the debris for each cropping cycle. Often, the end product is a fire-induced grassland climax (for example, mission grass, *Pennisetum polystachyon* in Fiji). Desert and desert steppe areas are little affected by fire because of their scattered vegetation.

Most susceptible are grassland and savannah zones where hot rainy seasons, followed by prolonged periods of drought, not only produce large quantities of highly combustible grassy fuel, but also provide ideal conditions for the spread of fire. In Africa, these conditions are made worse by the desiccating effects of the harmattan wind which blows out of the desert during the dry season. Short-grass communities that are kept short by the heavy grazing of livestock are less affected by fire (Lamprey, 1979).

Fires occurring annually throughout savannah areas over a long period tend to encourage tree species selectively that are fire-tolerant because of their corky bark (for example, Australian *Eucalyptus* spp., *Balanites aegyp-*

tiaca and *Acacia seyal* in Africa's Sudan zone, and *Acacia brevispica* and *Tarchonanthus camphoratus* in Kenya); because they reproduce vegetatively by means of suckers from ligno-tubers (for example, the Australian brigalow, *Acacia harpophylla*); or because they shed abundant seed, the germination of which is stimulated by the passage of fire (for example, *Acacia flavescens* in Australia). Fire also selectively influences grass survival. The annual burning of *Dichanthium-Bothriochloa* grasses in eucalyptus open forest in central coastal Queensland has led to the dominance of *Heteropogon contortus* (black or bunch spear grass) (Shaw & Bisset, 1955). In the Sudan frequent fires promoted the growth of *Andropogon gayanus, A. amplectans, Cymbopogon giganteus* and *Hyparrhenia* spp. Under protection these rapidly lost vigour and were invaded by an annual *Pennisetum* sp. which became dominant after another three years. Skovlin (1971) showed that the survival of important range grasses after burning in East Africa was high (up to 95 percent). Burning reduced the crown area, herbage weight and seed number. *Chloris roxburghiana, Digitaria milanjiana* and *Panicum maximum* were particularly affected, whereas *Pennisetum mezianum* and *Themeda triandra* were little damaged by burning and *Hyparrhenia lintonii* appeared to be improved by burning. However, more research on the reaction of species to fire is needed.

The basic facts upon which range management must be built concern the differential response of various grasses and tree species to burning at different times of the year, and the grazing requirements of the grazing animal at specific times during the year.

Observers are generally agreed about the gross effects of fire on West African vegetation (Rains, 1963; Rose-Innes, 1971).

- It is accepted that fire retards the natural development of vegetation toward a forest or woodland type of climax and may hold it more or less permanently in a tree savannah stage.
- Early burning at the end of the growing season tends to damage perennial grasses which are still partially green and have not yet returned all their food reserves from the leaves to storage in the roots. Should they be induced by fire to sprout again out of season, they must do so at the expense of partially replenished root reserves. Finally, if the resulting regrowth is grazed, root reserves are even further depleted; the most vigorous of plants die in the course of a very few years of such treatment.
- Moderate, "cool" fires occurring early in the dry season cause minimal damage, but if the regrowth is then grazed, plants are seriously weakened. Grasses seem to have a fairly constant "period of sprouting", ranging from six to ten days (Hopkins, 1963). Few long-term ill effects have been found in northern Australia (Perry, 1960).
- Savannah trees in Africa usually sprout well before grasses and often long before the beginning of the rains. They are most sensitive to fire damage at

this time, whereas grasses are dormant and escape harm. Fire promotes sprouting in woodland savannah trees and there is a fairly constant sprouting period, varying from 18 to 30 days for trees (Hopkins, 1963).
• Fierce, hot fires occurring late in the dry season are particularly destructive to woody vegetation, especially seedlings and suckers, except the most fire-resistant species. In the semi-arid woodland zone of the eastern Serengeti Plains (Tanzania), withholding of fire is necessary for the regeneration of the browse trees *Acacia drepanolobium, A. tortilis* and *A. xanthophloea* (Lamprey, 1979). However, the reaction of grass species to burning is important to maintain feed quality. In the north Guinea savannah *Andropogon gayanus* is encouraged by early, rather than late, dry season burning; the latter encourages the coarse, unpalatable *Loudetia acuminata*.

Hopkins (1965) showed that ground-level temperatures in both late and early dry-season burns invariably exceeded 538°C, but dropped at a height of 10 cm, quite sharply in the case of early burns, more gradually in late burns. Early burns did not generate temperatures of more than 66°C above a height of 3 m, but late burns exceeding 538°C were sometimes recorded up to 3 m above ground and temperatures over 100°C were consistently recorded to heights of more than 6 m. Wind speed was an important factor in increasing the severity of the fire.

Grass fires last only a short time and high temperatures are maintained for only a few seconds. Temperatures at soil level rise very steeply to between 100 and 850°C depending on wind, height and density, and usually return to ambient temperature within a few minutes. The soil temperature at a depth of about 2 cm changes little, varying at most by 14°C, but more often as little as 3-4°C or less (Pitot & Masson, 1951). The effect on subterranean portions of grasses is thus probably slight. Seedlings and saplings of all tree species are subject to heavy damage by fire (e.g. *Stoebe vulgaris* in the high veld of South Africa) (Krupko & Davidson, 1961) because they grow close to the ground in a sphere subject to the highest fire temperatures and thus are more likely to suffer the loss of apical buds and destruction of the cambium than well-grown trees of the same species (Hopkins, 1963).

One of the main problems in the use of fire in controlling vegetation is the control of the fire itself. Fire breaks are essential and these can be most conveniently provided by cultivating strips of ground to contain the fire within specified limits. The cultivation of two strips about 40 m apart gives added safety in grasslands, the centre grass strip being burnt under supervision. Where labour is not costly, hand cultivation can be carried out. The chemical desiccation of breaks, and firing while the surrounding vegetation is still green is effective but costly and is commonly used in developed countries for protecting power lines or forest areas.

Burning frequency is governed by climatic factors and the specific type of pasture and bush involved, and can be determined only by long experience with local conditions. Burning programmes in East Africa are based mainly on intervals of from one to seven years between burns, the frequency tending to decline with decreasing annual rainfall.

In northern Australia, Norman (1969) found that the frequency of burning and the rainfall of preceding years markedly affected botanical composition and could act as a guide to a burning schedule. He also found in Katherine that biennial burning was more beneficial than annual burning (Norman, 1963a).

Van Rensburg (1971) states that "herbage should be utilized for forage, bedding, compost, packing or other useful purpose. If it cannot be utilized purposefully, redundant herbage should be disposed of by controlled burning at the most propitious time. ... Controlled burning disposes of unpalatable old herbage, controls encroachment of undesirable woody plants, destroys parasites and produces a fresh, healthy sward. It also aids better distribution of animals, reduces fire hazards, stimulates seeding and helps to prepare a favourable seed bed." It is true that where there is a concentrated rainy season followed by a well-defined dry season, the wet season growth far exceeds livestock needs; this produces hay of low quality and presents a fire hazard in the dry season.

Tothill (1971) states that "in the central coastal area of Queensland, Shaw and Bisset (1955) have shown that 2 200-3 300 kg/ha of dry matter of less than 3 percent protein content remains on annually burnt grazed pasture at the end of the dry season." Normal grazing accounts for only from one-third to one-fifth of the total seasonal production of herbage. Miles (1949) reports 5 000 kg/ha of dry matter produced on unburnt pastures with a protein content of 2 percent. Norman and Wetselaar (1960) at Katherine, Northern Territory, report 1 320 kg/ha of dry matter of 2.5 percent protein from annually burnt pasture. This surplus poses a problem. In most privately owned ranches it is impractical to keep sufficient seasonal livestock numbers to use seasonal production effectively. In communal grazing lands, sufficient livestock may be present to utilize seasonal production, but these cannot or will not then be sold, which results in overgrazing and starvation in the subsequent dry season. Burning this accumulated dry matter is a waste. About two-thirds of the energy contained in non-woody aerial shoots is lost to the biological system through annual fires (Hopkins, 1963). Mowing stimulates growth better than burning (Steinke & Nel, 1967; Shaw, 1957), but mowing is costly, and often inefficient or impossible. Mowing and collection as conserved fodder may not prove economic because the product will often be of low quality and suffer further losses in storage. Central Queensland cattle ranchers used to say, "No burn, no fats", because the regrowth from areas of *Heteropogon contortus* burnt in the spring and responding to stored soil moisture or spring rains was of good quality and fattened cattle quickly.

More recently, as a result of basic work by Shaw (1957) in central coastal Queensland, and later by Stocker and Sturtz (1966) in the Northern Territory, Stobbs (1969a, b, c) in Uganda, Partridge (1975) in Fiji, and in numerous other areas, it has been found that native pastures can be oversown with pasture legumes to increase carrying capacity greatly and reduce the need to burn. Indeed, burning is detrimental to such improved pasture.

In grazing with sufficient numbers of cattle or goats, the young growth of most woody species in Queensland coastal areas (for example *Eucalyptus* and *Acacia*) is kept in check.

Tothill (1971) sums up the role of fire in range management in the future: "A system of strategic burning coupled with planned conservation of standing dry roughage is likely to be the best course of action for those lands not being developed to any significant extent. For such a system to be effective fire must be controlled on a regional basis by organized grazier cooperation", to which could be added, or under adequate supervision by government range management officers or local leaders where the individual ownership of land is not a practice.

5. The case for improved pastures to replace indigenous species

As discussed in Chapter 4, the live-weight performance of cattle grazing indigenous pastures is one of annual live-weight gains (LWG) during the growing season, with no gain (and sometimes substantial losses) during the dry season. Animals thus take a long time to reach a marketable weight for slaughter. Similarly, annual milk production shows uneven levels.

However, in subhumid and humid areas, rainfall is sufficient to support improved pastures that not only are adapted to higher rainfall, but will respond to applied fertilizers. In this way much higher yields and better animal performance can be obtained; with the selection of compatible and effectively nodulated legumes that will grow in association with grasses, more economical and highly nutritious pastures can be developed (Fig. 5.1).

With the decreasing area devoted to pastures in the wake of advancing cultivation in subhumid and humid areas, it is imperative that the productivity per unit area of pasture be improved. Table 5.1 shows such a response in Queensland in a subhumid environment. In Puerto Rico the yields from native pasture, a grass-legume mixture of *Pueraria phaseoloides* (puero or tropical kudzu), and Merker grass (*Pennisetum purpureum* var. *merkeri*) were compared. The results are shown in Table 5.2.

The unfertilized native grass yielded only one-third as much as *Pueraria* mixed with *Melinis minutiflora* (molasses grass), and just over one-fifth as much as the Merker-type elephant grass. With fertilization the ratio between the yield of native grass and the molasses grass/puero mixture remained the same, but the protein content of the grass-legume mixture was some 50 percent higher.

At Rodd's Bay, Queensland (latitude 23°50′S, rainfall 813 mm per year), the natural spear grass (*Heteropogon contortus*) dominant pasture averaged a live-weight gain of 25 kg/ha over a seven-year grazing period from 1959-66 (Shaw & 't Mannetje, 1970). With the addition of the annual legume *Stylosanthes humilis* and fertilization with phosphorus, potash and molybdenum, it produced 148 kg LWG/ha. The progressive improvement results from the control treatment are given in Table 5.3.

The LWG per head and per hectare from oversowing *S. humilis*

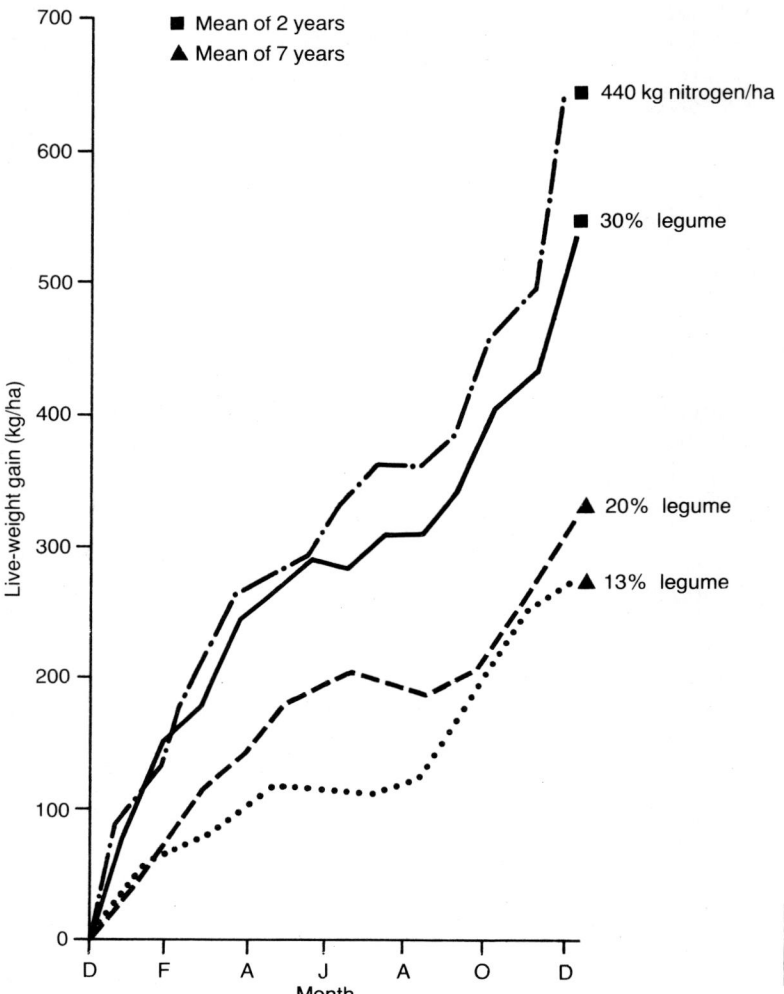

Figure 5.1. Cumulative live-weight changes in grazing pastures with varying percentages of legumes and a pure grass sward fertilized with nitrogen, Beerwah, southeastern Queensland, Australia (**Source:** Bryan & Evans, 1967)

(Townsville stylo) were considerably higher than the fertilizer treatment on natural pasture. The sowing of *S. humilis* was also a once-only operation, while fertilizer treatment is a recurrent annual expense. The *S. humilis* plus phosphorus, potassium and molybdenum treatment resulted in a six times higher live-weight gain per hectare than that from the control treatment. The

TABLE 5.1 **Comparative performance of native and improved pastures, Queensland, Australia**

Type of pasture	Fertilizer applied (kg/ha)	No. breeding cows on 6 ha	Change in body wt of cows (kg)	Av. wt of weaned calves (kg)	Total calf wt from 6 ha (kg)
1. Native	0	1	−23	117	177
2. Green panic + siratro	125	3	−12	210	630
3. Green panic + siratro	125	4	+11	219	876

Source: CSIRO, 1973

TABLE 5.2 **Comparative performance of native and improved pastures, Puerto Rico**

Treatment/ha (lime to pH 6.5)	Yields of dry forage/ha			Yields of crude protein/ha					
	Native pasture (*Axonopus* dominant) (kg)	Puero and molasses grass (kg)	Merker grass (kg)	Native (kg)	Native (%)	Puero and molasses grass (kg)	Puero and molasses grass (%)	Merker grass (kg)	Merker grass (%)
Control	4 581	13 877	20 418	353	7.7	1 607	11.6	1 635	8.0
Lime + 224 kg P_2O_5	5 986	19 270		504	8.4	2 408	12.5		
Lime + 224 kg P_2O_5 + 224 kg K_2O	7 028	20 238		594	8.4	2 861	13.9		
Lime + 224 kg P_2O_5 + 224 kg K_2O + 224 kg N	8 344				9.2				
Average	6 485	17 797	20 418	554		2 318		1 635	

TABLE 5.3 **Progressive live-weight gains with increasing intensification of stocking rate**

Treatment	Stocking rate (beasts/ha)	Live-weight gains	
		(kg/head)	(kg/ha)
Control	0.30	83	25
Control + P, K, Mo	0.62	100	62
Control + *S. humilis*	0.78	121	94
Control and *S. humilis* + P, K, Mo	1.00	148	148

great effect of the *S. humilis* treatment was due to the improvement of the available grazing material and the increased yield of pasture resulting from increasing nitrogen in the soil-plant-animal system (Dirven, 1970).

It is expected that in the 1980s and beyond fertilizer and seed prices will rise significantly. The cost input-output relationship will therefore have to be studied carefully to determine the most desirable amount of fertilizer to apply. A move toward the use of a more efficient species of grazing plants will be of paramount importance.

6. Pasture improvement by introducing new species

The decision to embark on improved pastures must be weighed according to a number of factors. Pasture establishment is generally costly in terms of land preparation, fertilizers, seed supplies and fencing. In terms of utilization, decisions must be arrived at carefully regarding the proposed animal production enterprise, the type of animals, and availability and cost.

A common fault, particularly in high rainfall areas, is to attempt to establish large areas of improved pastures too quickly, which leads to a high initial outlay, with high repayment costs and poor preparation likely, indifferent establishment and a timber regrowth problem. Teitzel, Abbott and Mellor (1974) have listed the following procedures for success.

- Attempt in each planting season only the area that finances and resources will allow to be done thoroughly.
- Consider all possible hazards, such as weeds, timber regrowth, competition from native grasses, dry spells, fire and pests, and prepare for them as soon as possible.
- Treat each type of country as a separate area for establishment.
- Aim at a fine, firm, weed-free seed-bed.
- Apply adequate fertilizer.
- Inoculate legume seed.
- Sow sufficient good, fresh seed of the recommended species for the class of country involved.
- Sow accurately.
- Sow into a moist seed-bed when follow-up rains are reasonably assured.
- Try, as far as possible, to cover the seed.
- Graze carefully during the first year, particularly in wet weather.

Land clearing

Types of country to be cleared. The types of country most suitable to be cleared are:

– dense tropical rain forest, which usually includes an undergrowth of vines, and lands that, although growing in a very wet environment with frequent rainy days, are moderately well drained;
– monsoon forest that is still dense, somewhat deciduous, and the soils well drained, although rainfall is less frequent;
– wet sclerophyll forest where there is little undergrowth but fairly dense tall timber, swampy areas of various sizes from large ponds and swamps to small gilgais, and frequent rainy days;
– wet heath country, also subject to waterlogging and frequent rain;
– dry sclerophyll forest, generally well-drained, with infrequent rain (not enough to hold up clearing operations) and varying tree populations;
– savannah woodland of variable rainfall and less tree density;
– low thornbush savannah in low- to medium-rainfall areas. It is dangerous to clear semi-arid shrub steppe because of the danger of wind erosion and loss of grazing.

Timber pulling. In most instances timber will have to be removed to prepare the land for seeding, although open grassland and old and new cultivations will need only ploughing and seed-bed preparation.

The removal of large trees in rain forest, monsoon forest and wet sclerophyll forest is now almost universally accomplished in areas larger than a few hectares using bulldozers. These large machines are essential for the time they save. The most economical method is to use two bulldozers abreast linked by a heavy ground chain which does the pulling. In some countries a third tree-dozer is brought into operation. It has a highly set blade and can give an extra push to hard-to-remove trees by following the two bulldozers pulling the chain.

In open forest country a "highball" (an old naval buoy attached to the middle of the chain by swivel lines) is necessary to keep the chain near the ground so that small trees do not escape being pulled. In a tropical rain forest a highball is not needed. In the humid tropics, where there is very little land still clothed with tropical rain forest, it is better to let it remain for national parks, wildlife sanctuaries and as a modifier of climate, than to destroy it. Even selective harvesting for building timber can cause damage. A vigorous reforestation programme should be in operation to grow the specific species required in plantations. There is sufficient other land that has been heavily overgrown with secondary vegetation to rehabilitate for permanent pastures. In subhumid areas receiving from 875 to 1 500 mm of annual rainfall with a well-defined dry season, there are still areas of savannah and forest that contribute little to timber resources but that can support productive improved pastures.

On land being cleared for pastures some timber should be left undisturbed either in strips, which is easier for bulldozers, or in strategic clumps for shade, shelter, wildlife refuges or for fencing materials. Stream banks and gullies should also be protected against erosion by an adequate gallery of timber. Aerial photographs of the area to be treated are useful in planning land clearing operations.

Contract specifications. Most large-scale land clearance schemes are based on contract arrangements with private operators. If sufficient operators are available, competition will favour arriving at a realistic cost.

Specifications for land clearing should be made carefully and clearly in terms of requirements, limitations, and exactly what is expected of the clearing operator, and should include *a*) a description of the present condition of the land; *b*) a description of the finished condition required, specifying maximum and minimum limits; *c*) the completion date and method of payment; and *d*) provision for a final inspection to ensure contract compliance.

General instructions, such as "removal of all vegetation", should be replaced with precise terms, as "removal of all material of *c*. 10 cm in diameter to a depth of 20 cm below the surface". Aerial photographs, maps and drawings will give locations; size of area; contours types and size of vegetation; soil conditions (types, stony land, gullies, drainage conditions and water table); the location of streams, rivers, present and planned roads and drainage works. A rainfall incidence chart indicating the average number of wet days is also helpful.

The most important detail is tree counts, which should be made for each type of vegetative growth present in the area. Besides the average number of trees per hectare, the count should include the size (diameter at breast height or immediately above buttress) and species of trees and other vegetation, especially vines. This tree count can best be obtained by a random sampling of perhaps three areas, each 100 m long by 10 m wide. The "finished condition required" stipulation should include the size and portion of trees to be left standing as individual trees and as belts of trees for shade, timber windbreaks and stream-bank protection and wildlife sanctuaries; the degree of stump and root removal required, and of debris to be removed; and the amount of submarginal land that may be left uncleared.

The contract should stipulate the starting time, time for completion of portions of the work, and for final completion of the job. The number of working days required for completion should be gauged in order that the contractor not be penalized for adverse weather conditions.

Other considerations. The following points should also be taken into account.

• The job should be of a size to permit the use of specialized equipment, efficient methods, competent management and supervision. A project large enough to write off a suite of machinery after completion is an advantage.

The distance from maintenance depots can also seriously affect costs and should be considered.
- The contract should be divided, if possible, into categories such as clearing, logging and road construction, assigning contractors according to their particular specialization.
- The experience, reputation and financial solvency of bidders should be investigated.
- Both vague and unnecessarily strict specifications should be avoided as these can cause a contractor to be less efficient.
- Unnecessary, submarginal land should not be cleared.
- Land should not be cleared until arrangements have been completed for effective utilization and management.
- Wherever possible, a pilot project should precede the work in order to determine the most efficient and economical methods; this will help with future budgeting.
- Arrangements should be made to harvest all millable timber before clearing commences.
- The roads used to initiate the clearing should be located so that some can be used for permanent roads, boundary fence roads, and fire-breaks.

Clearing rain forest should begin in the spring to allow the trunks and branches to dry sufficiently to take a burn, but not so soon that the leaf will be lost or that sucker regrowth occurs. When clearing open forest, the time of clearing is not very important as most forest species are climax species, most of which will be somewhat fire resistant and will not burn readily. Clearing is generally best done after the wet season when the ground is soft.

Wind-rowing. Where the timber is dense enough to take a running fire, it is usual to burn where it has fallen and as soon as it is sufficiently dry. However, since with some species, such as *Acacia harpophylla* (brigalow), an early burn without subsequent cultivation has often resulted in excessive regrowth, a recommendation has been made not to burn until nine months after pulling (Johnson, 1965). Thus, a knowledge of the species involved is necessary for intelligent management. Burning *in situ* prior to wind-rowing takes advantage of the abundant dead leaf material that may be lost in the wind-rowing operations, and fire consumes the great bulk of woody material so that there is less to deal with after the burn. However, waiting for a burn delays land preparation, a running fire destroys valuable humus, and some species (such as *Eucalyptus* spp.) are difficult to burn with a running fire. Where these factors are a consideration, wind-rowing can begin immediately and the material can be burnt in the wind-rows later.

The overriding factor favouring wind-rowing lies in the possibility of using a wheeled tractor, thus allowing easier control of subsequent weed invasion. Deaths of cattle from leg injuries and the difficulty of mustering for stock management can be costly and time-consuming if wind-rowing is not carried out.

Wind-rowing is best carried out with a heavy tine rake rather than with a bulldozer blade, as the rake allows the fertile topsoil to move between the tines and be left in place, and the weight of the debris moved is less (see Plate 3). These scrub or bush rakes are of varying strength and widths; some are designed to carry timber as well as to push timber together into wind-rows. However, in heavy, wet clay soils, rakes can clog up; although a bulldozer with an angle blade may ride more easily over the land, it is better to wait until the soil will pass through a rake. The distance between wind-rows affects costs and should not exceed 140 m. Wind-rows should be arranged on the contour of the land as protection against erosion until the pasture cover is established. Two scrub rakes working side by side usually accomplish wind-rowing more easily. Dead running can be avoided if wind-rowing is done in a forward step-wise direction, reversing the movement when the end of the field is reached (see Fig. 6.1).

Root-ripping. This operation is mainly for clearing land for cultivation. Tractors fitted with heavy tines on a toolbar rip the soil to a depth of 20-45 cm to remove shallow roots in advance of ploughing.

Root-raking and stick-picking. Following burning and ripping there will always be short pieces of timber that have escaped all previous treatment and present a hazard to the operations of ploughing, cultivation and to seeding machinery. Small pieces can be picked out by hand or raked together with root rakes and stick-pickers, and subsequently burned. The Wake root rake operates like a hay-tedding machine and sweeps the sticks into wind-rows with a side delivery action, to be picked up and burnt *in situ*.

Clearing savannah woodland. The population of trees in savannah woodland may not at first sight be high enough to warrant using two bulldozers and a chain; a single bulldozer operating against each tree may be all that is necessary with individual trees being burnt *in situ*. However, there is a lot of time wasted between trees when the tractor is not working, and it may be worthwhile to use two tractors and a chain and to wind-row the trees into heaps.

Clearing low thornbush and heath country. Where low thornbush and heath country is to be cleared for pasture improvement, a root-plough can be used; this machine shears off the roots 30-45 cm under the ground with a serrated shear blade of wide pitch and leaves the material above ground untouched. At King Ranch, Texas, a root-plough has been used to treat mesquite (*Prosopis glandulosa*) country. Two large track-laying tractors have been mounted abreast with about 1 m clearance and fitted with a tree pusher at the front end reaching 5 m high, with an angle-dozer blade at ground level. The

Tropical Grasses

Figure 6.1. Wind-rowing without dead running

machine straddles the brush, the angle dozer clears the brush ahead of the tracks so that they traverse flat ground, and the root blade is attached behind to cut the roots (see Plate 4). In heath country, ploughing with a heavy duty stump-jump plough (such as the Shearer "Majestic") may be sufficient (see Fig. 6.2).

Establishment of grass in pure and mixed swards in areas of adequate rainfall

Areas receiving less than 625 mm annual rainfall constitute a risk when introducing new species and will be dealt with in Chapter 10, "Reseeding the arid and semi-arid range".

Grasses may be established with varying (and usually limited) success in areas that receive about 625-650 mm of rain annually by oversowing into natural pastures without cultivation and without fertilizers. Rhodes grass (*Chloris gayana*) has been successfully established in ring-barked brigalow country in Queensland, Australia (>625 mm rainfall) simply by broadcasting it from horseback. While stocking reasonably heavily with cattle immediately afterwards helps to press some seed into the ground, the rate of establishment depends on the incidence of subsequent good rain. If seed is abundant it can be sown this way. However, the seed of improved species is usually too costly to be broadcast under conditions where success is rare.

Sod-seeding. Pasture legumes can usually be successfully sod-seeded into grass swards when supplied with adequate phosphorus. Gamba grass (*Andropogon gayanus*) was seeded in heavily grazed native pasture at Carimagua, Colombia, by row-seeding with hand-applied phosphorous fertilizer to beat weed competition. Low-density planting in fertilized hills was effective and cheap. *A. gayanus* developed from planting 1 000 plants per hectare to a population of >150 plants/m^2. The important factor is not to fertilize an area until the introduced grass is well established and then to apply fertilizer between hills (CIAT, 1978).

Establishment in the ashes after burning timber. Where the original timber was thick enough to provide a clean burn and where there is an accumulation of ashes, grasses can be successfully broadcast (generally by aerial seeding) and established if subsequent rains of sufficient volume are received soon after seeding.

Thousands of hectares of Queensland's brigalow (*Acacia harpophylla*) and gidgea (*Acacia cambagei*) forest and scrub woodland have been pulled and burnt, and grass species successfully established in the ashes (*Chloris gayana, Panicum maximum* var. *trichoglume, Cenchrus ciliaris* and *Sorghum almum* in the brigalow country, and *Cenchrus ciliaris* and *Panicum antidotale* in the gidgea country).

Specially adapted seed-boxes can now handle buffel grass seed alone. The base of the seed-box protrudes slightly below an aircraft's fuselage and can be lowered so that the slipstream will continuously sweep away thin layers of seed. The aircraft flies at 100 m during seeding. Seed of at least 90 percent purity is essential with this type of box. An aircraft can sow 1 200-1 600 hectares a day (Paull, 1973). Seed can also be broadcast by fertilizer distributors.

The soils are grey and brown alkaline clays, annual rainfall 625-750 mm in the brigalow area and 400-500 mm in the gidgea area. Fire destroys weed seed, releases nutrients previously mobilized in the trees and generally provides an ashy seed-bed suited to seedling emergence. Furthermore, the *Acacia* species have built up a supply of soil nitrogen from nodulation and leaf senescence; the initial soil nitrogen is high after a burn and stimulates young grass growth. In wetter subhumid and humid regions throughout the world, *Melinis minutiflora, Panicum maximum, Hyparrhenia rufa* and other species have been established in this way.

Seeding should take place just ahead of an expected rainy season. It is important to seed the ashes within a few days of the burn in case heavy rains fall before the seedlings emerge, causing a subsequent loss of both seed and ash by erosion from sloping land.

Weed growth, and often bush regrowth, is rapid shortly after a burn and it is therefore important to obtain a dense cover of improved pasture as soon as possible to suppress weeds. Rhodes grass (*Chloris gayana*) (Plate 32) in semi-arid areas and molasses grass (*Melinis minutiflora*) (Plate 41) in subhumid and humid regions are very effective early colonizers in these situations.

For small areas a Cahoon-type hand broadcaster can be effective and is inexpensive. A canvas bag containing the seed is slung from the shoulder and seed dropped on to a disc spinning horizontally, activated by gears operated manually with a handle.

Establishment on a prepared seed-bed. The seed of most pasture species is very small; it is necessary to prepare a fine seed-bed, free from weeds and with adequate fertilizer and soil moisture for successful establishment.

Cultivation must be sufficiently deep to destroy the original grass, weed and shrub vegetation, yet not so deep as to turn up the highly infertile subsoil. Generally the initial cultivation depth should be 8-10 cm deep to destroy *Imperata cylindrica* rhizomes and other weedy grasses, and to uproot small shrubs and seedlings. A heavy stump-jump disc plough is very effective with taller shrubs and young trees in all soils (including clays) when in a moist condition (see Fig. 6.2), while an offset disc cultivator will usually handle soils that are sandy, or friable loams. After this initial ploughing, the land should be allowed to rest to permit the roots and other vegetative material either to dry out or rot down; a second cultivation should then be given to chop up the material further, incorporating it in the soil, and the land levelled for sub-

Figure 6.2. Heavy duty stump-jump disc plough removing *Acacia harpophylla* (brigalow) regrowth before overseeding the pasture

sequent drilling or broadcasting. This "discing" can take place in late spring or early summer; wind-rows can be burnt during this period.

Seeding. Seeding is usually accomplished in early summer if the danger of weed competition is minimal so that the new pasture will have a long growing season in which to become established. If weed competition is expected, planting can be delayed until mid-summer to allow the weeds to germinate; subsequently they will be killed with cultivation at seeding. In some cases it is best to wait until late summer to escape heavy weed competition, high temperatures and heavy monsoonal rains which may cause the seedlings to be flooded. In any case seeding should be done when good establishment rains are expected to follow.

Seeding is usually carried out by broadcasting on the soil surface from the air or by using ground machines. However, the safest method is to use a combined seed and fertilizer drill for the accurate measurement and placement of seed and fertilizer. A disc drill is preferred to a tine machine in early establishment on newly cleared country because there is less blockage from small sticks and debris. On clean land, a tine drill can be used.

Broadcasting seed on to the soil surface without the seed-to-ground delivery chutes of a conventional drill is often carried out, but in such cases immediate rolling with a Cambridge-type roller is essential. On wet clay soils, a home-made roller (made from old automobile tyres packed together over a hardwood frame) will shed the wet soil better than a metal roller. Many grass "seeds" are difficult to plant with a seed drill because of the difficulty of separating the seed clusters, requiring special adaptations to the drill for satisfactory seeding (Paull, 1973). The plates above the fertilizer "stars" and the boxes are removed. The outside of each star-wheel is then tapped and a 10 cm by 9 mm bolt screwed in and locked in position with a locking nut. These act as agitators (see Fig. 6.3).

Some seeds (such as buffel [*Cenchrus ciliaris*] and gamba grass [*Andropogon gayanus*]) are hammer-milled before planting to remove the seed coat and break dormancy as well as to permit easy seeding, or to remove the beards from the seed. It is dangerous to plant only caryopses in semi-arid areas as a light rainfall after seeding may germinate the seed, while subsequent heat waves may destroy the seedlings. With the seed coat intact and dormancy gradually broken down over a period of two years, there is a better chance of good establishment if heavier seeding rates are employed. Where seed-harvesting ants are likely to be troublesome (see Plate 5), the seed can be treated before planting with 450 g of 20 percent lindane wettable powder per 100 kilograms of seed.

The optimal sowing depth for any species appears to depend on both soil type (particularly structure) and the soil moisture regime (Bogdan, 1964). Variations in soil moisture can influence the gaseous phase, which in turn influences germination.

If seed placed on the soil surface is to germinate, the water gain must exceed water loss. To do this the seed must be in close contact with available water, and either the soil moisture tension must be low or the rate of water lost from the seed to the atmosphere must be lowered by increased humidity (Harper & Benton, 1966). Harper and Benton found that seeds that have smooth testas, a small amount of contact, and that produce no mucilage (such

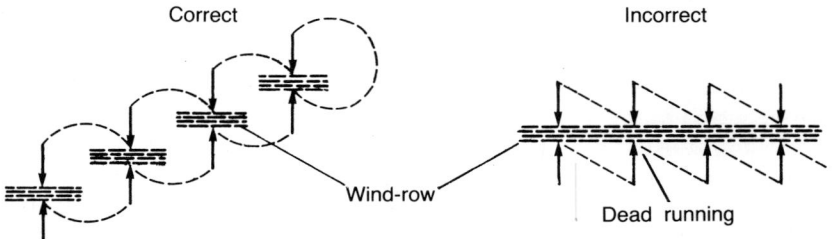

Figure 6.3. Agitators attached to the fertilizer plates in a grain/fertilizer drill sowing *Cenchrus ciliaris* (buffel grass) (**Source:** C.J. Paull, Queensland Department of Primary Industries)

as Kikuyu seeds), did not germinate at water tensions of 50 cm or greater when placed in sintered glass plates. The surface sowing of Kikuyu seed is therefore unlikely to succeed.

Nutrient reserves in the germinating seedling are depleted within about a week. For the plant to survive, nutrients must then be available in the soil solution (McWilliam, Clements & Dowling, 1970). Nutrient levels in the soil solution will depend upon both the soil type and the fertilizer added; the actual levels required will depend upon the species. Kikuyu is strongly responsive to high levels of nitrogen and phosphorus. Buffel and birdwood grasses show a good response to phosphorus, and to nitrogen in the presence of phosphorus (Humphreys, 1959). *Setaria sphacelata* cv. Kazungula has been found to be competitive for phosphorus, nitrogen and potassium and can establish itself on soils where deficiencies have been shown using other species (Mears, 1969).

Some grasses do not produce viable seed and are established only by using vegetative material. Pangola grass (*Digitaria decumbens*) is one such grass and can be easily established by mowing the grass, hand-loading the cuttings on a trailer and spreading them manually from the back of the trailer, followed by discing the cuttings into the soil. The material can also be harvested with a forage harvester, blown into a trailer and later spread by hand or a manure spreader and disced in. Para grass (*Brachiaria mutica*) is usually planted in furrows about 1 m apart, using a hoop of the stolon about 30 cm in diameter, and covered with soil. The hoop allows plant contact with the soil at two surfaces. The grass can also be planted from material chopped in lengths of about 30 cm, broadcast and disced in.

In many countries grasses are hand-planted from root splits. The root stocks are divided and planted in furrows or holes 60 cm apart in rows 90 cm apart. Guinea grass (*Panicum maximum*) is usually planted in this way in Puerto Rico because of its low seed viability (Vicente-Chandler *et al.*, 1953). *Pennisetum purpureum* is planted with stem cuttings of usually three nodes, placed in furrows 90 cm apart with the cuttings buried in the soil to about one-half their length every 20 cm, and covered. Alternatively, the whole stalk can be laid in the furrow and covered. Sugar cane species (*Saccharum officinarum* and *S. sinense*) are similarly planted.

Establishment under a nurse crop. At Entebbe, Uganda, *Chloris gayana* was sown with *Sorghum bicolor;* both germinated well and when the sorghum was cut, the Rhodes grass was 45-60 cm high. In Kenya, *Chloris gayana, Bothriochloa insculpta* and *Melinis minutiflora* have been established under a maize crop. In Tanzania, *Chloris gayana* has been established under maize crops (van Rensburg, 1969).

The International Centre for Agriculture (CIAT) in Colombia studied pasture establishment in growing crops. Pure *Stylosanthes guianensis* 184 and mixed swards of that legume with *Brachiaria decumbens* or *Panicum maximum* were tried under cassava, beans and rice at two soil fertility levels — "low" (0.5 t/ha of dolomitic lime and 100 kg P_2O_5/ha as triple superphosphate) and "high" (4 t/ha lime, 400 kg P_2O_5/ha and 100 kg K_2O/ha), all broadcast and incorporated with the top 15 cm of the soil. The crops were sprinkle-irrigated when required. The simultaneous planting of cassava and the legume *S. guianensis* was successful, although cassava yields were lower by 20 percent. The mixed pastures failed owing to a serious reduction in cassava growth. Pastures planted 60, 210 and 310 days after the cassava failed because of light competition.

With rice, simultaneous planting failed. However, pastures sown between 30 and 45 days after the rice (*Brachiaria decumbens* and *Desmodium ovalifolium*) established successfully; with vigorous, short-statured, high-yielding rice varieties, pasture establishment is feasible. With beans, pastures can be established about half-way through the bean life cycle and have no effect on bean yields. The pasture yield, however, is only about half the normal yield. Pasture establishment is highly desirable for this short-season crop (CIAT, 1978).

In Suriname *Hymenachne amplexicaulis* is planted in rice paddies and suppressed during the growth of the rice. After the rice harvest *Hymenachne* grows and provides excellent grazing between rice crops (Dirven, 1963a & b).

At Brian Pastures, Gayndah, Queensland, Rickert (1970) had no success in establishing *Panicum maximum* cv. Sabi under a nurse crop of oats as the competition for light, moisture and/or nutrients adversely affected germination.

Establishment under a surface mulch. Establishment of fine-seeded grasses is often disappointing on heavy soils in semi-arid and subhumid areas. Leslie (1965) showed that successful emergence often depends on the rate of emergence in relation to the rate of soil drying. Self-mulching black earth soils dry out readily and the establishment of grasses is difficult. Rickert (1970) successfully established *Panicum maximum* cv. Sabi on a basalt-derived self-mulching sandy clay at Brian Pastures, Gayndah, Queensland (lat. 25°39′S), under a chaffed straw mulch made from native pasture applied after surface sowing at rates of 2 500, 5 000 and 10 000 kg/ha. Soil moisture was adequate from rain and irrigation. The higher rate of germination under mulch was associated with lower soil temperatures and reduced rates of soil drying, which would make mulch effective under hot summer conditions where evaporation is usually a problem. A practical method of adapting this finding is needed. Meantime, the incorporation of crop residues such as wheat stubble would help. Success has been achieved by mixing small seeds with sieved cow manure and planting it with a seed drill.

7. Selection of pasture grass species, seed purchase and storage, and fertilizer needs

The following characteristics should be looked for in pasture grasses:

- **Nutritive value.** A high intake and digestibility of grasses by the animal results in good meat, milk and wool production. Stages of maturity, species and variety are all factors involved, and there are considerable seasonal fluctuations. Frost is an important reason for the decline in nutritional value of grasses grown in the subtropics. Maturity and desiccation cause nutrient decline and lead to insufficient nutritive value in winter and early spring. Grasses that seed late in the season and that hold their nutritive value well into maturity are valuable. High soluble carbohydrate content in the forage is an advantage. High palatability encourages high intake.

- **Compatibility with a legume.** Short, leafy grasses are preferred to reduce competition for light, and bunch grasses are preferable to strongly stoloniferous or rhizomatous ones. It is difficult to maintain legumes with *Brachiaria brizantha*, *B. decumbens*, *B. mutica*, *Digitaria decumbens* and *Pennisetum clandestinum* and a high standard of management is needed. Seasonal conditions exert an effect. Where the dry season is severe and long the most drought-tolerant grasses may not be the most desirable since they compete more strongly with a legume for limited water. A less competitive grass that tolerates the dry season well and recovers quickly in the following rainy season may be a more desirable companion grass for legumes. Soil type also plays a role; in northeast Thailand Para grass is less vigorous on a lateritic soil while *Centrosema pubescens* grows quite well. The Para grass is thus less competitive and a satisfactory grass-legume balance is maintained.

- **Persistence and regeneration.** These characteristics are affected by suitability to the environment and management practice. Grasses with horizontal stems growing close to or under the ground (stoloniferous and rhizomatous species) and with massive root systems favour resistance to grazing and regeneration after adverse conditions. Stocking rate is the most important factor determining the profitability of improved pasture, so tolerance to grazing is very important.

- **Drought tolerance.** Deep roots such as those of buffel grass (*Cenchrus ciliaris*), stem and leaf modifications to reduce transpiration (such as those of *Brachiaria dura*) and heavy early seeding all help a species resist or escape drought.

- **Tolerance to flooding.** The ability to survive typical flash flooding in the tropics and subtropics and to grow in slowly running or standing water is important to make use of swampy land and river valleys. Para grass (*Brachiaria mutica*) is valuable in such situations.

- **Tolerance to fire.** Fire is an almost universal feature of the annual picture of grasslands in subhumid and semi-arid grasslands. Species that will tolerate fires have a special niche to fill in these situations.

- **Tolerance to frost.** The ability to tolerate frosts, grow well into the autumn and preferably show some winter greenness is of utmost importance in the subtropics, and plant exploration has sought such species. Species and cultivars native to high mountain grasslands in the tropics are especially valuable, such as *Setaria sphacelata* cv. Nandi. The ability to survive heavy frosts and regenerate is also important.

- **High seed production and ease of harvest.** While vegetative reproduction is satisfactory for small areas, it is costly in time and labour for extensive ranching. High production of viable seed that ripens evenly over a restricted period without shattering, and that is easy to harvest mechanically, has tremendous advantages. The price of such seed can be reduced to encourage quicker and more widespread use of improved species. Seed that matures late, but ahead of frost damage, allows a grass to provide a long grazing season. Common Rhodes grass (*Chloris gayana*) seeds too early and too often, thus reducing its nutritive value.

- **Easy establishment from seed.** This is important especially where aerial seeding of large areas is involved, where the seed-bed may not be as well prepared as for drilling, or where a surface mulch dries out quickly, as in the self-mulching black clay soils.

- **Adaptability to a wide range of soils.** Such adaptability helps to simplify the choice of species and quantity of seed available, as demand will be kept high. However, if the species do not give good performances on these varying soil types, specially adapted species may be preferred.

- **Response to fertilizers.** As fertilizer prices increase it is becoming more important that species respond sufficiently to fertilizers to justify their application. Those that have the ability to extract nutrients from unfertilized soil are also being sought and widely sown.

- **Resistance to pests and diseases.** Species and cultivars that have resistance to certain pests and diseases are being sought or bred artificially. Pangola grass (*Digitaria decumbens*) has been severely attacked by rust (*Puccinia oahuensis*) and a search is being made for other *Digitaria* spp. that are resistant and that might replace it. *Sorghum* spp. that are resistant to the various leaf-destroying fungi are being bred. Resistance to the American sting nematodes (*Belmolaimus longicaudatus*) is being sought in *Hemarthria altissima*.

- **Toxicity.** Toxicity from the ingestion of grasses is usually not important except in the case of *Sorghum* spp., where hydrocyanic acid (prussic acid) poisoning is often serious. Low HCN cultivars are being bred and released. High oxalate content can be a problem.

The objective in species selection is to produce pastures that can maintain the growth of young stock and keep breeding animals in good condition throughout the year, while finishing off animals for slaughter and providing an adequate diet for lactating cows or production of wool by sheep. In a subtropical environment with a monsoon climate, emphasis should centre on species that will reduce weight or production losses, or both, during the winter and early spring.

Pure grass pastures or grass-legume mixtures

Natural grasslands composed of grass, either alone or in association with shrubs, sparse legumes and other forbs have been the mainstay of tropical livestock industries and of wildlife over the centuries, and still support the major proportion of the world's animals. But they do so at a low level of productivity and in most cases at a declining level. As livestock population has increased, overgrazing has led to deterioration in quality and a change in botanical composition to species of lower nutritive value. As human population has increased the demand for food has risen, and more land has been converted from grazing to cash crops. Thus more pressure is applied to the remaining pastoralists to raise productivity.

Productivity can be increased by better grassland and livestock management, but not at a level sufficient to keep pace with the increasing demand for livestock products.

It has become evident that if productivity from grassland is to be increased in areas with favourable rainfall, new pasture species able to respond to improved fertility must replace the existing swards.

The supply of nitrogen to the grass has been shown by Henzell (1970) to be the main limiting factor in increasing grassland productivity and nitrogen is the most effective single element in increasing grass production. When the needs for other elements are satisfied the addition of increasing levels of nitrogen can often lead to spectacular production.

Grass can obtain its nitrogen in a number of ways, but the most important sources are from fertilizers or associated legumes. Legumes vary in their ability to provide nitrogen, and for the most responsive grasses no legume can adequately supply the needs of the grass. Hence the simplest way to achieve maximum production from grass is to apply artificial fertilizer with a high nitrogen content. But this is expensive and will become more expensive as world fertilizer prices continue to rise. Therefore, for economical production, except in high-intensity livestock husbandry in the humid tropics or under irrigation it will be better to settle for an optimal level of production utilizing legumes as the main source of nitrogen.

Grasses take up nitrogen at the expense of legumes, and some vigorous stoloniferous grasses will compete so severely for the available nitrogen that they will eliminate the legume and create a monospecific sward. These grasses are best fertilized with nitrogen in pure swards. The main grasses in this category, used in the humid tropics where legumes may succumb to disease, include *Brachiaria decumbens* (signal grass), *Digitaria decumbens* (pangola grass), *Pennisetum clandestinum* (Kikuyu grass), *Brachiaria mutica* (Para grass), *Cynodon nlemfuensis* (star grass) and *Eriochloa punctata* (carib grass). *Pennisetum purpureum* (Napier or elephant grass) is used as a cut grass.

In Puerto Rico the pressure for land is so great that each hectare must be made to produce at its maximum. Combined with a high rainfall and responsive pasture species grassland productivity is among the highest in the world. Elephant or Napier grass fertilized with 897 kg N/ha yielded 84 800 kg DM/ha per year (Vicente-Chandler, Silva & Figarella, 1959).

At Orocovis, Puerto Rico (lat. 18°30'N, rainfall 1 735 mm), *Digitaria decumbens* + 350 kg N/ha yielded 1 000 kg/ha live-weight gain and a *Melinis minutiflora/Pueraria phaseoloides* pasture yielded 500 kg live-weight gain/ha (Vicente-Chandler et al., 1964). At the same station, Napier grass fertilized with 336 kg N/ha yearly produced 26 490 kg DM and 1 986 kg protein, compared with a Napier-kudzu pasture fertilized with all other necessary elements except nitrogen yield of 13 700 kg DM containing 1 081 kg protein. The difference would have been greater had the pure Napier grass sward received more nitrogen, as it can respond linearly to 1 792 kg N/ha in that environment. Vicente-Chandler and his colleagues in Puerto Rico believe that under their conditions grass pastures fertilized with nitrogen are both economic and the best solution in the humid tropics.

Most of the nitrogen-fertilized grass is utilized in intensive industries, particularly dairying, where continuity of production is desirable at a high level to maintain production quotas of fresh milk to meet the demands of population in the larger cities. It can also be justified in the following cases:

- In the early stages of farm development, when large amounts of feed can be produced on small areas while other parts are being developed;
- As strategic feed under irrigation in winter, when other grasses are becoming less productive;
- In areas of prolonged rainy periods or waterlogged soils where legumes are subject to severe disease hazards;
- To provide quick feed after frosts in wetter areas or under irrigation.

A recommendation that is frequently made for areas where there is sufficient rainfall or irrigation is to use one-quarter of the farm area for nitrogen-fertilized grass and the remainder for grass-legume mixtures.

The questions that should be asked when considering nitrogen-fertilized grass are as follows (Havilah & Mears, 1968).

- Under what conditions does nitrogen deficiency restrict growth?
- What is the response in terms of increased dry matter or nitrogen yield per kilogram of nitrogen supplied?
- What forms of nitrogen should be used?
- Will the pasture grown with applied nitrogen be of sufficient quality?
- Will the extra pasture be efficiently converted into animal products?
- Is the use of applied nitrogen economic?

Species recommended for sown grasses in the humid and subhumid zones

There are numerous grasses that have been sown in these areas and the characteristics and use of most of them are listed in the catalogue of grasses (see Chapter 15). Grasses for special purposes are listed in Chapter 12.

The more commonly used species are listed in Tables 7.1 and 7.3. Mixtures to provide a full year's grazing are desirable and are listed in Tables 7.2 and 7.4.

Seed purchase and storage

Pasture improvement is costly and is meant to provide long-term benefits. Hence only seed of the highest quality should be planted. Quality in seed means seed with high viability and high purity. Purity is important because contamination with weed species can cause weed problems for years to come,

TABLE 7.1 **Pure grass sward species for the humid tropics**

Species	Common name
Andropogon gayanus (P)	Gamba grass
Axonopus scoparius (F)	Imperial grass
Brachiaria decumbens (P)	Signal grass
B. humidicola (P)	Koronivia grass, creeping signal grass
B. mutica (P,F)	Para grass
B. ruziziensis (P)	Kennedy ruzi grass or Congo signal grass
Coix lacryma-jobi (A,F)	Job's tears
Cynodon nlemfuensis (P)	Star grass
Digitaria decumbens (P)	Pangola grass
Echinochloa polystachya (P)	German or Alemán grass
Eriochloa punctata (P,F)	Carib grass or Janeiro
Hemarthria altissima (P)	Limpo grass, red vlei grass, rooikweek
Ixophorus unisetus (P,F)	Honduras or Mexican grass
Melinis minutiflora (P)	Molasses grass or gordura
Panicum maximum (P)	Guinea grass, green panic, gramalotte
Paspalum plicatulum (P)	Brownseed paspalum or plicatulum
Pennisetum purpureum (P,F)	Napier or elephant grass
Tripsacum dactyloides (F)	Eastern gamma grass
Zea mays (A,F)	Maize or corn

Source: Teitzel, Abbott and Mellor, 1974
NOTE: A-annual; P-pasture; F-fodder.

and off-type seed of the desired species will give inferior performance. Most governments have regulations governing the quality of seed offered for sale, including minimum germination percentages, minimum purity percentages, and a list of prohibited seed contaminants. Some seed species are certified if the volume of production warrants it, especially if it is difficult to identify cultivars or varieties. Such seed is grown under supervision and seed packages officially sealed.

It is necessary to know the percentage of pure germinable seed to deter-

TABLE 7.2 **Grass-legume mixtures for the humid tropics**

Situation	Grass-legume mixture
Well drained fertile soils	Guinea-centro-puero
Well drained soils of moderate fertility	Guinea-centro-puero-stylo
Well drained soils of low fertility	Guinea-puero-stylo or signal-puero-stylo
Moderately drained soils	Guinea (cv. Hamil)-centro-puero-stylo
Poorly drained soils	Para-centro-puero-stylo

Source: Teitzel, Abbott and Mellor, 1974
NOTE: centro-*Centrosema pubescens*; puero-*Pueraria phaseoloides*; stylo-*Stylosanthes guianensis*.

TABLE 7.3 **Pure grass sward species for the subhumid tropics**

Species	Common name
Brachiaria brizantha (P)	Palisade grass
Cenchrus ciliaris (P)	Buffel grass
Cynodon aethiopicus (P)	Star grass
C. plectostachyus (P)	Naivasha star grass
Digitaria decumbens (P)	Pangola grass
Echinochloa frumentacea (A,P,F)	White panicum
E. utilis (A,P,F)	Japanese millet
Panicum coloratum var. *makarikariensis* (P)	Makarikari
Paspalum notatum (P)	Bahia grass
Pennisetum americanum (A,F)	Pearl or bulrush millet
P. clandestinum (P)	Kikuyu grass
Sorghum almum (A,P)	Colombus grass
S. bicolor (A,F)	Sorghum, jowar or dhurra
S. sudanense (A,P,F,)	Sudan grass
Tripsacum dactyloides (F)	Eastern gamma grass
Zea mays (A,F)	Maize or corn (USA)

Source: Teitzel, Abbott and Mellor, 1974
NOTE: A-annual; P-pasture; F-fodder.

mine planting rates. Most pasture seeds have a dormancy period directly after harvest due to post-harvest ripening and they gradually improve in germination under adequate storage. Some seeds need treatment before planting to break the dormancy, for example *Brachiaria decumbens* (signal grass)

TABLE 7.4 **Grass-legume mixtures for the subhumid tropics**

Andropogon gayanus - *Stylosanthes hamata*
Chloris gayana - *Macroptilium atropurpureum*
Digitaria decumbens cv. Pangola - *Lotononis bainesii*
D. decumbens cv. Transvala - *Centrosema pubescens*
Echinochloa frumentacea (A) - *Vigna unguiculata* (A) (Cowpea)
Panicum maximum - *Macroptilium atropurpureum* - *Centrosema pubescens* - *Leucaena leucocephala* - *Neonotonia wightii*
P. maximum var. *trichoglume* - *Macroptilium atropurpureum* - *Neonotonia wightii*
Paspalum dilatatum - *Trifolium repens* - *T. semipilosum*
P. plicatulum - *Desmodium intortum* - *Macroptilium atropurpureum* (Plate 64)
Pennisetum clandestinum - *Desmodium intortum, D. uncinatum* (Plate 68), *Trifolium repens, T. semipilosum*
Setaria sphacelata - *Desmodium intortum, Neonotonia wightii* - *Trifolium semipilosum* (with cv. Narok)
Urochloa mosambicensis - *Stylosanthes hamata* - *S. humilis*

Source: Teitzel, Abbott and Mellor, 1974
NOTE: Grasses for semi-arid areas are listed in Chapter 10.

may need treatment with sulphuric acid to break dormancy before planting. Details are given in the catalogue of grasses (Chapter 15).

Because tropical pasture seeds are generally difficult to grow in pure stands for seed production, fertilizers are needed to give high yields, and harvesting must be well timed and carefully carried out. There is often a shortage of seed and the costs are high. It is important therefore to order seed well ahead of planting, store it in a dry, well-aerated place, and test it for germination immediately before planting. The best method to store seed on a farm is to hang the seed in seedbags suspended from the rafters of a shed, allowing the air to circulate freely (Teitzel, Abbott & Mellor, 1974). If seed is to be stored in sealed containers or confined space, it should be artificially dried at temperatures of 32°, 37° and 43°C for initial seed moisture contents of more than 18, 10-18, and less than 10 percent moisture respectively (Humphreys, 1979).

Fertilizers for establishment and maintenance

Humid tropics. With the increasing cost of fertilizers it is important to know the minimum nutrient requirements as well as the optimum level for the most economical return. This knowledge should be available from local research or may be obtained from the samples taken from the soil in question at a depth of 10 cm. Some 30 borings to this depth mixed together to give a composite sample are needed. These can then be analysed and nutrient needs assessed, but field research must be carried out to verify the assessment from analyses. Teitzel, Standley and Wilson (1978), in the wet tropics (1 500-3 750 mm rainfall) of north Queensland have found that the local geology is a good guide to field fertilizer requirements when considered in relation to the vegetation. In the wet tropical area of Queensland the authors summarize the needs for a mixed grass-legume pasture as follows:

Soils derived from basalt. Phosphorus, molybdenum and sulphur deficiencies have been recorded. It has been difficult to get a reliable chemical test for phosphorus on these soils. Field trials show 200-400 kg of superphosphate per hectare are insufficient to produce marked increases in production. An initial dressing of 500 kg superphosphate plus 0.5 kg sodium molybdate per hectare is required for establishment, followed by 300 kg superphosphate plus 0.5 kg sodium molybdate every second year.

Soils derived from granite. Excellent pastures have been found where soil acid extractable phosphorus is about 30 parts per million (ppm) and exchangeable potassium about 120 ppm. When pasture potassium needs are satisfied, the treatment should last for several years. For soils initially carrying rain forest, 250 kg superphosphate per hectare is needed initially, but for other areas 500 kg is required at establishment. Soil phosphorus gradually declines, and 300 kg superphosphate per hectare is required every two years to maintain soil phosphorus at about 30 ppm. Other important deficiencies in granitic soils include sulphur, copper, zinc and molybdenum. Superphosphate (single) contains 10 percent sulphur and satisfies this requirement. Where copper, zinc and molybdenum deficiencies are suspected, 8 kg/ha each of copper and zinc sulphates and 0.5 kg of sodium molybdate are recommended for establishment and thereafter every fourth year.

Soils derived from metamorphic rocks. Deficiencies of phosphorus, potassium, molybdenum and sulphur are all found, and initial dressings of 500 kg superphosphate, 60-125 kg muriate of potash plus 0.5 kg/ha sodium molybdate are needed for establishment, with 300 kg/ha superphosphate plus 0.5 kg/ha sodium molybdate needed every second year. If the soil is high in aluminium an initial dressing of 500 kg/ha lime is required to reduce acidity.

Mixed alluvial soils. Phosphorus, potash, copper, zinc, molybdenum and sulphur deficiencies have been found and the above dressings can be adapted as required.

Soils derived from beach sands. These are extremely infertile and so fertilizer should be applied often in small quantities. On establishment, a dressing of 500 kg/ha superphosphate, 190 kg/ha muriate of potash, 8 kg/ha each of copper and zinc sulphate, and 0.5 kg/ha sodium molybdate is needed, followed by 150 kg/ha superphosphate, 50 kg/ha muriate of potash at the end of each wet season, and copper, zinc and molybdenum applied every fourth year. These dressings may appear to be quite heavy, but in a high rainfall regime the productivity of well-fertilized pastures is high. There is heavy removal of nutrients in the high yields of forage produced with intensive management, loss of nutrients by leaching and fixation, and low soil fertility. Applications of less than 175 kg N/ha per year to pastures are ineffective in the humid tropics. Apply the maintenance dressings in winter after the wet season when fertilizer can be most easily applied, as pastures are short, the ground is firm and grasses and legumes most need stimulation of growth by added nutrients.

In Puerto Rico in the humid tropics where land is limited, pure grass pastures are preferred and these are intensively managed. The predominant soil where experiments have been made is a deep red acid ultisol with a pH of 4.8 with 12 meq of exchangeable bases per 100 g of soil and an organic content of 3.4 percent. Clay minerals are predominantly koalinitic with high amounts of free iron and aluminium oxide contents. Liming is necessary to maintain soil pH at 6.0. At Orocovis, Puerto Rico, well-fertilized pastures harvested by cutting removed an average of 328 kg nitrogen, 54 kg phosphorus, 422 kg potassium, 128 kg calcium and 75 kg magnesium per hectare yearly in the harvested crop. The common fertilizer is a 15:5:10 (N:P:K) mix applied at 5 t/ha yearly to Napier grass (*Pennisetum purpureum*) as cut feed, and up to 3.75 t/ha to pangola grass (*Digitaria decumbens*), star grass (*Cynodon nlemfuensis*), Congo grass (*Brachiaria ruziziensis*), carib grass (*Eriochloa punctata*), Guinea grass (*Panicum maximum*) and Para grass (*Brachiaria mutica*). One tonne of lime is applied for each tonne of fertilizer used on these soils (Vicente-Chandler *et al.*, 1974). The initial application of lime should be made after the first ploughing, and ploughed in to mix with the soil. Liming should be used to prevent build-up of soil acidity, as once soil has become very acid deep in the profile it is difficult to correct. In Puerto Rico lime is added up to 70 percent soil base saturation. Less fertilizer is required if the manure is returned to the land and if all excreta and uneaten forage is returned to the land it is theoretically possible to reduce fertilization to about 20 percent of optimum rates (Vicente-Chandler *et al.*, 1974).

Subhumid tropics with 625-1 500 mm rainfall. In these areas a wide variety of soil also exists. Some coastal soils will be heavily leached while the black earths may be very fertile. Fertilizer needs will therefore have to be determined by soil analyses followed by field trials. In general the leached soils will require phosphorus, sulphur and molybdenum if legumes are included in the pasture mixtures (Ostrowski, 1978). Maintenance dressings of 100-200 kg/ha per year of superphosphate may be needed, substituting molybdenized superphosphate every three to four years. Potash may not be required for some time, and, as it is expensive, a close watch can be kept for the visible marginal chlorotic legume leaf symptoms of potash deficiency. A dressing of 50-100 kg/ha may be needed.

For pure grass pastures a basal dressing of superphosphate and potash can be given along with 110 kg/ha of nitrogen to initiate growth if needed. Additional nitrogen up to 200-400 kg/ha in split applications can be given during growth. The split applications are designed to offset any losses of nitrogen by leaching, volatilization or denitrification. With irrigated grass pastures dressings of 350-400 kg/ha are common.

It is usually worthwhile to use nitrogen up to 110 kg/ha in early spring to initiate growth until the associated legume can release nitrogen to the grass. A second application in autumn, when legume growth declines, is also useful (Jones, 1970). Even in the dry season in the wet tropics, a nitrogen response is possible in winter and spring (Teitzel, McTaggart & Hibberd, 1971). The application of nitrogen to pure grass pastures, such as pangola grass, late in the season in autumn can prolong the nutritive value and improve intake and live-weight gains (Minson, 1967).

Nitrogen is preferably supplied as urea or ammonium nitrate (Nitram). Sulphate of ammonia is not recommended because it lowers the soil pH, which has to be raised by adding lime, which is a costly operation. In all cases of nitrogen fertilization there must be adequate levels of phosphorus and potash in the soil.

Fertilizers for the small farmer

In many developing countries, landholders have very small areas of land from which to attempt to obtain sufficient food and fibre crops for subsistence. Subsistence is the primary aim using all available local manures with any surplus production sold for cash. There is therefore little time, labour or capital for development.

Power is provided by draught animals — usually buffalo or cattle, of which each farmer usually owns between two and seven. They are worked during the day and coralled at night, usually under the farmer's dwelling. Dairy cows are likewise housed nightly in small numbers and graze on waste land or roadsides during the day. A convenient source of good quality feed for these essential animals is a small plot of nutritious fodder, grown in the back yard, and cut and fed as a night ration. Gutteridge and Robertson (1979) have developed this concept in northeast Thailand employing *Panicum maximum* cvs. Common and Hamil, *Pennisetum purpureum* (Napier or elephant grass) and *Leucaena leucocephala* (leucaena) as forage species. The land nearest the house is usually highly fertile from the accumulation of animal excreta. This could be augmented by collection and application of human urine, diluted about 20 times with water. Where schistosomiasis is prevalent this practice should not be attempted unless there is adequate sanitary treatment to destroy any infective material. This is done in China (FAO, 1977). The algae *Azolla* spp., grown in ponds, can fix significant amounts of nitrogen. This is harvested and applied to rice fields, but there would still be no surplus for pasture improvement in a subsistence society.

Nitrogen fixation by grasses

Formerly it was thought by botanists, microbiologists and pasture agronomists that legumes were the only forage group that could fix nitrogen through associated growth of micro-organisms in the rhizosphere. Döbereiner in 1961 and 1968 (Day, Neves & Döbereiner, 1975), suggested that the rhizosphere nitrogen fixation by grasses may be of economic importance, and that it is more likely to occur in tropical environments.

Nitrogenase activity is found exclusively on the roots and is closely linked to plant photosynthesis. Tropical grasses possessing the efficient C_4 dicarboxylic acid photosynthetic pathway are involved. It has been shown that nitrogenous fertilizers applied to tropical grasses impose transitory limitations on rhizosphere nitrogen fixation. This observation may be of economic importance in pasture agronomy because of the possibility of making use simultaneously of mineral nitrogen and biological nitrogen fixation as long as only low fertilizer applications are used each time (Day, Neves and Döbereiner, 1975). It is suggested that plant breeders may be able to

TABLE 7.5 **Nitrogenase activity in intact cores of some native and introduced grass species from northern Australia**

Species	g N/ha/day	Area sampled
Rhynchelytrum repens	346	Townsville
Urochloa oligotricha	316	Rockhampton
Heteropogon contortus	232	Townsville
Chloris barbata	179	Townsville
Urochloa mosambicensis	157	Townsville
Bothriochloa insculpta	155	Rockhampton
Sorghum plumosum	136	Townsville

NOTE: *Azospirillum brasiliense* activity was also found with *Panicum maximum, Astrebla lappacea, Iseilema vaginiflorum, Brachiaria decumbens* and *Digitaria decumbens. Spirillum lipoferum* was found on the roots of *Iseilema vaginiflorum, Astrebla lappacea, Sorghum plumosum* and *Panicum maximum. Azotobacter* sp. was associated with *Brachiaria decumbens* and *Sorghum plumosum. Beyerinckia* sp. was found with *Astrebla lappacea* and *Sorghum plumosum. Enterobacter aerogenes* was found with *Astrebla lappacea*.

enhance nitrogen-fixation by plant genetics and that the use of selected strains of nitrogen fixing bacteria added as an inoculum at the time of sowing C_4 grasses may increase nitrogen fixation.

Soil temperature, soil moisture, light influence and soil type have an influence on nitrogen fixation, and a wide C/N ratio and ease of decomposition of organic matter favour nitrogen fixation. Heavier soils showed the highest activity, probably due to their high moisture-holding capacity. The portion of root closest to the root crown showed the greatest nitrogenase activity (Weier, McCrae & Allen, 1978).

The *Paspalum notatum/Azotobacter paspali* association has been the most intensely studied. There is a close relationship between the bacteria and the higher plant, *A. paspali,* not occurring only in the rhizosphere of certain cultivars, but everywhere this grass grows (Day, Neves & Döbereiner, 1975). Nitrogen fixation by this grass reaches 93 kg N/ha per year. *Spirillum lipoferum* has increased yields of *Pennisetum americanum, Panicum maximum* and *Cenchrus ciliaris* in Florida. The most noteworthy increase was from 480 kg/ha with no nitrogen to 1 690 kg/ha with 49 kg N/ha added (Quesenberry *et al.,* 1976). In Queensland *Spirillum lipoferum* has contributed 131.2 grams of N/ha per day to *Digitaria decumbens* pastures. *Azospirillum brasiliense* has been found to be present in the rhizosphere of 95 percent of the grasses collected in northern Australia. Weier, MacRae and Allen (1978) tested 28 species of tropical grasses in Queensland and all showed nitrogenase activity. Those showing significant amounts of nitrogen are shown in Table 7.5.

This source of nitrogen shows some promise and further research is under way to assess the extent of nitrogen fixation by grasses, the variation between species, and the likelihood of improving the process by selection of bacteria and grasses that will respond.

8. Pasture leys

A pasture ley is essentially a short-term pasture, usually of three years' duration, interposed within a pasture/crop cycle. Its functions are twofold: to build up fertility, crumb structure and soil condition, and to provide a crop of nutritious herbage for grazing, green chop (soilage), or conservation to meet the requirements of the livestock.

The end products of ley farming should be heavier crop yields after ploughing up of the ley, prevention of soil erosion, healthier and more productive livestock, reduction of crop diseases, reduction of weed population and flexibility in the choice of enterprise.

The practice involves mixed farming. Cultivation reduces the organic matter in the soil and also the elements nitrogen, phosphorus, sulphur, the bases potassium, calcium and magnesium, and there is a deterioration in the physical and microbiological status of the soil.

Many traditional rotation cycles have been evolved to help to restore soil fertility, such as shifting cultivation and the gum (*Acacia senegal*) crop cycle of the Sudan. These involve a legume or a diverse collection of shrubs, trees and weeds, with little grass component.

Pure grass leys are not common but have become traditional in certain areas.

In the Sudan, a method of periodically burning heavy stands of certain well established grasses, known as *hariq* cultivation, is used on rich clay plains to produce cereals and other annual crops (Semple, 1970). Normally, cultivators allow the grass to grow and accumulate for two to four years before burning, so that fertility is restored and the minimum of cultivation is required for seeded crops. This method requires great care to avoid accidental burning of the grass in an area where nearby grazing lands are burned annually. It is customary to make a fire-break by cultivating strips about 100 m wide, by clearing off the grass by hand or by burning early, when the fire can be easily controlled. The time allowed for the grass to accumulate varies with the time during which the land has been cultivated, the supply of land and the nature of the grass growth. Ideally, a certain climax in plant growth is awaited, though, occasionally, good results can be obtained after one year's growth, especially when the land has been cultivated for only one year and this is followed by an exceptionally heavy growth. A four-year accumulation is usually considered ideal, and from two to four years' cultivation may follow.

Anis (*Sorghum purpureosericeum* Schweinf. & Aschers.) is considered the best grass for *hariq* cultivation. Other useful grasses for this purpose are *Aristida mutabilis* Trin. & Rupr., Umm ritsu (*Hyparrhenia pseudocymbaria* (Steud.) Stapf.), Umm chir (*Brachiaria obtusiflora* Stapf.), dukn misi khat (*Pennisetum mollissimum* Hochst. in Flora), Umm belila (*Rottboellia exaltata* L.), and na'al (*Cymbopogon nervatus* (Hochst.) Chiov.). Tufted or bunch grasses are not satisfactory and trampling by cattle is objectionable. Burning takes place after the first rains, when new growth has appeared, so that the land is perfectly clean for the desired crop to be seeded in the ashes. Research in southern Nigeria has shown that burning mature vegetation is practically as beneficial to the fertility of the soil as turning the plant material under for green manure (Doyne, 1937).

The increased fertility is probably due to some fixation of nitrogen within the rhizosphere of the grasses, and improved organic matter content. Nitrogen fixation by grasses, especially on heavy black soils, is being discovered in Queensland, Australia (Weier, 1977; Weier, MacRae and Allen, 1978). The burning in the Sudan loosens the soil surface so that it can be easily cultivated with hand hoes.

Vegetable growers in coastal Queensland often use a pure grass ley of short duration as a green manure crop especially where table beans may be affected by bean fly (*Agromyza* sp.) if alternate legume hosts are grown as green manure. *Sorghum* spp. and maize, seeded heavily, provide fast growing and high-yielding crops that can be ploughed in.

Tobacco farmers in Zimbabwe use *Chloris gayana* cv. Katambora (Humphreys, 1978) and those in Tanzania use *Cynodon dactylon,* while tomato farmers in Florida use pangola grass (*Digitaria decumbens*) as rotation crops to reduce damage by soil nematodes.

In these cases the grass ley usually utilizes the residual fertility left as a result of fertilizing the main crop. An improvement in subsequent soil fertility is accomplished by fertilizing the grass, especially with nitrogen. In such cases it is important to select a grass species that will respond to this added fertilizer.

Where a pure grass pasture is used and fertilized with nitrogen it will be necessary to supply nutrients limiting plant growth, and provided the grass is well fertilized with nitrogen to produce near maximum dry matter, then the requirements of a pure grass pasture for phosphorus, sulphur and magnesium may be approximately equal to that of a legume. A grass normally takes up luxury amounts of potassium compared to its requirements.

Foster (1971) at Kawanda Research Sation, Uganda tested a three-year elephant grass (*Pennisetum purpureum*) ley followed by a three-year cropping cycle including cotton, maize, sweet potatoes, maize and beans. He found that crop yield declined markedly over a two-year cultivation period due mainly to the loss of soil nitrogen. The crops grown immediately after three years of unfertilized elephant grass gave much better yields than crops grown after three years of continuous cultivation. This was due to the greater availability of nitrogen in the soil after the grass rest. Where the ley phase was fertilized with nitrogen or grazed the following maize yields were higher, indicating the good response of maize to nitrogen. A significant decrease in the yield of cotton, sweet potatoes and beans, but not of maize, was recorded on plots where soil potassium had been depleted by a preceding elephant grass crop which had been fertilized with a large amount of nitrogen and potash and had been cut. Cutting and removal of the grass depleted the soil potash.

A ley of mixed grass-legume pasture, adequately fertilized to preserve the legume component, povides a much better pasture both for grazing and for building up soil fertility during the ley, especially with respect to nitrogen.

Jones (1967) at Samford, southeastern Queensland, compared the yields of forage sorghum following previous pasture treatments which included a pure *Paspalum plicatulum* cv. Hartley sward, the same grass plus 110 or 220 kg N/ha per year, and in association with the legumes *Macroptilium atropurpureum* and *Lotononis bainesii*, each separately. The pastures were grazed for a period of four years after establishment and then ploughed in and the land sown to the fodder sorghum, *Soghum bicolor* cv. Sugar Drip. The yields of nitrogen and carbon percentages are given in Table 8.1.

Both the fertilizer nitrogen and the grass with *M. atropurpureum* significantly improved yields, the legume contributing almost the equivalent of 200 kg N/ha/year.

In southern Africa yields of maize after one- to four-year leys of five perennial grasses (*Digitaria pentzii, Eragrostis curvula, Sorghum almum, Pennisetum purpureum, Cynodon dactylon*), a legume (*Medicago sativa*) and the annual *Pennisetum americanum* were compared with those of continuous culture. The yields were highest after *Medicago sativa, P. purpureum* and *Cynodon dactylon*. The annual yields were increased significantly in accordance with the duration of the ley (Du Plooy, Le Boux & Coetzee, 1965).

Bryan and Evans (1971) outlined the changes in fertility in a gleyed podzolic soil at Beerwah, southeast Queensland at a depth of 0-10 cm as a result of sheep grazing improved grass/legume pastures fertilized with a mixture of

TABLE 8.1 **Forage sorghum yields after previous pasture treatment**

Previous pasture treatment	Sorghum yield (kg/ha)	Soil nitrogen (%)	Soil carbon (%)
Grass alone	1 260	0.089	1.91
Grass + 110 kg/N/ha/yr	2 560	0.108	2.44
Grass + 220 kg/N/ha/yr	6 240	0.114	2.52
Grass + *M. atropurpureum*	5 520	0.125	2.59
Grass + *L. bainesii*	1 410	0.103	2.00

TABLE 8.2 **Changes in fertility in a gleyed podzolic soil from grazing a fertilized grass/legume mixture in southeast Queensland**

Mineral composition	Virgin soil	After 11 years' grazing	After 15 years' grazing
pH	5.2	—	5.0
Organic carbon (%)	0.84	2.60	1.56
Nitrogen (%)	0.047	0.15	0.103
Sulphur (ppm)	70	207	205
Total phosphorus (ppm)	18	150	160
Available phosphorus (ppm)	4	35	40
Total potash (ppm)	31	90	202
Total calcium (ppm)	30	550	320
Total magnesium (ppm)	29	29	—

625 kg superphosphate, 625 kg calcium sulphate, 125 kg potassium chloride, 8 kg each of copper sulphate and zinc sulphate and 280 g of ammonium molybdate per hectare (see Table 8.2).

    With increasing rainfall and decreasing temperature, soil organic matter tends to increase. Rainfall controls the kind and density of vegetation. Temperature mainly affects the rate of decomposition of plant residues either directly by accelerating the process of oxidation or through its influence on the microbial activity. This latter is the more important process. It is considered that 25°C is the critical temperature. In well-aerated soils of the humid warm tropics, the organic matter tends to accumulate below this temperature and to disappear above 25°C. Sunlight increases organic matter loss by radiation. The carbon nitrogen ratios for soils from unfertilized and fertilized areas with pasture growth are approximately the same and the organic matter from fertilized and unfertilized plots has approximately a constant proportion of carbon, nitrogen, sulphur and phosphorus in the ratio of 155:10:1.4:0.68. A grass-legume pasture will increase substantially the nitrogen content of the soil. The growth of the legume is dependent on the supply of phosphorus and work in temperate climates indicates a production of 26 kg of nitrogen in the surface 10 cm of soil for every 51 kg of superphosphate applied (Williams & Donald, 1957).

    Pasture leys may pose a problem with stock management involving fencing and with the manipulation of large-scale cultivation machinery. Grasses and legumes must be compatible in mixtures and easy to plough out when desired. Returns per hectare may not initially be as high from grazing animals as from commercial cropping, but in the long term the beneficial effect of pasture leys on overall fertility and stability of a mixed farming system will usually become evident.

9. Management of improved grassland in semi-intensive and intensive production systems

Most areas of impoved pastures in the tropics are concerned with the production of products such as milk and prime beef under semi-intensive or intensive conditions. Pasture management involving what is best for the pasture species must be considered in relation to herd management involving what is best for the grazing animal and the overall farm management programme and the financial situation.

Questions of maximum production, optimum production and pasture persistence arise. The establishment of improved pastures is costly and management will need to be aimed at persistence and maintenance of botanical composition as well as productivity.

Grazing methods

Although continuous grazing has proved most successful under extensive conditions with indigenous pastures, the integration with herd management and land use in intensive systems generally requires some form of rotational grazing.

Rotational grazing. In a dairying situation, separate provision must be made for milking cows, dry cows, bulls, heifers of mating age, younger heifers, steers (if kept) and young calves. In a beef situation a similar problem occurs, substituting cows with calves (breeders) for milking cows and eliminating the calf paddock, but segregation of steers into age groups may be required with a "bullock paddock" for final finishing of male castrates for slaughter.

Rotational grazing is also important in parasite control. In subhumid and humid areas endoparasites are often a major problem. The majority of endoparasites affecting stock require a period of at least three days between the time the eggs or early stage larvae are voided onto pasture and the time at which they reach an infective stage. Thus if animals are moved frequently to fresh pasture some worm control is possible. Effective reduction of ectoparasites such as the Queensland cattle tick (*Boophilus microplus*) is possible by using a grazing system involving rest periods of three months or more (Wheeler, 1962).

Rotational grazing involves a good deal of subdivision and usually costly fencing. Suspension and electric fencing can reduce costs.

In Uganda, Stobbs (1969c) compared continuous and rotational grazing on improved pastures of *Panicum maximum/Macroptilium atropurpureum* over a period of 1 218 days (29 grazing cycles). Three systems of grazing were imposed using zebu-type steers at a stocking rate of 5 beasts per hectare in the following ways:

- Continuous or free grazing using only one paddock;
- Rotational three-paddock grazing, moving cattle every two weeks;
- Rotational six-paddock grazing, moving cattle weekly.

Figure 9.1 shows the relationships.

Production from the three-paddock system over the whole 1 218 days was 1 544 kg/ha, from the continuously grazed pastures 1 462 kg/ha and from the six-paddock system 1 310 kg/ha. The continuously grazed pastures were

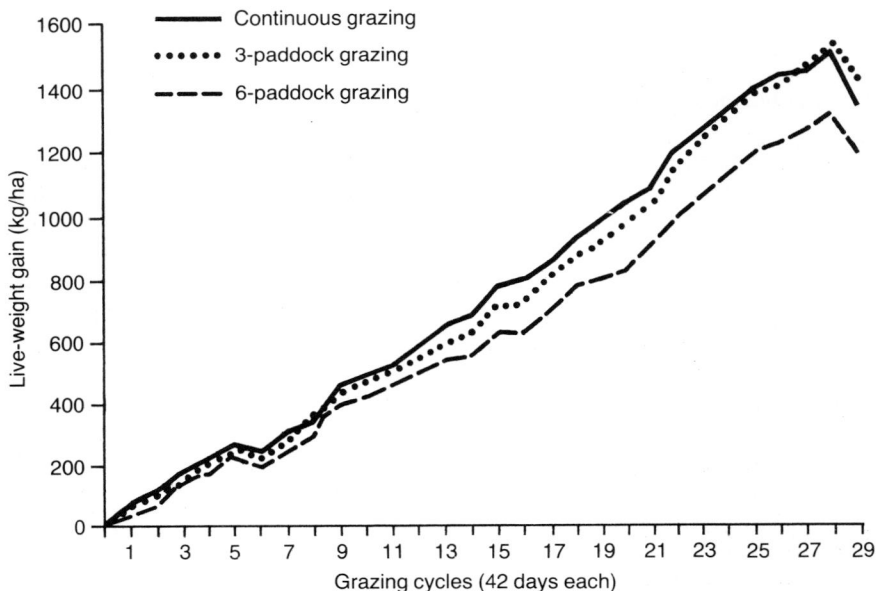

Figure 9.1. Cumulative live-weight gains from rotationally and continuously grazed pastures (**Source:** Stobbs, 1969c)

significantly higher yielding than those under the six-paddock system, but there was no significant difference between the final weight gains from the continuously grazed plot and the three-paddock system. There was however a deterioration in the botanical composition of the continuously grazed plots and Stobbs concluded that over long periods rotational grazing is necessary to maintain a satisfactory sward under Serere (Uganda) conditions, and the three-paddock system with two weeks' grazing and four weeks' rest appeared to be the most suitable practice.

CSIRO workers examined 12 grazing trials comparing continuous and rotational systems. Continuous grazing gave better live-weight gains in eight trials, rotational grazing in two, and in two other trials gains were equal.

Grazing frequency. The improved pasture species need a rest period between grazing cycles to produce new tillers and new leaf. The response to defoliation varies with species; the more prostrate stoloniferous species and those with buried crowns will stand closer grazing than the upright tussock types.

In Uganda, Stobbs (1969b) subjected a grass-legume mixture to two grazing cycles with results set out in Table 9.1.

In the dry season the heavier defoliation gave a much inferior performance as recovery was slower whereas during the wet season there was little difference.

Leader and follower system for lactating cows. Milking cows need a high plane of nutrition at all times and the quality of feed on offer is most important. When animals enter a new pasture they select the leafy and more digestible and nutritious parts of the plants first, and if grazing is prolonged the quality declines. This has led to a practice of letting the lactating cows (leaders) have first use of the pasture (top grazing) and non-lactating animals (followers) the pasture remaining (bottom grazing). This was the principle behind the Hohenheim system of grazing in Europe, also called the "shifting stable" method.

Stobbs (1973) tested this system on a Rhodes grass (*Chloris gayana*) cv. Pioneer and *Panicum maximum* cv. Gatton pasture fertilized with 200 kg/ha

TABLE 9.1 **Live-weight gain under two seasonal grazing cycles (kg/ha/day)**

Grazing cycle	Dry season	Wet season
3.5 days on, 14 days off	1.30	2.20
7 days on, 28 days off	0.95	2.28

Source: Stobbs, 1969b

superphosphate annually and 250 kg N/ha each summer in five applications. The pasture regrowth was grazed at three- to four-week intervals for 14 days by leader Jersey cows and then the followers grazed for a further 14 days. Both leader and follower cows were allocated 40 kg DM per cow per day. Leader cows produced 8.0 kg milk per cow per day compared with 5.8 kg milk per cow per day by the followers, which indicated a higher intake of digestible nutrients. Cows showed a preference for leaf, and after easily accessible leaf was removed by leader cows the follower cows had a small bite size. Follower cows partially compensated by increasing grazing time, mainly at night.

Stobbs and Minson (1978) showed that plant structure determines rate of intake. As the animal moves in a horizontal plane it selectively grazes in a vertical plane and the rate of intake is influenced by the spatial distribution of the leaf in the sward. Hence easily accessible nutritious digestible leaf can improve intake and production.

Strip grazing. Strip grazing of lactating cows is often used as a means to increase milk production. An electric fence ahead of the cows rations the grazing, and one behind them prevents grazing low-quality material. This allows followers to graze the stubble. On the Atherton Tableland in north Queensland, Ottosen, Brown and Maraske (1975) compared the milk production from non-pregnant Friesian cows under continuous grazing and strip grazing *Panicum maximum* var. *trichoglume/Glycine* pastures. They found milk yield was lower on the strip-grazed pasture because this group of cows was forced to reduce its selective grazing and eat pasture plants closer to the ground, thus eating a more fibrous diet of lower nutritional value. The free-ranging cows were able to select more grass and left more legume (38 percent of legume was left compared with 16 percent by the strip-grazed group).

At equal stocking rates there is generally little difference in yield per head under the two systems, but where more stock are carried, milk yields are higher under the strip-grazing method (Wheeler, 1962). However, the strip method is a good way to ration the available improved pasture when green feed is short.

Stocking rate. Manipulation of stocking rate is one of the most effective means of improving grassland productivity in terms of live-weight gain and milk production, but other factors such as water supply, labour and fertilizers are also important.

In general, the live-weight gain per animal decreases as stocking rate increases, but gain per unit area increases up to an optimum level and then decreases. Walshe (1975) and Matches and Mott (1975) have illustrated the relationships in Figures 9.2 and 9.3 in a temperate and a tropical environment. The optimum grazing pressure occurs just below the maximum gain per hectare. Below this optimum, pastures are undergrazed and beyond this

Management of Improved Grassland

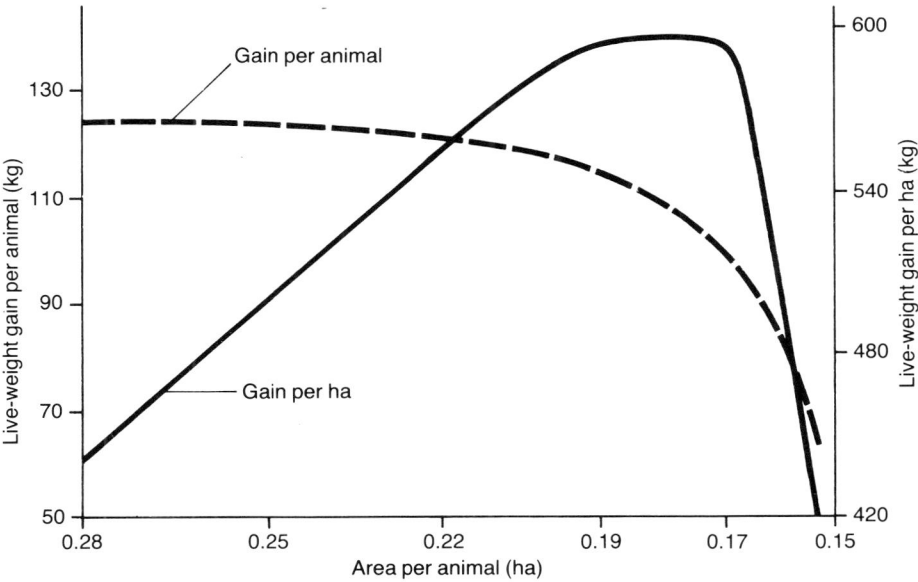

Figure 9.2. Effect of stocking rate on production per animal and per hectare (**Source:** Walshe, 1975)

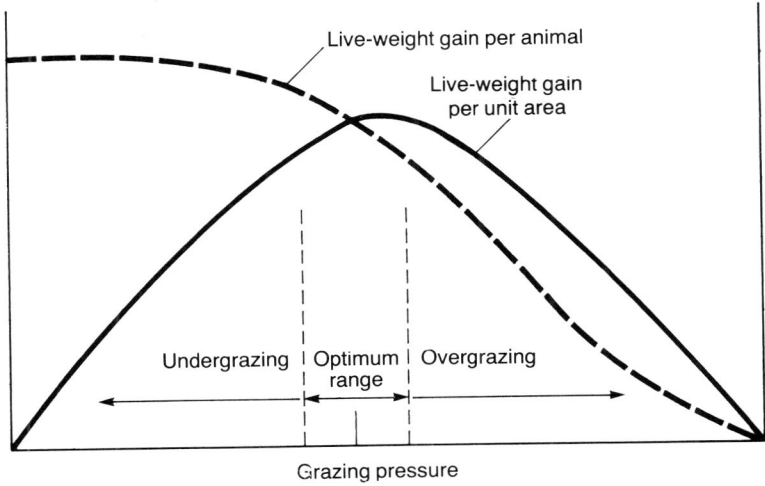

Figure 9.3. Relationship of grazing pressure to gain per animal and gain per unit area of land (**Source:** Matches & Mott, 1975)

81

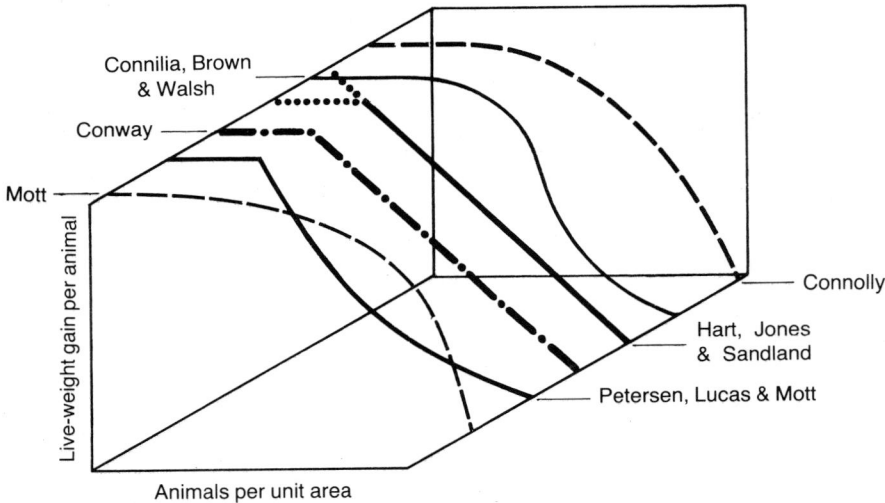

Figure 9.4. Proposed relationship between stocking rate and live-weight gain per animal (**Source:** Hart, 1978)

optimum, overgrazed. A careful watch must be kept on the pasture to ensure that overgrazing is not occurring, as once a pasture is overgrazed, rehabilitation is slow.

Various modifications of this curve have been devised and are shown by Hart (1978) in Figure 9.4. They introduce a plateau in which gains per animal remain constant below the optimum stocking rate and decline either hyperbolically or linearly thereafter. There is some disagreement as to the shape of the curve at greater than the optimum (critical) stocking rate, some studies arguing that although the seasonal result may well be represented by a linear decline, the actual monthly movement will vary because the forage supply will vary month by month and differential rates of herbage removal affect both quantity and quality of herbage (see Fig. 9.5). In May in Wyoming, USA, with plenty of high-quality forage, gain is independent of stocking rate because gain is then limited only by efficiency in conversion by the grazing animals. In June, quality and/or quantity of forage has been reduced at the highest stocking rates, as have gains, but at lower rates, animals are still gaining at the same rate as in May. By July, quality has been reduced at all stocking rates as plants mature so maximum gain is reduced even at those stocking rates less than the critical rate. Critical stocking rates continue to decrease, reflecting combined effects of plant maturity, stocking rate, and increasing animal weight and maintenance requirements. This pattern continues through October, by which time only the lowest stocking rate is less than the critical rate and maximum gains are only 40 percent of maximum gains at the start of the season.

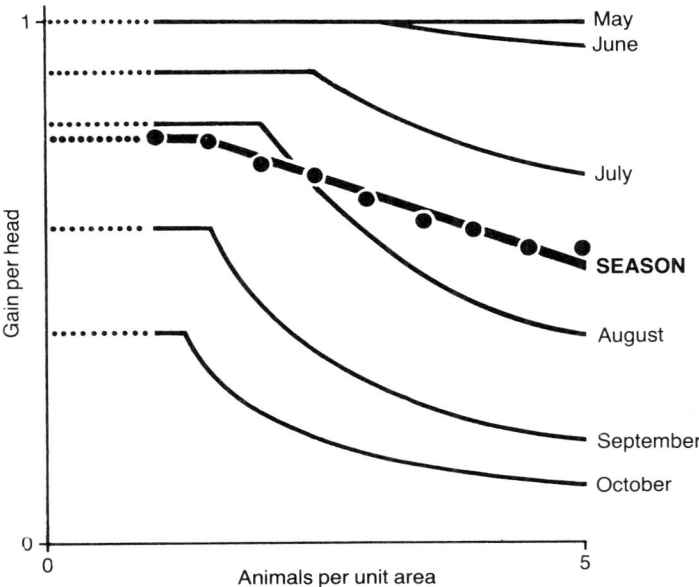

Figure 9.5. Hypothetical response of gain per animal to stocking rate for each month and for the entire season (**Source:** Hart, 1978)

The model produced by Jones and Sandland in 1974 (see Fig. 9.6) allows an accurate assessment of optimum stocking rate to give maximum gains per hectare. Maximum production occurs midway between zero and the stocking rate required to give no animal gain (maintenance only). The maximum gain per hectare is exactly half the stocking rate at which zero gain is predicted. It enables an estimate of "turn off" rate at different stocking rates from the mean rates of gain per day, and also of the stocking rate at which a desired weight gain within a year will be obtained.

The linear relationship at greater than the optimum stocking rate generally applies at all stocking rates likely to be employed experimentally and commercially.

Kind of grazing animal

Different animals have different grazing habits. Of the domestic animals, horses are selective grazers and crop the pasture close to the ground. They generally ignore browse plants. Cattle are less selective grazers and are less severe on pastures. They mainly eat grass, but will browse any edible shrubs which appear in their path. Sheep generally are close grazers, but will remove

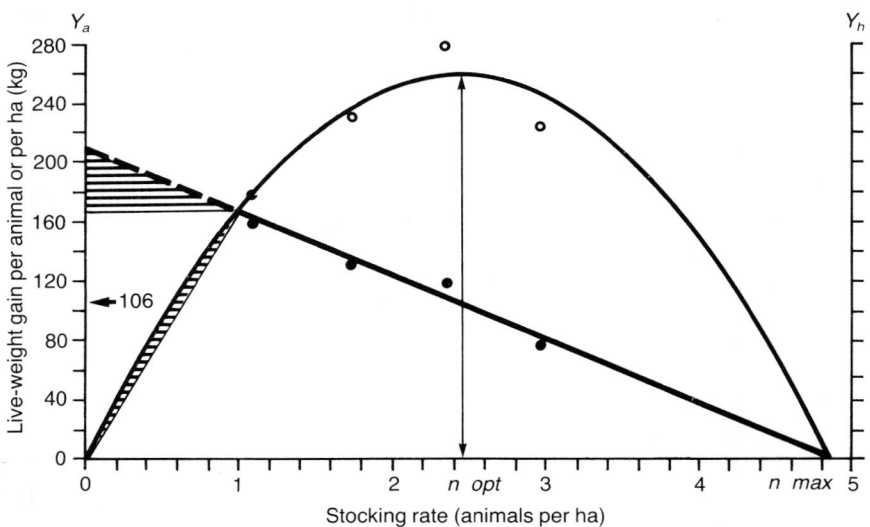

Figure 9.6. Relationship between gain per animal (Ya) and gain per hectare (Yh) in response to increasing stocking rate on a setaria/siratro pasture. The effect of holding animal gain constant below one animal per hectare is indicated (**Source:** Jones & Sandland, 1974)

the leaf of tall grasses from top to bottom. They usually select a diet much more nutritious than would be revealed by normal botanical sampling. Goats are mainly browsers. Grazed close to villages they can denude the areas immediately adjacent to the housing or keep the edible trees surrounding the site permanently pruned. An individual goat is not ordinarily devastating, but they usually graze in large numbers. Camels are normally browsers and can handle quite coarse, woody twigs, as well as leaf material. In areas where bush is invading grasslands, a mixed herd of cattle and goats can best control the problem, as neither animal individually would provide a satisfactory solution. Among the game animals there is also a wide variation in eating habits and surveillance of game parks requires a sound knowledge of the variations among species.

Large herds of zebra, wildebeest and Thomson's gazelle migrate in succession across the plains of Tanzania, the animals grazing selectively from herb levels of different chemical composition, according to their nutritional requirements. Zebras tend to feed on the upper parts of grasses and herbs, wildebeest on the middle parts, while Thomson's gazelle is the most selective, feeding on herbage of high protein content at the base of the sward. Migrations are synchronized with the availability of specific tissues of grasses. In the wet season, when grasses are longest, animals congregate in the short grass areas of the Serengeti Plain. In the early dry season when the grass stops growing and the short grass is removed, the amount of available feed declines and the animals move west in succession grazing on the longer grasses. In the late dry season the animals move more northerly and at each stage early members of the grazing succession prepare the structure of the vegetation for the followers (Bell, 1971). Buffalo and topi graze the long grasses, Grant's gazelle and eland are mainly browsers, while elephants, although eating clumps of grass, bark and knock down trees for their nourishment. Grass preferences have been studied and vary generally with the season.

Night kraaling

Throughout most of Africa, cattle, sheep and goats are housed in kraals or bomas at night, traditionally to protect them from the ravages of wild beasts, but also from burglary. There is also the advantage that they amass significant amounts of easily accessible manure for fertilizing fields or for crop growth in the manured area when the nomadic herds are on trek. The practice denies the animals any night grazing and often their release from the kraal is delayed until well into the morning.

Joblin (1960) studied the effect of restricted grazing compared with all-night grazing by zebu-type steers. He found that restriction of night grazing led to a significant decline of 30 percent in live-weight gain as the animals with a longer grazing period were able to select a diet of superior quality and quantity. Under very good or very bad grazing conditions the availability of night grazing did not appear to be critical.

10. Reseeding the arid and semi-arid range

Many of the tropical and subtropical rangelands have declined in fertility as a result of overgrazing, drought, wind and water erosion, and frequent fires. Even if visual evidence does not appear to confirm this, livestock production in real figures has fallen and botanical composition has changed as a result of continued selective grazing which has removed the more palatable and nutritious species.

There has always been a desire to reintroduce the original grass species or replace them with selected new species to attempt to restore the productivity. But this is not an easy task and often the cost of the reseeding programme has been lost. It is not economical to fertilize rangeland.

If the affected area is closed to stock the rangeland will gradually revegetate from seed lodged in protected areas or blown in by wind and, to a lesser extent, carried by water or animals. This is a very slow process, but has succeeded (see Plate 2) where there is some residual seed *in situ*.

In East Africa, Heady (1960a, b) gave little chance of success with reseeding areas receiving less than 375 mm of annual rainfall, but a high success rate in areas receiving more than 625 mm. However, any arable land receiving more than 625 mm will most surely be utilized for cropping so that reseeding to pasture will only be an option in difficult areas where special seeding techniques will be needed.

Heady also wisely suggests that the most suitable sites should be reseeded first. This would test the project and save a good deal of money if results were not satisfactory.

Selection of species

Experience throughout the world with reseeding has shown that the species most likely to be successful in a reseeding programme are native species found on sites similar to those being reseeded. Bogdan and Pratt (1967) state that

> The requirements of a grass for reseeding are that it must be sufficiently drought tolerant to survive, perpetuate itself and provide a good quality of herb-

age of fair or good grazing value. It should also produce an adequate amount of viable seed which can be easily harvested and it should be easy to establish.

This then brings up the matter of collecting seed in quantities sufficient to carry out the operation. It will involve some preliminary research on seed production, seed yield, dormancy, seed treatment for planting, seeding techniques and germination problems.

Bogdan and Pratt (1967) recommended 32 species for reseeding denuded rangeland in Kenya: *Aristida mutabilis, Bothriochloa insculpta, B. pertusa, Cenchrus ciliaris, C. pennisetiformis,*[1] *C. setigerus, Chloris gayana, C. mosambicensis,*[1] *C. roxburghiana, C. virgata, Chrysopogon aucheri, Cynodon dactylon, C. plectostachyus, Dactyloctenium* spp., *Echinochloa haploclada, Enteropogon macrostachyus, E. somalensis, Eragrostis caespitosa,*[1] *E. chloromelas, E. cilianensis, E. curvula, E. superba, Eriochloa fatmensis, Eustachys paspaloides,*[1] *Leptochloa obtusiflora, Leptothrium senegalense, Panicum coloratum, P. maximum, Paspalidium desertorum,*[1] *Sporobolus helvolus, S. marginatus,*[1] *Tetrapogon villosus,*[1] *Themeda triandra* and *Tricholaena teneriffae.*

In East Africa, Heady (1960a, b) found that *Cenchrus ciliaris, Bothriochloa insculpta* and *Eragrostis superba* had been most successful under a rainfall of about 625 mm per year, *Cynodon dactylon* (planted), *C. plectostachyus* (planted) and *Panicum maximum* had been successful with *Chloris gayana* succeeding for about four years in areas with rainfall above 750 mm.

In Arizona, United States, Humphrey (1960a) has used *Aristida adscensionis, Chloris virgata, Cynodon dactylon, Eragrostis chloromelas, E. curvula, E. superba, E. lehmanniana, Heteropogon contortus, Panicum antidotale* and *Sporobolus airoides* of which *Eragrostis lehmanniana* (Lehmann's love grass) has proved best adapted to reseeding Arizona ranges.

In India, *Cenchrus ciliaris, Sehima nervosum, Pennisetum polystachion* and *Dichanthium annulatum* are used in range reclamation (Jackobs, 1961). In Queensland, where little reseeding has been carried out, *Cenchrus ciliaris, Dactyloctenium giganteum* and *Anthephora pubescens* show promise.

There is scope for more autoecological studies of Australian grasses as several of these, such as reflexed panic (*Paractaenum novae-hollandiae*), woolly-butt grass (*Eragrostis eriopoda*), never-fail grass (*Eragrostis setifolia*) for fixing sand dunes in the 375 mm rainfall area; the perennial Mitchell grasses (*Astrebla* spp.); the Flinders grasses (*Iseilema membranaceum* and *I. vaginiflorum*) and button grass (*Dactyloctenium radulans*) as short-term annuals in the 400-500 mm area; and *Dichanthium tenuiculum* for winter greenness in depressions in the 800-900 mm heavy soil country, would seem to have use in several other tropical countries. There are several other promising species.

---

[1] Then untried.

Seeding

It is also important to plant the seed just ahead of the expected rains and double the usual seeding rate. Heady (1960a, b) recommends 1.12 kg/ha, if viable seed is needed, to give one established plant of each species 0.9 m^2. As hand-picked seed may contain a lot of chaff, it is necessary to know the content of viable seed in the harvest. Cook (1978) found that burning a native *Heteropogon contortus* pasture just prior to overseeding to reduce competition favoured seedling survival of sown species. Overseeding must be done in the wet season rather than in spring to reduce competition from native grass regrowth. Trees should also be killed just before seeding grasses.

Experience has shown that only under the most favourable conditions will uncovered broadcast seed produce acceptable stands. Bogdan and Pratt (1967) showed that the more complete the cultivation the better the stand. The main problems are the lack of seed cover and plant competition from any existing species in the higher rainfall areas. In sandy areas, seed may be able to bury itself with the help of rain. The passage of cattle grazing a grass sward carrying ripe seed will often result in satisfactory ground cover and this is often the aim of a deferred grazing system. A soil cover of 0.6-1.2 cm is satisfactory for small grass seed. With modern tree-pulling machinery in western Queensland where mulga (*Acacia aneura*) is pushed for drought feeding, the broadcasting of seed into the fallen tree canopy and also on to the tracks of the tracklaying tractor, and the scattering of buffel grass (*Cenchrus ciliaris*) seed around the bases of large woodland trees (such as *Eucalyptus populnea*) where fertility is higher (Ebersohn & Lucas, 1965; Christie, 1975b) has given good nucleus stands of grass.

Bogdan and Pratt (1967) list the following fundamental requirements for success in a reseeding programme:

- An appreciation of the ecological potential of the area concerned;
- Grasses suitable for reseeding purposes and sufficient seed of adequate quality;
- The integration of the reseeding operation into an overall land management policy, embracing grazing control and bush control where necessary;
- Some form of seed-bed preparation and a degree of seed protection in keeping with site requirements;
- A period of complete rest from grazing after seeding;
- Reasonable rains during the establishment seasons.

Machinery

Various types of machinery have been used for overseeding. Bogdan and Pratt (1967) used the Holt VIIIb weed breaker, cultivating the total area and leaving corrugations about 12.5 cm deep. It is similar in operation to the basin lister used in the United States. A home-made lister can be produced by removing two opposite quarters of disc plough. A set of eccentric discs can perform a similar pattern. The Kenya authors found this machine gave good germination but a poor final establishment, but in the United States such a pitting machine is highly regarded for its land preparation for moisture retention.

Good grass cover was obtained with the Rockland tiller, a heavy tine cultivator used by the United States Forest Service, giving complete cultivation with tines spaced 45 cm apart. Bishop (1973) had some success in revegetating clay-pans in northwest Queensland by making 15-22 cm high banks and then ripping the enclosed area to a depth of 15-22 cm with tines 1 m apart. He then ploughed with offset discs to make broad, spiralling banks. In other areas in the same region some success was achieved by seeding *Cenchrus ciliaris* behind a scalloped disc cultivator (see Fig. 10.1).

The covering of soil over the seed should be light and can be accomplished with a very light harrow or by dragging bushes across the seeded area.

Figure 10.1. A grass seeder mounted on a tandem offset scalloped disc cultivator for seeding rangeland. The planting mechanism is operated by cable from the rear ground wheel

Grazing

Reseeding rangeland is useless unless stock numbers can be controlled. It is especially important to protect the young seedlings from grazing for one or more years to allow the grasses to become fully established and to seed, thickening the stand. Subsequent grazing should also be controlled and the Arizona recommendation (Anderson *et al.*, 1957) is to remove livestock from reseeded areas when no more than 50 percent of the weight of the herbage has been consumed.

11. Handling difficult grasses

Axonopus affinis

Although *Axonopus affinis* originated in the Americas and has been used as a forage grass for a considerable time in those areas, its spread is indicative of declining fertility in the primary pastures established on newly cleared sub-humid coastal lands of tropical and subtropical Australia, Kenya and other countries (Cassidy, 1957).

In eastern coastal Australia the primary pasture established after land clearing for dairying was dominantly one of *Paspalum dilatatum* (paspalum or dallis grass) sometimes in association with *Trifolium repens* (white clover). On the upland areas of latosolic or red loam soils pure pastures of *Pennisetum clandestinum* (Kikuyu grass) were developed in the early dairying years.

There has been a decline in the productivity of these pastures and with the decline has come an invasion mainly by *Axonopus affinis* which in certain areas now covers up to 90 percent of the former paspalum sward, and the white clover component has virtually disappeared. *Axonopus affinis* flowers and sets seed early in the season and this is partly responsible for its decline in nutritive value, as after seed setting its crude protein content may fall as low as 4-5 percent. For a non-lactating cow a protein content of about 9 percent is needed and for milking cows a level of from 11-16 percent depending on the level of milk production. Phosphorus levels also fall at seed setting.

Axonopus spp. are used as permanent pastures in Florida, West Indies, Fiji, Malaysia, Guyana and Hawaii (McIlory, 1964) and a similar invasion of paspalum pastures has been recorded in Florida (Lovvorn, 1944) and Kenya. On the Atherton Tableland in north Queensland a Kikuyu-paspalum-carpet grass pasture has emerged after years of dairying.

The treatment of carpet grass dominated pastures aims at two goals: complete eradication and replacement by other grasses or crops, or manipulation of the botanical composition of the sward by mechanical renovation, fertilizer use, grazing management and supplementation.

Complete eradication is a drastic measure and removes the pasture from the production line, which may upset feed supplies. However, it can lead to the establishment of new pastures or fodder crops. It is usually effected by deep disc ploughing and subsequent chopping up with disc harrows or rotary hoeing to break up the sod to kill the grass more quickly.

Because of abundant seed production by *Axonopus* the pasture will quickly re-establish if cultivation is withdrawn. For this reason it is better to follow the ploughing out with a series of annual fodder or grain crops before planting a new pasture.

Declining yields from paspalum and Kikuyu pasture after years of extraction of soil nutrients mainly in milk and beef leads to a deficiency of soil nitrogen and phosphorus (Cassidy, 1957) and encourages the invasion by *Axonopus* spp. Cassidy showed improved performance by raising the fertility level with the addition of these elements through fertilizers in the Gympie district, Queensland. Gartner (1969) made specific measurements of the actual changes in botanical composition of the pasture with additions of nitrogen. He selected a pasture in the Millaa Millaa district of the Atherton Tableland, north Queensland (lat. 17°30'S, altitude 823 m, annual rainfall 2 570 mm) with an initial botanical composition in 1961 of 16 percent Kikuyu (*Pennisetum clandestinum*), 32 percent paspalum (*P. dilatatum*) and 58 percent narrow-leaved carpet grass (*Axonopus affinis*). He applied nitrogen as urea in split applications, half in late spring (October-November), and half in autumn (April-May) at rates of 56, 112, 224 and 448 kg per hectare per year over four years. An unfertilized control plot was included. The change in botanical composition is shown in Figure 11.1.

The crossover from the inferior carpet grass to the better paspalum-Kikuyu mixture occurred at an application of 168 kg N/ha but Kikuyu dominance was not achieved until 224 kg N/ha per year was applied. Carpet grass was virtually eliminated within a year from plots fertilized with 224 or 448 kg/ha, but it remained dominant in the unfertilized plot. Paspalum occupied an intermediate position in the sward, being more or less codominant with carpet grass and/or Kikuyu at the 56, 112 and 224 kg N/ha levels. Kikuyu dominated the 448 kg level.

Such heavy applications of nitrogen would have to be evaluated in terms of increased returns from animal products compared with the cost of fertilizer and its application. It has the advantage that the continuity of grazing is not interrupted.

Cassidy (1971), working in southeast Queensland (lat. 27°06'S) with a similar rainfall was unable to reduce the *Axonopus* component with 224 kg N/ha per year plus a basic complete fertilizer mixture including 49 kg P/ha per year. However, total paspalum yield increased sixfold and white clover went from an insignificant component to a major factor in increasing overall yield. The mat grass component was always 90 percent and there was no Kikuyu grass on the sward, as was the case with Gartner's trial. Cassidy (1971) suggested that a more competitive grass than *Paspalum dilatatum* was needed to oust the *Axonopus affinis,* such as Kikuyu.

Chemical treatment to reduce the dominance of *Axonopus* and encourage white clover growth was tried in Alabama, United States, by Searcy and Patterson (1961). They sprayed 2,2-DPA at 4.8 kg/ha a year to suppress the

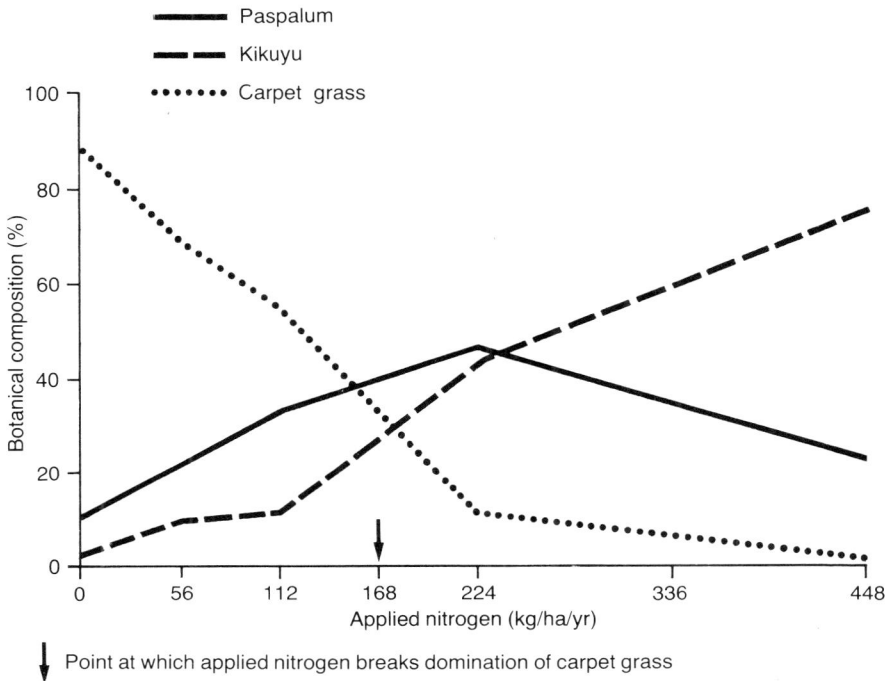

Figure 11.1. Effect of three years' nitrogen fertilization at different rates on the botanical composition of a Kikuyu/paspalum/carpet grass sward, November 1964 (**Source:** Gartner, 1969)

Axonopus and then sod-seeded white clover seed through the sod to the moist soil underneath. Murtagh (1977), on the north coast of New South Wales, confirmed the beneficial effect of 2,2-DPA in suppressing *Axonopus,* but the height of the grass and the time of application are significant factors in producing the desired effect. When grass leaves were 10 cm long and little white clover was exposed to the spray in March (autumn), increased clover yields up to 54 percent were obtained the following spring. An April spraying, with grass length at 5 cm and more white clover visible, significantly reduced clover yields. The most effective rate of herbicide application was 2.2 kg/ha per year. Murtagh suggested that the method be tested further.

In the absence of major soil fertility-building to replace *Axonopus,* supplementation of the ration for lactating animals grazing this grass is necessary to obtain satisfactory production. Supplementation with non-protein nitrogen (NPN) can lead to a marked improvement in production. It is, however, rarely possible to get a useful response in milk production quickly with simple NPN supplements without feeding extra protein and energy. Preformed protein (such as that in meat meal and legumes) is usually required for a quick milk production response where the level of protein is still too low when NPN is fed. Rather larger increases in energy are required because the cows are usually poor. Initially, grain or molasses must be fed at 5 kg or more per head per day, and must be introduced gradually to avoid scouring. A phosphorus supplement should be fed at all times. To combine the non-protein nitrogen, energy material and phosphorus, a daily intake of no less than 56 g of urea, 230 g molasses and 10 g phosphorus is needed. This can be provided by a mixture of 45 litres urea, 13.6 kg mono-ammonium phosphate, 17 kg molasses and 136 litres water.

This can be dispensed through a molasses-urea drum licker, or in the case of milking cows, can be fed in the feed trough in the bails (Edgley & Harle, 1974).

Cymbopogon nardus

Current ranch development in Uganda is being jeopardized by an apparent increase in *Acacia hockii* bushes, and in the tussock grass *Cymbopogon nardus* (previously *C. afronardus*). This grass is unpalatable to cattle (Harrington & Pratchett, 1972) and cattle have been known to die of starvation when an abundance of it, in green condition, was available (Harrington, 1974). The removal of *C. nardus* from fully stocked pastures improved cattle growth rates by about 30 percent, but the rate of recolonization can be extremely rapid, with all improvement eliminated within two years, under a management favourable to the species (Harrington, 1974).

Local pastoralists said that in 1907, *C. nardus* was uncommon. Tse-tse fly and its accompanying trypanosomiasis disease drove the cattle from the area and they were not reintroduced until 1960, after a successful bush-clearing scheme had eliminated the tse-tse fly. On the return of the cattle and the pastoralists in 1960, it was estimated that *C. nardus* made up 40 percent of the ground cover. Annual burning of the grassland had been accepted practice before 1907. Harrington (1974), after studying the effect of various factors came to the conclusion that reduced grazing pressure and reduced burning coupled with a high seed ascendency encouraged the spread of *C. nardus* to the detriment of the carrying capacity of the land.

Approximately 9 000 km^2 of east Ankole district in Uganda is potentially highly productive grassland. The altitude is 1 225-1 525 m, rainfall a bi-modal 750-1 000 mm, and the vegetation an *Acacia-Cymbopogon-Themeda* complex. The grass layer is widely dominated by *Themeda triandra* in the valleys and *Cymbopogon nardus* and *Loudetia kagarensis* on the hillsides. The whole area is usually burnt annually.

> *C. nardus* spreads only by seeds. The leaves contain bitter aromatic oils, the texture is rough and fibrous, and, except when very young, cattle avoid eating them,

thus allowing an annual build-up of coarse, rank vegetation which is difficult for cattle or humans to move through. The leaves attain a height of up to 1.5 m and one tussock can smother a circle up to 2 m in diameter, although only a fraction of this area is occupied by rooted stems. The smothering effect of the leaves, which do not stand upright but fall semi-prostrate around the plant is characteristic. The competitive advantage gained by the habit of growth is heightened by the production of fresh leaf immediately following burning in the dry season and thus, when the rains begin and associated grasses start to grow, *C. nardus* has already developed a smothering foliage. A further property of this species is the production of large numbers of flowering culms, up to 3 m high, after burning.

The difficulty of physical movement through a heavy *C. nardus* population results in cattle avoiding such pasturage, with a concomitant loss of large areas of potential grazing. When forced to graze these grasslands the cattle spend a disproportionate amount of time and effort in searching for edible grasses and this can cause a loss of up to one third of the potential growth of growing bullocks (Harrington & Thornton, 1969).

In 1967 it was reported that *C. nardus* had been eradicated from small paddocks in Kigezi District, Uganda, by night-kraaling cattle on infested paddocks.

Separation of the top-growth from the roots at ground level effectively kills the plant if no rain falls for four weeks after treatment. A team of labourers working in a line cutting off the tops took 72 worker-hours per hectare to clear a heavy stand of *C. nardus* completely.

Using fire and grazing management, Harrington and Thornton (1969) were able to convert a *C. nardus* dominated pasture, which was virtually useless to cattle, to a high-quality pasture. The *C. nardus* was burnt and the regrowth kept at 15 cm above the 10 cm stubble left after burning, by stocking until the new growth, unpalatable except to hungry stock, was eaten down by stocking at 25-30 bullocks/hectare. Early burning as soon as the grass would carry a fire, or burning twice a year, reduced the size of the *C. nardus* clumps and the dominance of the grass.

Application of 158 kg/ha of nitrogen per year to the pasture improved the nutritive value of the *Cymbopogon* and encouraged the associated *Brachiaria decumbens*. Stocking over $2^1/_2$ years at 4 beasts per hectare per year gave a better pasture than rotational grazing at 2 beasts per hectare on land cleared of *C. nardus*, and did not cause erosion or alter the botanical composition.

Digitaria abyssinica

This grass is widely distributed in the moister regions of East Africa from sea-level to 3 500 m, and is the most important of the rhizomatous grass weeds. *Digitaria abyssinica* (previously *D. scalarum*) occurs in a wide range of crops, including coffee, tea, sisal, pyrethrum, wattle, cotton and many other annuals and perennials. The growth and yield of crop plants is greatly reduced where the weed occurs and coffee bushes in particular can be completely killed by a bad infestation. It may fairly be regarded as the most troublesome of all East African weeds (Ivens, 1967).

It is a perennial grass which produces long slender rhizomes forming a dense mat beneath the soil surface. The rhizomes can grow to depths greater than 1 m and are extremely difficult to remove. Mechanical control by repeated cultivation in open fields is possible but in long-term crops it is almost impossible as the rhizomes penetrate among the crop roots where considerable damage can be done by attempts to dig out the weed (Ivens, 1967). Repeated digging with a forked hoe or digging fork will remove many rhizomes but total eradication of small viable fragments from the soil is virtually impossible and the method is both time consuming and costly (Terry, 1974).

Several chemicals have been tried. Richardson (1967) used Na-TCA (sodium trichloracetic acid) at 56 kg/ha to eradicate *D. abyssinica* in sisal almost completely. Ivens (1967) recommends the use of dalapon in coffee. Dalapon at 5.5 kg/ha sprayed at 550-1 100 litres/ha on young grass just emerging at the start of the rainy season is fairly effective, the spray being kept off the coffee foliage. Terry (1974) later recommended a follow-up spraying with paraquat five days after the dalapon treatment, to control annual weeds appearing. Parker (1970) achieved good control with asulam at 4.48 and 6.73 kg/ha but it did not last for long. Terry (1974) tested glyphosate (formulated as MON-1139), asulam and dalapon on African couch grass and found glyphosate — also sold as "Roundup", the most effective. One week after the application of 2 and 4 kg/ha glyphosate approximately 50 percent of the *D. abyssinica* foliage was chlorotic; by two weeks the chlorosis was severe. At 47 days there was almost complete control with the 2 kg and 4 kg/ha treatments. There was a prolific growth of young annual weeds where *D. abyssinica* had been controlled but retreatment with 1 and 2 kg/ha glyphosate gave excellent control of *D. abyssinica* and annual weeds were greatly suppressed.

At Nioka, Zaire, Risopoulos (1966) found that *Setaria sphacelata,* which is indigenous to the savannahs, can dominate couch grass if it is not burnt or grazed, and regenerates the soil. A *Setaria* cover for three to six years ensures an optimum build-up of fertility to up to 60 percent of the soil's initial fertility.

Eleusine jaegeri

Eleusine jaegeri is a grass of high altitudes in East Africa, occurring widely at altitudes above 2 000 m in Ethiopia, Kenya, Tanzania and Uganda. It is avoided by stock and its common name is derived from its frequent occurrence on the sites of abandoned Masai cattle enclosures (manyattas). It invades pastures of *Themeda triandra* (red-oat grass) and Kikuyu (*Pennisetum clandestinum*) (Ivens, 1967).

Ivens (1967) suggested that it could be controlled in *Themeda triandra* pastures by periodic burning. O'Rourke, Terry and Frame (1976) imposed the following treatments on an *Eleusine jaegeri* grassland on the rim of Empaki Crater in the Ngorongoro Highlands in Tanzania. The elevation of the area is 2 600 m and the mean annual rainfall 900 mm. The following treatments were imposed:

- Cut and left — *Eleusine* was cut close to ground level and cuttings left lying.
- Not cut — *Eleusine* was not cut.
- Cut and removed — *Eleusine* was cut close to ground level and cuttings were removed.

Chemicals were superimposed on the above cutting treatments at the following rates:

Chemical	Rate (kg AI/ha)
Dalapon ("Dowpon")	5.0
Dalapon	10.0
Glyphosate ("Roundup")	1.0
Glyphosate	2.0
Glyphosate	4.0
No chemical	—

All treatments were applied in water at a volume rate of 300 litres/ha with an Oxford Precision Sprayer operating at a pressure of 2.1 kg/cm^2 with size "0" fanjet nozzles. Applications were made at midday in warm, sunny, windless conditions and no rain fell within 24 hours. *Eleusine* was healthy and vigorous; mature plants had inflorescences and fresh green shoots were emerging from stumps cut eight weeks earlier.

- Destumped — *Eleusine* plants were individually dug out and removed.
- Burned — *Eleusine* foliage was destroyed by broadcast burning.
- Spot-burned — individual *Eleusine* plants were burned with a drip torch.

Destumping proved to be the most desirable method of control allowing desirable forage plants to take the place of *Eleusine*. Cutting and burning *E. jaegeri* is useful in *Pennisetum clandestinum* pastures. After cutting and removing cuttings, dalapon at 5 kg/ha reduced *Eleusine* and increased beneficial forage plants. Heavier doses did not lead to increased forage. All the chemicals reduced the *Eleusine* population, but dalapon at 5 kg/ha was the least costly.

Imperata cylindrica

Imperata cylindrica is a perennial grass with vigorous rhizomes that are difficult to eradicate. It also produces abundant seed, which is seated in one-flowered spikelets surrounded by long silky hairs from the obtuse basal callus (see Fig. 15.85). The spikelets fall entire from the pedicels and by means of the silky hairs are carried considerable distances. Fire is common in many tropical countries during the months when *Imperata* flowers, so that the seeds are blown far and wide to come to rest on areas that have been burned. When rain falls *Imperata* seedlings emerge.

The total area of *Imperata* grasslands in the world has been estimated at 500 million hectares, about 200 million of which are in Southeast Asia (Martoatmadjoun, 1976). Soerjani (1970) estimated that *Imperata* vegetation has covered more than 16 million hectares of waste land in Indonesia, with an annual increase of more than 150 000 hectares (1.25 percent). Another 23 million hectares of land are still used for shifting cultivation, which is the main source of the increase in the *Imperata* area. Total land being used for shifting cultivation increases by 100 000 hectares (1.05 percent) annually. In Malaysia there are an estimated 1.6 million hectares of *Imperata*, which are largely restricted to the understorey of rubber plantations.

In the past farmers cleared primary forest to obtain rich soil for cultivation. Because of insufficient fertility maintenance after crop removal, the soil became very poor in nutrients, and the farmers left this land to look for another piece of primary forest. The abandoned land was then rapidly occupied by *Imperata*, which, because of its competitive ability and resistance to fire, could become almost a pure stand. Farmers could only take one or two harvests of crops such as maize and upland rice before abandoning the area. Such shifting cultivation has dramatically increased the land under *Imperata*. Its competitiveness is due to its vigorous rhizome system, its high seed production and ability to colonize, plus some allelopathetic mechanism in the roots. The world distribution of five varieties of *Imperata* is shown in Figure 11.2 (Hubbard *et al.*, 1944).

Imperata has a number of different uses. It is widely used for thatch. Its extensive rhizomatous system effectively protects the soil from erosion. Indeed, without *Imperata* much land would be completely ruined (Semple, 1956). It is useful for packaging. It is readily burnt but survives fire to provide useful, if scant, grazing of young regrowth for village livestock (Lebrun, 1932; Holmes, Lemerle & Schottler, 1976). It can, with concentrates, provide an important roughage for a more adequate diet (Soewardi *et al.*, 1974). Hence, there is a tendency on the part of villagers to live with it, clearing more land in new or "fallowed" areas for cropping. It is a cheap system and in the absence of finance, heavy machinery, fertilizers and technology, little improvement is contemplated. However, the amount of cultivable land is decreasing and shifting cultivation must make way for a more permanent sys-

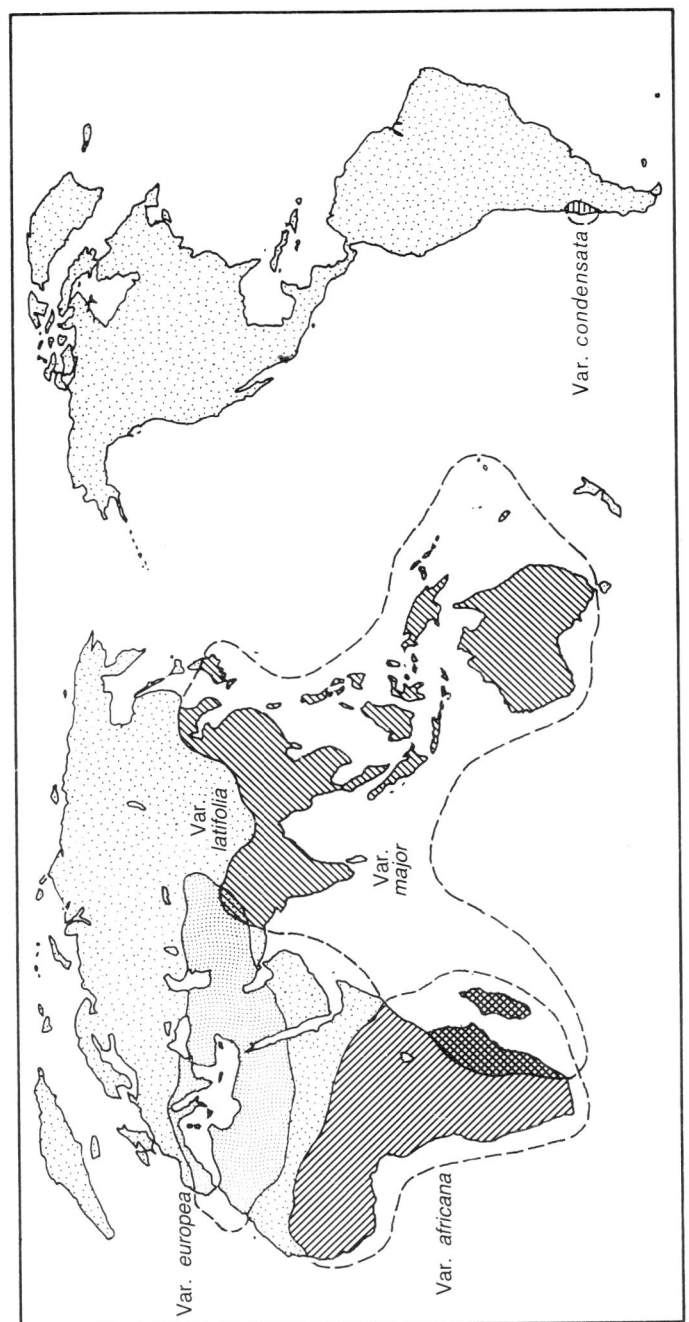

Figure 11.2. World distribution of *Imperata cylindrica* (**Source:** Soerjani, 1970)

tem of agriculture. *Imperata* is recorded as having seriously affected yields of rice (Medan, Indonesia), cotton (Uganda), teak forests (Indonesia), coffee (Uganda), cinchona (Indonesia), tea (Uganda, Indonesia), oil palm (Malaysia) and annual crops of maize, sweet potatoes, cassava, groundnuts and other crops throughout the tropical world. It therefore must be effectively controlled.

Methods of suppression or control

The main means of suppression and control can be dealt with under four headings — mechanical, chemical, ecological and biological.

Mechanical control. Before dealing with the actual mechanical methods it is necessary to consider the behaviour of the rhizomes. The above-ground growth annually replenishes the food reserves of the rhizomes. The rhizomes may penetrate to a depth of 15 cm in heavy soils and to about 60 cm (rarely up to 85 cm) in light soils. The roots may grow vertically downwards to a depth of 80-150 cm (Soerjani, 1970).

Ploughing. This is the most satisfactory way to destroy the *Imperata* completely. Deep ploughing is preferable for quick results. Deep ploughing to a depth below 15 cm should disturb more than 50 percent of the rhizomes. While it does not affect the sprouting ability of the chopped rhizomes, because chopping into short lengths breaks apical dominance and encourages shooting, deep covering reduces the length of shoots produced and weakens their survival ability after subsequent cultivations.

Burning before ploughing is common as it increases visibility and removes a great bulk of dead grass which may interfere with the operation of the machinery. One ploughing will usually not be sufficiently effective and is usually followed by a second (preferably at right angles) after a lapse of a month or so. Risopoulos (1966) used a single ploughing with a Rome plough in Zaire, and a single ploughing has been recommended for South Pacific islands. At Ingham, north Queensland, Larkin (1965) ploughed deeply with a "Majestic" disc plough, then with an offset disc cultivator, rolled the land level and disced it again before seeding. Tilley (1977) recommended repeated cultivation under dry conditions to eradicate *Imperata* for crop production. The rhizomes are desiccated by low humidity (Soerjani, 1970). Burial of the rhizomes by ploughing is more effective in the subtropics as 20°C is the minimum temperature at which the rhizome pieces sprout (Soerjani, 1970).

Slashing. This removes the top growth and therefore reduces the food reserves in the rhizomes. However, slashing by itself does not eliminate the rhizomes. Slashing every ten days over a three-year period still left a small number of rhizomes in the soil as a source of regrowth but the soluble carbohydrates in the rhizomes would have been severely reduced (Soerjani, 1970). One advantage with slashing is that the harvested material can be used as a mulch, for example in coffee plantations.

Chemical control. *Herbicides.* Most chemicals are expensive and their use for the treatment of the extensive areas of *Imperata* are out of the question. Treatment of small strategic areas may be practicable. In Malaysia sodium arsenite has been used extensively against *Imperata* in rubber plantations. It was banned in Malaysia in 1969 and much work has been done to develop less toxic chemicals. Special mineral oils have been marketed for *Imperata* treatment and mixtures of sodium chlorate and dalapon or aminotriazole have been found to give long-lasting control. By itself, dalapon appears to be somewhat variable. Under favourable conditions a dose of 11 kg/ha can be successful; sometimes as in East African trials, even 22 kg/ha is ineffective. In northeast Thailand, McLeod (1972) obtained 95 percent kill with 22.6 kg/ha in split applications six weeks apart. In Malaysia best results have been obtained with treatments applied to standing grass without preliminary cutting. Under the drier conditions of East Africa there is likely to be more chance of success by cutting to stimulate new growth, which would be more vulnerable to herbicides.

In rubber plantations it is used at 10-15 kg/ha with lower strength follow-up treatments. It is a slow-acting, systemic herbicide that can be absorbed by both the leaves and the roots, but quicker results are attained by spraying it on the leaves. At more than 20 kg/ha it may defoliate the rubber trees. A complete kill is not obtained but it severely retards growth.

In Colombia, *Imperata* has been controlled in oil-palm plantations by 2,4,5-T at 10 kg/ha with isolated regrowth treated with a 2,4,5-7 treated cloth.

In Nigeria, in oil-palm plantations, paraquat applied five times at fortnightly intervals exhausted the root reserves of the rhizomes to give a large measure of permanent control (Ivens, 1967). Repeated sprayings with paraquat at 0.2 to 0.7 kg a.c./ha in about 100 litres of water with a surfactant such as 0.1 percent teepol or agral is used in Indonesia, India and Malaysia in rubber and tea estates (Soerjani, 1970). In north Queensland Tilley (1977) used glyphosate at 2 litres of a 360 g AI per litre of product (e.g. Roundup) per 200 litres of water sprayed to wet the *Imperata* foliage but avoiding runoff for spot spraying, making sure to avoid non-target plants.

It is important when planning the use of herbicides to take full consideration of local environmental, edaphic and biological conditions, as each herbicide requires a specific ecological niche to exert a positive response.

Fertilizers. Raising soil fertility levels with phosphorus, which encourages smothering by legumes, is a promising method of suppression. *Imperata* responds to nitrogen and the nitrogen need is so high that one of its main competitive effects is on nitrogen utilization.

Ecological control. This can take many forms, such as flooding, burning and plant competition.

Flooding. Flooding with water retards the growth of *Imperata,* but does not eradicate it. It is, however, very effective in paddy rice cultivation where *Imperata* is not a problem.

Burning. Imperata burns readily, even in the green state, and millions of hectares are burnt each year, either intentionally or accidentally. Burning removes most of the above-ground growth but has little effect on the rhizomes, so burning really prepares a non-competitive situation in which *Imperata* can persist and quickly dominate. Repeated burning at short intervals has the same effect as slashing because the young new growth will not immediately burn.

Plant competition. Imperata is intolerant of shade and as its rhizomes extend only to a depth of 15 cm, deep-rooted plants can compete with it for water and nutrients, so that a deep-rooted plant that develops a canopy to overtop the *Imperata* can eventually suppress it. One such plant is *Eupatorium odoratum,* common in Southeast Asia as a somewhat weedy shrub. It is resistant and can survive fire along with *Imperata.* When *Imperata* ceases growth at flowering *Eupatorium* can overtop it and gradually suppress it by shading. The sequence of succession is described by Eussen and Wirjahardja (1973). *Imperata* at 90-100 percent population is burnt, this is followed by equal populations of *Imperata* and *Eupatorium* which is also burnt and the *Eupatorium* decreases in favour of taller-growing tree species to form secondary forest. This, if cleared, will still contain some *Imperata* that can regenerate. In Malaysia, *Cordia cylindristachya* produces similar effects (Gilliland *et al.*, 1971). Jagoe (1949) showed that *Leucaena leucocephala* (petai) could be established in *Imperata* pasture in Malaysia as a browse plant. Its subsequent canopy suppressed the *Imperata* and enabled better grasses to become established.

Such a system of ecological control can be made more effective if the species used to compete with *Imperata* by deep rooting and an effective canopy has commercial value, such as for timber, food or livestock forage. The following are selected from a list provided by Soerjani (1970) with some additions:

Crop
    Euphorbiaceae — cassava (*Manihot esculenta*)

Creepers

    Papilionaceae   — *Calopogonium mucunoides*
                              Pueraria phaseoloides
                              Vigna hosei
                              Centrosema pubescens
                              Desmodium intortum
                              Macroptilium atropurpureum

Bushes/shrubs

    Papilionaceae   — *Crotalaria* spp.
                              Stylosanthes spp.
                              Moghania macrophylla
                              Gliricidium sepium
                              Sesbania grandiflora
    Mimosaceae   — *Leucaena leucocephala*
                              Albizia falcata

Trees

    Myrtaceae   — *Syzygium aromaticum* (clove tree)
                              Eucalyptus grandis (eucalyptus)
    Pinaceae   — *Pinus elliottii*
                              Pinus merkusii
    Euphorbiaceae — *Aleurites moluccana*
    Bombaceae   — *Ceiba pentandra* (kapok tree)
    Anacardiaceae — *Anacardium occidentale* (cashew nut)
    Moraceae   — *Artocarpus integra* (jack fruit)
                              Morus australis (mulberry)
                              and several others

It is important when using such species that fire be prevented. Some species such as *Pinus merkusii* and *Eucalyptus grandis* do have some fire resistance.

Biological control. *Insects.* A Cecidomyed fly, *Orscotiella javanica*, attacks *Imperata* in Indonesia. The insect forms galls like an onion leaf on the plant. The larvae reach the plant through the lower leaves and penetrate the apical meristem. The leaf-sheaths become fused and the growth of the plant is affected. It is more noticeable in slashed plants. Although the control effected by the insect is not great, it is thought that in association with slashing, more suppression could be exerted (Soerjani, 1970).

Root parasites. Ivens (1967) reported the discovery in Nigeria of a plant (*Sopubia ramosa*) that parasitizes the roots of *Imperata,* but its controlling effect needs to be evaluated.

Livestock grazing. Cattle will readily graze the young, soft regrowth of *Imperata* after burning, but will avoid mature material. Elephants, however, are reported to graze the mature plant in the dry season around Lake Edward in Uganda (Field, 1971). In New Guinea, heavy continuous grazing of *Imperata* regrowth after burning around the villages has so weakened *Imperata* that it can be eradicated more easily for a cropping or improved pasture sequence. Similar reports have been received from Sri Lanka and most parts of the tropical world.

The huge area of *Imperata* would not permit such a practice to be used on a large scale, as there would be insufficient livestock to impose such a high grazing pressure over large areas. However, selected smaller areas could be so treated and the *Imperata*-dominated land reclaimed.

Recommended measures for control of *Imperata*

It is obvious that no one method will effectively control *Imperata* except complete deep cultivation for crops. Thus, a combination of methods must be used. In a village situation burning and grazing may be all that can be achieved. Burning is a cheap and effective preliminary operation to improve visibility and expose the extent of the improvement needed, but it must not be practised where it will damage associated commercial crops.

Plantation crops. Plantation crops will need protection from *Imperata* during the establishment phase until the crop canopy can suppress the grass. Ploughing and subsequent disc or rotary cultivation can take place to prepare a seedbed for the establishment of cover crops where grazing is impractical and for improved legume-based pastures where grazing can be permitted.

Suppression in forestry plantations. In forestry plantations it is only necessary to suppress the *Imperata* until the dense forest canopy can dominate the site. Individual trees planted in cultivated rows or cleared individual planting sites can be protected by periodically slashing the grass to keep it in check.

Grazing in non-cultivable land. In Erap, Papua New Guinea, Chadokar (1977) successfully oversowed siratro (*Macroptilium atropurpureum*) seed at 6 kg/ha with fertilizer onto burnt *Imperata* on a heavy soil. Sowing in the ashes of the burn gave a very good stand of siratro. A light discing of the burnt area and/or rolling after sowing further increased the stand. Stylo contributed 30 percent of the average dry-matter yield of 11 890 kg/ha from the discing treatment. This was equivalent to the yield of plots of *Imperata* fertilized with nitrogen at the rate of 240 kg/ha.

There is usually some bare space between *Imperata* tussocks, which can be exposed by burning. These can be colonized by vigorous creeping legumes such as *Centrosema pubescens, Macroptilium atropurpureum, Desmodium intortum* and the erect growing *Stylosanthes guianensis* if the soil fertility is improved with adequate supplies of phosphorus especially, and potash, sulphur and other minor element deficiencies such as molybdenum, copper and zinc. The comparatively large seeded legumes are stimulated into vigorous growth by the phosphorus and can dominate the *Imperata*. Blair, Pualillin and Samosir (1978) improved *Imperata* grassland in Sulawesi, Indonesia by burning the *Imperata,* applying such a nitrogen-free fertilizer mixture and oversowing the legume *Centrosema pubescens.* Responses on this glossudalf soil were recorded to sulphur and molybdenum and in one experiment elemental sulphur applied at 13 kg/ha increased the yield from 4.07 t/ha green weight to 10.49 t/ha over a 17-week period. The increase in *Centrosema* production led to a decline in the growth of *Imperata cylindrica* and an improvement in animal production. In northeast Thailand, *Desmodium intortum* developed a dense stand up to 50 cm high and overshadowed *Imperata.* In north Thailand, Gibson (1976) showed that legumes could be successfully introduced into *Imperata* grassland after cutting, burning, heavy grazing or light cultivation

Full cultivation, fertilizing and sowing improved pastures. This is the most effective way of dealing with *Imperata* grassland where the land is cultivable. It is costly, but if well done the increased return from animal production at higher stocking rates will usually recoup the cost.

At Ingham in north Queensland, oversown *Centrosema pubescens* at 2.5-3.5 kg/ha, molasses grass at 1-2.5 kg/ha and Guinea grass at 2.5-4.5 kg/ha, with Para grass hand-planted later, gave an excellent pasture. The land was deep ploughed, then disced and rolled to level it, with a second discing just before planting. It was sown mid-December and the pasture was well established before the main wet season (Larkin, 1965).

In Zaire, Risopoulos (1966) ploughed *Imperata* grassland with a Rome plough after burning at the beginning of the rainy season, sowed *Stylosanthes guianensis* at 3 kg/ha, kept the pasture closed to grazing for eight to nine months and then grazed it with cattle. A rotary slasher was used as needed there after. At the beginning of the experiment in 1956, the ratio of *Imperata* to *Stylosanthes* was 5.5:3.2, in 1957 it was 1:5, in 1958, 1:5, and in 1959, 1:5.

In the South Pacific region ploughing after an *Imperata* burn has been recommended. This can be followed by drilling in inoculated legume seed with Guinea grass or *Brachiaria brizantha* after fertilizing with 370-500 kg/ha of superphosphate.

At Tully, north Queensland the natural carrying capacity of *Imperata* pasture has been one beast per eight hectares, with an annual live-weight gain of 5-7 kg/ha. By ploughing the *Imperata* soon after the wet season, about May or June, at a depth of 10-15 cm, the *Imperata* roots are brought to the surface where they dry out and die. Three or four passages with a disc harrow follow to prevent regrowth and work the surface to a fine seed-bed. Seed of Guinea grass at 5 kg/ha and centro (*Centrosema pubescens*) at 5 kg/ha (the latter inoculated with effective *Rhizobium*) are broadcast and the seed rolled in during December-January when there is adequate moisture and follow-up rains are expected. A dressing of 450 kg/ha of superphosphate was needed originally to establish the pasture. Hand weeding was necessary to prevent hormone damage to the *Centrosema*. The first grazing was given toward the end of the wet season when the Guinea grass was commencing to seed. After grazing to about 30 cm every four to six weeks and slashing uneaten stalks to a similar height, the pasture was well established. The final carrying capacity was one beast per 0.4-0.8 hectares, with a live-weight gain of 440 kg/ha, compared with one beast per eight hectares, and an annual live-weight gain of 5-7 kg/ha (Sweeney, 1961).

Pennisetum polystachyon

Also known as mission grass, it occupies thousands of hectares of the dry and intermediate zones of Viti Levu, Fiji, mainly on steep hills with a highly sulphur-deficient black soil. It was introduced to Fiji in the 1920s (Partridge, 1975). It is also abundant in northeast Thailand where it is called khachornchob. In both cases it is regarded as a fire-climax, due to regular burning. It is also grown in India where it is known as thin Napier. Until recently it was mainly utilized as a cut-and-carry grass, and in India it is regarded as a useful fodder species and has been fertilized with nitrogen. It has also been made into silage.

It has long been regarded as a weed in Fiji and Thailand, but it has been observed to be palatable in its young growth and also remains green in the "dry" season in Fiji if grazed to prevent flowering. It flowers in April and seeds in May-June, after which the flower stems lignify to a completely inedible straw (Payne *et al.*, 1955). Its bulk and height (two metres) block both light and grazing animals from reaching the lower green leaves. Hence, little use has been made of it for grazing until recent years, a major factor being the lack of costly fencing to impose adequate stocking rates. It does not persist under heavy grazing, and set-stocked at 2.5 animal units per hectare it can be suppressed (Ellison & Henderson, 1973).

The poor tillering ability and tussocky nature of *P. polystachyon* allow legumes to grow in association with it. Several legumes have become naturalized, including *Desmodium heterophyllum* (hetero) (Parham, 1955), which responds to phosphate (Payne *et al.*, 1955). The grass will burn to ground level to facilitate the oversowing of legumes.

Partridge (1975) successfully established siratro (*Macroptilium atropurpureum*) on mown (because it was too wet to burn) mission grass. It was fertilized with 450 kg/ha single superphosphate with scarified and inoculated siratro seed broadcast at 5 kg/ha. The crude protein content of the grass improved considerably in association with the siratro, and it was concluded that mission grass could be converted to a good pasture by burning it to the ground and oversowing with siratro plus superphosphate. The naturalized legume *Desmodium heterophyllum* will automatically take its place if shading can be reduced by grazing management. *Stylosanthes guianensis* can replace siratro.

Such heavy application of fertilizer will be costly and Partridge (personal communication) has recently determined the commercial fertilizer recommendations shown in Table 11.1.

Sorghum halepense

Johnson grass was collected as early as 1696 at Aleppo, Syria. It is still known in Europe and north Africa as Aleppo grass or Sorgho d'Alep. It is regarded as a native of the Mediterranean coastal countries of Europe, Africa, and Asia, extending eastwards into India. It was introduced into South Carolina from Turkey in 1830. Ten years later it was popularized by a Colonel William A. Johnson of Selma, Alabama, which accounts for its present name. Despite its weedy habits Johnson grass was recommended as a crop by certain agricultural authorities as late as about 1890, and again about 1935 in the USA. It is such a noxious weed pest in the USA that its importance as a pasture and forage crop is often overlooked. It is probably the leading perennial hay crop of the southern states of the USA, and furnishes an appreciable portion of the pasture. In some years 60 000 hectares of Johnson grass are cut for hay in Texas alone and it is widely spread throughout the south and as far north as the thirty-eighth parallel from the Atlantic coast to the Colorado border. It is found even farther northward in the Potomac and Ohio valleys and in California, Oregon and Washington. Johnson grass seldom has been sown during the past [70] years (Martin & Leonard, 1959).

TABLE 11.1 **Fertilizer application schedule for the suppression of *P. polystachion***

Timing	Treatment
Initial dressing	250 kg/ha single superphosphate (22 kg P, 25 kg S)
Second year	Nil
Third year	125 kg/ha (11 kg P, 12 kg S)
Fifth and every subsequent two years	Sulphur reinforced fertilizer providing 4 kg P and 14 kg S

NOTE: A stocking rate of 2.5 beasts per hectare is suggested.

The grass has spread to other areas of the subtropics and is at present a most serious problem on the black earths of Queensland's Darling Downs, one of the main granaries of Australia.

Having learned too late of its invasive powers, persistence and competitiveness, farmers in the southern United States have learned to live with it, rather than attempt costly eradication. To improve the quality of its hay in the United States, soybeans or sweet clover are sown when the Johnson grass sod is being renovated and the combined crops are made into hay. Rough pea, a winter forage, is sometimes planted into frosted Johnson grass and the field used for winter forage, the pea providing green material during the winter. *Medicago* sp. crops infested with Johnson grass can also provide useful mixed hay.

Successive mowings, slashings or grazings can gradually reduce the vigour of the rhizomes, making them easier to destroy and allowing other grasses, such as *Paspalum dilatatum* or *Chloris gayana,* to compete more successfully. Mowing at eight-week intervals during two successive growing seasons greatly reduced the Johnson grass population on the Darling Downs, Queensland (Marley, 1978). Constant cultivation of "bare" fallows over 12 to 18 months will keep it in check.

Chemical treatments can also be effective. Seedlings of Johnson grass, if treated prior to rhizome formation, are susceptible to post-emergence applications of paraquat (1,1'-dimethyl-4,4'-bipyridylium dichloride), dalapon (sodium 2,2-dichloropropionate), MSMA or low rates of glyphosate. A number of herbicides are registered for control of established Johnson grass. These include Bromacil (5-bromo-3-sec-butyl-6-methyluracil), TCA sodium chlorate, dalapon, MSMA and glyphosate. Probably the most cost-efficient of these chemicals in eradication of Johnson grass is glyphosate. It is virtually non-selective, and will kill or severely thin associated vegetation. It has no residual soil activity and the application of this chemical must be followed, on death of the residual stand, with a residual herbicide. This is necessary to control the subsequent germination of seedlings which can quickly reinfest the area. Bromacil and diuron N'-(3,4-dichlorophenyl-NN-dimethylurea) are suitable residuals for the control of Johnson grass seedlings. Bromacil is preferred to diuron except in cotton areas.

The following programme of control is suggested by Marley (1978):

Burn or mow the standing vegetation in late December (summer) so that glyphosate can be applied to Johnson grass at the early flowering stage in autumn. Glyphosate is most effectively applied to Johnson grass at that growth stage and season. The dead material from the growth of a previous summer should support a fire in December. Deep ploughing can follow.

When the resultant Johnson grass growth commences flowering in late February or in March (autumn), apply glyphosate. Overall coverage of the plants including the seed-heads is essential. Rate of application is 6 litres per hectare of Roundup by boom spraying. The volume of water is not critical provided overall coverage is received. For spot spraying with high volume equipment, 1 part of Roundup in 100 parts of water is recommended. Rainfall within six hours of appli-

TABLE 11.2 **Commercial pesticide information**

Active ingredient	Product	Manufacturer
Glyphosate	Roundup (36% w/v)	Monsanto
Bromacil	Hyvar X (80% w/v) or	Du Pont
	Hyvar X-L (25% w/v)	Du Pont
Diuron	Several (80% w/v)	Du Pont, Bayer, ICI, Lane Chemical and Air Services
	Flowable diuron (50% w/v)	Ciba Geigy
Paraquat	Gramoxone (20% w/v)	ICI
MSMA	Daconate 8 (80% w/v)	Ag. Chem. Pty Ltd
	Nocweed MSMA (50% w/v)	Lane Ltd

cation will probably reduce effectiveness. When the aerial parts have died, and before the following spring, apply bromacil at 7 kg active ingredient (AI) per ha or diuron at 18 kg AI per hectare. Care must be taken that bromacil is not introduced into the area of water flow to irrigated crops. Both these chemicals should give season-long control of germination of annual broad-leaved weeds and grasses as well as those of Johnson grass and Columbus grass.

Any regrowth from original rhizomes of Johnson grass or Columbus grass during the following spring should be treated as previously with glyphosate. A close watch should be kept on the treated area for several seasons. Any indications of re-establishment by seedlings should be prevented by immediate retreatment with bromacil or diuron.

Where selective chemical control is needed repeated applications of MSMA are effective in the control of Johnson grass. Green couch (*Cynodon dactylon*) and Rhodes grass (*Chloris gayana*) tolerate repeated applications of MSMA. Apply the MSMA at 4 kg AI per hectare several times in each of two consecutive seasons at the early flowering stage of Johnson grass. Temperatures above 25°C at application are required for maximum effectiveness.

The chemical data for Australia as of January, 1978 are shown in Table 11.2.

12. Grasses for special purposes

Several grasses are adapted to ecological niches other than that of an ordinary pasture sward.

Tropical grasses exhibiting some tolerance to salinity and/or alkalinity

As the demand for arable land increases, pasture is increasingly being pushed on to soils of low fertility or which are otherwise unsuitable for cropping. These include soils variously affected by salinity. Grasses with some adaptation to saline conditions are required for pasture and to protect the soil against erosion or enlargement of the saline areas.

Salinity is common both in dry land and in tidal areas. The causes of dryland salinity have been summarized by Hughes (1979) in Table 12.1. Salinity in coastal areas occurs when flat land adjacent to the sea is periodically flooded by tidal water.

Salt-affected soils may be classified according to either chemical or morphological systems. According to the chemical classification, saline soils contain concentrations of soluble salts high enough to depress plant growth; sodic soils (formerly called "alkali" in the United States) contain enough exchangeable sodium to affect plant growth adversely. The salt tolerance of plants is usually determined by measuring the electrical conductivity of a soil saturation extract (EC_e) obtained from the active root zone. The lower limit for a saline soil has been defined as 4 mmhos/cm (electrical conductivity of the saturation extract at 25°C) and for sodic soils as 15 percent exchangeable sodium (Bernstein, 1962).

Maas and Hoffman (1977), evaluating irrigation water for crops, classified water suitability under the following system: 0-8, suitable for salt-sensitive crops; 8-16, suited to moderately sensitive crops; 16-24, suited to moderately tolerant crops; 24-32, suited to tolerant crops; over 32, unsuitable for any crops.

Salinity initially affects germination. Once germination has succeeded, subsequent growth may not be seriously retarded. Ryan, Miyamoto and Stroehlein (1975) showed germination of range grasses to be influenced not only by salt concentration (or osmotic pressure) but also by the nature of the

TABLE 12.1 **Summary of characteristics and causes of dry land salinity**

Process	Manifestations	Natural/Induced	Basic causes	Other determinants
Water-table	Bare, scalded areas Salt concentration at surface Adjoining areas with salt-tolerant vegetation Waterlogging Bare, flat areas	Induced	Clearing Saline ground water Poor and impeded drainage Soil erosion	Cultivation Capillary rise Evaporation Excessively wet season or seasons
Scalding	Deflated areas Surface crust or hard, impermeable surface	Partly induced, partly natural	Saline and/or sodic soil Surface sealing Overgrazing Wind and water erosion	Dry season Wet season (water erosion, run on waters) Evaporation
Saline springs (arid land)	Salt encrustations Soil salting Salt-tolerant vegetation Scalding	Mainly natural, partly induced	Saline aquifer Wet seasons	Clearing Grazing Evaporation

Source: Hughes, 1979

ions in the salt solution. Using solution concentrations of 50, 100, 150 and 200 meq/litre for each salt and germinating over 12 days at a temperature of 27°C they recorded the results shown in Table 12.2.

Aldon (1975) working with alkali sacaton (*Sporobolus airoides*) seed showed that it required almost zero moisture tension to germinate. However, large seeds about a year old were necessary, with an optimum temperature range for germination of 27-32°C. Dilution of the salt at planting and soon afterwards gave successful establishment, so the planting site should be saturated just before planting. At least 6 mm further rain is necessary within five days. Otherwise, a further irrigation is required (Knipe, 1967). Japanese workers (Okamoto, Kawatke & Horiuchi, 1975) also found that species with small seeds, such as *Panicum* spp. and *Chloris gayana,* were affected more severely than species with large seeds, such as *Sorghum* spp. In all, species germination rate decreased with increasing osmotic suction.

The salt concentration around the planted seed can be reduced by using a special planting technique. If a hill is formed by ploughing furrows in semi-arid areas affected by salt, evaporation of saline capillary water will leave a salt crust at the top of the furrow. If the seed is planted in a shallow furrow half way up the hill or in a shallow furrow on top of the hill (Price & Stokes, 1966) salinity will be lower. This concept has been used in the design of the

TABLE 12.2 **Salt concentrations needed to reduce germination (meq/litre)**

Species	NaCl	MgCl$_2$	CaCl$_2$	Na$_2$SO$_4$	MgSO$_4$
20% reduction					
Sporobolus airoides (alkali sacaton)	41	–	–	–	–
Panicum antidotale (blue panic)	62	54	105	106	150
Eragrostis lehmanniana (Lehmann's love grass)	53	54	100	64	200
Eragrostis curvula (weeping love grass)	165	100	>200	153	90
Eragrostis superba (Wilman love grass)	150	127	190	115	200
50% reduction					
Sporobolus airoides	113	–	–	–	–
Panicum antidotale	116	114	161	132	192
Eragrostis lehmanniana	81	97	184	83	>200
Eragrostis curvula	200	122	>200	171	133
Eragrostis superba	187	182	>200	164	>200

Source: Ryan, Miyamoto & Stroehlein, 1975.
NOTE: *E. curvula* and *E. superba* could be successfully germinated on soils having high levels of sodium.

Mallen Niche Seeder designed by Malcolm and Allen at the Western Australian Institute of Technology (see Fig. 12.1, Plate 6).

Tests for salt tolerance among grasses have been made in a number of countries but the range of plants has been rather restricted. Gusman, Cowley and Barton (1954) compared five grasses for total yield under 11 irrigations with salinized water (see Table 12.3).

The tolerance to salinity in descending order was Rhodes grass (*Chloris gayana*) > coastal Bermuda (*Cynodon dactylon*) > blue panic (*Panicum antidotale*) > buffel grass (*Cenchrus ciliaris*) > Angleton grass (*Dichanthium aristatum*). Treatment with the saline water seriously affected the phosphorus content, the effect on Angleton grass being most marked.

Graham and Humphreys (1970) compared the salt tolerance of cultivars of buffel grass (*Cenchrus ciliaris*) and found that cv. Biloela was the most tolerant with its yield remaining relatively higher up to 80 m.e. sodium chloride per litre.

Farnworth (1974) in trials at Hofuf, Saudi Arabia, tested a range of summer grasses on saline high calcium soils and Rhodes grass yielded 8.9 t DM/ha in 188 days, more than double the yield of any other species. After good establishment, most species suffered severe seedling mortality and many showed severe chlorosis.

Russell (1976) in Queensland tested some tropical grasses and legumes

Figure 12.1. The Mallen Niche Seeder, which plants seeds in a shallow trough atop the bank, thus reducing the impact of saline soils (see also Plate 6) (**Source:** C.V. Malcolm, Western Australia Department of Agriculture)

in pots of clay soil given 0-9.79 g NaCl per pot. He found the most tolerant grasses were *Chloris gayana, Panicum coloratum, Pennisetum clandestinum, Sorghum almum* and *Digitaria decumbens,* with *Setaria sphacelata* being the least tolerant. Both sodium and chlorine contents in plants generally increased with increasing salt level but no consistent relationship existed.

If certain salt-tolerant plants are established they will gradually bring about the necessary base exchange as a result of the carbon dioxide liberated from the roots. This forms carbonic acid and if calcium is present the calcium salt is formed and base exchange takes place with the sodium clay. Bermuda grass has been successfully used in this way as have *Chloris gayana* and *Sorghum sudanense*. Bermuda grass is able to shunt photosynthates from tops to roots to help survival under saline conditions. The weight of above-ground parts decreases while root weight to top weight ratios increase with increasing salinity (Youngner & Lunt, 1967).

Analyses of numerous ecological surveys of vegetation throughout the subtropical and tropical world show three tribes dominating the soils associated with salinity: the Chlorideae, Sporoboleae and Aeluropodeae (Lipschitz & Waisel, 1974). Salt glands were observed on both leaf surfaces

TABLE 12.3 **Effects of 11 irrigations with salinized water on the total yield of five grasses (Yields in tonnes/ha; percentage reduction in yield in parentheses)**

Total soluble salts in water (ppm)	Coastal Bermuda grass	Rhodes grass	Buffel grass	Blue panic	Angleton grass
900	15.62	12.45	14.72	12.85	13.02
1 500	13.32 (15%)	17.32	9.85 (33%)	11.42 (16%)	6.95 (46%)
3 000	11.67 (25%)	11.62 (6%)	9.55 (35%)	9.92 (27%)	7.25 (44%)
4 500	12.10 (23%)	10.02 (19%)	6.27 (57%)	8.35 (38%)	6.72 (48%)

Source: Gusman, Cowley & Barton, 1954.

of species belonging to genera of these tribes. The glands appear in longitudinal rows parallel to the veins. The concentration of solution around the roots is positively correlated with the salt content of the leaves as well as with the quantity of salt secreted. Secretion and Na content increased and leaf potassium decreased in solutions containing NaCl. In all species tested better growth occurred when about 50 M NaCl was added to the basic nutrient solution. The salt glands on leaves of *Chloris gayana* consist of a basal large cell and an upper small one. The secreted salts crystallized in the form of whiskers on top of the upper gland cell (Lipschitz *et al.*, 1974). From these findings the authors postulate that the Chlorideae, Sporoboleae and Aeluropodeae must have evolved from closely related or even the same ancestors which must have occupied a saline habitat.

Tropical and subtropical grass species that show tolerance to salinity and/or alkalinity include the following:

● Grasses with high salt tolerance

Cenchrus ciliaris (buffel grass). A climax species on the plains of Rajasthan, India (Magoon & Shankarnarayan, 1974);
— cv. Biloela is the most tolerant Australian cultivar, maintaining relatively high yields up to 80 meq NaCl/litre (Graham & Humphreys, 1970);
— it is not as tolerant as *Chloris gayana, Cynodon dactylon* or *Panicum antidotale.*
Cenchrus setigerus (birdwood grass). Also a climax species on the plains of Rajasthan, India (Magoon & Shankarnarayan, 1974);
— it occurs frequently on alkaline soils in the Rift Valley, Kenya.
Chloris barbata (purple top chloris). In Barbados occurred at a salt concentration of > 16 meq/100 g, which prevented the establishment of 30 other species (Eavis, Cumberbatch & Medford, 1974);

— it is an early colonizing annual in India (Magoon & Shankarnarayan, 1974).
Chloris gayana (Rhodes grass). C. Katambora yielded 8.9 tonnes DM/ha at Hofuf, Saudi Arabia in alkaline, high calcium soils — more than double the yield of other species (Farnworth, 1974);
— it is one of the most tolerant species in Queensland (Russell, 1976);
— it occurs on saline areas where water from hot springs drains on to the flood plains of the Kafue River in Zambia (Sayer & Lavieren, 1975);
— seed tolerated 0.4 M NaCl (16.6 atm) (Abd-el-Rahman & El-Monayeri, 1967);
— a salt pan strain comes from the salt pan near Hammanskraael, Pretoria, where it flourishes on the bottom of the pan, in granite soil impregnated with brine (Chippendall, 1955);
— in Italy it grows on land previously under saline water, and it grows on saline soils in Mississippi, United States. It is tolerant to $CaCl_2$ and Na_2SO_4, less so to $NaNO_3$ and NaCl, and is sensitive to $MgCl_2$ (Bogdan, 1969);
— it grows on saline seepage areas throughout Queensland and in Western Australia (Teakle, 1937).
Chloris montana (phulna). Grows on recent alluvial clay containing carbonates and to a lesser extent Na, Ca and Mg with pH up to 10.0.
Dichanthium aristatum (Angleton grass). Highly salt tolerant.
Digitaria decumbens (pangola grass). One of the most tolerant species in Queensland among commonly sown species (Russell, 1976).
Diplachne fusca (brown beetle grass). A climax species on the plains of Rajasthan, India (Magoon & Shankarnarayan, 1974);
— it occurs under saline conditions in the salt pans in the Siloana Plains, Western Province, Zambia (Verboom & Brunt, 1970);
— it occurs under saline conditions on the flood plain of the Kafue River in Zambia (Sayer & Lavieren, 1975);
— it makes better growth on alkaline than normal soils, and grows well on soil of pH 10. Fifty percent yield reduction occurs at $EC_e 20$ or at ESP_{72}. Growth is reasonable at $EC_2 25$ using NaCl in gravel culture. Sodium appears to be beneficial to its growth, and it is more tolerant of NaCl than of Na_2SO_4 (Malcolm, 1980);
— it is common on inundated heavy soils of the Murray River flood plains in South Australia; in artesian bore drains in central Australia. At Cooloota Springs and other bores in the area it inhabits the salt-encrusted overflow areas of the bore tanks (Lazarides, 1970);
— it occurs as a pure sward in the alkaline swamps of the Rukwa Valley in Central Africa (Vesey-Fitzgerald, 1963).
Echinochloa pyramidalis (antelope grass). The soils on which this species grows are often alkaline. Where drainage is closed the pH may be as high as 9.2 (Vesey-Fitzgerald, 1963).
Eragrostis curvula (weeping love grass). Least affected by increasing osmotic

suction at Kijushu Experimental Station in Japan (Okamoto, Kawatke & Horiuchi, 1975);
— it is very tolerant of salinity; seed germinated well under high levels of soil sodium (Ryan, Miyamoto & Stroehlein, 1975);
— in Western Australia, CPI14369 was moderately tolerant (Rogers & Bailey, 1963).

Eragrostis superba (Masai love grass). Has a high tolerance to salinity and alkalinity. Seed will germinate well (Ryan, Miyamoto & Stroehlein, 1975).

Oryza sativa (rice). Roots loosen the soil and render it more favourable to leaching. Rice also creates an acidic environment which reduces pH;
— cvs. IR8, IR-8-68 and Jaya have high salt tolerance among the "japonica" varieties and Jhona 349, Damodar and MCM among "indica" varieties (Yadav, 1975).

Panicum coloratum var. *makarikariense* (makarikari panic). Least affected by increasing osmotic pressure at Kijushu Experimental Station, Japan (Okamoto, Kawatke & Horiuchi, 1975);
— it is one of the most tolerant species in Queensland among sown grasses (Russell, 1976).

Panicum miliaceum (French or proso millet). Has high tolerance to alkaline (Na_2CO_3) soils (Chapko, 1977).

Panicum porphyrrhizos. Occurs in salt pans of Siloana Plains, Western Province of Zambia (Verboom & Brunt, 1970).

Panicum repens (torpedo grass). Occurs in saline sands in western Zambia (Verboom & Brunt, 1970).

Panicum virgatum (switch grass). Occurs in salt meadows in the United States (Chapman, 1974).

Paspalum distichum (formerly *P. vaginatum*) (seashore paspalum). Occurs on inland swamp saline areas of west Pakistan plains and has high grazing potential (Chaudri, Sheikh & Alam, 1969);
— in Western Australia it has grown successfully in salt seepage patches where the ground water just below the surface contained 3 000 mg of sodium chloride per litre. It can stand lawn irrigation with water containing up to 14 000 mg per litre total soluble salts where heavy watering can be applied. Sea water is too saline (Malcolm & Laing, 1976);
— it grows along the seafront in Suriname and is frequently flooded with sea water (Dirven, 1963a & b).

Paspalum paspaloides (formerly *P. distichum*) (water couch). Tolerates moderate salinity (Leithead, Yarlett & Shiflet, 1971).

Phragmites australis (common reed). Var. *pokornyi* grows in saline habitats (Chapman, 1974);
— while usually growing in fresh or brackish water, it will also grow in sea water. Growth was 86.25 cm in fresh water and only 11.25 cm in sea water. It will not tolerate over 2.25 percent salt in a water solution (Chapman, 1974).

Saccharum officinarum (sugar cane). An electrical conductivity of

4 mmhos/cm of soil saturation extract is the point above which sugar cane growth was markedly reduced (Shoji & Sund, 1967);
— it gave maximum yields at an EC_e of 1.8 mmhos/cm, 50 percent at 10 mmhos/cm and nil at 18.7 mmhos/cm (Maas & Hoffman, 1976);
— the varieties Co 975, Co 453, B 37172 and Co 1184 tolerate salinity (Yadav, 1975).
Sorghum almum (Columbus grass). One of the most tolerant of the species commonly sown in Queensland (Russell, 1976).
Sorghum bicolor (sorghum or jowar). Has high salinity tolerance after the plants have become well established. Seed planted in a small furrow in a hill will encounter less salinity than when planted at the top of a hill (Price & Stokes, 1966);
— Sorghum has high tolerance to sodium carbonate (Chapko, 1977).
Sorghum sudanense (Sudan grass). Sown in sodic soils after gypsum treatment in Karnal, Punjab (Yadav, 1975).
— it has high tolerance to alkali (Na_2CO_3) soils (Chapko, 1977);
— it gave its maximum yield at an electro-conductivity of the soil extract of 3 mmhos/cm, 50 percent of the maximum at 15 mmhos/cm, and nil at 26 mmhos/cm (Maas & Hoffman, 1977).
Spinifex hirsutus (beach spinifex). Continually exposed to salt water spray.
Sporobolus spp. Occur in saline soils around Pergamino and Sante Fe, Argentina (Chapman, 1974).
Sporobolus airoides (alkali sacaton). Occurs around the Great Salt Lake, Utah, United States (Chapman, 1974);
— it will tolerate salinities up to 1.0 percent;
— it can use drainage waters with salinities up to 10-15 mmhos/cm (Le Houérou, 1977a).
Sporobolus arabicus (synonym: *S. pallidus*) (usar). An early colonizer of saline land in India (Magoon & Shankarnarayan, 1974);
— it dominates the usar lands in Uttar Pradesh, India on recent clay alluvium containing chiefly carbonates and, to a lesser extent, Na, Ca and Mg with pH up to 10.0 (Seth, 1955);
— it is used on alkaline soils in Punjab (Yadav, 1975).
Sporobolus asperifolius. Occurs in the final beach community around the Salton Sea, California (Chapman, 1974).
Sporobolus coromandeleanus. An early colonizing annual in saline lands in India (Magoon & Shankarnarayan, 1974).
Sporobolus helvolus (okrich). Dominates saline rangeland in Rajasthan and is stocked at two- to four-month intervals by lambs (Ahuja & Vishwanatham 1976);
— it can tolerate drainage waters with salinities up to 10-15 mmhos/cm (Le Houérou, 1977a).

Sporobolus marginatus. A climax species on the plains of Rajasthan (Magoon & Shankarnarayan, 1974);
— it occurs on saline areas where saline waters from hot springs drain on to the flood plain of the Kafue River in Zambia (Sayer & Lavieren, 1975).
Sporobolus pyramidatus. Occurs on saline soils in Salinas Grandes, Argentina (Chapman, 1974).
Sporobolus spicatus. Occurs in the Sunda depression, Tanzania (lat. 8°30'S) on an old lake bed in a zone of alkaline soil (pH 9.6) as a covering sward (Vesey-Fitzgerald, 1963);
— it dominates the salt pans of the Siloana Plains in Western Province, Zambia (Verboom & Brunt, 1970);
— it occurs in saline areas where saline waters from hot springs drain on to the flood plain of the Kafue River in Zambia (Sayer & Lavieren, 1975).
Sporobolus virginicus (saltwater couch). Occurs in the Gulf of Mexico at 0.6-2.1 percent salinity (Chapman, 1974);
— it occurs on the Pacific coast of Peru (Chapman, 1974) and Australia (Moore, 1970), and in the Maritime marshes of South Africa in sandy areas (Chapman, 1974);
— it is stabilizing sandy soils between Sinamaica and Paraguaipoa near the coast southwest of Maracaibo, Venezuela (Tamayo, 1963).
Stenotaphrum secundatum (buffalo grass). A seashore grass, it will tolerate salt spray (Wheeler, 1950).
Urochloa mosambicensis (sabi grass). All species of *Urochloa* in India show high sodium tolerance.
Urochloa oligotricha (formerly *U. bulbodes*). Thrives on many alkaline soils in the Caribbean (Burt, personal communication).

- Grasses with moderate salt tolerance

Brachiaria mutica (Para grass). In southeast Queensland it grows on deep loamy soils overlying saline clays. It merges with saline grasses on marine plains in north Queensland.
Cenchrus pennisetiformis (Cloncurry buffel grass). Grows on slightly alkaline soils in northwest Queensland (Hall, 1978).
Eragrostis lehmanniana (Lehmann's love grass). Tolerates high pH caused by calcium and magnesium rather than sodium (Ryan, Miyamoto & Stroehlein, 1975).
Eriochloa fatmensis (formerly *E. nubica*). Tolerant to slightly saline soils on lagoon flats (Rose-Innes, 1977).
Hyparrhenia hirta (South African bluestem). In southwestern Australia clones of CPI 15786 were moderately tolerant of salinity (Rogers & Bailey, 1963).

Iseilema laxum (machuri). Mildly tolerant, associated with *Sporobolus marginatus* (Whyte, 1964).
Lasiurus hirsutus (sewan grass). Has moderate tolerance to salinity.
Panicum antidotale (blue panic grass). Has some tolerance to salinity but more to alkalinity by sodium and magnesium than to the chlorides (Ryan, Miyamoto & Stroehlein, 1975).
Pennisetum americanum (pearl millet). Tolerates salinity and was used for reclamation of salt lands in the Sind because of its ability to take up salts (Tamhane & Mulwani, 1937; Ravikovitch & Porath, 1967);
— concentrations of 1 400-2 600 ppm of soil salt produced only slight tip burn (Smith & Clark, 1968).
Pennisetum clandestinum (Kikuyu grass). One of the most tolerant species of grass commonly sown in Queensland (Russell, 1976);
— lawns in western Queensland will tolerate saline soil and saline bore water if watered heavily to keep the soil salts at depth (Everist, 1974).

- Grasses with low to negligible salt tolerance

Bothriochloa ischaemum (King Ranch bluestem). Shows some tolerance (Dalrymple, 1978).
Setaria sphacelata (setaria). Two hexaploid accesssions, CPI 32847 and CPI 32714, from near the Aberdare Mountains in Kenya, showed some tolerance in southeast Queensland (Evans, 1967) but generally the species has low salt tolerance (Russell, 1976).
Zea mays (maize). Gave maximum yields at EC_e of 2 mmhos/cm, 50 percent at EC_e 9 mmhos/cm and nil at 15 mmhos/cm (Maas & Hoffman, 1977);
— maximum salt concentration in the soil solution extract that does not reduce maize yields is about 1 100 mg/litre total dissolved salts (EC_e 1.7 dS/m). The maximum permissible salt concentration of irrigation water to sustain maize production is about 300 mg/l or an EC_w of 0.45 dS/m (Hoffman et al., 1979).

Pastures under plantation crops

As weed species growing beneath tree crops have to be removed at very considerable cost, Payne (1976) suggested that they be replaced by forage species whose growth can be controlled by grazing animals. The conversion of excess forage into meat and milk would increase production per unit area of land while diversifying production output and labour input. In the long term there might be an improvement in soil fertility. The grazing animal is an important pathway for nutrient transfer from forage to soil.
The main tree crops grown in the tropics are coconuts, rubber and oil

palm, but there are several minor ones such as kapok and cashew nut. Until quite recently there has been little integration of livestock with plantation crops. The following are some reasons for the lack of grazing beneath tree crops:

- There is little animal husbandry experience, especially among coconut growers who, because of the general ecology of coconut growing "within sound of the sea" obtain most of the protein in their diets from fish. They must be trained in animal husbandry, especially in parasite control and nutrition.
- The availability of forage varies widely within seasons and years. One of the main problems is the competition for light between the pasture and the tree crop. In the case of the coconut, there would be good grazing in a juvenile plantation for the first three years, then a decline for the next ten years. Grazing would remain at that low level for several years. This means that as pastures decline, cattle may have to be sold or their diet supplemented, and supply and sale price become important. Generally, a coconut plantation needs to be 20-25 years old before there is sufficient light to support a permanent pasture as a ground cover.
- Grazing young coconut plantations is likely to cause damage to the trees. In rubber plantations collection cups may be disturbed. The pastures and the tree crop compete for light, water and nutrients.
- Fencing is needed to control the animals.
- If there is no grazing, labour is required for cutting and carting the cut material. A creeping legume is especially difficult to cut.
- Tall grasses tend to hide the fallen coconuts, making them difficult to collect.
- Some heavy-textured soils and high water-tables associated with rubber and oil-palm are subject to pugging and soil compaction, damaging field drainage, bunds and terraces and causing erosion on sloping lands. In Indonesia oil-palm yield was depressed by 1.0-1.5 t/ha together with a loss of 0.2-0.3 t/ha of palm kernel (Thomas, 1978).
- In Southeast Asia there is always a danger of invasion of pastures by *Imperata cylindrica* until the tree canopy can suppress it. In young plantations there is thus the danger of fire as the *Imperata* burns readily with consequent damage to the plantation crop.

However, there is a huge potential for livestock production when properly integrated with plantation agriculture.

Pastures under coconuts. Payne (1974) has estimated the areas of coconut plantations in the humid tropical countries in 1970, as shown in Table 12.4. If 50 percent of this area were planted to pastures at a stocking rate of 1.25 beasts per hectare, an annual extraction rate of 10 percent could produce about 46 500 tonnes of beef (Guzman, 1974).

TABLE 12.4 **Estimated area planted to coconuts in humid tropical countries (1970)**

Country or region	Area planted to coconuts (thousands of ha)
Latin America and the Caribbean	275
Africa	310
Sri Lanka	410
India	780
Indonesia	785
Malaysia	205
Philippines	1 100
Other Asian countries	260
Oceania	290
Total	4 425

Source: Payne, 1974

If pastures are not planted until the trees are 20 years old or over, competition for light is reduced. At Laguna in the Philippines, where palms were planted in a triangular pattern with 8 m between them and with trees 8-9 m tall, light intensity was less than 50 percent in early morning and late afternoon; around noon the light intensity reaching the ground under the canopy was 60 percent (Guzman, 1974).

In Western Samoa in a 20-30-year-old coconut plantation with 50 percent light transmission, seasonal production of *Brachiaria mutica* and *B. decumbens* was affected by shade (Reynolds, 1978), but in Bali, Indonesia, palms spaced 10-12 m apart gave noon light transmission of 77-80 percent and very satisfactory pasture development (Steel & Humphreys, 1974).

Competition for nutrients depends upon the pasture species and the fertility of the soil. If added fertilizer is sufficient there should not be yield reduction, but the economics of fertilizer application must be assessed (Rodrigo, 1945). Competition for moisture will be apparent in dry years (Krishna-Marar, 1953, 1961).

In areas where rainfall exceeds 2 125-2 200 mm (Kannegieter, 1970), where the soils are light and where the farmer is able and willing to provide adequate fertilizer, ruminant livestock can be successfully managed in coconut plantations (Santhirasegaram, 1964) (see Plate 7). The major advantages of this system are that coconut plantations provide a desirable environment for highly productive livestock. The air currents generated by the almost continuous movement of the coconut fronds reduce the heat load on the animal, while the animal provides livestock products and assists in the control of weeds.

Improved pastures for livestock are now used in Tanzania, Sri Lanka, the Philippines (22 percent of coconut groves carry livestock), Papua New Guinea, Samoa, Fiji (10 percent of beef cattle come from coconut estates), Malaysia and several other countries.

Fernandez (1972) has shown that pastures of *Brachiaria subquadripara* (formerly *B. miliiformis*) or *B. brizantha* can increase coconut yields by 30-40 percent after 14 years, but inadequately fertilized *Panicum maximum* can reduce coconut yield. Combined grass/legume mixtures may be used. Santhirasegaram (1964) recommended *Brachiaria brizantha* and *B. subquadripara* in conjunction with *Pueraria phaseoloides* on lighter soils and *Centrosema pubescens* on heavier soils. Fernandez (1973) preferred *B. subquadripara* and *Digitaria decumbens* from a group of *Brachiaria mutica*, *Paspalum notatum*, *P. plicatulum*, *P. scrobiculatum*, *Setaria sphacelata* and *Panicum coloratum*. *Brachiaria humidicola* is used in the southern high rainfall zone and *S. sphacelata* cv. Kazungula did well at Lunuwila, Sri Lanka. *Panicum maximum* cultivars have also performed well. Fernandez (1972) selected *Brachiaria brizantha* and *B. ruziziensis* as the best for grazing.

In Papua New Guinea *Brachiaria mutica*, *B. ruziziensis*, *Pennisetum purpureum* and *Panicum maximum* are preferred (Hill, 1969). Few records are available for actual live-weight gains of livestock under coconuts, but some carrying capacity figures are recorded.

In Sri Lanka, 0.5-2.0 head per hectare of Sinhala cattle were carried on a pasture under coconut palms but were supplemented with 1.4 kg/head per day of coconut press cake (Rajaratnam & Santhirasegaram, 1963). In Papua New Guinea, Brahman cattle graze a *Paspalum conjugatum/Calopogonium mucunoides* pasture under coconuts at a stocking rate of 2.5 beasts per hectare (Hill, 1969).

In the Solomon Islands, two to three cattle per hectare are grazed under coconuts (Eden, 1953). Management includes controlling growth to keep the pasture at a height that will not impair visibility of fallen coconuts and can be fertilized adequately. Rodrigo (1945) found that *Pennisetum purpureum* caused a 39 percent reduction in copra yield, but with adequate NPK fertilizer, copra yield was increased by 59 percent and fodder yield increased eightfold. At Los Baños, the Philippines, dry-matter yields of 23-33 tonnes/ha per year have been obtained from *Brachiaria mutica* when fertilized with 150:75:150 kg/ha per year of NPK and cut at 60-day intervals. Use of a legume reduces fertilizer need; thus the legume component of a mixed pasture should be carefully maintained.

Pastures under oil palm. Cover crops in oil-palm plantations were traditionally composed of pure legumes. In 1975 the MARDI/CSIRO pasture team established a mixture of common Guinea grass (*Panicum maximum*) and

Schofield stylo (*Stylosanthes guianensis*) in a six-year-old plantation. The pasture was grazed by Kedah-Kelantan cattle at two and four beasts per hectare. After seven months of grazing, the sown species had disappeared and the pastures consisted of mainly native grasses such as *Paspalum conjugatum, Axonopus compressus* and *Ottochloa nodosa*. When the stocking rate was reduced to one and two beasts per hectare, the animals gained 116 kg per head at one beast per hectare, and 62 kg per head on the two-beasts-per-hectare pasture in the subsequent 12 months. The botanical composition of the pasture grazed continuously at one beast per hectare was stabilized but at two beasts per hectare there was a substantial increase in non-palatable species, comprising mainly *Melastoma malabaricum, Eupatorium odoratum* and *Ageratum conyzoides* (Ajij Singh Sidhu *et al.*, 1977).

Grasses suitable for erosion control and their rainfall ranges

Andropogon gayanus. Has been used in Kano, Nigeria, for reclaiming badly overgrazed and eroded land (Bowden, 1963a & b). 400-1 400 mm.
Axonopus affinis. Widely used in the United Sates to prevent erosion and stabilize road banks (Bennett, 1962). In excess of 750 mm.
Axonopus compressus. Used for stabilizing slopes and banks of dams. 775-2 000 mm.
Bothriochloa glabra. Has proved to be the most useful grass for soil erosion control. 500-800 mm.
Bothriochloa insculpta. Useful for erosion control on self-mulching black clays if seeded heavily (Bissett & Graham, 1978). 500-800 mm.
Bothriochloa ischaemum. Can form a sod almost as dense as *Cynodon dactylon* and can be used for erosion control. Cultivar King Ranch Bluestem is recommended (Srinivasan, Bonde & Tejwani, 1962). Used in Gujarat, India. 375-1 000 mm.
Bothriochloa pertusa. Used for reseeding eroded land in India and is promising for revegetating spoil from central Queensland mines (Whyte, 1968). 500-1 375 mm.
Brachiaria brizantha. Has been used successfully for erosion control in southeast Queensland and India (Gandhi, 1957; Patil & Ghosh, 1963). 500-2 000 mm.
Brachiaria decumbens. A valuable grass for erosion control as it covers the ground well. 1 000-3 000 mm.
Brachiaria humidicola. A very effective grass for erosion control (Chippendall & Crook, 1976). 875-1 000 mm.
Brachiaria mutica. Useful for stream bank and stream erosion control. 900-3 000 mm.
Cenchrus ciliaris. One of the best grasses for erosion control in semi-arid areas, the stoloniferous varieties being preferred. 375-750 mm.

Cenchrus pennisetiformis. A useful stream bank stabilizer in northwestern Queensland (Hall, 1978). 250-600 mm.
Chloris gayana. Useful for erosion control if well grazed. Cv. Katambora establishes and covers the ground rapidly. 600-1 600 mm.
Chloris roxburghiana. Has been used successfully in reseeding eroded rangeland in Kitui and Baringo districts, Kenya (Jordan, 1957; Pratt, 1963, 1964). 500-625 mm.
Chloris virgata. Although short-lived, it is one of the first grasses to colonize bare ground. 375-750 mm.
Cynodon aethiopicus. An excellent stoloniferous species. 500-875 mm.
Cynodon dactylon. Excellent for erosion control, but may be difficult to eradicate. Spreads by stolons and rhizomes. 500-875 mm.
Cynodon nlemfuensis. An excellent stoloniferous species for erosion control. 500-875 mm.
Cynodon plectostachyus. An excellent stoloniferous grass for erosion control on the black clay soils of the Darling Downs, Queensland, and on sandy soils of the coast. 500-875 mm.
Dichanthium annulatum. Has an elaborate root system and is one of the best grasses for stabilizing the bunds in the ravine lands of Gujarat (Srinivasan, Bonde & Tejwani, 1962) and for erosion control on 20° slopes in India (Misra, Ambasht & Singh, 1977). 500-900 mm.
Dichanthium aristatum. Useful in high rainfall areas with a rainfall in excess of 875 mm.
Dichanthium caricosum. Excellent for erosion control. It colonizes bare areas quickly both by seed and runners (Roberts, 1970a, b). 600-1 000 mm.
Digitaria abyssinica (scalorum). Has been planted on the slopes of Cape Peninsula in South Africa to control erosion (Chippendall, 1955) but it is the most troublesome of all African weeds (Ivens, 1967). 500-1 000 mm.
Digitaria decumbens. Excellent in rainfalls from 650-1 200 mm.
Digitaria didactyla. Excellent on sandy soils with a rainfall of 700-1 250 mm.
Digitaria smutzii. Has been recommended for revegetating abandoned crop land in Orange Free State and Transvaal, but is usually propagated vegetatively.
Echinochloa utilis. A very rapidly colonizing annual fodder crop that quickly stabilizes newly cleared sandy coastal areas. 500-1 200 mm.
Eragrostis curvula. Widely used in Kenya, Sri Lanka and the United States for stabilizing terraces, water discharge areas and banks of earth tanks. In Japan it has helped stabilize mountain slopes for at least three years (Endo, 1978). 300-1 000 mm.
Eragrostis lehmanniana. Has been successful in reseeding the range in the southwestern United States and gives a rapid soil cover (Humphrey, 1960a). 250-375 mm.
Eragrostis tef. Useful in erosion control (Narayanan & Dabadghao, 1972). 950-1 500 mm.

Heteropogon contortus. Has proved useful on 20° slopes in India (Misra, Ambasht & Singh, 1977). 500-1 500 mm.

Lasiurus hirsutus. Used for stabilizing sand dunes in Iraq (Dougrameji & Kaul, 1972). Below 250 mm.

Melinis minutiflora. Excellent in high-rainfall areas of cleared rain forest and for stabilizing road banks in the wet tropics. 750-1 750 mm.

Panicum antidotale. Used extensively for erosion control of flood plains in the United States, mainly to protect against wind erosion. It is sown at right angles to the prevailing wind. 500-750 mm.

Panicum coloratum var. *makarikariense* Cv. Pollock is useful in erosion control because of its large crown. 650-1 700 mm.

Panicum repens. Used for binding coastal sands and lake shore lines, for stabilizing steeper slopes of ponds in Zambia where cattle approach to drink, and to fix mine dumps in Zimbabwe (Chippendall & Crook, 1976) and Gujarat, India (Srinivasan, Bonde & Tejwani, 1962). 500-1 500 mm.

Panicum turgidum. Valuable for fixing sand dunes in the 100-400 mm rainfall areas.

Paspalum dilatatum. Where well established it will control water erosion. It is used to fix mine dumps in South Africa (Chippendall & Crook, 1976). 750-1 250 mm.

Paspalum distichum. Useful in erosion control on salted lands and lands subject to tidal influences (Cameron, 1959). 400-750 mm.

Paspalum nicorae. Excellent for stabilizing waterways and airfields. Rainfall the same as *Paspalum notatum.*

Paspalum notatum. Used to stabilize terraces against erosion in the United States (Stephens & Marchant, 1960). 750-1 500 mm.

Pennisetum clandestinum. Excellent for stabilizing waterways in cultivation, for stabilizing irrigation channels and for sloping lawns (Read, 1975). 650-1 300 mm.

Pennisetum polystachyon. Quickly covers the ashes of a forest burn forming a dense tussock grassland in Thailand and Fiji (Partridge, 1975). 1 000-1 250 mm.

Pennisetum purpureum. When established it will effectively control erosion, but its height may be an inconvenience. 1 000-1 750 mm.

Phragmites australis. A reed that effectively prevents water erosion along drains and river banks. 750-3 000 mm.

Phragmites karka. Like *P. australis* it withstands heavy floods and stabilizes river banks (Rose-Innes, 1977). 1 500-3 000 mm.

Setaria porphyrantha. A valuable grass for strip cropping in black clay soils and is easily established (Watt, 1976). 500-700 mm.

Setaria sphacelata. Cv. Kazungula, with its large crowns, gives very good erosion control. 750-1 250 mm.

Stenotaphrum secundatum. Excellent for erosion control on flat land on "muck" soils in Florida, United States. 750-1 500 mm.

Urochloa mosambicensis. Has been used successfully in India. 600-1 200 mm.

Grasses suited to aquatic, semi-aquatic and moist areas (hydrophilous grasses)

There are numerous grasses that thrive in or near water and many have been mentioned in the literature. Rains (1963) and Verboom and Brunt (1970) have attempted to classify rather broadly the extent of inundation some grasses will tolerate, but no close attention has been given to depth or duration of inundation on a wide scale. The following lists have been drawn up from the papers by the above authors, together with Colman and Wilson (1960), Anderson (1970a, b, 1972, 1974) and others. There is some overlapping where grasses have a wide range. Only the species dealt with in this book are listed, with their main references.

- Grasses flourishing in perennially wet places (lakes, marshes, swamps)

Acroceras macrum (Verboom & Brunt, 1970)
Brachiaria mutica
Echinochloa crus-galli (rice-fields)
Echinochloa pyramidalis (Rains, 1963; Pursglove, 1976)
Hymenachne acutigluma (Henty, 1969)
Hymenachne amplexicaulis (Dirven, 1963)
Leersia hexandra (Rains, 1963; Dirven, 1963; Verboom & Brunt, 1970)
Vossia cuspidata (Rains, 1963; Verboom & Brunt, 1970)

- Grasses growing on the banks of perennial streams

Panicum repens (Sayer & Lavieren, 1975)
Pennisetum purpureum
Phragmites australis (Linedale, 1974)
Phragmites karka
Saccharum spontaneum

- Grasses growing in seasonally wet places

Acroceras macrum (Verboom & Brunt, 1970)
Andropogon gayanus var. *gayanus* (Rains, 1963)
Axonopus affinis
Axonopus scoparius (Gonzalez & Pacheco, 1970)
Bothriochloa pertusa (Whyte, 1964)
Brachiaria humidicola

Brachiaria mutica
Coix lacryma-jobi
Cynodon dactylon
Cynodon plectostachyus
Dichanthium aristatum
Dichanthium caricosum (Partridge, 1979b)
Digitaria abyssinica (previously *D. scalarum*) (Verboom & Brunt, 1970)
Diplachne fusca (Lazarides, 1970)
Echinochloa colona
Echinochloa crus-galli (in rice)
Echinochloa haploclada
Echinochloa polystachya (Semple, 1970)
Echinochloa turneriana (Skerman, 1947)
Entolasia imbricata (Mwakha, 1971)
Eriochloa fatmensis (previously *E. nubica*) (Rose-Innes, 1977)
Eriochloa punctata (previously *E. polystachya*)
Hemarthria altissima (Rattray, 1973)
Ischaemum indicum (Dabadghao & Shankarnarayan, 1973)
Ischaemum magnum
Iseilema laxum (Whyte, 1964)
Ixophorus unisetus (Gonzalez & Pacheco, 1970)
Oryza sativa
Panicum antidotale
Panicum coloratum var. *makarikarienses* cv. Bambatsi (Anderson, 1972)
Panicum repens (Sayer & Lavieren, 1975; Verboom & Brunt, 1970)
Panicum whitei (Skerman, 1947)
Paspalum distichum (previously *P. vaginatum*) (Colman & Wilson, 1960)
Paspalum paspaloides (Colman & Wilson, 1960)
Paspalum plicatulum (Colman & Wilson, 1960)
Paspalum scrobiculatum (Colman & Wilson, 1960)
Paspalum urvillei
Paspalum wettsteinii (Leggett, 1968)
Phragmites australis (Linedale, 1974)
Setaria sphacelata cv. Kazungula (Colman & Wilson, 1960)
Sporobolus helvolus (Gupta, Saxena & Sharma, 1972)
Sporobolus virginicus (Leithead, Yarlett & Shiflet, 1971)
Stenotaphrum secundatum cv. Roselawn (Haines *et al.*, 1965)
Vetiveria zizanioides

Grasses that tolerate high levels of soil aluminium

During 1977, 38 grass and legume species were tested in the field. 0, 0.5, 2 and 6 t/ha of lime were applied to give an aluminium saturation level of approximately 90, 85, 60 and 15 percent, respectively. Several species, including *Brachiaria decumbens, B. humidicola, B. mutica, Andropogon gayanus, Panicum maximum* and *Digitaria decumbens* showed excellent tolerance to aluminium, and all reached near maximum yield with application rates of 0 to 0.5 t. On the other hand, *Hyparrhenia rufa,* a grass that is quite common in the tropics, responded very satisfactorily up to a level of 2 t/ha.

Cenchrus ciliaris was most affected by aluminium concentrations up to 4 ppm, followed by *Hyparrhenia rufa* and *Panicum maximum* (Spain, 1979).

Other grasses remain to be tested. *Hemarthria altissima* (CIAT, 1978) and *Tripsacum dactyloides* have been reported to tolerate acid soils.

13. Utilization and conservation of forage

When faced with choice in the use of a green forage crop, it is important to consider the potential uses and ultimate yield to livestock. Table 13.1 shows the potential losses associated with various utilization schemes for tropical grasses.

Grazing

Grazing forage grasses is the usual method of utilization and species are selected on their ability to tiller profusely, stand a series of defoliations during the grazing, withstand trampling and respond to fertilizers, as well as for their palatability, accessibility and nutritive value.

Problems in grazing a forage crop include losses of material through trampling, faecal contamination, selective grazing and early maturity, when a crop grows too quickly for the number of grazing animals and becomes stemmy and unattractive. In clay soils there is an additional problem of pug-

TABLE 13.1 **Utilizable yields of tropical grasses under various usages**

Losses/ utilizable yield	Grazing	Green chop (cut-and-carry)	Hay	Silage	Sealed storage (Harvestore)
Harvesting losses (%)	50	2	25	5	10
Storage losses (%)	—	—	5	30	2
Feeding out losses (%)	—	15	15	15	1
Total dry matter losses (%)	50	17	45	50	13
Balance of fodder to animal (%)	50	83	55	50	87
Estimated actual maximum yield of dry matter (%)	60	60	92	92	80
Utilizable yield of potential dry matter (%)	30	50	51	46	70

ging, which causes excessive plant damage and eventual crusting, leaving hard, uneven clods on the soil surface.

Cloudy days reduce the soluble carbohydrate content of the forage and subsequent production, and toxicity such as prussic acid (HCN), nitrite poisoning, or magnesium tetany may take its toll.

With separate night paddocks for a dairy herd there is usually a transfer of nutrients by greatly increased defecation and urination at these sites, resulting from a transfer of nutrients from the paddocks grazed during the day to the paddocks grazed and camped in during the night.

Grazing requires some measure of control in semi-intensive and intensive enterprises and subdivision by permanent fences, temporary fences (both of which can be electrified) or shepherding are needed.

On the credit side, grazing requires less labour and is less time-consuming than other feeding methods. The animal selects its diet in both quality and quantity and nutrients are returned to the soil via the excreta.

Green chop

Cutting green forage in the field and transporting it to the livestock is a common practice of village dwellers in tropical countries. Those who own little land for grazing have to keep draught or milk animals, be they cattle, sheep or goats, horses or donkeys, within the confines of their compounds during the night, and in some cases both day and night. In this context, the feeding system is called "cut-and-carry". It is also practised by commercial livestock farmers such as dairymen and feedlot cattle finishers who mechanize the operation in an intensive feeding system.

The operation is called "soilage" when its aim is to collect animal excreta in a confined space and return it to the land to build fertility. In Bukoka, Tanzania, *Hyparrhenia* spp. are fed to one or two cattle in small stalls from which the bedding, plus excreta, is later placed among the plants in a banana plantation.

Green chopping, cut-and-carry and soilage are not systems of fodder conservation, but involve immediate use of the cut fodder. The advantages of such systems are that there is a much higher recovery of plant material as it is not lost by trampling or faecal contamination; there is little selectivity because the animal intake is usually rationed and the system allows for this rationing; and the feeding can be arranged at a determined convenient site. The excreta can be recovered and can be utilized as desired and where desired, generally as farmyard manure applied to more intensive crops.

Continued removal of plant material without return of the nutrients to the field in which it was grown can lead to deficiencies of soil nutrients, particularly of potash, which will then have to be replaced by fertilizers or organic manures.

Using and Conserving Forage

For a large number of livestock, green chopping involves a significant expenditure of machinery to cut and cart (see Fig. 13.1). In tropical and subtropical countries, heavy storms may immobilize the machinery, requiring a supplementary storage of hay or other fodder at the feeding site to tide the herd over the wet days. Another problem is the advancing age of the crop from the beginning to the end of the cutting period. However, the intake of green leaf is a vital determinant of animal production and green chopping as an overall system has many advantages.

Forage can be conserved in the field uncut, as standing hay, or as stubble remaining from a harvested grain crop; it can be cut and dried in the field or artificially indoors to become hay, which itself can be pelleted or converted to meal; it can be made into silage or haylage or allowed to mature to be harvested for grain.

Conservation as standing hay or deferred feed

In arid and semi-arid areas this is practically the only way to conserve forage because of low dry-matter yields, high harvesting costs and the generally inferior quality of the herbage. Figure 13.2 shows that with maturity both

Figure 13.1. A combined chopper/catcher forage harvester for green chop

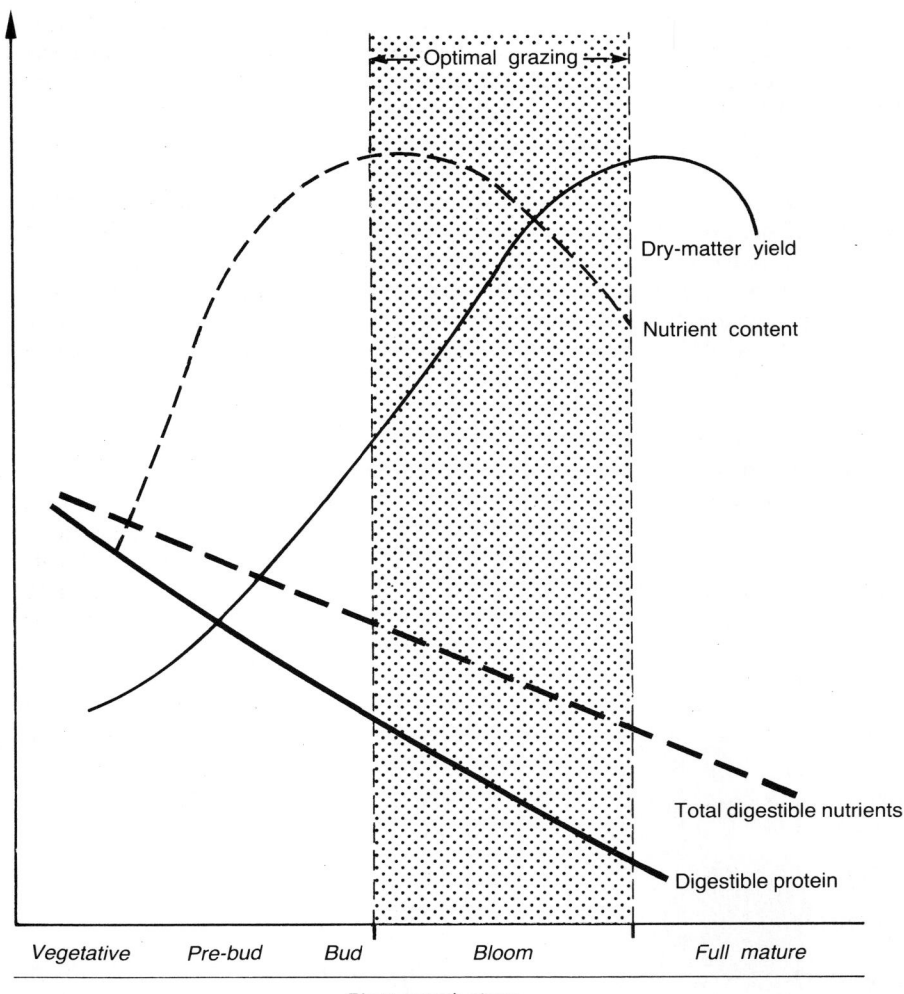

Figure 13.2. Growth cycle of crops (**Source:** R.H. Skerman, personal communication)

total digestible nutrients and digestible protein contents decline. The arid and semi-arid grasses have been subjected to natural selection for survival. They are mostly early-maturing annuals that seed quickly and ensure regeneration, or perennials with special morphological or physiological adaptation to dry conditions.

Some species, such as *Paspalum scrobiculatum,* can maintain a reasonably high level of nutrient content right up to maturity. If the rainy season ends abruptly, grasses hay off before maturity, providing a better quality forage. Standing hay is generally the cheapest and easiest way of conserving forage as it does not require machinery or physical handling. It is practised by large ranch-holders in the drier areas. Occasionally it is practised by village communities by common consent in communal grazing lands. This is to be encouraged.

Two major problems arise with such a practice. The more dangerous is the risk of fire, which can destroy the whole conserved area in a matter of minutes or hours. An efficient system of fire control is essential. Fire-breaks about 10 m wide can be cultivated by hoes in the hands of local people in a village effort, by ploughing the whole area or by grading two strips about 10 m apart and 2 m wide and burning out the centre section. Chemicals can also be used. A 3-m-wide break about 3.5 km long requires 25-30 kg of diuron per hectare.

The second major problem with standing hay is the occurrence of light falls of rain when the culms and leaves are dead. Mould growth results, "blackening" the dead grass and making it useless as forage. This problem is experienced in some years in the Mitchell grass areas in central Queensland.

Standing hay becomes important when some of the supplementary feeds such as urea-molasses licks are used. Such supplements require an adequate amount of roughage to stimulate the ruminal flora of the grazing animal to utilize the non-protein nitrogen. Without such roughage, the ration gives little nutrition in a maintenance situation.

A urea-molasses mixture has been sprayed on standing pastures, but with generally disappointing results. In the case of a tussock grassland, such as Mitchell grass grassland, too much of the nutrient spray is lost on bare ground between the tussocks. In addition, spraying is needed almost daily to provide fresh mixture as the animals require it.

Stored fodder

The need for fodder during the severe winters of the Northern Hemisphere led to the conservation of hay and later silage. This practice has been followed, but less ardently, by livestock owners who faced periodic droughts in the warmer countries. On snow-bound Northern Hemisphere farms the very continuance of livestock enterprises demanded storage of fodder. In areas where snow is not usual or is never experienced, drought periods throughout the year or over a series of years cause decline in production and often death of livestock. The compulsion to conserve fodder is not so evident where the predictability of a drought is low, compared with the high predictability of snow in the Northern Hemisphere.

To provide raw material for conservation in the Northern Hemisphere it became necessary to grow specific areas of grass and fodder crops. When

Townshend, in 1730, introduced a grass ley into England's four-course rotation to improve soil fertility, the dual benefit of pasture extended the practice gradually to other continents.

Fodder is usually conserved as hay, chaff, pellets, haylage or silage. The aim is usually to harvest the crop at its maximum nutrient content, but the time of harvest may be earlier if a higher protein content is required or later if maximum dry matter is desired (see Fig. 13.2).

Hay

Hay is the most commonly stored fodder on the farm and is one of the oldest feed systems used to level out the feed supply throughout the year. It is generally the most convenient processed form of storage.

The aim in making hay is to conserve the maximum of dry matter and nutrients at the lowest cost. "Good hay — the cheapest form of feed nutrients during the non-grazing season, is weed-free forage, dried without loss of leaves from handling or deterioration in dry matter and nutrients, but with its natural colour and sweetness" (USDA Yearbook, 1948).

In tropical areas where there is the likelihood of heavy afternoon storms or continuous rain during the time the hay is on the ground, the decision to make hay rather than silage is often critical. The reliability of haymaking weather could be assessed by a statistical analysis of existing meteorological weather records with a knowledge of the length of time needed to cut, cure and store the particular type of hay. However, at Gualaca, Panama, Medling (1972) found that even when two to three non-rainy days occurred, air humidity remained too high and there was insufficient air movement and sunshine to dry the hay. Hence other factors must be taken into consideration.

Hay should be made at the optimum date to maximize yield and still have the percentage of digestible dry matter necessary to meet the nutrient needs of the livestock. It is best cut early in the flowering stage. When cut earlier, the nutritive value is higher but yield is lower and the moisture content is too high for easy curing. If cut after flowering, the increased yield does not compensate for decreased palatability and nutritive value. The first cut of hay from a crop is usually of better quality than subsequent ones.

Cutting interval definitely affects both yield and quality of grass hay crops. Table 13.2 shows the results of a three-year study with coastal Bermuda grass (*Cynodon dactylon*) in Georgia, United States (Altom, 1978).

Curing of hay. To make good hay it is imperative that the grass is dried quickly and not unduly exposed to the sun. Raking hay into wind-rows when it is dry and brittle, or mowing and raking at high speed can reduce hay quality by large losses of leaf. Poor weather conditions after mowing can lower hay quality. Rain can cause leaf losses and leaching of nutrients. The loss of nu-

TABLE 13.2 **Effect of cutting frequency on yield, crude protein and crude fibre of** *Cynodon dactylon* **fertilized with 672 kg N/ha per year (1959-61)**

Interval between cuttings (weeks)	Forage yield (tonnes/ha)	Crude protein (kg/ha)	Crude protein (%)	Crude fibre (%)
3	8.50	2 735	18.5	27.0
4		2 595	16.4	29.1
5		2 608	15.4	30.6
6		2 567	13.3	31.6
8		2 123	10.7	32.9
12		1 805	9.0	33.4
24		1 239	8.4	33.9

Source: Altom, 1978

trients in hay-making is about 25 percent under normal conditions (Göhl, 1975).

The moisture content to which the hay, to prevent degradation, should be reduced in the field needs to be researched for tropical pastures. A water content of 25 percent is safe for temperate pastures and 22 percent has proved safe for coastal Bermuda grass at Tifton, Georgia (Hellwig, 1965). The Tifton group used a rotary slasher, and 28 hours after harvest the water content of the slashed grass was 14 percent, while that of the mown grass was 20.9 percent. Catchpoole (1969) investigated the methods of making hay from *Setaria sphacelata* cv. Nandi at Samford, Queensland (lat. 27°22'S, altitude 50 m, rainfall 1 150 mm with summer dominance). Pasture yield was 4 480 kg DM/ha and the water content of the hay was near 25 percent. He found that mechanical tedding was better than slashing and crushing (conditioning). The slashed and crushed grass lost more water during the first day than tedded grass, but this advantage was lost when the hay took up more water at night and during rain than tedded grass. Gordon (1974) supported this finding. Catchpoole suggests frequent tedding of the hay on the first day and, if the weather is fine, once or twice on the following days. The dry matter losses were usually well below 10 percent and the field losses of nitrogen were 6-14 percent. The hay was cured in 50-55 hours in good weather and 70-75 hours in dull weather.

Collection, transport and storage of hay. Hay may be taken from the field in the loose form, being loaded by hand fork or by a pick-up or buck rake fitted either in front or behind a wheel type tractor. In the northern United States where the winters are severe, the hay is often left in the field for overwintering by beef cattle. In such cases loose hay can be put up in small stacks with a loose hay loader which elevates the hay into a mobile frame. When filled, the frame is removed and the hay released to remain in the field while the mobile unit moves on to the next site (see Plate 8).

More generally, the hay will be baled. For overwintering a roller baler has been developed to leave large bales (see Plate 9) in the field for the livestock. The bales can also be stored in the open near the feeding sheds. Deterioration of such bales in the field is relatively slow, so less covered storage space is needed.

In Gualaca, Panama, Medling (1972) found that though hay-making and storage of grass hay was difficult under the prevailing 3 997 mm/year rainfall, rice straw could be successfully baled and stored using the large rotary-compressed bale.

The normal rectangular bale of about 30 kg capacity is widely used. The pasture is cut, wind-rowed and cured in the field, then a pick-up baler bales the material, which can be loaded manually or mechanically into following trailers. The rectangular bale packs easily into current shed designs, and the hay can be fed easily either to individual animals or whole herds. Round bales of about the same capacity are left in the field for overwintering cattle. Baling machines tie the hay with either twine or wire. Care must be taken with wire-tied bales to prevent short pieces of wire being left around feeding areas and being ingested by animals. Numerous cases of rumen puncture from loose wire have been reported.

Roller-baler systems. In this system the hay is rolled up from the wind-rows like rolling up a carpet or is lifted into a confined bale chamber where pressure is applied to produce the bale. There are three common sizes of bales: a 227-kg bale approximately 1 m long and 1.2 m in diameter; a 364-kg bale measuring approximately 1.2 m long and 1.5 m in diameter; and a 545-kg bale 1.5 m long and 1.9 m in diameter. The 545-kg bale is the most common. A bale mover attached to a tractor's three-point hitch handles the transport and feeding operation.

Barn storage is slightly more effective in preserving protein and also reduces the dry-matter loss by 5-10 percent when compared to outside storage, but the savings in dry matter usually do not warrant the erection of permanent storage structures for roll bales.

To minimize storage losses from outside storage:

● Remove the bales from the field to reduce spoilage and prevent moisture accumulation on the bottom of the bale. This also aids feeding, as the field frequently becomes impassable in wet weather.
● Store bales on a well-drained site to prevent moisture concentration in the lower portion of the bale.
● Select a site near the feeding area to reduce labour.

- Leave a minimum of 30 cm space between each bale to ensure proper drainage.
- Do not store all the bales in the same yard because it could mean total loss in a fire.

In feeding out, three systems can be used: free choice, daily feeding and controlled feeding with panels. In the free choice system the bale is available to stock at all times. It is a cheap way to feed and need not be attended to daily, but up to 36 percent of hay may be lost from a sorghum-Sudan grass bale by trampling. Losses can be reduced by feeding one bale at a time and ensuring this is all used before feeding another.

In daily feeding the bale is opened and distributed in cut sections on the ground or in a trough. It means more labour, but there is little loss when the stock are exposed to it for short periods of 30 minutes to one hour per day. Losses can be kept to less than 2 percent. Controlled feeding by adjusting regulating wooden panels through which the stock gain access to the feed reduces the labour and minimizes hay waste, as the stock cannot trample the hay. Losses may be less than 3 percent (Rider, 1979).

An effective protection for hay stacked in the field or in the yard, without the expense of building a shed, is to cover the stack with movable sheets of curved galvanized iron which can be used over and over again as the hay is used for feeding (see Fig. 13.3). A stack is built in the shape of a rectangle over a floor of 30-cm round hardwood timber laid crosswise at 1.5 m apart. Logs are placed lengthwise atop these hardwood poles at an average of 15 cm apart.

The stack is roofed with curved, 24-gauge, corrugated galvanized iron. The curve is formed by bolting two 3-m sheets together end to end. Each is curved to a 30-cm spring: the centre is 30 cm higher than the ends when standing on its two ends. This forms an arch with a span of approximately 4.5 m and a height of 1.07 m (see Fig. 13.4). A stack 4.25 m wide across the short base, 4.5 m across the top, and 15 m long, with the eaves 4.5 m high and the top of the centre of the stack 5.5 m will hold approximately 50 tonnes, or one tonne for each 30 cm of length.

The pairs of iron roofing sheets are assembled on the ground by erecting a stand 1.07 m high of one stout piece of timber level with the ground and long enough to carry three pairs of curved, galvanized iron sheets together leaving a metre at each end for working room. The pairs of sheets are then bolted together with roofing bolts 0.6 cm × 3 cm, which must also be galvanized. Use three to a pair of sheets, allowing for an overlap of 15 cm at the top of the arch. Two additional bolts are inserted down each side of each 3-m sheet so that each complete arch is joined to its neighbour by five bolts. The necessary bolt holes can be punched in the galvanized iron a little larger than the bolt size for ease in assembling. When the roof is bolted in place it can be weighted down with No. 8 plain fencing wire. It must pass completely over the roof, to

Tropical Grasses

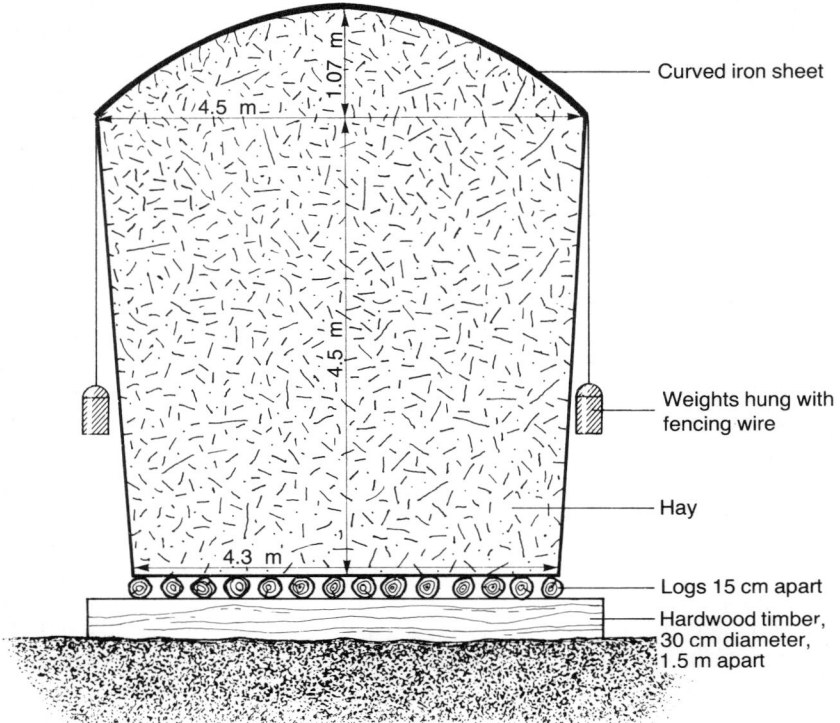

Figure 13.3. Elevation of haystack, showing foundations, stack and weights

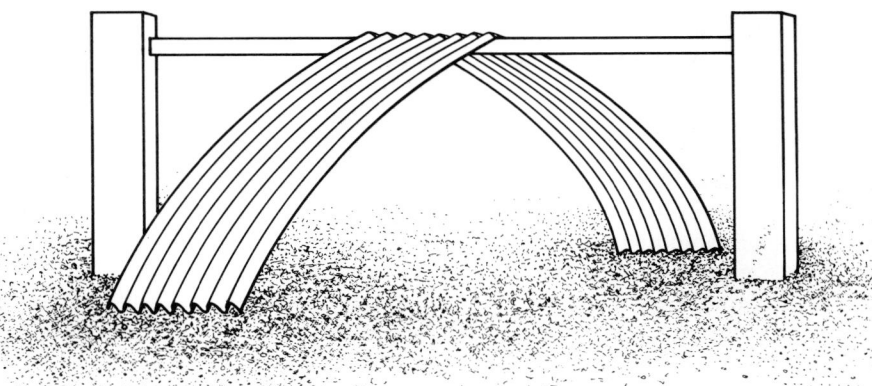

Figure 13.4. A frame for joining the iron sheets

which logs are hung to weigh it down. An extra wire can be passed over the end sheets and tied down to the ground logs and tightened as the hay settles down.

When opening the stack, one pair of sheets can be removed at a time as needed. If feeding ceases, the remainder of the stack will then still be covered. The sheets can be stored to be used again. It is useful to number the sheet pairs 1, 1A, 2, 2A, etc., for ease in reassembling.

Hay-making for the small operator. Where the green pasture is cut by hand, it is best to harvest after the dew has evaporated from the green material later in the morning. It can then be cut and placed into small heaps about 20-30 cm high and turned frequently in the sun to encourage quick drying. When the initial moisture has evaporated, the material can then be placed under the roof of any dwelling and allowed to dry completely away from the sun. This will conserve the colour and the nutritive value of the hay.

If the weather is humid or rainy, the material can be placed off the ground on home-made tripods, made by cutting six poles each 2-2.5 m long and 5-10 cm in diameter. The ends of three poles are then joined together to form a tripod. The legs of the tripod can then be stabilized by lacing the remaining three poles around the original three, forming a bar about 50 cm from the ground. The grass is stacked around the tripod, building upwards from the cross-arms. In this way the grass is kept off the ground and a hollow cone in the centre of the tripod allows the air to circulate freely and hasten the drying process (Thorp, 1979).

Artificial drying of grass. The artificial drying of grass is generally likely to be too expensive for most farmers in the tropical countries, although it could be a means of conserving a good deal of wet-season forage which is now wasted. Furthermore, most drying equipment is based on oil burning and oil is becoming an expensive commodity.

Artificial drying of grass can take place in any weather and at any time. It gives a product of high feeding value as nutrient losses are minimal. Protein, vitamin A and vitamin B are preserved. Storage space is also less than for naturally dried hay.

Driers are usually classified by their capacity to evaporate water. Initial moisture contents are assumed to be 85 percent, 80 percent and 75 percent, with a final dried-crop moisture content of 10 percent. If it is possible to wilt the grass to a moisture content below 75 percent, output will be increased and costs reduced. A tonne of hay at 40 percent moisture requires the removal of twice as much water as does a tonne at 30 percent moisture. Thorp (1979) gives a table of the evaporative capacities of green-crop driers in terms of output of dried material in kilograms per hour (see Table 13.3).

Among types of driers are high temperature rotary drum driers (costly), medium temperature rotary drum driers, medium temperature conveyor

TABLE 13.3 **Capacities of green-crop driers**

Drier rating, water evaporation (kg/hour)	Output of dried material (kg/hr)		
	85 percent Initial moisture content	80 percent Initial moisture content	75 percent Initial moisture content
2 500	500	700	1 000
3 000	600	850	1 150
4 000	800	1 150	1 550
5 000	1 000	1 450	1 950
6 000	1 200	1 700	2 300
10 000	2 000	2 840	3 850
15 000	3 000	4 275	5 800
20 000	4 000	5 700	7 700

Source: Thorp, 1979

driers, and low temperature batch driers. The Lister machine is one of the latter type and is more frequently found on the commercial farm. During the dry season heating of the air is not essential, all that is necessary is that the crop be ventilated. Artificial drying of grass is expensive and should only be practised on high quality material. The dried material can subsequently be milled or pelleted.

The percentages of harvested dry matter conserved by various curing methods are shown in Table 13.4.

The densities (kg/m^3) of various forms of hay are shown in Table 13.5.

TABLE 13.4 **Net dry-matter conservation of various curing methods**

Curing method	Net dry-matter conservation (%)
Quick drying in an oven	90
Barn drying with heated air	87
Barn drying with unheated air	85
Field curing without rain	70
Field curing with rain	50-70

Source: Reid, 1973

TABLE 13.5 **Densities of various fodder forms**

Form of fodder	Density (kg/m³)
Long hay	80
Chopped (2- to 10-cm lengths)	144
Chopped and wafered	256-480
Fine ground meal	336
Finely ground and pelleted	705-961

Source: Reid, 1973

Pelleting

Artificially dried forage can be made more dense for transport, storage and feeding by compressing it into bales or into wafers or cobs (chopped forage) extruded from a ram-press or rotary die press respectively, or it may be milled and pelleted. Mechanical treatment in machines that produce wafers or cobs will break down the dry, brittle herbage to a varying extent. This breakdown, or hammer milling, has a profound effect on the potential nutritive value of the dried feed, mainly through voluntary intake. Intake depends a great deal on the structure and density of the package. If the pellet is too hard (indicated by a density in excess of 1 gram per cc) young cattle may eat less even than the same material in chopped form. If the package is too easily broken and is dusty, the full potential intake may not be achieved. When dried forage is supplied *ad lib.* as the sole feed, and excessive hardness or dustiness are avoided, the highest intakes by ruminants have been obtained when the herbage has been ground in a hammer mill with screens of about 2-4 mm. Milling gave an additional live-weight gain of 2-78 percent (mean 37 percent) over that obtained from the same grass in chopped form. Beardsley (1964) gave a mean increase of 98 percent in the United States.

Wafering without milling gave an average live-weight gain of 70 g per day; 110 g a day was obtained from milling without wafering. The potential role of dried forage in beef production is as a concentrate feed with grazing or silage. The major requirements of the forage used are high digestibility, adequate protein content, with a modulus of fineness close to 1. A proportion of long forage or roughage is desirable as a means of maintaining good health in an animal (Tayler, 1970).

Voluntary intake of ground, pelleted hay is generally 1-30 percent greater than that of the same hay fed in the long or chopped form. The intake of finely ground hay is higher because the small particle size gives an increased rate of digestion and passage of the residue through the gastro-intestinal tract, and wetting or pelleting the material gives a further increase because it reduces dust. Table 13.6 compares the feeding value of ground, pelleted hay with that of long or chopped hay from the same source.

Silage

The manufacture of silage is of great antiquity. "Silage" is the name given to the product, "ensilage" is the process.

When a green crop is cut and put together in a heap, it continues to respire and in the process heat is developed. Bacteria are present on the cut material and several aerobic types continue to increase in number until oxygen is used up in one to four hours. Enzymes use readily available carbohydrates to produce heat and carbon dioxide and when the oxygen is used up the anaerobic condition, the heat, and available sugars in the material favour the development of lactic-acid-producing bacteria, mainly *Lactobacilli,* and in the matter of a few days they are completely dominant. These *Lactobacilli* produce mainly lactic acid with small amounts of acetic, propionic, formic and succinic acids. When the lactic acid concentration reaches 8-13 percent of the dry matter, the *Lactobacilli* are inactivated by their own excretory prod-

TABLE 13.6 **Comparison of feeding value of ground, pelleted hay with that of long or chopped hay from the same source**

Form of hay	Manner of feeding	Hay DM intake (kg/day)	Live-weight gain (kg/day)	Fat-corrected milk (kg/day)	Fat in carcass or milk
Fattening lambs					
Chopped	*ad lib.*	2.48	0.27	—	30.8
Ground, pelleted	controlled	2.54	0.26	—	30.8
Ground, pelleted	*ad lib.*	3.25	0.40	—	28.4
Fattening steers					
Long	*ad lib.*	14.0	1.80	—	28.4
Ground, pelleted	*ad lib.*	17.1	2.17	—	25.7
Lactating cows					
Chopped	*ad lib.*	26.4	−0.11	23.7	4.09
Ground, pelleted	*ad lib.*	33.3	1.66	28.1	4.01
Chopped	controlled	29.1	0.09	33.7	3.8
Ground	controlled	29.8	0.63	29.8	2.2

Source: Reid, 1973

uct and so the original material is really "pickled" in lactic acid and is known as "silage". In the absence of air it should keep indefinitely. At this stage the pH has been reduced to about 4.2 or below, inhibiting further bacterial growth and enzyme action. However, if air enters the silo, mouldy, rotten, very hot silage results.

The production of lactic and acetic acids depends a good deal on the amount of sugar in the original material from which these acids are derived by fermentation. A sugar content of about 6 percent is necessary for successful silage (Wieringa, 1966). Thus, material like sweet sorghum is ideal because of its high sugar content.

In freshly cut material, approximately 15-25 percent of the total nitrogen will be in the non-protein nitrogen form, of which 7-80 percent is present as amino-acids. Breakdown of proteins by proteolytic organisms is inhibited by a pH below 4.2 or a dry-matter content above about 43 percent, due to osmotic pressure and increasing fermentable sugar concentration.

If very young material is being ensiled it may not contain enough sugar and its dry matter will be too high to effect a lactic acid fermentation. The butyric-acid-forming bacteria, Clostridia (mainly *Clostridium butyricum*) will proliferate and produce butyric acid, ammonia, cadaverine, histamine, putrescine, tryptamine and tyramine. The resulting silage is foul smelling and of very poor quality. Heavy use of nitrogen fertilizer can affect the fermentation of grass in the young stage. The nitrate may be reduced to nitrite under the anaerobic conditions, and nitrite inhibits butyric acid formation (Wieringa, 1966). At nitrate levels below 1 percent the silage fermentation is influenced by factors other than nitrate (for example dry matter, sugar and protein). At 1-2 percent nitrate, no butyric acid is produced, being inhibited by nitrite, but above 2 percent nitrate, butyric acid occurs again. High nitrate is correlated with low sugar content and as nitrate in the silage disappears in a few days, if there is insufficient sugar to bring the pH below 4.2, butyric acid will build up again and spoil the silage. Hence, silage from low-nitrate grass is poor, that from 1-2 percent nitrate grass is good, and that from higher-nitrate grass is poor. Quality standards for grass silage produced from temperate pastures in the Netherlands are shown in Table 13.7.

TABLE 13.7 **Quality standards for silage**

Standard	Quality of silage		
	Good	Medium	Poor
pH	<4.2	4.3-4.5	>4.5
butyric acid (%)	<0.2	0.2-0.5	>0.5
ammonia N (%)	<8.0	9-15	>15

It is better to ensile high-nitrate grass when it is older and has a higher dry matter content. Young green material often produces unsatisfactory silage due to the buffering effects from protein and organic acids (Playne, 1966) and the effluent from young material when ensiled leads to significant nutrient losses in the effluent.

Wilted silage. Because of the unsatisfactory results in ensiling high moisture material the practice of pre-wilting the material in the field is often adopted in the United States. The cut material is wilted to a dry-matter content of approximately 65 percent and then ensiled. With such material, a lactic acid silage is produced but there is less fermentation and a pH of 4.5 will still produce good silage. Wilting in the field under the vagaries of tropical weather is always hazardous, and compaction of wilted material is always difficult. Silage made from grazed green panic/glycine pasture on the Atherton Tableland which contained 44 percent of dead and senescent leaf made good silage, but it was not a dominantly lactic acid silage. When there was no senescent material, the silage quality was poor.

The use of additives. The use of additives is advised if the material being ensiled in non-sealed, airtight storage contains over 70 percent moisture and is known to be low in sugar content. Materials such as maize, sorghum and elephant grass do not need additives.

The feed additives furnish a readily available source of carbohydrates for bacterial fermentation and some absorb some of the moisture from high moisture material. Approximately 75-80 percent of the feed nutrients added will be recovered in the field. Additives include:

- Ground maize, barley, oats, maize grain and cobs or meal in amounts from 50-100 kg per tonne, depending on the moisture content of the crop;
- Products such as citrus pulp, maize cobs and chopped hay to reduce seepage losses from the silo (materials such as fruit cannery waste and brewer's grains can be ensiled alone);
- Molasses.

The chemical additives include mineral acids such as hydrochloric and sulphuric acids which, added to the fresh material in sufficient amounts, reduce the pH to below 4.0 immediately, thus preserving the silage in fresh condition. The concentrated acids are diluted with water to a strength of 2 normal and sprinkled on the ensiled grass at the rate of 18-27 litres per tonne. This is the basis of A.I.V. silage introduced by A.I. Vertanen of Sweden. Good silage results and it indicates that heat is not necessary for the manufacture of good silage. The method is costly but less so in industrialized countries where the acids may be obtained more cheaply. Formic acid at 0.5 percent has been used more recently and is said to give good preservation. Animal intake of the silage and animal production are good. Formic acid must be handled with care.

Sodium metabisulphite at rates of 3.5 and 5.4 kg/t has been used as an additive. It has a bacteriostatic action but is little used now. It had little effect on *Paspalum dilatatum* silage in Queensland (Levitt, Taylor & Hegarty, 1962), probably due to its volatility at temperatures above 30°C.

The antibiotic bacitracin has been used at 5.5 g zinc bacitracin per tonne of forage with some advantage in ensiling *Digitaria decumbens, Cynodon dactylon* and *Paspalum notatum* (Catchpoole & Henzell, 1971).

Molasses has long been the most common additive to silage in tropical countries where sugar cane is grown. However, its use for industrial processing and as a component in licks has reduced its availability in many areas. It does not entirely reduce the dry-matter losses but gives a satisfactory fermentation. The quantity added has been subject to experimentation. In temperate regions too little molasses can give worse results than none at all, and may even increase butyric acid levels (Catchpoole & Henzell, 1971). The addition of 4 percent (approximately 32 kg/t) of molasses to the green material is usual, but this has been found insufficient with tropical grasses by Levitt, Taylor & Hegarty (1962), Catchpoole (1962) and Medling (1972). Levitt and his colleagues found 7 percent (approximately 71 kg/t) was needed to make satisfactory silage. Medling (1972) used 10 percent molasses to ensile most tropical species successfully in plastic bags at Gualaca, Panama.

Levitt and his colleagues found that salt at 1.2 kg/t plus molasses at 48.5 kg/t gave the smallest losses of dry matter. When 80 kg/t of molasses was used, some of the dry-matter losses were from the molasses. While salt will act as a preservative, the cost of salt at 1.2 kg/t in northern Australia is four times that of 80 kg of molasses/t.

The type of silage produced depends a good deal on the temperature, and the U.K. Ministry of Agriculture recognizes four types of silage.

Sweet, dark brown silage. Made when the material ferments too quickly and the temperature rises above 45°C. This usually results from the use of comparatively dry material, either that which was mature at harvest, or which has been allowed to dry somewhat after being cut. Such dry crops facilitate fermentation in two ways. Because they do not pack so tightly, they allow air to penetrate the silo readily, and because the heat generated by fermentation has comparatively less moisture to heat, the temperature rises more.

Acid, light brown or yellow-brown silage. Made with less air than sweet, dark brown silage, the material does not heat up so much. This type commonly occurs at temperatures of 30-40°C. As a rule there is not much juice expressed from the silo when this type is being made. Acid brown silage is commonly made in pit and trench silos. This silage has a yellow-brown to brown colour and an acid though pleasant smell, largely due to the presence of acetic acid (the acid of vinegar). The yellowish types have the more pleasant smell. It is readily eaten by stock. This is the most common form of silage and it is much superior to the dark brown variety.

Green, fruity silage. Usually this quality is only made by chance, as it is hard to control conditions to make it with certainty. It is made by rapidly ensiling fresh, lush, leafy grass and clover at a temperature of about 30°C, but no higher. This type of silage has a green to olive-green colour, and a smell that is delicious — neither sweet nor sour — and is best described as "fresh" and "fruity". It is greedily eaten by stock, and its digestibility is high, but the relatively large amount of effluent carries away valuable nutrients.

Sour silage. Generally has a dark brown or olive-brown colour, and a pungent and very unpleasant smell, due largely to the presence of butyric acid. It is commonly made when a very immature and succulent crop is ensiled. In this case the watery fodder packs down very closely in the silo and excludes the air to such an extent that little heating is possible. It is frequently found at the bottom of trench silos, especially if made in wet weather, because of limited fermentation. Slower filling will help prevent this condition. The sour silage has a high feeding value and is quite palatable despite its unpleasant smell. Since all silage may give rise to objectionable odours which are readily absorbed, it should be kept well away from milk or cream.

Several types of silos are available for preserving green materials.

The silage trench. This is the best type of silo for tropical conditions because it is cheap to construct with local resources. It can be located near the cultivation areas to reduce the time and cost of transport. It can be dug by hand or, more easily, by a bulldozer.

The hillside trench is preferable, as it can be filled and emptied without lifting heavy materials. Animals can feed themselves and any effluent can drain away. It requires a fixed site, but can be used again and again. The earth trench may need some reshaping but if it is lined with concrete at the sides and bottom it provides continuous use and storage (see Plate 14).

The dimensions vary according to the usual amount of material to be ensiled. A convenient size holds about 100 t of material which will be provided by a trench 30 m long, 7 m wide and 2 m deep. The sides of the trench slope slightly outward from the bottom to allow the silage to wedge itself into the trench, thus increasing compaction (see Fig. 13.5). The upper end of the trench should have a slope of 20-30° to allow a vehicle to drive in with its load, discharge it and drive out the other end of the trench (see Plate 10). This also compacts the material during each passage.

If there is no natural site on a suitable slope, a double-ended trench can be excavated on flat land to suit the purpose. Such a trench is more difficult

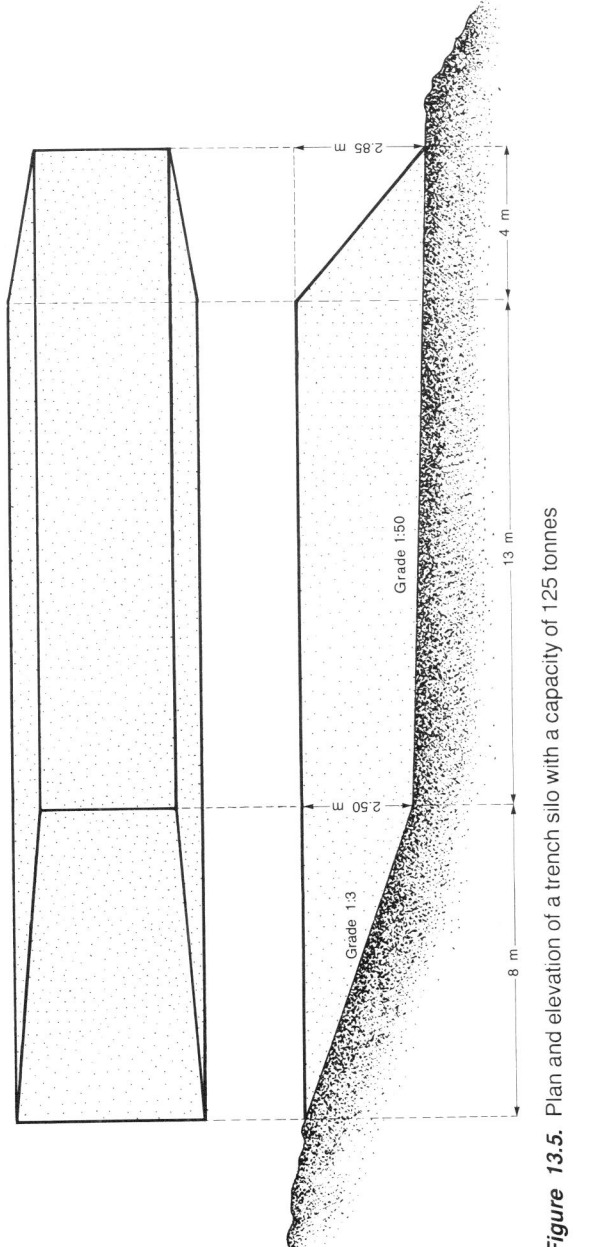

Figure 13.5. Plan and elevation of a trench silo with a capacity of 125 tonnes

to empty and does not provide good drainage, but if it is made of earth it will usually absorb the effluent.

The trench silo can be filled with uncut material, but it is difficult to remove for feeding, as it must be cut out in blocks with a hay knife or similar tool. It is preferable to ensile chaffed or slashed material. Chaffed material is the easiest to remove for feeding.

Fill the trench as quickly as possible, compacting the material by trampling or with a heavy vehicle after each load is discharged. A wheeled tractor gives more pressure than a track-laying tractor. A road grader helps distribute the load across the trench while compressing it. Fill the trench to a height of about 1 m above ground, allow the material to settle for two days, and then cover the material with about 50 cm of earth, or with a plastic sheet first and then earth, or old rubber tyres. Building up the material (see Plate 11) above ground gives it extra pressure for wedging itself into the silo and provides a camber that will shed rain easily. The silage to be made should keep for many years. If a clay soil is used for cover, it may dry out and crack during the dry season. It should be lightly cultivated or worked over to close the cracks and keep the material airtight. Losses in such well prepared trenches should be only a few centimetres of material from the top and sides.

When the silage is needed, excavation should start at the lower end, removing only sufficient material for each day's use. Then when there is no more need for silage, the remaining material is still covered. Excavators are illustrated in Plates 12 and 13.

A frame of wood or pipe can be fitted to a hillside trench, allowing the animals to feed directly from the silage face. If a limited amount is uncovered each day and the ration controlled, losses from trampling will not be unnecessarily large.

The bunker silo. This type of silo is essentially the same as the trench silo but is erected above ground, using wooden or concrete slabs or earth for sides. It has good drainage and animals can feed themselves. There is some danger of collapse if a tractor is used to compress the material and to cover the bunker as it reaches the final stages of filling. Research is under way at the National Institute of Agricultural Engineering, in the United Kingdom, into the use of flexible-wall bunker silos to reduce costs.

The bun silo. This is a simple and inexpensive type where the green material is placed in a heap on the ground, compressed by a tractor to the desired height and covered with plastic and car tyres or earth. It can be located anywhere, preferably on a well-drained site. Silage can be made in wet weather with this system. Tractor drivers must be careful in compressing the heap, as the material may move and overturn the tractor. The bun silo is a temporary system to be used until a more permanent system is adopted. Losses can be high.

The stack silo. This is a common method of storing grass as silage. With coarse-stemmed material such as maize or sorghum it is impossible to obtain sufficient compaction and the ingress of air causes large losses. Stacks of grass silage can be successful where the grasses are thin-stemmed. Stacks 4 m high by 5 m diameter should contain 30-35 t of silage. It is most essential that enough material be used to give sufficient height to provide a stack that, when settled, will be at least 3 m high. The height is needed to give enough pressure to compact the material and exclude air. The amount of shrinkage is considerable. It will be necessary to ensile about 50 t of green pasture to produce 40 t of silage. Stacks should be located on high, well-drained sites.

Cut enough material to build the stack 2 m high the first day. No grass is added to the stack for a few days to allow it to settle, providing a solid foundation on which to build the stack. This is most important, as a stack erected on a faulty foundation is sure to slip as it settles.

Once the temperature has reached 43-49°C stacking can be continued. More material should be added at once if the heat gives indications of rising above 49°C. Usually, material should be added every second day until the desired heat is obtained. Cover with 25-30 cm of soil at the edge and 45-60 cm in the centre. The soil can be held in place by a perimeter of bags filled with soil.

The tower silo. This method involves very expensive equipment and demands a permanent silo, usually located at the feeding site for dairy or beef cattle. Such expenditure is usually beyond the financial capacity of many tropical farmers and ranchers. A temporary and cheaper silo can be constructed from wire mesh lined with plastic or bitumenized paper and filled by hand with pitchforks or by a chopper-blower or chopper-elevator. However, the plastic or paper lining is easily punctured by sharp ends of stems, by wire or by the trampling process.

For large herds a concrete silo is commonly used. Its height must be at least twice the diameter to give enough weight to compress the silage and exclude the air (see Plate 14). The crop is cut with a forage harvester and then fed into the top of the silo with an elevator or blower, the material being compacted by trampling as the silo is filled. Small doors at suitable intervals allow the silo to be entered and emptied as needed. The silage can also be taken out by a top-unloader that sits on top of the silage and is centred by castor wheels against the internal wall. A screw auger moves the silage into a central blower which discharges the silage into trucks or hoppers for feeding (see Fig. 13.6).

Tower silos provide good silage with very little waste.

The vacuum silo. In this system a heap of silage is made on a plastic sheet and lightly compressed to shape by a tractor, much like the system for a bunker silo. When the correct shape has been attained a complete plastic cover is placed over the top and sealed to the bottom sheet. Air is then evacuated

Figure 13.6. A top-unloading device for tower silos which excavates the silage and blows it into a trailer. Castor wheels centre and stabilize the excavator

using a vacuum pump until the stack is hard, usually reducing its volume to one-half. The top sheet is then removed and filling continued until it is big enough for further evacuation.

It is a mobile system requiring little initial consolidation, but the cost of the plastic bunker is high and it seldom lasts more than two or three years. It requires a vacuum pump. Sharp points of stem tend to puncture the plastic and allow air to enter the system, and the plastic is also open to puncture by the hazards of general farming operations.

Choice of species for ensilage

The production of silage from tropical forage grasses is rarely practised because of climate, physical and quality problems. The conclusions reached by Catchpoole and Henzell (1971) in relation to silage made from tropical herbage species were as follows:

- The densities of these silages can be low, and special precautions are needed to exclude air from them.

- Several tropical species produce stable silage without additives. The factors responsible for this preservation are not known and certainly do not include the production of high concentrations of lactic acid.
- Addition of molasses improves preservation of silage in the tropics, but only when large amounts (up to 80 kg/t wet weight of plant material) are added.
- Wilting prior to ensiling also improves silage preservation, but mould growth may be a problem. In some tropical countries it is difficult to achieve the desired degree of wilting.
- Most tropical grasses (cut plant material) have a relatively low feeding value, and changes during ensiling may reduce it even further.
- Feeding of tropical grass silages to supplement the diet of beef cattle on standing winter or dry-season feed has given little or no live-weight response. High levels of concentrates are necessary with tropical grass silages to obtain satisfactory milk yields or beef production in feedlots.

However, tropical fodder crops such as maize, the sorghums and elephant grass are frequently ensiled because of their high yield of good quality silage without additives. Elephant grass is usually preferred for silage in the humid tropics, maize for the wetter subhumid regions, and the *Sorghum* spp. for the drier subhumid and wetter semi-arid areas (see Plate 15).

At Gualaca, Panama, with an average annual rainfall of almost 4 000 mm, Medling (1972) found that maize was unsatisfactory because of the many pests associated with its growth, despite a spraying schedule. He also found that the sorghums universally suffered from rust and made poor growth. Pearl millet (*Pennisetum americanum*) gave poor germination and poor growth. Best results were obtained with elephant grass (*Pennisetum purpureum*). It produced 16.1 t of green matter per hectare per month, containing 10 percent crude protein compared with 11.7 t and 7 percent crude protein from the best sorghums. Moreover, the elephant grass had a useful life of three years, while maize and sorghum are annuals.

It is generally recommended to harvest maize for silage when 50-75 percent of the grains have become dented at the tip. At this stage the contents are turning from "milky" to the "dough" stage. Sheldrick (1975), working in Kenya with the maize hybrid 611C, found the greatest yield of dry matter occurred 187 days from planting. Its crude protein content was 6-7 percent. He recommended that cutting commence when the grain reaches the early dough stage (crop moisture 75 percent, about 176 days from planting, for this cultivar) and continue till the hard dough stage (crop moisture 65 percent, 198 days old) to give the optimum harvest of digestible organic matter.

Catchpoole (1962) studied the time of cutting and its effect on moisture, protein and sugar content of *Sorghum bicolor* cv. Saccaline in Queensland. His results are shown in Table 13.8 (see Plate 16).

At ten weeks (early milk stage), yields of silage and silage protein were

TABLE 13.8 **Moisture, protein and sugar contents of fresh sorghums cut at seven different stages**

Age of crop (weeks)	Stage of growth	Moisture (% fresh wt)	Crude protein (%)	Crude protein (kg/ha)	Reducing sugars (% glucose)	Total sugars (dry wt)
4	Ear not initiated	90.6	22.6	436.8	—	—
6	Ear initiated	91.6	14.0	963.2	6.1	8.6
8	Boot stage	88.4	9.5	1 064.0	7.5	9.2
10	Early milk stage	81.9	7.3	1 220.8	12.2	38.1
12	Early dough stage	78.7	5.6	1 064.0	11.1	14.7
14	Hard dough stage	72.1	4.2	974.4	5.0	7.6
16	Bottom leaves dead	68.4	4.8	1 176.0	4.0	6.1

Source: Catchpoole, 1962

at maximum levels. The crude protein content of silage was 6.7 percent of the dry matter. Delaying harvest to the dough stage reduced crude protein contents. Harvesting before the milk stage reduced silage yields, but increased crude protein contents. Only a very early harvest (four weeks) significantly reduced yields of silage protein. The four- and six-week harvests produced silage of poor quality, shown by pH values well above 4.2; ratios of non-volatile acid to volatile acids of less than unity; and low ratios of amino-acids to volatile bases. All harvests after eight weeks (boot stage) produced excellent acid silage showing no evidence of protein degradation. The yields and losses are shown in Table 13.9.

The four- and six-week material suffered high effluent losses, with associated high protein losses, on ensilage. Effluent losses were reduced markedly at eight weeks and further at ten weeks, when there was virtually none.

Later Catchpoole (1972) studied the ensilage of *Sorghum almum* cv. Crooble, a forage sorghum, and had similar results. Good silage without additives was made from material harvested after seven weeks. One result of these trials was that material that was wet during harvest made very poor silage, even though material at that age usually produced excellent silage. Thus, ensiling during rainy weather should be avoided. Young and predominantly vegetative *S. almum* decomposed badly and lost large amounts of effluent during storage. The practice of wilting or the use of additives would probably be needed to produce successful silage from this material.

After the initial immature stage (up to six weeks) the time to cut sorghum for silage will depend on whether a lower-yielding, high-quality product is required for direct feeding, or whether a large yield of lower-quality material is required for roughage. The latter must be supplemented with protein con-

TABLE 13.9 **Moisture and protein contents, silage yields and silage losses at varying age of crop**

Age of crop (weeks)	Moisture (% wet wt)	Crude protein (% dry wt)	Silage yield (tonnes/ha)	Silage protein (kg/ha)	Percentage lost during ensilage		
					Total loss (% orig. wt)	Effluent loss	CP loss (% CP)
4	80.8	17.2	8.8	246.4	63	56	43
6	82.9	12.6	40.5	761.6	57	53	21
8	82.7	9.1	63.6	896.0	42	34	16
10	80.5	6.7	97.8	1 108.8	14	11	10
12	78.3	4.8	102.6	952.0	4	1	7
14	73.5	5.4	86.9	1 086.4	—	0	—
16	71.9	5.1	88.6	1 108.8	4	0	4

Source: Catchpoole, 1962

centrates, so costs and freight charges on protein supplements may resolve this question.

Silage for the small farmer

Most of the literature on the feeding of livestock deals with the problems of commercial farmers, large-scale ranchers or feedlot operators. In the developing countries, livestock production is usually in the hands of the villagers who can handle only one to six animals. They are usually kept in stalls under or adjacent to the house or are managed as part of a communal herd. These animals are shepherded during the day on communal land and returned to their village compound at night. Fodder is best conserved as hay rather than silage under these conditions.

However, silage can be made in cheap plastic bags if available. Heavyweight polythene bags of 25-kg capacity are suitable for such storage. Chopping the material to improve compaction, tramping it into a bag and tying tightly will generally exclude enough air. In laboratory studies the air is replaced with nitrogen gas.

The quality of the silage will be determined by the amount of sugar in the ensiled material. It must be high enough to give a quick fermentation. When Donefer, James and Laurie (1973) ensiled sugar cane tops and fith with no special precautions to exclude oxygen, no observable spoilage resulted even after long storage.

In Gualaca, Panama, Medling (1972) experimented with plastic bags to make silage. The pasture was chopped with a stationary forage harvester and the chopped material caught in plastic bags of 1.5 mm thickness. The air was removed by vacuum pump until the material was rock hard, then the mouth of the bag was doubled over, tied with twine, and placed in a second bag which was also evacuated and tied. The double bag was then hung on the rafters until required. With pangola grass, *Cynodon plectostachyus, Pennisetum purpureum* and some other grasses this was successful. However, with young leafy material some effluent collected in the bottom of the bag. Sixty-six species of grasses and legumes were tried, of which only six gave satisfactory silage without the addition of molasses. Adding 10 percent molasses invariably gave satisfactory silage. Catchpoole (1965) showed that with 4 percent molasses added lactic acid fermentation did not always take place. He suggested that higher levels were needed with grass silage. It would seem that 4-10 percent molasses is satisfactory for most grasses. If moisture content is kept below 70 percent, little effluent should develop. This procedure deserves more research for small-scale silage manufacture. With satisfactory moisture, adequate sugar and effective compaction, a practical method could be developed for most tropical species.

Haylage

Silage is a high-moisture plant product that requires a comparatively large amount of lactic and acetic fermentation to ensure sufficient acidity (about pH 4.2) to prevent adverse fermentation and putrefaction by *Clostridium butyricum*. The high moisture of silage is necessary when there is no airtight system available to give good compaction and prevent oxidation.

Haylage, on the other hand, is a low-moisture (45-50 percent), short, chopped plant product that undergoes considerably less fermentation. It is spontaneously combustible and hence must be stored in oxygen-limiting conditions. The Harvestore was designed to meet these conditions.

The Harvestore creates an oxygen-limiting environment by three features which are also engineered to ensure maximum flexibility and long life.

- The structure, erected on site, is made of glass fused to steel sheets. The fusing process provides the corrosion resistance of glass with the normal elasticity and strength of steel. The sheets are belted and sealed with mastic to prevent the ingress of air. Porosity, and hence diffusion of air, is nil.
- There are "breather bags" in the roof. They allow the expansion and contraction of gases to take place inside the tower with the necessary volume of air to prevent a partial vacuum occurring. It is drawn into the large triple-laminated butyl bags, which have a capacity of over 30 m^3 of air movement. This prevents air from reaching the stored material.
- The bottom unloader allows the structure to be used as a processing unit, being fed at the top and unloaded from the bottom on the principle of "first

in — first out." It also obviates the need to open the structure fully to install, service or unload the material. It is therefore critical to oxygen-limiting storage.

The Harvestore system eliminates the persistent odour of silage, and effluent is no problem because of the comparative dryness of the material. Physical work is minimal since the system is operated mechanically.

The material that can be converted into haylage is diverse. Any common forage crop or pasture can be processed, as well as miscellaneous crops such as sunflowers, potatoes and rape. Factory waste such as coffee grounds, fruit pulp, brewers' grains and stover can be stored, as well as chicken litter.

For forage crops and pastures the material is preferably cut in the boot or pre-boot stage (see Fig. 13.2). With maize and sorghum the crop can be cut a little later (dent stage for maize and hard dough stage for sorghum) if it is intended to feed the grain and make haylage of the stubble. For hay, stover and dry stubble the material can be reconstituted with water to 40 percent moisture and stored.

Harvesting and storage. The material is cut in one operation with a mower/conditioner/wind-rower machine, then wilted in the field to 55 percent moisture. When a sample of fresh material has lost half its original weight in sun-drying, the crop is ready to be chopped. The material is cut into 40-mm pieces with a fine chop machine and blown into the Harvestore using a swan-neck delivery pipe. The material is cut and stored as quickly as possible (3-4 m height a day with a common type Harvestore). No tramping or other compaction is used because there is no access; anyway, the finely chopped material consolidates well and excludes air. As soon as the available oxygen is used up and replaced with carbon dioxide by respiration (generally two to four hours), the fermentation is arrested.

Utilization. The material can be used at any time, but it is usually left for at least three weeks to allow full fermentation. The advantage of the haylage system is that material can be added to the Harvestore at any time, regardless of its composition. There are usually no problems. If the material is too wet when added, there will be some nutrient loss through effluent. Wet material is very hard to move with the unloader. It causes premature wear on the machine and raises maintenance costs.

If the material is too dry there are problems of consolidation and a danger of spontaneous combustion. This latter problem should not arise if the Harvestore hatches are kept closed when not in use. If the level of haylage in the Harvestore is getting low, the unloader should be modified with the short-arm attachment or removed.

The Harvestore is expensive to install. It is essentially a unit for continuous feeding and not for sole use as a storage of fodder for times of drought.

Only about 2 percent of Australian dairy farmers would be in a position to finance such a system, but once established in a large-scale operation (more than 70 milking cows) it can be economical. It is very convenient and can be incorporated in a zero-grazing situation. Cow production performances with haylage feeding are usually better than with other types of fodder (Skerman, R.H., personal communication).

The economics of fodder conservation

Hutchinson (1966), working with temperate pasture species at Armidale, New South Wales, showed that animal production was depressed by the reduced grazing area caused by hay-making. Grass was cut in the spring, but with a high stocking rate, low production continued into the following summer. He suggested that, under the local conditions, only fodder in excess of current needs should be harvested.

In the tropical and subtropical areas where there is more than four months of dry season, however, fodder conservation does have advantages in year-long production and is a valuable standby in times of drought. The main problem is to have sufficient feed conserved to maintain essential animals throughout the drought. Sale of surplus stock after a predetermined date if rain has not fallen is a sound practice in areas of erratic rainfall. One pastoral company in Queensland ordered all surplus stock to be marketed immediately after the first of April if no worthwhile wet season (November-March) rain had fallen by that date.

At Samford, southeast Queensland (lat. 27°22′S, alt. 50 m, annual rainfall 1 150 mm), Jones (1976) compared grazing of two pastures, Nandi setaria and Samford Rhodes grass. Each was fertilized with 336 kg N/ha in four applications annually and grazed continuously at two stocking rates (3.75 and 5 beasts per hectare after the first year). Surplus feed was cut in summer, conserved, and fed back in winter in the same paddock, over a four-year period at two sites. The paddocks also received 250 kg superphosphate and 125 kg KCl/ha in the spring. Hay was cut on 16 December, 12 March, 22 February, 11 January and 4 April. Conservation and feeding back improved mean gains per animal by 12.3 percent and mean gains per hectare by 13.6 percent. There was no difference between the grasses. The "grazers" were ahead of the "conservers" until July and the commencement of hay feeding, when they lost weight for three to four months before gaining weight again in spring and early summer. The conservers maintained or gained weight in winter and their spring and summer gains were similar to those of the grazers.

Nandi hay averaged 1.46 percent nitrogen and 0.25 percent phosphorus; Rhodes hay averaged 1.61 percent nitrogen and 0.26 percent phosphorus. Conservers did much better than grazers in winter when severe frosts and heavy winter rain caused standing herbage to deteriorate rapidly.

TABLE 13.10 **Mean cumulative live-weight increase per hectare (kg)**

Year	Stocking rate (beasts/ha)	Live-weight increase/ha		
		Hay conservation	No hay conservation	Mean
1963	2.5	108.4	81.4	94.9
	1.5	118.2	94.1	106.2
1964	2.5	165.6	160.3	163.0
	1.5	170.9	122.1	146.5

At Brian Pastures, Queensland, Scateni (1966) compared two stocking rates of natural pastures of *Heteropogon contortus,* one weaner steer per 0.4 ha and one weaner steer per 0.66 ha, with and without hay conservation. The grasses were from improved pastures, grown in midsummer and fed back in the winter, spring and early summer. He found that production per head was greater at the lighter stocking rate and that the practice of conserving hay from improved pastures of *Panicum maximum* var. *trichoglume* (green panic) and *Medicago sativa* (alfalfa or lucerne) and feeding it back gave superior gains to grazing throughout on native pasture. Thus, by a complementary use of sown pastures for winter to early summer (dry season) grazing and natural pastures for summer and autumn (wet season) grazing, annual cattle performance can be greatly increased in terms of both production per head and overall production per hectare. The comparative figures are given in Table 13.10.

14. The chemical composition and nutritive value of tropical grasses

By D.J. Minson

The productive value of any feed depends on the quantity eaten and the extent to which the feed consumed supplies the animal with the required energy, protein, minerals and vitamins. Many tropical grasses have been studied in both the laboratory and animal house, and this chapter attempts to bring together the results of these studies, which are widely dispersed in the scientific literature. There are differences among grass species and varieties, but composition of a grass is also influenced by climate, soil fertility, stage of maturity and method of feeding, so one must exercise caution when comparing grass species grown in different environments.

Voluntary intake

The quantity of dry matter eaten each day by an animal is the most important factor controlling the productive value of a feed. Clearly, if an animal can consume only a small quantity of grass, production of meat or milk will be low no matter how high the protein, digestible energy or mineral content of the grass.

The gross, or total, energy content of tropical grasses is relatively constant, varying between 17.2 and 18.4 kilojoules per gram of dry matter (Minson & Milford, 1966). Because of this constancy, the results of most intake studies are expressed in terms of dry matter. Temperate pastures also have a relatively constant energy content of a similar value, varying between 18 and 19.1 kJ/g of dry matter (Hutton, 1961). Any variation in the gross energy content of grasses is usually associated with differences in the proportion of protein or ash.

Ideally, the voluntary intake of pasture plant should be determined in

---

D.J. Minson, Division of Tropical Crops and Pastures, CSIRO, Cunningham Laboratory, Brisbane, Australia.

the field, where selective grazing can operate. Unfortunately, there are no accurate methods for measuring the voluntary intake of grazing animals. Intake data for tropical grasses have therefore been mainly obtained from indoor feeding studies. Sheep have generally been used, although a few results have also been obtained with cattle and goats. The quantity of grass dry matter eaten depends not only on the quality of the grass but also on the size of the animal. To eliminate the effect of differences in body size within animal species on the intake values of feeds, most results of intake studies are quoted in terms of grams of feed dry matter eaten per unit of metabolic weight, where metabolic weight is the 0.75 power of body weight in kilograms. Feed intake expressed in this way varies from 24 g/kg $W^{0.75}$ per day for a mature tropical grass to 100 g/kg $W^{0.75}$ per day for immature temperate pasture.

Figure 14.1 shows the histogram for the voluntary intake of 450 fresh or dried tropical grasses fed to sheep in different parts of the world. The mean intake of the tropical grasses by sheep was 50.8 g/kg $W^{0.75}$ per day, compared with a mean intake of 67.3 g/kg $W^{0.75}$ per day for temperate grasses. The distribution of the intake values about the temperate grass mean is similar to that for tropical grasses. Relatively few intake values have been published for tropical grasses fed to cattle. In a comparison of cattle and sheep fed leaf and

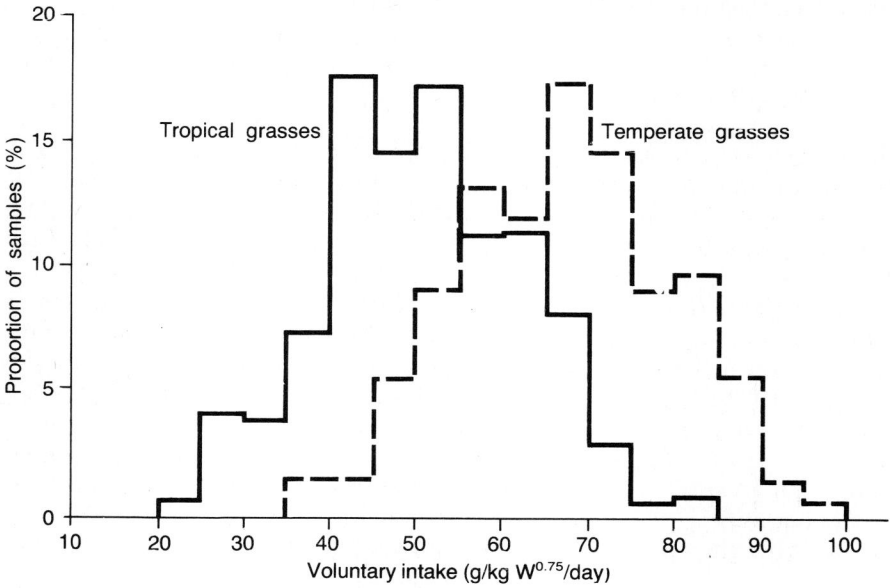

Figure 14.1 Frequency distribution of voluntary intake observations of tropical and temperate grasses

Chemical Composition and Nutritive Value

TABLE 14.1 **Voluntary intake and digestibility of some tropical grasses**

Species	Voluntary intake[1]	Correlation coefficient	Residual standard deviation
Chloris gayana	0.74 D + 3.4	0.73	± 3.7
Panicum spp.	1.42 D − 20.6	0.76	± 6.8
Three tropical species	1.19 D − 0.25	0.73	± 4.8

[1] D - dry-matter digestibility.

stem fractions of grasses, the mean voluntary intakes were 62.3 and 46.0 g/kg $W^{0.75}$ respectively (Ternouth, Poppi & Minson, 1979). This 35 percent higher intake by the cattle, even when the conventional 0.75 power of body weight is used to express intake, illustrates the danger of applying intake values to species different from those used to test the feed. Despite this difference between sheep and cattle in the actual quantity of grass eaten, it has been found that sheep rank tropical grasses in the same order of intake as do cattle (Playne, 1970b). This is why sheep are used in many pasture evaluation studies.

Stage of growth. For sheep there is a general trend for the voluntary intake to decrease with increasing age of the regrowth and declining digestibility of the dry matter. The relations between voluntary intake (V) and dry-matter digestibility (D) are shown in Table 14.1 for *Chloris gayana* (Milford & Minson, 1968), *Panicum* spp. (Minson, 1971a, b), and three tropical grasses used in a fertilizer nitrogen study (Minson, 1973).

Leafiness. Despite highly significant correlations between voluntary intake and digestibility, there can still be large differences in voluntary intake between grasses with the same digestibility. In a study of six *Panicum* varieties, intake differences of up to 50 percent were found between varieties, although there were no differences in digestibility (Minson, 1971a, b) (see Fig. 14.2). The variety with the highest intake had the highest proportion of leaf. Subsequent work with separated leaf and stem fractions of grasses has shown that sheep consumed larger quantities of leaf, even when there was no difference in digestibility between leaf and stem (see Table 14.2). These studies have recently been extended to cattle, which were also found to eat larger quantities of leaf (see Table 14.3). With both sheep and cattle, the lower intake of stem appears to be due to the longer time required for the stem to be broken down in the rumen.

Deficiencies of essential nutrients. The physical structure of the plant appears to be the main factor limiting the intake of animals grazing tropical pastures.

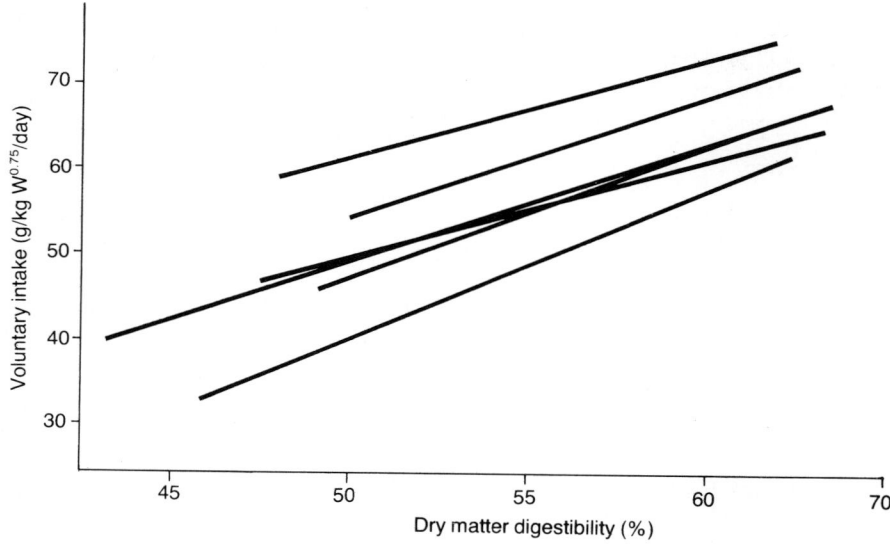

Figure 14.2. Relation of voluntary intake to digestibility for six varieties of *Panicum* species

However, if the diet is deficient in protein or an essential mineral element, intake will be depressed below the level set by the fibre in the diet. This will only occur when the animal's reserve of the essential nutrient has been depleted. With some minerals (for example, sodium) this takes many months.

TABLE 14.2 **Voluntary intake and digestibility by sheep of separated leaf and stem fractions of five grasses, each cut at three stages of growth**

Species	Voluntary intake (g/kg $W^{0.75}$/day)		DM digestibility (%)	
	Leaf	Stem	Leaf	Stem
Digitaria decumbens	58	40	53	54
Pennisetum clandestinum	50	35	51	52
Chloris gayana	57	45	53	58
Setaria splendida	59	32	56	59
Panicum maximum	64	47	51	56
Mean	58	40	53	56

Source: Laredo & Minson, 1973

Chemical Composition and Nutritive Value

TABLE 14.3 **Voluntary intake and digestibility by cattle of separated leaf and stem fractions**

Grass species	Regrowth (weeks)	Voluntary intake (g/kg $W^{0.75}$/day)		DM digestibility (%)	
		Leaf	Stem	Leaf	Stem
Digitaria decumbens	6	29	20	57	57
	12	26	19	48	50
Chloris gayana	6	31	25	60	59
	12	28	20	50	49
Mean		28	21	54	54

Source: Poppi, Minson & Ternouth, 1981

Protein. When the crude protein content of tropical grasses falls below 6-8 percent, appetite will be depressed by a crude protein deficiency in the animal. Since the appetite is limited by a protein deficiency and not by excess fibre, changing the physical structure of the feed by grinding and pelleting has no effect on the intake of low-protein tropical grass (Minson, 1967).

The protein content of tropical grasses may be increased by applying fertilizer nitrogen. With *Digitaria decumbens,* a late application of fertilizer nitrogen increased the crude protein content from 4.1 to 9.9 percent and lifted dry-matter intake of beef animals from 4.3 to 7.7 kg per day (Chapman & Kretschmer, 1964). When fed the unfertilized grass, cattle lost 0.22 kg per day, but when fed the nitrogen-fertilized grass they gained 0.69 kg per day.

Digestibility

The most desirable way of expressing the energy value of a feed is as net energy, since this takes into account the feed energy lost as gas, as heat and in the excreta. Elaborate equipment is required for the measurement of net energy, and only two cuts of tropical grasses have been studied in this way (Graham, 1967). In digestion trials, only the energy lost in the faeces is measured and the digestibilities of many tropical forages have been determined with sheep and cattle. The results of digestion trials are usually expressed as digestibility percentages of the dry matter (DMD) or organic matter (OMD) and these are closely related to the digestibility percentage of the energy (DE) (Minson & Milford, 1966). Table 14.4 shows digestible energy percentages of two grasses.

The mean energy contents of one gram of digestible dry matter or digestible organic matter are 17.22 and 17.89 kJ, respectively (Minson & Milford, 1966). Many feed rationing systems are now based on metabolizable energy

TABLE 14.4 **Digestibility percentages of the energy of *Digitaria decumbens* and *Sorghum almum***

Species	Digestible percentage			Relation coefficient
	of energy	of dry matter	of organic matter	
Digitaria decumbens	0.996	− 3.36		0.998
	0.994		− 4.45	0.998
Sorghum almum	0.961	1.23		0.999
	0.967		− 1.54	0.998

(ME). For tropical grasses ME is approximately 0.81 times the digestible energy content (Minson, 1979).

Temperature. The dry-matter digestibility percentage determined for 543 samples of tropical grass ranged from 30-75 percent with a mean of 54 percent. This is 13 percent lower than the mean dry-matter digestibility of 592 samples of temperate grass, which have a digestibility range from 45-85 percent (Minson & McLeod, 1970) (see Fig. 14.3). The lower digestibility of tropical grasses is associated with the high temperature at which they are grown. When temperate and tropical grasses were grown at different temperatures, the dry-matter digestibility of monthly regrowths was negatively correlated with mean daily temperature (r = − 0.76, Minson & McLeod, 1970) (see Fig. 14.4). Studies in controlled environment rooms have confirmed the

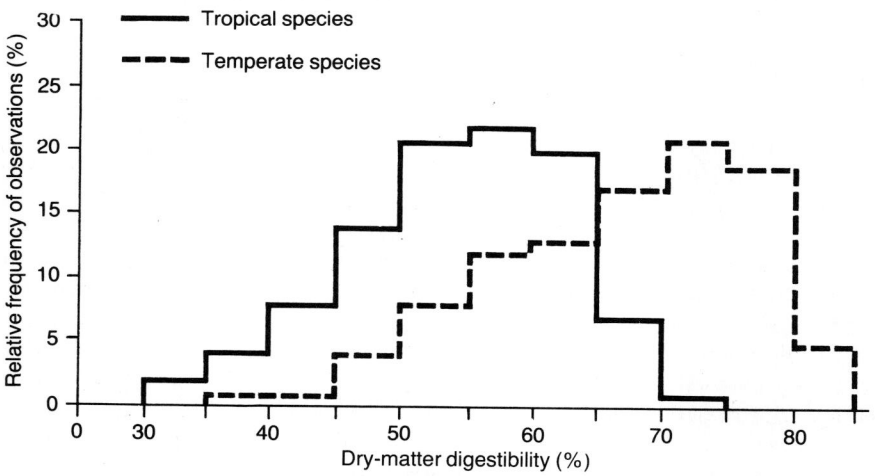

Figure 14.3. Frequency distribution of digestibilities for 543 cuts of tropical grasses and 592 cuts of temperate grasses

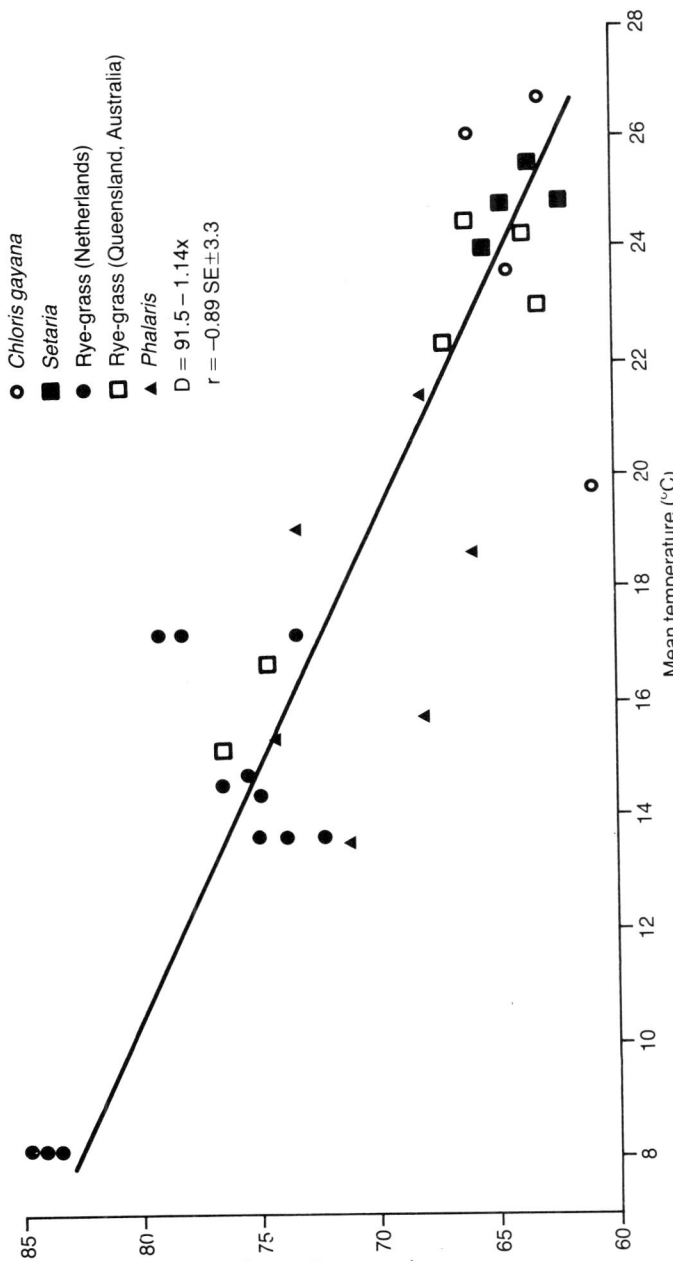

Figure 14.4. Relation of mean temperature to dry-matter digestibility

depressing effect of high temperature on digestibility (Wilson, Taylor & Dolby, 1976). When tropical grasses are grown under wet, low-temperature conditions *in vivo* digestibility is similar to that normally found for temperate grasses (Van Wyk *et al.*, 1955).

Fertilizers. The application of fertilizer nitrogen to tropical grasses will usually increase the level of crude protein, but has no consistent effect on dry-matter digestibility (see Table 14.5). In some studies fertilizer nitrogen increased dry-matter digestibility by 9 percent, while in others there was a 3 percent depression. The cause of this variable response to fertilizer nitrogen is not known, but it does not appear to be caused by differences in yield, leafiness or flowering (Minson, 1973).

Sulphur fertilizer application has been found to increase by 5 percent the *in vivo* dry-matter digestibility of a seven-week regrowth of *Digitaria decumbens* when fed to sheep receiving no sulphur supplement. When a sulphur supplement was given, the digestibility of the sulphur-deficient grass was increased to the same level as the sulphur-fertilized *D. decumbens* (Rees, Minson & Smith, 1974). Subsequent work with *Digitaria decumbens* cut at seven stages of growth confirmed the beneficial effect of feeding supplementary sulphur on the digestibility of sulphur-deficient grass (Rees & Minson, 1978).

Fertilizer containing calcium increased the dry-matter digestibility by two percentage units for *Digitaria decumbens* grown on sandy soil (Rees & Minson, 1976). However, the application of superphosphate had no consistent effect on the dry-matter digestibility of *Digitaria pentzii*, *Digitaria* sp. variety *Pretoria*, *Panicum phragmitoides* and depressed by 10 percent the dry-matter digestibility of *Panicum coloratum* var. *makarikariense* (Myburgh, 1937).

Stage of growth. It is generally recognized that the dry-matter digestibility of grasses falls as they mature. As the leaf and stem fractions of tropical grasses have similar dry-matter digestibilities (Hacker, 1971; Laredo & Minson, 1973), the fall in digestibility of tropical grasses as they mature is mainly due to a drop in the digestibility of both the leaf and stem fractions, not to the increase in the proportion of stem.

As pastures mature, digestibility generally falls at 0.1 to 0.2 percent per day (Minson, 1971b) although larger short-term rises and falls have been reported (Grieve & Osbourn, 1965). In a study of eleven tropical grasses, dry-matter digestibility of summer regrowths decreased by 0.27 percent per day between days 28 and 67, and by 0.14 percent per day between days 67 and 94 (Minson, 1971a, b, 1972). The rate of decrease in digestibility of autumn regrowths was much slower, with mean values of 0.14 percent per day between days 42 and 70 and only 0.04 percent per day between days 70 and 102, a difference possibly due to the lower yield of autumn regrowths.

TABLE 14.5 **Effect of fertilizer nitrogen on the crude protein and DM digestibility percentage**

Species	Stage of regrowth (days)	Crude Protein (%) Control	Crude Protein (%) N-fertilized	DM digestibility (%) Control	DM digestibility (%) N-fertilized	Difference		Reference
Chloris gayana	Young	8.1	13.4	60	66	+6		Milford, 1960
	28	10.7	14.1	59	62	+3	n.s.	Minson, 1973
	Mature	6.9	7.8	50	49	−1		Milford, 1960
Cynodon dactylon	18[a]	9.6	16.6	47	50	+3		Carver et al., 1975
	21[a]	11.2	21.1	65	65	0		Webster et al., 1965
	35[a]	—	—	37	46[b]	+9		Fribourg et al., 1971
	50[a]	7.9	9.3	56	56	0		Alexander et al., 1961
	100[a]	6.0	7.4	54	57	3		Alexander et al., 1961
Digitaria decumbens	28	8.1	13.2	60	62	+2	n.s.	Minson, 1973
	Not stated	4.7	6.8	65	65	0		Chapman & Kretschmer, 1964
	Not stated	4.4	8.1	61	58	−3		Chapman & Kretschmer, 1964
	82	4.9	8.7	53	52	−1	n.s.	Minson, 1967
	84	3.7	7.2	48	52	+4	n.s.	Minson, 1967
	Mature	3.1	3.6	34	41	+7		Chapman & Kretschmer, 1964
Panicum maximum	9	9.0	11.4	59	57	−2		Owen, 1964
	45	11.7	12.9	60	58	−2	n.s.	Chacon et al., 1971
	65	7.8	8.9	55	55	0		Chacon et al., 1971
Pennisetum clandestinum	28	10.6	14.7	58	59	+1	n.s.	Minson, 1973
Mean		7.6	10.9	54	56	2		

[a] Approximate.
[b] Also higher levels of fertilizer P and K.

Protein

The published concentration of crude protein of 560 tropical grass samples, grown and determined in different parts of the world, ranges from 2 to 27 percent of the dry matter according to stage of growth and level of soil fertility, with a mean of 10.6 percent. These levels of crude protein are much lower than those found in 340 tropical legume samples (mean 16.7 percent) and 470 temperate grass samples (mean 13.3 percent). The range of crude protein and the proportion of samples of tropical grasses, tropical legumes and temperate grasses with different levels of crude protein are shown in Figure 14.5.

As grasses mature, there is usually a decrease in the crude protein percentage. This decrease is caused by an increase in the proportion of stem, which has a lower protein percentage than the leaf fraction. The crude protein percentage of both the leaf and stem fractions decreases with age (see Fig. 14.6). When excess pasture is offered, cattle selectively graze the leaf fraction, thus increasing the crude protein content of their diet above that of the average value for the grass offered (Stobbs, 1973). This advantage of selective grazing is particularly important with mature tropical grasses. However, it is obvious that although selective grazing will improve the diet when the animals start to graze a fresh pasture, it will automatically reduce the quality of the grass left for subsequent grazing.

The protein content of tropical grasses also depends on the availability of soil nitrogen. Fertilizer nitrogen application usually increases the crude protein percentage in tropical grasses (see Table 14.5). The level of soil nitrogen is also increased by growing a legume in association with the grass. This has increased the crude protein percentage of the grass component of the sward (Jones, Davies & Waite, 1967).

The apparent digestibility of the crude protein in tropical grasses varies from 20 to 80 percent, with most grasses in the 30-70 percent range (Milford & Haydock, 1965). This variation is mainly associated with the level of crude protein in the grass; low digestibilities are associated with low crude protein percentages in the grass and vice versa. More important than the digestibility percentage of the crude protein is the quantity of crude protein that will be digested in each 100 grams of feed. For tropical grasses an equation has been published relating the quantity of digestible crude protein (DCP) in 100 grams of dry matter to the crude protein (CP) percentage of the dry matter (Milford & Minson, 1965). This equation is similar to the equation for predicting the DCP of tropical legumes. Both equations are shown in Table 14.6.

Although the equations for predicting DCP are very similar for grasses and legumes, it should be remembered that the quantity of DCP in grasses is much lower than that in legumes, since legumes usually have a higher level of crude protein. Thus, the mean level of DCP in grasses and legumes is 6.3 and 11.5 g/100 g feed, respectively.

Mineral composition

Phosphorus. The published phosphorus content of 586 samples of tropical grasses grown throughout the world ranges from 0.02 to 0.58 percent in the

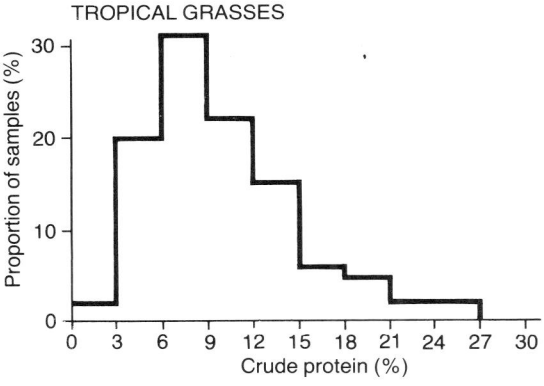
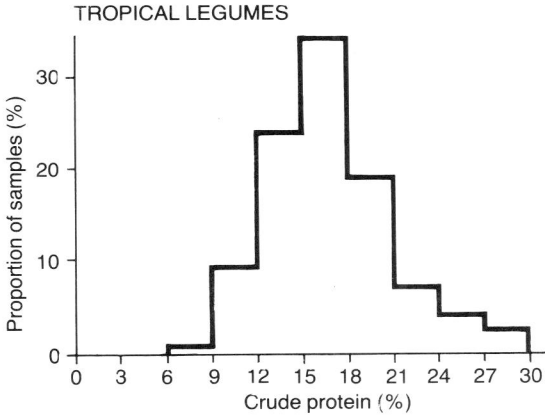
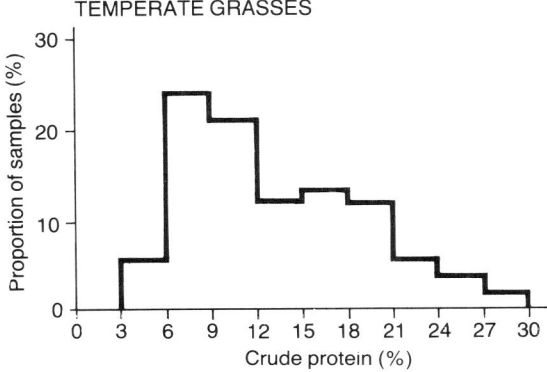

Figure 14.5. Frequency distribution of crude protein for tropical grasses, tropical legumes and temperate grasses

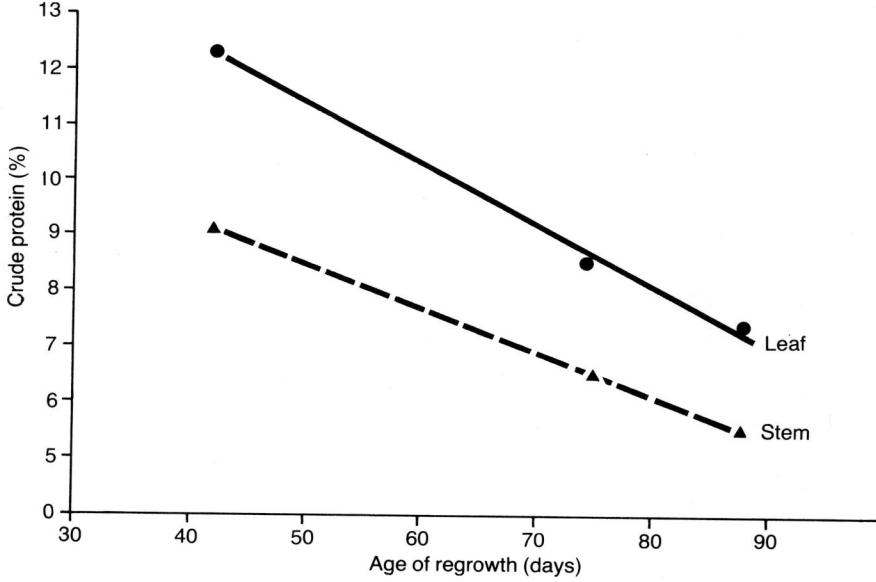

Figure 14.6. Relation of the age of regrowth to the crude protein contents of stem and leaf fractions of tropical grasses (**Source:** Laredo & Minson, 1973)

dry matter, with a mean of 0.22 percent. The histogram showing the variation in the phosphorus content is shown in Figure 14.7. The recommended level of phosphorus in the diet of cattle weighing 450 kg and gaining 0.5 kg each day is 0.17 percent of the diet (National Research Council, 1968). About one-third of all the published values for phosphorus in tropical grasses are below this level. Lactating cows and young animals require higher levels of phosphorus in their diet, so the problem of phosphorus deficiency is potentially more serious than for the fattening animal and can lead to a depression in

TABLE 14.6 **Equations to predict the digestibility of crude protein (DCP)**

Species	Equation[1]	Residual standard deviation	Correlation coefficient
Grasses	DCP = 0.90 CP − 3.25	± 0.84	0.98
Legumes	DCP = 0.93 CP − 3.99	± 1.17	0.96

[1] CP - crude protein percentage of the dry matter.
Source: Minson, 1977

Chemical Composition and Nutritive Value

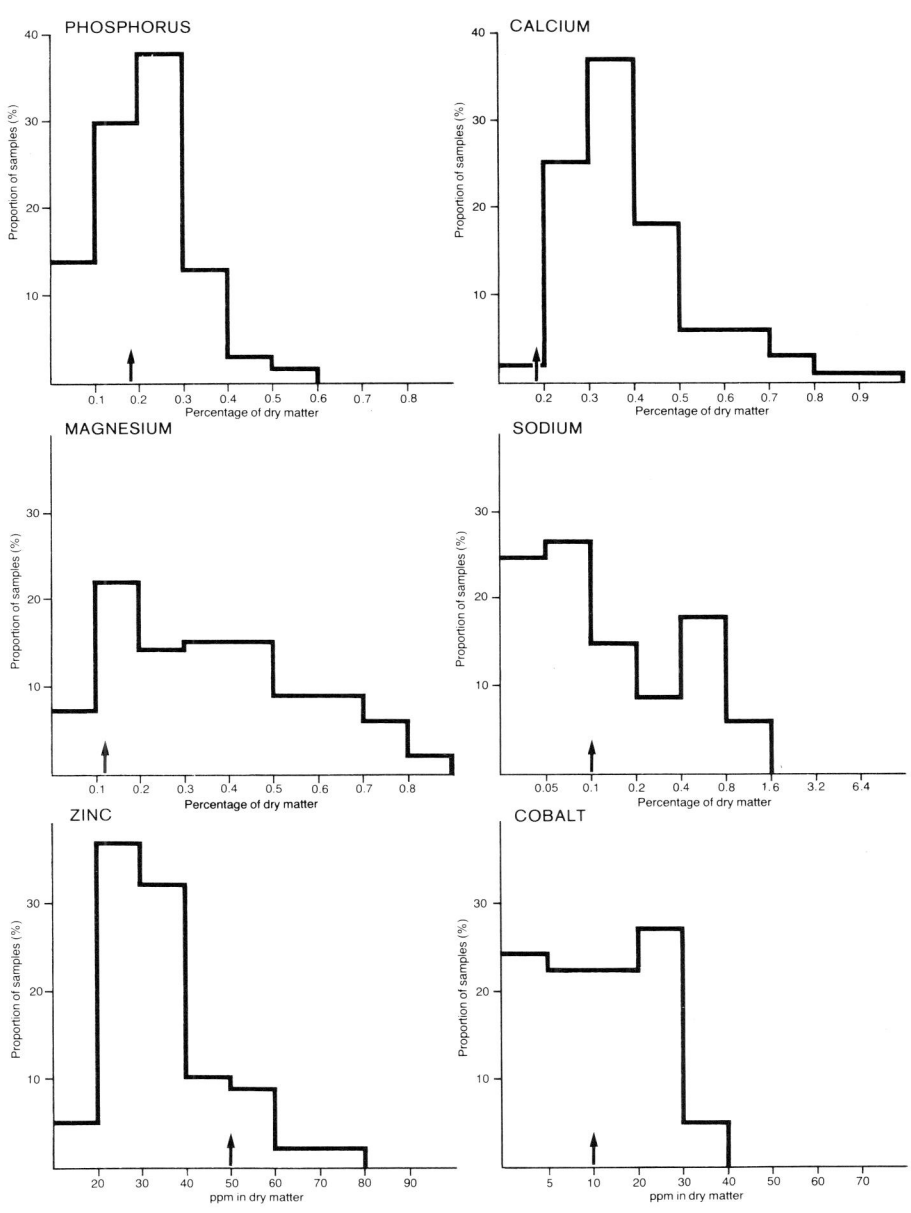

Arrow indicates recommended levels

Figure 14.7. Frequency distribution of six elements in tropical grasses

food intake and the eventual development of the classic symptoms of phosphorus deficiency.

Animals grazing pastures low in phosphorus usually take a long time to become phosphorus deficient. This is because they usually have large reserves of phosphorus in their bones which can be drawn on to supplement the diet. There is even some evidence that the recommended levels of phosphorus are unnecessarily high (Little, 1980). Animals can also selectively graze the younger fraction of the pasture, which contains a higher concentration of phosphorus.

A major factor causing differences in phosphorus content of pasture is the stage of growth. In a study of eleven tropical grasses, the mean phosphorus level was 0.31 percent at the beginning of the wet season. It fell to 0.08 percent during the following eight months as the pasture matured (Jones, 1964). However, there is no difference in the percentage of phosphorus in leaf and stem fractions (see Fig. 14.8). The level of phosphorus in grasses also depends on the level of available phosphorus in the soil. This will vary with soil type and the quantity of phosphorus fertilizer applied.

Calcium. The published calcium content of 390 samples of tropical grass grown throughout the world varied from 0.14 to 1.46 percent in the dry matter with a mean of 0.4 percent. The variation in calcium content of these samples is shown in Figure 14.7. The recommended level of calcium in the diet of cattle weighing 450 kg and gaining 0.5 kg each day is 0.17 percent (National Research Council, 1968). Nearly all tropical grass samples contained sufficient calcium to meet the animals' requirements. Weaner calves require up to 0.3 percent calcium in their diet. One-quarter of the tropical grass samples contained less than this level.

The percentage of calcium in the leaf fraction is twice that in the stem fraction, so selective grazing of the leaf fraction leads to a diet with a higher calcium percentage. With increasing maturity there is a decrease in the calcium percentage of both leaf and stem fractions (see Fig. 14.8).

Magnesium. The published magnesium content of 280 tropical grasses grown in various parts of the world varied from 0.04 to 0.9 percent in the dry matter, with a mean of 0.36 percent. The variation in the magnesium content of these samples is shown in Figure 14.7. The recommended level of magnesium in the diet for cattle weighing 450 kg and gaining 0.5 kg a day is approximately 0.11 percent (Agricultural Research Council, 1965). Less than 10 percent of the samples of tropical grasses were deficient in magnesium. Lactating cows require a diet containing 0.14 percent magnesium. For these animals, 15 percent of the tropical grass samples were deficient in magnesium.

As pastures mature there is a fall in the percentage of magnesium in both

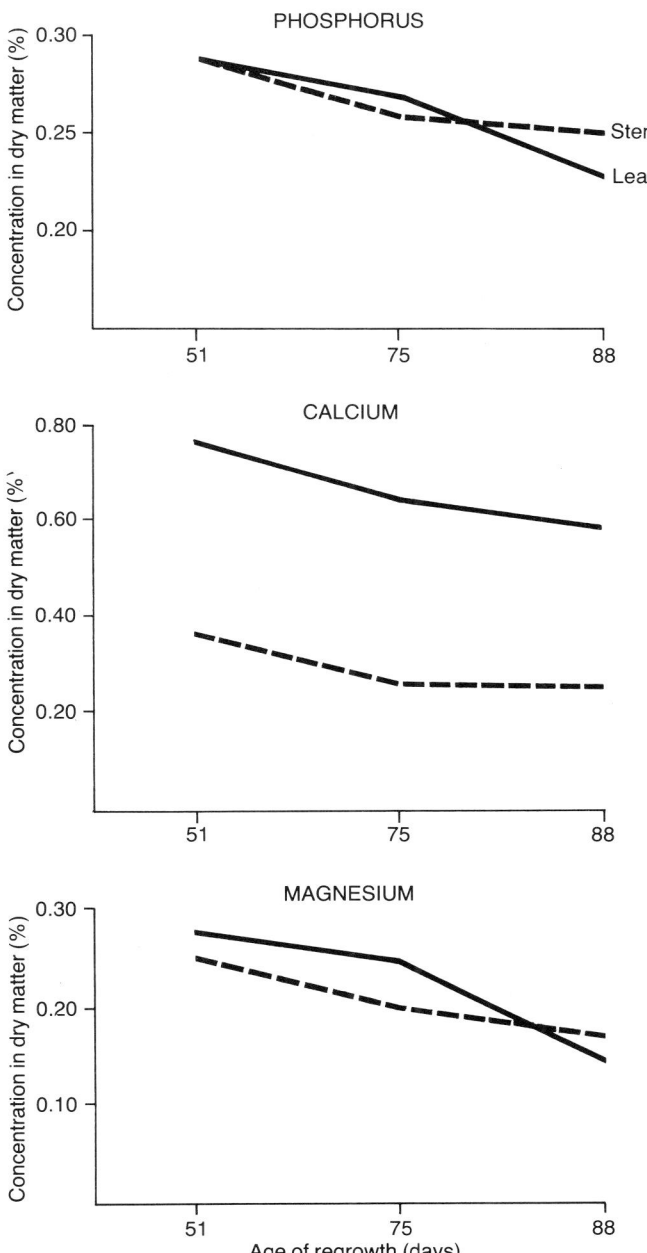

Figure 14.8. Effect of age and plant part on mineral composition of five tropical grasses

the leaf and stem fraction (see Fig. 14.8). Unlike calcium, magnesium is present in equal proportions in leaf and stem fractions.

Sodium. The published sodium content of 192 samples of tropical grasses grown in various parts of the world varied from 0.01 to 1.8 percent in the dry matter, with a mean of 0.26 percent. The distribution of the sodium levels was not balanced. Fifty-two percent of the samples had less than 0.1 percent sodium and another 18 percent were within the range 0.4 to 0.8 percent (see Fig. 14.7). The recommended level of sodium in the diet of 450-kg cattle gaining 0.5 kg each day and eating 81 kg DM per day is about 0.08 percent (Agricultural Research Council, 1965). Forty-three percent of the grass samples described in the literature were deficient in sodium. However, Morris and Gartner (1971) found no adverse effects when cattle were fed a diet containing 0.05 percent sodium, and suggested that the true requirement is considerably less than that suggested by the Agricultural Research Council (1965). Another problem is that cattle usually have large reserves of sodium and are very efficient at conserving sodium when fed sodium-deficient diets. Non-lactating cattle must be fed a sodium-deficient pasture for at least six months before they become sodium-deficient. With lactating animals, however, a decrease in milk production can occur after two months' feeding of a low sodium diet, since the animal cannot reduce the secretion of sodium in the milk.

Differences in the sodium concentration of tropical grasses are mainly associated with genetic factors. Six varieties of *Panicum* grown on the same soil varied from 0.14 percent for *P. maximum* to 1.33 percent for *P. coloratum* var. *kabulabula* (Minson, 1975). In a comparison of *Setaria* spp., Hacker (1974) found that cultivars from Kenya were consistently very low in sodium, while those from southern Africa and Zaire had high levels of sodium when all were grown on the same soil.

Zinc. The published zinc concentration in 119 samples of tropical grasses grown in various parts of the world varied from 15 to 120 ppm in the dry matter, with a mean of 36 ppm. The distribution of the zinc concentration in these samples is shown in Figure 14.7. The recommended level of zinc in the diet is 50 ppm (Agricultural Research Council, 1965). Eighty-four percent of the samples of tropical grasses studied would be considered deficient in zinc. In one review of zinc requirements of cattle it was concluded that although 9 ppm was adequate for calves, there were a few instances where 20 to 40 ppm were not sufficient for optimum performance. Until more information is available it is probably wise to consider any zinc values below 30 ppm indicative of a potential deficiency.

Cobalt. The cobalt concentration in 45 samples of tropical grasses grown in many parts of the world varied from 0.02 to 0.91 ppm in the dry matter, with

TABLE 14.7 **The soluble oxalate content of various tropical grasses**

Grass	Oxalic acid (%)	
	Summer	Autumn
Paspalum commersonii	0.05	0.09
Paspalum plicatulum	0.08	0.13
Chloris gayana	0.07	0.08
Dicanthium aristatum	0.07	0.08
Bracharia ruziziensis	0.13	0.18
Panicum coloratum (Bambatsi)	0.20	0.17
Brachiaria decumbens	0.27	—
Digitaria smutsii	0.36	0.57
Digitaria decumbens (leaves)	0.35-0.65	
Urochloa mosambicensis	0.44	0.67
Panicum coloratum (Kabulabula)	0.50	0.39
Panicum maximum (Petrie)	0.52	0.80
Cenchrus ciliaris (Molopo)	1.50	1.40
Setaria sphacelata (Nadi)	3.1	2.8
Setaria sphacelata (Kazungula)	4.2	3.7

Source: Jones & Ford, 1972 a, b

a mean of 0.16 ppm. The distribution of these concentrations is shown in Figure 14.7. The recommended level of cobalt in the diet is 0.1 ppm (Agricultural Research Council, 1965). Of the samples, 46 percent were considered deficient. Cobalt is required for the synthesis of vitamin B_{12}. Cattle only become cobalt deficient when liver reserves of vitamin B_{12} have been depleted. Cobalt deficiency has recently been reported in cattle grazing pastures in Malaysia ('t Mannetje, Sidhu & Murugaiah, 1976) and Australia (Winter, Siebert & Kuchel, 1977) that were grown on cobalt-deficient soils.

Copper. The copper concentration of 94 samples of tropical grasses grown in various parts of the world varied from 3 to 100 ppm in the dry matter, with a mean of 15 ppm. The recommended level of copper in diets for cattle is 10 ppm (Agricultural Research Council, 1965) and 26 percent of the tropical grass samples appear to be deficient in copper. However, this value may be too high. With grazing animals no growth response was found when feeding copper, although the copper levels in the pasture were considerably less (3-8 ppm) than the recommended 10 ppm (Winter, Siebert & Kuchel, 1977).

Undesirable factors

Goitrogens. Goitrogens have not been detected in tropical grasses, but goitre has been reported with nitrogen-fertilized *Cynodon aethiopicus* (Rudert & O'Donovan, 1974). Nitrogen-fertilized *C. aethiopicus* contains cyanogenic glycosides which release hydrocyanic acids after digestion. This is converted in the liver to thiocyanate, which inhibits iodine uptake by the thyroid gland. This problem is readily overcome by feeding an iodine supplement (Rudert, 1975).

Oxalate. Oxalate is found in most tropical grasses and varies between 0.05 percent for *Paspalum commersonii* to 4.2 percent in *Setaria sphacelata* cv. Kazungula (see Table 14.7). The oxalate is rapidly metabolized by ruminants, and adverse effects of oxalate on calcium metabolism have only once been reported in cattle (Jones, Seawright & Little, 1970). With horses, the oxalate leads to calcium deficiency, lameness, swelling of the jaw bones and loss of condition (McKenzie, 1978). The problem is directly related to the type of pasture being eaten, and the horses normally recover when transferred to pasture species low in oxalate.

Conclusion

In this chapter many chemical and nutritional factors that are known to influence animal production have been described. In practice it is a difficult problem to decide which of these factors is limiting production. There is no hard and fast rule that can be applied to all situations, but there are a few guidelines that can help diagnose limiting factors in grazing studies.

A rapid loss of production occurring soon after animals are transferred to a new pasture indicates the presence of undesirable factors in the pasture or an animal disease. Mineral deficiencies are usually less dramatic in their effects (except deficiencies of calcium and magnesium), and take many months to affect production since the animals usually have large reserves of most elements. This may be one factor that accounts for the lower incidence of mineral deficiencies than would be expected from the data presented in Figure 14.7. Deficiencies of phosphorus, zinc and cobalt are limited to heavily leached soils, and are unlikely on alluvial soils or well-fertilized pastures. Deficiencies of sodium are related to particular pasture species and may be found in any area, provided the drinking water is also low in sodium. Once detrimental factors and mineral deficiencies have been eliminated, the most common problem is the protein content of the diet. In many grazing studies, the level of production is related to the level of legume in the diet. Once fully developed, well-fertilized, grass/legume pastures are established, animal production will be limited mainly by a low digestibility or intake of the pasture. In this case improving the level of minerals or protein content of the diet will have no beneficial effect on animal production.

Plate 1. A severely overgrazed area of thornbush (*Acacia* sp.) country in Riwa, Kenya. The area has been divided into quadrants using a bulldozer and each quadrant removed from grazing for 12 months by arrangement with tribal elders

Plate 2. The effect of exclosure from grazing animals for 12 months, Riwa Grazing Scheme, Kenya. The grasses have regenerated and the trees have put on foliage

Plate 3. A scrub rake which pushes fallen timber into wind-rows. The spaced tines allow the soil to pass through

Plate 4. The King Ranch root plough. Note the scrub pusher in front and serrated blade on the rear root plough

Plate 5. Seed-harvesting ants at work in the Sahel, Kordofan Province, the Sudan

Plate 6. The shallow trough or niche left on top of the bank by the Mallen Niche Seeder (Figure 12.1), for seeding saline soil (***Photo:*** C.V. Malcolm, Dept Agriculture, W. Aust.)

Plate 7. Santa Gertrudis heifers grazing Para grass under coconuts at Lais, Mindanao, the Philippines (**Photo:** R.E. Harrison)

Plate 8. Mobile field hay-stacking frame, Nebraska, United States

Plate 9. Large bales of rolled hay left in the field for winter feeding, United States

Plate 10. Consolidating silage in a trench with a tractor during the day's filling. It is important to compact the material thoroughly, preferably several times during the day

Plate 11. Silage trenches sealed by excavated soil to a depth of 45 cm and finished to provide a camber to ensure run-off from the silo surface

Plate 12. A grab attached to the end-loader of a tractor excavates silage from a trench and loads it into a truck for distribution

Plate 13. An excavator-elevator for removing silage from a trench and loading a following truck

Plate 14. A hay barn (left), tower silo with external unloading chute (rear) and a concrete-lined hillside trench silo (foreground)

Plate 15. Feeding young Merino sheep on sorghum silage, central Queensland, Australia. The silage is fed on hard ground near to water. The provision of troughs or other feed equipment is too expensive in this situation

Plate 16. A crawler tractor pulling a forage harvester and side-tipping trailer harvesting sweet sorghum (*Sorghum bicolor* cv. Saccaline) for silage. The full trailer is emptied by tipping it sideways hydraulically into a trench silo. At the end of each day the material is well compacted in the trench by the tractor

Plate 17. *Andropogon gayanus* (gamba grass) (***Photo:*** J.G. McIvor, CSIRO, Australia)

Plate 18. *Anthephora pubescens* (***Photo:*** F. Smith, Queensland Department of Primary Industries)

Plate 19. A 10 000-bale stack of *Astrebla* spp. hay

Plate 20. Emus stroll across Mitchell grass, Queensland, Australia

Plate 21. Mitchell grass grassland in the early wet season, Queensland, Australia

Plate 22. Mitchell grass grassland in early autumn, Queensland, Australia

Plate 23. A mixture of *Heteropogon triticeus, H. contortus* and *Themeda australis* (left), and heavily grazed *Bothriochloa pertusa* (right) (**Photo:** J.G. McIvor, CSIRO, Australia)

Plate 24. *Brachiaria decumbens* and *Centrosema pubescens* (**Photo:** J.G. McIvor, CSIRO, Australia)

Plate 25. *Brachiaria mutica* and *B. humidicola,* Fiji

Plate 26. Para grass (*Brachiaria mutica*), centro and calopo, the Philippines (***Photo:*** R.E. Harrison)

Plate 27. The annual *Cenchrus biflorus* growing on the deep Goz sands in an *Acacia senegal* (gum arabic) plantation in Kordofan Province, the Sudan. Note the improved fertility resulting from mineralization of the deciduous leaf, and possibly from nodulation breakdown

Plate 28. *Cenchrus ciliaris* (buffel grass) can often be established in arid areas with poor soils by scattering seed around the base of large shade trees, where soil fertility is higher owing to the accumulation of fallen leaves and livestock excreta. Trees such as this *Eucalyptus populnea* are used by livestock for shade and camping

Plate 29. *Cenchrus ciliaris* cv. Gayndah (Gayndah buffel grass) established in the ashes at Charleville, Queensland, Australia (lat. 26° 25' S), after an *Acacia cambagei* (gidgea) scrub was pulled and burnt. The soil is a brown grumusol clay and the average annual rainfall is about 500 mm

Plate 30. A home-made *Cenchrus ciliaris* (buffel grass) seed harvester. The height of the hydraulic end-loader frame on the tractor can be adjusted. The beater reel in the seed box has an 18-cm clearance from the floor; when the tractor is driven at 2.5 km/hour it beats off only ripe seed into the seed box. Seed is removed through the door on the bottom left of the seed box into bags

Plate 31. *Cenchrus ciliaris* cv. Gayndah (Gayndah buffel grass) stabilizing the surface of a heavy cracking clay soil denuded from heavy livestock traffic around a permanent water-point in Mitchell grass tussock grassland, Longreach, Queensland, Australia

Plate 32. *Chloris gayana* (Rhodes grass) suppressing regrowth of *Acacia harpophylla* (brigalow) and providing a useful pasture. The Rhodes grass was aerially sown on to the ashes of pulled and burnt brigalow scrub. Untouched brigalow can be seen in the background and to the right

Plate 33. Boran steers grazing a star grass (*Cynodon* sp.) pasture in cleared thorn-bush (*Acacia* sp.) and leleswha woodland at Coles' Ranch, Gilgil, Kenya. The woodland was hand cleared for charcoal production and then planted to *Cynodon*

Plate 34. Shorthorn steers being finished on irrigated *Digitaria decumbens* (pangola grass) at Parada, north Queensland, Australia (lat. 17° 06' S, annual rainfall 900 mm). A live-weight gain of 2 000 kg/ha was obtained when the grass was fertilized with 672 kg N/ha and stocked at 10 beasts/ha

Plate 35. *Echinochloa pyramidalis* (antelope grass) (**Photo:** J.G. McIvor, CSIRO, Australia)

Plate 36. *Eleusine jaegeri* (manyatta grass) invading Kikuyu grass pastures on the Kenya highlands. Improving the fertility of the soil helps to control its spread

Plate 37. A "cut-and-carry" system in Kordofan Province, the Sudan. The donkey is loaded with the annual grass *Eragrostis tremula,* used for overnight feeding of village livestock

Plate 38. Awned seed clusters of *Heteropogon contortus* (bunch spear grass). The awned seeds can pierce the skin of sheep and are troublesome in wool processing. During alternating periods of moisture and dryness the awned seeds can bury themselves beneath the surface of the soil to survive fire (***Photo:*** J.G. McIvor, CSIRO, Australia)

Plate 39. *Heteropogon contortus* seeds have fallen from cattle rail wagons to invade this railway enclosure after annual burning. This land normally supports a *Chloris-Paspalidium* association

Plate 40. A fire-induced *Imperata cylindrica* (lalang or blady grass) grassland on Sulawesi, Indonesia. There are 16 million ha of such grasslands in Indonesia, and they expand by 150 000 ha each year

Plate 41. *Melinis minutiflora* (molasses grass) stabilizing a newly cleared and burnt rain-forest slope at Millaa Millaa, north Queensland, Australia (annual rainfall 2 625 mm). The slope is being grazed by Hereford bullocks

Plate 42. *Panicum antidotale* (blue panic) established by aerial sowing into ashes of pulled and burnt *Acacia cambagei* (gidgea) scrub at Yalleroi, central Queensland, Australia (lat. 24°S, annual rainfall 520 mm). The soil is a brown cracking clay

Plate 43. New pasture of *Panicum maximum* (Guinea grass) and *Stylosanthes guianensis,* Tully, northern Queensland, Australia

Plate 44. A *Paspalum dilatatum/Trifolium repens* pasture, renovated (left) and untreated (right)

Plate 45. *Pennisetum americanum* (pearl or bulrush millet)

Plate 46. *Pennisetum americanum,* Kordofan Province, the Sudan

Plate 47. Breeding plots of *Pennisetum americanum,* Tifton, Georgia, United States

Plate 48. Kikuyu grass, *Desmodium intortum* and *D. uncinatum*

Plate 49. *Pennisetum purpureum* — one year's growth from a single stem piece

Plate 50. A Mexican variety of *Pennisetum purpureum*, South Kalimantan, Indonesia (***Photo:*** R.E. Harrison)

Plate 51. Sugar cane and Para grass, Mt Bartle Frere, northern Queensland, Australia

Plate 52. *Sorghum almum* (Columbus grass)

Plate 53. A crop of *Sorghum bicolor* and cowpea (*Vigna unguiculata*) for ensilage

Plate 54. The root parasite *Striga hermonthica* on sorghum Province, Kordofan, the Sudan

Plate 55. Johnson grass (*Sorghum halepense*) invading a peach orchard, Georgia, United States

Plate 56. *Triodia pungens*, Charleville, Queensland, Australia

Plate 57. Young regrowth of *Triodia pungens* after burning, northwest Queensland, Australia

Plate 58. Teosinte × maize hybrids, United States

15. The Tropical Grasses Catalogue

Acroceras macrum Stapf

Common name. Nile grass (South Africa).
Natural habitat. Swamps and seasonally-flooded damp grassland.
Distribution. Widespread in northeastern and southern tropical Africa.
Description. A perennial with extensively creeping, rather wiry rhizomes. Culms mostly 40-110 cm high, simple, slender, geniculate and sometimes prostrate in the lower part. Leaves bright green, glabrous or scantily hairy. Inflorescence mostly 15-25 cm long, usually made up of two to five spikelike racemes, solitary and widely spaced on a central slender axis. The lower racemes are 6-9 cm long, the upper shorter. Spikelets arranged singly on distinct pedicels, or in pairs, with one pair almost sessile. Spikelets 5 mm long, awnless, glabrous. It has the C_3 photosynthesis (Oliveira *et al.*, 1973) (see Fig. 15.1).
Altitude range. 1 000-2 000 m.
Rainfall requirements. It occurs on valley bottoms where moisture accumulates in regions with a rainfall of 625-1 500 mm.
Drought tolerance. As it occupies valley bottoms and vlei soils there is usually sufficient soil moisture to allow it to survive, but it has little drought tolerance.
Tolerance to flooding. It tolerates seasonal flooding.
Soil requirements. It inhabits topographic bottom land with a wide soil range from sandy to black clay soils.
Fertilizer requirements. It will respond to fertilizers.
Ability to spread naturally. Good. It spreads by its creeping rhizomes.
Sowing methods. It is established vegetatively.
Response to photoperiod. It is indifferent to day length for flowering (Evans, Wardlaw & Williams, 1964).
Compatibility with other grasses and legumes. Clatworthy (1970) successfully grew *Trifolium semipilosum* and *Lotononis bainesii* with *A. macrum* with added nitrogen to 74.6 kg N/ha per cut. In the third-year no-N plots,

Figure 15.1. Acroceras macrum. **A**-Habit **B**-Inflorescence **C**-Raceme

increased yield due to legumes was 92 percent for *T. semipilosum* and 73 percent for *Lotononis bainesii* (of the yield of pure grass plus N at 112 kg N/ha).
Ability to compete with weeds. Good. It can become a weed itself and is difficult to plough out. Thus it should not be used as a short-term ley (Chippendall & Crook, 1976).
Response to fire. It should be protected from hot veld fires (Chippendall & Crook, 1976).
Genetics and reproduction. 2n=36 (Fedorov, 1974).
Dry- and green-matter yields. On old wattle plantations near Ermelo, eastern Transvaal, of seven grasses tried (Botha, 1953), Nile grass gave best performance yielding an average for the first three years of 13 200 kg/ha per year of hay.
Suitability for hay and silage. It makes most palatable and nutritious hay (Chippendall & Crook, 1976) and is used for silage in South Africa (Semple, 1970).
Chemical analysis and digestibility. Göhl (1975) gives the analysis in Table 15.1.
Palatability. It is very palatable (Verboom & Brunt, 1970). In the wildlife areas it is heavily grazed in summer.
Toxicity. In Taiwan some minor phytotoxicity has been shown by the roots of *A. macrum* towards lettuce seedlings (Chou, 1977). Ferulic, syringic, p-coumaric, vanillic, p-hydroxybenzoic and (O-hydroxyphenyl)-acetic acids were identified as active factors. In Zambia scouring occurs when cattle move from the fibrous forest grazing to the rich plains' grasses consisting of *Echinochloa pyramidalis, Acroceras macrum, Hemarthria altissima, Leersia*

TABLE 15.1 **Acroceras macrum**

	DM	As % of dry matter				
		CP	CF	Ash	EE	NFE
Fresh, early bloom, Kenya		8.7	30.8	9.9	5.9	44.7
Fresh, whole aerial part, Suriname		14.0	34.6	8.7	2.6	40.1
Fresh stems, Suriname		7.9	38.5	9.0	1.5	43.1
Fresh leaves, Suriname		21.3	30.0	8.4	3.8	36.5
Hay, South Africa		8.5	31.8	6.1	1.8	51.8

	Animal	Digestibility (%)				
		CP	CF	EE	NFE	ME[1]
Hay, South Africa	Sheep	57.5	66.2	47.3	69.9	2.35

[1] ME (metabolizable energy) is always given in megacalories per kilogram of dry matter.

hexandra, Vossia cuspidata and *Echinochloa scabra (stagnina)* and it may be three to four months before they regain condition (Verboom & Brunt, 1970).
Seed production and harvesting. It does not produce viable seed. Strains which yield viable seed and are rust resistant are being sought.
Diseases. It is susceptible to rust and other leaf diseases.
Economics. It has been widely cultivated as a planted pasture in vleis and wet soils, always being established vegetatively and requiring fertility and responding to fertilizer. Common on shallowly flooded levees and margins of the flood plain on the Kapic River in Zambia (Sayer & Lavieren, 1975) and in northern Zimbabwe along the Chambeshi River.
Animal production. No figures have been cited.
Main attributes. Its ability to vegetate moist valley bottoms and stand seasonal flooding, its palatability and good hay quality.
Main deficiency. Its ability to become a weed.
Further reading. Vesey-Fitzgerald, 1963.

Andropogon gayanus **Kunth**

Common names. Gamba grass (Africa), Sadabahar (India), Rhodesian andropogon (southern Africa), Rhodesian blue grass (Zimbabwe), onaga (northwest Africa).
Natural habitat. Open woodland and savannah.
Distribution. Native to tropical Africa and now introduced to many countries.
Description. A perennial species which in Africa grows in large tufts up to 2 m high with a reedlike habit. It is blue, with a waxy bloom. The leaves are distinctive because the many blades appear to be petioled. In the lowest leaves the laminar tissue is greatly reduced, almost to the midrib, above the ligule. The leaf widens gradually and then narrows again, so that the blade is finely pointed (Chippendall & Crook, 1976). Fifty percent of its roots are fibrous, and less than 0.5 mm in diameter, growing laterally at an angle of 10-15° to the surface for the first 50 cm, and then growing parallel to the surface for 1 m. These collect the early falls of rain. Forty percent are cord roots up to 2 mm in diameter which grow downwards at 30-40° and rarely measure 50 cm long. Ten percent of the roots are 0.5 mm in diameter, branch rarely, and form ropes. These grow vertically down to 80 cm and give drought tolerance (Bowden, 1963a) (see Fig. 15.2, Plate 17).
Season of growth. Summer.
Optimum temperature for growth. The optimum temperature for flowering is 25°C (Tompsett, 1976).
Frost tolerance. It is probably not tolerant of frost.
Latitudinal limits. Probably about 20°N and S.
Altitude range. Sea level to 2 000 m, but grows best below 1 000 m.
Rainfall requirements. It grows in the 400-1 400 mm rainfall regime but prefers a rainfall of 750-1 260 mm and more than three months dry season — up to six months.
Drought tolerance. Excellent. Ten percent of its roots form ropes 0.5 mm thick which go down to more than 80 cm (Bowden, 1963a). In an oxisol at Carimagua, Colombia it dried the profile to 120 cm (CIAT, 1978).
Tolerance to flooding. Rains (1963) lists *A. gayanus* var. *gayanus* among grasses growing in seasonally flooded places.
Soil requirements. It will grow on a wide range of soils including those of low fertility, from sands to black cracking clays, but prefers sandy clays of medium to high fertility. In South America it has shown outstanding results in oxisol-ultisol soils (CIAT, 1978). In the Sudan it is common on sandy loams to loamy sands.
Fertilizer requirements. A. gayanus performs very well without fertilizer nitrogen or phosphorus (CIAT, 1978) and thus is a valuable grass for low-input

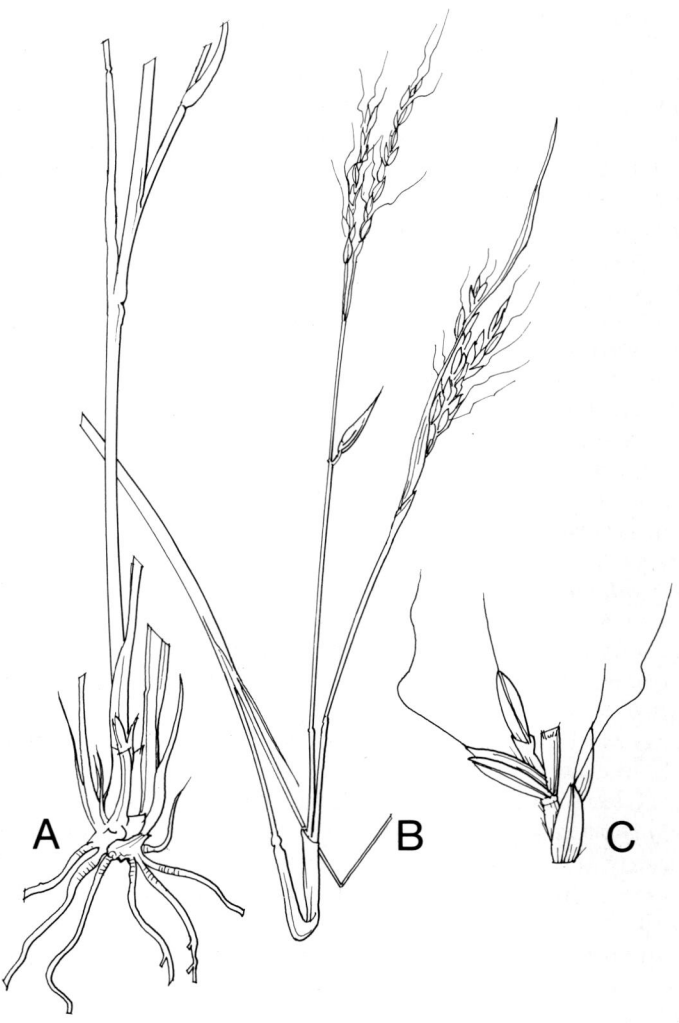

Figure 15.2. *Andropogon gayanus.* **A**-Habit **B**-Inflorescence **C**-Spikelet

pasture production. Haggar (1975) found the highest return of dry matter per unit of fertilizer (14.4 kg DM/kg N) was from only 28 kg N/ha and there was only a modest increase in crude protein up to 10.5 percent of the dry matter with increasing nitrogen. However, Haggar (1975) obtained almost linear increases in dry-matter production with increasing amounts of sulphate of ammonia up to 112 kg N/ha, but from 112 kg to 896 kg N/ha yields were curvilinear with maximum yield at 500 kg/ha. Falade (1975) determined the optimum phosphorus content of the dry matter for growth was 0.19 percent in a sandy loam soil at Ibadan, Nigeria when fully fertilized. Andrew and Robins (1971) obtained a critical percentage of 0.185 in Queensland. Some basic potassium may be required. It tolerates high aluminium (Spain, 1979). On the Quilichao ultisol in Colombia, *A. gayanus* gave maximum yield in the establishment year without fertilizer, and on the Carimagua oxisol the maximum yield was obtained with 50 kg P_2O_5/ha. Phosphorus fixation in the Carimagua oxisol is about half that of the Quilichao ultisol (CIAT, 1978).

Ability to spread naturally. It spreads slowly by seed.

Land preparation for establishment. A clean, firm seed-bed is required.

Sowing methods. Cleaned and de-bearded seed is drilled in shallow rows or broadcast and rolled. It can be planted also from root-stocks (splits), the best being mature woody stumps.

Sowing depth and cover. Sow seed at 1-2.5 cm below the surface.

Sowing time and rate. Sow at the beginning of the rainy season at 5 kg/ha (35-70 kg/ha uncleaned).

Dormancy. Germination improves for a time with storage.

Seed treatment before planting. De-beard the seed — a machine has been developed at CIAT, Colombia (CIAT, 1978).

Seedling vigour. Good.

Vigour of growth and growth rhythm. Dry-matter yield increased during the wet season from June to October in Nigeria, reaching a maximum of about 3 800 kg/ha in October, declining then until February. Cutting in early October gave best balance of bulk and quality (Haggar, 1970).

Response to photoperiod. It is a short-day plant with a critical day length for flowering between 12 and 14 hours. Flowering increased as day length shortened from 12 to eight hours (Tompsett, 1976).

Response to light. It prefers to grow in full sunlight.

Compatibility with other grasses and legumes. It combines naturally with *Stylosanthes fruticosa* in the Sudan and the United Republic of Tanzania and with *Stylosanthes* spp. and *Desmodium ovalifolium* and other legumes in Colombia (CIAT, 1978).

Ability to compete with weeds. It could suppress weeds by shading and root competition in dry areas.

Response to defoliation. At Fashola Livestock Farm, Nigeria, *A. gayanus* required intervals of more than six weeks between cuttings, and a cutting height of about 4 cm to maintain productivity and a good stand (Ahlgren *et al.*, 1959). It cannot stand heavy grazing until it is well established, but requires high stocking rates to maintain reasonable height.

Grazing management. It should be utilized when young, as once flowering stems appear it becomes harsh and of little nutritional value. Burning during the dry season is universal. However, it is important to maintain some residual dry matter and leaf area after grazing in such erect grasses (CIAT, 1978).

Response to fire. It tolerates fire and in Ghana and elsewhere it is burnt every year. Early dry-season burning promotes its growth, whereas late burning promotes the unpalatable *Loudetia acuminata* (Ramsay & Rose-Innes, 1963).

Genetics and reproduction. 2n=40 (20, 35, 40, 42, 43, 44; Fedorov, 1974).

Dry- and green-matter yields. Adegbola (1964) recorded 14 800 kg DM/ha per year at Agege (Lagos), Nigeria. In India 3 300 kg/ha fresh grass was obtained. Hendy (1975) obtained a production of 40 000 kg DM/ha per year at the Livestock Research Station, Tanga, United Republic of Tanzania, from a fertilizer application of 44 kg P_2O_5, 30 kg K_2O and 50 kg N/ha per year. A selection of *A. gayanus*, No. 621 from Shika, Nigeria, yielded 4 000 kg DM/ha at Quilichao, Colombia without fertilizer nitrogen but with adequate phosphorus (CIAT, 1978).

Suitability for hay and silage. It has been conserved as silage and hay, but its low nutritive value (Ademosun, 1973) does not justify the work involved (Miller, Rains & Thorpe, 1964).

Value as a standover or deferred feed. It is coarse and of low nutritive value after maturity, with only 1.5 percent crude protein (Miller, Rains & Thorpe, 1964).

Chemical analysis and digestibility. Göhl (1975) lists its feeding analysis in Table 15.2, to which Boudet's (1970) figures are added. Crude protein content in all categories of leaf and stem rose to a maximum at ear emergence. Maximum yields of digestible nutrients can be obtained by cutting at that time (Haggar & Ahmed, 1971). *A. gayanus* selection 621 gave low digestibility but high nitrogen levels in Colombia (CIAT, 1978).

Palatability. It is palatable when young and cattle will eat it up to flowering. Palatability ranking was *A. gayanus* > *Panicum coloratum* > *P. maximum* > *Pennisetum purpureum* (Bowden, 1963b).

Toxicity. No toxicity has been reported by Everist (1974).

Seed production and harvesting. Generally it is a good seed producer. Manual harvest produces more than 100 percent more than mechanical methods. In India, Mishra and Chatterjee (1968) obtained the highest yield by cutting twice a year (in mid-January and early July) and fertilizing with 38.9 kg N/ha plus 22.2 P_2O_5/ha. It maintained its seed yield with a third cut late in August.

Seed yield. Haggar (1966) recorded 21-86 kg/ha. Caryopses constitute 10 percent of this figure. At CIAT, Colombia, 34 kg/ha of pure live seed was harvested manually. Collection of shattered seed from the ground increased

TABLE 15.2 *Andropogon gayanus*

	DM	As % of dry matter				
		CP	CF	Ash	EE	NFE
Fresh, early vegetative, Nigeria	27.8	6.5	29.6	9.9	3.1	50.9
Fresh, mature, Nigeria	27.5	4.1	30.3	8.3	0.6	56.7
Fresh, 4 weeks, Ghana	21.0	7.7	21.5	9.3		
Fresh, 8 weeks, Ghana	19.9	12.9	25.6	8.5		
Fresh, 12 weeks, Ghana	19.1	12.1	26.5	8.5		
Fresh, 16 weeks, Ghana	30.9	8.4	29.2	6.6		
Fresh, 24 weeks, Ghana	59.4	5.4	29.9	5.5		
45-day-old growth, wet season, West Africa	22.7	8.5	37.2	8.5		
Leaves 40 days old, cool dry season, West Africa	44.0	6.4	33.3	9.0		
Old leaves and tillers, cool dry season, West Africa	44.8	1.9	35.1	6.9		
Hay, Nigeria	88.5	6.1	35.1	7.9	1.7	49.2
Silage, Nigeria	25.0	5.8	37.4	7.4	1.9	47.5

	Animal	Digestibility (%)				ME
		CP	CF	EE	NFE	
Fresh, early vegetative, Nigeria	Cattle	52.3	74.0	64.5	67.4	2.33
Mature, Nigeria	Cattle	41.5	59.4	0.0	63.1	2.02
Hay, Nigeria	Cattle	11.5	54.1	35.3	50.0	1.65
Silage, Nigeria	Cattle	20.7	63.9	42.1	42.9	1.71

Source: Göhl, 1975

yield and germination capacity, when not affected by the weather (CIAT, 1978).

Cultivars. There are several different types of plants and four varieties:

- Var. *bisquamulatus* (Hochst.) Hack — common in the savannahs from Senegal to the Sudan, colonizing denuded and waste land. It is very palatable to livestock.
- Var. *gayanus* — in periodic swamps in the same region. Good for erosion control in damp places.
- Var. *squamulatus* (Hochst.) Stapf — occurring also in this area and extending to the United Republic of Tanzania, Zimbabwe, Mozambique, Transvaal and Angola (Bowden, 1963b).
- Var. *tridentatus* — confined to Ghana (Rose-Innes, 1977).

Bisquamulatus and *squamulatus* grow best on well-drained sandy clays of medium to high fertility.

Pasto Carimagua cv. 621 was to be released by CIAT in 1980.

Value for erosion control. It has been used in Kano, Nigeria for reclaiming badly overgrazed and eroded land (Bowden, 1963b).

Diseases. It has no disease problems.

Pests. It may be attacked by the Brazilian spittle bug (CIAT, 1978).

Economics. *A. gayanus* (gamba grass) is a dominant constituent of large areas of natural and sown grasslands in Nigeria and other savannah areas of tropical Africa. It especially suits low-lying tropical areas which have moderate to low rainfall and a long dry season. When incorporated into a rotation it has been found to be a fertility builder. The stems are used for weaving grass mats and for thatching (Bowden, 1963b). It is a promising grass in northern Australia.

Animal production. In Nigeria, natural grassland containing 60 percent of *A. gayanus* resulted in a weight gain of 0.31 kg per day when grazed by N'Dama and Keteku cattle, but when consumed as silage the weight gain was 0.11 kg/day (Adegbola, Onayinka & Eweje, 1968).

Main attributes. Its excellent growth and dry-matter production in acid, infertile soils with minimum inputs, exceptional tolerance to drought stress, burning, and high levels of aluminium saturation; low P and N requirements; no known disease or major insect attacks; excellent seed-producing ability; compatibility with legumes; adaptability to low-cost pasture establishment systems, acceptable nutritional quality and high intake due to high palatability; high animal production levels during the first year (CIAT, 1978). Its adaptability to low-lying tropical areas which have low to moderate rainfall and its ability not only to survive a long dry season of several months but also to remain green for much of it and begin regrowth very early in the following rains to provide an early bite (Bowden, 1963b).

Main deficiencies. As it approaches and reaches maturity, it coarsens and is also of low nutritional value.

Further reading. Bowden, 1963b; CIAT, 1978.

Anthephora pubescens Nees

Common name. Wool grass (Australia).
Natural habitat. Widely distributed in Africa on sandy subtropical soils.
Distribution. Africa and Australia.
Description. It is a leafy, palatable, perennial tussock grass (see Plate 18).
Season of growth. Summer.
Rainfall requirements. In South Africa it occurs in the 250-650 mm rainfall areas, most common in the 350 mm area.
Drought tolerance. Excellent.
Tolerance to flooding. It will not tolerate flooding.
Soil requirements. Suited to the lateritic red earth soils and sandy soils with a pH range of 6-7. It will not grow on heavy soils (Smith, personal communication).
Fertilizer requirements. It responded to superphosphate at 250 kg/ha on a lateritic red earth soil at Charleville, Queensland (lat. 26°4'S, altitude 980 m and rainfall 450 mm) (Silcock, 1976). In South Africa it responds to lime or superphosphate (at 140 kg/ha) on acid sandy soils.
Ability to spread naturally. Poor.
Land preparation for establishment. It requires some rough soil disturbance.
Sowing methods. Broadcast the seed.
Sowing depth and cover. It is surface sown and uncovered in western Queensland.
Sowing time and rate. Sow just before the wet season at 4 kg/ha.
Number of seeds per kg. It varies considerably, with an average of about one million caryopses or 125 000 spikelets.
Dormancy. There is some post-harvest dormancy for about nine months. Dehusking greatly improves germination (Smith, personal communication).
Response to defoliation. It can stand some grazing, but during severe grazing tillers can be pulled from the ground.
Response to fire. Fire has little adverse effect on regrowth; it stimulates seed production in South Africa and at Charleville, southwestern Queensland.
Genetics and reproduction. It is apomictic.

TABLE 15.3 ***Anthephora pubescens***

	DM	As % of dry matter				
		CP	CF	Ash	EE	NFE
Fresh, mid-bloom, Niger		14.1	38.0	8.4	1.8	41.4

Dry- and green-matter yields. On sandy, lateritic red earths at Charleville its yield over four years ranged from 1 555-3 980 kg/ha with a mean of 2 898 kg/ha. From one harvest on heavy soil the yield was 1 330 kg/ha per year.

Chemical analysis and digestibility. At Charleville, material growing on sandy lateritic red earth contained 1.08 percent N and 0.07 percent P. Göhl (1975) gives the analysis in Table 15.3.

Palatability. It is very palatable and is sought by domestic animals and wildlife.

Seed production and harvesting. Severe chipping or burning prior to growing the seed crop increases seed production. The seed-heads are clearly exserted well above the leaf growth, making mechanical harvesting attractive. The seed on each head ripens fairly evenly and can all be harvested at once. Emergence of seed-heads continues for about a month, with a peak seven days after the first heads emerge, followed by a long tailing off period. Ripe seed can usually be harvested 28-32 days after emergence of the seed-head (33-39 days after first head emergence). With adequate water and fertilizer, seed can be harvested twice a year, once in early summer and once in autumn.

Seed yield. Varies enormously. From one hand harvest at Charleville the spikelet yield was 175 kg/ha (Smith, personal communication).

Economics. It is a useful grazing grass in South Africa in low rainfall sandy areas, and shows promise for sandy lateritic mulga (*Acacia aneura*) soils in southwestern Queensland (altitude 325 m; lat. 26°4'S and average annual rainfall 450 mm).

Further reading. Silcock, 1976.

Aristida adscensionis L.

Synonym. A. sub-micronata Schumach.
Common names. Six-weeks three-awn (United States), common needle grass (Kenya).
Natural habitat and distribution. Occurs in any disturbed poor soils in grassland and open bush throughout tropical Africa. Introduced to the United States.
Description. An annual growing up to 90 cm high. Culms many-noded, few. Leaf-sheath tight. Leaf-blade linear, long-pointed. Upper glume always distinctly emarginate or two-fid at the summit and with a very distinct mucro or short awn between the acute or obtuse lobes. Awns scabrous, about 1.5 cm long (Andrews, 1956). The culms are yellow to bright green, curing to a straw colour. The seed-heads may be purple (see Fig. 15.3).
Season of growth. Summer, but can grow out of season if temperature and moisture are suitable.
Altitude range. Sea-level to 2 250 m. It is dominant in Zimbabwe at altitudes below 600 m.
Rainfall requirement. It is dominant in Zimbabwe in a rainfall regime of 300-380 mm falling mainly between November and March. It requires a rainfall in excess of 250 mm.
Drought tolerance. Being an annual, it usually escapes drought once it has had sufficient rain to germinate.
Soil requirements. It prefers sandy soils but has a wide tolerance.
Land preparation for establishment. It will establish on roughly prepared ground and can be oversown into unprepared land.
Sowing methods. Broadcast the seed on to the surface and, where possible, cover lightly.
Genetics and reproduction. 2n=22 (Fedorov, 1974).
Chemical analysis and digestibility. Göhl (1975) has listed its chemical analysis in Table 15.4.

TABLE 15.4 *Aristida adscensionis*

	DM	As % of dry matter				
		CP	CF	Ash	EE	NFE
Fresh, early bloom, Kenya		8.9	36.8	11.4	1.7	41.2
Fresh, early vegetative, Kenya	25.8	10.5	33.6	9.5	1.6	44.8
Fresh, mature, Kenya		5.2	39.1	9.6	1.5	44.6
Standing hay, Kenya	91.0	2.0	45.7	7.9	0.9	43.5

Figure 15.3. *Aristida adscensionis.* **A**-Plant **B**-Spikelet **C**-Stamens, ovary and lodicules **D**-Lower glume **E**-Upper glume **F**-Lemma

Palatability. It is a late-flowering, unpalatable annual in the United Republic of Tanzania (Wigg, Owen & Makurasi, 1973). The sharp seeds and stiff awns repel animals when mature. It provides an abundance of feed for a short time in Arizona.

Seed production and harvesting. It will grow and set seed at any time of the year in Arizona when moisture and temperature are favourable, but is predominantly summer-growing.

Value for erosion control. It easily colonizes bare ground.

Economics. It is one of the better annual grasses in Arizona ranges, but provides poorer forage than most perennials.

Animal production. It is used for semi-arid cattle ranching in Zimbabwe. At Tuli station in southern Zimbabwe, the natural veld was estimated by Rattray (1962) to carry one animal per ten hectares. When the carrying capacity was increased to one animal per eight hectares an almost pure stand of *Aristida adscensionis* was induced, which is unpalatable and provides little ground cover during the dry season.

Main attributes. A common pioneering species on disturbed soil, waste land, rocky places and fallows.

Main deficiencies. Its short life and severe decline in nutritive value soon after it dries.

Further reading. Humphrey, 1960a.

Aristida latifolia Domin

Common names. Feather-top wire grass, curly spear grass (Australia).
Natural habitat. Semi-arid open grassland.
Distribution. Throughout central and northern Australia in Mitchell grass grassland and on heavy soils.
Description. Densely caespitose, the culms robust, simple, 40-50 cm tall, glabrous and glaucous; leaf-sheaths ciliate at the orifice, blades long, flat except for the setaceous-convolute apex, prominently nerved, scabrous above, 2-3 mm broad, the older leaves becoming curved or twisted. Panicle long and narrow, 12-15 cm long, not dense, pale; the branches long and erect, the lower branches distant. Spikelets pale. Glumes subequal, glabrous, one-nerved or keeled, narrow, scarious, acuminate or mucronate, 11-15 mm long, the lower scabrous on the keel. Lemma narrow, convolute, 5-6 mm long (including the callus), glabrous or scabrous towards the apex, the callus densely silky — villous; awns arising from the simple, spirally twisted, non-articulate column continuous with the summit of the lemma and 5-7 mm long, the awns very slender, subequal, 15-30 mm long (Gardner, 1952) (see Figs. 15.4, 15.5).
Latitudinal limits. 23-25°S.
Altitude range. 250-300 m.
Rainfall requirements. It occurs in the under 500 mm rainfall belt in western Queensland associated with *Astrebla* spp. in grassland. It responds to light falls of rain better than *Astrebla* spp.
Drought tolerance. It has a shallow root system and is less drought resistant than *Astrebla* spp., but better than *Dichanthium sericeum* and *Digitaria* spp. (Purcell & Lee, 1970).
Soil requirements. It thrives on grey-brown and black alkaline cracking clays.
Seedling vigour. Its seeds germinate and establish earlier and later than Mitchell grass (*Astrebla* spp.), but both species have the same temperature range for germination.
Vigour of growth and growth rhythm. It remains greener into the autumn than Mitchell grasses.
Response to defoliation. It will not stand heavy grazing and is not common in pastures stocked with horses unless the grazing is exceptionally light.
Grazing management. It should be heavily stocked if possible, or burnt or slashed to prevent it seeding.
Response to fire. Burning and heavy stocking have no differential effect on the persistence of the grass (Purcell & Lee, 1970).
Economics. Aristida latifolia is regarded as an undesirable species in the Mitchell grass (*Astrebla*) grasslands of western Queensland. Its three-awned seed when present in wool increases processing costs and results in lower wool prices. The seeds also work their way into the muscle of sheep. Growth

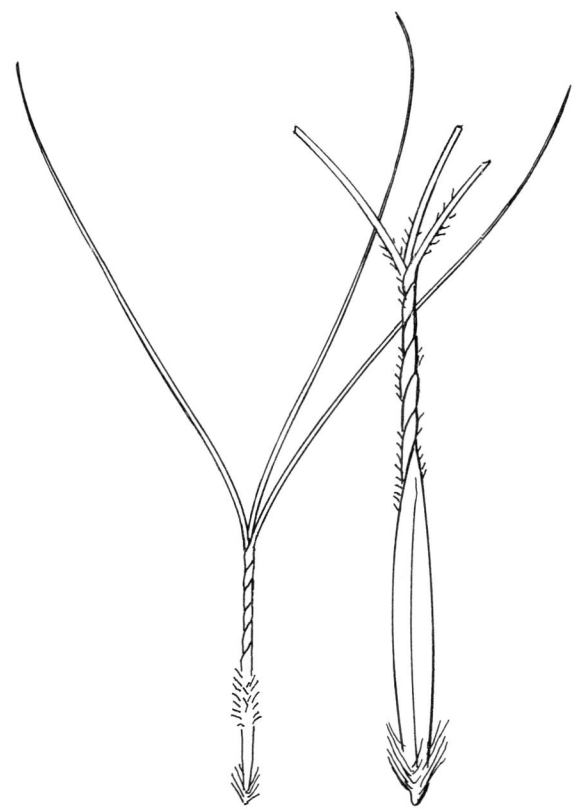

Figure 15.4. Aristida latifolia, plant, lemmas

Figure 15.5. Aristida latifolia invading Astrebla spp. at Blackall, central Queensland, Australia. The soil is a grey-brown cracking clay

and production of young sheep are impaired markedly. *A. latifolia* became increasingly important in the late 1950s when there was a succession of dry years. There is need for research into the autoecology of this grass and the general grazing management. Ploughing will destroy it, but in areas where it is most abundant, ploughing is a costly operation.

Animal performance. The young growth is lightly grazed by sheep, but it is generally ignored in favour of more palatable grasses.

Further reading. Purcell & Lee, 1970.

Astrebla spp.

Common name. Mitchell grasses.
Natural habitat. Open black and brown clay plains in subtropical areas of Australia within the annual rainfall isohyets of about 200 and 500 mm.
Distribution. They occupy an area of some 450 000 km^2, predominantly in western Queensland with outliers into the Northern Territory extending sparsely into Western Australia (see Fig. 15.6).
Description. As the four species usually occur together with only minor variations in ecological environment they will be treated together. They are perennial, summer-growing tussock grasses whose persistence and extraordinary resistance to drought and continuous grazing are due largely to the root-stock, which consists of many short, stout, thick, branched, scaly rhizomes. In *A. lappacea* numerous wiry roots spread outward from the bottom of the rhizome into the surrounding surface soil horizon and then turn vertically downward, continuing unbranched through the layer of columnar clay. The roots branch into fine rootlets where soil cracks cease and a gypsum layer begins. Main tillers arise from the root-stock and axillary tillers develop from axils on the main tillers (Orr, 1975). A key to the species (Gardner, 1952) is as follows:

A. Spikelet or raceme dense, 8-20 mm broad; internodes of the rachilla less than 1 mm long; spikelets three- to nine-flowered.
   1. Lobes of lemma all similar, symmetrically attenuated into rigid and tough bristle-like points, sometimes hooked; lemmas densely silky-villous on and around the lateral nerves from the base upwards: *A. squarrosa* C.E. Hubb. (bull Mitchell grass) (see Fig. 15.7A).
   2. Middle lobe of the lemma alone symmetrically tapering into a tough bristle-like unhooked point; lateral lobes semi-lanceolate to semi-ovate; lemmas villous all over the entire portion of the dorsal surface.
      a. Spikelets densely imbricate; lower glume five- to nine-nerved; glumes much shorter than the lemmas: *A. pectinata* (Lindl.) F. Muell. (barley Mitchell grass) (see Fig. 15.7B).
      b. Spikelets loosely imbricate and alternate; lower glume one- to three-nerved; upper glume as long as the spikelet: *A. lappacea* (Lindl.) Domin (curly Mitchell grass) (see Fig. 15.7C).
B. Spikelets or racemes slender, 2-3 mm broad, the spikelets distant or loosely imbricate; internodes of the rachilla 2-4 mm long, lower glume one-nerved; upper glume seven- to 11-nerved, half as long as the spikelet; spikelets two- to four-flowered: *A. elymoides* F. Muell. (hoop or weeping Mitchell grass) (see Fig. 15.7D).

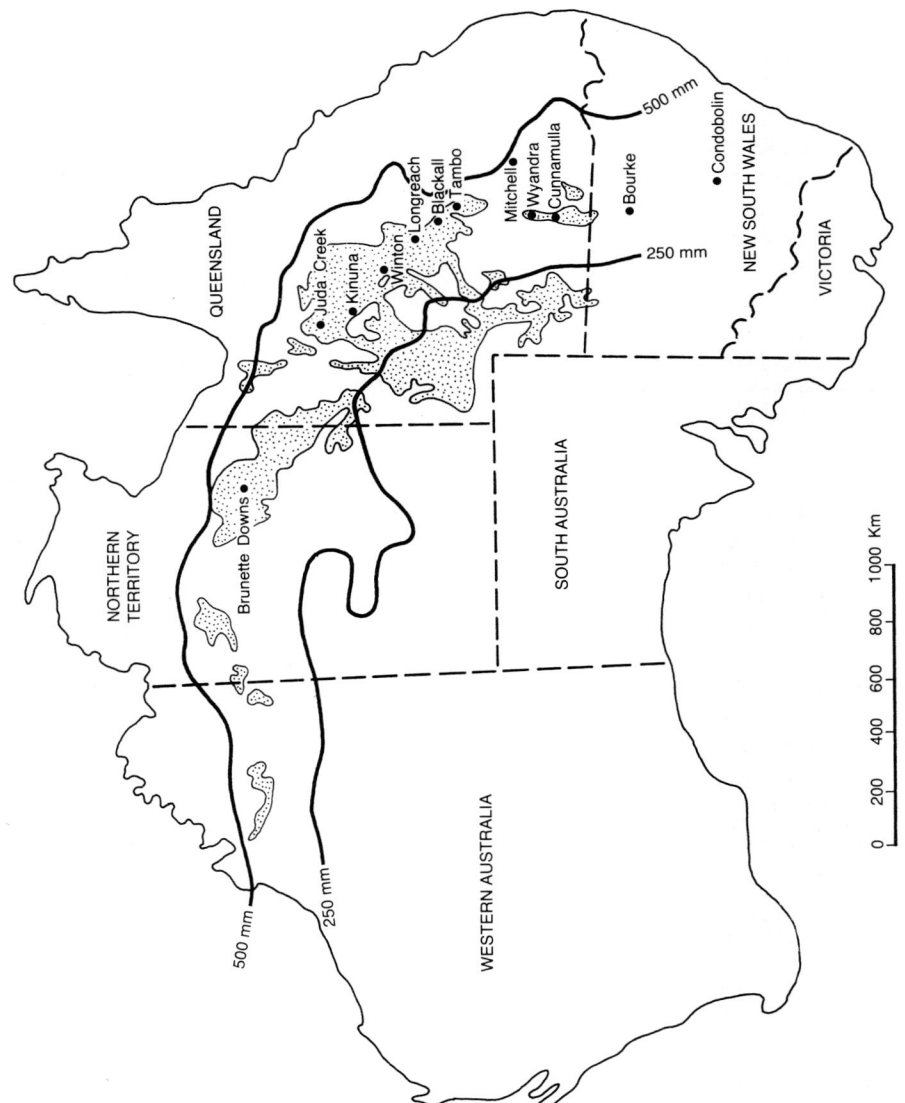

Figure 15.6. Distribution of *Astrebla* spp. pastures in Australia

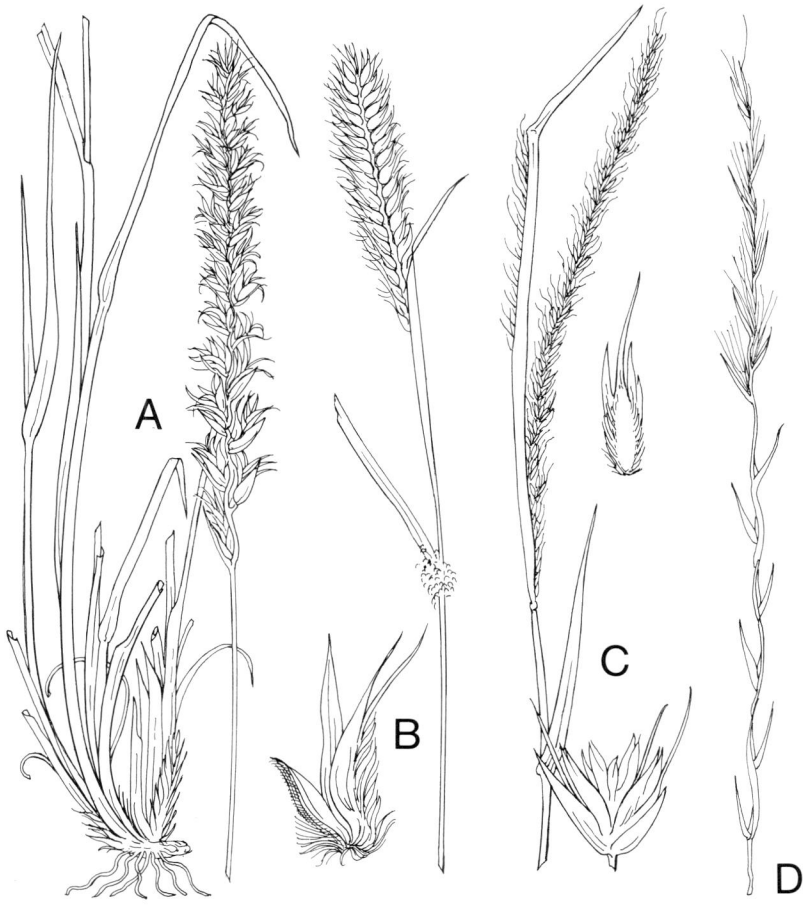

Figure 15.7. Astrebla spp. **A**-*A. squarrosa* **B**-*A. pectinata* **C**-*A. lappacea* **D**-*A. elymoides*

Season of growth. Summer.
Optimum temperature for growth. Mean monthly maximum temperature of 35°C or over. Optimum tillering in *A. pectinata* occurs at day/night temperatures of 28/23°C and growth and leaf production increased with temperature up to 30/25°C (Jozwik, 1970).
Frost tolerance. It tolerates frosts, as the southern area of its distribution receives an average of 50 frosts per year, but the vegetative parts are frosted.
Latitudinal limits. From 18-28°30'S, approximately.
Altitude range. 300-1 000 m.
Rainfall requirements. The Mitchell grasses reach their best development between the 250 and 550 mm annual rainfall isohyets in regions with pronounced summer rains. Roe and Allen (1945) decided that the growth of *Astrebla* spp. depends on the current season's precipitation rather than on stored soil moisture. A second rain following germination by about six weeks is necessary to enable tillers and secondary roots to develop (Everist, 1951). Old plants respond to light falls of rain by producing axillary tillers from the lower nodes, heavier rains initiate new basal tillers (Jozwik, Nicholls & Perry, 1970).
Drought tolerance. The Mitchell grasses have exceptional drought tolerance. They will "hay off" if no rain falls after maturity and grow again after good rains in the next summer. The plants do not break up when dry, and while in that state are still acceptable to stock, but light rain at this time will cause mould growth and blackening, making them inedible.
Tolerance to flooding. The species *A. lappacea* and *A. pectinata* are killed by flooding but *A. elymoides* and *A. squarrosa* are tolerant. *A. squarrosa* occupies the lower and wetter drainage lines in Mitchell grass pastures.
Soil requirements. The Mitchell grasses are restricted to grey, brown and red alkaline cracking clay soils with a minimum of 40 percent clay. The surface is self-mulching clay to sandy clay loam overlying a massive cracking clay. Lime is usually present in the upper profile, with gypsum below.
Tolerance to salinity. They are tolerant of mild salinity; many of the soils on which they grow have a pH approaching 8.5.
Fertilizer requirements. The Mitchell grasses are never fertilized. The potash and phosphorus status in the soils is usually adequate and the birch effect of nitrogen release on wetting of the cracking clays helps provide additional nitrogen for early wet season growth.
Ability to spread naturally. The Mitchell grasses produce abundant seed in a good season. In years of exceptional wet season rains, seedlings will appear in abundance but there is usually some seed emergence in a normal wet season.
Land preparation for establishment. Either prepare a good seed-bed with ploughing and cultivation or merely broadcast seed over the self-mulching clays and trample it in with livestock (Breakwell, 1923) during wet weather.
Ability to compete with weeds. Their very drought-tolerance allows them to outlast weeds, and not many common "weeds" are weeds where grazing

merino sheep are concerned, as they do well on numerous edible plants.
Tolerance to herbicides. No information is available.
Response to defoliation. The Mitchell grasses stand a good deal of defoliation; but overstocking removes seedlings, and heavy grazing during active growth depletes root reserves for recovery.
Grazing management. Mitchell grasses are normally continuously grazed, and rotational grazing at Cunnamulla (Roe & Allen, 1945) proved no benefit in a long-term trial. Vagaries of seasons outweigh stocking practices over a series of years. Normal stocking rates are one sheep per 1.2-2 hectares. Grazing should not be heavy during the period of active growth and the grasses should be allowed to seed freely to recoup the paddock seed supply. Heavy stocking by horses can destroy a Mitchell grass pasture because they graze closely and trample and dig up tussocks with their hooves in dry times.
Response to fire. Burning is unnecessary and undesirable in most years, and only when excessive foliage canopy excludes the inter-tussock herbage is a fire warranted. Mitchell grass recovers well from occasional accidental fires.
Genetics and reproduction. 2n=40 (Jozwik, 1969; Fedorov, 1974).
Dry- and green-matter yields. Dry matter yields and rainfall during the growing period have been recorded as 400 kg/ha (163 mm) at Brunette Downs, Northern Territory, 2 250 kg/ha at Claverton, Wyandra, Queensland.
Suitability for hay and silage. A good deal of hay has been made over the years in Queensland. The tussocky nature of the country makes hay-making difficult; the black clay soils are difficult to negotiate after rain and unless the grasses are cut early the quality of the hay will deteriorate. It is really low-quality roughage. In open-air storage, oxidation lowers its nutritive value, but in its usual habitat, protection of stored hay is often costly (see Plate 19).
Sowing methods. Sow on a well-prepared seed-bed with suitable drills, or broadcast into loosely mulched soils.
Sowing depth and cover. Sow no deeper than 1 cm and lightly cover or roll.
Sowing time and rate. Sow in spring at 3.5-4.5 kg/ha.
Dormancy. Post-harvest dormancy exists in all species for a period of up to 12 months to reach 88 percent germination and maintains this for at least another year (Myers, 1942b). Germination occurs in a temperature range of 15-42°C but mainly 22-38°C.
Seedling vigour. Seedling growth is good when heavy rains fall to germinate the grasses.
Vigour of growth and growth rhythm. Growth responses are determined by rainfall. If rainfall of only about 40 mm occurs, fine rootlets developed from the main roots stimulate some culm growth using stored starch. The main root system is stimulated by 75 mm or more. Once the season breaks dry matter increases fourfold in one month, 20-fold in two months and 35-fold in three months.
Response to light. Mitchell grasses do not produce well in shade. Scattered

trees (shady downs) with sparse foliage do not interfere with growth.
Compatibility with other grasses and legumes. Mitchell grasses form a tussock grassland. When the wet season arrives the inter-tussock space is occupied by annual grasses and herbs, e.g. *Iseilema* spp., *Dactyloctenium radulans, Brachyachne convergens, Rhynchosia minima, Boerhaavia diffusa* and a short-term perennial, *Dichanthium sericeum,* in very wet years. In the winter rainfall zone *Medicago* spp. are common associates (see Plates 20, 21, 22).
Value as standover or deferred feed. It is quite valuable as standover feed. Though its nutritive value has deteriorated as a whole to submaintenance levels, sheep selectively graze the more nutritious portions. Light rain falling on mature Mitchell grasses causes mould growth and blackening, which spoils its value as roughage.
Chemical analysis and digestibility. Crude protein levels recorded vary from a low 3.5 percent in midwinter, through 5-6 percent for fair quality, 8 percent for good quality, and 18.4 percent for young three- to four-week-old leafy material; carbohydrates 39-51 percent, crude fibre 26-33 percent (Orr, 1975), ether extract 1.4-2 percent (dry material), ash 8.8-12 percent (Siebert, Newman & Nelson, 1968).
Palatability. Mitchell grasses are generally shunned in favour of annual grass and herb associates during the growing season, but are eaten when other feed is not available.
Toxicity. No toxicity has been reported by Everist (1974).
Seed production and harvesting. Mitchell grasses appear to have abundant seed, but "seed" consists of spikelets with a variable number of caryopses. Some 3 percent of spikelets have no seed, 18 percent one seed, 18 percent four seeds (Myers, 1942b). Seed is generally hand picked as demand is low.
Minimum germination and quality required for commercial sale. 35 percent germinable seeds; 75 percent purity (Queensland).
Cultivars. There are no registered cultivars but four species are recognized: *A. lappacea* (curly Mitchell grass); *A. pectinata* (barley Mitchell grass); *A. elymoides* (hoop Mitchell grass); and *A. squarrosa* (bull Mitchell grass).
Value for erosion control. Owing to their tussocky nature, Mitchell grasses do not give very effective control of run-off, but their extensive roots intertwine to give some protection against soil erosion.
Diseases. Of little importance.
Pests. Of little importance. Kangaroos can be competitive with other grazing animals in dry times, and on young shoots appearing after fires have passed through.
Economics. A most important natural tussock grassland occupying some 450 000 km^2 across the northern half of Australia.
Animal production. The carrying capacity of the Mitchell grass tussock grassland is rated at one sheep to one to two hectares in the 600-mm rainfall zone and one sheep to two hectares in the 300-mm region. Wool production per hectare from merino sheep ranges from 2.3-3 kg. Cattle carrying capacity is

rated at one per 15 hectares. With beef shorthorn cattle specially selected for heat tolerance, live-weight gains of 0.7 kg per head per day were obtained from Mitchell grass pastures at Muttaburra, Queensland, over a whole year, reaching an average live-weight gain of over 1 kg per head per day during the winter-spring period (Dowling, 1960).

Main attributes. Their drought resistance, adaptability to a harsh environment, and quick response to rainfall in the presence of the grazing animal.

Main deficiencies. Low nutritive value during the winter and lack of growth. Difficulties with germination of seed.

Further reading. Davidson, 1954; Davies, Scott & Kennedy, 1938; Orr, 1975; Roe & Allen, 1945.

Axonopus affinis Chase

Common names. Mat grass (Queensland), compressum, Durrington grass (New South Wales), narrow-leaved carpet grass (Australia).

Natural habitat. Subhumid and humid woodland and savannah.

Distribution. *A. affinis* is believed to have originated either in the southern United States, the West Indies or Central America. It is now found in the tropical and subtropical regions of America, Africa, Asia, Australia and the Pacific islands.

Description. A glabrous perennial, 25-75 cm high, having short rhizomes and slender to moderately stout arching stolons with short internodes. The grass often forms a dense mat. Leaf-blades are 5-20 cm long, 2-6 mm wide, flat or folded with a rounded or obtuse apex. The inflorescence consists of two to three slender, sessile, erect spikes, the upper pair approximate and the third a little remote. The spikes are straight and the rachis is 0.5 to 0.75 mm wide. The spikelets, each 2 mm long, are arranged in two neat rows. The fertile floret almost equals the spikelet in length, and is white to pale yellow in colour. It differs from *A. compressus* in having more slender culms and stolons, narrower leaves and shorter (2 mm) spikelets which are more obtuse; it has wavy or flexuous peduncles and fertile lemmas with glabrous apex or only a very reduced tuft of apical hairs (Barnard, 1969) (see Fig. 15.8).

Season of growth. Summer growing, with peak growth in late summer.

Optimum temperature for growth. Top growth was maximum at a temperature of 27-32°C with a day length of 15 hours.

Minimum temperature for growth. Temperatures below 12.8°C inhibited flowering (Lovvorn, 1945). In Australia it is found further south than *A. compressus*.

Frost tolerance. It has some degree of tolerance to frost.

Rainfall requirements. Usually in excess of 750 mm per annum.

Drought tolerance. It is more drought-resistant than *A. compressus* and will invade hilly, as well as flat, country.

Tolerance to flooding. It prefers moist soil but does not withstand prolonged flooding or permanently swampy conditions very well, though better than *Paspalum dilatatum*.

Soil requirements. It prefers a sandy, moist soil and will flourish in sandy and heath soils too infertile for *Paspalum dilatatum*. It will respond however to more fertile conditions.

Fertilizer requirements. It has a low fertility requirement but will respond to fertilizers, particularly nitrogen, but its efficiency of use of applied nitrogen is low compared with other pastures. Its shallow root system (96 percent of roots in the 0-5 cm layer), lack of root response to fertilizer, and distribution, suit carpet grass to infertile, moist sandy soils. Apply 224 kg/ha superphosphate at sowing and annually (McLennan, 1936).

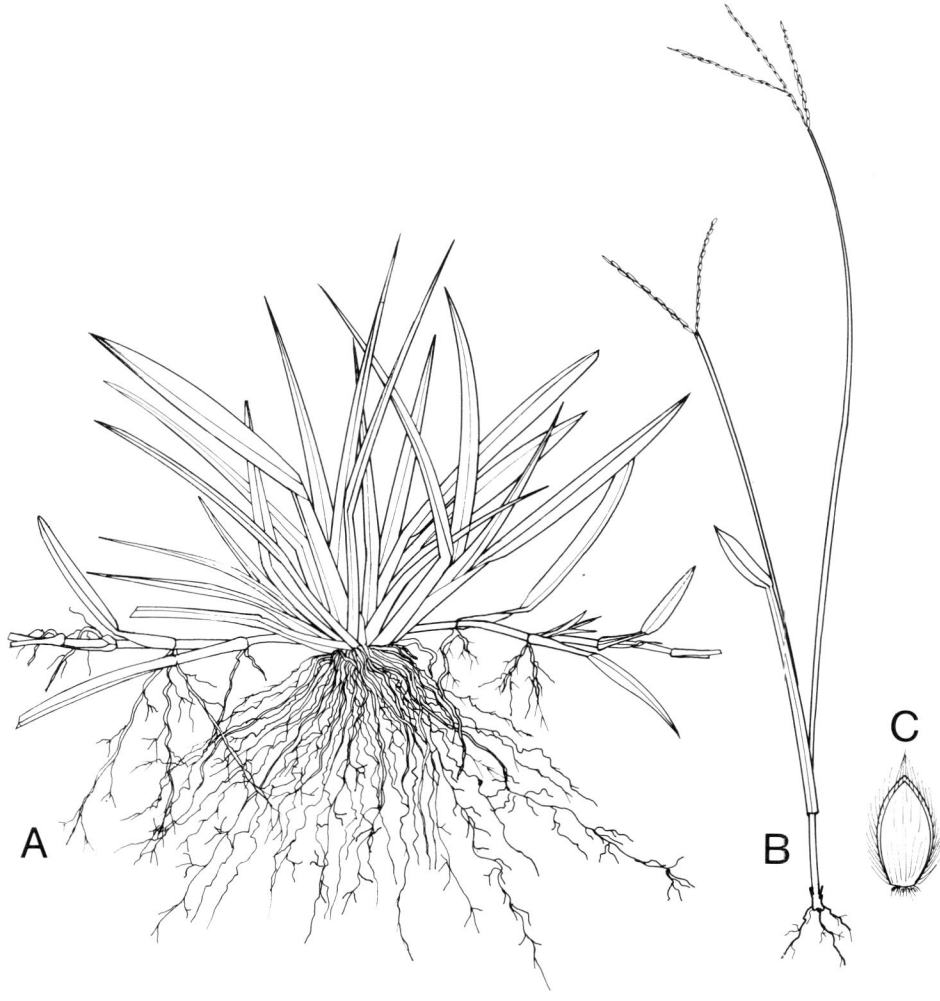

Figure 15.8. Axonopus affinis. **A**-Habit **B**-Inflorescence **C**-Spikelet

Ability to spread naturally. Under favourable conditions this grass will rapidly spread by stolons and rhizomes, and the abundance of light seed enables it to spread quickly also from seed spread by grazing animals.

Land preparation for establishment. It can be established on a well-prepared seed-bed or broadcast into ashes after a scrub burn.

Sowing methods. Usually broadcast on the surface and harrowed or rolled in.

Sowing depth and cover. Surface sowing with the minimum of cover.

Sowing time and rate. In the United States it is sown at 6-12 kg/ha from spring to midsummer.

Number of seeds per kg. Approximately 2 860 000.

Seedling vigour. It has a very vigorous seedling stage.

Vigour of growth and growth rhythm. It has a short growing season, extended by early nitrogen application.

Response to photoperiod. It flowers in all day lengths from eight to 16 hours, but seed production is greatest under 12- to 14-hour day lengths (Knight & Bennett, 1953).

Compatibility with other grasses and legumes. It is a low-fertility grass; as soil fertility declines it can successfully invade *Paspalum dilatatum* pastures, as well as *Cynodon dactylon* lawns. Few legumes will grow with it to compete with its dense sod.

Ability to compete with weeds. It dominates annual weeds in a pasture. It is often a weed itself (see Chapter 11).

Tolerance to herbicides. Bromacil at 1.8 kg of an 800 g AI/kg product (e.g. Hyvar X) per 200 litres water plus 250 ml surfactant sprayed at a minimum of 450 litres water per hectare will control it (Tilley, 1977). 2,2-DPA at 4.8 kg/ha sufficiently restrained the growth of carpet grass to allow sod-seeding of white clover in Alabama, United States (Searcy & Patterson, 1961).

Response to defoliation. It stands heavy defoliation. Under high cutting height its yield is reduced.

Grazing management. It should be kept in the vegetative state by frequent grazing and periodically renovated and fertilized with nitrogen, especially in spring to prolong its feeding value.

Response to fire. It recovers quickly from fire.

Genetics and reproduction. 2n=80, 54 (Fedorov, 1974).

Dry- and green-matter yields. Yields of fertilized carpet grass ranged from 812-5 197 kg/DM/ha per year in the southeastern United States with nitrogen applications from 56-370 kg N/ha per year (Martin, 1975).

Suitability for hay and silage. It makes poor-quality hay, since when it is high enough to harvest it is low in nutritive value.

Value as standover or deferred feed. In its mature state it is really only poor-quality roughage.

Chemical analysis and digestibility. Its forage is generally of much poorer quality and lower feeding value than *Paspalum dilatatum*. Its copper levels are much lower than those of the latter. After seed set, the crude protein level in mat grass may fall as low as 4 or 5 percent. Grain or urea-molasses supplements need to be fed to lactating dairy cattle in this case, and phosphorus should be available. Chemical analysis is set out in Table 15.5.

Palatability. It is fairly palatable till flowering, after which its palatability declines.

Toxicity. No toxicity has been reported with this grass.

Seed production and harvesting. A. affinis is a prolific seeder and seed is shed easily. If seed were required it could be beaten off with revolving beaters on to a tray. It is harvested mechanically in Mississippi and Louisiana, United States.

Minimum germination and quality required for commercial sale. 60 percent germinable seeds, 97 percent purity (Queensland) germinated at 20-35°C with KNO$_3$ moistening agent.

Cultivars. No cultivars are recorded.

Value for erosion control. In the United States it is widely used to prevent erosion and stabilize road banks (Bennett, 1962).

Diseases. No major diseases occur.

Pests. No major pests are evident.

Economics. It is a low-quality pasture which indicates declining fertility throughout the tropical world.

Animal production. Animal live-weight gains from carpet grass are low compared with other pasture species, and live-weight losses occur in winter. Live-weight gains from pure pasture alone have been 84-98 kg/ha per year unfertilized; with white clover and fertilized a peak of 693 kg/ha per year (Martin, 1975). When *A. affinis* replaces *Paspalum dilatatum* as a pasture butter production will drop 33 percent in northern New South Wales (McLennan, 1936).

Main attributes. It grows on poor soil and covers the ground against erosion,

TABLE 15.5 **Axonopus affinis**

	DM	As % of dry matter				
		CP	CF	Ash	EE	NFE
Fresh, cut at 4-week intervals, Malaysia	28.9	8.3	30.4	5.9	1.7	53.7
Fresh, cut at 6-week intervals, Malaysia	30.7	7.5	30.9	5.9	1.3	54.4
Fresh, mid-bloom, Colombia	34.2	6.2	37.2	5.1	1.4	50.1
Cut at 7.5-10 cm, Wollongbar, New South Wales	–	9.2	20.2	7.5	2.7	60.3

Source: Göhl, 1975

and it tolerates overgrazing. It is considered a good horse feed, as horses eat the masses of seed-heads avoided by cattle (McLennan, 1936).

Main deficiencies. It has very low nutritive value, especially after seeding, low dry-matter yield and response to fertilizer, short growing season, poor root development and low animal production.

Further reading. Barnard, 1969; Martin, 1975.

Axonopus compressus (Swartz) Beauv.

Common names. Broad-leaf carpet grass (Australia), savannah grass (West Indies), nudillo (Peru), bes-chaitgras (Suriname), cañamazo (Cuba).
Natural habitat. Subhumid and humid woodland and savannah, flourishing in moist soils.
Distribution. Native to the southern United States, Mexico and Brazil. Now introduced into most tropical and subtropical countries, especially west tropical Africa, South Africa, India, the Philippines, Indonesia, Australia and the Pacific islands.
Description. A. compressus is similar to A. affinis in most of its botanical characters but it is more robust and stoloniferous. It has stouter culms and stolons, wider leaves and longer spikelets which are more acute. Its leaves are 9-12 mm wide and it forms a dense mat over the surface of the ground, seldom reaching a height of more than 15 cm. The spikelets are 2.2-2.5 mm compared with 2 mm for A. affinis. There is a more pronounced tuft of hairs at the apex of the lemma than in A. affinis (Barnard, 1969). There is a good deal of variation in size and stolon and rhizome formation with environment and management (see Fig. 15.9).
Latitudinal limits. About 27°N and S.
Season of growth. A summer-growing perennial very similar in needs to A. affinis.
Frost tolerance. Not so tolerant of frost as A. affinis.
Rainfall requirements. Higher than for A. affinis, with about 775 mm minimum. In Trinidad it needs 40 cm of rain between January and May (Gumbs & Shastry, 1978).
Drought tolerance. Less tolerant of dry conditions than A. affinis.
Tolerance to flooding. It is not tolerant of swampy conditions.
Soil requirements. It prefers moist sandy soil throughout the year, a little more moisture than required for A. affinis, and it also has a lower soil fertility requirement than *Paspalum dilatatum*, but will respond to moderate fertilizer treatment.
Fertilizer requirements. CSIRO workers (Weier, 1976) have shown that A. compressus has an active nitrogenase system and over a 12-week summer growing period fixed 13 kg N/ha per day. A complete fertilizer is applied unless soil tests indicate otherwise.
Ability to spread naturally. It spreads more quickly by stolons and rhizomes under favourable conditions than A. affinis. It also spreads by seed, but does not produce such abundance of seed as A. affinis.
Land preparation for establishment. For a pure grass stand prepare a fine, weed-free seed-bed.
Sowing methods. It can be surface sown through a drill or broadcast.

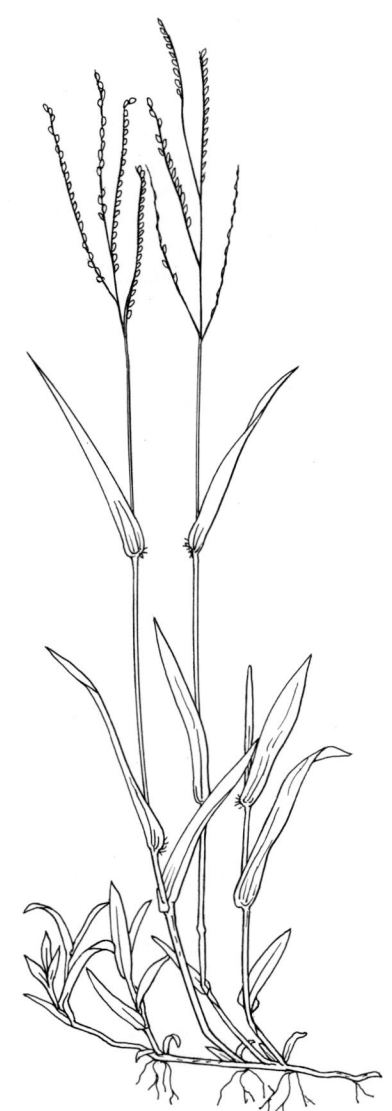

Figure 15.9. Axonopus compressus

Sowing depth and cover. Sow on the surface and roll after planting.
Sowing time and rate. Sow at 6 kg/ha in early to late summer.
Number of seeds per kg. 2 970 000.
Seedling vigour. Excellent.
Response to light. It grows well in the shade. In Malaysia the yield of carpet grass under the leguminous rain tree *Pithecolobium saman* was 20 percent higher and the protein level 14 percent of the dry matter and 11 percent under non-leguminous shade (Jagoe, 1949).
Compatibility with other grasses and legumes. It will gradually invade *Cynodon dactylon* in lawns and will grow in association with white clover (*Trifolium repens*) and *Desmodium triflorum*.
Ability to compete with weeds. It will compete with weeds under favourable conditions.
Tolerance to herbicides. It can be controlled by similar treatments to *A. affinis* (Tilley, 1977).
Response to defoliation. Carpet grass stands a good deal of defoliation and in lawns has to be mown frequently.
Grazing management. Heavy grazing to maintain the grass in a vegetative condition is essential.
Response to fire. It will survive a fire but, owing to its habitat, fires are infrequent.
Genetics and reproduction. 2n=40, 50, 60 (Fedorov, 1974).
Suitability for hay and silage. It rarely provides enough material for conservation, and the stems are coarse.
Value as standover or deferred feed. Very poor — once it has seeded its nutritive value is very low.
Chemical analysis and digestibility. The forage quality of broad-leaf carpet grass is poor, but higher than for *A. affinis*. Harrison (1942) recorded the dry-matter content of carpet grass (no stage mentioned) as 39 percent, the digestible crude protein at 1.8 percent, and the starch equivalent as 21.8 percent. Göhl (1975) has listed percent digestibility of six-week-old, 20-cm-tall growth in Table 15.6.
Palatability. It is fairly palatable.
Seed harvesting methods. The seed can be easily harvested with a stripper type harvester.
Minimum germination and quality required for commercial sale. 60 percent (Queensland) germinable seeds, germinated at 20-35°C moistened with KNO_3.
Value for erosion control. Good for stabilizing slopes against erosion where conditions suit it, and also for stabilizing banks of dams.
Economics. Used widely as a lawn grass in the tropics and subtropics, it invades as a low-quality pasture grass in these areas.
Animal production. In São Paulo, Brazil, zebu steers gained an average daily liveweight of 0.175 kg over 672 days which included two dry seasons (Rocha

TABLE 15.6 *Axonopus compressus*

	DM	As % of dry matter				
		CP	CF	Ash	EE	NFE
Fresh, dry season, 4 weeks, Trinidad	28.6	9.0	29.2	10.5	1.5	49.8
Fresh, dry season, 8 weeks, Trinidad	35.6	7.6	28.8	3.1	1.1	54.4
Fresh, wet season, 4 weeks, Trinidad	23.8	10.5	43.1	12.4	1.2	32.8
Fresh, wet season, 8 weeks, Trinidad	24.9	11.4	42.4	10.4	1.8	34.0
Fresh, wet season, 6 weeks, 20 cm, Thailand	25.6	8.2	30.5	10.2	2.0	49.1
Fresh, 7.5-10 cm, Wollongbar, New South Wales		13.0	19.4	9.6	2.9	55.0

	Animal	Digestibility (%)				
		CP	CF	EE	NFE	ME
Fresh, 6 weeks, Thailand	Sheep	48.0	31.0	37.0	30.0	1.10

Source: Göhl, 1975

et al., 1962). In Fiji it is considered a useful feed, especially if sensitive plant (*Mimosa pudica*) is growing with it (Parham, 1955).
Main attributes. It grows quickly and stabilizes erosive soils in the higher rainfall tropics.
Main deficiencies. Its tendency to invade better pastures, its low quality after seeding, and the frequency of cutting needed in lawns.
Further reading. Barnard, 1969; Gledhill, 1965; McLennan, 1936.

Axonopus scoparius (Flügge) Hitch.

Synonym. Paspalum scoparium Flügge.
Common names. Imperial (South America), maicillo (Peru), cachi (Central America).
Natural habitat. Moist ground.
Distribution. Tropical America — Colombia to Brazil.
Description. An upright perennial bunch grass 1.5-2 m tall with solid and succulent stems, leaf-blades 5-20 mm wide, and numerous slender erect or ascending racemes 10-20 cm long, aggregated toward the summit of the stem, spikelets 2.5-3 mm long, appressed, short-villous. The leaves are blunt-ended and hairy.
Season of growth. Summer.
Minimum temperature for growth. 0°C.
Frost tolerance. It will tolerate low temperatures to 0°C (Göhl, 1975).
Altitude range. It thrives best at 600-1 200 m in Panama, 1 200-2 000 m in Colombia and 2 500 m in Costa Rica, but grows at 300 m in the eastern Llanos where nights are cool.
Rainfall requirements. It grows best in areas of high rainfall, well distributed (Rattray, 1973).
Drought tolerance. It will tolerate drought on deep soils.
Soil requirements. It prefers well-drained soils, such as sandy and alluvial soils well supplied with organic matter.
Fertilizer requirements. With nitrogen and potassium, yields as high as 44 tonnes per hectare were obtained, the effect of phosphorus was minor. Liming from pH 4.7 to pH 5.2 affected the yield (de Alba, Basadre & Mason, 1956). It responds favourably to applied nitrogen but less than many grasses.
Land preparation for establishment. A well-prepared seed-bed ploughed to 15-20 cm.
Sowing methods. It is propagated by division of clumps at a recommended spacing of 60 × 60 cm in furrows. It requires about 2 tonnes per hectare of clumps (Whyte, Moir & Cooper, 1959).
Response to defoliation. It does not persist well under grazing. It is basically a "cut-and-carry" grass.
Grazing management. Introduce the cattle when the grass is 80-100 cm high and remove them when the grass is grazed down to 20-25 cm. Renovate to break up compacted soil and reduce weeds.
Genetics and reproduction. 2n=20 (Fedorov, 1974).
Dry- and green-matter yields. At Turrialba, Costa Rica, imperial grass gave a yield of up to 42 000 kg DM/ha with an application of nitrogen and potash (de Alba, Basadre & Mason, 1956). Dry forage yields range from 10 000-14 000 kg/ha without fertilizer, but are doubled with 50 kg N/ha after each harvest (four to six years) (Crowder, Chaverra & Lotero, 1970).

TABLE 15.7 **Axonopus scoparius**

	DM	As % of dry matter				
		CP	CF	Ash	EE	NFE
Flowering, Costa Rica	10.0	7.2	25.4	6.6	1.7	49.2
Fresh, 4 weeks, Costa Rica	14.6	11.4	28.6	12.2	3.4	44.4
Fresh, 6 weeks, Costa Rica	15.0	7.1	30.0	9.5	2.7	50.8
Fresh, 8 weeks, Costa Rica	15.3	6.2	29.7	14.2	1.4	48.5
Fresh, late bloom, Brazil	18.3	7.3	30.5	7.8	1.9	52.5

	Animal	Digestibility (%)				
		CP	CF	EE	NFE	ME
Fresh, 8 weeks, Costa Rica	Zebu	48.6	75.7	53.5	58.4	2.02

Source: Göhl, 1975

Suitability for hay and silage. It is widely used for silage in Colombia (Crowder, Chaverra & Lotero, 1970) and is persistent only when cut (Göhl, 1975).
Chemical analysis and digestibility. Figures from Costa Rica and Brazil have been recorded in Table 15.7.
Palatability. It is quite palatable.
Cultivars. Vegetative material of two selections, Imperial ICA Clone 60 and Clone 72, is distributed by Tulio Ospina Station (Crowder, Chaverra & Lotero, 1970).
Diseases. A bacterial disease caused by *Xanthomonas axonoperis* reduces or eliminates the pasture. It is spread by farmers' machetes or grazing animals. Disease-free fields are maintained with clean stem pieces for establishment and by cutting with a machete dipped in disinfectant (Crowder, Chaverra & Lotero, 1970).
Economics. Cultivated as a soilage crop in Central and South America.
Animal production. No figures have been cited.
Further reading. Gonzalez & Pacheco, 1970.

Bothriochloa bladhii (Retz.) S.T. Blake

Synonyms. *B. intermedia* (R. Br.) A. Camus; *Andropogon intermedius* R. Br.

Common names. Forest blue grass (Australia), lautoka grass (Fiji), Australian blue-stem (United States).

Natural habitat. Open forest on heavier soils.

Distribution. Africa, India to Australia and the Pacific, introduced to United States.

Description. An erect to geniculately ascending branched perennial bunch grass to about 1 m high; the culms, often rooting at the lower nodes, or, rarely, producing stolons bearing pale green to purplish racemes, simple or occasionally divided, arranged fairly densely about a central axis; the sessile spikelets with short, fine, bent hairlike awns and lower glumes conspicuously pitted with a single tiny hole like a pin-prick. Leaves strongly aromatic when crushed, smelling and tasting like turpentine. Blade linear-lanceolate and tapering gradually from the base to a fine hairlike point. Ligule inconspicuous, short membranous and backed by sparse white hairs. Sheath glabrous and slightly compressed. Culms round, slender, glabrous. Roots coarse, aromatic (Rose-Innes, 1977) (see Fig. 15.10).

Season of growth. Summer.

Frost tolerance. It survives seasonal frosts though the culms may be frosted. It is not as tolerant as *D. sericeum*.

Drought tolerance. In Queensland, Australia, it occurs in the 700-800 mm rainfall belt. It is fairly tolerant of drought conditions.

Tolerance to flooding. It will tolerate short-term flooding.

Soil requirements. It occurs mainly on heavy clay loam to clay soils often derived from basalt in Queensland, and on heavier alluvial soils.

Tolerance to salinity. Not recorded.

Fertilizer requirements. It is not usually fertilized but will respond to nitrogen (Gill, Rana & Negi, 1970).

Ability to spread naturally. It will spread slowly by seed.

Land preparation for establishment. The land can be fully prepared for seeding or roughly prepared for rooted slips.

Sowing methods. It is propagated by rooted slip in India.

Response to defoliation. It will stand heavy grazing up to one beast to 0.4 ha at Gayndah, Queensland (700 mm rainfall with summer dominance) (Scateni, 1966).

Grazing management. Under experimental conditions at Gayndah, Queensland, a stocking system of two weeks' grazing, six weeks' rest was adopted (Scateni, 1966). In this environment it was shown that utilizing the native pasture (e.g. *Bothriochloa bladhii*, *Heteropogon contortus* dominant) for

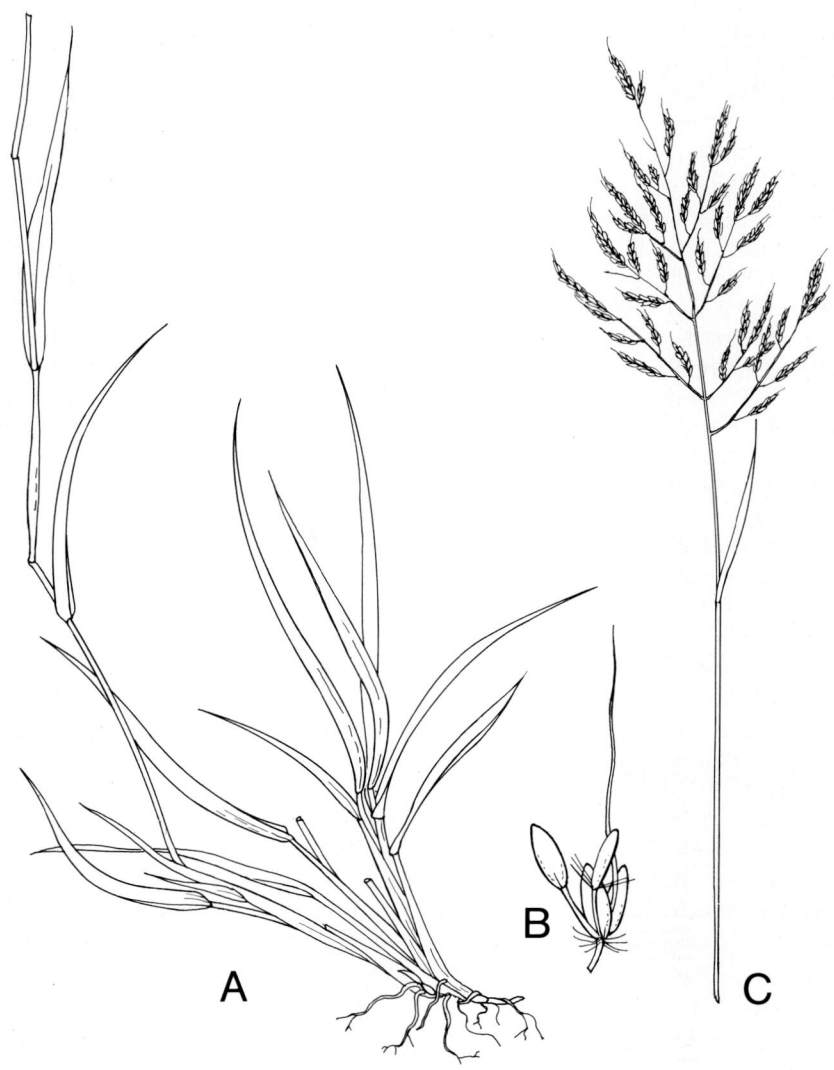

Figure 15.10. Bothriochloa bladhii. **A**-Rachis **B**-Spikelet **C**-Inflorescence

summer and autumn grazing and sown pasture (e.g. green panic and lucerne) for winter and early summer grazing improved annual production (Scateni, 1966). Without sown pasture, winter/spring grazing of 0.74, 1.24 or 2.47 animals per hectare increased the basal cover of *B. bladhii* and *H. contortus* at the expense of other species and improved the sward. Continuous summer grazing reduced the percentage of *B. bladhii* at the expense of *Dichanthium* spp. Rotational grazing resulted in an increase in *B. bladhii*.
Genetics and reproduction. 2n=40, 50, 60, 80 (Fedorov, 1974).
Suitability for hay and silage. It makes quite useful hay in Australia, the United Republic of Tanzania and India. In the Kangra district in Punjab, *B. bladhii* top-dressed with 28 kg N/ha in June and harvested in early September gave the highest yields and the best quality hay (Gill, Rana & Negi, 1970).
Value as standover or deferred feed. It will provide useful roughage in the winter to be supplemented with licks or concentrates.
Chemical analysis and digestibility. Göhl (1975) gives figures in Table 15.8.
Palatability. It is a palatable grass in Queensland. In Ghana it is regarded as unacceptable to livestock. At Richmond, New South Wales, dairy cattle refused to eat it both in the young stage and later. It is somewhat aromatic.
Seed production and harvesting. In Queensland it seeds from midsummer (December) to late autumn. No seed is harvested commercially.
Cultivars. No cultivars are registered but hybrids exist. A key has been drawn up of the *Bothriochloa intermedia* complex by Faruqi (1969).
Value for erosion control. It is useful, but there are better grasses, e.g. *B. insculpta*, suited to similar soils.
Diseases. It is often attacked by smut.

TABLE 15.8 ***Bothriochloa bladhii***

	DM	As % of dry matter				
		CP	CF	Ash	EE	NFE
Fresh, 4 weeks, Ghana	30.0	13.6	25.0	13.4		
Fresh, 8 weeks, Ghana	21.8	9.2	26.2	8.1		
Fresh, 12 weeks, Ghana	23.4	8.3	29.9	9.0		
Fresh, 16 weeks, Ghana	30.5	5.6	31.6	6.6		
Fresh, 36 weeks, Ghana	59.7	9.3	28.7	5.7		
Fresh, 1st cut, India		3.9	38.9	9.5	1.3	46.4
Fresh, 2nd cut, India		3.7	37.7	9.2	1.1	48.3
Fresh, 3rd cut, India		2.1	37.8	8.5	1.3	50.3
Hay, flowering stage, India		4.2	43.2	9.2	1.5	41.9

Source: Göhl, 1975

Economics. A very useful native grass for beef cattle in central and southern coastal Queensland, Australia.

Animal production. At Brian Pastures Research Station, Queensland, a pasture growing on a heavy soil derived from basalt containing 43 percent *B. bladhii,* 17 percent *Dichanthium* sp., 17 percent *Heteropogon contortus,* 7 percent *Chloris divaricata,* 4 percent *Aristida* sp. and 12 percent other species stocked at 1.24 steers per hectare during the dry season (winter) from June to November gave an average live-weight gain of 8.5 kg/ha over three years; when supplemented with 0.75 kg/ha per day of cotton-seed meal (40 percent crude protein) they gained 69.5 kg/ha. When grazed at 0.62 steers per hectare during the wet season (December-May) in summer, steers gained 69.2 kg per head (Addison, 1970). It was found useful in reseeding rangeland in the southern United States.

Main attributes. A palatable native grass which will tolerate drought conditions and survive annual burning. It utilizes heavy soils well.

Main deficiencies. In Ghana it is unpalatable and an indicator of overgrazing on the Accra plains. It also invades lawns in the coastal savannahs (Rose-Innes, 1977).

Further reading. Scateni, 1966.

Bothriochloa caucasica (Trin.) C.E. Hubbard

Synonym. *Andropogon caucasicus* Trin.
Common name. Caucasian blue-stem (United States).
Distribution. Native of the southern USSR and introduced into the United States.
Description. A tufted perennial, 0.6-1 m tall, leafy at and near the base with rather fine stems. It differs from King Ranch blue-stem (*B. ischaemum*) in the upright stiff stems, which seldom dry to a straw colour, the normally shorter basal leaves containing some red pigment and the branched purple panicle having a longer axis.
Minimum temperature for growth. It survived −26 to −28°C in Oklahoma in 1950-51.
Frost tolerance. It is more cold resistant than King Ranch blue-stem and has survived as far north as southern Ohio, Colorado and Nebraska, United States, with no winter kill.
Rainfall requirements. It generally grows in the 500-700 mm rainfall regimes.
Drought tolerance. It is drought resistant, more so than King Ranch blue-stem.
Soil requirements. It grows well on poor, shallow soils.
Sowing methods. Seed is broadcast on to a fine seed-bed and rolled, or it can be broadcast into sorghum stubble.
Sowing depth and cover. Sow on the surface, or no deeper than 2.5 cm.
Sowing time and rate. Sow in January to May in Texas, United States at 1 kg/ha in 1-m rows or 2 kg/ha in a solid stand.
Number of seeds per kg. 2.2 million.
Dormancy. Fresh seed has a low germination rate, but after four months germination may reach 60 percent.
Vigour of growth and growth rhythm. It makes rapid growth.
Ability to compete with weeds. It competes well with weeds.
Suitability for hay and silage. It makes useful hay.
Seed production and harvesting. The seeds are difficult to harvest as they shatter readily. It is harvested when most of the seed is ripe with a combine at 1 000-1 200 rpm. Turn the fresh seed two to three times daily to dry.
Cultivars. 'Medio' is an improved variety with finer stems and leaves that produce a thicker turf than the common type.
Value for erosion control. It is used for stabilizing waterways in Texas.
Further reading. Archer & Bunch, 1953.

Bothriochloa insculpta (A. Rich.) A. Camus

Synonym. Amphilophis insculpta (A. Rich.) Stapf.
Common names. Sweet pitted grass (East Africa), pinhole grass (Zimbabwe and South Africa), stippel grass (South Africa), creeping blue grass (Australia).
Natural habitat. Open bush and grassland, mainly in heavy textured black soils.
Distribution. Kenya, Uganda, Zaire, southern and northeastern tropical Africa. Common in Madras, the Nilgiris and Pulneys in India. Introduced to Australia as CPI 2695 in 1931 as *Amphilopis* (syn. *Bothriochloa*) *glabra* (Bisset, 1978).
Description. A stoloniferous perennial, 30-100 cm high. Sessile spikelets shiny, shallowly grooved below the pit and glabrous, rarely with the margins finely hairy, pedicelled spikelets with three pits, rarely with one (Napper, 1965). It resembles *Bothriochloa bladhii* and *B. ewartiana* — all three have scented leaves and the scent persists in the hay. Creeping blue grass can be distinguished by its conspicuously hairy nodes, the reddish-purple colouring of the exposed stem portions and the development of creeping, as well as upright, stems. It differs from *Dichanthium aristatum* and *D. annulatum* in being scented, and having a pit on one side of the seed hull, and from *D. aristatum* in also having no hairs at the base of the branches (Bisset & Graham, 1978). Leaves, stems and seed-heads are aromatic and the aroma persists in stored hay (see Figs. 15.11, 15.12).
Season of growth. Summer.
Frost tolerance. Good. It has some winter hardiness.
Altitude range. Sea-level to 2 000 m.
Rainfall requirements. The lower yearly rainfall limit is about 500 mm and its range generally 500-800 mm.
Drought tolerance. It is drought tolerant.
Tolerance to flooding. It will not thrive under waterlogged conditions.
Soil requirements. It does very well on black clay soils but also on loams and clay loams, and will vegetate on eroded clays, scalded areas and puggy soils.
Fertilizer requirements. Although able to persist in unfertilized ground it does respond well to nitrogen fertilizer.
Ability to spread naturally. It spreads gradually — in one case 1 m on each side of a planted row in two years (Bisset & Graham, 1978).
Land preparation for establishment. Establish on a well-prepared seed-bed, or in the ashes of a burn. Oversowing may give reasonable results.
Sowing methods. Surface sow and lightly roll or harrow; broadcast on to a scrub burn; broadcasting into barley stubble and brigalow is effective. In aerial sowing push the seed out of the hopper by hand (Bisset & Graham, 1978).
Sowing depth and cover. Not deeper than 1 cm.

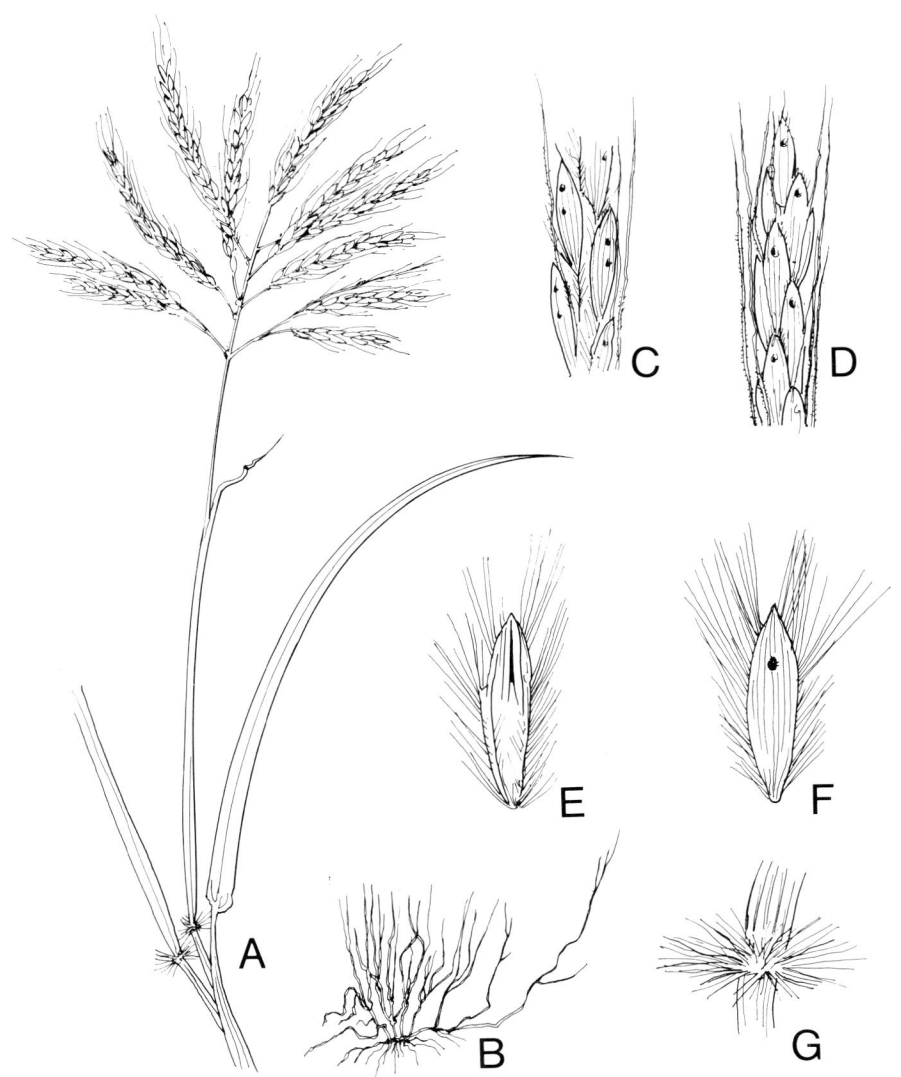

Figure 15.11. *Bothriochloa insculpta.* **A**-Habit **B**-Creeping stem **C**-Upper sterile spikelets **D**-Lower sterile spikelets **E,F**-Seed **G**-Node

Figure 15.12. *Bothriochloa insculpta* growing with *Macroptilium atropurpureum* (siratro) at Rockhampton, Queensland, Australia (**Source:** K. Kelly, Queensland Department of Primary Industries)

Sowing time and rate. Sow from November to January (summer) at 2 kg/ha. For soil conservation sow at 9 kg/ha.
Number of seeds per kg. About 1 210 000 fertile spikelets with one seed in each.
Dormancy. There is little dormancy.
Seed treatment before planting. A treatment to remove awns from the seeds would assist sowing. A "de-awner" used for *Stylosanthes humilis* is successful with *B. insculpta* seed.
Seedling vigour. Good.
Vigour of growth and growth rhythm. A summer-growing perennial, it continues to grow into winter until cut by frost, but is slow in coming away in the spring in comparison with Rhodes grass and green panic.
Response to photoperiod. It flowers in late April and in November in central Queensland (Bisset & Graham, 1978).
Response to light. It does not grow well in shade.
Compatibility with other grasses and legumes. It will grow with native grasses and gradually dominate them. Where tried, it combined well with siratro and *Stylosanthes humilis* (Bisset & Graham, 1978).
Ability to compete with weeds. Good.

Response to defoliation. Heavy grazing by hippopotomuses leads to dominance of *Heteropogon contortus* around Lake Edward, Uganda, and still heavier use will lead to poor *Sporobolus* and *Aristida* weeds (Heady, 1966).
Grazing management. Graze when the runners are well developed. Trampling by stock will develop the ground cover. Graze well during summer to prevent if from becoming stemmy. In pure stands or with *Stylosanthes humilis* keep it grazed to 10-15 cm; with siratro let the pasture reach 30 cm (Bisset & Graham, 1978).
Response to fire. Good. An annual burn is necessary if grazing is not heavy.
Genetics and reproduction. 2n=50, 60 (Fedorov, 1974).
Dry- and green-matter yields. No records of yields have been found.
Suitability for hay and silage. It has been made into hay before the grass becomes stemmy. The hay has been eaten by horses and beef cattle but rejected by dairy cattle (used to better material) and sheep. With siratro its palatability is better. The hay is aromatic.
Value as standover or deferred feed. It is useful as low quality roughage but is rather stemmy.
Chemical analysis and digestibility. Figures from various sources are given in Table 15.9.
Palatability. It is reasonably palatable, much better than *Heteropogon contortus,* and is improved in association with siratro.

TABLE 15.9 ***Bothriochloa insculpta***

	DM	As % of dry matter				
		CP	CF	Ash	EE	NFE
Fresh, early vegetative, 20 cm, Kenya	25.0	9.7	32.5	12.7	2.6	42.5
Fresh, early bloom, 30 cm, Kenya	20.0	9.2	34.9	12.0	2.7	41.2
Fresh, mid-bloom, 30 cm, Kenya	40.0	8.4	32.4	14.2	2.3	42.7
Hay, mature, Kenya	92.5	5.5	37.2	11.2	1.7	44.4
Standing hay, Kenya	91.0	4.6	35.2	9.9	1.8	48.5

	Animal	Digestibility (%)				ME
		CP	CF	EE	NFE	
Early vegetative, Kenya	Sheep	62.7	73.3	55.6	64.3	2.22
Early bloom, Kenya	Sheep	63.5	74.9	46.1	61.1	2.20
Mid-bloom, Kenya	Sheep	54.3	69.5	54.8	57.3	1.99
Hay, mature, Kenya	Sheep	39.6	60.6	48.0	47.4	1.73
Standing hay, Kenya	Sheep	33.6	56.8	45.6	55.3	1.82

Source: Göhl, 1975

Toxicity. The scent of the flowerheads will not cause taint in milk if eaten, though it deters cattle when more palatable feed is available.

Seed production and harvesting. It flowers twice a year in Rockhampton, Queensland, on the Tropic of Capricorn, in autumn and late spring. For seed the pasture is grazed till late summer and shut till seed ripens in late autumn. The crop is direct headed. Clean seed plots have been established in Queensland to prevent contamination with seed of *Dichanthium aristatum* (Angleton grass). Seed yield: 10-30 kg/ha with up to 80 kg/ha per crop with nitrogen fertilizer and irrigation.

Minimum germination and quality required for commercial sale. It is difficult to get more than 50 percent purity, though germinations up to 70 percent can be obtained. Seed germination declines rapidly after two years.

Cultivar. The only registered cultivar is cv. Hatch — after Garney Hatch of The Caves, Rockhampton, Queensland, who sponsored its early acceptance. It is described above.

Value for erosion control. It is useful for erosion control on self-mulching black clay slopes. It is one of the few suitable grasses as it establishes quickly and forms a good ground cover. The runners do not take root readily, but this is improved by prolonged wet weather, trampling or by soil cover, and using high seeding rates. The pastures must be grazed or burnt periodically.

Diseases. In a wet autumn a leaf rust (*Puccinia duthiae*) can reduce seed yields.

Pests. There are no major pests.

Economics. One of the few grasses suited to sowing in black self-mulching clays. It has been used with fair success in reseeding the medium rainfall areas of Baringo, Kenya.

Animal production. Around Rockhampton (rainfall 1 495 mm) on the Tropic of Capricorn, carrying capacity is one animal to 1.2 hectares.

Main attributes. Its ability to compete with native grasses on forest country without nitrogen fertilizer and to invade *Heteropogon contortus* pastures. Its hardiness and ease of establishment.

Main deficiencies. It lacks a little in palatability because of aromatic leaves.

Further reading. Bisset & Graham, 1978.

Bothriochloa ischaemum (L.) Keng

Synonym. Andropogon ischaemum (L.).
Common names. Plains blue-stem, King Ranch blue-stem, Turkistan blue-stem, yellow blue-stem (United States).
Natural habitat. Dry stony places, borders of fields and slopes.
Distribution. Afghanistan, India, Iraq, Pakistan, USSR, Turkey, southwestern United States (Oklahoma, Texas, Kansas, Missouri, Arkansas).
Description. It is a warm-season perennial bunch grass, sometimes forming a sod. It grows to 0.3-0.5 m with creeping root-stock, erect culms, simple or sparingly branched above, glabrous or pubescent at nodes. Leaves glaucous, hairy at base. Three to ten digitate spikes, linear, 4-6 cm long, greyish (see Fig. 15.13).
Rainfall requirements. It is adapted to 375-1 000 mm rainfall.
Drought tolerance. It is drought resistant.
Tolerance to flooding. It has no tolerance to flooding.
Soil requirements. 'Plains Blue-stem' is adapted to a wide range of soils from well-drained good sandy soils to loam and clay loam soils, but not to deep sands. It has some tolerance to soils which produce iron chlorosis. It prefers limestone soils.
Tolerance to salinity. 'King Ranch Blue-stem' has some salt tolerance.
Fertilizer requirements. In the United States a planting mixture of NPK (16-20-0, 18-46-0 or 10-20-40) is used. It responds well to fertilizer, but gives good quality forage at low fertility rates. A yield of 30 kg of dry grass can be expected for every kilogram of nitrogen.
Ability to spread naturally. It volunteers readily from seed.
Land preparation for establishment. A good, clean, firm seed-bed is needed.
Sowing methods. The fluffy seed is mixed with fertilizer and sown through a drill or fertilizer spreader using a coarsely ground fertilizer.
Sowing depth and cover. Sow on surface and roll or cover lightly.
Sowing time and rate. Sow in the early summer at 11-17 kg/ha.
Number of seeds per kg. 3.1 million.
Dormancy. There is some post-harvest dormancy.
Seed treatment before planting. The seed requires hammer-milling before planting.
Seedling vigour. Excellent.
Vigour of growth and growth rhythm. It has a growth similar to Bermuda grass (*Cynodon dactylon*) in Oklahoma. It can be grazed throughout the winter there. It grows well in late summer and autumn in the United States, flowering from June to October.
Genetics and reproduction. $2n=40, 50, 60$. It is a pseudogamous apomict.
Dry- and green-matter yields. The average annual yield of green matter on a

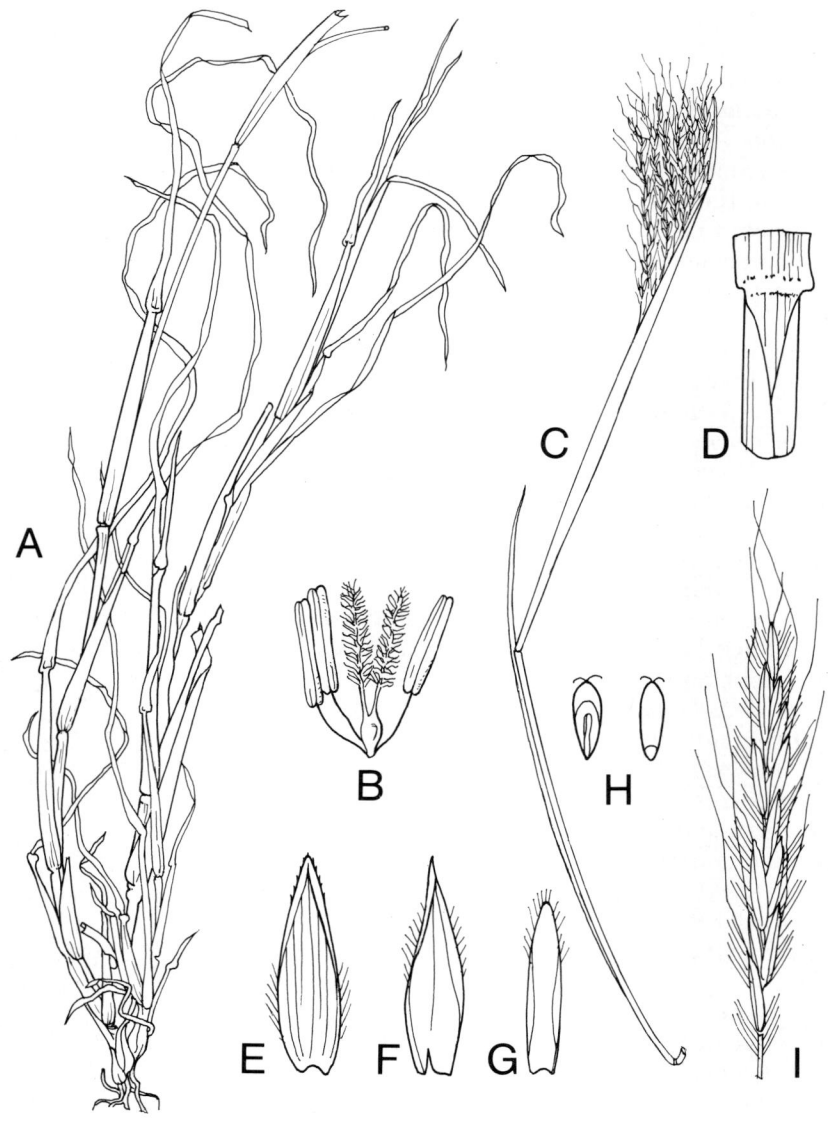

Figure 15.13 Bothriochloa ischaemum. **A**-Habit **B**-Flower **C**-Inflorescence **D**-Ligule **E,F,G**-Glumes of sessile spikelet **H**-Grain **I**-Portion of false raceme

black clay soil in Maharashtra State, India, with light irrigation during the hot months December to May was 8 724 kg/ha from three cuts each for three years and four cuts one year (Whyte, 1964). In Oklahoma, United States, yields of dry matter per hectare were 3 288 kg with no nitrogen fertilizer, 6 545 kg with 110 kg N/ha, and 10 039 kg with 220 kg N/ha (Dalrymple, 1978). In a seven-year study at the Noble Foundation, Oklahoma, it outyielded Bermuda grass cv. Midland by 23 percent. In Gujarat, India, it yielded 15 680 kg/ha green matter — better than *C. ciliaris, Dichanthium annulatum, Eragrostis superba, Panicum antidotale* and *P. repens*.

Suitability for hay and silage. It has made good hay in the United Republic of Tanzania and in the drier zones of Hawaii.

Chemical analysis and digestibility. Advanced plains blue-stem with stemmy growth gave an average digestibility of 48.6 percent. The crude protein content averaged 4.7 percent during the winter of 1978 with a phosphorus content of 0.08 percent (Dalrymple, 1978).

Palatability. 'Plains Blue-stem' is quite palatable and leafy, but is extremely prostrate in Kenya.

Toxicity. No toxicity has been reported with this grass.

Cultivars.

● 'Plains Blue-stem' — a blend of morphologically similar selections of the same pasture species. The selections come from Afghanistan, India, Iraq, Pakistan, the USSR and Turkey.

● 'King Ranch Blue-stem' — has been used for several decades in the United States. It is a high-quality grass that can produce greater yields than native grass, but less than 'Plains' and 'Caucasian' blue-stems. It can be used well as a range grass but as there are more productive grasses, very little 'King Ranch Blue-stem' is planted today. It can form a sod almost as dense as Bermuda grass and can be used for erosion control.

● 'Elkan Blue-stem' — more cold resistant than other strains.

Value for erosion control. Valuable for reseeding eroded soils in the central and southern Great Plains of the United States. It was found to be one of the most effective soil stabilizers to maintain the bunds in the ravine lands of Gujarat, India, with an elaborate root system and excellent ground cover (Srinivasan, Bonde & Tejwani, 1962).

Animal production. Newly purchased 159 kg steers gained 0.45 kg per day grazing 'Plains Blue-stem' during late summer and autumn in Oklahoma. Weaner calves gained 0.66 kg per day for the first 35 days and 0.24 kg/day for the last 34 days for an overall daily gain of 0.45 kg per head per day for 69 days (Dalrymple, 1978).

Main attributes. Its winter hardiness and palatability.

Further reading. Dalrymple, 1978; Harlan *et al.*, 1961.

Bothriochloa pertusa (L.) A. Camus

Synonyms. Amphilophis pertusa (L.) Stapf; *Andropogon pertusus* (L.) Willd.
Common names. Seymour grass, hurricane grass (Africa), camagueyana (Cuba), Indian blue grass (Australia), Barbados sour grass.
Natural habitat. Grassland on clay soils and open woodland.
Distribution. Kenya, Uganda, southern and northeastern tropical Africa, Southeast Asia, Caribbean archipelago.
Description. Tufted perennial 30-70 cm high, rarely stoloniferous. Sessile spikelets finely hairy on the back and sides; pedicelled spikelets, usually with one pit but occasionally with three or none (Napper, 1965). Very close to *Bothriochloa insculpta,* the latter having a glabrous lower glume and one to three pits. It is smaller with slightly smaller panicles and shorter spikelets. In some specimens of *B. pertusa* the culms creep about the surface of the soil and root at the nodes (Bor, 1960). Roots penetrate to 55 cm with production of 6 356 kg/ha of root on alluvial soil at Varanasi, India (Ramam, 1970). Rhizomes constitute 80 percent of the total below-ground dry matter (see Fig. 15.14).
Altitude range. Sea-level to 2 100 m in India.
Rainfall requirements. In the areas of 500-1 375 mm in India (Whyte, 1968) and in the dry zone of Burma with 600-1 000 mm of rainfall, up to 1 800 mm in Timor.
Drought tolerance. Withstands short dry spells. Shankarnarayan (1963) lists it as a perennial drought-evading plant.
Tolerance to flooding. Whyte (1964) lists it as an associated species in loamy soils subject to waterlogging at Ballasar colony, North Gujarat, India.
Soil requirements. It will grow on poor soils and succeeds *H. rufa* on poor soils in Panama. It is common on the black cotton soils of India (Chinnamani, 1968) and Timor (Whyte, 1968) and also grows on a variety of soils from coarse to fine-textured with a pH range from pH 5.8-7.5 and on lateritic soils (Whyte, 1968).
Ability to spread naturally. It is a vigorous weed in the pastures of the United States Virgin Islands, invading Guinea grass, molasses grass and Bermuda grass (Oakes, 1968).
Land preparation for establishment. A fine seed-bed is preferable, but it will establish slowly in a rough seed-bed.
Sowing methods. Broadcast the seed on the surface of the soil and roll or cover lightly.
Sowing time and rate. Sow in early summer at 1.5 kg/ha.
Number of seeds per kg. 1 210 000 pure seed (Bogdan & Pratt, 1967).
Ability to compete with weeds. Excellent. It can invade also Guinea, molasses and Bermuda grasses, although can be suppressed by pangola grass (Oakes, 1968).

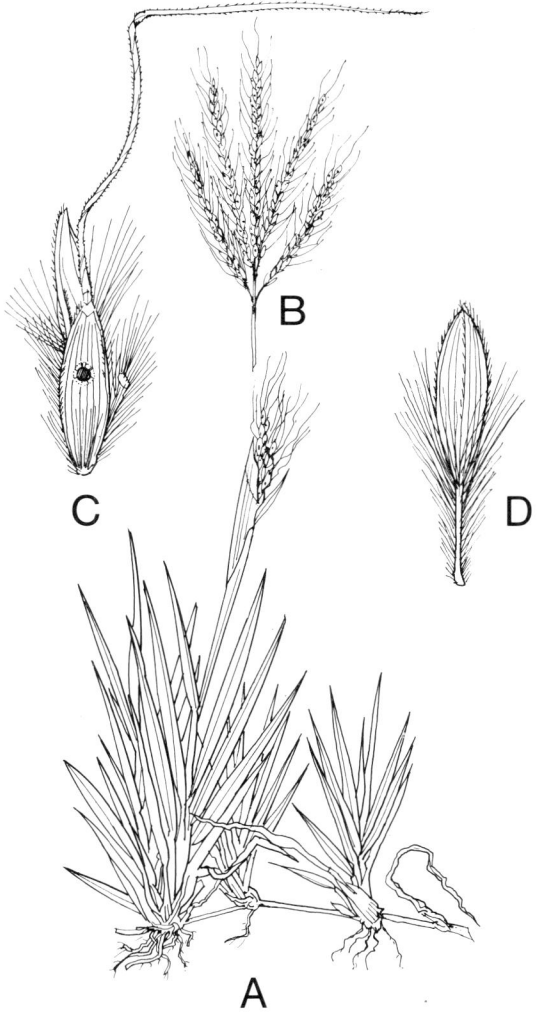

Figure 15.14. Bothriochloa pertusa. **A**-Stolon **B**-Inflorescence **C**-Sessile spikelet **D**-Pedicelled spikelet

Response to defoliation. It withstands heavy grazing and trampling (Bor, 1960). It represents the first stage in degradation of the *Sehima/Dichanthium* cover in India (Dabadghao & Shankarnarayan, 1973) (see Plate 23).
Response to fire. It survives fire (Whyte, 1968).
Genetics and reproduction. 2n=40, 60 (Fedorov, 1974). In India it behaves both as an apomict and also undergoes sexual reproduction (Gupta, 1969-70).
Dry- and green-matter yields. In Cuba, Pérez Infante (1970) obtained average annual yields of 15 000 kg DM/ha, of which 40 percent was produced in the dry season under sprinkler irrigation (Bogdan, 1977).
Suitability for hay and silage. Silage was made successfully in pits 7.62 × 0.45 × 1.83 m deep in India from 1940-45 from this grass (Whyte, 1964). It makes useful hay (Bor, 1960).
Chemical analysis and digestibility. See Table 15.10.
Palatability. It is quite palatable.
Seed production and harvesting. The seed is a hairy spikelet with an attached joint and pedicel.

TABLE 15.10 **Bothriochloa pertusa**

	DM	As % of dry matter							Reference
		CP	CF	Ash	EE	NFE	C_aO	P_2O_5	
Fresh, mature, India		3.9	37.9	10.0	2.3	45.9			Patel, 1966
		6.0	33.2	11.7	1.4	47.7			Sen & Ray, 1964
Hay, India	94.3	3.0	39.2	10.7	1.2	45.9			Lander & Dharmani, 1932
1st cut, Bareilly		5.44	36.49	12.31	1.23	44.63	0.55	0.52	Sen, 1957
2nd cut, Bareilly		3.86	35.68	11.32	1.13	48.01	0.47	0.41	Sen, 1957
3rd cut, Bareilly		3.19	36.50	11.28	1.27	47.76	0.60	0.69	Sen, 1957
Fresh, early bloom, Venezuela	34.7	5.7	34.9	11.7	1.0	46.7			International Network of Feed Information Centres, 1978

	Animal	Digestibility (%)				ME	Reference
		CP	CF	EE	NFE		
Hay	Zebu	1.8	59.7	21.4	47.7	1.66	Lander & Dharmani, 1932

Source: Göhl, 1975

Cultivars. In Maharashtra State, India, Oke (1971) recommends cv. Ghana Marvel 20 which gives 40-100 percent more dry matter than the local strain. The Queensland Department of Primary Industries is evaluating six types of this species.

Value for erosion control. It has been recommended for reseeding eroded land in India and shows promise for erosion control in central Queensland. It is a good grass for revegetating spoil from central Queensland coal mines.

Diseases. There are no major diseases.

Pests. There are no major pests.

Economics. A bad weed of pastures in the Virgin Islands (Oakes, 1968). It is low-yielding — used for fodder, hay and mulching in India. Considered one of the better grasses in Barbados, Jamaica, Uganda and India.

Animal production. No figures have been found.

Main attributes. It will stand up to trampling and constant grazing.

Main deficiencies. Tends to become a weed.

Further reading. Oakes, 1968; Pérez Infante, 1970; Rao, 1970.

Brachiaria brizantha (A. Rich.) Stapf

Common names. Palisade grass (Samoa), signal grass (East Africa), St Lucia grass (Queensland), Ceylon sheep grass (Sri Lanka), upright brachiaria (Zimbabwe), bread grass (South Africa), estrella de Africa, pasto alambe (Latin America).
Natural habitat. Grassland valleys and open woodlands.
Distribution. Native to tropical Africa but now introduced into most tropical countries.
Description. Perennial up to 120 cm high, with stout erect culms and broadly lanceolate leaf-blades. Two to five racemes, up to 15 cm long, with two rows of almost sessile, overlapping, rounded spikelets, 4-6 mm long on the underside. It differs from *B. decumbens* in that the spikelets have a sub-apical fringe of long purplish hairs, and the spikelets are longer than those of *B. decumbens* (see Fig. 15.15).
Season of growth. Spring to autumn.
Optimum temperature for growth. About 30-35°C.
Minimum temperature for growth. It grows well into the winter, being green when other tropical grasses are brown and dry.
Frost tolerance. It will survive frosts.
Latitudinal limits. It can be used in pastures at high altitudes in Burundi (Scaillet, 1965).
Altitude range. Sea-level to 3 000 m.
Rainfall requirements. It requires a rainfall generally in excess of 500 mm.
Tolerance to drought. It is fairly drought tolerant (Bor, 1960; CIAT, 1978).
Tolerance to flooding. It will not tolerate flooding (Bor, 1960; CIAT, 1978).
Soil requirements. It tolerates a wide range of soils and is tolerant of acid conditions.
Fertilizer requirements. It performs quite well without fertilizer and was found to be one of the best grasses (*B. brizantha* 665) under low nitrogen and low phosphorus at CIAT, Quilichao, Colombia. However, it responds well to fertilizer, producing over 4 000 kg DM/ha with no nitrogen but adequate phosphorus, and high yields with complete fertilizer.
Ability to spread naturally. It can spread slowly by seed as the seed ages to break its dormancy.
Land preparation for establishment. A well-prepared seed-bed is preferable.
Sowing methods. It can be propagated vegetatively by sods, root pieces and stems.
Response to light. It tolerates shade under coconuts well in Sri Lanka (Bor, 1960).
Compatibility with other grasses and legumes. In Sri Lanka, Fernando (1961) grew it in association with *Alysicarpus vaginalis* and *Centrosema pubescens*.

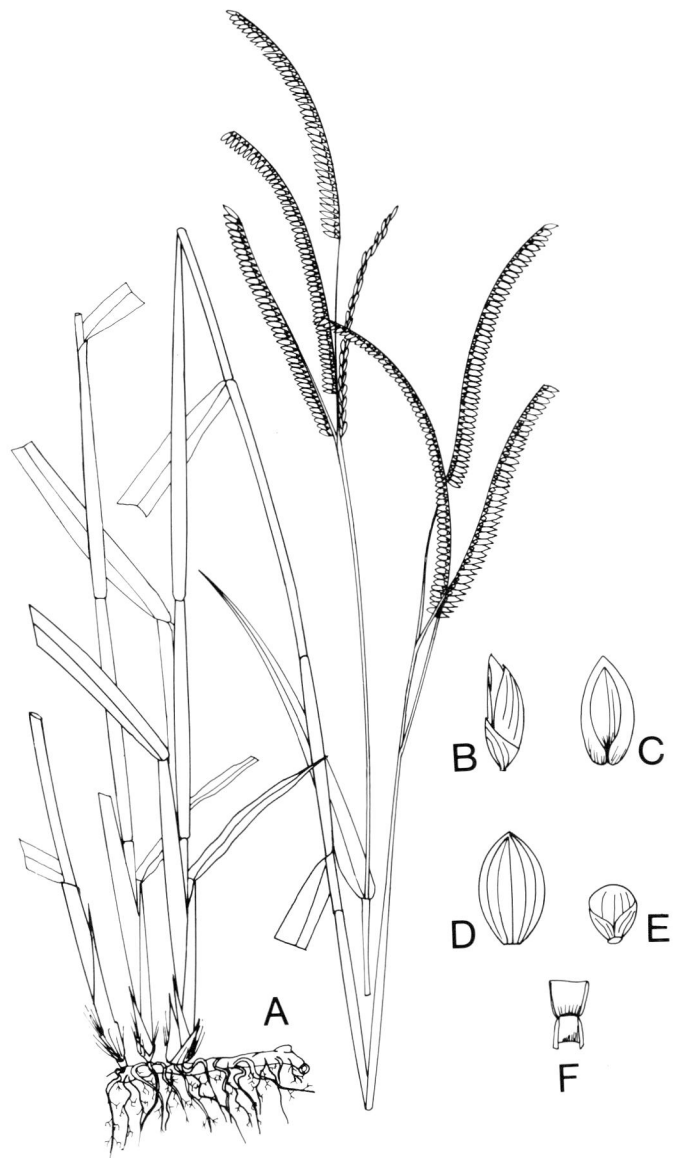

Figure 15.15. Brachiaria brizantha. **A**-Habit **B**-Spikelet **C**-Palea **D**-Upper glume
E-Lower glume **F**-Ligule

High nitrogen applications decreased the proportion of legumes, while phosphorus applications increased the proportion. In Malaysia *Stylosanthes guianensis* and centro have been successful (Vendargon, 1964). However, if fertilizer mixtures are not balanced in favour of legumes the grass will dominate.

Response to defoliation. It will stand heavy grazing. In Sri Lanka, Sivalingam (1964) recommended a cutting interval of 30 days when fertilized with nitrogen at 0, 45, 132 and 396 kg/ha.

Grazing management. It can be heavily grazed if used as a monospecific sward and regularly fertilized with nitrogen. If grown with a legume the grazing system must favour the legume and adequate phosphorus must be maintained.

Response to fire. *B. brizantha* will not tolerate fire. In Zambia annual burning of dominantly *Hyparrhenia* grassland for three years reduced the *B. brizantha* cover from 0.38 to 0.09 percent (Brockington, 1961). Selection 665 at CIAT, Quilichao, Colombia has good resistance to burning, but burning is not recommended (CIAT, 1978).

Dry- and green-matter yields. With *Alysicarpus vaginalis,* fertilized with 88 kg N and 88 kg P_2O_5/ha it yielded 6 750 kg DM/ha from three cuts (Fernando, 1961). In Fiji over a three-year period it yielded an average of 7 850 kg DM/ha with a crude protein content of 7.6 percent (Roberts, 1970). It yielded 16 800 kg/ha green matter in Tanzania. In Sri Lanka *B. brizantha* gave herbage dry-matter yields per year ranging from 10 368 to 17 377 kg/ha at nitrogen applications ranging from 56 to 280 kg/ha (Appadurai, 1975).

Suitability for hay and silage. In the United Republic of Tanzania it has made useful silage (van Rensburg, 1952) and in Sri Lanka it is useful both for hay and silage (Bor, 1960). In Burundi a mixture of *Eragrostis curvula* (at least 50 percent), *B. brizantha* and *Setaria sphacelata* is recommended for silage (Scaillet, 1965).

Chemical analysis and digestibility. It has good nutritional value. Göhl's (1975) analyses and digestibility figures are shown in Table 15.11.

Palatability. It is very palatable, with a good leaf/stem ratio.

Toxicity. At the Queensland Agricultural College, Lawes, Queensland, Australia, crossbred wether sheep grazing on a vigorous sward of *Brachiaria brizantha* growing on a black clay soil developed severe photosensitization and icterus, marked by drooping ears, swelling of the subcutis of the face and eyelids, and congested, yellowish mucous membranes. The sheep rapidly lost condition and died. In these animals the skin over the muzzle, ears, and eyelids was necrotic and the conjunctival sac filled with purulent exudate with consequent blindness (Briton & Paltridge, 1941).

Seed production and harvesting. Selection 665 at CIAT, Colombia has good seed production. Seed remains viable for about three years (Jones, 1973).

Value for erosion control. It has been used successfully in Australia (Queensland) and India (Gandhi, 1957; Patil & Ghosh, 1963).

Disease resistance. Good.

TABLE 15.11 *Brachiaria brizantha*

	DM	As % of dry matter				
		CP	CF	Ash	EE	NFE
Fresh, mature, Nigeria	30.4	4.0	31.1	11.1	1.3	52.5
Hay, Kenya	91.4	8.6	31.3	10.6	1.9	47.6
Stem cured, Kenya	90.6	8.7	28.0	10.8	1.6	50.9

	Animal	Digestibility (%)				ME
		CP	CF	EE	NFE	
Fresh, mature, Nigeria	Cattle	20.0	56.6	15.4	61.0	1.84
Hay, Kenya	Sheep	63.3	73.3	62.4	72.9	2.41
Stem cured, Kenya	Sheep	65.2	57.0	0.0	61.7	1.96

Source: Göhl, 1975

Pests. It is attacked by some insect pests.

Economics. B. brizantha has shown to be an outstanding pasture for leys in Kenya (Bogdan, 1959), Madagascar (Birie-Habas, 1959; Granier & Lahore, 1961), the United Republic of Tanzania, Fiji (Payne *et al.*, 1955), Sri Lanka (Panabokke, 1959), Nigeria (Foster & Munday, 1961), Uganda (Bredon & Horrell, 1962), Burundi, Zaire (Scaillet, 1965) and other countries.

Animal production. In a grazing trial in Sri Lanka, Fernando (1961) obtained a live-weight gain of 464 kg/ha from grass alone, 647 kg/ha from a *Pueraria phaseoloides*/*Brachiaria brizantha* sward and 631 kg/ha from a *Centrosema pubescens*/*B. brizantha* mixture over a 260-day season.

Main attributes. Its productiveness, drought resistance, ability to spread and suppress weeds and its ability to grow in shade.

Main deficiencies. Its tendency to produce monospecific swards. Its low seed production.

Further reading. Fernando, 1961.

Brachiaria decumbens Stapf

Synonym. B. eminii (Mez) Robyns.
Common names. Signal grass (Australia), Suriname grass (Jamaica), Kenya sheep grass.
Natural habitat. Open grasslands and partial shade on the Great Lakes Plateau in Uganda and adjoining countries of East and Central Africa (Loch, 1978).
Distribution. Native to Africa and now widespread in the tropics and sub-tropics.
Description. A trailing perennial 30-60 cm high with heavy lanceolate leaf-blades 8-10 mm wide. Two to five racemes, 2-5 cm long with a broad ciliate rachis and 4-mm-long spikelets (Napper, 1965). The erect stems arise from a long, stoloniferous base and root down from the lower nodes producing a dense sward (Loch, 1978) (see Fig. 15.16).
Season of growth. A perennial with a long growing season from spring to late autumn.
Optimum temperature for growth. 30-35°C.
Frost tolerance. It is readily frosted, but its winter production is better than pangola grass in frost-free areas.
Latitudinal limits. About 27°N and S.
Altitude range. Sea-level to 1 750 m.
Rainfall requirements. It is essentially a grass of the wet tropics, but still has good drought tolerance adapted to a dry season of four to five months. It prefers 1 500 mm or more of rain.
Drought tolerance. It does not do well where the dry season is more than five months, but is more productive than *B. mutica, Digitaria decumbens, Panicum maximum, Hyparrhenia rufa* and *Andropogon gayanus* in the late dry season (CIAT, 1978).
Soil requirements. It is tolerant of a wide range of soils and is little affected by high aluminium soils (Spain & Andrew, 1977) or shallow soils. It needs good drainage and fertile conditions for its maximum growth but persists on poor soils.
Fertilizer requirements. On an oxisol at Carimagua, Colombia, it gave maximum yields at 50 kg P_2O_5/ha, responding much more to phosphorus than *Panicum maximum* and *Andropogon gayanus*. On an ultisol at Quilichao, Colombia, it gave a linear response to more than 400 kg N/ha (CIAT, 1978). It has excellent tolerance to aluminium (Spain, 1979).
Ability to spread naturally. It spreads gradually from seed.
Land preparation for establishment. It will establish in rough seed-beds but gives better results on a well-prepared seed-bed.
Sowing methods. Initially planted from vegetative material, it is now established mostly from seed, the dormancy of which can be broken.

Figure 15.16. Brachiaria decumbens. **A**-Habit **B**-Inflorescence **C**-Portion of raceme

Sowing depth and cover. Sow no deeper than 1 cm and roll after sowing.
Sowing time and rate. Sow during the wet season at 2-5 kg seed per hectare.
Number of seeds per kg. 220 000 to 225 000.
Dormancy. The seed needs after-ripening for ten to 12 months to break dormancy (Harding, 1972). Treating the seed for ten to 15 minutes with commercial sulphuric acid improved germination of recently harvested seed from 0-33 percent (Grof, 1968).
Seedling vigour. Excellent — greater than buffel grass and green panic (Cook, 1977).
Vigour of growth and growth rhythm. Within three months a complete ground cover should be obtained. It has a long growing season producing active growth from spring to late autumn, but is frost-tender (Cook, 1978).
Response to photoperiod. It is a short-day plant.
Response to light. It grows very well in the reduced light under coconuts (Gutteridge & Whiteman, 1978).
Compatibility with other grasses and legumes. It is compatible with legumes such as *Stylosanthes*, *Centrosema* and *Pueraria* for a short time but these are soon suppressed by the vigorous grass to leave a pure grass sward (see Plate 24). Hetero (*Desmodium heterophyllum*) has some tolerance (Loch, 1978; Gutteridge & Whiteman, 1978).
Ability to compete with weeds. When established it will suppress weeds very effectively (Harding, 1972).
Tolerance to herbicides. It tolerates pre-emergence application of atrazine at 2.5 kg of 80 percent product per hectare which gives good control of a wide range of weeds when establishing the grass in red latosolic soils on the Atherton Tableland, Queensland (Loch, 1978; Hawton, 1979) — see also *Panicum maximum*.
Response to defoliation. It withstands heavy grazing (Harrington & Thornton, 1969) and trampling (Harding, 1972). CIAT (1978) recommend waiting for a month after the opening rains before establishing rainy-season stocking rates.
Grazing management. It needs to be stocked heavily. Added nitrogen is required to keep it in leafy active growth. Frequent applications of nitrogen — up to six times per season — keep the grass in a very nutritious condition and improve live-weight gain especially under high rainfall conditions. Grazing at 42-day rotations gave the best balance with *B. decumbens* and *Pueraria phaseoloides* at Carimagua, Colombia. At 56-day rotations the *Brachiaria* seeded and new seedlings invaded the *Pueraria*. At 28-day rotation the *Brachiaria* maintained good stands (CIAT, 1978).
Response to fire. It is affected by burning but if the environment is dry enough, *B. decumbens* will take a fire and recovery after fire is usually satisfactory. It colonizes bare patches in *Cymbopogon nardus* grassland in Ankole, Uganda, which is usually burnt annually (Harrington, 1974).
Genetics and reproduction. 2n=36 (Fedorov, 1974). It is a tetraploid (Com-

monwealth Plant Introduction number 1694) and it is an obligate aposporous apomict (Barnard, 1969).

Dry- and green-matter yields. A selection from *B. decumbens,* no. 666 yielded over 4 000 kg DM/ha without nitrogen but with adequate phosphorus at Quilichao, Colombia (CIAT, 1978) and it was one of the better grasses at low nitrogen and low phosphorus. In a cutting experiment its annual dry matter production was 36 300 kg/ha which significantly exceeded that of pangola (*Digitaria decumbens*), Para grass (*Brachiaria mutica*) and Guinea grass (*Panicum maximum*) (Barnard, 1969). In Honduras, mean annual production was 23 072 kg DM/ha with added nitrogen as 450 kg sulphate of ammonia per hectare (Romney, 1961). At Koronivia, in Fiji's wet zone, it yielded 34 126 kg DM/ha over an 11-month period (Roberts, 1970a, b) while in the Solomon Islands it produced up to 30 000 kg DM/ha per year under coconuts.

Suitability for hay and silage. It makes excellent hay.

Value as standover or deferred feed. It is a useful standover feed if it has not been weathered.

Chemical analysis and digestibility. See Table 15.12. As the protein content decreases, so does the digestibility of protein, crude fibre and nitrogen-free extract (Butterworth, 1971).

Palatability. It is excellent in the young leafy stage, but as the stems become hard palatability declines.

Toxicity. No toxicity has been reported by Everist (1974). Garcia-Rivera and Morris (1955) recorded only 1.0 percent of oxalates in the dry matter and Ndyanabo (1974) 1.10 percent.

Seed production and harvesting. On the Atherton Tableland, north Queensland, in years with an early start to the wet season, two seed crops are possible — the first and major one early in January and the second in May. The seed is harvested with an "all-crop" harvester. At CIAT, Colombia, flowering

TABLE 15.12 *Brachiaria decumbens*

	DM	As % of dry matter				
		CP	CF	Ash	EE	NFE
Fresh, early bloom, Kenya		11.2	28.0	9.9	2.8	48.1
Fresh, mid-bloom, Trinidad	19.5	8.2	33.4	8.4	2.5	47.5

	Animal	Digestibility (%)				ME
		CP	CF	EE	NFE	
Fresh, mid-bloom, Trinidad	Wethers	46.9	72.9	66.1		2.21

Source: Göhl, 1975

commenced in early July and peak seed production occurred on 9 August. Combine harvesting recovered only 59 percent of the seed obtained by manual harvest.

Seed yield. CIAT obtained 100 kg pure seed per hectare. Yields range up to 200 kg/ha.

Minimum germination and quality required for commercial sale. 15 percent germinable seeds with 50 percent purity in Queensland.

Cultivars. Cv. Basilisk is the only one available in Queensland. It was introduced into Australia from Uganda in 1930. CIAT, Colombia, is making selections.

Value for erosion control. It is a valuable grass for erosion control as it covers the ground well, stands heavy grazing and establishes on poor and rocky soils.

Diseases. It is relatively disease free.

Pests. Various leaf-eating insects may cause problems.

Economics. A valuable high-rainfall pasture grass. In the wet tropics it is useful to have 25 percent of the farm sown to a vigorous stoloniferous grass. If adequately fertilized with nitrogen, such a grass can support large numbers of cattle in late winter and spring when feed is in short supply. In this way, the stocking pressure on more vulnerable grass/legume mixtures can be reduced during their period of slow growth. On well-drained soils 'Basilisk' signal grass is excellent for such nitrogen-fertilized pastures. It makes very efficient use of fertilizer nitrogen (Loch, 1978).

Animal production. At Turpina Station, Colombia (altitude 50 m) two steers per hectare grazed on *B. decumbens* and gained 0.60 kg per head per day. In the Cauca Valley a live-weight gain of 1 700 kg/ha was obtained from steers grazing an irrigated signal grass pasture fertilized initially with 100 kg N/ha, then 50 kg/ha after each grazing period (Crowder, Chaverra & Lotero, 1970). At Carimagua, Colombia, *B. decumbens* is capable of producing 260-280 kg LWG/ha per year (CIAT, 1978).

Main attributes. High productivity under intensive use, tolerance of low fertility and relative freedom from pests and diseases. Its good performance under coconuts (Loch, 1978).

Main deficiencies. It can dominate a pasture and not allow associated legume growth. It may become a weed but can be easily ploughed out.

Further reading. Loch, 1978.

Brachiaria dura Stapf

Natural habitat. Dry, upland woodlands. After cropping for two years to *Pennisetum americanum* and *Manihot utilissima* it invades the old cultivations.
Distribution. It is recorded from Western Province, Zambia.
Description. A perennial up to 0.8 m high, compactly caespitose on a short, oblique rhizome. Stems wiry, erect and simple, more or less geniculate and branched, smooth. Leaf-sheaths hard, hairy, up to 7.5 cm long; ligule a narrow ciliolate rim, leaf-blades narrow, linear, convolute, very wiry, up to 30 cm long, 27 cm wide when flattened out but only 1.25 cm wide when rolled. Inflorescence a solitary terminal upright spikelike raceme (Verboom, 1966).
Season of growth. Summer.
Optimum temperature for growth. It grows well at 30-35°C.
Minimum temperature for growth. The minimum temperature in June is 10°C.
Frost tolerance. It tolerates the frosts between June and July down to 3°C.
Latitudinal limits. It occurs between latitudes 14 and 17°45'S and longitudes 22 and 25°30'.
Altitude range. In Western Province, Zambia, it occurs at an average altitude of 1 000 m.
Rainfall requirements. In Western Province, Zambia, it occurs in a rainfall area receiving 875-1 000 mm between November and March, with the remainder of the year dry.
Drought tolerance. It is exceedingly drought resistant. *B. dura* has masses of root hairs 1-3 mm long from tip to base, intermingling with sand particles. A vegetable glue (polysaccharides) is extruded by the roots, encasing them with sand particles, thus creating its own soil complex and forming an absorbent mantle of soil. The lack of water retention and ion absorption in the loose sand is thus overcome. This important characteristic promotes forage production.

The lower part of the shoot is covered also with a dense mass of felty hairs. The tufty ligule catches dew from the rolled-up leaves. In this way, moisture finds its way to the root base. During the dry season, the leaves are rolled up to expose a minimum evaporation surface. In the wet season, the leaves flatten out to allow a maximum surface for photosynthesis (Verboom, 1966).
Tolerance to flooding. Unknown.
Soil requirements. It is adapted to sandy soils, usually coarse, of low mineral status, low organic matter and absorption complex, of high leaching potential and poor water retention capacity.
Tolerance to salinity. Unknown.
Sowing methods. Transplanting new shoots is successful.

TABLE 15.13 **Brachiaria dura**

Stage	CP	CF	EE	Ash	NFE	Ca	Mg	Na	P	K	Cl
Young shoots	7.6	41.3	2.4	2.8	46.0	0.16	0.05	0.04	0.25	0.23	0.05
Flowering	6.4	43.3	2.2	2.4	45.7	0.36	0.05	0.04	0.19	0.28	0.25

Dormancy. The seed is dormant after ripening.

Vigour of growth and growth rhythm. It stays green and sets seed in Western Province, Zambia the whole year round.

Response to defoliation. It stands heavy grazing.

Chemical analysis and digestibility. Verboom and Brunt (1970) recorded values as percentages of the dry matter (see Table 15.13).

Palatability. Its palatability is very high and it remains green year round.

Economics. The rumen contents of game from the Mulonga and Siloana Plains in western Zambia have high percentages of *Brachiaria dura*.

Main attributes. Its adaptation to sandy soils of low fertility, its relatively high protein, its long period of succulence and its root adaptation to drought.

Further reading. Verboom, 1966.

Brachiaria humidicola (Rendle) Schweick.

Synonym. B. dictyoneura (Fig. and De Not.) Stapf.
Common names. Koronivia grass (Fiji), creeping signal grass (southern Africa).
Natural habitat. Valley grassland in moist situations, road verges, and vleis.
Distribution. Tropical Africa.
Description. A procumbent stoloniferous perennial with lanceolate leaf blades; three to four racemes with hairy spikelets 3.5-4 mm long (Napper, 1965). It is distinguished from other species of the genus by its creeping habit. The rachis is narrow and angled. The lower glume is three-fourths as long as the spikelet, and conspicuous because it is purple or purple-brown in the upper part, contrasting with the light green colour of the rest of the spikelet. The leaf-blades are flat, and at least 5 mm wide, often more (Chippendall & Crook, 1976) (see Fig. 15.17).
Season of growth. Summer.
Optimum temperature for growth. About 32-35°C in Fiji.
Minimum temperature for growth. It has some cool-season productivity (Roberts, 1970a, b).
Altitude range. 1 000-2 000 m.
Rainfall requirements. It is adapted to the wetter zone of Fiji.
Drought tolerance. It has good drought tolerance (CIAT, 1978) and remains green better than other species with better dry-season production in Fiji (Partridge, 1979b).
Tolerance to flooding. Good, but not as great as *Brachiaria mutica*.
Soil requirements. While it does well on deep fertile soils, it proved to be well suited to coral rubble soil of high pH in less-shaded old coconut plantations in the Solomon Islands (Gutteridge & Whiteman, 1978).
Fertilizer requirements. It responds well to high nitrogen applications and is much more productive than Batiki blue grass (*Ischaemum indicum*) (Roberts, 1970a, b). It has excellent tolerance to aluminium (Spain, 1979) and has low phosphorus requirements.
Ability to spread naturally. Being intensely stoloniferous, it covers the ground well on red soils but is slow to colonize heavy black soils in Fiji.
Land preparation for establishment. A rough ploughing is sufficient to accommodate the vegetative material planted.
Sowing methods. It is usually established vegetatively, small bundles of stolons being planted at spacings of 1-2 m (Roberts, 1970a, b; Partridge, 1979a).
Sowing depth and cover. The cuttings are planted at 15-20 cm with a spade, and covered.
Sowing time and rate. Sow in the wet season (November to March) in the dry zone, and at any time in the wet zone, in Fiji.

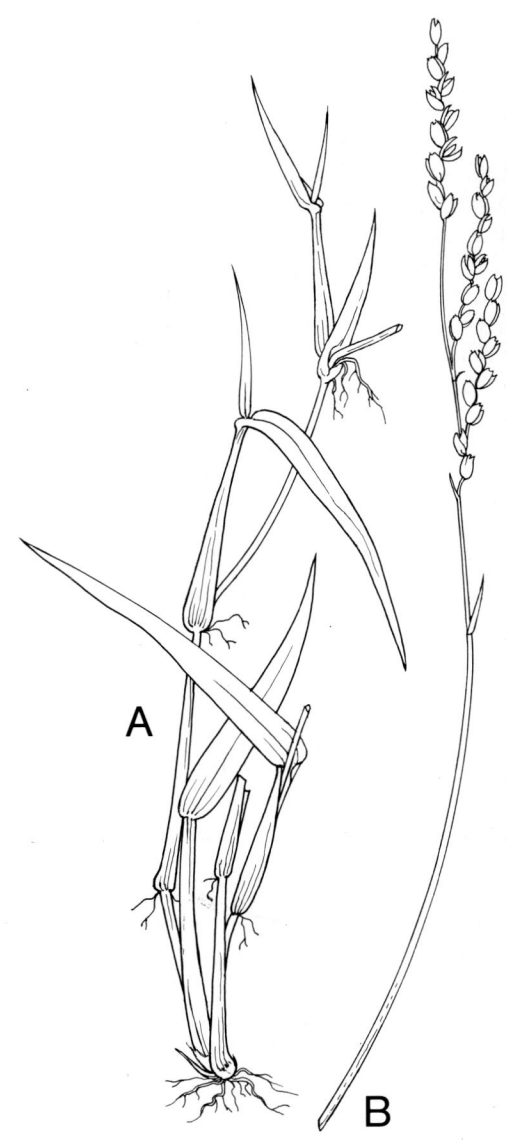

***Figure* 15.17.** *Brachiaria humidicola.* **A**-Habit **B**-Inflorescence

Seedling vigour. Good (CIAT, 1978).
Vigour of growth and growth rhythm. It grows well in summer, and flowering does not normally commence till November in Fiji.
Response to light. In less-shaded coconut plantations it performs very well (Gutteridge & Whiteman, 1978).
Compatibility with other grasses and legumes. Because of its dense growth it is difficult to establish legumes with it. Sod-seeding siratro and centro into it has had some success if the grass is mown closely, but weed invasion by *Mimosa pudica* is a problem in Fiji (Roberts, 1970a, b). It combines with *Desmodium heterophyllum* in Fiji (Partridge, personal communication). In Zimbabwe it combines well with *Trifolium semipilosum* and *Lotononis bainesii* (Clatworthy, 1970).
Ability to compete with weeds. It forms a dense, somewhat woody mat layer beneath grazing level and effectively suppresses weeds (Roberts, 1970a, b).
Response to defoliation. It will withstand heavy defoliation.
Grazing management. It withstands heavy grazing except in areas where Navua sedge (*Cyperus aromaticus* is a threat in Fiji (Partridge, personal communication).
Response to fire. It will not tolerate burning (CIAT, 1978).
Genetics and reproduction. 2n=72 (Fedorov, 1974).
Dry- and green-matter yields. In Fiji it produced 10 929 kg DM/ha unfertilized and achieved a yield of 34 018 kg/ha with an application of 452 kg N/ha with a linear response to nitrogen up to that peak yield (Roberts, 1970a, b). In 1972 at Sigatoka, Fiji, when fertilized with 450 kg/ha superphosphate, it produced 17 500 kg DM/ha, of which 49 percent was produced in the dry (winter) season (Partridge, 1979a). At CIAT, Quilichao, Colombia, selection 679 yielded 2 500 kg DM/ha during an 18-week period without phosphorus, and over 5 000 kg DM/ha with the application of 50 kg P_2O_5/ha (CIAT, 1978).
Chemical analysis and digestibility. See Table 15.14.
Palatability. It is palatable when young, but of low palatability at maximum productivity (Roberts, 1970a, b). Selection 679 has low dry-matter digestibility and low nitrogen content at CIAT, Quilichao, Colombia (CIAT, 1978).

TABLE 15.14 ***Brachiaria humidicola***

	As % of dry matter				
	CP	CF	Ash	EE	NFE
Fresh, early bloom, Kenya	5.1	37.4	9.8	1.6	46.1
Fresh, full bloom, Kenya	7.9	35.5	14.7	2.0	39.9

Source: Göhl, 1975

Toxicity. No toxicity has been reported.
Seed production and harvesting. It is a shy seeder, and this factor alone contributes to its slow spread (Roberts, 1970a, b). The seed also has low viability.
Seed yield. In Colombia, Ferguson (1979) obtained 10-50 kg/ha.
Cultivars. Selections are being made at CIAT, Colombia.
Value for erosion control. It would prove a very effective grass for erosion control (Chippendall & Crook, 1976).
Diseases. It is resistant to diseases.
Pests. It has no serious insect pests. It is resistant to spittle bug (*Deois incompleta*) in the Brazilian humid tropics (Serrão *et al.*, 1979).
Economics. It is used for grazing in Brazil, Fiji and other parts of the world.
Animal production. Selection 679 was to be subjected to grazing trials at CIAT, Colombia during 1979 and subsequently.
Main attributes. Its strongly stoloniferous habit, with ability to root at the stolon nodes, covers the ground rapidly and competes particularly well with weeds. Its low phosphorus requirement and resistance to spittle bug.
Main deficiency. Its low seed production.
Further reading. CIAT, 1978; Clatworthy, 1970.

Brachiaria mutica (Forsk.) Stapf

Synonyms. Panicum muticum Forsk.; *P. purpurascens* Raddi.
Common names. Para grass (Africa, Australia, United States), Mauritius signal grass (South Africa), pasto para and malojilla (South America), gramalote (Peru), parana (Cuba).
Natural habitat. Swampy places and stream banks.
Distribution. Tropical areas of Africa and America, now introduced into most tropical countries.
Description. A short-culmed, stoloniferous perennial up to 200 cm high with long, hairy leaf-blades about 16 mm wide. Panicle 10-20 cm long with solitary racemose or compound branches and glabrous, acute, irregularly multi-seriate spikelets 3-3.5 mm long (Napper, 1965) (see Fig. 15.18).
Season of growth. A summer perennial.
Optimum temperature for growth. Mean annual temperature 21°C (Russell & Webb, 1976).
Minimum temperature for growth. 15°C (Allen & Cowdry, 1961).
Frost tolerance. Para grass is frost sensitive and severely affected by it, but persists well. It will not stand a lower temperature than 8°C (Wheeler, 1950).
Latitudinal limits. It extends to about 27°S in selected areas. Viable seed was collected at latitude 13°S in the Northern Territory of Australia (Wesley-Smith, 1973, Russell & Webb, 1976).
Altitude range. Sea-level to 1 000 m.
Rainfall requirements. Adapted to high-rainfall tropical and subtropical conditions, but in protected areas it can persist with rainfall as low as 900 mm per year.
Drought tolerance. It usually tolerates general drought by reason of its specific swampy environment, being maintained by the residual moisture from the wet season.
Tolerance to flooding. Para grass is semi-aquatic and can persist in standing and running water. It inhabits wetter areas than *B. humidicola* (see Plate 25).
Soil requirements. It prefers alluvial and hydromorphic soils but will grow on a wide range of moist soil types.
Tolerance to salinity. In southeast Queensland it grows on deep loamy soils overlying saline clays and merges with saline grasses on marine flood plains.
Fertilizer requirements. It responds readily to nitrogen (Roberts, 1970a, b). On phosphorus-deficient soils a dressing of 500 kg/ha should be applied prior to planting with subsequent top-dressings of 120-250 kg/ha per year for a few years. Nitrogen applied toward the end of summer or in autumn will give better winter growth (Currie, 1975). It tolerates high aluminium (Spain, 1979).

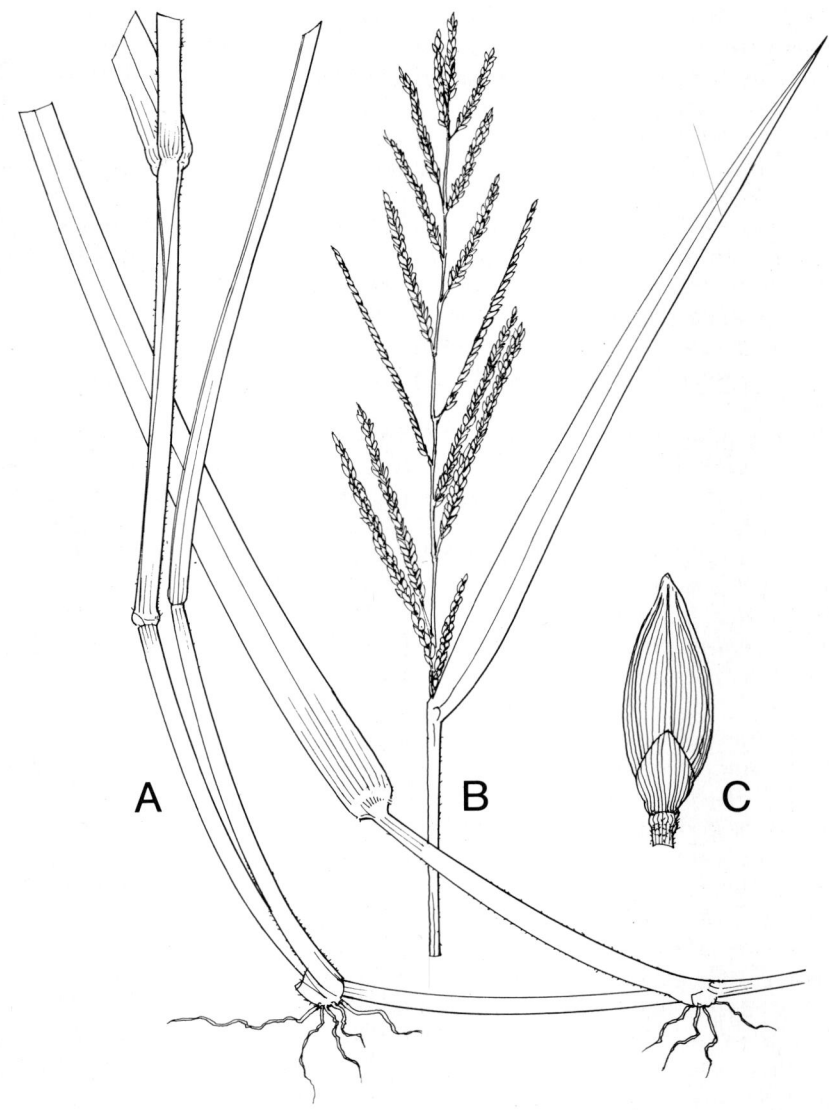

Figure 15.18. Brachiaria mutica. **A**-Rooting culms **B**-Inflorescence **C**-Spikelet

Ability to spread naturally. Under suitable conditions Para grass sends out long stolons and branches bend over and both root at the nodes. It can spread some 5 m in a growing season (Cameron & Kelly, 1970).
Land preparation for establishment. An initial ploughing may be necessary for a rough seed-bed in cleared land. Sprigs can be hand planted in the ashes of burnt wet sclerophyll forest or rain forest, or can be directly planted into swampy land.
Sowing methods. Para grass is usually sown by partly burying sections of the stem in loose, moist earth. Stem cuttings about 25 cm long are planted, preferably as two to three-noded loops at 1-m spacing, or cuttings can be broadcast and disc-harrowed in. Where seed is used, it can be sown in the ashes of a burn or well-prepared seed-bed, or directly into wet areas.
Sowing depth and cover. Sow seed no deeper than 1 cm and roll after seeding.
Sowing time and rate. In spring, summer or late winter, depending on soil moisture, at 2.5-4.5 kg/ha where seed is available.
Number of seeds per kg. About 1 000 000.
Seedling vigour. Seedlings are very robust and vigorous.
Response to photoperiod. It is a short-day plant (Wang, 1961).
Response to light. It will grow in partial shade but prefers full sunlight.
Compatibility with other grasses and legumes. Para grass generally dominates a sward where conditions are favourable. Legumes suitable to moist conditions such as hetero (*Desmodium heterophyllum*), puero (*Pueraria phaseoloides*), centro (*Centrosema pubescens*) and calopo (*Calopogonium mucunoides*) can persist for varying periods with it (see Plate 26). If conditions favour the legume and suppress the grass, the mixture can be maintained longer. In northeast Thailand, Para grass sown on a lateritic red earth soil is not vigorous enough to suppress *Centrosema pubescens,* and a good legume/grass balance is possible.
Ability to compete with weeds. In its natural habitat, Para grass competes successfully with weeds. On Navua plantation in Fiji there is severe competition with *Cyperus aromaticus*. The Para grass is disc-harrowed to chop the stems and submerge them partly in the mud to encourage more rooting and a heavier population of plants to give more competition.
Tolerance to herbicides. Para grass can become troublesome in irrigation ditches, drains and earth tanks. It can be killed by spraying with 2,2-DPA at 2.3 kg of a 740 g AI/kg product (e.g. Dowpon) plus paraquat at 85 ml of a 200 g AI/litre product (e.g. Gramoxone) plus wetting agent at 250 ml per 200 litres of water. Spray till the solution runs off the leaves. Spray when the plant is growing vigorously. Glyphosate at 2 litres of a 360 g AI/litre product (e.g. Roundup) per 200 litres of water will also kill it (Tilley, 1977). Undiluted diesel distillate and sodium chlorate at 100 g/litre of water will also kill it (Kleinschmidt & Johnson, 1977).
Response to defoliation. Low grazing and cutting can seriously deplete Para grass stands. Livestock preferentially eat the leaves and young shoots, and if

animals are forced to eat the stems the damage to new growing points is severe.

Grazing management. It should not be grazed too closely, and the first grazing should be deferred till the grass is 30-60 cm high and well-established: about 12 months. Regular, controlled light grazing will ensure rapid regrowth (Currie, 1975).

Response to fire. Para grass usually grows in such moist situations that fire is unusual. Burning is not recommended. If the plants survive, fire damage is usually superficial but regrowth is slow (Currie, 1975).

Genetics and reproduction. Reproduction is usually by vegetative means, though seed production does take place at low latitudes. The chromosome number is 2n=36 (Fedorov, 1974).

Dry- and green-matter yields. Three to four cuttings a season each of 2 500-7 500 kg/ha can be obtained. At Laguna, Philippines, Furoc and Javier (1976) obtained 84 300 kg/ha of green matter from an irrigated crop, equivalent to 24 000 kg DM/ha, sufficient to feed eight steers for one year. In Fiji, green weights per hectare vary between 83 000-91 000 kg/ha with a crude protein percentage of 5.5 to 15 percent of the dry matter (Roberts, 1970a, b). At South Johnstone in north Queensland 29 818 kg DM/ha per year were obtained.

Suitability for hay and silage. The waterlogged nature of moist soils growing Para grass prevents the passage of machinery for hay and silage making. In a "cut-and-carry" system, hay and silage can be made. The grass should not be cut too low, as subsequent regrowth will be retarded. It makes good silage with dry-matter losses of 10 percent (Paterson, 1945).

Value as standover or deferred feed. Para grass is very valuable in this regard, as its environment is such that soil moisture persists well into the dry season so green growth is usually available for livestock at a crucial time.

Chemical analysis and digestibility. Available figures are given in Table 15.15. Göhl (1975) records 17 other analyses.

Palatability. Young Para grass stems and leaves are very palatable. The coarse old stems are eaten only if grazing pressure is applied.

TABLE 15.15 ***Brachiaria mutica***

	DM	As % of dry matter					Reference
		CP	CF	Ash	EE	NFE	
Floral initiation, Costa Rica		6.36	30.34	8.41	1.35	43.54	Gonzalez & Pacheco, 1970
Flowering, Nov., Africa	29	6.50	35.20	7.00			Boudet, 1975
Regrowth, March, Africa	23	8.90	32.30	15.40			Boudet, 1975
30 days' regrowth, Africa	19	18.20	29.20	12.21			Boudet, 1975
Silage, old forage	26.6	5.80					Paterson, 1945

Toxicity. It is usually non-toxic. One case was reported in 1913 in Queensland as being cyanogenetic (Everist, 1974).

Seed production and harvesting. Application of 125 kg/ha of urea at the onset of the wet season significantly increases both the number of reproductive tillers and also the yield of harvestable seed from about 20 kg/ha to over 25 kg/ha (Currie, 1975), while Grof (1969) increased seed yield from 13 to 31 kg/ha with 112 kg N/ha. Harvest within the week in which general anthesis is completed (Grof, 1969). It can be harvested mechanically by direct heading or it can be hand collected.

Seed yield. Up to 31 kg/ha.

Minimum germination and quality required for commercial sale. 15 percent germinable seed, 40 percent purity in Queensland, Australia.

Cultivars. In Zaire, var. Lopori gives high yields and is recommended for permanent pastures (Risopoulos, 1966). There are no registered cultivars in Australia.

Value for erosion control. Para grass can be quite useful for stream bank and stream erosion control, and on steep slopes where rainfall and soil permit vigorous growth.

Diseases. It is relatively free of diseases. Coccid bug attack associated with sooty mould fungus (*Capnodium* sp.) causes damage to young leafy shoots (Cameron & Kelly, 1970). Blast (*Piricularia* sp.) and sheath blight (*Rhizoctonia* sp.) occur in Thailand (Vinijsanond, 1978).

Pests. Navua sedge (*Cyperus aromaticus*) infests Para grass pastures in Fiji. Some relief is obtained by chopping the Para grass pastures with a fluted roller, burying more nodes under the mud and thickening the stand.

Economics. It has been used as a pasture in Australia since its introduction in 1849. It is now used extensively for grazing and cut fodder in most tropical and subtropical countries with adequate rainfall (Cameron & Kelly, 1970).

Animal production. At Palmira Station, Colombia, crossbred Brahman steers grazed Para grass at 2.5 beasts per hectare on unfertilized plots and gained 0.6 kg per head per day. On nearby pastures fertilized with nitrogen and irrigated, daily gains were 0.78 kg per head (Crowder, Chaverra & Lotero, 1970). In alluvial swampy areas it will carry more than three animals per hectare (Currie, 1975). At Parada (north Queensland) Evans (1969), using a Para grass/centro mixture under irrigation, recorded a live-weight gain of 0.96 kg per head per day with shorthorn bullocks.

Main attributes. A well-adapted grass for growing in swamps and shallow streams in both running water and temporary waterholes, providing grazing in both wet and dry seasons. It is still the best grass for this purpose in the tropics.

Main deficiencies. A pest in irrigation ditches, headlands, drains and earth tanks. When this blockage leads to waterlogging, adjacent crops may be lost. It is a coarse fodder when mature. It does not combine easily with legumes.

Further reading. Cameron & Kelly, 1970; Currie, 1975.

Brachiaria radicans Napper

Synonym. B. latifolia.
Common name. Tanner grass (United States).
Natural habitat. Swampy and seasonally flooded grassland.
Distribution. Kenya, Tanzania and Uganda, introduced to the Americas.
Description. A stoloniferous perennial, 100-150 cm high, with glabrous, geniculate culms, rooting at the lower nodes, and lanceolate leaf-blades up to 15 cm long by 12 mm wide. Six to ten racemes with a wavy rachis 1.5 mm wide and glabrous, biseriate, subacute spikelets 4 mm long (Napper, 1965).
Season of growth. Summer.
Altitude range. 1 300-2 000 m in Tanzania.
Rainfall requirements. In Puerto Rico it grows in a rainfall range of 1 500-2 000 mm of evenly distributed rain per year.
Drought tolerance. It does not tolerate drought.
Soil requirements. In the humid tropics of Puerto Rico, deep red acid ultisols with a pH of 4.8 are common with high free iron and aluminium oxides.
Fertilizer requirements. In the eastern Llanos of Colombia on the oxisols at Carimagua, Spain (1979) obtained major increases in stolon length per plant from additions of phosphorus, and in the number of stolons produced by *B. radicans* with the addition of potash. It is able to fix nitrogen. Three hundred and thirty kg/ha of a 15:5:10 fertilizer is applied two weeks after planting in Puerto Rico (Vicente-Chandler *et al.*, 1974). Liming to reduce acidity and prevent aluminium and manganese toxicity may require 2 250-4 500 kg/ha.
Land preparation for establishment. A well-prepared, deep seed-bed is required.
Sowing methods. The cuttings are planted in rows 1 m apart and covered lightly with soil to a depth of 5-8 cm.
Sowing time and rate. Sow at the beginning of the rainy season in spring in Puerto Rico, using 2 500-4 500 kg/ha of mature stem-cuttings.
Dry- and green-matter yields. At Orocovis, Puerto Rico, Tanner grass yielded 12 880 kg/ha of green matter three months after planting, compared with 15 120 kg from star grass (*Cynodon nlemfuensis*), 9 520 kg from *Brachiaria ruziziensis* and 6 720 kg from pangola grass (*Digitaria decumbens*) (Vicente-Chandler *et al.*, 1974), showing that it is quick to establish.
Toxicity. Some anaemia has been reported in horses and sheep grazing Tanner grass during the winter in Brazil (Rosenfeld *et al.*, 1976).
Further reading. Spain, 1979; Vicente-Chandler *et al.*, 1974.

Brachiaria ruziziensis Germain and Everard

Synonym. Brachiaria eminii Mez.
Common names. Kennedy ruzi grass (Australia), Congo signal grass (Africa), prostrate signal grass (Kenya).
Natural habitat. A pioneer species of cleared rain forest in Africa.
Distribution. Lake Edward and Lake Kivu districts, Rwanda, Burundi, and the Ruzizi plains in Zaire, now widely distributed in the tropics.
Description. A spreading perennial with short rhizomes, similar in habit to Para grass. The inflorescence consists of dense and spikelike racemes. The spikelets are all sessile and close together, the rachis of the racemes winged, broad and over 3 mm wide. The spikelets are hairy and the lower glume under half the length of the spikelet (Harker & Napper, 1960). It has softer leaves than *B. brizantha* (see Fig. 15.19).
Season of growth. Summer.
Optimum temperature for growth. 33°C day, 28°C night (Dienum & Dirven, 1972).
Minimum temperature for growth. Low yields resulted from a 24/19°C regime. Ludlow (1976) found this species, of several tropical grasses, the most affected by low temperatures.
Frost tolerance. It is killed by heavy frosts, and spring regrowth is very slow after light frosts.
Altitude range. 1 000-2 000 m in Kenya, up to 1 200 m in Panama (Rattray, 1973).
Rainfall requirements. It requires a reasonably high rainfall, but can endure hot dry spells. A rainfall of 1 000 mm or more is best.
Drought tolerance. It has good drought tolerance.
Tolerance to flooding. It does not tolerate flooding.
Soil requirements. It requires a soil of high fertility, such as latosols carrying mesophyll rain forest. It will tolerate acid soils. It needs good drainage.
Fertilizer requirements. It needs high phosphorus in the early growth on a wide range of soils. It responds well to nitrogen, either inorganic or from legumes, but its nitrogen requirement exceeds that of Guinea grass, which makes the latter more attractive (Mellor, Hibberd & Grof, 1973b). Risopoulos (1966) recorded an increased yield of 10 739 kg/ha from nitrogen application in Zaire.
Ability to spread naturally. It spreads well from rhizomes.
Land preparation for establishment. A well-prepared seed-bed is recommended, but light disc-harrowing gives good results.
Sowing methods. Drill the seed into a well-prepared seed-bed. In Zaire it has been sown in rows 60 cm apart, or broadcast over the land after scarification of the soil with a disc harrow or brushcutter, without burning the native pastures, and grazed as soon as it is ready (Risopoulos, 1966).

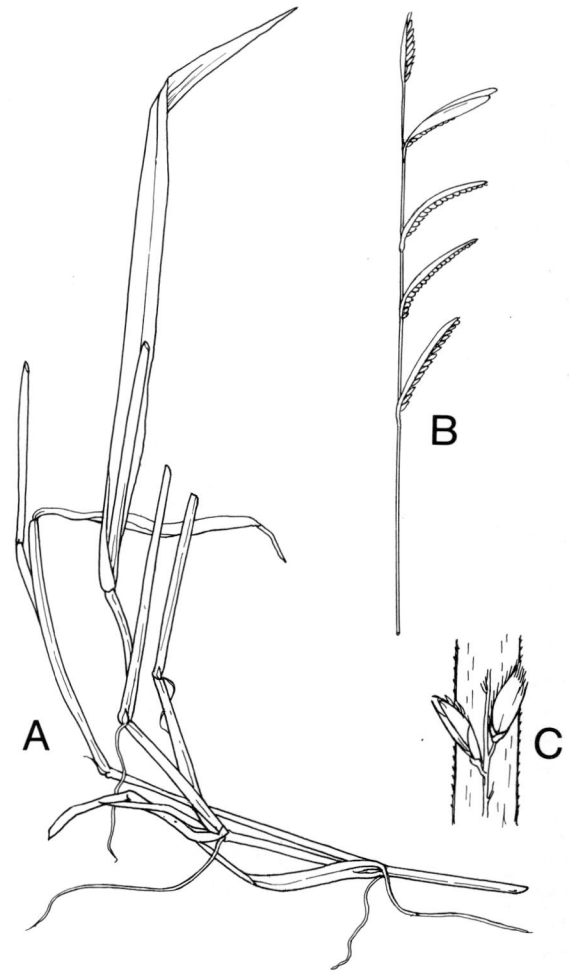

Figure 15.19. Brachiaria ruziziensis. **A**-Habit **B**-Inflorescence **C**-Portion of raceme

Sowing depth and cover. Surface sow in moist soil, and sow no deeper than 2 cm in dry soil (Bogdan, 1964). In Zaire it is recommended to sow at a depth of 1-2 cm. Under humid conditions seeds lose their vitality after one year (Risopoulos, 1966).
Sowing time and rate. In Zaire the seed rate recommended is 30 kg/ha.
Number of seeds per kg. About 250 000.
Dormancy. Fresh seed shows post-harvest dormancy and delayed germination. Fresh seed gave 20 percent germination in Queensland, and after 12 months' storage this increased to 40 percent (Davidson, 1966). Dormancy can also be broken by treating the seed with concentrated sulphuric acid for 15 minutes (Barnard, 1969) or mechanical scarification (Jones, 1973).
Seedling vigour. Excellent (Davidson, 1966).
Vigour of growth and growth rhythm. It gives good early wet season growth for eight weeks after the opening rains (Falvey, 1976) and it seeds heavily in April at South Johnstone, north Queensland (lat. 17°36'S).
Response to light. Yields increase with increasing light intensity (Dienum & Dirven, 1972).
Compatibility with other grasses and legumes. Ruzi grass combines well with legumes such as *Centrosema pubescens* or *Pueraria phaseoloides* if the mixture is leniently grazed. In Zaire it has combined well with *Setaria sphacelata* and *Stylosanthes guianensis*. In northern Australia *Stylosanthes humilis* and *S. hamata* can be introduced by cultivating the grass and oversowing the legumes (Falvey, 1979).
Ability to compete with weeds. It successfully suppresses weeds.
Response to defoliation. It forms a dense mat under grazing which withstands grazing well (Davidson, 1966). The yields of dry matter did not vary very significantly in Sri Lanka with monthly cutting at 2.5 cm or 7.6 cm but bi-monthly cuts yielded a little higher (Appadurai & Goonawardene, 1973).
Grazing management. In combination with *Stylosanthes humilis* in northern Australia it must be grazed heavily to maintain this legume in the sward (Falvey, 1976).
Response to fire. Selection 6019 at CIAT, Colombia, does not tolerate fire (CIAT, 1978).
Genetics and reproduction. It appears to be apomictic.
Green- and dry-matter yields. In Tanzania, ruzi grass yielded 21 159 kg DM/ha (Naveh & Anderson, 1967). At South Johnstone, north Queensland it yielded 19 500 kg DM/ha under a six-week cutting interval and an input of 220 kg N/ha/year (Grof & Harding, 1970). In Sri Lanka yields of 16 807, 22 031 and 25 585 kg DM/ha per year with nitrogen applications of 112, 224 and 366 kg N/ha (Appadurai, 1975). In French Guiana the yield was 20 574 kg DM/ha and 1 180 kg/ha crude protein (Borget, 1966) and in Zaire yields of 31 352 kg and 21 468 kg green matter per hectare per year were obtained in successive years, 1958-59, with 100 kg nitrogen and 100 kg superphosphate per hectare per year (Risopoulos, 1966). At Gualaca, Panama, it

produced 11 000 kg DM/ha without fertilizer and 27 000 kg DM/ha when fertilized with 600 kg N/ha per year in a rainfall area of 3 997 mm per year.

Suitability for hay and silage. It made very good silage with *Stylosanthes guianensis* in Zaire with 1 percent molasses and without additive (Risopoulos, 1966) and made good hay in Zambia (van Rensburg, 1969).

Chemical analysis and digestibility. High temperatures have an adverse effect on digestibility (Dienum & Dirven, 1972). Digestibility decreased in 18-day material from 78.4 percent at day/night temperatures of 24/18°C to 72.7 percent at 29/30°C and 69.5 percent at 34/30°C (Dirven, 1973). Feeding value declined when it seeded heavily at South Johnstone, Queensland, in April (Mellor, Hibberd & Grof, 1973a). Scaut (1959) in Zaire found fresh grass to contain 13.9 percent crude protein, 27.2 percent crude fibre, 9.0 percent ash, 2.3 percent ether extract and 47.6 percent nitrogen-free extract in the dry matter.

Palatability. It is very palatable. At the Cerrado Centre, Brazil, it was preferentially grazed ahead of *Stylosanthes guianensis* during the rainy season.

Toxicity. No toxicity has been recorded by Everist (1974).

Seed production and harvesting. It seeds heavily at Gandijika, Zaire (lat. 6°45'S with four months dry season) (Risopoulos, 1966). Seed can be harvested in May in Queensland, either by hand or with a tractor-mounted buffel type seed harvester but yields are lower by this method (Davidson, 1966).

Seed yield. A seed yield of 125 kg/ha has been recorded in Queensland and 200 kg/ha in Zaire (Risopoulos, 1966).

Minimum germination and quality required for commercial sale. The seed is germinated at 20-35°C after treatment with sulphuric acid for ten minutes. Fifteen percent germinable seed and 40 percent purity in Queensland.

Cultivars. 'Kennedy', described above, is the only present cultivar. Selection 6019 has been tested at CIAT, Colombia.

Value for erosion control. It is useful for erosion control in areas where it grows well.

Diseases. It is comparatively free from diseases.

Pests. There are few pests.

Economics. An important grazing species in the wetter tropics.

Animal production. Brahman steers grazing Kennedy ruzi grass at Utchee Creek, north Queensland, gave a poor winter performance on grass in association with legumes compared with the performance on Guinea grass, but nearly equalled Guinea when 27 kg N/ha was applied. Ruzi grass/legume pastures produced 1 157 kg/ha live-weight gain, while ruzi grass plus 200 kg N/ha per year gave 1 513 kg/ha live-weight gain (Mellor, Hibberd & Grof, 1973a). Live-weight gains from the Cerrado Centre in Brazil are shown in Table 15.16.

Main attributes. Its fast growth early in the wet season, its compatibility with *Stylosanthes humilis* and *S. hamata*, its good seed production and ease of establishment.

TABLE 15.16 **Brachiaria ruziziensis**

Stocking rate (steers/ha)	1.5	1.9	1.9	2.3	2.3	2.7
kg P_2O_5/ha	120	120	240	200	240	240
Days on experiment	238	238	238	224	224	224
Daily gain per animal (g)	582	492	481	514	519	343
Daily gain per ha (g)	873	934	913	1174	1188	929
Total wt gain during expt (kg/ha)	209	222	217	264	266	208

Main deficiencies. Its winter growth is slow. It needs well-drained fertile soils.
Further reading. Davidson, 1966; Dienum & Dirven, 1972; Mellor, Hibberd & Grof, 1973.

Brachiaria subquadripara (Trin.) Hitchc.

Synonyms. B. *miliiformis* (Presl.) Chase; *Panicum distachyum* (L.).
Common names. Cori grass (Sri Lanka), green summer grass, two-spiked panic, two-finger grass (Australia), Thurston grass (Fiji).
Natural habitat. Under coconuts, on disturbed soil.
Distribution. Southern Asia, Malaysia and Australia.
Description. A vigorously growing semi-erect annual grass. The spikelets are more than 3 mm long, hairless or almost so and the lower lemma is without a membranous palea (Tothill & Hacker, 1973) (see Fig. 15.20).
Rainfall requirements. It should be sown only in areas receiving more than 1 750 mm so that the grass will not compete unfavourably with coconuts for moisture.
Soil requirements. It will grow on a variety of soils, especially sandy ones.
Fertilizer requirements. It will respond to a complete fertilizer mixture, especially on sandy soils under coconuts.
Ability to spread naturally. It spreads well by stolons when trampled in by light grazing.
Land preparation for establishment. Under coconuts little land preparation is necessary; digging holes by hand to insert cuttings is satisfactory.
Sowing methods. Broadcast stem cuttings, to be harrowed in, or dig holes a metre apart, drop in a few cuttings and cover.
Sowing depth and cover. Up to 15 cm, leaving a portion of the cuttings above ground level; cover.
Sowing time and rate. The grass should be sown only in coconut stands more than 30 years old. The best time to plant in Sri Lanka is at the beginning of the southwest and northeast monsoon rains, as the soil must be wet at planting.
Vigour of growth and growth rhythm. It grows vigorously once established and adequately fertilized.
Response to light. It stands a good deal of shade, but coconut trees less than 30 years old may shade out the grass on the ground (Coconut Research Institute, 1966).
Tolerance to herbicides. This mat-forming grass competes with young sugar cane for light and moisture in Queensland. To control, spray with a long-term pre-emergent spray of Trifluralin at 2.8 l/ha of a 400 g AI/l product (e.g. Treflan EC). It must be well incorporated into the soil immediately following application. It is a pre-emergent treatment for grass only; it will not control broad-leaved weeds. For a knock-down treatment in canefields spray with Diuron at 6 g of an 800 g AI/kg product (e.g. Karmex, Diurex) per litre of water plus 6 ml surfactant. Apply with a mister to the point where the spray runs off the leaves. For general control in non-cane situations use Diuron at 5 kg of an 800 g AI/kg product (e.g. Diurex, Karmex) plus paraquat at 285 ml

Figure 15.20. Brachiaria subquadripara. **A**-Plant **B**-Spikelet

of a 200 g AI/l product per 200 l water. Use a minimum of 340 l water per ha (Tilley, 1977).

Grazing management. Early light grazing a month after planting will encourage more rooting of the stolons. Eat down to 15 cm rotationally and fertilize with nitrogenous fertilizer every three to six months.

Genetics and reproduction. 2n=54-56 and 72 (Fedorov, 1974).

Suitability for hay and silage. It makes quite good hay where sufficiently abundant.

Palatability. It is highly palatable.

Seed production and harvesting. It sets only a small amount of seed and is propagated by cuttings.

Main attributes. Its ability to grow in partial shade and its palatability.

Main deficiencies. Lack of seed production and low bulk.

Further reading. Coconut Research Institute, 1966.

Cenchrus biflorus Roxb.

Synonym. C. barbatus Schumach.
Common names. Haskanit, abu sha'ar (the Sudan), wezzeg (Nigerian Sahel), cram-cram (Sahel), initi (Mauritania), Indian sandbur (India).
Natural habitat. In bush, and as a weed on disturbed land.
Distribution. Throughout tropical Africa and India, usually on sandy soils.
Description. A tufted annual up to 90 cm high with flat, rigid leaf blades. False spikes 3-15 cm long, dense; spikelets 3.5-5 mm long, surrounded by a rigid involucre 4-7 mm long, the outer bristles short and spiny, retrorsely scabrid becoming spreading, the inner flattened with ciliate margins (Napper, 1965) (see Fig. 15.21).
Season of growth. Summer.
Optimum temperature for growth. Adapted to hot, dry tropical areas with a short growing season.
Frost tolerance. It generally occurs in areas free from frost.
Altitude range. Sea-level to 500 m.
Rainfall requirements. It occurs in the 400-500 mm rainfall area in Kordofan Province, the Sudan, with an eight-month dry season.
Drought tolerance. Being an annual it escapes drought by early seeding.
Tolerance to flooding. It does not have any flood tolerance.
Soil requirements. It is usually found on sandy soils in arid and semi-arid regions such as Mauritania (Boudet & Duverger, 1961) and the Sudan (Skerman, 1966) and is often secondary after cultivation.
Sowing methods. Seeds are usually distributed by grazing animals, to which the mature burs adhere.
Vigour of growth and growth rhythm. It grows vigorously during the wet season and flowers in August (Mauritania) and is standing hay from February to June (Boudet & Duverger, 1961).
Compatibility with other grasses and legumes. It is associated with the leguminous gum arabic tree (*Acacia senegal*) in the Sahel of northern Africa (see Plate 27).
Grazing management. It is rarely managed. Grazing before seeding is preferable because of higher nutritive value.
Genetics and reproduction. 2n=30, 34 (Fedorov, 1974).
Dry- and green-matter yields. An above-ground biomass of 590 kg/ha from this and associated grasses has been reported from Senegal (Morel & Bourlière, 1962) and of 600 kg/ha at Kanem, Chad in 1961 (Gillet, 1967).
Suitability for hay and silage. It makes good hay if cut before the burs harden.
Value for standover or deferred feed. It is one of the best available standover feeds in the Sahel after its seed bur has shed. Being an annual, it does not last long.

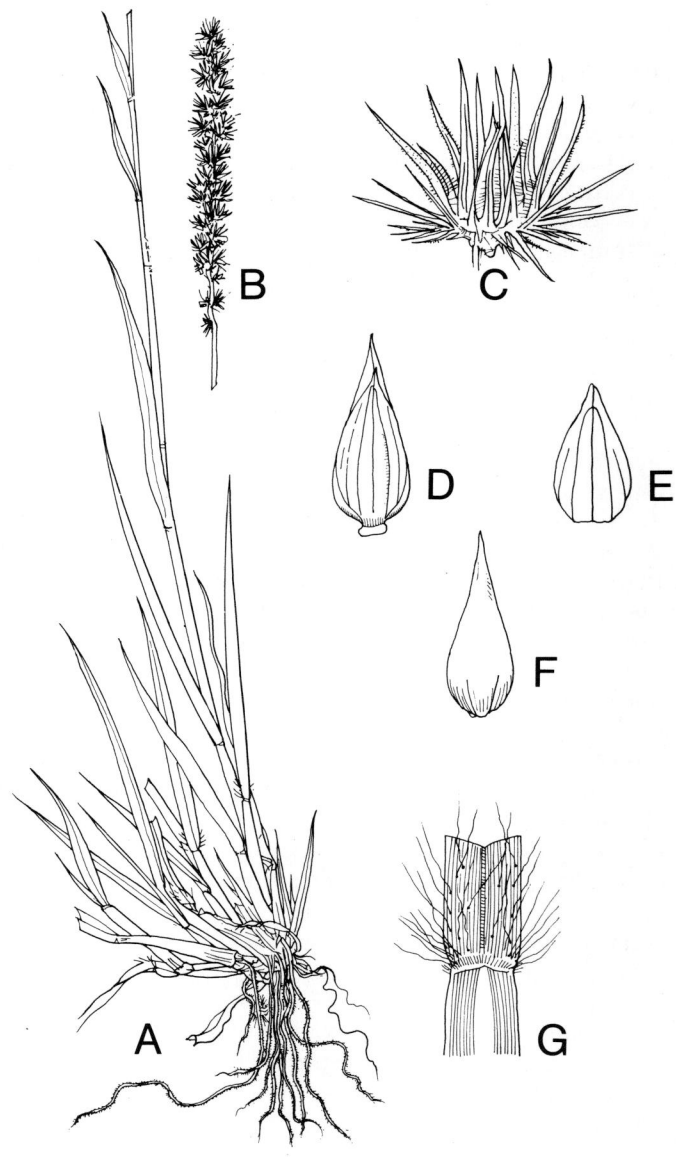

Figure 15.21. *Cenchrus biflorus.* **A**-Base of plant **B**-Inflorescence **C**-Involucre **D**-Spikelet **E**-Upper glume **F**-Upper fertile floret **G**-Ligule

TABLE 15.17 *Cenchrus biflorus*

	DM	As % of dry matter					References
		CP	CF	Ash	EE	NFE	
Fresh, early bloom, Tanzania	20.0	10.3	37.8	6.7	2.0	43.2	French, 1941
Fresh, late bloom, Tanzania		11.6	33.9	7.7	2.8	44.0	French, 1941
Fresh, mid-bloom, Niger		10.0	34.6	11.1	1.5	42.8	Bartha, 1970
Hay, early bloom, India	94.9	9.0	32.9	11.6	1.0	45.5	Sen, 1938
Growing, Aug., the Sahel	27.0	8.6	34.1	13.5			Boudet, 1975
Flowering, September (under grazing), the Sahel	23.0	16.0	30.3	11.8			Boudet, 1975
Dry stems, Oct. to Feb., the Sahel	94.0	3.1	38.8	9.0			Boudet, 1975
Dry stems, Mar. to June, the Sahel	94.0	2.6	39.1	11.1			Boudet, 1975

	Animal	Digestibility (%)				ME	References
		CP	CF	EE	NFE		
Fresh, early bloom, Tanzania	Sheep	71.0	75.0	64.0	69.0	2.51	French, 1941
Fresh, late bloom, India	Sheep	70.0	72.0	78.0	68.0	2.48	French, 1941
Hay, early bloom, India	Zebu	58.4	58.0	36.9	50.2	1.77	Sen, 1938

Chemical analysis and digestibility. Göhl (1975) gives figures in Table 15.17.
Palatability. It is readily eaten in the preflowering stage and after the seed bur has shed.
Economics. It is abundant in the African Sahel in low-rainfall woodland savannah on sand and as "ephemeral prairie", providing rich and valuable wet season grazing from June to August and standing hay later. It is being used in de-desertification on the Thar Desert of Rajasthan (Aggarwal & Lahini, 1977). In times of scarcity the seeds are eaten by humans (Bor, 1960).
Animal performance. No records of animal performance have been located.
Further reading. Boudet & Duverger, 1961; Whyte, 1968.

Cenchrus ciliaris L.

Synonyms. Pennisetum cenchroides Rich.; *P. ciliare* (L.) Link.
Common names. Buffel grass (Australia), African foxtail (United States, Kenya), dhaman grass, anjan grass, koluk katai (India).
Natural habitat. Open bush and grassland.
Distribution. Hotter and drier parts of India, Mediterranean region, tropical and southern Africa, now widely introduced.
Description. The genus *Cenchrus* belongs to the tribe Paniceae, in which the two-flowered spikelets fall when ripe, leaving no glumes. The spikelets are solitary and the pedicels never swollen. *Cenchrus* resembles the genus *Pennisetum,* except that the bristles are wavy and the inner ones flattened at the base. *Cenchrus ciliaris* has slightly hairy inner bristles, connate at the base only, fine and only slightly flattened at the base. It is a tufted or spreading perennial 12-120 cm tall (Harker & Napper, 1960). It is deep rooting (see Figs. 15.22, 15.23).
Season of growth. Summer.
Optimum temperature for growth. Ludlow and Wilson (1970b) found growth at 30°C was 12.5 times that at 20° for cv. Biloela.
Minimum temperature for growth. 5-16°C (Russell & Webb, 1976).
Frost tolerance. It is affected by frosts, but killed only in areas of prolonged frosting. The tall-growing varieties are less affected.
Altitude range. Sea-level to 2 000 m.
Rainfall requirements. In Queensland it grows in the 375-750 mm rainfall regime with 60 percent or more of the annual rainfall falling in summer. It does not do well in high-rainfall areas. For good establishment, buffel seed needs to be moist for about five days and a minimum of 30 mm of rainfall is needed for a *C. ciliaris* pasture to respond to nitrogen fertilizer. For high growth rates soil moisture must be adequate for 30 consecutive days (Henzell et al., 1975).
Drought tolerance. The buffel grasses are very drought resistant.
Tolerance to flooding. Anderson (1974) showed that buffel grasses are killed after six days' flooding in the field. The tall, rhizomatous cultivars ('Tarewinnabar', 'Nunbank', 'Boorara', 'Biloela' and 'Molopo') are more tolerant than the short, non-rhizomatous 'Gayndah' and 'American'. Rhizome development is not the reason, as 'Molopo', with the best rhizome development, was the least flood tolerant of the tall cultivars. At short flood durations, plants were more severely affected if completely covered by water than if some parts of the leaves were exposed. Where flooding in the field occurs irregularly and does not last more than six days, buffel grasses can be used in pasture mixtures. FAO No. 8693, collected at 680 m, 8 km south of game post, Ewaso, Ngiro, came from seasonally flooded ground.
Soil requirements. Lighter textured soils of high phosphorus status are best,

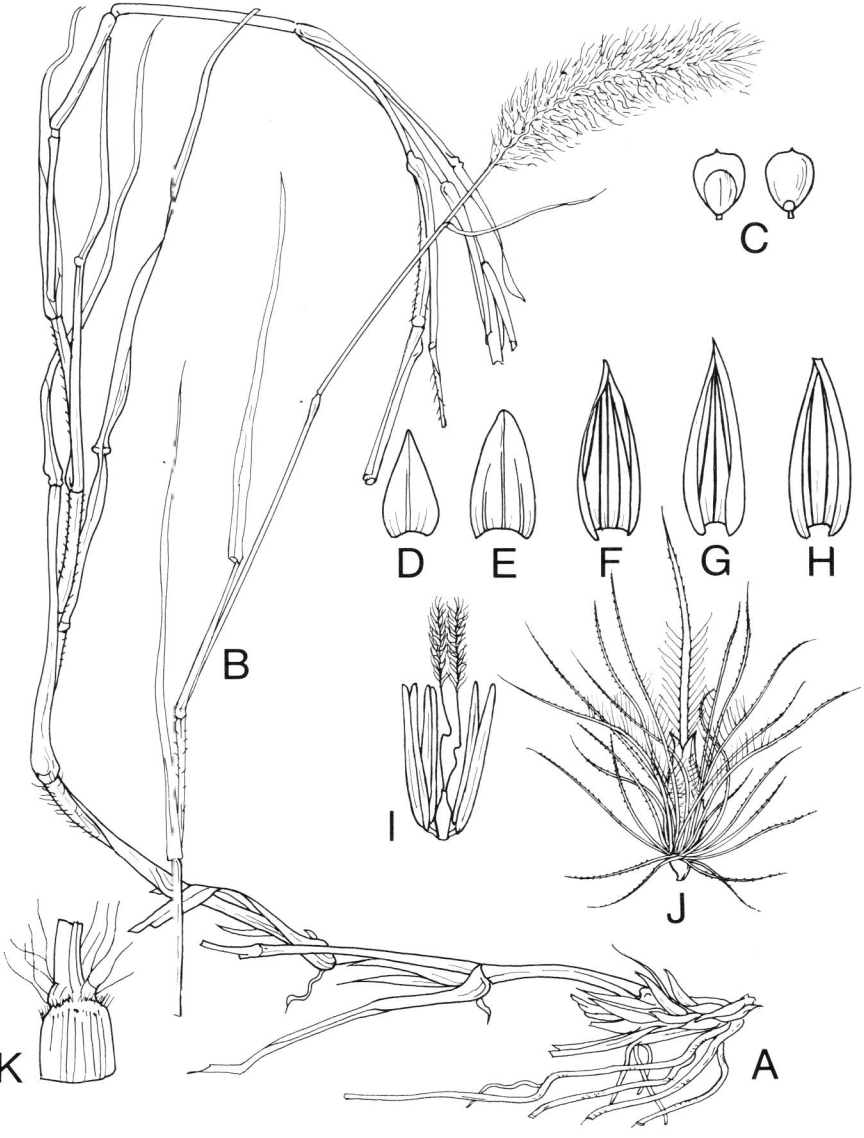

Figure 15.22. Cenchrus ciliaris. **A**-Habit **B**-Inflorescence **C**-Grain **D**-Lower glume **E**-Upper glume **F**-Lower lemma **G**-Upper lemma **H**-Upper palea **I**-Flower **J**-Spikelet with involucre **K**-Ligule

Tropical Grasses

but it thrives on the clay loams and clays of the cleared brigalow (*Acacia harpophylla*) and gidgea (*A. cambagei*) scrubs of inland Queensland. It is slow to establish on the black cracking clays, but when established it does well, for example in central Tanzania. The optimum soil reaction is pH 7 to 8, but it grows on soil with a pH as low as 5.5. Ramia and Fernandez (1974) and Spain and Andrew (1977) showed buffel grass to be very sensitive to soils containing high levels of aluminium.

Tolerance to salinity. The buffel grasses are less tolerant to salt than Rhodes grass, Bermuda grass and blue panic. Cv. Biloela is more tolerant than other

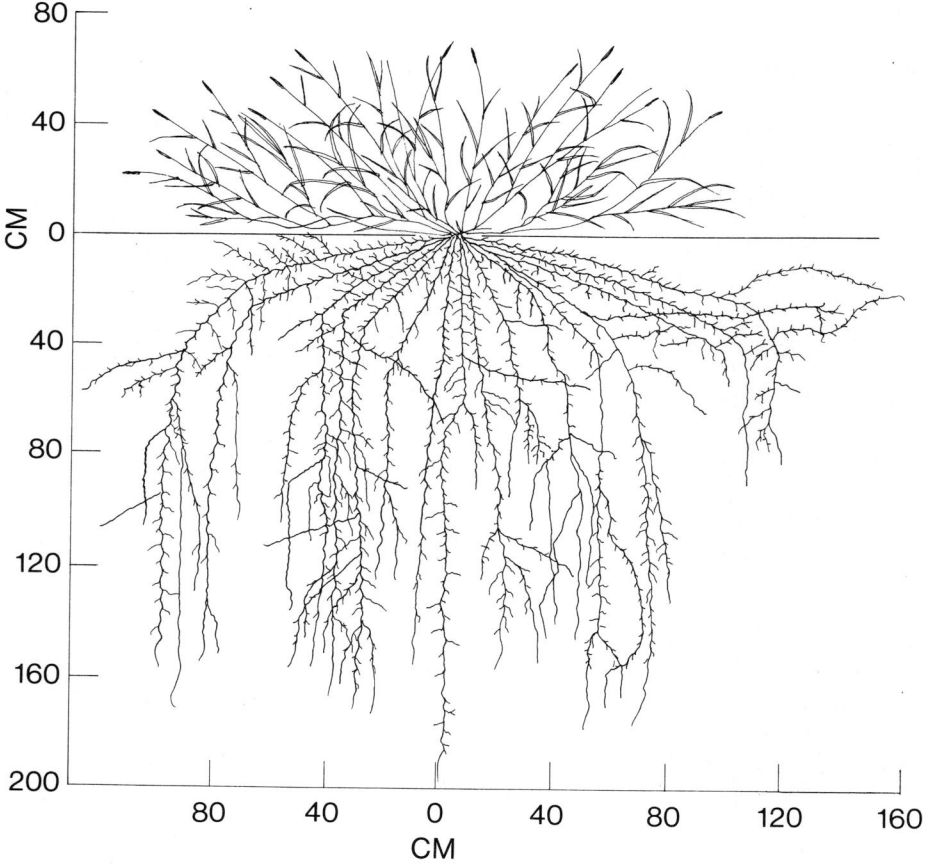

Figure 15.23. Vertical section of the root system of *Cenchrus ciliaris* cv. Biloela, seven months after planting

cultivars, maintaining relatively high yields up to 80 meq sodium chloride per litre (Graham & Humphreys, 1970).

Fertilizer requirements. Buffel grass is not often fertilized in a semi-arid environment. Henzell (1976-77) showed that application of 167 kg N/ha increased the annual yield of buffel grass from 9 780-15 250 kg DM/ha at Narayen, Queensland, in 1975-76 (average rainfall 710 mm). In Ghana, Asare (1970) obtained highest yield with a complete NPK mixture. The critical value for phosphorus is 0.26 percent of the dry matter.

Ability to spread naturally. C. ciliaris spreads well by seed where the soil pH ranged from pH 7-8 in Tanzania (Brzostowski, 1962).

Land preparation for establishment. Most of the buffel grass in Queensland has been aerially sown into the ashes left after a scrub burn. However, a rough seed-bed can be prepared by ploughing or disc-harrowing, and hollows left by uprooting tree stumps and by the tracks of a bulldozer when clearing scrub are effective seed-beds. Soil disturbance is generally essential for initial establishment.

Sowing methods. Aerial sowing is common and under ideal conditions an aircraft can sow up to 1 600 ha per day at rates as low as 200 g/ha, in strips. Costs vary according to distance from the landing airfield. Because the seed is light, wind direction is important. Seed can be sown through special buffel seed drills fitted with agitators in the seedbox to separate the seed, or with an augur device to perform the same function (Paull & Lee, 1978). It can be hand-broadcast into fallen timber, under big trees such as *Eucalyptus populnea* where fertility is higher (Ebersohn & Lucas, 1965; Christie, 1975b) due to mineralization of fallen vegetation and livestock camping (see Plate 28) and into stump holes. Further methods of planting can be found in a paper by Paull (1973). It will not establish under *E. crebra* (Humphreys, 1978).

Sowing depth and cover. It is surface sown and lightly harrowed or rolled where possible.

Sowing time and rate. It is best sown just ahead of the expected rainy season at 0.5-4 kg/ha, according to seed supplies, costs, and expected rate of full ground cover.

Number of seeds per kg. 450 000 to 703 000.

Dormancy. Removal of glumes improves germination, but is risky in arid areas (Pandeya & Pathak, 1978). Anthocyanins and phenolics in the seed coat inhibit germination (Pandeya & Pathak, 1978). Freshly harvested seed has poor germination and after-ripening for three to 12 months is desirable. Seed older than two years should be tested for germination. Scarification will also break dormancy (Humphreys, 1978).

Seed treatment before planting. Where seed-harvesting ants are prevalent, dusting with 20 percent lindane dust at 1 kg per 80 kg of seed is recommended to reduce seed removal. Pelleting with lime to increase seed weight and reduce soil acidity has been successful in lateritic red earths. Germinate for ten days at 40°C. Seed treatment by hammer milling for semi-arid planting is

not recommended, as it is more attractive to ants and a good initial germination may end in the total loss of seedlings (Paull & Lee, 1978).
Seedling vigour. Buffel grass seedlings usually have good vigour.
Vigour of growth and growth rhythm. New, fully-expanded leaves developed at the rate of every nine to ten days in summer, 11 days in spring and 14 to 12 days in autumn, with senescence highest in the summer growing season and slowest during the cooler dry autumn (Wilson & 't Mannetje, 1978).
Compatibility with other grasses and legumes. Buffel grasses are frequently sown with Columbus grass (*Sorghum almum*) at 4 kg/ha to give a quick, temporary pasture to recoup establishment costs. The *S. almum* lasts for one or two years, and then the buffel grass and its other associated grasses can take over. Buffel grass at 1.5-2 kg/ha can also be sown with Rhodes grass at 0.5-1 kg/ha, Bambatsi panic, green panic or *Urochloa mosambicensis*. Eventually the buffel grass is likely to become a pure stand. Few legumes will persist with it for longer than 12 months unless well managed. In subtropical areas, *Medicago sativa, Macroptilium atropurpureum* and some annual *Medicago* spp. may last a little longer.
Ability to compete with weeds. In the semi-arid areas it usually dominates weeds (see Plate 29).
Tolerance to herbicides. Buffel grass often invades urban areas in a semi-arid environment and becomes a weed. Control is possible with 2,2 DPA or with paraquat (Gramoxone) applied before flowering when the soil is moist. Diuron can be used as a pre-emergence spray to destroy germinating buffel. In establishing buffel grass pasture, any broad-leaved weeds can be controlled, if necessary, with 2,4-D at 1 kg active constituent per hectare, if the buffel seedlings have developed at least four leaves (Paull & Lee, 1978). Atrazine at 4 kg/ha used as a post-emergence spray affected survival of *C. ciliaris* but at 1 kg/ha it was less damaging (Scateni, 1978).
Response to defoliation. Buffel grass will stand considerable grazing once it is established. Newly established pastures can be used during the winter and spring following planting and, if necessary, locked up during the following summer to seed and increase plant density. Frequent grazing improves nitrogen content. Cuttings at 10 cm at intervals of 20 days in a wet year and 30 days in a dry year gave highest yields in Rajasthan (Dabadghao, Roy & Marwaha, 1973).
Grazing management. Where the area of buffel grass is small, it is best to graze the native pastures during the summer growing period and keep the buffel grass, especially the cold-tolerant varieties, for use during the winter. On hard setting soils some renovation with a tined implement will promote plant growth if done after rains in spring and summer. Do not overdo the renovation. An occasional fire to retard the development of woody plants and destroy old grass is generally necessary. Graze according to herd needs (Humphreys, 1967). Nitrogen fertilizers will be required to maintain productivity or increase it.

Response to fire. Buffel grass will stand the burning of the old vegetative grass. The crowns will not be adversely affected and nutritive value may improve.

Genetics and reproduction. The chromosome numbers are 2n = 32, 36, 40, 54. It is apomictic in general but Dr E.C. Bradshaw in 1961 in Texas discovered sexual reproduction in his plots. The method of reproduction is controlled by two genes and epistasis (Bashaw, Hovin & Holt, 1970). 2n = 32, 34, 36, 40, 52, 54 (Fedorov, 1974).

Dry- and green-matter yields. At Narayan, Queensland, Henzell (1976-77) recorded an increased annual dry matter production by Biloela buffel grass from 4.80 t/ha to 9.08 t/ha with the application of 168 kg N/ha each year over a six-year period. In the forest region of Ghana, Asare (1970) cut 10-cm buffel grass at six-week intervals. With a complete fertilizer (135 kg N/ha, 73 kg P/ha and 11.7 kg K/ha), the yield was 24 200 kg DM/ha; unfertilized it was 18 800 kg DM/ha. In India, a yield of 2 010 kg DM/ha was obtained by cutting at 60-day intervals (Shankarnarayan *et al.*, 1977).

Suitability for hay and silage. Buffel grass makes reasonable-quality hay when cut in the early flowering stage, yielding up to 2 500 kg/ha per cut with a protein content of 6-10 percent of dry matter. Old grass, after the seed has been harvested, can give low-quality roughage for drought feeding with supplements. This old grass will have a protein content of 4-6 percent. Little buffel grass has been made into silage, as the moisture content in the semi-arid areas is usually low.

Value as standover or deferred feed. It is a valuable standover feed for winter grazing and for roughage along with supplements such as urea-molasses.

Chemical analysis and digestibility. See Table 15.18. For Biloela buffel grass unfertilized, buffel grass + 84 kg N/ha per year, buffel grass + 168 kg/N per year, and buffel grass + siratro, t' Mannetje's (1977) figures for green matter are given in Table 15.19.

Palatability. It is very palatable when young, and remains fairly palatable at maturity.

Toxicity. Buffel grass has caused bighead disease in horses due to a high oxalate content. The symptoms are ill thrift, lameness, and swelling of the skull. Disease is most prevalent during the wet season when the buffel is young and lush. To control it, vary the diet away from pure buffel grass, or feed a supplement of ground limestone or dolomite (Walthall, 1977). Playne (1976) found oxalate levels in buffel grass ranged from 1.2 to 2.8 percent total oxalate and 1.2 to 2.2 percent water-soluble oxalate.

Seed production and harvesting. Seed plots can be grazed to late spring or early summer when good rains are received. They are then shut up and, if necessary, slashed to obtain an even height when seed is ready to harvest. The seed remains mature on the heads for 14-20 days and can be harvested up to three times a season if rainfall is adequate. Cv. Molopo will retain ripe seed for only six to eight days. Seed can be hand-harvested and such seed should

TABLE 15.18 *Cenchrus ciliaris*

	DM	As % of dry matter				
		CP	CF	Ash	EE	NFE
Fresh, early vegetative, Pakistan	41.4	9.8	38.4	9.8	5.4	36.6
Fresh, mature, Pakistan	21.9	7.3	41.9	8.8	4.8	37.2
Flowering, India		6.2	34.4	13.3	1.5	44.6
Fresh, early bloom, Tanzania	20.0	11.0	31.9	13.2	2.6	41.3
Hay, 1st cut, Tanzania	87.0	7.4	35.2	11.7	1.7	44.0

	Animal	Digestibility (%)				
		CP	CF	EE	NFE	ME
Fresh, early bloom, Tanzania	Sheep	76.2	76.2	85.0	72.9	2.50
Hay, 1st cut, Tanzania	Sheep	54.0	71.6	47.0	67.5	2.22

Source: Göhl, 1975

not need cleaning. Most is direct-harvested by machinery either home-built (see Plate 30) or purchased. Basically, the ripe seed is beaten off on to trays which collect the seed. Refinements include removal of most of the trash and bagging the seed.

Seed yield. 10-60 kg/ha of clean seed per harvest. Seed remains viable for two to three years (Jones, 1973).

Minimum germination and quality required for commercial sale. 20 percent germinable seeds, 90 percent purity (Queensland). Pre-dry fresh seed and germinate at 20-35°C, moistened with water (Prodonoff, 1966).

Cultivars.
* 'Biloela' — introduced from Dodoma in Tanzania in 1937 by CSIRO,

TABLE 15.19 *Cenchrus ciliaris*, green matter

	In vitro digestibility %	%N	%P
Buffel alone	63.4	1.60	0.49
Buffel + 84 kg/N	65.0	2.24	0.33
Buffel + 168 kg/N	65.0	2.25	0.33
Buffel + siratro	66.5	2.07	0.41

Source: t'Mannetje, 1977

Australia and tested at Rockhampton, Queensland. It is a tall-growing type named after the Biloela Research Station, Queensland, where it was further developed. It grows in a wide range of soils and will adapt to poorer soils. 703 000 seeds per kg.
- 'Boorara' — introduced from Kenya and extensively propagated and proved by W.H. Rich of "Boorara" cattle ranch in central Queensland. It will grow on poorer soils than the other cultivars.
- 'Nunbank' — introduced to Australia from Uganda and tested by CSIRO, Australia chiefly on the property of B.C. Clark of "Nunbank", Taroom, Queensland. It will grow on poorer soils as well as on fertile ones.
- 'Manzimnyarna' and 'Sebungwe' — dwarf strains suitable for semi-arid conditions in Africa.
- 'Tarewinnabar' — seed came from Kenya and was tested by the Queensland Department of Agriculture and Stock and developed principally on the property of Sir William Gunn at Tarewinnabar, Goondiwindi. It has greater frost tolerance than most of the other cultivars except 'Molopo'. Seed is scarce and seed production low.
- 'Molopo' — seed came from near the Molopo River, western Transvaal, and was developed by the New South Wales and Queensland Departments of Agriculture. It has good frost tolerance along with 'Tarewinnabar'. It gives poor seed production and prices of seed are high. 535 000 seeds per kg.
- 'Lawes' — seed came from the Department of Agriculture, Pretoria, South Africa. Commercial seed is not available. It is identical with the American cultivar T3782, blue buffel, and is very similar to 'Molopo'.
- 'Zeerust' — a tall, leafy cultivar adapted to the 500-625 mm rainfall area in South Africa.
- 'Gayndah' — seed came originally to Australia from Kenya, was grown at the Gayndah state school and developed mainly by a local grazier, C.J. Pinwell. This shorter variety is very suitable for sheep grazing. 479 000 seeds per kg.
- 'Chipinga' — a fine, leafy variety from Zimbabwe.
- 'American' — introduced to Australia from America and is identical with the American material T.4464. It is a short variety, very suitable for sheep. Its palatability may lead to its overgrazing and disappearance. 454 000 seeds per kg.
- 'Higgins' — a true-breeding apomictic variety developed from the sexual plant found in Texas by a rancher, Pat Higgins, and used by Dr E.C. Bashaw (Burton, 1970).
- 'West Australian' — believed to have arrived in Australia as seed and hay in an Afghan camel harness between 1870 and 1880. A very short variety and one of the first cultivars to be used in western Queensland on cleared gidgea (*Acacia cambagei*) country. It is the least drought-tolerant of them all. Milford (1960a, b) found this cultivar the most nutritionally valuable of all the cultivars.

- 'Kongwa 531' — has given the best results on the red earth soils at Kongwa, Tanzania under an annual rainfall of 561 mm. It is a fine-leaved, erect type of medium height which produces ample seed and has excellent drought resistance (Wigg, 1973).
- 'Edwards Tall' — (Kitale introduction no. K.5148, Kenya) does not seed well but has a robust habit which gives it added resistance to overgrazing. There are also several hybrids in existence which are being tested in the United States and Australia (Paull & Lee, 1978).

Value for erosion control. Buffel grass is valuable in that it is one of the best adapted grasses to semi-arid conditions. Its tussocky nature, however, does not allow for complete ground cover (Robinson, 1978).

Diseases. In Kenya the seed is often destroyed by attacks of smut in wetter areas (Bogdan & Pratt, 1967). Blast (*Piricularia* sp.) and rust (*Uredo* sp.) occur in Thailand (Vinijsanond, 1978).

Pests. It has few pests.

Economics. It is one of the best forage grasses for semi-arid areas in the subtropics and tropics.

Animal production. At Narayan Research Station, Queensland, Biloela buffel grass fertilized with 84 and 168 kg N/ha per year gave an annual live-weight gain of 160 kg/ha at a stocking rate of one beast per hectare, where unfertilized buffel grass produced a weight loss between the months of May and November. The unfertilized native pasture gave an annual live-weight gain of 30 kg/ha ('t Mannetje, 1977). Siratro with buffel grass at an equivalent stocking rate yielded 80 kg more live-weight gain than did buffel grass on its own, and buffel grass with siratro and buffel grass with 168 kg N/ha grazed at the same stocking rate gave similar gains (162 kg/ha at 1.1 steers per hectare) ('t Mannetje, 1972).

Main attributes. Its hardiness, deep-rooting, ability to grow in semi-arid conditions, and generally free-seeding habit. Its persistence and resistance to trampling (e.g. around stock-watering points), and drought tolerance (see Plate 31).

Main deficiencies. Its difficulty in establishment on heavy soils and its ultimate dominance under conditions which suit its persistence. The difficulty of removing it for cultivation and its depressing effect on a following crop (Humphreys, 1967).

Further reading. Humphreys, 1967; Paull, 1973; Paull & Lee, 1978.

Cenchrus pennisetiformis Hochst. and Steud. ex Steud.

Common names. Cloncurry, white, or slender buffel grass (Australia).
Distribution. Mediterranean region to the hotter and drier parts of India, Burma, Sri Lanka and northeast Africa.
Description. In *Cenchrus pennisetiformis* the inner bristles are glabrous, connate for 1-3 mm from the base; flat and rigid throughout. It is a tufted perennial up to 60 cm high (Harker & Napper, 1960). It has paler involucres (Gardner, 1952) and wider spacing of the spikelets on the rachis (Humphreys, 1978) than *C. ciliaris* (see Fig. 15.24).
Season of growth. Summer.
Optimum temperature for growth. It grows in hotter areas than *C. ciliaris* (average temperature range: 10-30°C) in Queensland, Australia and Kenya.
Frost tolerance. Its natural habitat is relatively frost free.
Latitudinal limits. Probably 10°N and 20°S latitudes (Kenya and northern Queensland).
Altitude range. Sea-level to 300 m in Queensland.
Rainfall requirements. It is adapted to arid conditions with an annual rainfall of 250 mm or less in Kenya (Bogdan & Pratt, 1967). In northwest Queensland it has spread mostly into the 370-560 mm annual rainfall regime with a summer maximum (Hall, 1978).
Drought tolerance. Excellent. It remains green during the dry season in India and cattle eat it avidly (Bor, 1960).
Tolerance to flooding. It will survive seasonal flooding.
Soil requirements. It established along fertile river alluvium in northwest Queensland but has since spread across frontage woodlands to stony, undulating country. It does not spread on to heavy cracking clays but does prefer high phosphorus and calcium soils of alkaline reaction. In Kenya it grows on sandy soils, loams and alluvial silts (Bogdan & Pratt, 1967).
Tolerance to salinity. It grows on slightly alkaline soils in northwest Queensland (Hall, 1978).
Fertilizer requirements. It appears to spread more rapidly on the alluvia of streams where the phosphorus status is higher in northwest Queensland. It does not appear to be sensitive to potassium. It also grows around the bases of trees of *Eucalyptus* spp. where fertility is higher.
Ability to spread naturally. In Queensland it spreads naturally by seed along the banks of watercourses where soil phosphorus levels are higher than surrounding land and the soil surface of a lighter texture, mostly in above-average seasons. It is now gradually spreading into poorer soils.
Land preparation for establishment. Minimum land preparation should be some soil disturbance with a disc cultivator or a rigid tine implement in strips across the range.

Figure 15.24. Cenchrus pennisetiformis

Sowing methods. It has normally been broadcast on top of a single light cultivation and around large trees, edges of roads and cattle tracks.
Sowing depth and cover. It is surface-sown and covered lightly with harrows or bushes dragged over the area, or left uncovered.
Sowing time and rate. It is best planted just ahead of the expected wet season at 0.75-3 kg/ha according to seed supply.
Number of seeds per kg. About 400 000 (Bogdan & Pratt, 1967) of pure seed.
Dormancy. It would appear to have some post-harvest dormancy.
Seed treatment before planting. It may be necessary to treat the seed with a deterrent against seed-harvesting ants, such as lindane dust.
Seedling vigour. Very good.
Vigour of growth and growth rhythm. It begins growth ahead of native perennials such as *Chrysopogon fallax, Dichanthium* and *Bothriochloa* spp.
Response to light. It will grow in partial shade along river banks and under larger trees, as well as in open country.
Compatibility with other grasses and legumes. It usually grows as a monospecific sward when established.
Ability to compete with weeds. Good. It gradually occupies the whole area and weed competition is not great in the arid and semi-arid areas.
Response to defoliation. It will stand heavy grazing.
Grazing management. On the large ranches on which it is established in northwest Queensland it is not usually managed. It is most useful when stocked early in the growing season, allowed to seed, and then stocked again to help spread the seed by trampling and adherence to animals' coats.
Response to fire. It recovers well from fire.
Genetics and reproduction. 2n=35, 42, 54 (Fedorov, 1974).
Value as standover or deferred feed. It is excellent as standover feed in the areas where it grows.
Chemical analysis and digestibility. No analyses have been found.
Palatability. The stems are soft and the whole herbage is well grazed.
Toxicity. No toxicity has been reported.
Seed production and harvesting. It seeds heavily and has mostly been hand picked. A buffel seed harvester should perform the operation in areas where it can manoeuvre. The "seed" is a cluster of spikelets surrounded by hairy bristles.
Value for erosion control. It is valuable for stream bank protection of the rivers and creeks where it has become established in northwest Queensland.
Diseases. There are no major diseases of this grass.
Pests. There are no major pests.
Economics. This grass is currently the only introduced pasture grass with the ability to increase carrying capacity and stabilize beef cattle numbers in the Mt Isa highlands region of northwest Queensland (Hall, 1978).
Animal production. There are some 860 000 ha of grazing country containing large, dense areas of Cloncurry buffel grass along the main river and creek

frontages surrounding Cloncurry, northwest Queensland (Hall, 1978).
Main attributes. Its ability to colonize the banks of streams in the dry tropics, its palatability compared with most other grasses, and its persistence.
Main deficiencies. It does not grow on heavy cracking clays and stony ridges colonized by *Aristida contorta* and *Enneapogon polyphyllus* in northwest Queensland (Hall, 1978).
Further reading. Hall, 1978.

Cenchrus setigerus Vahl

Synonym. C. setiger Vahl.
Common names. Birdwood grass (Australia), moda dhaman grass (India).
Natural habitat. Open dry bush and grassland, usually on alkaline soils.
Distribution. Northwest India and northeast tropical Africa.
Description. Cenchrus setigerus is a tufted perennial up to 60 cm tall with flat or folded leaf-blades. False spike dense, 1.5-9 cm long; spikelets 3-4.5 mm long, surrounded by a rigid involucre 3.5-5 mm long, the outer bristles minute or absent, the inner flattened, grooved on the back. Each cluster of spikelets contains one to three caryopses (see Fig. 15.25). The inner bristles, in contrast to *C. ciliaris,* are glabrous connate for 1-3 mm from the base, flat and rigid throughout (Harker & Napper, 1960).
Season of growth. Summer.
Optimum temperature for growth. Probably 30-35°C. It is extremely tolerant of heat and drought.
Minimum temperature for growth. It does not respond well to winter rain.
Frost tolerance. It survives frost.
Latitudinal limits. 30°N and S.
Altitude range. 500-800 m.
Rainfall requirements. It is adapted to arid and semi-arid climates with a long dry season and responds very quickly to light rains.
Drought tolerance. It is very tolerant of drought and will grow in areas of annual rainfall as low as 200 mm, making it excellent for improvement of low-rainfall grazing land. It is more tolerant than *C. ciliaris.*
Tolerance to flooding. FAO Nos 8688 and 8703 at altitude 670-680 m, 8 and 13 km south of game post Ewaso Ngiro, Kenya came from flooded ground.
Soil requirements. It prefers light-textured sandy soils but does well over a wider range of soils than *C. ciliaris* (Suijendorp, 1953).
Tolerance to salinity. It occurs frequently on alkaline soils in the Rift Valley, Kenya.
Fertilizer requirements. In areas where birdwood grass finds its most valuable niche it is uneconomical to fertilize, but it will respond to nitrogen and phosphorus.
Ability to spread naturally. Birdwood grass spreads only slowly where seed is incorporated in the soil by animal treading or cultivation.
Land preparation for establishment. The land needs some cultivation or the seed can be sown aerially or by hand in the ashes of a burn, for example after felling and burning *Acacia cambagei* scrub in brown clay soils of central Queensland, Australia.
Sowing methods. Sow aerially in ashes or through modified cereal drills. Hand broadcasting is suitable for small areas.

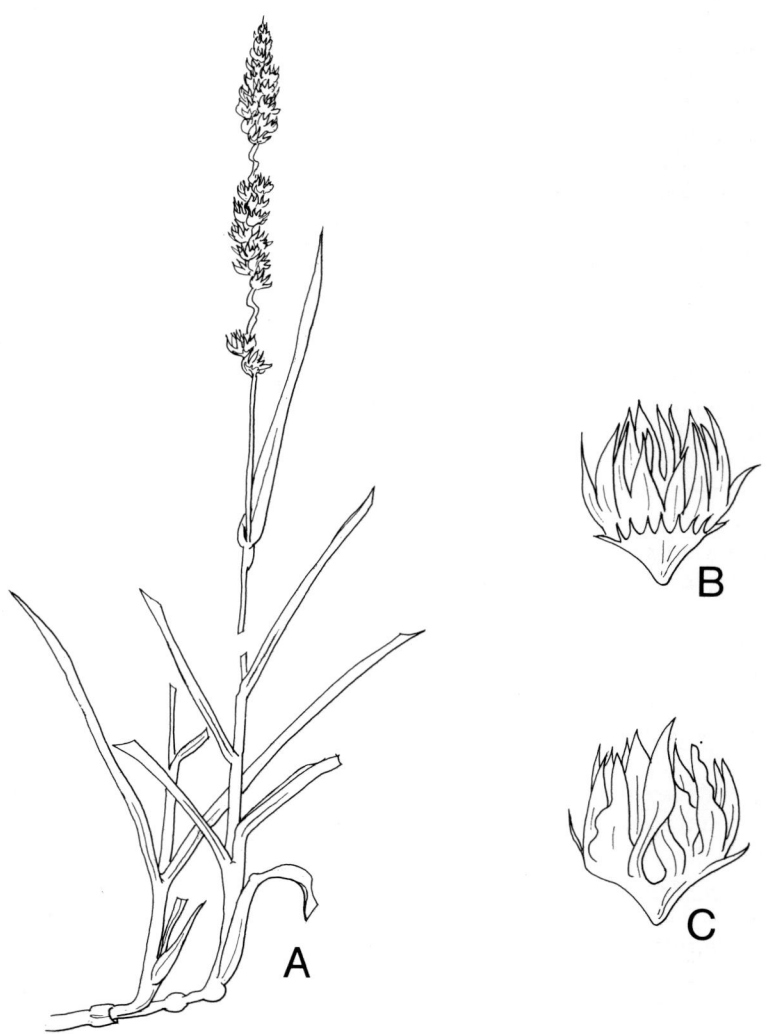

Figure 15.25. *Cenchrus setigerus.* **A**-Habit **B,C**-Spikelets

Sowing depth and cover. Seed is usually sown on the surface and covered with harrows or bushes dragged across the sown area. Seed should not be covered to a depth greater than 2 cm.

Sowing time and rate. Sow just before the usual rainy periods in summer at the rate of 1.5-3.0 kg/ha depending on seed supplies, cost and rapidity of cover desired.

Number of seeds per kg. 350 000.

Dormancy. Seed has some physiological dormancy, the germination rate increasing up to two years after harvest.

Seed treatment before planting. It can be treated with lindane dust to deter removal by seed-harvesting ants.

Seedling vigour. The young seedling is quite vigorous.

Vigour of growth and growth rhythm. It is not a vigorous grower and its tussocky nature does not provide a great deal of bulk.

Response to photoperiod. It flowers in shortening days.

Compatibility with other grasses and legumes. It will grow with *S. humilis*, but tends to dominate (Norman, 1962b).

Response to defoliation. Birdwood grass, once established, withstands grazing well. Cutting at 30-day intervals gave the highest yields in Rajasthan, and cutting height (5, 10 or 15 cm) made no difference.

Grazing management. Once established, birdwood grass can stand heavy grazing. To thicken the stand it should be allowed to seed every two to three years.

Response to fire. Fire will remove the dry top growth, but unless the fire is very severe the plants will recover well after rain.

Genetics and reproduction. Birdwood grass is apomictic but pollination is necessary to endosperm formation and seed set. The chromosome number is $2n=34, 36, 37$ (Fedorov, 1974). Haploids with *C. ciliaris* have been found in Kenya.

Dry- and green-matter yields. In India, cutting at ten-day intervals yielded 400 kg DM/ha and cutting at 60-day intervals (during 1970-72) yielded 2 120 kg DM/ha (Shankarnarayan, Dabadghao & Kumar, 1977). At Jodhpur, Rajasthan, average monthly yields per plant were 45 g, 3.2 g and 11.9 g in 1968-69 (rainfall 178.8 mm), 1969-70 (92.7 mm) and 1970-71 (504.8 mm) (Gupta, Saxena & Sharma, 1972). At Victoria River Downs, Northern Territory, Australia, it yielded 679 kg DM/ha per year over three years (Pearson, Hill & Allen, 1979).

Suitability for hay and silage. Birdwood grass, if in areas which can be mown, makes useful hay, though the yield is not high. No record is available of its use for silage.

Value as standover or deferred feed. Because of its hardy nature and ability to grow in low-rainfall areas, birdwood grass is a valuable standover feed in these areas. Care must be taken to prevent grass fires in these arid climates.

Chemical analysis and digestibility. Göhl (1975) notes one record from

Dougall and Bogdan in Kenya. Fresh material at the early-bloom stage contained 18.6 percent crude protein, 28.3 percent crude fibre, 11.9 percent ash, 1.9 percent ether extract and 39.3 percent nitrogen-free extract.

Palatability. Birdwood grass is quite palatable and readily accepted by stock. It is grazed in preference to *Cynodon plectostachyus* in Kenya.

Toxicity. No toxicity has been reported.

Seed harvesting methods. Birdwood grass matures in eight weeks and seeds heavily. The seed can be hand picked or harvested mechanically with cereal harvesters or special grass-seed harvesters. The seed is a cluster of spikelets surrounded by hard scales.

Minimum germination and quality required for commercial sale. Germination is poor. Germinate at 20-35°C moistened with water. 20 percent germination and 60 percent purity in Queensland.

Cultivars. Introductions into Australia exhibit two distinct seed clusters — one is light straw-coloured and the other brown or black (Barnard, 1969).

Value for erosion control. Its tussocky nature, lack of vigour, bulk and persistence makes it of minor value in erosion control, but it would help to set up some barrier against moving sand. It is useful in the Northern Territory of Australia (Robinson, 1978).

Diseases. There are no major diseases.

Pests. There are no major pests.

Economics. A valuable grass throughout the arid and semi-arid tropics.

Animal production. In the Rajasthan desert area, *C. setigerus* carries one sheep to 2.6 hectares in Jodhpur, and to 6.01 hectares in Pali. In the Pindan country in the West Kimberleys, Western Australia, characterized by sand plains and sand dunes, birdwood grass introduction with 50 kg double superphosphate per hectare every two years has lifted live-weight gain from 30-50 kg/ha to 72-116 kg/ha, and with mineral supplements live-weight gain reached 146 kg/ha in the wet season, January to May (Holm & Payne, personal communication, 1975).

Main attributes. Its drought resistance, hardiness and palatability, and its non-fluffy seed, making the seed easier to sow than that of *C. ciliaris*.

Main deficiencies. Its tussocky nature and lack of bulk.

Further reading. Suijendorp, 1953.

Chloris gayana **Kunth**

Common names. Rhodes grass (Australia, United States, Africa), pasto Rhodes (Peru).

Natural habitat. Open woodland and grassland on a wide range of soils.

Distribution. Native to Africa, introduced to the United States in 1902 and now widely grown in tropical countries.

Description. A glabrous, usually stoloniferous perennial up to 90 cm high, but very variable. Inflorescence up to 15 spikes, occasionally in two whorls, but usually one. Its roots descend to 4.7 m; 47 m of roots occur in the first 30 cm^3 of soil, but they are sparse beyond 2.4 m (Hosegood, 1963) (see Fig. 15.26).

Season of growth. Spring and summer.

Optimum temperature for growth. Ivory (1976) recorded 30/26°C to 40/29°C day/night temperatures, Russell and Webb (1976) suggested a range of 16.9 to 22.3°C, and Bogdan (1969) suggested 35°C. Ludlow (1970b) found growth at 30°C was 5.85 times greater than at 20°C for cv. Samford.

Minimum temperature for growth. Ivory (1976) determined the critical mean temperature for growth as 8°C. Russell and Webb (1976) gave the mean temperature for the coldest month for Rhodes grass at 2.6 to 12.4°C. In the USSR minimum temperatures of –10°C and in Texas, United States, of –9.4°C have been recorded, but there is usually no growth below 0°C. Ivory (1976) recorded –2.6 to –3.5°C.

Frost tolerance. It can survive sub-zero temperatures. In Uruguay it survived the intense and repeated frost of 1942 (Bogdan, 1969). Plants usually survive and grow the next season. On the Darling Downs, Queensland, Australia, it was one of the most frost-tolerant grass species during its first winter, with cv. Pioneer having 97 percent survival (Jones, 1969).

Latitudinal limits. A range of 18-33.4°N and S with a mean of 25.7° (Russell & Webb, 1976).

Altitude range. 600-2 000 m in the equatorial zone, lower to the north and south (Bogdan, 1969); sea-level to 500 m in Queensland.

Rainfall requirements. It grows best in the 600-750 mm rainfall area and Russell and Webb (1976) gave its range as 691-1 597 mm. It is widely used in irrigated pastures in Israel and the United States.

Drought tolerance. Good. Rhodes grass roots can extract water to a depth of 4.25 m.

Tolerance to flooding. It tolerates seasonal waterlogging, but is killed by root submergence over 30 cm (Colman & Wilson, 1960).

Soil requirements. It grows on a wide range of soils, but may have some establishment problems on acid soils. It prefers loose-textured loams of volcanic origin.

Tolerance to salinity. Excellent. A "salt-pan" strain comes from the salt-pan near Hammanskraal, Pretoria, where it flourishes on the bottom of the pan, in granite soil impregnated with brine (Chippendall, 1955). In Italy it grows on land previously under saline water, and it grows on saline soils in Mississippi, United States. It is tolerant to $CaCl_2$ and $NaSO_4$, less so to $NaNO_3$ and

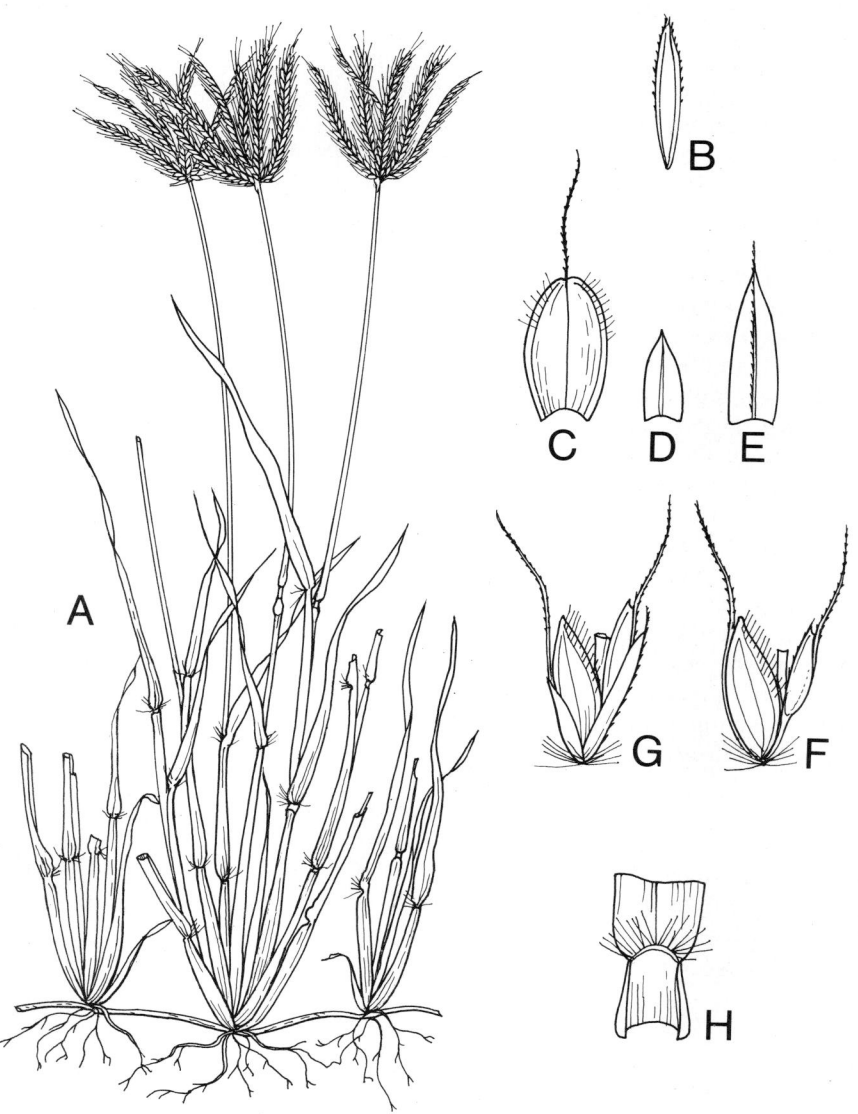

Figure 15.26. Chloris gayana. **A**-Habit **B**-Palea **C**-Lower lemma **D**-Lower glume **E**-Upper glume **F**-Spikelet **G**-Spikelet without glumes **H**-Ligule

NaCl, and sensitive to $MgCl_2$ (Bogdan, 1969). In the Sudan *C. gayana* seed germinated in soil containing 0.4M NaCl (Abd el-Rahman & El-Monayeri, 1967).

Fertilizer requirements. Rhodes grass rarely gives a response to potash, gives an increased response to phosphorus in some areas and usually a spectacular linear response to nitrogen in the presence of adequate phosphorus and potassium, both in yield and in crude protein content. Split applications after each cut or after grazing cycles are better than one basic application with the usual rate of 275-400 kg/ha. The critical value for phosphorus as a percentage of the dry matter at the immediate pre-flowering stage is 0.23.

Ability to spread naturally. Excellent. It produces stolons which creep over the ground, rooting at the nodes, and also produces abundant seed to give rise to new plants.

Land preparation for establishment. The better the seed-bed preparation the better the establishment, although a rough ploughing will help provide some plants from which to slowly build a sward. In Australia most seed is aerially sown into ashes left after a scrub burn.

Sowing methods. Aerial sowing into ashes one week after a fire is usual. Drilling seed through a modified seed drill mixed with sawdust and sown through the fertilizer box with the delivery chutes removed is another method. Seed can also be undersown in maize, thus giving good grazing after the maize is harvested.

Sowing depth and cover. It is sown on the surface, or no deeper than 2 cm, and covered by rolling or with a bush or a light metal harrow. Mulching after sowing with up to 5 000 kg/ha of hay has significantly increased herbage production in Zimbabwe (Smith, 1966).

Sowing time and rate. It should be sown into ashes as soon as the ashes have cooled and just ahead of the normal wet season. Severe erosion of ashes and loss of seed will result if a storm occurs over sloping sown land before the seedlings stabilize the soil. Seeding rate is 1-4 kg/ha.

Number of seeds per kg. 7 250 000 to 9 500 000 most cultivars, 4 250 000 for cv. Katambora.

Seed treatment before planting. Treat the seed with lindane dust, 20 percent dust at 1 kg per 80 kg of seed, if seed-harvesting ants are prevalent.

Seedling vigour. Excellent — better than Makarikari grass (Lloyd, 1970).

Vigour of growth and growth rhythm. Growth commences early in the spring. A newly sown field will be ready to graze four to six months after planting and reaches its highest production in the second year. Feeding value is low at flowering.

Response to photoperiod. Optimum day-lengths are ten to 13 hours (Bogdan, 1969).

Response to light. It does not grow well in shade.

Compatibility with other grasses and legumes. Rhodes grass does grow with legumes, though experiences have been erratic. In Zambia, when grown with

Stylosanthes guianensis, yields were increased by about 20 percent and with *Neonotonia wightii* the increase was nearly 100 percent. In Kenya *Medicago sativa* and *Trifolium semipilosum* also stimulated yields (Bogdan, 1969). In Australia there has been a tendency to grow Rhodes grass and *Medicago sativa* separately but Christian and Shaw (1952) showed that *M. sativa* at two to four plants per m^2 enhanced Rhodes grass production on a black clay soil.

Ability to compete with weeds. Rhodes grass is of prime importance in pasture mixtures for the brigalow (*Acacia harpophylla*) scrub soils in Queensland because of its vigorous early growth after the scrub has been burnt. Rhodes grass seed sown in the ashes helps suppress sucker regrowth unless suckers are well established before the Rhodes grass, in which case it cannot compete (see Plate 32).

Tolerance to herbicides. It can be killed by cultivation or a heavy spraying with paraquat at 570 ml of a 200 g AI/l product (e.g. Gramoxone) per 200 l water plus surfactant at 250 ml/200 l water. Spray until the solution runs off the leaves (Tilley, 1977). Pre-emergence treatment with atrazine at 1 and 4 kg/ha severely affected emergence of *Chloris gayana* and the 4 kg/ha treatment also adversely affected survival after emergence on Mywybilla clay on the Darling Downs, Queensland (Scateni, 1978).

Response to defoliation. Rhodes grass stands a good deal of defoliation, but in Israel, Dovrat and Cohen (1970) showed that in irrigated and fertilized fields, dry-matter production was some 50 percent higher at 28-day cutting intervals than at 14-day intervals.

Grazing management. It should be allowed to establish and then grazed to prevent flowering, as the nutritive value declines rapidly toward maturity. Fertilizer nitrogen should be added as necessary, or farmyard manure applied.

Response to fire. Rhodes grass usually recovers well after a fire.

Genetics and reproduction. 2n=20, 30, 40 (Fedorov, 1974). The diploids (2n=20) include cvs. Pioneer and Katambora and the tetraploids (2n=40) include cvs. Callide and Samford. Breeding and selection aim at plants that are leafier and late flowering. Rhodes grass is cross-pollinating.

Dry- and green-matter yields. In Zambia, Rhodes grass alone yielded 58 000 kg DM/ha. Under irrigation in Texas, a yield of air-dried herbage of 15 775 kg/ha was recorded, cv. Callide yielded 6 084 kg/ha cut at 50 days, 11 350 kg at 90 days, 11 817 kg at 153 days and 14 157 kg at 188 days. The leaf percentages were 52, 28, 30 and 20 percent respectively and nitrogen in the dry matter 1.4, 0.78, 0.52 and 0.50 percent. In southwest Australia, a yield of 23 639 kg/ha was obtained from an irrigated Rhodes grass pasture treated with three dressings of fertilizer at eight-week intervals during the summer (November to April), each dressing providing 56, 22 and 45 kg/ha of nitrogen, phosphorus and potassium, respectively (Roberts & Carbon, 1969).

Suitability for hay and silage. It makes quite good hay if cut just as it begins to flower or a little earlier. Old stands give low-quality hay. Silage has been made successfully in Nigeria, Zambia and northern Australia, but generally it does not give satisfactory silage (Catchpoole, 1965).

Value as standover or deferred feed. It can be used as low-quality roughage in conjunction with urea-molasses licks.

Chemical analysis and digestibility. In Rhodes grass the contents of organic compounds usually vary as follows: crude protein 4-13 percent, crude fibre 30-40 percent, ether extract 0.8-1.5 percent, nitrogen-free extract 42-48 percent (Bogdan, 1969). In Australia crude protein increased from 6.3 percent unfertilized to 9.5-9.8 percent when fertilized with 440 kg N/ha. Sodium chloride content may be high. Digestibility is usually 40-60 percent of the dry matter. Göhl (1975) recorded seven sets of analyses.

Palatability. Young growth is very palatable, but after the plants have seeded they are less attractive.

Toxicity. No toxicity has been recorded in Australia. Ndyanabo (1974) recorded 0.44 percent total oxalic acid in the dry matter, but no toxicity in India.

Seed production and harvesting. Rhodes grass seed matures 23-25 days after flowering. It is a short-day plant and may flower over a long period making seed harvesting difficult and also reducing its nutritive value. A harvesting cycle of 60-70 days gives seed of better viability and spikelets with a higher caryopsis content (Humphreys, 1973). In cv. Mbarara anthesis is completed in three weeks. Seed is harvested with a combine with a threshing drum operating at 1 100-1 400 rpm and the blast reduced to prevent loss of seed. Harvest ten to 14 days after "smoking" (pollination) has ceased. The crop can also be cut with a reaper and binder a little earlier and the seed cured in the field before threshing, or the heads stripped off with a stripper harvester. The seed must be subsequently cleaned. Hand picking is satisfactory for small areas, and this seed needs minimal cleaning. The seed should be stored carefully. Its viability lasts for up to two years with mature seed (Jones, 1973) but rarely past the first year for immature seed. Pure seed plots should be isolated by 30-60 m.

Seed yield. 100-650 kg/ha.

Minimum germination and quality required for commercial sale. 60 percent germinable seeds and 50 percent purity in Queensland. Germinate at 20-30°C moistened with water. Germination is increased by exposure to light.

Cultivars.
- 'Giant Rhodes Grass' — robust with thick stems, long awns (6-9 mm) and a long tuft of hairs at the awn base. Drought resistant and productive. In Kenya it is known as 'Mpwapwa', in Tanzania as 'Kongwa', and in Australia as cv. Callide. It is a tetraploid. It flowers from January to May in Queensland.
- 'Katambora' — originating from the banks of the Zambesi River, Zimbabwe, it is leafy, dense-growing and a good seed producer. It suppresses nematodes in soil. It is a diploid and flowers from January to May in Queensland.

- 'Mbarara' — from Uganda. Somewhat stemmy, very productive, and gives the highest seed yield and has high seedling vigour.
- 'Pioneer' — the first Rhodes grass introduced to Australia. It is early flowering and hence of lower nutritive value. It flowers from November to May in Queensland.
- 'Rongai' — not a cultivar but a regional group of types in the Rongai area near Nakuru, Kenya. Drought-resistant and stemmy.
- 'Nzoia' — of African origin. Very leafy and has good seed yields but is susceptible to *Helminthosporium* infection and hence has low persistence.
- 'Samford' — introduced to Australia as CPI 16144 in 1952 from Musaia, Sierra Leone. A tetraploid (2n=40) with vigorous stolon development, it flowers late in Australia (April-May), is less frost-tolerant than 'Pioneer' and responds well to nitrogen. Its seed production is good and its palatability outstanding, even in the mature dry state (Barnard, 1972).
- 'Pokot' — a single plant selection from the Pokot district in Kenya. Leafy and vigorous, high yielding and late-flowering.
- 'Masaba' — first introduced into Kenya as 'Endebess' but later renamed 'Masaba', it is leafy and productive but seed production is affected by smut.
- 'Karpedo' — suited to the drier areas of Kenya.

Value for erosion control. It effectively controls erosion when well established. It is rather tall and needs management by light grazing or hay production. Cv. Katambora establishes and covers rapidly and persists well even at low fertility.

Diseases. The cultivar Nzoia is susceptible to *Helminthosporium* and cv. Masaba is affected by a smut.

Pests. In Uganda thrips have damaged Rhodes grass seed.

Economics. Rhodes grass has been used as a short-term pasture ley in East Africa, the United States and Australia where it restores soil structure and provides organic matter.

Animal production. In Kenya, live-weight gains in cattle were 382 kg/ha in the first year of Rhodes grass growth, 228 kg/ha in the second, and 167 kg/ha in the third. In Zimbabwe, Rhodes grass fertilized with 220 kg/ha superphosphate + 440 kg/ha ammonium sulphate supported five steers per hectare for four months of peak growth in summer while they gained 117 kg/head, or 234 kg/ha. When only 220 kg/ha of sulphate of ammonia was used, the stocking rate had to be reduced to 2.5 steers/ha for the same level of live-weight gain (Bogdan, 1969). In Zimbabwe, Rhodes grass fertilized with 270 kg and 38 kg P/ha gave live-weight gains of 230 kg/ha from a stocking rate of 12.4 heifers per hectare (Rodel, 1970).

Main attributes. Its wide adaptability, ease of establishment, persistence and early nutritive value.

Main deficiency. The short season of nutritive peak in many cultivars.

Further reading. Bogdan, 1969.

Chloris mosambicensis K. Schum.

Synonym. Tetrapogon mosambicensis (K. Schum.)
Natural habitat. Grassland and open bush, usually on wet, dark, clay soils.
Distribution. In Africa, in Swaziland, eastern Transvaal and on the coastal lands of Kenya.
Description. Tufted, stoloniferous perennial with the culms up to 30 cm high, compressed and glaucous-based. Inflorescence of three to four digitate spikes 4-7 cm long, straw-coloured, variegated with purple; spikelets 3-4 mm long, lemmas with an inconspicuous fringe above (Napper, 1965; Meredith, 1955). It resembles Rhodes grass but has shorter stolons (Bogdan & Pratt, 1967) (see Fig. 15.27).
Altitude range. Sea-level to 300 m. Specific to low altitudes.
Rainfall requirements. It occurs in areas with rainfall around 575 mm (Bogdan & Pratt, 1967).
Drought tolerance. Extremely drought tolerant.
Soil requirements. It is adapted to loose sandy soils (Bogdan & Pratt, 1967) but has been collected in Kenya from heavy clay soils, also.
Land preparation for establishment. Some soil disturbance is necessary, preferably with a rigid tine cultivator or chisel plough or basin lister type machine.
Sowing methods. Drill where possible, otherwise broadcast.
Sowing depth and cover. Sow into a roughly prepared seed-bed and lightly cover with harrows or brushes.
Sowing time and rate. Sow in summer at 0.3 kg/ha.
Number of seeds per kg. 1 650 000 pure seed (Bogdan & Pratt, 1967).
Response to fire. It survives annual fires in Kenya.
Chemical analysis and digestibility. At an early flowering stage analysis showed almost 14 percent crude protein and 32 percent crude fibre, 9.5 percent ash, 1.8 percent ether extract, 43.3 percent nitrogen-free extract (Dougall & Bogdan, 1960).
Palatability. Its palatability is satisfactory.
Seed production and harvesting. It is a good seeder and the seed can be collected readily from natural stands in Kenya. It is hard to distinguish the seed from that of Rhodes grass. The "seed" consists of a spikelet and one caryopsis (Bogdan & Pratt, 1967). So far it has been hand harvested, but could be handled with a stripper or beater harvester.
Economics. It deserves attention for reseeding semi-arid rangeland (Bogdan & Pratt, 1967).
Further reading. Bogdan & Pratt, 1967.

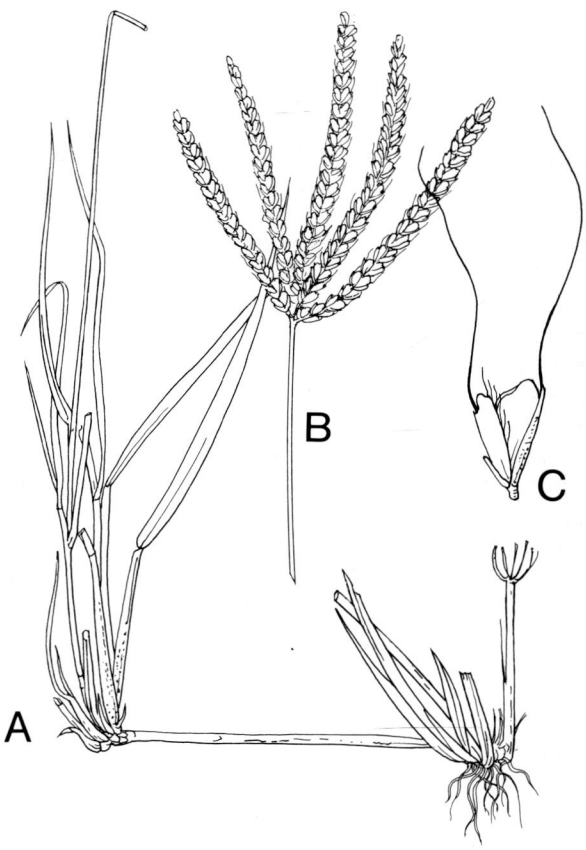

Figure 15.27. Chloris mosambicensis. **A**-Habit **B**-Inflorescence **C**-Spikelet

Chloris roxburghiana Schult.

Synonym. C. *myriostachya* Hochst.
Common names. Horsetail grass and plume chloris (Zimbabwe).
Distribution. In Africa, in Botswana and north and northeast Transvaal, often on old termite nesting grounds.
Description. A tufted perennial with culms up to about 120 cm high; lowest leaf-sheaths usually white or straw-coloured; panicle 5-15 cm long, straw-coloured or purple; spikelets long-awned (Meredith, 1955). It has characteristically flat shoot bases and dense, feathery panicles which are pale green or purple when young (Bogdan & Pratt, 1967) (see Fig. 15.28).
Season of growth. Summer.
Altitude range. Sea-level to 1 500 m.
Rainfall requirements. A single-season rainfall of 500-625 mm (Bogdan & Pratt, 1967).
Drought tolerance. It is drought tolerant (Bogdan & Pratt, 1967).
Soil requirements. Loose sandy soils, loams and alluvial silts (Bogdan & Pratt, 1967).
Sowing time and rate. In summer, at 100-200 g/ha of scarified naked caryopses.
Number of seeds per kg. 6.6 million naked caryopses (Bogdan & Pratt, 1967).
Seed treatment before planting. Free the caryopses from the spikelets. Treat the seed with an insecticide such as Aldrin (Bogdan & Pratt, 1967).
Response to fire. It is severely affected by fire (Skovlin, 1971).
Genetics and reproduction. 2n=20.
Chemical analysis and digestibility. It has up to 16 percent crude protein in the dry matter at the early flowering stage. The crude fibre content at this stage is around 30 percent of the dry matter (Bogdan & Pratt, 1967).
Palatability. Very succulent early spring growth.
Seed production and harvesting. It seeds well and has been harvested by hand. Bogdan and Pratt (1967) recorded that the spikelets are not easily detached from the panicles due to the matting of the long, fine awns and so it is more convenient to cut the panicles and thresh the seed later by rubbing the panicles between two pieces of rubber. They suggest that one of the rubber surfaces should be a section of automobile tyre, the groove being adaptable to hold the seed during abrasion.
Economics. It has been used successfully in reseeding eroded rangeland in Kitui and Baringo districts in Kenya (Jordan, 1957; Pratt, 1963; 1964). Could be used as a garden ornamental.
Further reading. Bogdan & Pratt, 1967.

Figure 15.28. Chloris roxburghiana. **A**-Habit **B**-Portion of raceme **C**-Spikelet

Chloris virgata Sw.

Common names. Woolly-top or feather-top Rhodes grass (Australia), black-seed (Kenya), feather finger grass (United States), feather-top chloris (South Africa).

Natural habitat. Roadsides and grasslands as a weed, secondary in cultivation.

Distribution. Widely distributed throughout the tropics, common on roadsides.

Description. A variable annual; culms 15-90 cm high, often decumbent and rooting from the lower nodes; leaf-blades acute or acuminate; seven to 15 spikes, usually erect and forming an almost spike-like inflorescence; spikelets 3-5 mm long, with one bisexual floret and one, more rarely two, empty lemmas (Chippendall, 1955). The ripe spikelets are black and have longer awns than Rhodes grass (Bogdan & Pratt, 1967) (see Fig. 15.29).

Season of growth. Summer.

Optimum temperature for growth. About 25-30°C.

Frost tolerance. It is killed by frost.

Altitude range. Sea-level to 2 000 m.

Rainfall requirements. It does well in the 500-750 mm rainfall zone with a dominantly summer incidence. The minimum rainfall is 375 mm.

Drought tolerance. Being an annual it is not regarded as drought tolerant.

Tolerance to flooding. It will not withstand flooding.

Soil requirements. It prefers heavy soils and does well on black cracking clays in Queensland, Australia, but has a wide soil range.

Tolerance to salinity. It is a common grass of the saline or usar tracts of north-west India (Bor, 1960) and forms 20 percent of the vegetation in the Mehesana district of the Rann of Kutch, India, on alkaline land (Whyte, 1964).

Ability to spread naturally. It produces abundant seed, which sheds and, being light, is easily transported by wind and water. A little is spread by livestock movements. In Queensland it soon occupies vacant spots in lucerne fields under 650-700 mm rain-fed conditions.

Land preparation for establishment. A rough seed-bed is all that is needed for oversowing.

Sowing methods. Broadcast the seed.

Sowing depth and cover. Sow on surface and roll if moisture is low.

Sowing time and rate. In summer at 0.5 kg/ha of pure seed.

Number of seeds per kg. About 20 000 000 spikelets containing one caryopsis each (Bogdan & Pratt, 1967).

Seedling vigour. Good.

Vigour of growth and growth rhythm. It has a short growing season.

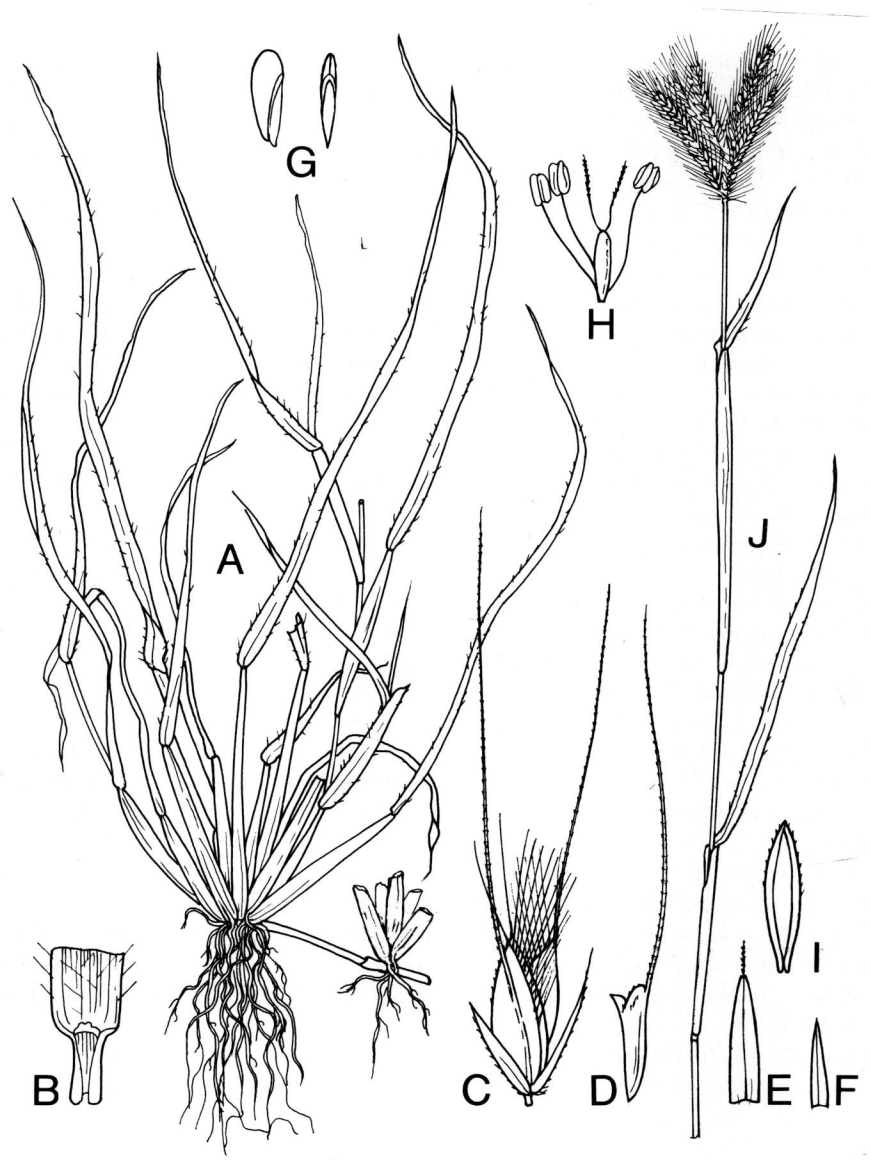

Figure 15.29. Chloris virgata. **A**-Habit **B**-Ligule **C**-Spikelet **D**-Empty lemmas **E**-Lower glume **F**-Upper glume **G**-Grain **H**-Flower **I**-Palea **J**-Inflorescence

Response to light. It will grow in shade, but prefers open country (Whyte, 1964).

Compatibility with other grasses and legumes. Being an annual, it is not persistent but quick regeneration from seed allows it to establish quickly in vacant situations in crops and along roadsides.

Ability to compete with weeds. Good in an annual situation, but it is regarded as a weed itself in Queensland's agricultural areas.

Tolerance to herbicides. It has been removed from lucerne fields by an application of Dalapon at about 6 kg/ha, sprayed at a height of 7.5 cm. Thorough land preparation and cultivation of fallows helps to reduce its dominance.

Response to defoliation. Being an annual, it has little persistence under grazing, but was found as the final result of overgrazing a perennial sward at Matapos, Zimbabwe (Rattray, 1960a).

Grazing management. When used as a reseeding grass in rangeland it should be allowed to seed periodically to provide a thicker stand.

Response to fire. It is easily destroyed by fire. An intense fire could destroy a lot of shed seed.

Genetics and reproduction. $2n = 14, 20, 26, 30$ (Fedorov, 1974).

Suitability for hay and silage. It makes somewhat inferior hay. Cut at flowering.

Value as standover or deferred feed. Being an annual it soon breaks down after maturity.

Chemical analysis and digestibility. It has 12-13 percent crude protein in the dry matter at fresh, full bloom stage (Bogdan & Pratt, 1967). Hay in Zimbabwe contained 10.3 percent crude protein in the dry matter on a 10 percent dry-matter basis (Göhl, 1975).

Palatability. Its palatability is fair only.

Seed production and harvesting. It seeds heavily, and the seed can be harvested as for Rhodes grass (*Chloris gayana*). The seed consists of a spikelet and one caryopsis.

Seed yield. Similar to Rhodes grass (about 100-650 kg/ha).

Value for erosion control. It is one of the first grasses to colonize bare ground, and has been used for reseeding denuded rangeland.

Economics. It has been used with some success by L.H. Brown in reseeding the Mwea Plains south of Embu (Kenya) (Bogdan & Pratt, 1967). In Queensland, Australia, it is regarded as a weed in cultivated and waste places. It can be troublesome in wool.

Further reading. Humphrey, 1960a.

Chrysopogon aciculatus (Retz.) Trin.

Synonym. Andropogon aciculatus Retz.
Common names. Mackie's pest, grass seed (Australia), love grass (Malaysia), manienie-ula (Hawaii), kase, seed grass (Fiji).
Natural habitat. It is common on abandoned cultivations on poor sandy soils.
Distribution. Widely distributed in the tropics of Asia, Polynesia and Australia at low elevations.
Description. A vigorous creeping grass with stout, tough rhizomes, the culms ascending to 45 cm. Inflorescence a small panicle, 7.5-10 cm long, with numerous slender branches. Spikelets narrow. Awn bristly, short and fine. The branches at first ascend almost vertically, spread obliquely at flowering and then bend upward again at fruiting. Each branch has three spikelets at its tip, one sessile and two pedicelled (see Fig. 15.30).
Drought tolerance. It is fairly drought tolerant.
Soil requirement. Although occurring on neutral soils, it favours sandy acidic loams with pH 5.1-6.1. It prefers moist soils.
Response to defoliation. One of the few grasses which can stand heavy grazing in India (Bor, 1960). Under heavy grazing it replaced *Arundinaria ciliata* at Khon Kaen, Thailand (Robertson, Humphreys & Edwards, 1976).
Genetics and reproduction. 2n=20 (Fedorov, 1974).
Value for erosion control. Its creeping rhizome and its capacity to resist hard grazing makes it useful for stabilizing embankments and similar sites.
Economics. An extremely common grass in village pasture in the plains of Asia because the prostrate, creeping stems resist overgrazing and trampling. Grazing animals suffer severely from the ripe fruits becoming attached to their hair by the sharp basal callus. By this means the fruit works its way into the flesh and causes extensive ulceration. Dogs frequently develop abscesses between the toes from the same cause, and germinating seeds of this grass can sometimes be pressed out of large bags of pus in the dog's flesh (Bor, 1960). Useful for rough lawns, forming a dense, hard-wearing turf, but a troublesome weed when uncontrolled because of the sharp-pointed seeds (Henty, 1969). A serious pest in north Queensland. The seeds work through clothing and cause irritating sores. It used to be used as a cover for coconut plantations in the Philippines, and in Guam the straw was used for making hats and mats. Probably represents the final stage in deterioration of the *Phragmites/Saccharum/Imperata* swamp grasslands in India (Dabadghao & Shankarnarayan, 1970).
Further reading. Dabadghao & Shankarnarayan, 1970.

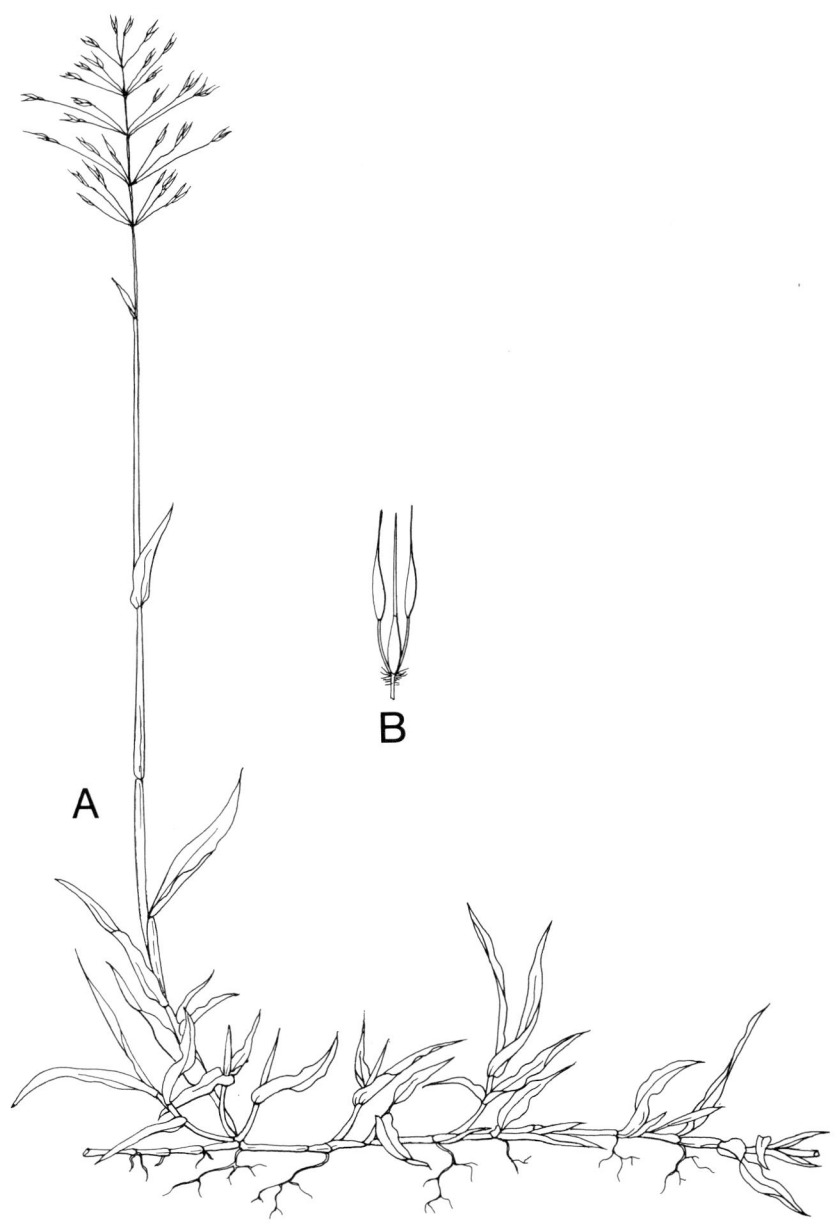

Figure 15.30. Chrysopogon aciculatus. **A**-Habit **B**-Triad of spikelets

Chrysopogon aucheri (Boiss.) Stapf var. *quinqueplumis* (A. Rich.) Stapf

Common name. Aucher's grass (Kenya).
Natural habitat. Dry, gravelly, often rather alkaline soils with grass or semi-desert cover.
Distribution. Afghanistan, Pakistan (Sind, Baluchistan), Iraq, Iran, Somalia, Kenya, Uganda.
Description. Tufted perennial up to 50 cm high with slender, wiry culms. Panicle spreading, with orange-haired spikelets in triads on slender peduncles; sessile spikelets 5-6 mm long with a geniculate awn from the lemma and a straight plumose bristle from the upper glume; pedicelled spikelets green, about as long as the sessile ones, with a plumose bristle from each glume (Napper, 1965). It forms low cushions (see Fig. 15.31).
Season of growth. Summer.
Optimum temperature for growth. Probably about 30-35°C.
Frost tolerance. It probably has little frost tolerance.
Latitudinal limits. About 30°N to 10°S latitude.
Altitude range. Sea-level to 2 000 m.
Rainfall requirements. This grass will grow in rainfalls of 250 mm to 625 mm a year.
Drought tolerance. Excellent, extending into the low-rainfall, annual-grass, semi-desert area of northern Kenya (Bogdan & Pratt, 1967).
Soil requirements. It prefers sandy soils or lava ash, but grows on rocky ground as well.
Tolerance to salinity. Occurs on dry, gravelly, often rather alkaline soils.
Sowing time and rate. Summer at 2.5 kg/ha.
Number of seeds per kg. About 450 000 pure seeds.
Genetics and reproduction. 2n=40 (Fedorov, 1974).
Dry- and green-matter yields. It has not been tested for these details.
Cultivars. None have been registered. An unusually large form from Karpeddo, Baringo, is bulk-seeded at Marigat, Kenya.
Palatability. It is a leafy grass, highly palatable to animals.
Seed production and harvesting. It is a poor seeder and this is accentuated by shedding of the ripe spikelets. The "seed" consists of a raceme with three spikelets, only one of them with a caryopsis (Bogdan & Pratt, 1967).
Economics. It is an important grazing grass in arid regions. It is the dominant grass growing on aridisols high in gypsum in Somalia, and is important in the arid rangeland in Kenya, and in dry areas of Uganda.
Animal production. It is a leafy grass of high nutritive value, well liked by grazing animals.
Further reading. Bogdan & Pratt, 1967.

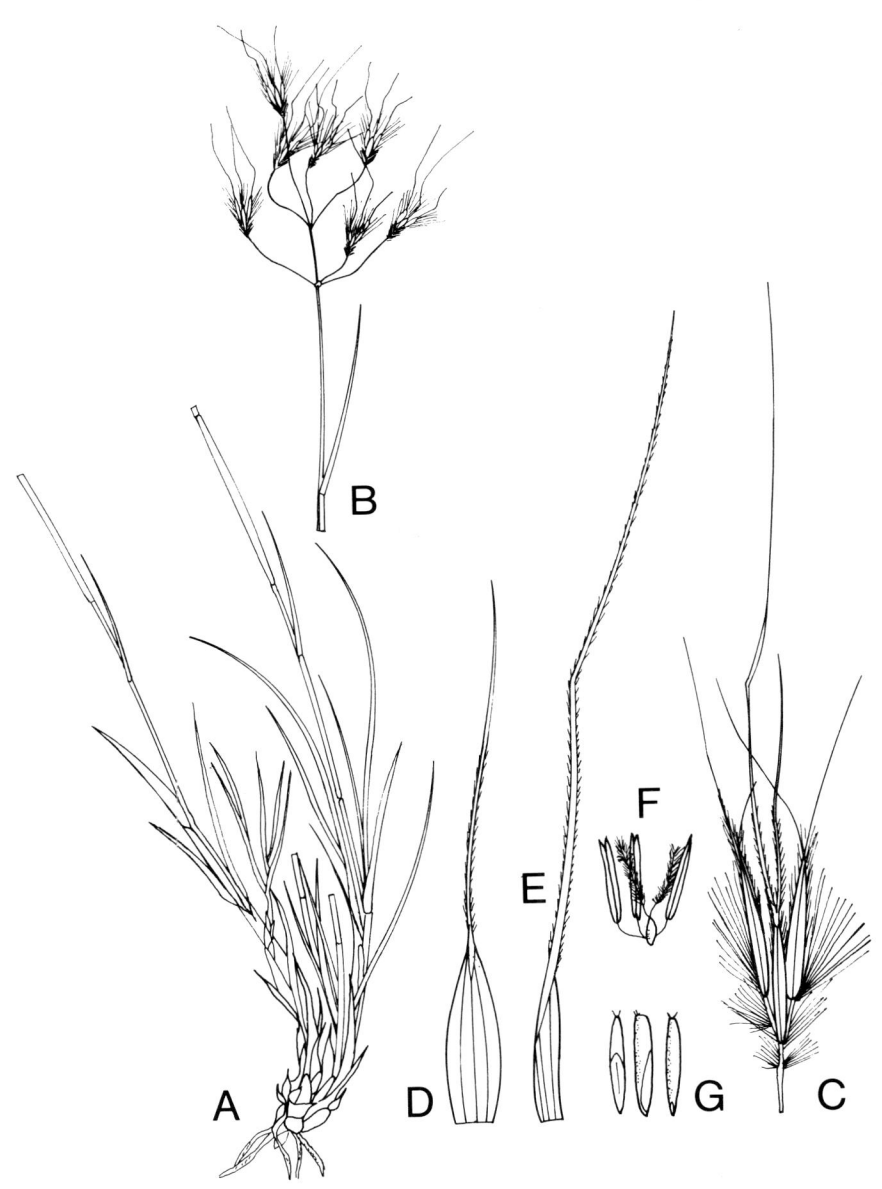

Figure 15.31. Chrysopogon aucheri. **A**-Habit **B**-Inflorescence **C**-Cluster of spikelets **D**-Upper glume of sessile spikelet **E**-Upper lemma and awn of sessile spikelet **F**-Flower **G**-Grain

Coix lacryma-jobi L.

Common names. Job's tears (Australia), adlay (Philippines), sila (Fiji), ma yuen (China).
Natural habitat. Swampy places and near streams, cultivated.
Distribution. Native to tropical Asia, but now widely distributed in the tropics.
Description. An annual, 1-2 m tall, stem erect, with brace-roots from the lower nodes. Inflorescences prolific; the first glume of the male spikelet narrowly winged, the wings not covering the raceme (Henty, 1969). Seeds yellow, purple, white or brown. The grass is monoecious with separate male and female flowers. There are soft-shelled forms for eating and hard-shelled ones for ornamentation (see Fig. 15.32).
Season of growth. Summer.
Latitudinal limits. About 22°N and S.
Altitude range. Sea-level to 2 000 m.
Rainfall requirements. It needs reasonably high rainfall in areas which grow maize and upland rice, usually in excess of 1 500 mm.
Drought tolerance. It is not tolerant of drought.
Tolerance to flooding. It is tolerant to flooding and is found occurring naturally in swamps.
Soil requirements. It requires a fertile soil for its best growth. In poor soils many of the fruits are hollow.
Fertilizer requirements. It needs a fertile soil or a dressing with complete fertilizer.
Ability to spread naturally. It spreads very slowly in favourable environments.
Land preparation for establishment. It prefers a well-prepared seed-bed.
Sowing methods. It is sown by seed. Drill it in with a machine able to handle large seeds or, by hand, dig holes with a hoe and plant at distances of 40-60 cm.
Sowing depth and cover. The seed is sown 5 cm deep and covered.
Sowing time and rate. It is sown at the beginning of the wet season at 10-15 kg/ha.
Response to photoperiod. Its flowering is accelerated by short days (Evans, Wardlaw & Williams, 1964).
Genetics and reproduction. 2n=10, 20 (Fedorov, 1974).
Cultivars.
- 'Stenocarpa' Stapf Hook. — the false fruits are more or less cylindrical.
- 'Agrestis' (Lour.) Backer — the false fruits are ovoid-globose (Henty, 1969).
- 'Lacryma-jobi' — the involucres are ovoid, hard and polished.
- 'Ma-yuen' — in Malaysia the involucres are ovoid, soft, shell-like and striate.

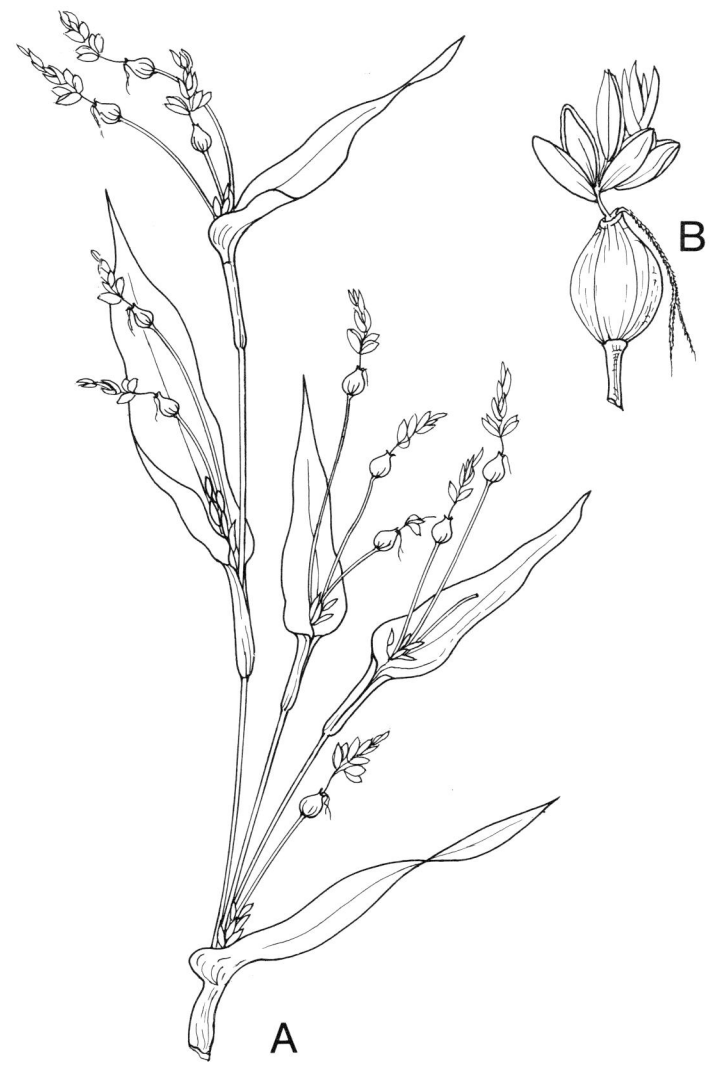

Figure 15.32. Coix lacryma-jobi. **A**-Inflorescence **B**-Male raceme

A third Malaysian variety, cv. Monilifer, has longer often globose involucres 7-10 mm in diameter, stony and flattened on one side (Gilliland *et al.*, 1971).

Dry- and green-matter yields. In India it is grown as a fodder plant in low-lying areas, and its average yield of green material is about 13.9 t/ha (Bor, 1960). For grain it yields 2 000-4 000 kg husked grain per hectare. The hulling percentage is 30-50 percent.

Chemical analysis and digestibility. Fresh early vegetative growth in India showed 29.9 percent dry matter, 8.5 percent crude protein, 27.9 percent crude fibre, 8.96 percent ash, 2.7 percent ether extract and 51.9 percent nitrogen-free extract (Sharma *et al.*, 1968). The husked grain contains 10.8 percent moisture, 13.6 percent protein, 60 percent fat, (ether extract), 58.5 percent carbohydrate (nitrogen-free extract), 8.4 percent fibre and 2.6 percent ash.

Palatability. The green material is very palatable.

Economics. A tall grass which is cultivated in many parts of the tropics, particularly among the hill-tribes who make a porridge and also brew beer from it. The soft-shelled races are cultivated for these purposes but the hill tribes also cultivate several hard-shelled varieties which are used for beads and other ornaments (Bor, 1960). The ground grain is used as feed for poultry. It is grown to a limited extent in southeastern Asia. Although it has been tried elsewhere, production is still very small.

Further reading. Schaffhausen, 1952.

Cymbopogon nardus (L.) Rendle

Synonym. Cymbopogon afronardus Stapf.
Common names. False citronella (Zaire), citronella grass (Taiwan), blue citronella grass (Kenya), naid grass (India).
Natural habitat. Common in grassland and open woodland of *Acacia* and *Combretum* on the hills in Uganda.
Distribution. Throughout southern and northeastern tropical Africa, Uganda, Tanzania and Kenya.
Description. Tall tufted perennial with narrow leaf-blades. Panicle narrow, 15-30 cm long with racemes 8-10 mm long, often rather villous; sessile spikelets flat or concave on the back with winged keels, awn 5-6 cm long (Napper, 1965) (see Fig. 15.33).
Season of growth. Summer.
Altitude range. 2 000-3 000 m.
Rainfall requirements. 750 mm or over.
Sowing methods. C. nardus establishes naturally from seed in the highly grazed areas beneath bushes.
Ability to compete with weeds. It is very competitive, and where overgrazing takes place in useful pastures it tends to increase.
Tolerance to herbicides. Dalapon and paraquat would probably control it.
Response to defoliation. Generally *C. nardus* is avoided in grazing. Light grazing encourages it, but heavy grazing pressure of one bullock per hectare prevented recolonization of the species (Harrington, 1974). Periodic very heavy stocking converted a slope pasture of 47 percent *C. nardus* to a mixed pasture dominated by *Brachiaria decumbens*. The application of 158 kg N/ha increased the content of *B. decumbens* still further (Harrington & Thornton, 1969).
Response to fire. It is very resistant to fire and too-frequent burning is one of the main causes of its increase. Harrington (1974) found that a late burn in the long dry season (usually late August in Uganda) carried out every third year reduced the biomass of *C. nardus* and encouraged the somewhat better, associated grasses of *Brachiaria decumbens, Themeda triandra* and *Hyparrhenia filipendula*. The burn should be against the wind and in weather which would minimize fire temperatures. This would prune the undesirable associated shrub *Acacia hockii*. Annual burning reduces the size of the *C. nardus* plants, but does not improve the sward.
Genetics and reproduction. 2n=20 (Fedorov, 1974).
Chemical analysis and digestibility. Its feeding value is low.
Palatability. The grass is unpalatable to cattle and cattle have been known to die of starvation when an abundance of it, in green condition, was available (Harrington, 1974). Buffalo will eat it sparingly and elephants will accept it during the dry season (Field, 1971).

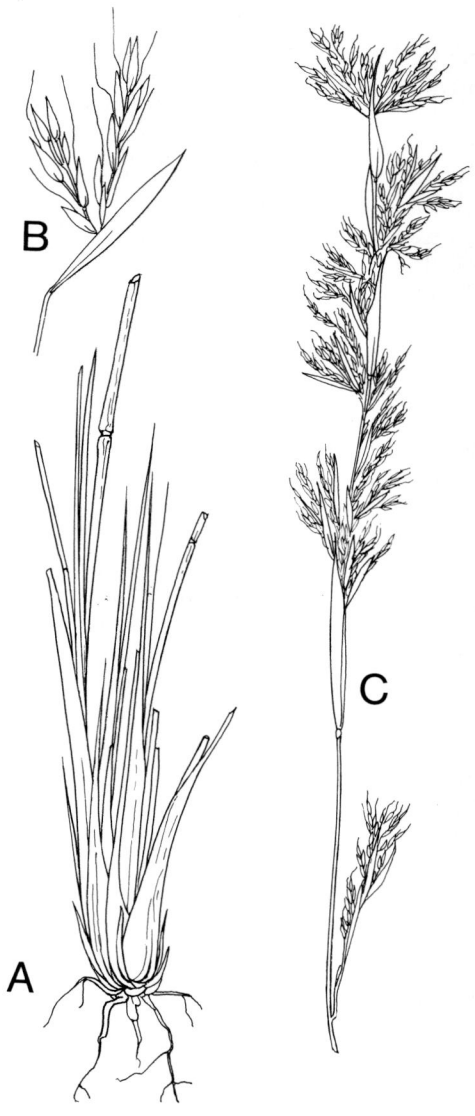

Figure 15.33. *Cymbopogon nardus.* **A**-Stem **B**-Raceme pair **C**-Inflorescence

Control. In addition to a late burn in the long dry season, van Rensburg (1971) showed that, during the wet season, a soil tillage treatment combined with the sowing of *E. curvula* at 2.64 kg/ha revegetated severely damaged *Themeda/Cymbopogon* veld and increased dry-matter production. Hand hoeing is the most effective method of elimination where labour is not expensive.

Economics. *Cymbopogon nardus* is an unpalatable, unwanted invader of Ankole district pastures in Uganda. Removal of *C. nardus* from fully stocked pastures improved growth rates by about 30 percent, but the rate of recolonization can be extremely rapid. The knowledge of the ecology of the grass is supremely important in the development of the area (Harrington, 1974). It is a good thatching and mulching material, and the grass produces citronellal, an aromatic oil.

Animal production. The invasion of a pasture by *C. nardus* always leads to a reduction in animal production.

Further reading. Harrington, 1974; Harrington & Pratchett, 1974b.

Cynodon spp.

Until recently the only two *Cynodon* species important agronomically were reported as *Cynodon dactylon* and *Cynodon plectostachyus*. A recent revision by Clayton and Harlan (1970) has shown that the true *Cynodon dactylon* rarely occurs in the tropics and that the majority of tropical African plants previously thought to belong to *C. dactylon* are in fact either *C. nlemfuensis* or *C. aethiopicus,* both stoloniferous perennials without underground rhizomes, whereas rhizomatous *C. dactylon* occurs mainly in the subtropics and in warm temperate countries. Bogdan (1977) says that it can be accepted with a high degree of probability that reports on the grazing of *C. dactylon* in tropical Africa both in natural stands and in cultivation actually refer to *C. nlemfuensis* and to a lesser extent *C. aethiopicus* (see Fig. 15.34).

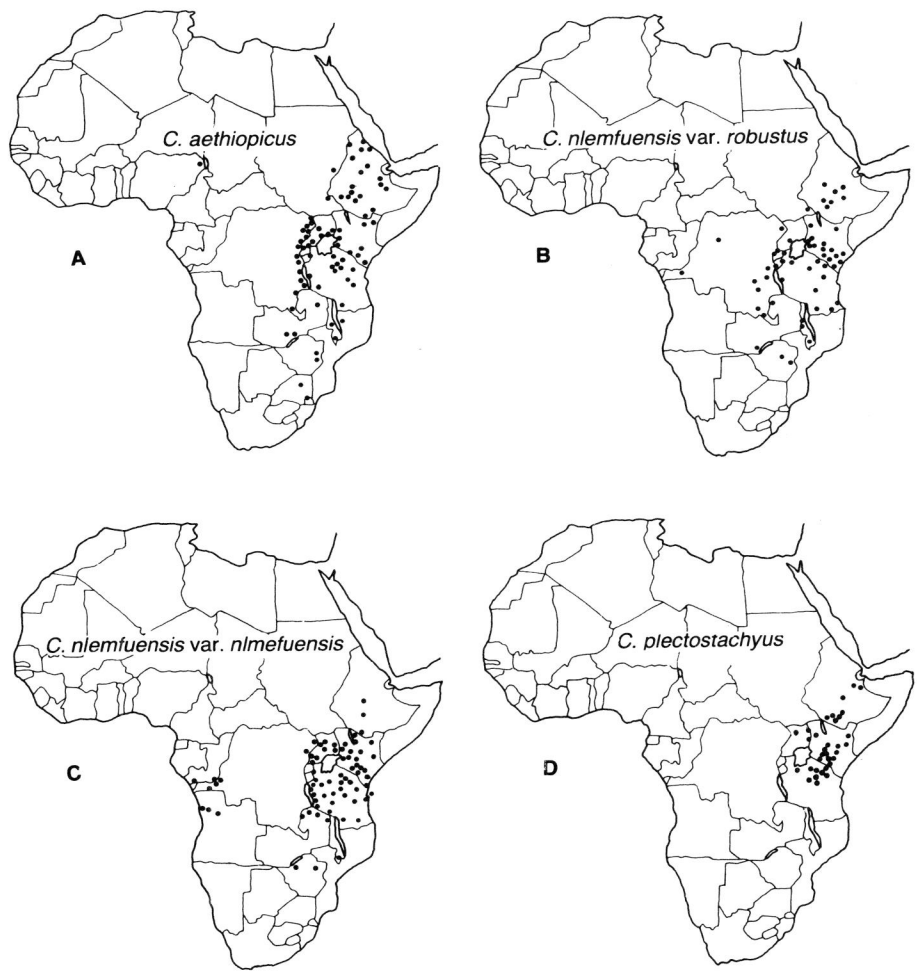

Figure 15.34. Distribution of *Cynodon* spp. in Africa. (**Source:** Harlan *et al.*, 1970)

Cynodon aethiopicus **Clayton and Harlan**

Common name. Giant star grass.
Distribution. Occurs down the eastern coast of Africa from the Red Sea to the Transvaal, most densely at the Uganda-Zaire border.
Description. A large, robust, non-rhizomatous grass. *C. aethiopicus* has a somewhat similar inflorescence to *C. plectostachyus,* but the glumes are about three-fourths the length of the spikelet and the foliage is stiff and harsh, the stems woody and the stolon internodes lie flat (Harlan, de Wet & Rawal, 1970). Racemes in two to five whorls (rarely one), stiff, red or purple (see Fig. 15.35).
Dry- and green-matter yields. Strickland (1976-77) recorded a range of yields from 1 000-2 000 kg/ha per month in summer and 165-500 kg/ha per month in winter from three accessions tested at Samford, Queensland (lat. 23°50′S, rainfall 813 mm).
Seed yield. Strickland (1976-77) obtained 7-15 kg/ha from one harvest in January 1977. There could have been an additional harvest.
Further reading. Strickland, 1976-77.

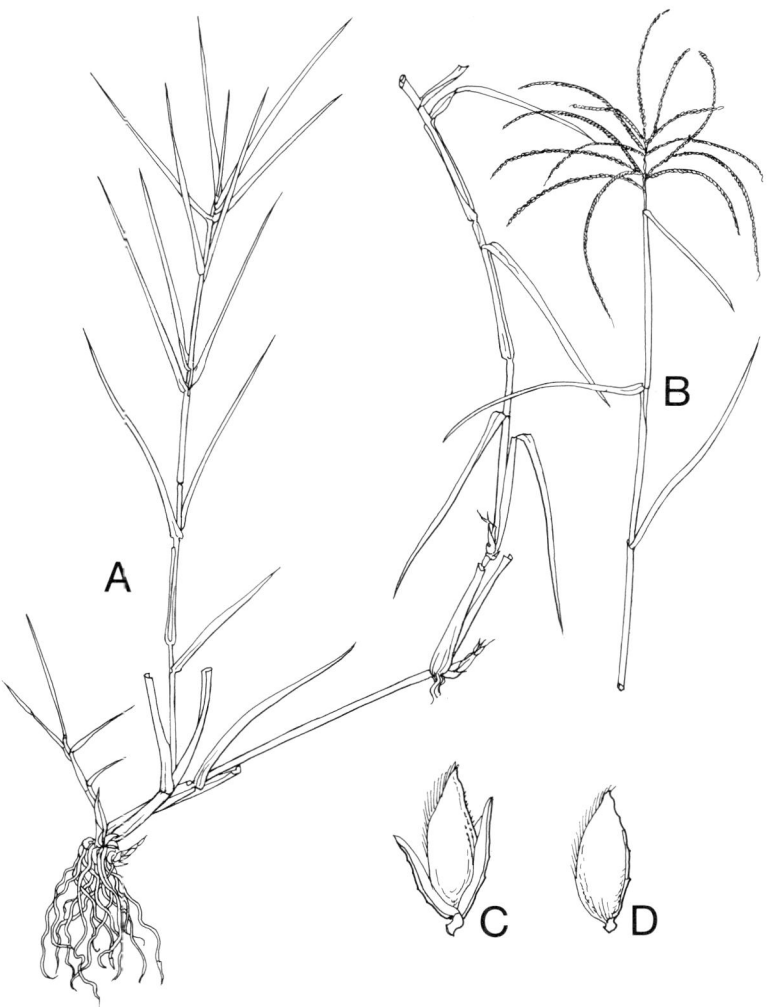

Figure 15.35. Cynodon aethiopicus. **A**-Habit **B**-Inflorescence **C**-Spikelet **D**-Floret

Cynodon dactylon (L.) Pers.

Common names. Couch grass, green couch (Australia), Bermuda grass (United States), kabuta (Fiji), dhoub grass (Bangladesh), Bahama grass, quick grass (South Africa), chepica brava, cama de niño, pata de perdiz, gramilla blanca (Peru), hierba-fina (Cuba), grinting, tigriston (Suriname).
Natural habitat. Grassland, lawns and pastures and as a weed in cultivation.
Distribution. Wheeler (1950) says the best evidence is that is originated in Asia, particularly India, and has now become pan-tropical.
Description. A variable perennial, creeping by means of stolons and rhizomes, eight to 40 culms, (rarely) to 90 cm high: leaves hairy or glabrous, three to seven spikes (rarely two), usually 3-6 cm long and in one whorl, or in robust forms up to ten spikes, sometimes in two whorls: spikelets 2-3 mm long, rachilla often bearing a reduced floret (Chippendall, 1955). It differs from *Digitaria scalarum* (African couch) in the vegetative stage in that there is no obvious membranous ligule where the leaf-blade joins the sheath (Ivens, 1967) (see Fig. 15.36).
Optimum temperature for growth. 35°C (Evans, Wardlaw & Williams, 1964), 37.5°C (Lovvorn, 1945).
Minimum temperature for growth. Grows very slowly at 15°C (Evans, Wardlaw & Williams, 1964). Day temperatures must exceed 10°C. The minimum temperature regime for growth consists of an eight-hour day at 15°C and a 16-hour night at 5°C (Youngner, 1959).
Frost tolerance. It frosts but recovers.
Latitudinal limits. 30°N and 31.4° ± 7.5° S (Russell & Webb, 1976).
Altitude range. Sea-level to 2 300 m.
Rainfall requirements. It usually occurs over a range of 625-1 750 mm of annual rainfall.
Drought tolerance. Good. The rhizomes survive drought well. Coastal Bermuda grass has proved very drought resistant in Georgia, United States.
Tolerance to flooding. In Bangladesh couch grass survives the annual flooding of the Ganges-Brahmaputra rivers to a depth of 6 m or more for several weeks. It is then oversown with *Lathyrus sativus* (Khesari) and used for dairy cattle grazing.
Soil requirements. There are varieties adapted for a wide range of soils. Coastal Bermuda prefers well-drained, fertile soils, especially heavier clay and silt soils not subject to flooding, well supplied with lime and high-nitrogen mixed fertilizers. Lawn couch grass is most frequently grown for sale on sandy loams easy to dig and rebuild.
Tolerance to salinity. Common couch has good tolerance to salinity, but makes only slow growth. It is able to shunt its photosynthate from the tops to the roots to enable it to survive under saline conditions (Youngner & Lunt,

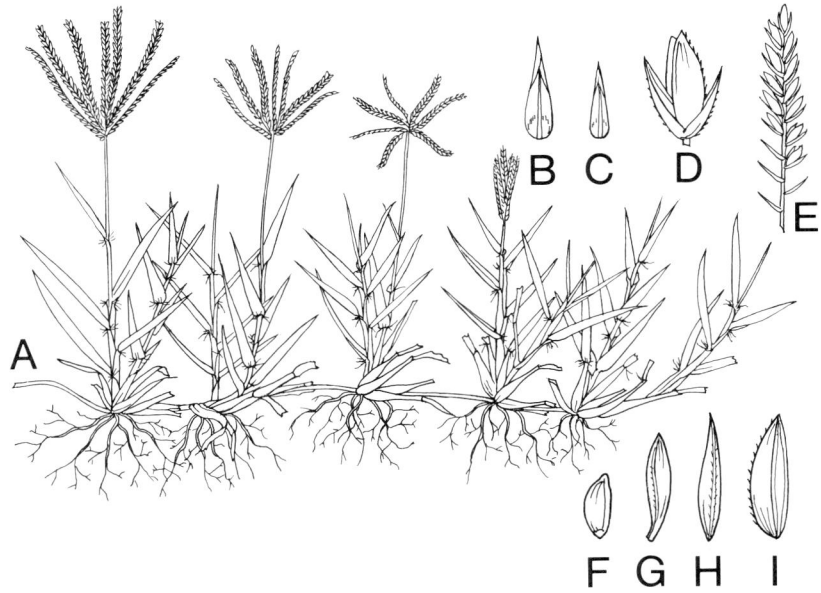

Figure 15.36. Cynodon dactylon. **A**-Habit **B**-Upper glume **C**-Lower glume **D**-Spikelet **E**-Spike **F**-Grain **G**-Palea **H**-Lemma, back **I**-Lemma, side

1967). It gave maximum yields up to EC_e 7 mmhos/cm, 50 percent at 15 mmhos/cm and nil at 22.5 mmhos/cm (Mass & Hoffman, 1976).

Fertilizer requirements. A good basic fertilizer with additional levels of nitrogen according to purpose. With 'Coastal Bermuda Grass', the efficiency of nitrogen utilization begins to decline with 220 kg/ha for hay production and 450 kg/ha for protein production. Application of farmyard manure and sulphate of ammonia to mixed pasture at Kongwa, Tanzania, caused an invasion by *Cynodon dactylon* and suppression of *Chloris gayana* and *Cenchrus ciliaris* (Wigg, Owen & Mukurasi, 1973). The average rainfall at Kongwa is 562 mm per year. In southern Texas, United States, *C. dactylon* fixed 30 kg N/ha in 100 days (Wright, Weaver & Holt, 1976).

Ability to spread naturally. *C. dactylon* spreads quickly by rhizomes and stolons, and less obviously by seed.

Land preparation for establishment. If planted by turf, rough ploughing will be sufficient, but if for a lawn grass sown from seed, a very well-prepared, fine, weed-free seed-bed is needed.

Sowing methods. It is usually sown as turfs or as seed for lawns. 'Coastal Bermuda Grass' is sown by seed or, more often, with sprigs, as it produces few viable seeds.

Sowing depth and cover. It is surface sown and rolled in.
Sowing time and rate. Sow in summer at 9-11 kg/ha.
Number of seeds per kg. 4 489 000.
Dormancy. No seed dormancy has been reported.
Seed treatment before planting. Treat with lindane dust if seed-harvesting ants are about.
Seedling vigour. Seedlings usually root down quickly.
Vigour of growth and growth rhythm. It grows vigorously once established.
Response to photoperiod. It is indifferent to day length for flowering (Evans, Wardlaw & Williams, 1964).
Response to light. It usually dies out under medium to dense shade.
Tolerance to herbicides. Dalapon at 6-12 kg/ha applied to young growth can give a high degree of control. Repeated cultivations will kill the plant, but repeated spraying with herbicides are effective. Spray young, vigorously growing plants with paraquat at 2.8 l/ha of a 200 g AI/l product (e.g. Gramoxone) plus surfactant at 250 ml/200 l of water, using a minimum of 400 ml water per hectare. TCA 2,2,-DPA and glyphosate (Round up) can also be used (Tilley, 1977).
Compatibility with other grasses and legumes. 'Coastal Bermuda' combines with lespedeza and white clover well. For northeast Thailand it is combined with *Stylosanthes humilis* (McLeod, 1972).
Ability to compete with weeds. It suppresses weeds well if kept mown or grazed closely and fertilized.
Response to defoliation. It stands close grazing probably better than any other grass.
Management. Graze closely to keep the feeding value high, and fertilize with nitrogen. Renovate by ploughing or discing when sod-bound.
Response to fire. It will stand severe fires due to the extensive rhizome development in most varieties and cultivars.
Genetics and reproduction. 2n=18, 27, 30, 36, 40 (Fedorov, 1974).
Harlan, de Wit and Rawal (1970) recognize six varieties:
- *dactylon* (4X) — Bermuda grass. A cosmopolitan weed, turf grass and forage grass.
- *aridus* (2X) — giant Bermuda grass. Southern India to Israel and the Sinai, and sparingly southward in dry areas to the Karoo of South Africa. Introduced to Hawaii and Arizona.
- *afghanicus* (2X, 4X) — Afghanistan.
- *coursii* (4X) — Madagascar.
- *elegans* (4X) — Africa south of 12°S latitude.
- *polevansii* (4X) — near Barkerspan, South Africa.

Dry- and green-matter yields. Coastal Bermuda grass receiving 550 kg/ha of complete 4-8-4 fertilizer plus 520 kg/ha of nitrate of soda produced 6 tonnes of air-dried hay in four cuttings in Georgia. Strickland (1976-77) recorded a range of dry-matter yields of 1 000-3 000 kg/ha per month in summer and 100-

TABLE 15.20 *Cynodon dactylon*

	CP	CF	NFE	EE	TA	DCP	TDN
Dry season	6.5	30.4	50.8	2.6	9.7	2.8	56.9
Wet season	8.0	30.8	50.2	1.7	9.3	3.9	37.2

Source: Karue, 1974

1 200 kg in winter at Samford, Queensland, from 20 accessions of *Cynodon dactylon*.

Suitability for hay and silage. Coastal Bermuda grass gives excellent hay, very quickly cured, and, if fertilized, of excellent nutritive value. It is frequently pelleted in the United States. Harvesting at eight weeks increased dry matter but reduced crude protein in comparison with a four-week cut (Utley *et al.*, 1978). It makes good silage, but not of the lactic acid type when ensiled with 41 kg maize grain per tonne. The pH was 5.0, volatile acid content was only 2-4 percent of the dry matter and it had the appearance of haylage (Miller, Clifton & Cameron, 1963).

Value as standover or deferred feed. It will provide standover or deferred feed if closed for grazing.

Chemical analysis and digestibility. Karue (1974) recorded the chemical composition as percent of the dry matter of grass receiving no special treatment in Kenya in Table 15.20. Göhl (1975) lists 11 analyses. Crude protein varies from 8.3 percent in mature to 14.0 percent in young grass. Burton, in Georgia, United States, has been able to reach 22 percent in nitrogen-fertilized grass for pelleting for poultry food.

Palatability. It is very palatable if kept short in growth and fertilized.

Toxicity. Most of the *Cynodon dactylon* types are non-toxic but an occasional case of HCN poisoning may occur. In the United States, frosted Bermuda grass can cause photosensitization. Kidder, Beardsley and Erwin (1961) and Ndyanabo (1974) recorded 1.10 percent total oxalic acid in the dry matter but no toxicity.

Seed production and harvesting. In the United States two seed harvests of 'Coastal Bermuda' are made — July and November. It is mowed into windrows, picked up and threshed by combines and subsequently cleared.

Seed yield. Cv. Coastal Bermuda, 275-350 kg/ha.

Minimum germination and quality required for commercial sale. 60 percent germinable seed, 97 percent purity (Queensland). Germinate at 20-30°C, moisten with KNO_3 solution.

Cultivars.
- 'Common Bermuda Grass', or *C. dactylon* var. *dactylon* — a tetraploid

(2n=36) originating in the Near East and is the common weed of arable land. It is excellent for erosion control and gives valuable feed especially in winter, though limited in quantity.
- 'Coastal Bermuda Grass' — bred by Dr Glen Burton of Tifton, Georgia, United States, from a cross between *C. dactylon* var. *dactylon* and *C. dactylon* var. *elegans*. It is outstanding for hay and pasture and has a wide soil range. It is larger than 'Common Bermuda Grass' with longer internodes. The leaves have a characteristic light green colour. It is almost seedless, but seed can be obtained. It is larger and more erect in habit than 'Tift', and its lighter green and more flexible leaves droop more than those of 'Tift'. More frost resistant than 'Common Bermuda'. It responds remarkably to nitrogen fertilizer. It is best planted as pieces of sod. It produced a yearly average of 130 kg/ha more live-weight gain than common couch.
- 'Tift Bermuda' — found in a cotton field near Tifton, Georgia, United States. It has long decumbent stems, few seed-heads and an abundance of large stolons and rhizomes. It is superior to 'Common Bermuda' or couch grass for both hay and pasture.
- 'St Lucia Bermuda Grass' — establishes and spreads by surface runners. It has no rhizomes. It is adapted to the muck and sandy muck soils underlain by lime on the lower east coast of Florida, United States (Wheeler, 1950).
- 'Alicia' — does not produce high-quality forage and is not very winter hardy. It spreads rapidly, established by cuttings or rhizomes. Forage digestibility is lower than for 'Coastal Bermuda' (Bates, 1978).
- 'Callie' — adapted to the area south of the line from North Carolina to Texas, United States. It is not winter hardy. It spreads and establishes very fast, gives hay yields 10-15 percent higher than 'Coastal' and is more digestible (Bates, 1978).
- 'Oklan' — not as winter hardy as 'Hardie' or 'Midland'. Propagated by stolons. Has high digestibility and good forage quality. Starts growth later in spring than 'Midland'.
- 'Suwannee Bermuda' — developed at the Coastal Plains Experiment Station, Tifton, Georgia. Is better adapted to soils of low fertility than 'Coastal' and is used in poor soils in Florida. It will not withstand close grazing.
- 'Coast Cross-1' — adapted to the southern United States, not as winter hardy as 'Coastal' or 'Oklan' and has lower forage production. A good hay type with high digestibility and better animal performance than other varieties. Starts late in spring.
- 'Midland' — the standard cultivar for Oklahoma, United States. The most winter hardy of the improved, upright, high-producing cultivars. Adapted to shallow, drought-prone soils. Starts off early in spring but is slow spreading.
- 'Hardie' — more winter hardy than 'Coastal' or 'Oklan'. Grows best on deep, fertile soils, has high digestibility and gives good daily gains. Suitable for hay. Starts growth early in spring. Produces a lot of forage (Bates, 1978).
- 'Greenfield' — best for erosion control with heavy growth of rhizomes and

TABLE 15.21 *Cynodon dactylon*, live-weight gains

Rate of nitrogen application (kg/ha)	Beef produced (kg/ha)
0	290
56	339
112	538
224	767

Source: Johnson, McGill and Gurley, 1960

stolons, fast spreading and grows well on thin, eroded soils. Winter hardy and produces early spring growth (Bates, 1978).
Value for erosion control. It has saved untold areas of soil from erosion by wind and water. It is a hardy pioneer which colonizes bare ground and holds and accumulates soil. It helps to bind the edges of roads and provides excellent grazing for village geese, ducks, goats, cattle and buffaloes if not trampled too much by these latter heavy beasts.
Diseases. C. dactylon is attacked by *Helminthosporium* leaf diseases in some areas.
Pests. Root knot nematodes attack common couch grass in sandy soils. Coastal Bermuda grass is more resistant to attack.
Main attributes. Cynodon dactylon has wide adaptability to soils and climate. It is palatable, nutritious, and stabilizes soil against erosion, stands heavy grazing and makes useful hay and silage.
Main deficiencies. It can become a weed in cultivation and it does not provide much bulk unless well fertilized.
Economics. A valuable lawn grass of wide adaptability. It produces excellent forage when adequately fertilized.
Animal production. In Zimbabwe, *C. dactylon* fertilized with 270 kg nitrogen and 38 kg phosphorus per hectare gave a live-weight gain of 480 kg/ha from grazing 12.4 heifers per hectare (Rodel, 1970). Live-weight gains on 'Coastal Bermuda Grass' in Georgia over the period 1950-52, with varying rates of nitrogen, are shown in Table 15.21 (Johnson, McGill & Gurley, 1960).
Further reading. Johnson, McGill & Gurley, 1960.

Cynodon nlemfuensis Vanderyst

Common names. Giant star grass (Nigeria), African star grass (Australia), robust star grass.

Natural habitat. Disturbed areas in grassland, cattle paddocks, verges, on moist alluvium.

Distribution. Var. *nlemfuensis* occurs mainly in Kenya, Uganda and Tanzania with small outliers in Zaire and Ethiopia. Var. *robusta* has a similar base with more representation in Ethiopia and eastern Zaire (see Fig. 15.34).

Description. A large, robust, non-rhizomatous grass, deep-rooted. *C. nlemfuensis* has two distinct varieties: var. *nlemfuensis* is somewhat finer and less robust than var. *robusta* and resembles a very large *C. dactylon* except that it has no rhizomes (Harlan, de Wet & Rawal, 1970). The inflorescence has four to nine racemes, each 4-7 cm long. Var. *robusta* has seven to 12 racemes, each 6-13 cm long (Chippendall & Crook, 1976). Inflorescence is one whorl (occasionally two), slender, green or pigmented (see Fig. 15.37).

Soil requirements. It has a wide range of soil fertility and can mobilize and recycle subsoil nutrients, especially calcium, to offset the increasing acidity from high levels of sulphate of ammonia application (Mohamed Saleem, Chheda & Crowder, 1975).

Fertilizer requirements. Cultivar 1B.8 requires 0.22 percent phosphorus in herbage for optimum production. It responds to heavy dressings of nitrogen.

Sowing methods. It is propagated by rooted runners.

Response to defoliation. At Fashola Livestock Farm in Nigeria in the derived savannah zone, *C. nlemfuensis* maintained a good stand and productivity when cut fortnightly to a height of 2-3 cm (Ahlgren *et al.*, 1959). Later, Chheda and Mohamed Saleem (1972) showed that cultivar 1B.8 should be cut to a height of 18 cm for optimum herbage production.

Genetics and reproduction. Var. *nlemfuensis* is mostly a diploid, but a tetraploid race has been encountered under cultivation and it crosses rather easily with the diploid variety of *C. dactylon*. Var. *robusta* is a diploid with tetraploid races under cultivation.

Dry- and green-matter yields. Strickland (1976-77) recorded an average dry-matter production of 1 600-2 000 kg DM/ha per month in summer and 400-1 000 kg/ha per month in winter from three accessions tested at Samford, Queensland (lat. 27°22'S, 1 050 mm rainfall, humid subtropical climate). In Nigeria the cultivar 1B.8 yielded 25 600 kg DM/ha over two harvests between 19 August and 26 October 1971 (Mohamed Saleem, Chheda & Crowder, 1975).

Chemical analysis and digestibility. Mohamed Saleem, Chheda and Crowder (1975) recorded a crude protein of 7.8-8.0 percent, potassium content of 2.8 percent, a range in phosphorus of 0.14 to 0.26 percent and of calcium from

0.16 to 0.31 percent, the latter from an application of 2.5 t/ha of calcitic limestone. Ademosun and Kolade (1973) compared two varieties of *C. nlemfuensis* var. *robusta* (local or 1B.1 and var. *nlemfuensis* (1B.8) and found that when harvested at six weeks both varieties yielded about 3 t/ha. Digestibility of dry matter by goats was about 60 percent for both varieties.

Palatability. It is extremely palatable, especially when young.

Seed production and harvesting. Strickland (1976-77) could only obtain a

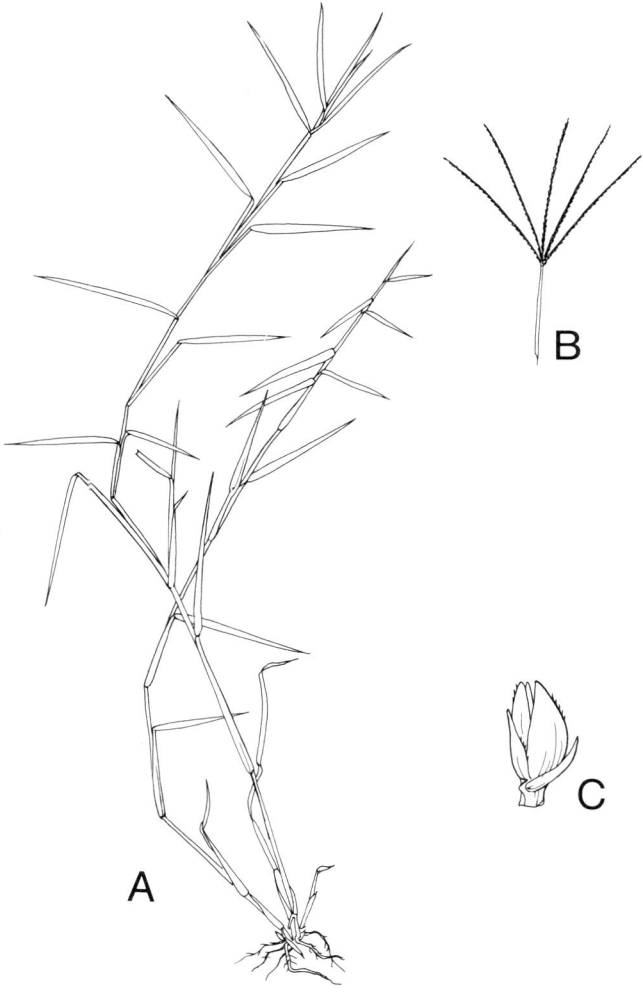

Figure 15.37. Cynodon nlemfuensis. **A**-Habit **B**-Inflorescence **C**-Spikelet

trace of seed from three accessions tested at Samford, Queensland, in 1976-77.

Cultivars. Cultivar 1B.8 is a deep-rooted, non-rhizomatous variety and 1B.1 is a local variety in Nigeria, the local variety being consumed to a greater extent than 1B.8 (Ademosun, 1973). Dry-matter intake values were 1.59, 1.36 and 1.51 kg per 100 kg body weight for the 1B.8 variety harvested at four, seven and ten weeks of age and fed to goats. The corresponding values for the local 1B.1 variety were 2.09, 2.26 and 2.20 kg per 100 kg body weight (Ademosun & Kolade, 1973).

Animal production. Owing to the confusion of nomenclature, figures are difficult to verify. Plate 33 shows a herd of Boran cattle grazing what is probably a *C. nlemfuensis* pasture in Kenya.

Further reading. Harlan, de Wet & Rawal, 1970; Strickland, 1976-77; Vicente-Chandler *et al.,* 1974.

Cynodon plectostachyus (K. Schum.) Pilger

Common names. Naivasha star grass (eastern Africa), estrella (South America), Bermuda mejorado, Hawaiiano (Costa Rica).
Natural habitat. Dry lake beds.
Distribution. C. *plectostachyus* has a fairly restricted natural distribution along the Rift Valley through Ethiopia, Kenya, northern Uganda and northern Tanzania.
Description. A large, robust, non-rhizomatous grass. True *C. plectostachyus* is a diploid which can easily be identified by the small glumes, rarely as long as one-third of the spikelet; soft foliage; racemes in two or more whorls and arching stolon internodes (Harlan, de Wet & Rawal, 1970) (see Fig. 15.38).
Season of growth. Summer.
Frost tolerance. It survives frost.
Altitude range. Sea-level to 2 000 m.
Rainfall requirements. It is adapted to semi-arid areas with rainfalls from 500-875 mm.
Drought tolerance. Very good.
Tolerance to flooding. Tolerates temporary flooding.
Soil requirements. It has a wide range of tolerance from sandy loams to alluvial silts and clays, and black cracking clay soils, but prefers soil of high fertility.
Tolerance to salinity. It is tolerant to alkaline soils and is always found in what appear to be alkaline areas in Kenya.
Fertilizer requirements. The optimum phosphorus content of the dry matter for growth was determined by Falade (1975) as 0.305. It responds well to nitrogen.
Ability to spread naturally. Excellent — under good conditions its stoloniferous habit allows it to spread rapidly.
Land preparation for establishment. Full land preparation is required for establishment from seed.
Sowing methods. It can be established from seed or by splits dug into the soil, this latter requiring less land preparation.
Sowing depth and cover. Seed should be surface sown and lightly covered and rolled.
Sowing time and rate. In the wet season at 6.5 kg/ha.
Number of seeds per kg. 25-40 million florets with one caryopsis (Bogdan & Pratt, 1967).
Seed treatment before planting. Treat with an insecticide.
Seedling vigour. Good.
Response to light. It prefers to grow in full sunlight.
Compatibility with other grasses and legumes. It tends to form a monospecific

sward, but will grow with *Medicago* species, *Trifolium semipilosum* and *Lotononis bainesii* (Clatworthy, 1970).
Ability to compete with weeds. Excellent.
Response to defoliation. Excellent. It can stand heavy grazing.
Grazing management. It should be grazed fairly heavily and top-dressed with nitrogen as required.

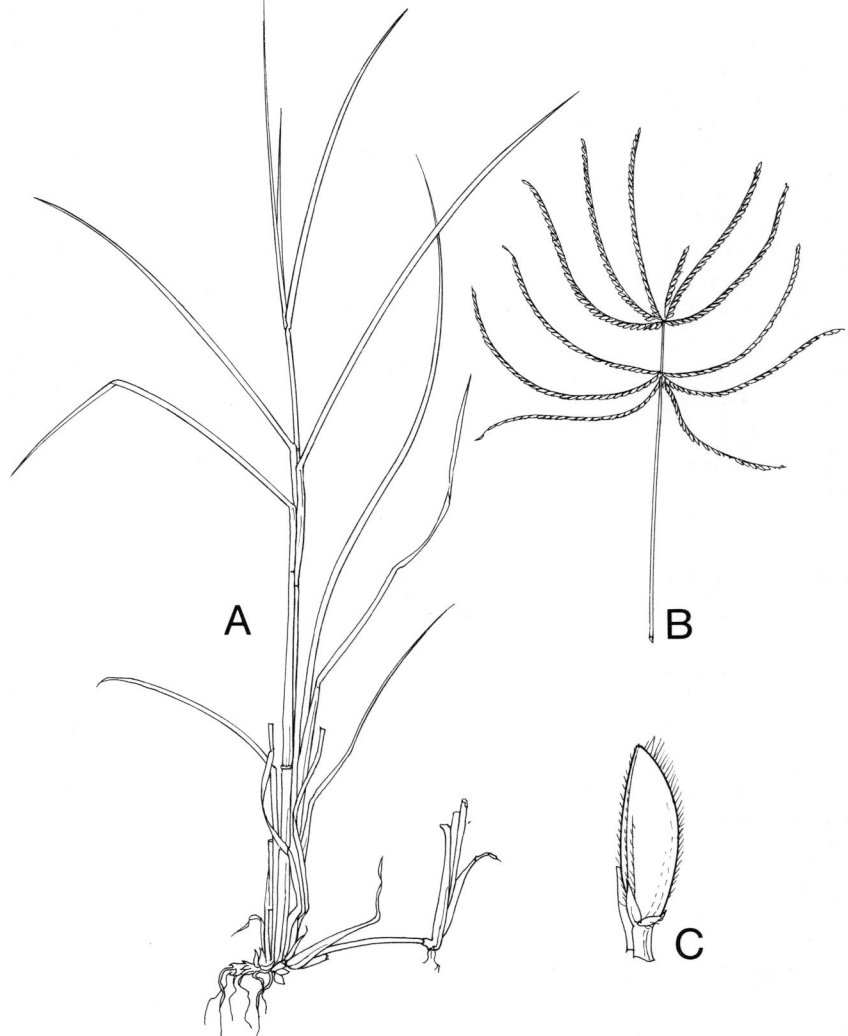

Figure 15.38. Cynodon plectostachyus. **A**-Plant **B**-Inflorescence **C**-Spikelet

Response to fire. It survives fire very well and quickly responds to subsequent rain.

Genetics and reproduction. A diploid — 2n=18 (Fedorov, 1974). Crossbreeding studies have shown it to be completely isolated genetically (Harlan, de Wet & Rawal, 1970).

Dry- and green-matter yields. Strickland (1976-77) recorded a range of dry-matter yields of 1 300 kg/ha per month in summer to 300-1 100 kg/ha per month in winter from three accessions tested at Samford, Queensland. In Nigeria, Moore (1965) cut 3 300 kg/ha of *C. plectostachyus*/centro hay from cutting at 5-10 cm at eight-week intervals. At Gualaca, Panama it produced 6 000 kg DM/ha without fertilizer and 32 000 kg DM/ha with 600 kg N/ha in a rainfall regime of 3 997 mm a year (Rattray, 1973).

Suitability for hay and silage. It makes quite good hay, and with the addition of 10 percent molasses makes good silage (Medling, 1972).

Value as standover or deferred feed. It makes quite good deferred grazing.

Chemical analysis and digestibility. In Costa Rica analysis of material at floral initiation revealed 14.98 percent crude protein, 26.20 percent crude fibre, 37.22 percent nitrogen-free extract, 1.93 percent ether extract, and 9.67 percent ash in the dry matter on a 10 percent moisture basis (Gonzalez & Pacheco, 1970). At Gualaca, Panama, it contained 20-25 percent dry matter in the wet season and 47 percent dry matter in the dry season (Rattray, 1973).

Palatability. It is extremely palatable.

Toxicity. No toxicity can be attributed to this grass in Queensland (Everist, 1974).

Seed yield. 25 kg/ha from one harvest at Samford, Queensland (Strickland, 1976-77).

Value for erosion control. Excellent. It has been used on the black cracking clay soils on sloping cultivated land on the Darling Downs, Queensland, with success.

Animal production. Because of taxonomic confusion, until recently the literature references to the role of *C. plectostachyus* may actually refer to other species. It is, however, an excellent grazing grass and one which stabilizes soil against wind and water erosion. In Zimbabwe, *C. plectostachyus* (?) pastures fertilized with 270 kg nitrogen and 38 kg phosphorus per hectare gave a liveweight gain of 830 kg/ha from the grazing of 12.4 heifers per hectare (Rodel, 1970).

Main attributes. Its rapid colonization of bare land and invasion of overgrazed land.

Main deficiency. It may become a little aggressive in cultivations.

Further reading. Harlan, de Wet & Rawal, 1970.

Dactyloctenium aegyptium (L.) Beauv.

Common names. Crowfoot grass (Africa), beach wire-grass (Hawaii), kra lekrab (Mauritania), giant button grass.
Natural habitat. Usually occurs on disturbed areas, especially in sandy soils.
Distribution. Throughout tropical Africa, introduced into America.
Description. A glaucous annual with culms up to 50 cm high, not stoloniferous, but often rooting from the lower nodes; leaves usually hairy on the margins and midrib, the hairs tubercle-based; usually four to eight spikes, rarely one to three, 1.5 to 6.5 cm long; spikelets 4 mm long, usually three-flowered. The stout spikes and rigid awns are rather distinctive. *Eleusine* can be distinguished from *Dactyloctenium* because it is awnless (see Fig. 15.39).
Altitude limits. Sea-level to 2 000 m.
Rainfall requirements. It generally occurs in a rainfall regime of 400-1 500 mm.
Drought tolerance. One of the most drought-resistant grasses because of its rapid growth and seeding each wet season, even if of short duration.
Soil requirements. It is adapted to soils of a wide range of textures.
Tolerance to salinity. It is tolerant of alkaline soils (Bogdan & Pratt, 1967).
Land preparation for establishment. Some soil disturbance such as discing is generally necessary.
Sowing methods. For reseeding rangeland in Kenya the seed has been broadcast on to uncultivated land among cut branches, and on to harrowed land.
Sowing depth and cover. It is surface sown and protected by branches, mulch or a light harrowing.
Sowing time and rate. It is sown just ahead of the wet season at 0.25 kg/ha of spikelets (Bogdan & Pratt, 1967).
Number of seeds per kg. About 1.25 million spikelets or 2.7 million loose caryopses (Bogdan & Pratt, 1967).
Vigour of growth and growth rhythm. It flowers in August in the Sahel and remains as standing hay into February (Boudet & Duverger, 1961).
Genetics and reproduction. $2n=20, 34, 36, 48$ (Fedorov, 1974).
Suitability for hay and silage. It makes excellent hay.
Chemical analysis and digestibility. See Table 15.22.
Palatability. It is quite a palatable grass.
Toxicity. It is rich in cyanogenetic glucosides at times and may be a danger to grazing stock.
Seed production and harvesting. It seeds heavily. Most of the seed to date has been hand harvested, but it lends itself to combine heading.
Economics. Sometimes used as a food grain in times of scarcity in India and Africa but is said to have an unpleasant taste and to cause internal disorders (Bor, 1960).

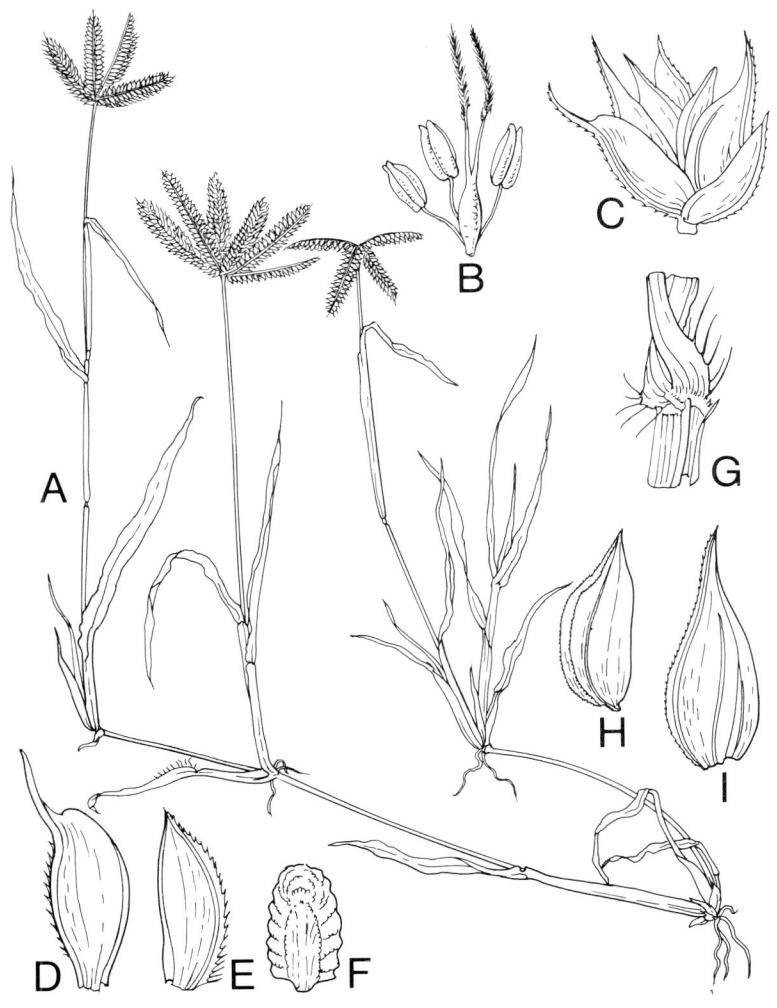

Figure 15.39. Dactyloctenium aegyptium. **A**-Habit **B**-Flower **C**-Spikelet **D**-Upper glume **E**-Lower glume **F**-Grain **G**-Ligule **H**-Palea **I**-Lemma

TABLE 15.22 *Dactyloctenium aegyptium*

	DM	As % of dry matter				
		CP	CF	Ash	EE	NFE
Fresh, early bloom, Kenya		15.8	26.8	10.0	1.8	45.6
Fresh, tall bloom, Kenya		15.6	27.9	13.6	1.5	41.4
Fresh, mid-bloom, Niger		8.7	32.4	10.6	1.5	46.8
Fresh, milk stage, India		7.3	33.7	12.5	1.2	45.3
Hay, Zimbabwe	91.3	8.3	36.9	6.5	0.8	47.5

	Animal	Digestibility (%)				
		CP	CF	EE	NFE	ME
Hay, Zimbabwe	Cattle	44.3	57.6	38.3	50.0	1.81

Source: Göhl, 1975

Animal production. No figures have been cited, but it is a useful, spontaneously growing component of pastures for grazing stock.
Main attributes. A quick-growing, short-term grazing plant which colonizes disturbed land.
Main deficiency. Its annual nature.
Further reading. Bogdan & Pratt, 1967.

Dactyloctenium giganteum Fischer and Schweick.

Common name. Giant button grass (Australia).
Natural habitat. On disturbed or sandy areas.
Distribution. Found throughout tropical Africa, especially southern Africa. The Queensland introduction Q.10091 was collected by Dr J.P. Ebersohn from southern Zimbabwe.
Description. It is similar to *D. aegyptium* but stouter and taller. It is an erect annual with broad leaves and a tendency to root at the nodes under favourable moisture conditions. The plant produces a profusion of tillers which lead to numerous inflorescences. The spikes are 4.5-10 cm long in groups of four to eight, often unequal (see Fig. 15.40).
Season of growth. Predominantly summer growing, but will overwinter if frosts are light.
Optimum temperature for growth. Germination is highest at 40-42°C.
Frost tolerance. It is killed by heavy frosts (–6°C) during June and July in Zimbabwe, but it retains its leaf.
Latitudinal limits. It occurs commonly between latitudes 25°N and S.
Rainfall requirements. The seed was collected in southern Zimbabwe in a rainfall regime of 450 mm spread over the months of October to April.
Drought tolerance. Once it germinates it generally escapes drought because of its short growing season.
Tolerance to flooding. It does not tolerate flooding.
Soil requirements. It prefers sandy soils and is a strong pioneer on the Kalahari sands in Africa. Under good moisture conditions it will establish on lateritic red earths and heavy self-mulching soils.
Fertilizer requirements. It is not usually fertilized but good yields were obtained by Bishop (1973) at Normanton, Queensland, in trial plots receiving 500 kg/ha of superphosphate and 250 kg urea per hectare.
Ability to spread naturally. It spreads well by seed under favourable conditions of soil, moisture and nutrients.
Land preparation for establishment. A rough seed-bed is usually sufficient, especially with sandy soils.
Sowing methods. The seed is broadcast on the surface and uncovered.
Sowing time and rate. At the beginning of the wet season at about 5 kg/ha.
Number of seeds per kg. About 2 million.
Dormancy. There is some after-ripening, the maximum germination in Queensland being obtained with two-year-old seed germinated at 40-42°C, with a figure of 68 percent. The germination figures for 1969 are shown in Table 15.23. Germination is stimulated by a submergence in water up to eight days from 11.3 percent after one day to 15.2 percent after eight days. Treatment with 1 percent Thiourea for 15 hours gave a germination of 40 percent; the control was only 9.2 percent (Batianoff, personal communication).

Figure 15.40. Dactyloctenium giganteum (**Source:** G. Batianoff, Queensland Department of Primary Industries)

Vigour of growth and growth rhythm. It flowers in 30-45 days and ripens seed in 60-80 days in latitudes 26-25°S in Queensland.

Dry- and green-matter yields. Yields of air-dried matter ranged from 650-5 500 kg/ha, the latter figure when fully fertilized.

Suitability for hay and silage. It makes good, soft, leafy hay.

Chemical analysis and digestibility. The percentage of nitrogen in the tops of *D. giganteum* was 0.3-0.35 without fertilizer and 0.3 and 0.4 with 500 kg/ha superphosphate. The percentage of phosphorus was 0.03 without and 0.05-0.08 with superphosphate.

TABLE 15.23 *Dactyloctenium giganteum*, germination

Age of seed	Germination (%)
3 weeks	1
8 months	18
1 year	34.7
2 years	56
3 years	64

Palatability. It is very palatable.
Toxicity. No nitrate poisoning has been reported.
Seed production and harvesting. It seeds heavily and is easily harvested.
Seed yield. 190 kg per hectare seed were harvested in four harvests in south-eastern Queensland in one year.
Economics. In Africa it is useful in quickly providing ground cover and grazing on cultivated land going out of production. In Kordofan Province, Sudan, it is used as a cut and carry grass for feeding livestock in the villages.

Dichanthium annulatum (Forsk.) Stapf

Synonym. *Andropogon annulatus* (Forsk.).
Common names. Sheda grass (Australia), lindi (the Philippines), karad, marvel grass, Delhi grass (India), Kleberg blue-stem (United States), pitilla (Cuba).
Distribution. Tropical Africa to Southeast Asia, New Guinea and northern Australia.
Description. Tufted perennial to 60 cm; the nodes bearded; leaves papillose-pilose at least on the upper surface; first glume of the sessile spikelet not indurate, or slightly indurate. Two to six racemes, sometimes more. Lower glume of sessile spikelet with tubercle-based hairs toward the tip (Tothill & Hacker, 1973), oblong, obtuse or truncate, keel not winged. Median nerve present, sheaths terete, ligule longish (Bor, 1960). It differs from *D. caricosum* in having the first glume keeled, not winged, a medial nerve, and large membranous ligule (Dabadghao & Shankarnarayan, 1973). Ninety-six percent of its roots end within a depth of 1 m. It differs from *Bothriochloa pertusa* in having no pitting on the glumes (Narayanan & Dabadghao, 1972) and from *Dichanthium sericeum* by the spikelets having a naked appearance due to the hairs being few or almost absent. The spikelets are also very blunt at the top (White, personal communication). Roots penetrate to 100 cm in alluvial soil at Varanasi, India, with a yield of 11 275 kg/ha of oven-dried roots (Ramam, 1970) (see Fig. 15.41).
Minimum temperature for growth. In Bihar, India, it produced 3 214 kg/ha in summer and 367 kg/ha in winter with a dressing of 44 kg N and 34 kg P_2O_5/ha per year (Singh & Chatterjee, 1968). There was some winter greenness.
Latitudinal limits. 8-28°N in India.
Altitude range. It has a range of 250-1 375 m in India.
Rainfall requirements. Tropical and subtropical rainfall patterns. It is found mostly in India in the 500-900 mm rainfall regime (Dabadghao & Shankarnarayan, 1973). It persisted poorly in arid zone trials at Alice Springs, Australia (Millington & Winkworth, 1970). The strain was from India.
Drought tolerance. It evades or endures drought well (Whyte, 1968).
Tolerance to flooding. It survives short-term flooding.
Soil requirements. It tolerates a wide range of soils but prefers black cotton soils in India and will not thrive in acidic soils.
Tolerance to salinity. It tolerates saline soils well and occurs on such soils in India in association with *Sporobolus marginatus*.
Fertilizer requirements. In India, the application of 22.75 kg N/ha to a natural pasture of predominantly *D. annulatum* increased the content of *D. annulatum* and decreased *Heteropogon contortus* and *Eremopogon* (Erasmus & Sud, 1976).

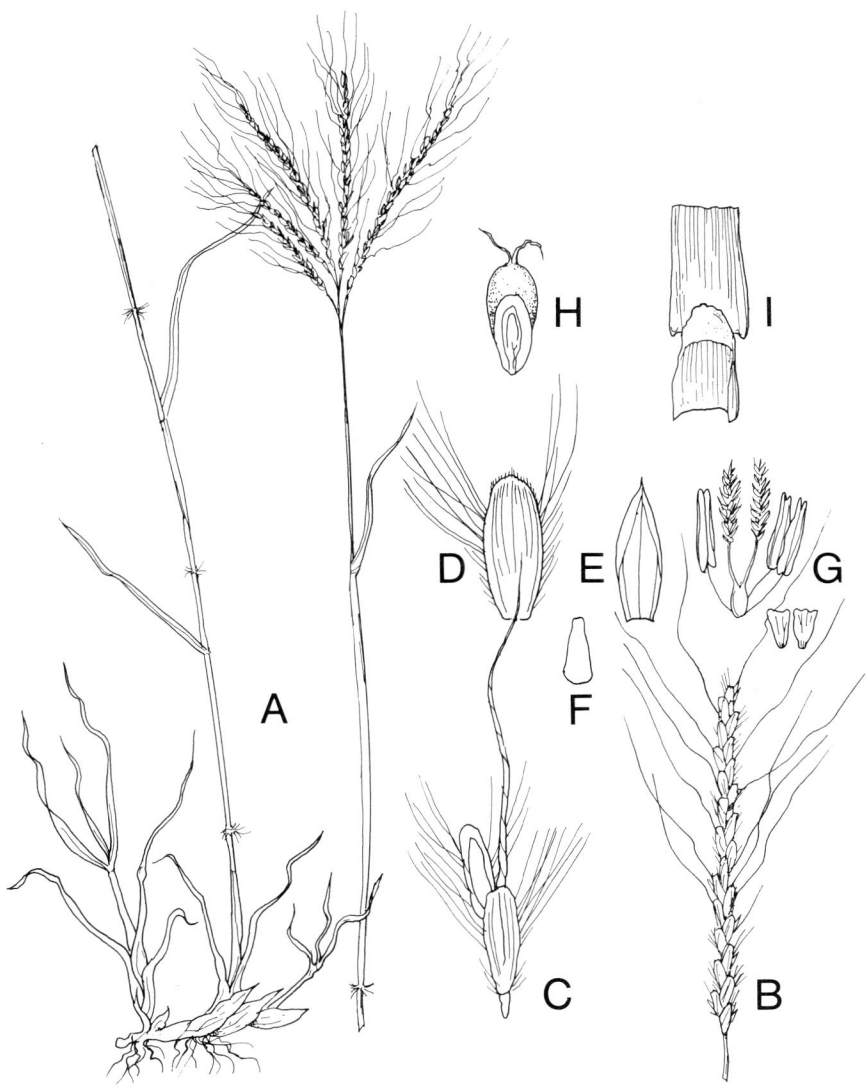

Figure 15.41. Dichanthium annulatum. **A**-Habit **B**-False raceme **C**-Pedicelled spikelet **D**-Lower glume of sessile spikelet **E**-Upper glume **F**-Lower lemma **G**-Flower **H**-Grain **I**-Ligule

Ability to spread naturally. Good.
Land preparation for establishment. A good seed-bed is required for early establishment, but it will gradually colonize a rough seed-bed.
Sowing methods. Usually established from root slips in India, as seed collection is laborious and expensive. It is sown in rows 60 cm apart with a similar distance between the plants, as they form large tussocks.
Sowing time and rate. Sow at the commencement of the wet season.
Dormancy. In India, Leelavathy (1969) found the filtrate of the rhizosphere fungus *Trichoderma viride* reduced the germination of *D. annulatum* seed from 90 to 77 percent.
Growth rhythm. It grows during the wet season from June to November in India and after harvest in November for hay. It provides spring growth from February to March, but this growth is stemmy (Dabadghao & Shankarnarayan, 1973).
Compatibility with other grasses and legumes. It does not grow well in mixtures as it crowds out other grasses.
Response to defoliation. It forms an open turf under grazing and stands very heavy grazing (Dabadghao & Shankarnarayan, 1973).
Grazing management. It is not usually managed, but, if overgrazed, a *Dichanthium-Iseilema* grassland in Bellary, Mysore, India deteriorates first to *Bothriochloa* sp., then *Eremopogon* sp. and *Andropogon* sp. and finally to an inferior *Aristida* sp./*Andropogon* sp./*Eragrostis* sp. sward (Chinnamani, 1968). Green-matter production fell from 6 000-10 000 kg/ha with *Dichanthium-Iseilema* to 200-1 500 kg/ha with the poor *Aristida/Andropogon/Eragrostis* association.
Genetics and reproduction. 2n=20, 40, 60 (Fedorov, 1974). It is quite a variable species.
Dry- and green-matter yields. At Bellary, Mysore, India, on black cotton soils a mixture of *D. annulatum* and *Iseilema anthephoroides* yields 6 000-10 000 kg/ha of green herbage (Chinnamani, 1968). An average hay production of 3 300 kg/ha can be expected from a good *D. annulatum* stand (Dabadghao & Shankarnarayan, 1973; Srinivasan, Bonde & Tejwani, 1962).
Suitability for hay and silage. It is widely used for hay in India (Narayanan & Dabadghao, 1972).
Chemical analysis and digestibility. See Table 15.24.
Palatability. Good. Preferred to *Cenchrus ciliaris* in India.
Seed production and harvesting. It seeds heavily and in India seeds are hand collected.
Cultivars. In Maharashtra State, India, Oke (1971) recommends 'Marvel 8', a strain which gives 40-100 percent better yields of dry matter than the local strain, while Chakravarty and Kackar (1971) selected Nos. 485, 487 and 490, selections which produced annual mean yields in the 1965-69 seasons of 1 450, 2 010 and 1 760 kg DM/ha respectively. There are two varieties in Queensland: var. *grandispiculatum,* which is taller and more robust than the

main type, and var. *monostachym,* which differs in having a single spike to the seed-head in place of the usual several.

Value for erosion control. Of numerous grasses tested by Srinivasan, Bonde and Tejwani (1962) for stabilizing the bunds in the ravine lands of Gujarat, India, *D. annulatum* proved one of the best because of its elaborate root system and excellent ground cover. It has also proved useful for erosion control on 20° slopes (Misra, Ambasht & Singh, 1977).

Economics. On the black cotton soils of Bellary, Mysore, India, it is a climax

TABLE 15.24 *Dichanthium annulatum*

	DM	As % of dry matter				
		CP	CF	Ash	EE	NFE
Fresh, 6 weeks, India	25.0	10.4	34.9	12.1	1.7	40.9
Fresh, 10 weeks, India	35.3	6.1	38.8	9.8	1.3	44.0
Fresh, 14 weeks, India	37.6	5.9	41.4	9.8	1.1	41.8
Fresh, early bloom, Kenya		8.5	39.0	6.9	1.1	44.5
Fresh, milk stage, India		4.7	40.7	10.9	1.3	42.4
Fresh, 1st cut, India		5.2	31.4	11.9	1.6	49.9
Fresh, 2nd cut, India		3.8	35.0	9.9	1.2	50.1
Fresh, 3rd cut, India		2.7	33.3	10.1	1.0	52.9
Fresh grown in saline soil, India		2.6	37.7	10.3	1.7	47.7
Hay, late vegetative, India		4.6	38.9	9.5	0.9	46.1
Hay, mid-bloom, India		4.1	39.9	10.6	1.0	44.4
Hay, mature, India		2.7	39.1	11.5	1.2	45.5

	Animal	Digestibility (%)				ME
		CP	CF	EE	NFE	
Fresh, 10 weeks, India	Cattle	47.4	70.3	44.6	46.2	1.89
Hay, late vegetative, India	Zebu	28.0	60.4	29.2	42.4	1.63

Source: Göhl, 1975

species with *Iseilema anthephoroides* and is a palatable and nutritious species (Chinnamani, 1968). It is a climax species along with *Sehima nervosum* over practically the whole of peninsular India and one of the most important grasses in the *Dichanthium/Cenchrus/Lasiurus* cover in semi-arid northern India (Dabadghao & Shankarnarayan, 1973). It is widely used in the Philippines for pasture improvement.

Animal production. In the Rajasthan desert area of India, *D. annulatum* has supported 6.93 sheep per hectare at Pali (Das, 1973).
Main attributes. Its wide adaptability, tolerance of alkaline soils and its effective erosion control.
Main deficiencies. Its variability and its dominance of other grasses.
Further reading. Harlan *et al.,* 1961; Whyte, 1964.

Dichanthium aristatum (Poir.) C.E. Hubbard

Common name. Angleton grass (Australia, Cuba), alabang X (the Philippines), Angleton blue-stem (United States), wildergrass (Hawaii).
Distribution. Originated in India and introduced into Australia, Africa and America. Well distributed over the Philippines (Farinas, 1970).
Description. A perennial with slender erect culms to 90 cm; nodes usually bearded, two to four racemes, erect and rather close, pedunculate, first glume of the spikelet not indurate (Henty, 1969). Stalks of the racemes hairy, pedicellate spikelet usually male or bisexual, sometimes neuter, but with both glumes well developed and often with lemmas (Tothill & Hacker, 1973). Var. *heteropogonoides* produces stolons up to 3 m long in the wet season and roots at the nodes — called alabang X in the Philippines (Farinas, 1970) (see Fig. 15.42).
Season of growth. Summer.
Minimum temperature for growth. It makes slow growth during winter and spring.
Frost tolerance. It is not frost resistant, but is acceptable to stock after frosting.
Rainfall requirements. It has a high water requirement and annual rainfall over 875 mm is recommended.
Drought tolerance. It is a very drought-resistant perennial (Farinas, 1970).
Tolerance to flooding. Very good.
Soil requirements. It tolerates a wide range of soils.
Tolerance to salinity. Highly salt tolerant.
Fertilizer requirements. It has a very low phosphorus requirement, but does respond to improved soil fertility.
Sowing methods. It has been broadcast into sorghum stubble in northern Queensland, the cattle trampling in the seed as the sorghum is grazed, and it has been successfully sown under a cover crop of oats, on the Darling Downs in Queensland, on a well-prepared seed-bed. It is also propagated by runners, 30-45 cm apart in rows 1 m apart (Archer & Bunch, 1953).
Seedling vigour. Excellent. It establishes easily.
Vigour of growth and growth rhythm. It is a vigorous grass with a tendency to dominate and become a weed in some areas.
Response to photoperiod. Short days obligate (Evans, Wardlaw & Williams, 1964; Knox, 1967). There is a negative relationship between day length and percentage apomixis (Knox, 1967).
Compatibility with other grasses and legumes. In northern Queensland, strip planted into *Heteropogon contortus* pastures at about 20-m intervals, it invaded and nearly eliminated the spear grass over a seven-year period. It combines well with Townsville stylo (*S. humilis*) (Onley & Sillar, 1965).

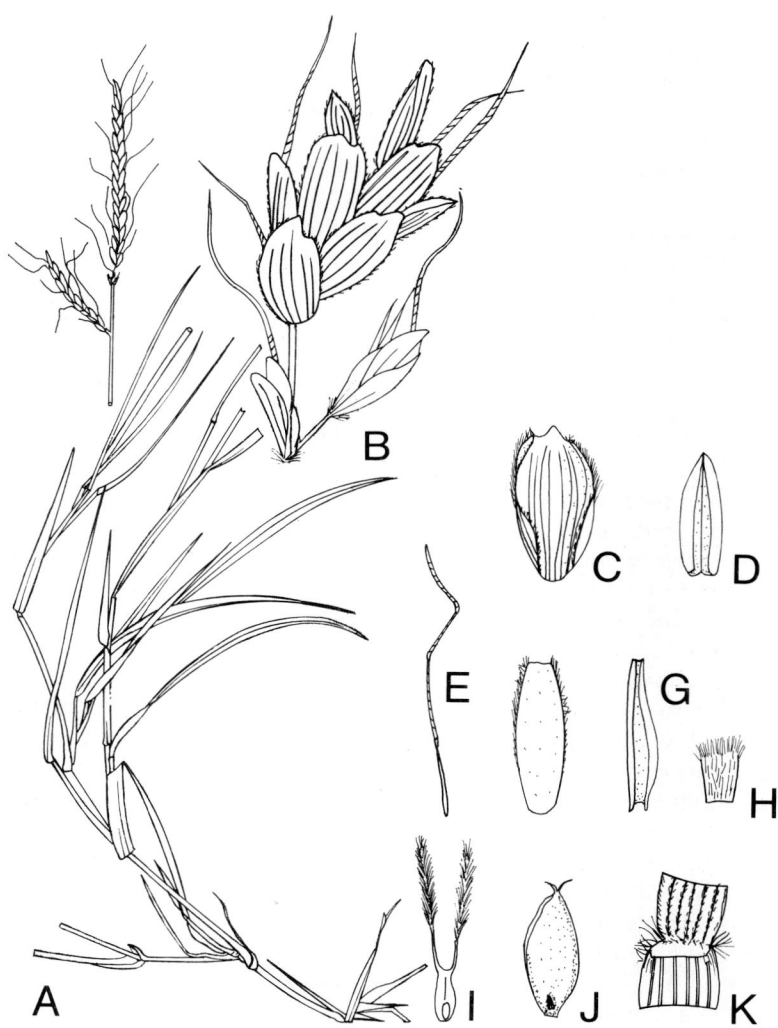

Figure 15.42. Dichanthium aristatum. **A**-Habit **B**-Inflorescence **C**-Upper glume **D**-Lower glume **E**-Upper lemma **F**-Lower lemma **G**-Upper palea **H**-Lower palea **I**-Flower **J**-Grain **K**-Ligule

Response to defoliation. It can stand heavy grazing along with its companion legume *S. humilis* during the summer.

Grazing management. In order to maintain a high proportion of the associated legume, *S. humilis*, it is necessary to graze it heavily during summer.

Genetics and reproduction. 2n=20, 40, 60 (Fedorov, 1974). It is a facultative apomict. In day lengths of less than 13 hours reproduction is mostly apomictic, while in longer days it becomes more than 50 percent sexual (Knox, 1967 quoted by Tothill, 1970).

Dry- and green-matter yields. It yields an average of 10 tonnes of hay per hectare in the United States (Archer & Bunch, 1953).

Suitability for hay and silage. It is cut for hay in the United States just before flowering.

Value as standover or deferred feed. Excellent.

Palatability. It is very palatable and is also acceptable to cattle after it has seeded and matured (Onley & Sillar, 1965).

Seed production and harvesting. At Taranga, Bloomsbury, northern Queensland, the pasture is closed in April and harvested for seed in May-June, by direct heading.

Seed production and harvesting. It is a prolific seeder.

Seed yield. 75-150 kg/ha from one or two cuts per year, when irrigated (Ferguson, 1979).

Minimum germination and quality required for commercial sale. Germinate at 20-30°C, moistened with water (Prodonoff, 1966).

Cultivars. There are numerous ecotypes but no cultivars have been released. Those ecotypes occurring naturally in the 600-750 mm rainfall zone in Queensland appear to be poor producers, whereas those in higher rainfall areas up to 1 800 mm have given good production in association with *Stylosanthes humilis* (Onley & Sillar, 1965).

Economics. An excellent fodder grass, widely used in the Philippines. At Lawes, Queensland, it provided excellent grazing over the short period from February to April (autumn).

Animal production. At Taranga, Bloomsbury, even after heavy summer grazing a Townsville stylo/Angleton grass pasture is capable of maintaining breeders in good condition for the remainder of the year at a stocking rate of one beast to 1.2 hectares.

Further reading. Onley & Sillar, 1965.

Dichanthium caricosum (L.) A. Camus

Common names. Nadi (pronounced "nandi") blue grass, nawai grass (Fiji), jiribilla (Cuba), Antigua hay grass (West Indies).
Natural habitat. Swampy places, open humid woodland, black cotton soils.
Distribution. Native to India, Burma, Sri Lanka, Malaysia, China, New Guinea and Fiji.
Description. Tufted perennial with slender culms to 45 cm; nodes bearded, leaves and sheaths glabrous; one raceme, sessile, first glume of the sessile spikelet more or less indurate; first glume of the pedicelled spikelet obovate, the upper margin purple, ciliate. In the strain Nadi blue grass, the nodes are glabrous (Henty, 1969). It has blue-tinged leaves. When the plants are young the stems are a bright mauve-blue colour (Parham, 1955). It differs from *D. annulatum* in having the first glume of the sessile spikelet winged, with no medial nerve and a short, ciliate ligule (Dabadghao & Shankarnarayan, 1973) (see Fig. 15.43).
Season of growth. Summer.
Optimum temperature for growth. It makes most growth between 32-35°C in Fiji.
Minimum temperature for growth. About 12-14°C in July in Fiji.
Altitude range. 600-1 000 m.
Rainfall requirement. It is an important grass in the "dry" zone of Fiji where rainfall ranges from 1 500-2 500 mm a year with a dry season from May to November. However, it does not produce well in dry weather.
Drought tolerance. It tolerates a rather dry winter-spring well, but makes little growth in dry weather.
Tolerance to flooding. It is quite tolerant of waterlogging (Partridge, 1979b) and is suited to the waterlogged black clays.
Soil requirements. It prefers dry, sandy habitats in Kerala, India.
Fertilizer requirements. It has a low fertility requirement.
Ability to spread naturally. It spreads well by vigorous stolons under moderate grazing.
Land preparation for establishment. In Fiji, burn the mission grass (*Pennisetum polystachyon* and, without cultivation, broadcast seed on the surface. (Partridge, personal communication).
Sowing methods. It is propagated by seed or by division of the roots. Broadcast the seed by hand or from aircraft on to burnt mission grass.
Sowing depth. Good germination results from surface sowing.
Sowing time and rate. Sow at a minimum rate of 2 kg/ha at the beginning of the wet season in November-December in Fiji (Partridge, personal communication).
Dormancy. Germination improves up to nine months in storage (Parham, 1960).

Figure 15.43. Dichanthium caricosum. **A**-Inflorescence **B**-Pair of spikelets

Seed treatment before planting. De-awning makes seed distribution more even (Partridge, personal communication).
Seedling vigour. Very good.
Vigour of growth and growth rhythm. It flowers in May and June at the beginning of the dry season in Fiji, after which the quality and quantity of growth decline rapidly (Partridge & Ranacou, 1974).
Response to photoperiod. It is a short-day plant. Seed ripens in early June in Fiji.
Response to light. It grows well under coconuts and *Pinus caribea* in Fiji.
Compatibility with other grasses and legumes. It combines well with *Stylosanthes hamata* and siratro (*Macroptilium atropurpureum*) and the naturalized legumes *Desmodium heterophyllum, D. triflorum, Alysicarpus vaginalis* and *Mimosa pudica* in Fiji (Partridge, 1979b).
Ability to compete with weeds. It is a strong competitor with weeds on the unfertilized hill soils of the dry zone of Fiji (Roberts, 1970a, b).
Response to defoliation. It forms a close turf when grazed.
Grazing management. It is very resistant to heavy grazing.
Response to fire. If rested to provide a body of grass it burns well, but if burnt when grazed the results may be patchy.
Genetics and reproduction. 2n= 40, 60 (Fedorov, 1974).
Dry- and green-matter yields. From 1950-52, the average green-matter yield of Nadi blue grass at Sigatoka was 22 725 kg/ha (Roberts, 1970a, b). At Sigatoka, Fiji, Nadi blue grass yielded an average of 11 500 kg DM/ha per year in 1971-72 with 31 percent of the yield in 1972 in the dry season (Partridge, 1979a).
Suitability for hay and silage. It makes a high-quality hay.
Value as standover or deferred feed. It holds its palatability to maturity.
Chemical analysis and digestibility. Crude protein contents range from 6.3 percent of the dry matter in the wet seasons to less than 2.5 percent in the dry season in Fiji (Partridge & Ranacou, 1974) (see Table 15.25).
Palatability. Excellent, even to maturity (Parham, 1955).
Toxicity. None reported.
Seed production and harvesting. Seed is produced in April and May in Fiji.
Cultivars. In Maharashtra State, India, Oke (1971) recommends two new cultivars, Marvel 40 and Marvel 93, which produce 40-100 percent more dry matter than the naturally occurring strain.
Value for erosion control. Excellent. It colonizes bare areas quickly by both seed and runners (Roberts, 1970a, b).
Economics. One of the best pasture grasses in the dry areas of Fiji and the West Indies.
Animal production. On Nadi blue grass alone, the annual live-weight gain of steers was 150 kg/ha in 1970/71, 105 kg/ha in 1971/72, and 95 kg/ha in 1972/73, to give a mean annual live-weight gain of 110 kg/ha (Partridge & Ranacou, 1974) (see Fig. 15.44). With 10 percent *Leucaena leucocephala* (vaivai) in the

TABLE 15.25 *Dichanthium caricosum*

	DM	As % of dry matter				
		CP	CF	Ash	EE	NFE
Fresh, mid-bloom, Tanzania	28.0	7.0	36.7	11.7	1.5	43.8
Hay, 1st cut, Tanzania	88.0	6.1	40.3	11.8	1.7	40.1

	Animal	Digestibility (%)				ME
		CP	CF	EE	NFE	
Fresh, mid-bloom, Tanzania	Sheep	54.3	77.9	64.0	66.8	2.31
Hay, 1st cut Tanzania	Sheep	45.9	61.3	65.0	47.6	1.79

Source: Göhl, 1975

pasture the mean live-weight gain was 170 kg/ha and, with 20 percent, 270 kg/ha. With siratro and superphosphate at one beast to 0.4 hectare, the annual live-weight gain was 364 kg/ha. At higher stocking rates, siratro declined but naturalized legumes increased and a live-weight gain of 380 kg/ha at a stocking rate of 3.5 beasts per hectare was optimum (Partridge, 1979b).

Main attributes. Its ability to colonize and spread by both seed and runners where fertilizers are not applied (Roberts, 1970a, b) in the "dry" zone of Fiji with 2 000 mm of rain a year. Its low fertility requirement, tolerance to waterlogging, tolerance of heavy grazing and good ground cover for the control of erosion and weeds (Partridge, 1979b).

Main deficiencies. It grows poorly in dry weather and establishment of legume in the sward without cultivation is difficult.

Further reading. Parham, 1955; Partridge, 1979b.

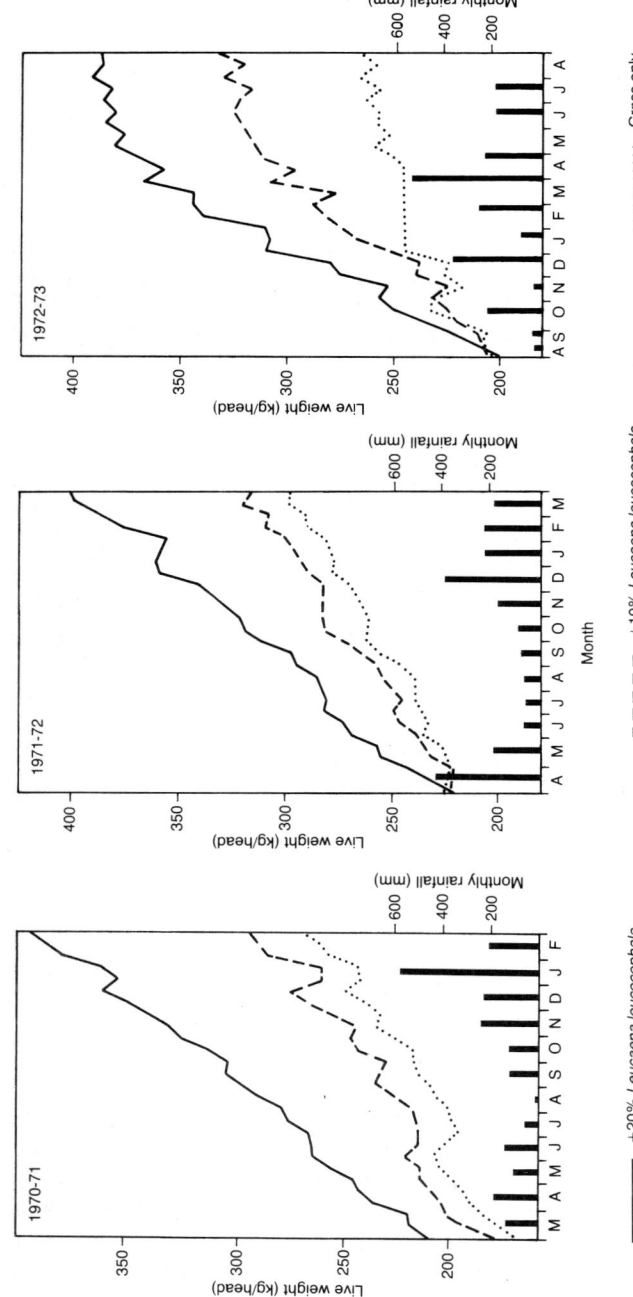

Figure 15.44. Effects of supplementary fertilized *Leucaena* spp. on cumulative annual live-weight gain (kilograms per head) of steers grazing *Dichanthium caricosum* in Fiji (**Source:** Partridge & Ranacou, 1974)

Dichanthium sericeum (R. Br.) A. Camus

Synonym. Andropogon sericeus R. Br.
Common names. Queensland blue grass (Australia), silky blue-stem (United States).
Natural habitat. Open grassland on heavy black clay soils.
Description. Erect perennial, often very finely stemmed and not over 30 cm high; nodes bearded, one or two racemes, sessile, the basal, imperfect spikelets often remaining as an involucre after the rest have fallen; first glume of the sessile spikelet more or less indurated. There is a good deal of variation within the species — var. *mollis* is a softly hairy form (Blake, 1944), var. *polystachyus* has a large head and ten to 30 spikes 3-5 cm long (Turner, 1891). Queensland blue grass is distinguished by its bluish-green colour, and soft silky seed-heads (see Fig. 15.45).
Season of growth. Spring, summer and autumn.
Frost tolerance. Extremely sensitive to frost, but rapidly comes into leaf on the approach of warm weather.
Latitudinal limits. 30°N to below 32°S.
Altitude range. About 200-300 m in Queensland.
Rainfall requirements. It generally grows in the 500-700 mm rainfall zone. If rainfall in excess of the higher figure occurs it becomes unpalatable in western Queensland.
Drought tolerance. It is not as tolerant of drought as the Mitchell (*Astrebla* spp.) grasses in western Queensland.
Soil requirements. It grows best on friable black earths on the Darling Downs and Central Highlands in Queensland where the soils are derived from basalt, but throughout Australia (except Tasmania) it occurs on a wide variety of soils. They must, however, be fertile, as the grass is easily pulled out on poor soils where rooting is poor.
Fertilizer requirements. D. sericeum responded to phosphorus, sulphur and nitrogen on a red-brown earth soil in northwestern New South Wales (Lodge, 1979). The response was linear up to 90 kg P/ha, 90 percent of maximum yield was obtained with 10 kg S/ha and response was linear up to 150 kg N/ha. The grass is not usually fertilized, and it may not respond to fertilizers when growing on black clay soils.
Vigour of growth and growth rhythm. It grows vigorously when conditions are suitable. It is one of the earliest grasses to shoot in the spring and flowers February to May in Queensland.
Response to defoliation. Scateni (1966) showed that cutting in January at Gayndah, Queensland (lat. 27.5°S, rainfall 745 mm) caused a decrease of 2.9 percent in the basal cover of Queensland blue grass whereas a February cut gave a slight increase, as did a May cut. The species flowers in early

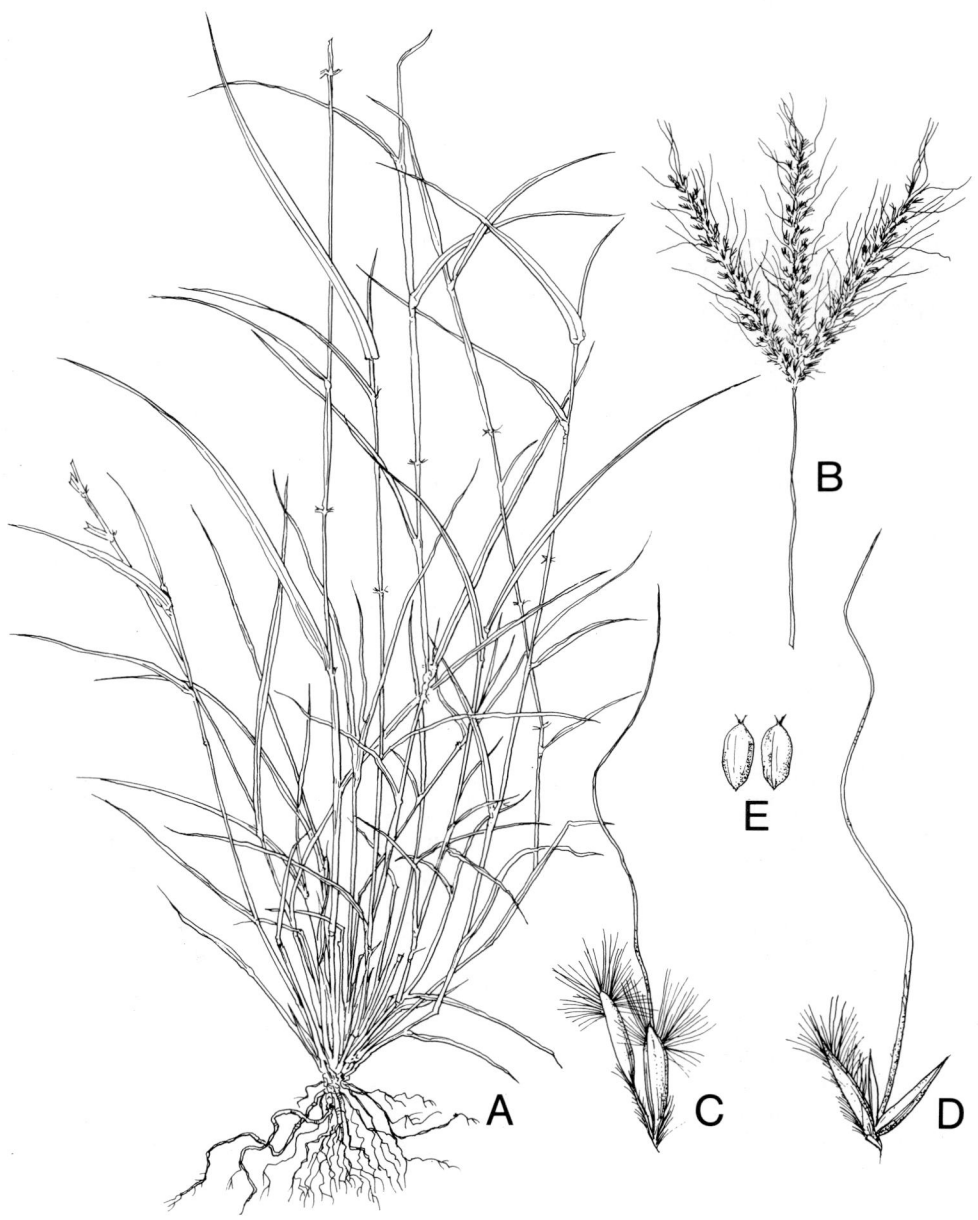

Figure 15.45. *Dichanthium sericeum.* **A**-Habit **B**-Inflorescence **C**-Pair of spikelets **D**-Open spikelet **E**-Grain

January. It stands a good deal of trampling. It disappears under heavy stocking (Everist, 1935).
Grazing management. It should be moderately stocked or it can be eaten out. It should be allowed to seed and thicken its stand periodically.
Genetics and reproduction. 2n=20 (Fedorov, 1974).
Suitability for hay and silage. It makes good soft hay.
Chemical analysis and digestibility. The Queensland Department of Primary Industries recorded 10 percent crude protein, 33.1 percent crude fibre, 1.1 percent ether extract, 0.54 percent CaO and 0.55 percent P_2O_5 in the dry matter of green material in seed (*Paspalum*, 1954).
Palatability. It is very palatable in the green state, but not attractive when high rainfall leaches its nutrients.
Seed production and harvesting. It is a heavy seeder. No mechanical harvesting is done in Queensland.
Minimum germination and quality required for commercial sale. 50 percent germinable seeds and 95 percent purity in Queensland.
Cultivars. Strain 64 has been selected by CSIRO (Downes, 1969).
Value for erosion control. Its tussocky nature and not very extensive root system do not make it effective for erosion control.
Economics. One of the best natural pastures for sheep and cattle in inland Australia.
Animal production. In Queensland the carrying capacity of *D. sericeum* pastures is rated at about one animal to five hectares in the 600-700 mm rainfall area of the Darling Downs.
Main attributes. Its palatability and early spring growth.
Main deficiencies. Its frost susceptibility and its extreme palatability tend to lead to its disappearance. Under heavy rainfall it becomes rank and unpalatable in central Queensland.
Further reading. White, 1935.

Dichanthium tenuiculum (Steud.) S.T. Blake

<u>Synonym.</u> *D. superciliatum* (Hack.) A. Camus.
<u>Common name.</u> Tassel blue grass (Australia).
<u>Distribution.</u> Northern Australia, Papua New Guinea, Timor, and the Philippines.
<u>Description.</u> A perennial tussock grass differing from *D. affine* and *D. sericeum* in having more than 15 racemes where the latter have two to 12 (Simon, 1980). It is a robust grass growing to 80 cm with bearded nodes; the first glume of the sessile spikelet not indurate, prominently seven-nerved; the first glumes of both spikelets bearing long silky hairs on the margins; a light fringe of hairs on the rachis-joints and pedicels (Henty, 1969).
<u>Economics.</u> It is a very valuable grass for the wetter heavy soil "melonholes" or gilgais in the Burdekin Valley in north Queensland (lat. 19°36'S, rainfall 850 mm) and gives dry season grazing (Kinsey, personal communication). It will grow on cracking clays, and flood plains with medium-textured red earths or finely cultivated soils.

Digitaria abyssinica (A. Rich.) Stapf

Synonym. *D. vestita* Fig. & De Not. var. *scalarum* Schweinf.; *D. scalarum* (Schweinf.) Chiov.
Common names. African couch grass (Tanzania), couch finger grass, Dunn's finger grass (South Africa), thangari (Kenya).
Natural habitat. Grassland, and as a weed in plantation crops.
Distribution. Native to Zaire and eastern tropical Africa.
Description. Perennial with slender long rhizomes and erect culms up to 30 cm high; leaf-blades 3-5 mm wide, but on occasions up to 7-8 mm. Panicle of two to nine racemes, often whorled and sub-erect, 2.5-8 cm long with broadly elliptic, completely glabrous obtuse spikelets 2 mm long (Napper, 1965). The rhizomes form a dense mat beneath the soil surface, extending to depths greater than 1 m, and may twine around the roots of perennial crops (Terry, 1974). It differs from *Cynodon dactylon* in the vegetative stage in having an obvious membranous ligule where the leaf-blade joins the sheath (Ivens, 1967) (see Fig. 15.46).
Altitude range. Sea-level to 3 000 m.
Rainfall requirements. It prefers more humid areas. Common near Bukoba on shores of Lake Victoria, west Tanzania. Rainfall should be in excess of 500 mm.
Tolerance to herbicides. Terry (1974) found glyphosate (Round-up) at 2 and 4 kg/ha and at split applications of 1+1 and 2+2 kg/ha gave very good control of *D. abyssinica* foliage but 1 kg/ha was less effective. Asulam at 2,4 and 8 kg/ha gave poor control. Dalapon at 5 kg/ha gave moderate control which was improved by 0.5 kg/ha paraquat applied 51 days after treatment with dalapon. Prolific growth of annual broad-leaved weeds occurred where *D. abyssinica* had been controlled with glyphosate and dalapon and these would need subsequent control. Richardson (1967) obtained success in Kenya sisal plantations with heavy applications of sodium trichloro-acetate (Na-TCA). It was successfully controlled in Zambia coffee plantations by running poultry intensively (Chippendall & Crook, 1976).
Genetics and reproduction. 2n=36 (Fedorov, 1974).
Dry- and green-matter yields. Richardson (1967) reported fresh weight yields of *D. abyssinica* from sisal land of 36 t/ha which caused substantial fibre losses in the sisal.
Chemical analysis and digestibility. Dougall and Bogdan (1960) record 14.7 percent crude protein, 29 percent crude fibre, 9.5 percent ash, 3.8 percent ether extract and 43 percent nitrogen-free extract in fresh material at the early bloom stage. Verboom and Brunt (1970) recorded 8.69 percent crude protein, 31.48 percent crude fibre, 3 percent ether extract, 6.17 percent ash and 50.66 percent nitrogen-free-extract from flowering material in Zambia.

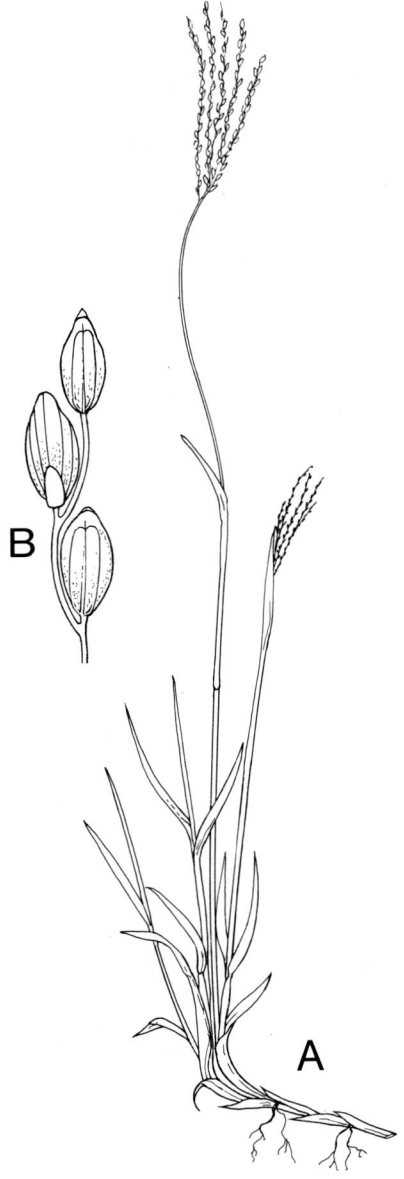

Figure 15.46. Digitaria abyssinica. **A**-Plant **B**-Spikelet

Palatability. Fairly palatable when young, but unproductive. It was accepted by Ankole bullocks in the wet season in Uganda, but not in the dry season (Harrington & Pratchett, 1972).
Value for erosion control. It has been planted on the slopes of the Cape Peninsula in Africa to control erosion (Chippendall, 1955).
Economics. The grass is often a troublesome weed in cultivations, in plantations and in orchards. It has been planted on the slopes of the Cape Peninsula in Africa to control erosion (Chippendall, 1955). The most troublesome of all African weeds (Ivens, 1967). It is, however, used in leys at Nemalonge, Uganda, in cotton rotations.
Further reading. Ivens, 1967.

Digitaria ciliaris (Retz.) Koeler

Synonym. *D. adscendens* (H.B.K.) Henr.; *D. marginata* Link; *D. sanguinalis* (L.) Scop.

Common names. Summer grass (Australia), pasto colchón, pata de gallo, gramilla (Peru), pata de gallina (Cuba), hairy crabgrass (Oklahoma), wild crabgrass (Hawaii).

Natural habitat. Sandy soils and loams, as a weed in cultivation.

Distribution. A widely distributed tropical weed.

Description. An annual, caespitose with branching culms; nodes pilose; leaves linear, acuminate, the sheaths pilose; ligule elongate, obtuse, glabrous; up to ten racemes on a triquetrous rachis; spikelets unilateral, geminate, one sessile, one pedicelled, ovate; lower glume small, upper three-nerved, pilose-ciliate; sterile lemma as long as the upper glume, three-nerved, margin ciliate; fertile lemma as long as the sterile lemmas, glabrous (Henty, 1969) (see Fig. 15.47).

Altitude range. Sea-level to 1 800 m.

Tolerance to herbicides. Cultivation will usually kill the grass. If a herbicide is needed, a pre-emergence spray with diuron at 6 g of an 800 g AI/kg product (e.g. Karmex) per litre of water plus 6 ml surfactant. Apply with a mister till solution runs off the leaves (Tilley, 1977). It can be treated with PCP either as a pre-emergence spray or when the seedlings are young. Oil emulsions fortified with PCP will kill or reduce the vitality of older plants (Kleinschmidt & Johnson, 1977).

Genetics and reproduction. 2n=54 (Fedorov, 1974).

Dry- and green-matter yields. In Oklahoma, United States, Dalrymple (1978) obtained yields of 9 520-11 760 kg DM/ha and 2 240-3 360 kg DM/ha when double-cropped behind small grains.

Chemical analysis and digestibility. Dewald's (1978) figures for Starr pearl millet, Tift Sudan grass, crabgrass and oats under similar conditions in Oklahoma are shown in Table 15.26.

Palatability. It is quite palatable.

Economics. A common weed of cultivated and waste or disturbed land, particularly on sandy soils. Germination is stimulated by exposure to light. In subtropical and dry tropical environments it affects the quality of *S. humilis* hay in northern Australia and *Medicago sativa* (lucerne) hay in southern Australia. In Oklahoma, United States, it is used as a forage species (Dewald, 1978).

Animal production. Average daily gains of steers on crabgrass pastures for short periods in Oklahoma averaged 0.98 kg/day during four grazing years: 1971, 1972, 1974 and 1976 (Dewald, 1978).

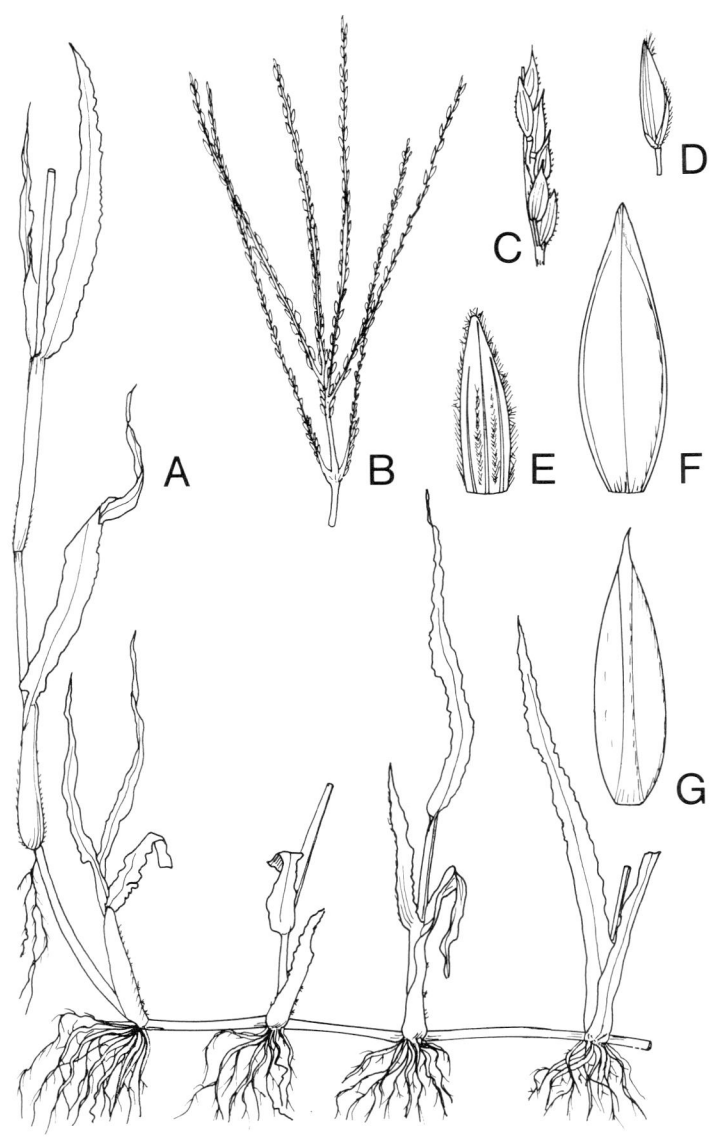

Figure 15.47. Digitaria ciliaris. **A**-Habit **B**-Inflorescence **C**-Portion of raceme **D**-Spikelet **E**-Upper glume **F**-Upper lemma **G**-Palea

TABLE 15.26 *Digitaria ciliaris*

	DM	CP	Fibre	TDN
Star pearl millet	91.7	8.0	32.2	5.1
Tift Sudan grass	92.2	7.6	33.3	4.5
Crabgrass	92.0	13.8	29.4	8.6
Oats	89.4	24.2	13.6	8.2

	Digestibility (%) (3 lambs)			
	DM	CP	Fibre	TDN
Star pearl millet	61.0	52.2	75.3	60.5
Tift Sudan grass	64.5	55.7	67.7	63.4
Crabgrass	66.4	70.6	73.0	64.4
Oats	81.3	82.3	84.1	81.7

Source: Dewald, 1978

Main attributes. It grows on poor soils and gives some useful forage in this situation.

Main deficiencies. It is a vigorous, stoloniferous, weedy species invading crops and pastures, especially on sandy soils.

Further reading. Dewald, 1978.

Digitaria decumbens Stent

Common names. Pangola grass (United States, Australia), pasto pangola (Peru), pangola digit grass (Florida).

Distribution. Originated in the Transvaal and now introduced into most tropical countries.

Description. A stoloniferous perennial which differs from *D. pentzii* mainly in having the culms much branched, usually decumbent, and often rooting from the lower nodes, the spikelets 2.5-3 mm long and quite glabrous, the hairs on the upper glume and lower lemma being short, fine and inconspicuous. *D. decumbens* is based on a plant from the Nelspruit district of the Transvaal. Subsequently plants were referred to it that were cultivated in Pretoria as the "Pangola River" strain of woolly finger grass (Chippendall, 1955) (see Fig. 15.48).

Season of growth. Summer, with a growth period of up to seven months. It commences growth in November and slows down in March in southeast Queensland.

Optimum temperature for growth. The mean temperature where it grows is 19-24°C (Russel & Webb, 1976). In Hawaii it grows well between 25 and 40°C (Whitney & Greeen, 1969). It grows actively above 26.5°C (Bryan & Sharp, 1965).

Minimum temperature for growth. The mean temperature of the coldest month in areas where it normally grows is 7-15°C (Russell & Webb, 1976). Blue, Gammor and Lundy (1961) suggested a 10°C minimum in Florida, and Bryan and Sharpe (1965) about 15°C in southeast Queensland.

Frost tolerance It is susceptible to frosts, but generally recovers during subsequent warm weather.

Latitudinal limits. The mean latitude is given as 21-30°N or S by Russell and Webb (1976).

Altitude range. In Hawaii, sea-level to 1 500 m.

Rainfall requirements. The mean rainfall is 900-1 975 mm (Russell & Webb, 1976). In South Africa it grows well under the 625-750 mm summer dominant rainfall on moist, fertile, well-drained soils but is better suited to rainfalls of 1 000-1 200 mm in coastal regions.

Drought tolerance. It will survive droughts fairly well if established, but will not be productive.

Tolerance to flooding. It can withstand temporary, but not sustained, flooding.

Soil requirements. Pangola grass will grow over a wide range of soils on wet sands or heavy clays and at low fertility levels. It will grow on poorer soil than *Paspalum dilatatum* and *Pennisetum clandestinum* but in low fertility soils is not very attractive to hungry stock.

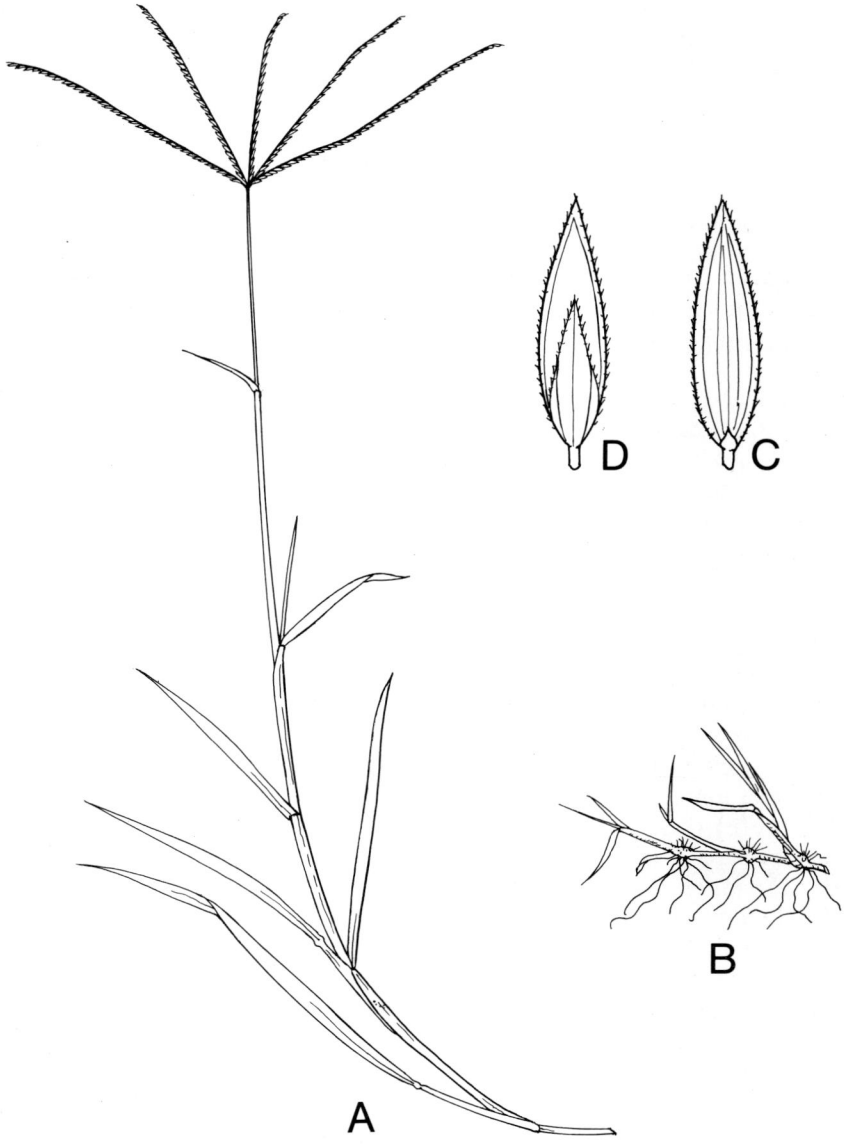

Figure 15.48. Digitaria decumbens. **A**-Decumbent stem **B**-Creeping stem **C**-First glume **D**-Second glume

Tolerance to salinity. Fair.

Fertilizer requirements. Pangola grass responds markedly to nitrogen when matched with adequate phosphorus, potash and trace elements. Normal applications of nitrogen under irrigation are about 300 kg/ha per year. At Palmira Experiment Station in the Cauca Valley, Colombia, dry-matter production increased in linear fashion with each additional increment of nitrogen up to 224 kg/ha and in a curvilinear manner with higher applications. The critical value for phosphorus in the dry matter at the immediate pre-flowering stage is 0.16 percent. It can fix up to 65 kg N/ha per day (Weier, 1976). It tolerates high aluminium (Spain, 1979).

Ability to spread naturally. Once pangola grass is established it spreads very rapidly by stolons. It does not produce viable seeds.

Land preparation for establishment. An initial ploughing or other soil disturbance is required so that sprigs can be incorporated in a loose soil.

Sowing methods. As pangola grass does not produce viable seed, propagation is effected by sprigs or roots. For large areas, the quickest method is to cut the vegetative plant material in a chaff-cutter or forage harvester adjusted to provide plant pieces with a few nodes, broadcast this material over the ploughed field and disc-harrow it in, or simply broadcast whole runners and disc the area. Hand planting of roots about 1 m apart will soon give a good cover. A small nursery area established ahead to provide material for future enlarged planting is a sound approach.

Sowing time. Preferably during the rainy season so that the planting material will root down quickly.

Vigour of growth and growth rhythm. Pangola grass grows vigorously when established.

Response to photoperiod. It is a long-day plant and flowers in 14 hours of daylight (Degras, Mathurin & Félicité, 1974).

Response to light. It showed low tolerance to 50 percent light under coconuts (Reynolds, 1978).

Compatibility with other grasses and legumes. Under suitable conditions for its own development, pangola grass dominates all other species. It combines well with the legume *Lotononis bainesii* (Bryan, 1961), as both stand heavy grazing. It can also grow with *Centrosema pubescens*. In Florida, Kretschmer (1965, 1966) has grown it satisfactorily with *Stylosanthes humilis* and *Macroptilium atropurpureum* (siratro) where addition of nitrogenous fertilizer is low.

Ability to compete with weeds. Pangola grass quickly suppresses weeds.

Response to defoliation. Pangola grass tolerates heavy grazing. Extremely heavy grazing is not harmful if the grass is allowed to grow again to a height of 30-45 cm afterwards (Bennett, 1973).

Grazing management. In Jamaica, Creek and Nestel (1965) found that pangola grass grazed at 32-day intervals produced more dry matter and more liveweight gain than when grazed at 40-day intervals. In the United States it is suggested that it be grazed rotationally, allowing one week's rest during grazings in midsummer and two to three weeks during the remainder of the growing season (Bennett, 1973). In Colombia, Crowder, Michelin and Bastidas (1964) recommended applying 84-114 kg N/ha after every cut at approximately two-monthly intervals. Recovery rates for nitrogen were 50-75 percent.

Genetics and reproduction. Pangola grass is highly male sterile because lagging chromosomes produce unbalanced nuclei in the pollen, and female sterile because meiosis fails to progress beyond the leptolene stage. Only one plant which has produced seed has been observed (Sheth, Yu & Edwardson, 1956). It is an aneuploid — 2n=17 (Hutton, 1970).

Dry- and green-matter yields. At Beerwah, southeast Queensland, under an annual summer dominant rainfall or 1 075 mm, it produced a mean annual yield of 10 565 kg/ha per year, fully fertilized (Evans, 1967a) and grows at the rate of 113 kg DM/ha per day in summer, but only produces 2.25 kg DM/ha per day in winter (Evans & Hacker, 1980). At South Johnstone in north Queensland (lat. 17°6'S, altitude 580 m and rainfall 900 mm) 28 282 kg DM/ha per year were obtained from five cuttings from a pasture fertilized with 220 kg nitrogen, 22 kg phosphorus and 55 kg potassium per hectare per year. It outyielded *Panicum maximum* cv. Hamil and *Brachiaria ruziziensis*, but yielded less than *Brachiaria decumbens* and *B. mutica* (Grof & Harding, 1970).

Suitability for hay and silage. It cures rapidly in dry weather and makes excellent hay when cut and cured before it gets too stemmy (Wheeler, 1950). It makes successful silage in Florida.

Value as standover or deferred feed. When pangola grass matures, its feeding value declines and standover feed, while providing roughage, is not very nutritious.

Chemical analysis and digestibility. The feeding value of pangola grass is high, but animal intake near maturity can be limited by protein deficiency (Minson, 1967) unless a late application of nitrogen fertilizer is given or it is grown with a legume. Bryan and Sharpe (1965) reported a range of 3.9 to 11.6 percent crude protein in the dry matter. In Costa Rica, analysis of material at floral initiation revealed 11.81 percent crude protein, 30.2 percent crude fibre, 36.3 percent nitrogen-free extract, 2.5 percent ether extract and 9.2 percent ash in 10 percent moisture material. Sulphur fertilization of pangola grass through superphosphate increased the nutritive value of pangola grass by overcoming a simple sulphur deficiency when diets contained less than 0.17 percent sulphur (Rees & Minson, 1976).

Palatability. Excellent when the material is young and vigorous. It is usually neglected in favour of other grasses when it becomes old and stemmy.

Toxicity. No toxicity has been recorded with this grass.

Cultivars. There is only one cultivar in Australia. In Florida, United States, a slender-stemmed type known as 'Leesburg No. 5' has achieved some success (Nestel & Creek, 1962) and also cv. Transvala (Gaskins & Sleper, 1974).

Value for erosion control. Wherever pangola grass can be successfully grown it will effectively control erosion.

Diseases. Pangola grass has been seriously attacked by a rust (*Puccinia oahuensis*) in many areas causing agronomists to look for a replacement grass among the *Digitaria* species (Evans, 1972). In Fiji it is affected by a stunt virus which is also serious in Florida.

Pests. In Florida the yellow sugar-cane aphid (*Sipha flava*) is a major pest (Bennett, 1973; Oakes, 1978). Another aphis (*Schizaphis hypersiphata*) is also troublesome, and the crabgrass leaf beetle (*Lemma rufotincta*) less damaging. The sting nematode may be serious. Minor pests are the lawn armyworm (*Spodoptera mauritia*) and the sod webworm (*Herpetogramma licarsisalis*) (Broadley & Rogers, 1978). Heavy aphis populations have caused severe damage in the wet tropics (Teitzel & Middleton, 1979). Natural predators such as ladybirds and hover flies exert some control.

Animal production. With dressings of 448 kg and 896 kg N/ha on the moist Beerwah (southeast Queensland) environment, pangola grass gave annual live-weight gains in cattle of 1 220-1 340 kg/ha (Bryan & Evans, 1967). At Parada, in north Queensland, a live-weight gain of 2 990 kg/ha/year was obtained from grazing irrigated pangola grass fertilized with 672 kg N/ha per year (Ebersohn & Lee, 1972) (see Plate 34). In the Ord River valley in northern Australia, a maximum live-weight gain of 1 330 kg/ha a year was obtained from 11.5 weaner steers per hectare and there was no increase with increasing dressings of nitrogen from 300 kg to 800 kg/ha per year (Blunt, 1978). A figure for live-weight gains of beef cattle of about 1 300 kg/ha per year has also been obtained in Florida (Motta, 1968), Puerto Rico (Caro-Costas, Vicente-Chandler & Burleigh, 1961; Caro-Costas, Vicente-Chandler & Figarella, 1965), Beerwah, Queensland (Evans, 1969) and Ayr, north Queensland (Deans *et al.*, 1976). In Brazil, Aronovich, Serpa and Ribeiro (1970) obtained an average live-weight gain per hectare per year over four years of 349 kg with pangola alone, 410 kg with a mixture of pangola grass and the legume *Centrosema pubescens* and 531 kg/ha from a pangola grass pasture fertilized with 100 kg N/ha per year.

At Turrialba, Costa Rica (605 m altitude, 2 688 mm rainfall) a herd of 50 cows grazing a pangola grass pasture at intervals of 21 days at 2.67 beasts per hectare produced 6 014 kg milk per hectare. The pasture was fertilized with 100 kg urea per hectare, 150 kg potassium chloride and 80 kg/ha triple superphosphate in April, followed by two additional applications of 200 kg urea per hectare at intervals of four months (Blydenstein *et al.*, 1969). In the West Indies, pangola grass fertilized with 330 kg N/ha produced an annual live-weight gain of 1 157 kg/ha, compared with unfertilized pasture production of 357 kg/ha (Oakes, 1960).

Main attributes. Its rapid cover, nutritious growth in the presence of nitrogen, ability to establish on poor soils and withstand heavy stocking, and its ability to combine with *Lotononis bainesii*.

Main deficiencies. Its lack of seed production and its susceptibility to rust, its short growing season and its aggressiveness in relation to legumes other than *Lotononis bainesii*.

Further reading. Aronovich, Serpa & Ribeiro, 1970; Bryan & Evans, 1967; Bryan & Sharpe, 1965; Nestel & Creek, 1962.

Digitaria didactyla **Willd.**

Synonym. *Panicum tennuissimum* Benth.
Common names. Queensland blue couch grass (Australia), seragoon grass (Malaysia).
Distribution. Native to the Mascarene Islands and Madagascar, and now introduced to many tropical countries, primarily as a lawn grass.
Description. A small, creeping grass, blades narrow about 2.5 cm long, with a fine setaceous tip and usually two racemes, conjugate and sessile. It is close to *D. ciliaris,* but differs in its perenniality, fine leaf-blades, small number of racemes and their slender build. It differs from *Cynodon dactylon* in its shorter, broader leaf and its distinctive bluish colour (see Fig. 15.49).
Season of growth. Summer.
Optimum temperature for growth. 25-40°C.
Minimum temperature for growth. 15°C.
Frost tolerance. It will become frosted but the frosted material is of good quality as forage. Frosted lawns recover.
Altitude range. Sea-level to 1 500 m.
Rainfall requirements. It generally grows within the rainfall limits of 700-1 250 mm.
Drought tolerance. It survives droughts very well.
Tolerance to flooding. It will survive temporary flooding.
Soil requirements. It has wide soil tolerance, but grows best on sandy loams and loams.
Fertilizer requirements. Blue couch grass can survive with little fertilizer, but responds to added fertilizer, especially nitrogen. At 110 kg N/ha it invaded Samford Rhodes grass pastures at Samford, Queensland (Jones, 1970). Hegarty (1958) recorded a linear response in yield up to 105 kg N/ha.
Ability to spread naturally. Blue couch spreads rapidly by runners and from seed.
Land preparation for establishment. It requires a very fine, well-prepared and level seed-bed for seed planting of lawns. When planting turfs a level area is also required, usually underlying the turfs with sand.
Sowing methods. Lawns are usually laid as turfs. Where a pasture is desired it can be planted by runners, turfs or seed.
Sowing time and rate. Sow in the wet season as turfs or slips.
Dormancy. There is some post-harvest dormancy. Seeds harvested in January, 1971, gave significantly higher germination after 13.5 months' storage than those from a November, 1971, harvest used after 3.5 months' storage. Tetrazolium tests showed 84 percent and 65 percent viability respectively.
Seed treatment before planting. Dehulling markedly inhibited germination of seed.

Compatibility with other grasses and legumes. In its natural state it will combine with the legume *Desmodium triflorum* in Queensland, but yields are low unless adequate fertilizer is applied.

Ability to compete with weeds. It generally grows vigorously enough to suppress weeds, though in southeast Queensland the annual *Ambrosia artemisifolia* is able to germinate and dominate local patches by strong root competition and shading.

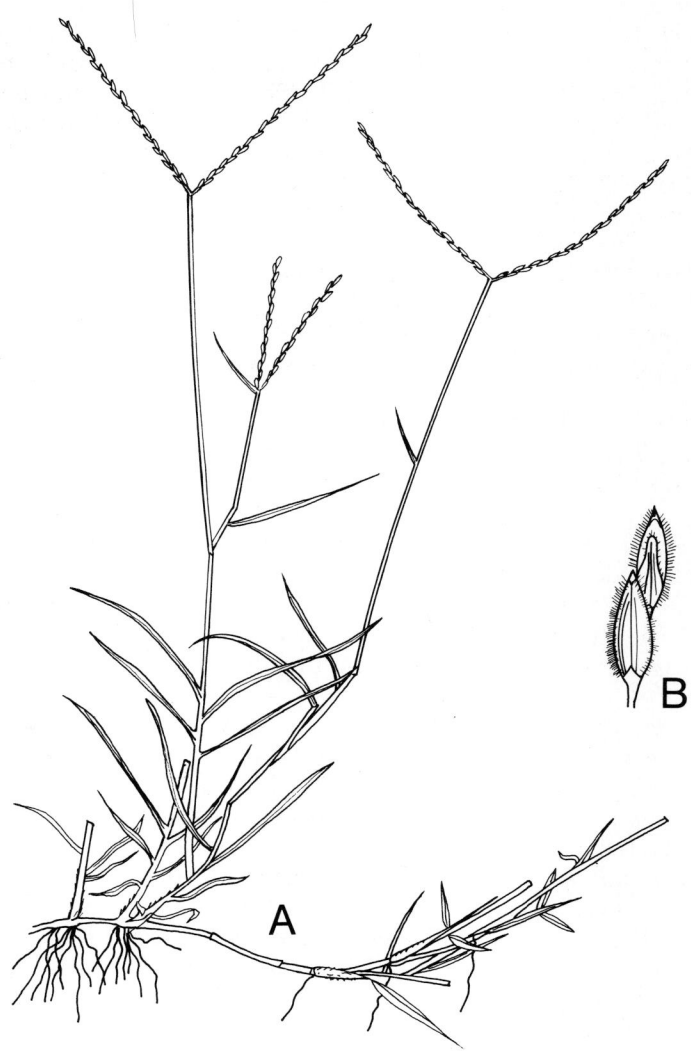

Figure 15.49. Digitaria didactyla. **A**-Plant **B**-Spikelets

Response to defoliation. It can stand heavy defoliation.
Grazing management. It requires little management, apart from preventing seeding, and fertilizing with nitrogen to keep it productive.
Genetics and reproduction. 2n=18, 36 (Fedorov, 1974).
Dry- and green-matter yields. Henzell (1963) recorded dry-matter yields of 11 200 kg/ha with up to 225 kg N/ha per year with little response above this nitrogen level.
Seed production and harvesting. Commercial seed production has not been undertaken in Queensland, but adequate seed is produced. For seed testing, dormancy is broken by treating pre-chilled seeds with 0.2 percent KNO_3 solution and holding them at 50°C for seven days.
Seed yield. Febles-Perez, Whiteman and Harty (1974) harvested 103 kg/ha in January 1971, and 98 kg/ha in November 1971, by mechanical harvesting.
Minimum germination and quality required for commercial sale. 60 percent germinable seed and 97 percent purity in Queensland.
Value for erosion control. Excellent.
Economics. One of the most popular lawn grasses in Queensland.
Animal production. No figures have been cited but it is a valuable pasture grass in southeast Queensland. It does not produce much bulk unless well fertilized whit nitrogen.
Main attributes. Its dense turf, response to nitrogen, suitability for erosion control and its palatability.
Main deficiency. Its lack of bulk unless heavily fertilized.
Further reading. Febles-Perez, Whiteman & Harty, 1974.

Digitaria milanjiana (Rendle) Stapf

Synonym. *D. swynnertonii* Rendle.
Common name. Milanje finger grass (South America, southern Africa), woolly finger grass (Fiji).
Natural habitat. Woodland and thicket. Common in disturbed areas and abandoned cultivations.
Distribution. Originated in the Zontpansberg district of northern Transvaal.
Description. A stoloniferous perennial 60-120 cm high with erect culms and long leaf-blades. Four to 11 racemes, digitate or subdigitate with pubescent spikelets 2.5-3 cm long, stout yellow bristles and long hairs occasionally present (Napper, 1965). It differs from *D. smutsii* in being stoloniferous and having the nerves of the lower lemma scabrid from minute spines (Meredith, 1955). Its characteristic rough spikelets can be felt by stroking the racemes downwards (Chippendall & Crook, 1976). It is a tall plant with bright green leaves which, in the young stages, form a tuft on the ground. The flower is very tall and has six to 12 long racemes given off from the apex of the stalk (Parham, 1955) (see Fig. 15.50).
Altitude range. Sea-level to 1 700 m.
Rainfall requirements. It requires a rainfall in excess of 375 mm.
Drought resistance. Excellent.
Soil requirements. It prefers sandy soils in Zambia (Verboom & Brunt, 1970) and also occurs on red soil stream banks in Kenya. On heavy black seasonally waterlogged soils it occurs as a tufted type, on red soil as a creeping, stoloniferous variety.
Fertilizer requirements. It responds quite well to nitrogen (Clatworthy, 1970).
Response to photoperiod. At Florida University, with a temperature of 25-30°C, *D. milanjiana* produced twice as much dry matter at eight hours with 1.5 hours light interruption of the dark period, than at 9.5 hours (Gaskins & Sleper, 1974). It flowered in 79-96 days at 14 hours and 132-151 days at 11 hours (Degras, Mathurin & Félicité, 1974).
Compatibility with other grasses and legumes. It has been grown successfully in combination with *Trifolium semipilosum* and *Lotononis bainesii* in Zimbabwe (Clatworthy, 1970).
Response to defoliation. In an experiment at Serdang, Malaysia, a mixture of *Brachiaria decumbens, Digitaria milanjiana* and *Panicum maximum* receiving 150 and 300 kg/ha per year of nitrogen as urea was grazed at six, eight and ten beasts per hectare. After 26 months of grazing, *D. milanjiana* remained as only 3 percent of the pasture grazed at ten beasts per hectare, and after 29 months it had died out of the eight- and ten-beast-per-hectare plots (Ajij Singh Sidhu *et al.*, 1977).
Response to fire. It is severely affected by fire (Skovlin, 1971).

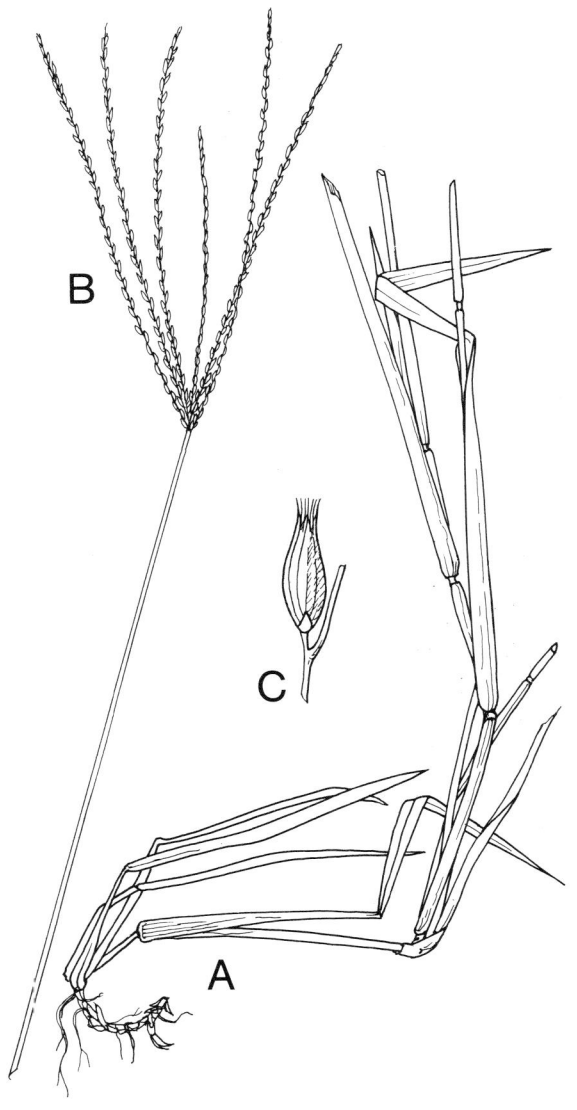

Figure 15.50. Digitaria milanjiana. **A**-Plant **B**-Inflorescence **C**-Spikelet

TABLE 15.27 *Digitaria milanjiana*

	DM	As % of dry matter				
		CP	CF	Ash	EE	NFE
Fresh, early bloom, Kenya		8.1	36.8	9.7	2.3	43.1
Hay, Malawi	89.6	12.5	29.2	9.2	2.3	46.8

	Animal	Digestibility (%)				ME
		CP	CF	EE	NFE	
Hay, Malawi	Cattle	56.4	68.1	41.0	60.1	2.11

Source: Göhl, 1975

Genetics and reproduction. 2n=18 (Fedorov, 1974).
Dry- and green-matter yields. In Fiji, an average yield of 6 328 kg DM/ha with a crude protein content of 8.7 percent was obtained over a three-year period (Roberts, 1970a, b).
Suitability for hay and silage. Successful silage was made at Gualaca, Panama by Medling (1972) in plastic bags with 10 percent molasses added.
Chemical analysis and digestibility. See Table 15.27.
Palatability. It is extremely palatable.
Cultivar. 'MRD.1. Digit'.
Main attributes. One of the very first species to recover after extreme drought. It remains green throughout the year in Kenya.
Main deficiencies. A poor seed producer.

Digitaria pentzii Stent

Synonym. *D. eriantha* Steud. var. *stolonifera* Stapf.
Natural habitat. In dry-land areas on fertile clay.
Distribution. Widely distributed in southern Africa. Introduced in India, Australia and other countries.
Description. A densely-tufted perennial that is strongly stoloniferous, the numerous runners shooting at all the nodes and rooting from some or all of them. Culms up to 120 cm high, simple or branched at the base, straight or bent at the nodes. Lowest leaf-sheaths densely hairy at the base, the lower with long fine hairs, the upper almost glabrous. Leaf-blades glabrous or hairy, up to 30 cm long and 6 mm wide; expanded, but often much less; ligule up to 5 mm long. Three to 14 racemes, up to 18 cm long, arranged digitately or on a central axis up to 3 cm long; spikelets 3-3.5 mm long, fairly conspicuously hairy (Chippendall, 1955) (see Fig. 15.51).
Season of growth. Summer.
Frost tolerance. It becomes frosted and brown when exposed to cold but retains its nutritive value.
Altitude range. It occurs at altitudes of 450-1 100 m in Zimbabwe.
Rainfall requirements. In Zimbabwe it occurs with *Eragrostis rigidior*, *Brachiaria nigropedata*, *Cenchrus ciliaris*, *Schmidtia bulbosa*, *Panicum maximum*, *Urochloa* spp. and *Heteropogon contortus* in the 380-640 mm annual rainfall regime with summer dominance (Barnes, 1972).
Drought tolerance. It can tolerate drought well.
Soil requirements. It occurs on fertile clay in its natural habitat but can adapt to a wide range of soils, including granite sands (Rattray, 1960b).
Sowing methods. It rarely sets seed, and so is propagated by stolons (Bor, 1960).
Response to photoperiod. It flowered in 79-96 days at 14 hours and 138-151 days at 11 hours daylight (Degras, Mathurin & Félicité, 1974).
Ability to compete with weeds. Good, its rapid ground cover quickly suppresses weeds.
Response to defoliation. It is very resistant to heavy grazing and possesses exceptional powers of recuperation (Bor, 1960).
Genetics and reproduction. 2n=18, 27, 35, 36, 45, 54 (Fedorov, 1974). It is cross pollinated.
Suitability for hay and silage. It makes excellent hay (Bor, 1960).
Chemical analysis and digestibility. See Table 15.28.
Palatability. Not very palatable during the grazing season, but eaten readily in autumn. Maintains palatability and nutritive value into winter.
Value for erosion control. It is good for stabilizing soil and improving structure (Göhl, 1975), but has to be established vegetatively.

Figure 15.51. *Digitaria pentzii.* **A**-Plant **B**-Inflorescence **C**-Spikelet

TABLE 15.28 *Digitaria pentzii*

	DM	As % of dry matter				
		CP	CF	Ash	EE	NFE
Fresh, vegetative, 12 cm, Tanzania	20.0	14.0	28.4	14.3	3.0	40.3
Fresh, early bloom, 20 cm, Tanzania	28.0	8.7	31.7	11.8	1.9	45.9
Fresh, early bloom, 30 cm, Tanzania	30.0	6.7	35.2	11.5	2.3	44.3
Hay, South Africa	93.4	10.1	32.9	7.8	2.0	47.2

	Animal	Digestibility (%)				
		CP	CF	EE	NFE	ME
Fresh, 12 cm, Tanzania	Sheep	83.6	82.4	77.0	80.6	2.70
Fresh, 20 cm, Tanzania	Sheep	60.9	77.0	37.0	72.8	2.37
Fresh, 30 cm, Tanzania	Sheep	55.2	71.3	57.0	60.3	2.13
Hay, South Africa	Sheep	57.3	68.6	40.2	62.5	2.19

Source: Göhl, 1975

Economics. It is a constituent of woodland savannah in Zimbabwe where cattle ranching is practised. Cattle weights increase through June and July and reach their peaks in September. Carrying capacity is about one beast per ten hectares (Rattray, 1960b).

Animal production. Over a period of 12 months, cattle grazing on three strains of *Digitaria pentzii* produced more live-weight gain on a per-hectare basis than comparable cattle grazing pangola grass (*D. decumbens*). Cattle grazing on *D. pentzii* CQ 911 produced twice as much as those on pangola, and continued to gain throughout the colder part of the year (May to September). The gains were positively correlated with the content of sown species, the bulk density of the species and the greenness of the species (Peake, Strickland & Hacker, 1976).

Further reading. Peake, Strickland & Hacker, 1976.

Digitaria smutsii Stent

Common name. Woolly finger grass.
Distribution. It occurs in the Transvaal, Orange Free State, northern Cape Province and parts of the Kalahari thornveld in southern Africa. The species was described from plants growing at Doornkloof, Field Marshal J.C. Smuts's home at Irene in the Transvaal (Chippendall, 1955).
Description. A robust, tufted, non-stoloniferous perennial with culms up to 150 cm high, usually branched; lowest leaf-sheaths densely hairy at the base, the leaves otherwise glabrous or with scattered tubercle-based hairs on the lower sheaths, blades up to 60 cm long, 6-12 mm or more wide, expanded; ligule 2-3.5 mm long; eight to ten, sometimes four to six racemes up to 15 cm long, arranged digitately or, more often, alone or in whorls on a central axis up to 7 cm long, the lower often divided and compound in the lower half; spikelets 3.5-4 mm long, fairly conspicuously hairy (Chippendall, 1955).
Season of growth. Summer.
Drought tolerance. It is very drought resistant, surviving a severe drought at Moree in northwest New South Wales (Darley, 1967).
Soil requirements. It does well on sandy soils.
Fertilizer requirements. It requires a balanced fertilizer as determined by soil tests. It responds readily to nitrogen.
Ability to spread naturally. It spreads rapidly from stolons.
Land preparation for establishment. A well-prepared seed-bed is preferred, but root-stocks can be established in a rough seed-bed.
Sowing methods. Propagated by division of root-stocks or by seed.
Vigour of growth and growth rhythm. D. smutsii CPI16778A has excellent spring growth characteristics (Hacker, 1976).
Ability to compete with weeds. It can compete successfully with weeds.
Response to defoliation. It will stand heavy defoliation.
Grazing management. It will stand heavy grazing and can be managed by short-term grazing at high stocking followed by top-dressing with nitrogen after grazing.
Genetics and reproduction. D. smutsii CPI16778A is almost completely sterile, but others, e.g. CPI38869, are fertile. Hybridization is in progress (Hacker, 1976). 2n=18, 36 (Fedorov, 1974).
Dry- and green-matter yields. In Sri Lanka an annual yield of 21.46 t/ha DM was obtained from a fully fertilized pasture (Pathirana & Siriwardene, 1973).
Chemical analysis and digestibility. In mid-country Sri Lanka, analyses of *D. smutsii* at four weeks showed 17.2 percent dry matter and 13.35 percent crude protein and at six weeks 17.64 percent dry matter with 11.44 percent crude protein, when fully fertilized with 140 kg N, 196 kg P_2O_5 and 252 kg K_2O/ha (Pathirana & Siriwardene, 1973).

Palatability. It is closely grazed at the Veterinary Research Farm at Entebbe, Uganda (van Rensburg, 1969).
Seed production and harvesting. It does not produce much seed at Moree, northwest New South Wales (Darley, 1967).
Value for erosion control. It has been recommended for revegetating abandoned cropland in southern Africa.
Main attributes. Its ability to spread rapidly; its tolerance to heavy grazing and its response to fertilizers.
Main deficiencies. Its poor seed production.
Further reading. Pathirana & Siriwardene, 1973.

Diplachne fusca (L.) Beauv.

Synonyms. Festuca fusca (L.); *Leptochloa fusca* Kunth; *Trioda ambigua* R. Br.; *Uralepsis fusca* Steud.; *U. drummondii.*
Common names. Brown beetle grass, brown-flowered swamp grass (Australia).
Natural habitat. Low, wet ground and brackish swamps, and in artesian water bore drains.
Distribution. Asia, Africa, Australia.
Description. A glabrous, erect perennial or biennial grass growing to 150 cm, with narrow, convolute leaves when dry, with long sheaths and a jagged ligule. Panicle narrow, 15-30 cm long with erect branches, the lower ones long (Turner, 1891). It forms well-rooted tussocks. There appears to be some variability, some are stoloniferous (Chippendall, 1955). Leaf-blades are 15-30 cm long and 1-2 mm wide when flattened out, usually more or less inrolled (see Fig. 15.52).
Season of growth. Summer.
Rainfall requirements. It occurs over a wide range from 300-450 mm in western New South Wales, and in areas of up to 1 000 mm or more in coastal areas. It is always found in or near water and the seed germinates in it.
Tolerance to flooding. Good.
Soil requirements. It grows in a wide range of soils, but prefers clay loams or clays.
Tolerance to salinity. Excellent. It is one of the most salt-tolerant grasses. It is common on inundated heavy soils of the Murray River flood plain in southern Australia and in alkaline artesian bore drains in central Australia. At Cooloota Springs and other bores in the area, it inhabits the salt-encrusted overflow areas of the bore tanks (Lazarides, 1970). In Central Africa the beds and verges of alkaline lakes are extremely flat, therefore when the basin is full, extensive areas tend to be shallowly flooded. This alkaline swamp is colonized by a single species of grass, *Diplachne fusca* (L.) Beauv., which extends for many kilometres. Growing as a water grass *D. fusca* remains green and flowers during the rainless months, provided the ground is flooded. If the swamp dries up, even for a period of years, as so often happens, the *D. fusca* mat remains in occupation and becomes burnt during the dry season. If man-made fires do not consume the dry mat during the dry season, lightning frequently ignites the tinder-dry herbage at the onset of the rains. Even under the stress of a clean burn every year the *Diplachne* mat yields to no other association on alkaline soils (Vesey-Fitzgerald, 1963).
Vigour of growth and growth rhythm. In Australia it grows rapidly in summer, generally flowers from March to May but also in September-October, and dries off till the next summer rain.

Palatability. It is a soft, herbaceous grass, very palatable to stock.
Further reading. Vesey-Fitzgerald, 1963.

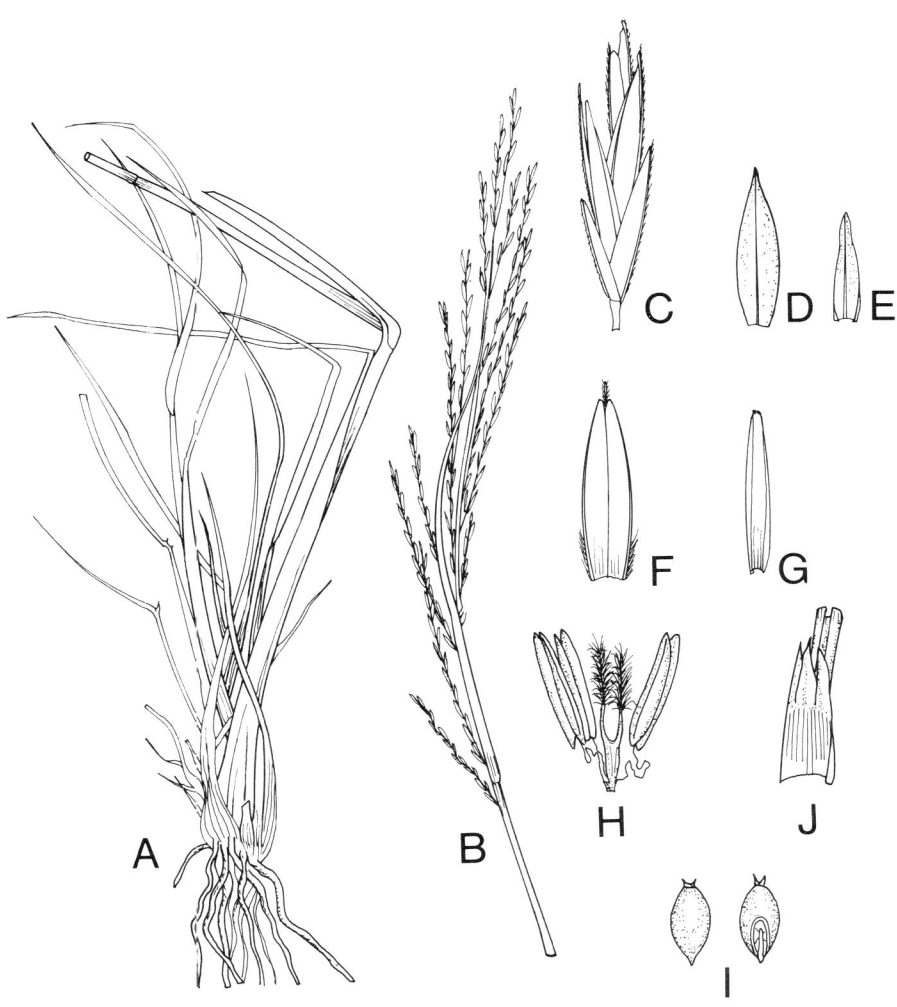

Figure 15.52. Diplachne fusca. **A**-Habit **B**-Inflorescence **C**-Spikelet **D**-Upper glume **E**-Lower glume **F**-Lemma **G**-Palea **H**-Flower **I**-Grain **J**-Ligule

Echinochloa colona (L.) Link

Synonym. Panicum colonum L.
Common names. Awnless barnyard grass (Australia), jungle rice (United States), grama de agua (Cuba), pasto colorado (Peru), azz (Mauritania).
Natural habitat. Swampy places and seasonally flooded grassland.
Distribution. Widely spread in tropical Africa, Asia and Australia.
Description. A tufted annual up to 60 mm high with geniculate culms. Panicle erect, 5-13 cm high, racemes rather distant with crowded, green or purplish apiculate spikelets 2.5-3 mm long. Ligule absent (Henty, 1969) (see Fig. 15.53).
Season of growth. Summer.
Altitude range. Sea-level to 1 800 m.
Rainfall requirements. It grows in environments ranging from 400 mm to about 1 200 mm. In the arid areas it grows in ponds and swamps while the water lasts, and usually seeds before it dies.
Soil requirements. It grows in a fairly wide range of soils, but is most common in loams, silts and clays in low places.
Vigour of growth and growth rhythm. In the Sahel it flowers in August and has dried off by February (Boudet & Duverger, 1961).
Tolerance to herbicides. If control is needed, a pre-emergence spray of 2,4-D sodium salt at 4.5 kg/ha of an 850 g AI/kg product (e.g. Hormicide) can be used. No wetting agent is required. Use a minimum of 340 litres of water per hectare. This gives short-term protection. For long-term control use trifluralin at 2.8 l/ha of a 400 g AI product (e.g. Treflan E.C.). Seedlings can be killed by paraquat at 1.4 l/ha of a 200 g AI/l product (e.g. Gramoxone) plus surfactant at 250 ml per 200 litres water (Tilley, 1977).
Genetics and reproduction. 2n=36, 48, 54, 72 (Fedorov, 1974).
Suitability for hay and silage. It makes very palatable hay.
Chemical analysis and digestibility. See Table 15.29.
Palatability. Extremely palatable.
Economics. It is a valuable grazing plant in its short season of growth. The seed is eaten by humans in times of stress. It can be an important weed of rice.
Animal production. No figures have been cited but it is a valuable short season grazing and hay grass throughout the tropical and subtropical world.

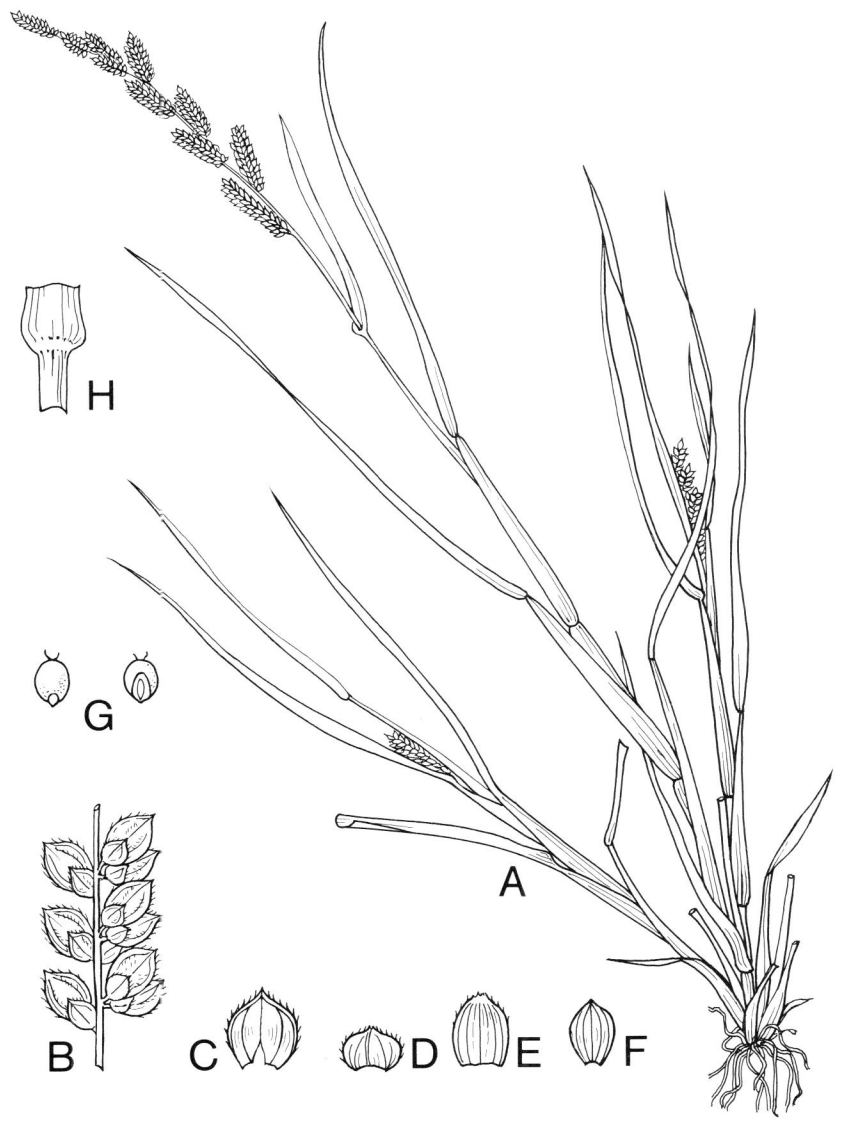

Figure 15.53. Echinochloa colona. **A**-Habit **B**-Portion of spike **C**-Upper glume **D**-Lower glume **E**-Lower lemma **F**-Upper lemma **G**-Grain **H**-Ligule

TABLE 15.29 *Echinochloa colona*

	DM	As % of dry matter				
		CP	CF	Ash	EE	NFE
Fresh, mid-bloom, Niger		12.5	28.6	12.4	2.4	44.1
Fresh, 6 weeks, India	14.3	15.8	28.6	15.2	2.3	38.1
Fresh, 10 weeks, India	21.1	9.9	31.3	17.4	1.9	39.5
Fresh, 14 weeks, India	27.7	7.2	31.7	16.2	1.5	43.4

Source: Göhl, 1975

Echinochloa crus-galli (L.) Beauv.

Synonym. Panicum crus-galli (L.).
Common names. Barnyard millet (Australia), water grass (United States), cockspur grass (South Australia), song chang (Viet Nam).
Natural habitat. Freshwater swamps.
Distribution. A worldwide weed of cultivation in the tropics.
Description. An annual with more or less robust culms ascending to 105 cm from a geniculate base; spikelets 3-4 mm, crowded in the racemes, which are often branched; panicle dense and stiffly erect (Henty, 1969). The basal sheaths are commonly purplish, owing to the folding together of the sides of the sheaths (Burbidge, 1968) (see Fig. 15.54).
Season of growth. Summer.
Altitude range. Sea-level to 2 500 m.
Tolerance to flooding. Excellent. It often grows in standing water and is the main weed in rice paddies throughout the world.
Soil requirements. It will grow in a variety of soils in wet situations, but is usually found on silts and clays in ponds and depressions.
Dormancy. The seed germinates with the sown rice, but cannot germinate in water deeper than 15 cm, so flooding the rice field with water to this depth will give the rice seedlings an advantage.
Response to photoperiod. Flowering is accelerated by short days (Evans, Wardlaw & Williams, 1964).
Tolerance to herbicides. To control chemically, use a pre-emergence spray of 2,4-D sodium salt at 4.5 kg/ha of an 840 g AI/kg product (e.g. Hormicide). No wetting agent is required when used as a pre-emergent spray. Use a minimum of 340 litres of water per hectare. To kill young plants, spray with paraquat at 570 ml of a 200 g AI/l product (e.g. Gramoxone) per 200 litres water plus surfactant at 250 ml/200 l water. Spray to a point of run-off. For advanced plants give one spraying with 2.2-DPA at 2.3 kg of 740 g AI/kg product (e.g. Shirpon, Dowpon) plus 250 ml wetting agent per 200 litres water. Thoroughly wet the plants (Tilley, 1977). In the Ord River valley, northern Australia, *E. colona* and *E. crus-galli* were controlled with Stam F-34 (3,4-dichloropropionanilide) at 3.25-7.5 kg active ingredient per hectare, two to three weeks after emergence.
Genetics and reproduction. $2n = 36, 42, 48, 54, 72$ (Fedorov, 1974).
Dry- and green-matter yields. At Laguna, the Philippines, Furoc and Javier (1976) gathered 3.88-11.07 t/ha green weeds (chiefly *E. crus-galli*) from a rice field from the first to the last weeding.
Economics. It is the world's worst weed in paddy rice. It is a useful forage plant for all herbivorous animals and the grain is eaten by humans in times of want.
Animal production. It is relished by stock everywhere, and where it does not

Figure 15.54. Echinochloa crus-galli. **A**-Habit **B**-Portion of inflorescence **C**-Upper lemma **D**-Lower lemma **E**-Grain

interfere with cropping it is valued for grazing, especially for dry-season forage when other grasses on dry land have matured. It is an important grazing plant for buffaloes in Viet Nam. In the Philippines, Furoc and Javier (1976) considered that one farmer could collect sufficient weeds to feed four steers daily from 0.43 hectares of weed-infested rice.

Main attributes. Its rapid germination and growth provide nutritious and succulent forage.

Main deficiency. Its problem as a weed in rice.

Further reading. Seaman *et al.*, 1968.

Echinochloa frumentacea (Roxb.) Link

Synonym. Echinochloa colonum var. *frumentacea* Ridley.
Common names. White panicum, Siberian millet (Australia).
Natural habitat. In cultivations and naturalized in wet grassland.
Distribution. Occurs widely in the tropics. Is cultivated in tropical Asia, Africa, Australia and the western United States and Canada as a fodder grass and cereal.
Description. A stout annual, 90-150 cm high, leaf-blades flat, wide, ligules absent. Panicle 10-25 cm long with dense racemes of awnless 3-4 mm long spikelets (Napper, 1965). The seed is 2-3 mm long and 1.5-2 mm wide. It differs from *Echinochloa crus-galli* mainly in having glabrous, awnless spikelets and heavier, thicker, more compact racemes (Barnard, 1969) (see Fig. 15.55).
Season of growth. Summer.
Optimum temperature for growth. 25-30°C.
Frost tolerance. It does not tolerate frosts.
Altitude range. Sea-level to 1 500 m.
Rainfall requirements. The millets have a lower rainfall requirement than the sorghums. They are usually grown in an annual rainfall regime of 500-750 mm with a summer dominance.
Drought tolerance. They are more drought tolerant than maize, and because of their early maturity they often escape droughts.
Fertilizer requirements. A basic treatment of a complete fertilizer may be required if soil tests reveal this. Excess nitrogen may cause the crop to lodge, but this is not very important where the crop is grazed. Application of 55-70 kg N/ha may be desirable for grazing crops.
Ability to spread naturally. Scattered seed will usually germinate, but it is generally sown as a crop on prepared land.
Land preparation for establishment. A well-prepared, firm seed-bed is preferable, but in early development a rough seed-bed will usually provide enough crop for grazing.
Sowing methods. Drilling the seed into a well-prepared seed-bed is usual, but the seed can be broadcast and harrowed in.
Sowing depth and cover. As millet seed is small, it should be sown no deeper than 2.5 cm and rolled to compact the soil around the seed.
Sowing time and rate. Sow in spring to late summer, depending on frost incidence, at 8 kg/ha drilled and 10 kg/ha broadcast.
Number of seeds per kg. 367 000 (Siberian millet).
Dormancy. For germination tests seed is pre-dried at 40°C for seven days.
Seed treatment before planting. If seed-harvesting ants are troublesome, treat the seed with lindane before planting.

Response to photoperiod. Flowering is accelerated by short days (Evans, Wardlaw & Williams, 1964).
Seedling vigour. Good.
Vigour of growth and growth rhythm. It grows vigorously in a semi-prostrate habit, flowers in 62 days and matures in 120 days.
Compatibility with other grasses and legumes. It grows well with other grasses, but is usually combined with an annual legume such as cowpea (*Vigna unguiculata*) for grazing, hay and silage, sown at the rate of 6-8 kg/ha of millet and 11-12 kg/ha of cowpea.

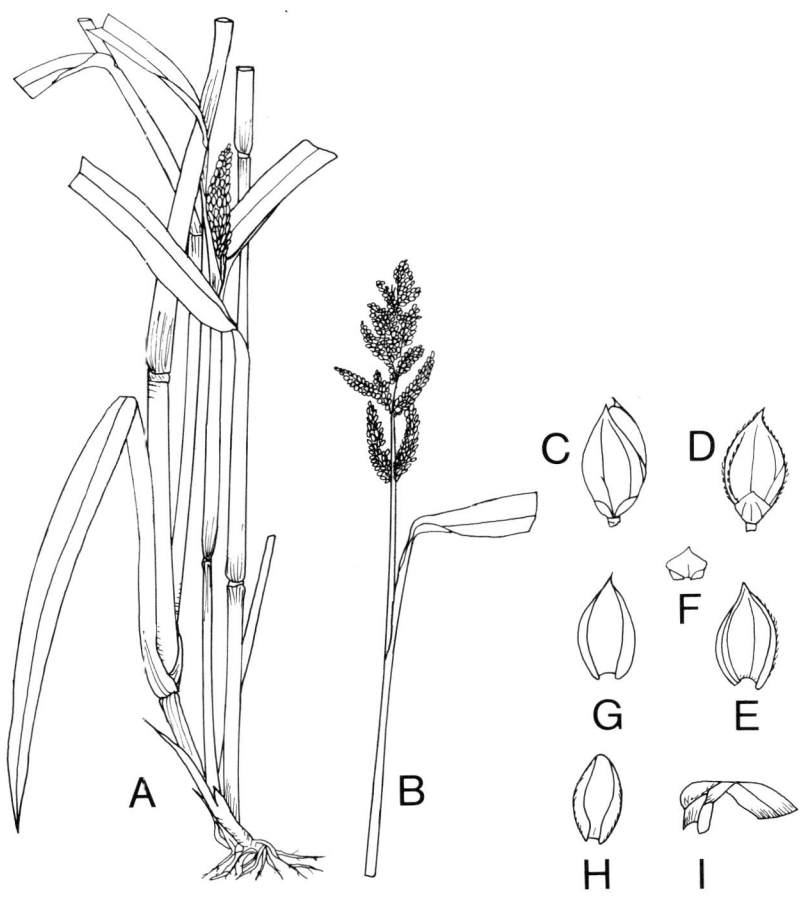

Figure 15.55. Echinochloa frumentacea. **A**-Habit **B**-Inflorescence **C,D**-Spikelets **E**-Lower glume **F**-Upper glume **G**-Lower lemma **H**-Lower palea **I**-Ligule

Ability to compete with weeds. Fairly good when established.
Tolerance to herbicides. To control weeds in grain crops, spray with MCPA (2-methyl 4-chlorophenoxy-acetic acid) at 1.3 kg acid equivalent per hectare at tillering stage, before heads start to form in the sheath.
Response to defoliation. If grazed quickly, a number of regrowths in a season can be utilized, depending on rainfall and soil fertility.
Grazing management. Graze heavily and then cease until the regrowth is ready to graze again.
Genetics and reproduction. 2n=36, 54, 56 (Fedorov, 1974).
Dry- and green-matter yields. Yields of up to 35 000 kg/ha of green material can be obtained.
Suitability for hay and silage. While white panicum has coarser stems than other short millets and takes longer to cure for hay, it makes quite good hay and silage, especially if combined with cowpea. For hay the millet is cut in the early heading stage and for silage when the grain is at the firm dough stage.
Chemical analysis and digestibility. No figures have been found.
Palatability. Although the stems of white panicum are fairly coarse, the grass is extremely palatable.
Seed production and harvesting. White panicum seed quite heavily. It can be harvested by combines using a small seed box. The seed should be dried thoroughly, as it will heat in storage if too moist. Store below 13 percent moisture.
Minimum germination and quality required for commercial sale. 75 percent germinable seed and 97.3 percent purity in Queensland, Australia.
Cultivars. No official cultivars are recognized, though in Queensland two varieties of *E. frumentacea* are sown: white panicum and Siberian millet.
Value for erosion control. The millets are frequently used for temporary control of erosion in newly cleared and ploughed sandy soils because they grow rapidly and seed is cheap.
Diseases. Covered smut of the seed-heads is caused by *Ustilago tricophora* and is controlled by treating the seed before planting with thiram.
Pests. No major pests affect the plants except periodical invasion by grasshoppers.
Economics. A widely used forage and hay crop throughout the developed tropical countries.
Animal production. No figures for animal performance have been found, but it is used a good deal for dairy-herd grazing in coastal southeast Queensland.
Main attributes. A quick-growing crop which seeds heavily, is very palatable in the young stage, makes good hay and fits into gaps in the feed year.
Main deficiencies. Its annual nature.
Further reading. Douglas, 1970.

Echinochloa haploclada (Stapf) Stapf

Natural habitat. It grows in seasonally flooded lowlands and ditches.
Description. A very variable rhizomatous perennial from 15-200 cm high, leaves usually without ligules but sometimes with a ciliate fringe on the lower ones. Inflorescence 10-23 cm long, usually dense with short racemes of acuminate or minutely awned spikelets 2-3 mm long (Napper, 1965) (see Fig. 15.56).
Altitude range. Sea-level to 1 750 m.
Rainfall requirements. It occurs in the 650-700 mm annual rainfall regime, generally in wet areas. Bogdan and Pratt (1967) record one form found near Mt Marsabit (Kenya) labelled K.53542, which is adapted to moderately dry conditions.
Tolerance to flooding. It is adapted to swampy conditions.
Soil requirements. It has a range from sandy loams to alluvial silts with a preference to loams (Bogdan & Pratt, 1967).
Ability to spread naturally. The grass is easily established and established plants self-seed readily (Bogdan & Pratt, 1967).
Sowing methods. Broadcast on to a well-prepared or roughly disc-harrowed seed-bed.
Sowing depth and cover. Surface sow and roll if possible.
Sowing time and rate. In summer at 0.5 kg/ha.
Number of seeds per kg. About 850 000 spikelets with one caryopsis each.
Genetics and reproduction. 2n=18, 36 (Fedorov, 1974).
Chemical analysis and digestibility. Herbage analysed at an early flowering stage yielded 14 percent crude protein in the dry matter (Bogdan & Pratt, 1967).
Palatability. It is readily grazed.
Seed production and harvesting. It is a prolific seeder and the seeds (one plump caryopsis per spikelet) are easy to harvest and handle (Bogdan & Pratt, 1967).
Further reading. Bogdan & Pratt, 1967.

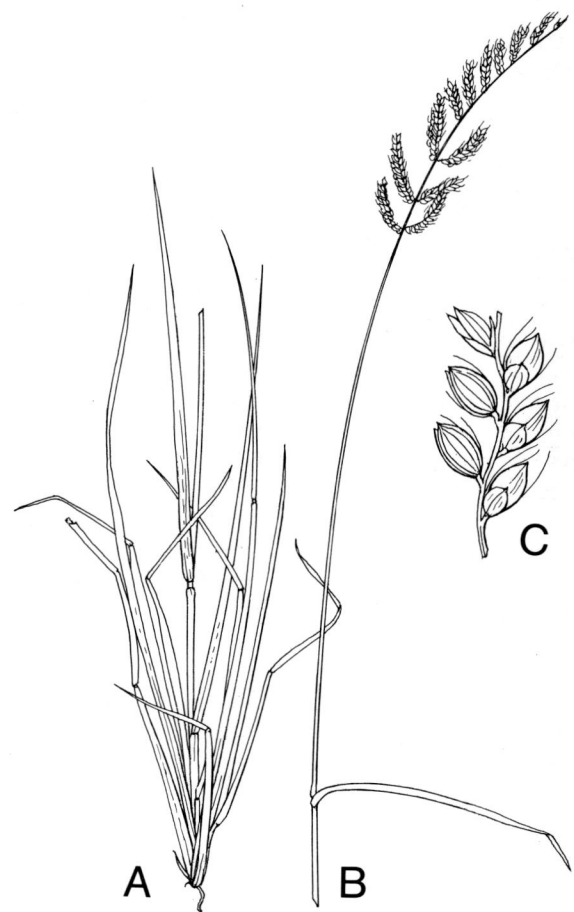

Figure 15.56. Echinochloa haploclada. **A**-Habit **B**-Inflorescence **C**-Portion of raceme

Echinochloa polystachya (Kunth) Hitchc.

Synonyms. Panicum spectabile Nees *Echinochloa spectabilis* (Nees) Link.
Common names. German or Alemán grass (Panama), pasto alemain (Venezuela), pardegrao, prasi-grasi (Suriname).
Natural habitat. Swamps and ditches near the coast.
Distribution. Mexico and West Indies to Argentina.
Description. Perennial, usually in colonies, culms coarse, 1 to 2 m tall, from a long creeping root base, glabrous; nodes densely hispid with appressed yellowish hairs; ligule a dense line of stiff, yellowish hairs; blades up to 2-5 cm wide, panicle dense, 10-30 cm long, racemes ascending; rachis scabrous, spikelets closely set, nearly sessile, about 5 mm long; sterile floret staminate, the awn 2-10 mm long; fruit rather soft, 4 mm long extending to a point 0.5 mm long (Hitchcock, 1930).
Season of growth. Summer.
Frost tolerance. It is not tolerant of frost.
Latitudinal limits. Between 30°N and S.
Altitude range. It prefers low elevations.
Rainfall requirements. It is a high-rainfall grass.
Drought tolerance. It does not tolerate drought.
Tolerance to flooding. It will grow well in standing water, and generally requires moist conditions (Semple, 1970).
Soil requirements. It is adapted to wet to very wet soils (Rattray, 1973).
Chemical analysis and digestibility. Göhl (1975) records one analysis from Cuba. Fresh vegetative material yielded 29.7 percent dry matter, 13.1 percent crude protein, 26.2 percent ash, 3.1 percent ether extract and 46 percent nitrogen-free extract (Calvino, 1952). Digestibility figures are recorded by Jiménez and Parra (1973) as obtained from the giant rodent, the capybara, in Venezuela. The intake (g DM/kg body weight × 0.75) was 83. The total digestibility of the dry matter was 58.8 percent, containing 11.7 percent crude protein and 36.2 percent crude fibre, with digestibility ratios of 63 and 52.1 percent, respectively. In Suriname, Dirven (1936b) found an average of 9.6 percent crude protein in the dry matter, with a range up to 18.9 percent.
Palatability. It is very palatable.
Economics. A useful summer pasture under high-rainfall conditions with some seasonal waterlogging, widely used in South American tropics and subtropics.
Animal production. In Vera Cruz, Mexico, with a rainfall of 1 060 mm from May to October and an altitude of 10-16 m, zebu steers grazed on a put-and-take system gained 280 kg/ha on *E. polystachya* compared with 406 kg/ha on *Digitaria decumbens*, 223 kg/ha on *Brachiaria mutica,* 190 kg/ha on *Hypar-*

rhenia rufa and 157 kg/ha on *Panicum maximum* (Arroyo & Teunissen, 1964).
Main attributes. Its adaptability to wet conditions and good productivity.
Main deficiencies. Its intolerance of frosts.
Further reading. Jiménez & Parra, 1975.

Echinochloa pyramidalis (Lam.) Hitchc. and Chase

Common name. Antelope grass (southern Africa).
Natural habitat. Seasonally flooded grassland and lake shores, floating meadows.
Distribution. Throughout tropical Africa and America and introduced to other countries. It forms a major part of the sudd of the upper Nile.
Description. A reedlike perennial up to 300 cm high, with solid stems, rarely to 450 cm, ligules of the lower leaves a fringe of hairs often absent on the upper. Inflorescence 15-30 cm long with racemes up to 8 cm long having purplish, acute, awnless spikelets 3-4 mm long (Napper, 1965). It forms a dense pure stand with a leaf table at 120-200 cm (see Fig. 15.57, Plate 35).
Season of growth. Summer.
Frost tolerance. It is not frost hardy.
Altitude range. Sea-level to 300-1 500 m.
Rainfall requirements. It generally grows in swamps, where rainfall accumulates.
Drought tolerance. It is drought resistant.
Tolerance to flooding. Excellent. It is native to seasonally flooded grassland and lake shores. Kuri cattle graze on the inundated antelope grass around Lake Chad (Pursglove, 1976).
Soil requirements. Usually associated with badly drained black clays ("black cotton" soils) of illuvial nature which become very sticky when wet but when dry turn very hard and deeply fissured or of a blocky structure. Normally these are deep, with lower horizons mottled with yellow or brown.
Tolerance to salinity. The soils on which this species grows are often alkaline. Where the drainage is closed the pH may be as high as 9.2, but if drained the soil is normally slightly acid (pH 6) (Vesey-Fitzgerald, 1963).
Sowing methods. It is propagated by cuttings.
Vigour of growth and growth rhythm. E. pyramidalis is characteristic of flood plain grasslands such as the Congo Basin of Tanzania and northern Zimbabwe. Growth starts at the onset of the rains, depending on the extent of flooding. Under optimum conditions the previous season's accumulation of dry matter rots away in the water and the new growth is very vigorous. Flowering occurs about half way through the wet season and seeds are shed before the end of the rains. Translocation of nutrients below ground then starts and the subaerial parts dry off although the site may still be flooded. However, node shoots remain green and there is some secondary flowering later. Early fires may not penetrate the *Echinochloa* stand. However, later in the season the whole stand becomes straw, and fierce fires, resulting in a clean burn, occur. Subsequently vigorous growth from ground level occurs without the incidence of rain, and this provides a green dry-season pasture which may

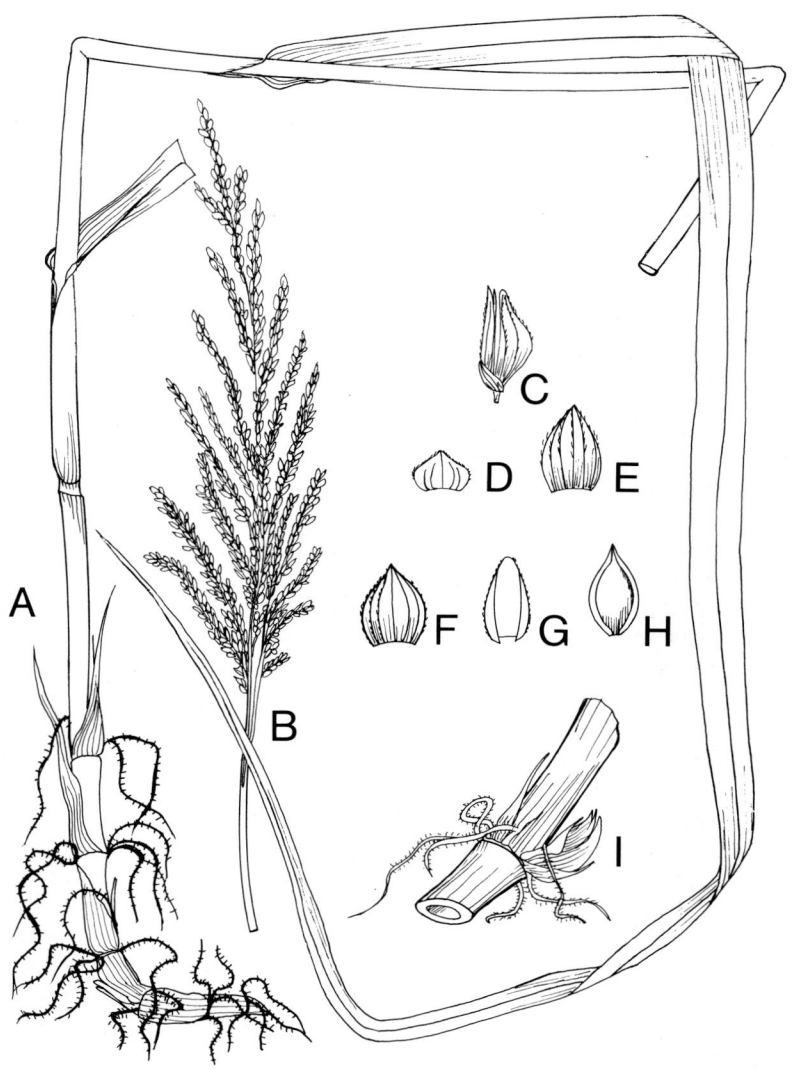

Figure 15.57. Echinochloa pyramidalis. **A**-Habit **B**-Inflorescence **C**-Spikelet **D**-Lower glume **E**-Upper glume **F**-Lemma of upper floret **G**-Palea **H**-Upper floret **I**-Rooting culm

TABLE 15.30 *Echinochloa pyramidalis*

	DM	As % of dry matter				
		CP	CF	Ash	EE	NFE
Fresh, full bloom, Kenya		7.00	31.40	8.60	1.10	51.90
Flowering, Zambia		2.94	32.82	5.92	2.49	55.83
Hay, Zimbabwe	82.5	15.60	33.50	10.20	2.40	38.30
Regrowth, after 30 days, March, West Africa	18.0	16.80	32.50	14.20		

	Animal	Digestibility (%)				ME
		CP	CF	EE	NFE	
Hay, Zimbabwe	Cattle	56.50	54.80	41.20	64.00	2.01

Source: Göhl, 1975

remain available until the commencement of the next rainy season (Vezey-Fitzgerald, 1963).
Response to photoperiod. It is indifferent to day length (Evans, Wardlaw & Williams, 1964), i.e. day neutral.
Tolerance to fire. It is burnt in the dry season in the Lake Kyoga swamps in Uganda and nutritious regrowth results.
Genetics and reproduction. 2n=36, 54, 72 (Fedorov, 1974).
Dry- and green-matter yields. In Malawi several varieties yield about 25 000 kg green matter per hectare in February, rather less at a second cut in July and before flowering.
Suitability for hay and silage. It makes useful hay and silage in South Africa. The types with glabrous or smooth leaf-sheaths should be used for hay: those with hairy leaf-sheaths are unpleasant to handle.
Value as standover or deferred feed. Excellent dry-season grazing after old growth is burnt off — a common African grazing cycle.
Chemical analysis and digestibility. See Table 15.30.
Palatability. Although extremely coarse, indigenous animals graze it readily to ground level at the end of the dry season. The young growth is very palatable after the old material has been burnt off.
Toxicity. No toxicity has been reported.
Seed production and harvesting. It is a heavy seed producer but sometimes there is low seed germination, and seed is shed during the rains.
Value for erosion control. Its dense, tangled, floating stems, rooting at the nodes, provide efficient protection against wave action on the walls of earth dams, or flood-induced erosion of river banks (Rose-Innes, 1977).

Cultivars. In Malawi, cv. Chirundu is an upright variety and cv. Parfuri a creeping type.

Economics. An excellent fodder grass. This grass and *Echinochloa scabra* (previously *stagnina*) are the dominant species in the great floating meadows of the Niger and Lake Chad and from the major part of the sudd at the headwaters of the Nile. The grain is used as human food in some parts of Africa (Chippendall & Crook, 1976). In Nigeria an impure salt or carbonate of soda is made by burning the grass.

Animal production. No quantitative figures have been cited, but it is an important dry-season grazing fodder throughout tropical Africa.

Further reading. Vesey-Fitzgerald, 1963.

Echinochloa scabra (Lam.) Roem. and Schult.

Synonym. E. stagnina (auct.).
Common names. Bourgou, gamarawal (West Africa), banti (India), hippo grass (Zimbabwe), long-awned water grass (South Africa).
Natural habitat. Seasonally flooded grassland, lake shores and swamps in water up to 3 m deep.
Distribution. Tropical Africa and Asia, in Sepik and Western districts of Papua New Guinea, tropical Australia.
Description. A perennial or sometimes annual grass, usually growing in deep water, the culms rooted on the bottom and floating (lengths up to 10 m are reported); panicle resembling *E. crus-galli* but the racemes simple or only slightly compound, and often up to 5 cm long; spikelets 4-6 mm long, 3-6 mm wide and distant. Ligules of the lower leaves a fringe of long hairs, absent on the upper leaves (see Fig. 15.58).
Season of growth. Summer.
Altitude range. 1 000-2 000 m in Tanzania.
Rainfall requirements. It is aquatic, which presupposes plenty of available water.
Drought tolerance. By reason of its adaptation to swamps it escapes the ravages of all but the most severe droughts in which the soil moisture disappears.
Tolerance to flooding. Excellent.
Soil requirements. It prefers clay soils of high lime content.
Land preparation for establishment. In India it is grown as a crop known as banti, on fully prepared land.
Sowing methods. It is drilled in rows 30 cm apart and thinned later (Solomon, 1953). It can also be sown by cuttings in prepared soil (Boudet, 1975).
Sowing depth and cover. It is sown at about 1-1.5 cm deep and lightly covered.
Sowing time and rate. In India it is sown in June and July at 7 kg/ha.
Vigour of growth and growth rhythm. In India it is planted in June and July and is ready for harvest in October.
Ability to compete with weeds. In India it is grown as a crop and inter-cultivated and weeded. It competes well with weeds in swamps.
Grazing management. In the dry season the pasture is grazed as the waters recede, cattle feeding in the deeper water followed later by sheep as the waters dry up.
Genetics and reproduction. 2n=36, 54, 108, 126 (Fedorov, 1974).
Dry- and green-matter yields. 4 000 kg DM/ha in young growth, 13 000 kg DM/ha at complete maturity, 150 kg DM/ha in 30 days' regrowth in the dry season and 2 500 kg/ha in 30 days' regrowth after irrigation (Boudet, 1975).
Suitability for hay and silage. The long trailing leafy stems have a high sugar content. If dried they make coarse, though palatable, hay (Göhl, 1975).

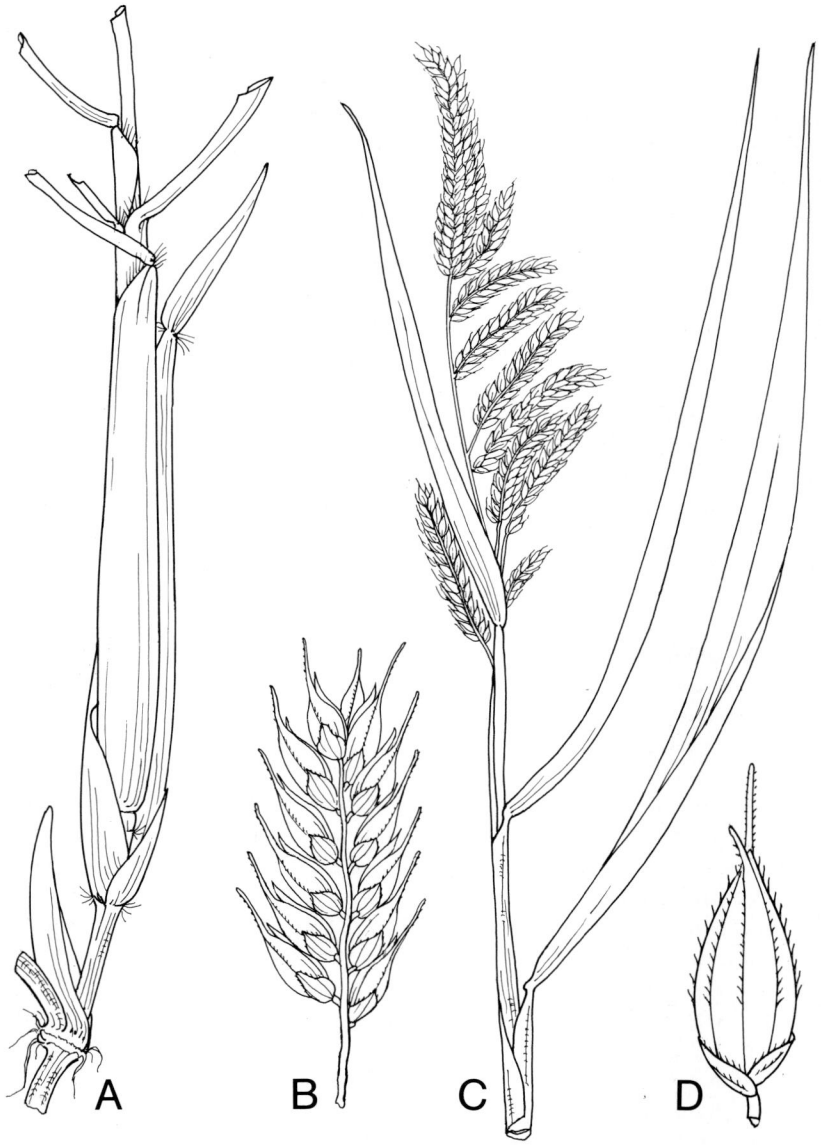

Figure 15.58. Echinochloa scabra. **A**-Culm and shoot **B**-Raceme **C**-Inflorescence **D**- Spikelet

Value as standover or deferred feed. When the water recedes the stems root at the nodes and produce excellent regrowth for grazing during the dry season.

Chemical analysis and digestibility. See Table 15.31.

Palatability. Excellent. The long trailing stems floating on water have a high sugar content. It is still palatable when dry.

Toxicity. In Zambia, scouring occurs when cattle move from the fibrous forest grazing to the rich plains grasses consisting of *Echinochloa scabra, E. pyramidalis, Acroceras macrum, Hemarthria altissima, Leersia hexandra* and *Vossia cuspidata,* and it may be three to four months before they regain condition (Verboom & Brunt, 1970).

Seed production and harvesting. It ripens in October in India and the grain is separated from the husk by pounding. It does not produce much seed.

Diseases. No diseases have been reported for this crop (Solomon, 1953).

Economics. It is mostly a dry-season reserve for animal grazing because the swamps dry out gradually at that time and livestock gain access. In India it is cultivated mainly by the poor and the seed is usually boiled and eaten like rice. It is wholesome (Solomon, 1953). In the Niger area of West Africa it yields excellent fodder and is used for thatching and caulking, and is burnt to produce a "salt" for making soap and indigo. It is also used to extract a sugary

TABLE 15.31 **Echinochloa scabra**

	DM	As % of dry matter				
		CP	CF	Ash	EE	NFE
Fresh, mid-bloom, Niger		11.30	32.50	9.9	2.20	44.10
Flowering, Zambia		12.19	36.24	8.4	1.63	50.54
Fresh, mature, Philippines	19.3	6.70	33.70	11.9	2.10	45.60
Flowering, Oct. - Nov., West Africa	24.0	9.30	35.80	12.5		
Submerged stand, Nov., West Africa	15.0	4.60	42.30	8.4		
Dry stems, Apr., West Africa	92.0	2.90	37.90	7.4		
30 days' regrowth, May, West Africa	27.0	14.40	27.00	16.7		
50 days' regrowth, irrigated, West Africa	19.0	16.90	29.90	14.8		

	Animal	Digestibility (%)				
		CP	CF	Ash	NFE	ME
Hay	Sheep	59.3	62.50	80.5	51.80	1.92

Source: Göhl, 1975

NOTE: Mineral analysis of DM—Ca, 0.39 g/kg; Mg, 0.12 g/kg; Na, 0.13 g/kg; P. 0.05 g/kg; K, 0.06 g/kg; Cl, 1.047 g/kg.

sap for making vinegar (Chippendall & Crook, 1976). It is abundant on the shores of Lake Volta in Ghana along with *Brachiaria mutica*.

Animal production. In the interior delta of the Niger it provides a most important source of green grazing for livestock during the dry season. The livestock graze the fodder as the waters recede under high evaporation, the cattle grazing first and then the sheep as the waters become more shallow. The grass is similarly utilized where it occurs in moist areas in other semi-arid and arid countries.

Main attributes. Its quick growth and adaptability to clay depressions and lake shores in the dry Sahel for dry-season grazing.

Further reading. Boudet, 1975; Solomon, 1953.

Echinochloa turneriana (Domin) J.M. Black.

Common names. It is officially called channel millet in Australia. Other names are native, swamp, western or wild millet, and wild or native sorghum.
Natural habitat. Seasonally flooded heavy clay soils.
Distribution. Occurs naturally in New South Wales, Northern Territory, South Australia and Queensland.
Description. A robust, tall-growing annual grass closely allied to Japanese millet and white panicum. It has rather stout stems from less than 1 m to over 2 m high (see Fig. 15.59).
Season of growth. In summer after seasonal flooding.
Latitudinal limits. It is usually found from latitudes 17 to 29°S in Australia.
Altitude range. 100-200 m in Australia.
Rainfall requirements. Although it is a most important cattle-finishing grass in low-rainfall areas of central Australia receiving from 250 to 375 mm of rain a year, this grass is essentially a flood-plain grass and the seed will not germinate unless flooded. Light rains do not germinate the seed in the heavy cracking clay soils. Flooding occurs when the catchment area, mainly of heavy, self-mulching grey and brown cracking clays, receives heavy rainfalls of the order of 75 mm or more over a period of a few days to a week.
Drought tolerance. It will not tolerate drought.
Tolerance to flooding. It tolerates and depends on seasonal flooding but not permanent water.
Soil requirements. It usually grows in heavy cracking clays and silty clays which will hold moisture for long periods.
Tolerance to salinity. It grows on soils of relatively high pH, from pH 7.0-8.0, but not on saline soils.
Fertilizer requirements. No figures have been obtained, but soil analyses from sites where it grows well reveal very high available phosphorus figures from 200-400 ppm with up to 1 000 ppm at depth. It is also high in exchangeable calcium and magnesium.
Seedling vigour. Excellent.
Vigour of growth and growth rhythm. It grows rapidly and seeds in a matter of three months or so.
Response to defoliation. It does not recuperate from heavy grazing, but in the natural habitat in western Queensland it is seldom grazed to its full capacity because there is usually excess feed for the cattle available.
Suitability for hay and silage. Good if it is accessible.
Value as standover or deferred feed. Excellent. It allows cattle to be finished for market for three to four months after the coastal and near coastal season is finished in Queensland and the cattle bring good prices.
Palatability. It is extremely palatable both green and in the ripened state with a heavy seed-head.

Figure 15.59.
Echinochloa turneriana
(**Source:** Department of Public Lands, Queensland)

Seed production and harvesting. It seeds very heavily in a good season and could be harvested with a combine if seed were in demand. The terrain would be rough to negotiate and would require well-sprung vehicles.

Economics. One of the most important cattle-finishing grasses on the irregularly flooded 2-3 million hectare Channel Country of the Georgina, Diamantina and Bulloo rivers and Cooper Creek in southwest Queensland. It is well worth trying in similar climates outside Australia. The grain would be a valuable human food in dry times.

Animal production. No quantitative figures have been recorded.

Main attributes. Its fast growth, palatability and nutritive value and its availability when other fodders are deteriorating.

Main deficiency. Its lack of readily available seed supplies.

Further reading. Everist, 1975; Skerman, 1947.

Echinochloa utilis Ohwi-Yabuno

Synonym. E. *crus-galli* var. *frumentacea* (Roxb.) Wight.
Common names. Japanese millet, Shirohie millet (Australia).
Natural habitat. Cultivated.
Distribution. Cultivated widely in the tropics and subtropics.
Description. A tall, robust annual 60 to 120 cm high. The inflorescence is a panicle made up of from five to 15 sessile erect branches. The spikelets are brownish to purple in colour, being crowded on one side of the rachis. The spikelet is subtended by two glumes within which are two florets. The lower floret is staminate while the upper one is perfect (see Fig. 15.60).
Season of growth. Summer.
Optimum temperature for growth. There was little difference between 15/20°C and 20/25°C in germination.
Minimum temperature for growth. Germination was depressed at 10/15°C.
Frost tolerance. It is intolerant of frost.
Rainfall requirements. It is grown generally within a rainfall range of 500-1 000 mm for grazing and fodder, with a summer dominance.
Drought tolerance. It is fairly drought tolerant.
Soil requirements. It prefers sandy loams to clay loams. Germination may be difficult in self-mulching heavy clays.
Fertilizer requirements. A basic complete fertilizer may be needed which would be determined by soil tests. A dressing of 55-70 kg N/ha will generally improve grazing performance.
Ability to spread naturally. It germinates readily from scattered seed but is usually planted.
Land preparation for establishment. A fully prepared seed-bed is required for a good crop. A rough ploughing may be sufficient for quick ground cover in a developing area.
Sowing methods. The seed is usually drilled into a prepared seed-bed. It can be broadcast and harrowed in.
Sowing depth and cover. Plant at a depth of 2-2.5 cm, harrow and roll.
Sowing time and rate. Sow spring to late summer at 9 kg/ha.
Number of seeds per kg. 345 000 (Japanese millet), 272 000 (Shirohie millet).
Dormancy. For germination tests seed is pre-dried at 40°C for seven days.
Seed treatment before planting. If there is a seed-harvesting ant problem, treat the seed with lindane before planting.
Seedling vigour. Excellent — used for quick cover on newly cultivated land.
Vigour of growth and growth rhythm. A very vigorous grower on fertile soils, reaches maturity in about six weeks. Grain crops mature in 100 days.
Response to light. It prefers to grow in full sunlight.
Compatibility with other grasses and legumes. It is usually the sole grass com-

Figure 15.60. Echinochloa utilis (**Source:** N.J. Douglas, Queensland Department of Primary Industries)

ponent in mixed swards. Used widely in association with *Vigna unguiculata* (cowpea) in fodder crops for hay and silage.

Ability to compete with weeds. Its vigorous growth tends to suppress weeds but it must be sown on a fine seed-bed to ensure dense stands.

Tolerance to herbicides. For weed control in Japanese millet, spray with NCPA at a rate not to exceed 250 g acid equivalent per hectare at tillering stage and before the heads start to form in the sheath.

Response to defoliation. It stands up to grazing several times in a season, but does not ratoon as well as white panicum.

Grazing management. It can be grazed as early as three weeks from sowing. Start grazing when the crop is 30-40 cm high, preferably by strip grazing to prevent unnecessary trampling. It can be grazed up to five times in a season.

Response to fire. It will not survive fire when dry enough to burn.

Genetics and reproduction. 2n=54 (Fedorov, 1974).

Dry- and green-matter yields. Yields of up to 35 000 kg/ha of green material can be obtained.

Suitability for hay and silage. It makes good hay and silage, especially when sown in association with cowpea. Sow at 6-7 kg/ha of millet seed with 11 kg/ha of cowpea.

Chemical analysis and digestibility. See Table 15.32.

Palatability. Excellent.

Toxicity. No toxicity has been recorded by Everist (1974).

Seed production and harvesting. It produces a heavy seed crop which can be harvested by a combine fitted with a small seed box. The grain should be dried to less than 13 percent moisture for storage, as it heats and spoils quickly if above this figure.

Minimum germination and quality required for commercial sale. 75 percent germinable seed, 97.3 percent purity in Queensland.

Cultivars. No official cultivars are registered, but Japanese millet and Shirohie millet are grown in Queensland.

Value for erosion control. Japanese millet is used extensively by urban development companies to hold soil from erosion between timber clearing and home construction. The seed is cheap and germinates well in most soils. The plant grows rapidly to hold the soil and is easy to eradicate when gardens are being established.

Diseases. A bacterial leaf spot caused by *Xanthomonas translucens* has been recorded from isolated areas in the Burnett district, Queensland (Douglas, 1974).

Pests. Bird damage is likely if the crop is in an area isolated from other grain crops.

Economics. One of the early millet species grown for both food and fodder.

TABLE 15.32 *Echinochloa utilis*

	DM	As % of dry matter				
		CP	CF	Ash	EE	NFE
Fresh, late bloom, United States	24.4	7.4	31.1	8.6	2.9	50.0
Hay, United States	89.1	13.5	32.7	10.4	2.5	50.9

	Animal	Digestibility (%)				ME
		CP	CF	EE	NFE	
Late bloom, United States	Sheep	57.0	59.0	60.0	64.0	2.14
Hay, United States	Cattle	60.0	75.0	69.0	76.0	2.50

Source: Göhl, 1975

Animal production. Little quantitative data have been recorded, but it is a common forage crop in the United States and Australia.
Main attributes. It is a fast-growing, short-term summer crop which can be used as a catch crop between seasons and is a valuable grazing and hay crop.
Main deficiencies. It is an annual which seeds heavily and may be a weed in some areas but is easily cultivated out.
Further reading. Douglas, 1970.

Eleusine coracana (L.) Gaertn.

Common names. Finger millet (Mali), kurrakan millet or koracan millet, ragi, nachni (India), African millet, rapoko (South Africa), dagusa (Ethiopia).
Natural habitat. Roadsides and banks, naturalized from cultivation.
Distribution. Native to Africa and Asia.
Description. A more robust annual grass than *E. indica* with thicker, heavier spikes, sometimes curved at the tips and assuming a brownish colour when mature. It differs from *E. indica* in that the rachilla is tough, whereas in *E. indica* it disarticulates, at least above the glumes. The lemmas are also obliquely ovate and obtuse whereas in *E. indica* they are lanceolate, oblong and acute (Gardner, 1952). The grain is globose, dark brown, smooth in some varieties and at other times somewhat rugose, with a depressed black hilum and slightly flattened on one side (see Fig. 15.61).
Season of growth. Summer.
Altitude range. Sea-level to 2 000 m but mostly at 900 m in East Africa.
Rainfall requirements. It is a low-rainfall plant, often interplanted with maize and sorghum in Africa so that if the normal rains fail a crop of finger millet has a good chance of maturing. It can tolerate a rainfall as low as 130 mm if it is well distributed (Göhl, 1975), but prefers 900 mm.
Drought tolerance. It tolerates dry spells in the early stages of growth and then grows rapidly.
Tolerance to flooding. It will not tolerate flooding.
Fertilizer requirements. 125 kg N/ha is recommended in Uganda, broadcast when the plants are 15 cm high.
Ability to compete with weeds. Poor. It is mostly hand weeded to remove *Eleusine indica* and *E. africana* in Uganda. These are hard to distinguish from finger millet in the young stage.
Land preparation for establishment. Seed-bed preparation should be thorough because of the small seed, and because it cannot stand weed competition. Often brush is burned on the site to provide ash.
Sowing methods. In India seedlings are raised in nurseries and transplanted when 10-12 cm tall into rows 40-45 cm apart, with 15-20 cm between plants.
Sowing depth and cover. Being a small seed it is usually surface sown. Rolling after sowing would improve germination.
Sowing time and rate. It is sown early in the season to spread the labour over various crops in East Africa. Use 35 kg/ha broadcast and 6-9 kg/ha in rows 30-33 cm apart (plants thinned to 5 cm).
Vigour of growth and growth rhythm. It matures in four to five months. Harvested in October and November in India.
Response to photoperiod. Short to medium day lengths are necessary for flowering (Evans, Wardlaw & Williams, 1964).

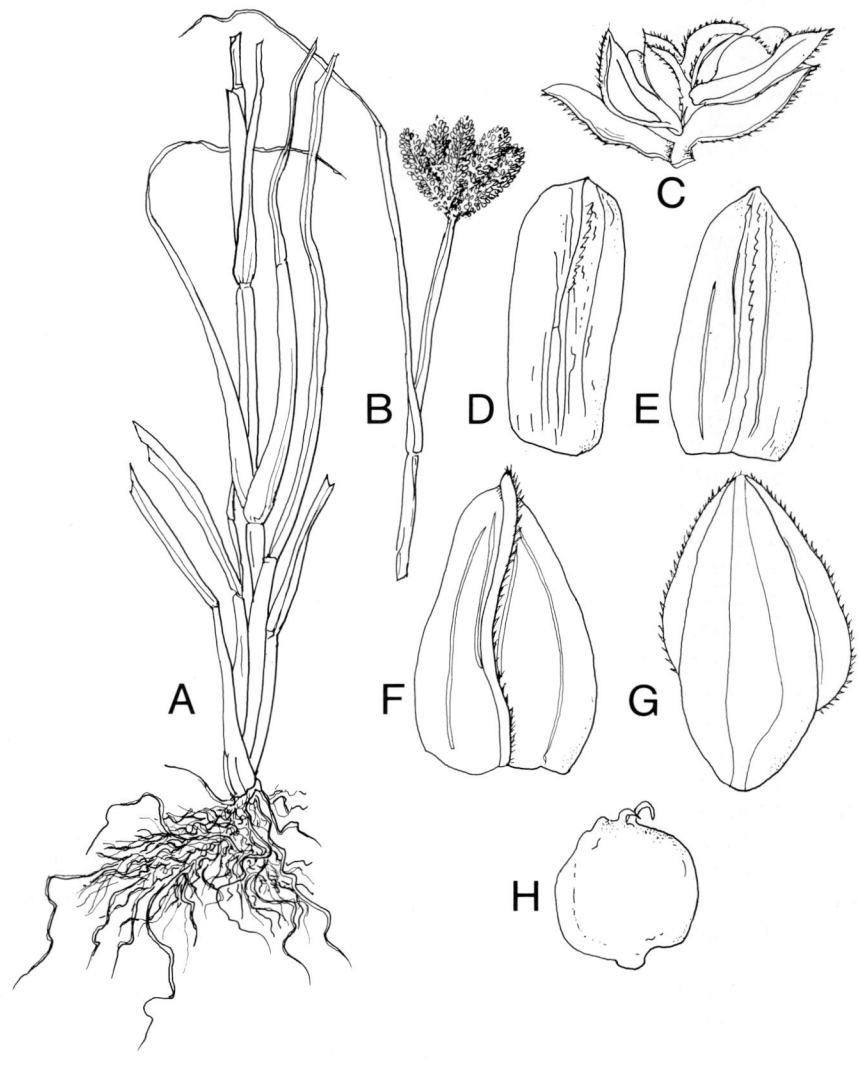

Figure 15.61. Eleusine coracana. **A**-Habit **B**-Inflorescence **C**-Spikelet cluster **D**-Upper glume **E**-Lower glume **F**-Lemma **G**-Palea **H**-Grain

TABLE 15.33 *Eleusine coracana*

	DM	As % of dry matter				
		CP	CF	Ash	EE	NFE
Fresh, late vegetative, India		7.6	33.6	15.1	1.1	42.6
Fresh, dough stage, India		7.1	28.8	12.5	1.7	49.9
Straw, India	95.0	3.2	34.2	7.9	1.3	53.4
Silage from straw, India		3.6	38.8	9.6	1.5	46.5

	Animal	Digestibility (%)				
		CP	CF	EE	NFE	ME
Straw, India	Cattle	16.0	79.6	47.0	59.3	2.20
Straw silage, India	Cattle	8.0	69.0	44.0	52.0	1.90

Source: Göhl, 1975

Genetics and reproduction. 2n=36 (Fedorov, 1974). A breeding unit to serve the whole of East Africa is located at the East African Agricultural and Forestry Research Organization (EAAFRO), Serere, Uganda.
Chemical analysis and digestibility. See Table 15.33.
Palatability. The straw is used as low-quality roughage in India and Uganda.
Seed production and harvesting. It seeds well, and as the straw is very tough the heads are cut off by hand mower in Uganda, but combine harvesters can be used in large fields. In India it is threshed under bullocks' feet.
Seed yield. Usually 450-900 kg/ha, but up to 4 500 kg/ha have been obtained.
Seed germination. The grain can be preserved for 50 years in dry grain pits (Solomon, 1953).
Cultivars. The variety Engeryi was released from the breeding station located at EAAFRO in 1969. It has good malting characteristics and foodgrain qualities (Peters, 1973).
Diseases. Head blast caused by *Piricularia oryzae,* which also attacks rice, can cause damage. Resistant varieties are being developed.
Pests. There are few pests, but birds may cause damage when the grain is in the soft dough stage.
Economics. In Uganda and west Kenya the grass is usually fried a little before being mixed with dried cassava chips. The two are then ground into a flour which is used for making "ugali" or "uji". The grass is also used for brewing. Straw is used for thatching, granaries and food containers (Acland, 1971). It forms the staple food for 50 percent of the population of Uganda, especially in areas of poor soil and low rainfall. In India it is grown in high-rainfall areas where the soil is too light for rice and too steep for terracing. In Fiji it is used

as food and in India the flour is made into a cooling drink called "ambli" and the green heads parched and eaten as "hurda".

Main attributes. It can be stored as grain for long periods without insecticides. The seeds are small, they dry out quickly, and insects cannot live inside them. This is important in humid Uganda where maize is difficult to store. Called a "famine" crop because it could be stored for lean years. Used as a first crop in new land in Kenya and Tanzania.

Main disadvantages. It has a low yield capacity and requires much labour at all stages — for seed-bed preparation, weeding, bird scaring, harvesting and threshing.

Further reading. Peters, 1973; Solomon, 1953.

Eleusine indica (L.) Gaertn. ssp. *indica*

Common names. Crow's foot grass (Australia), goose grass (United States), grama de caballo (Cuba), pata de gallina (Peru), rapoka grass, crab grass (South Africa), Indian goose grass, kavoronaisivi (Fiji), mangrasi (Suriname).
Natural habitat. Widespread weed of disturbed land.
Distribution. Tropical and subtropical regions.
Description. Coarse, caespitose annual, branching at the base, 30-60 cm tall, the culms ascending or prostrate, smooth, compressed; leaf-sheaths smooth, blades linear, flat or folded, 3-8 mm wide. Two to six spikes, digitate, sessile, 4-15 cm long, with usually one inserted lower on the culm, the rachis prominently flattened with the spikelets loosely imbricate and secund. Spikelets sessile with three to 15 flowers, 3-4 mm long. Glumes rather unequal, the lower narrow, oblong, obtuse, one-nerved, the upper lanceolate or ovate-lanceolate, three-nerved, rather acute. Lemmas lanceolate, rather acute, sometimes keeled, 3 mm long. Pericarp persistant, very loose and membranous, enclosing the rugose seed (Gardner, 1952). Distinguished from *Chloris* and *Dactyloctenium* by having awnless spikes. It has a particularly tough root system and is hard to pull out. Subspecies *africana* is a tetraploid, larger, with larger spikelets and a ligule that has a definite ciliate fringe (Chippendall & Crook, 1976). At low densities it can compensate by producing more tillers (Jones & Aliyu, 1976) (see Fig. 15.62).
Season of growth. Summer.
Optimum temperature for growth. The optimum mean temperature for seed germination was 23°C at Kairi (lat. 17°18′S, altitude 700 m) on the Atherton Tableland, Queensland (Hawton, 1979).
Minimum temperature for growth. Hawton (1979) found mean temperatures below 23°C restricted germination, but it could occur at temperatures as low as 20°C.
Altitude range. Sea-level to 2 000 m.
Rainfall requirements. It commonly grows in the 500-1 200 mm rainfall range.
Drought tolerance. Its extensive root system allows it to forage for moisture well during its annual growth.
Seedling vigour. Seedlings have exceptional vigour and quickly establish themselves.
Vigour of growth and growth rhythm. On the Atherton Tableland vigorous growth occurs between 8 September and 16 January at the expense of *Setaria sphacelata* seed crops (Hawton, 1979).
Tolerance to herbicides. To control this grass use cultivation, but if chemical control is needed use a pre-emergent spray of 2,4-D sodium salt at 9.5 kg/ha of an 840 g AI/kg product (e.g. Hormicide). No wetting agent is required

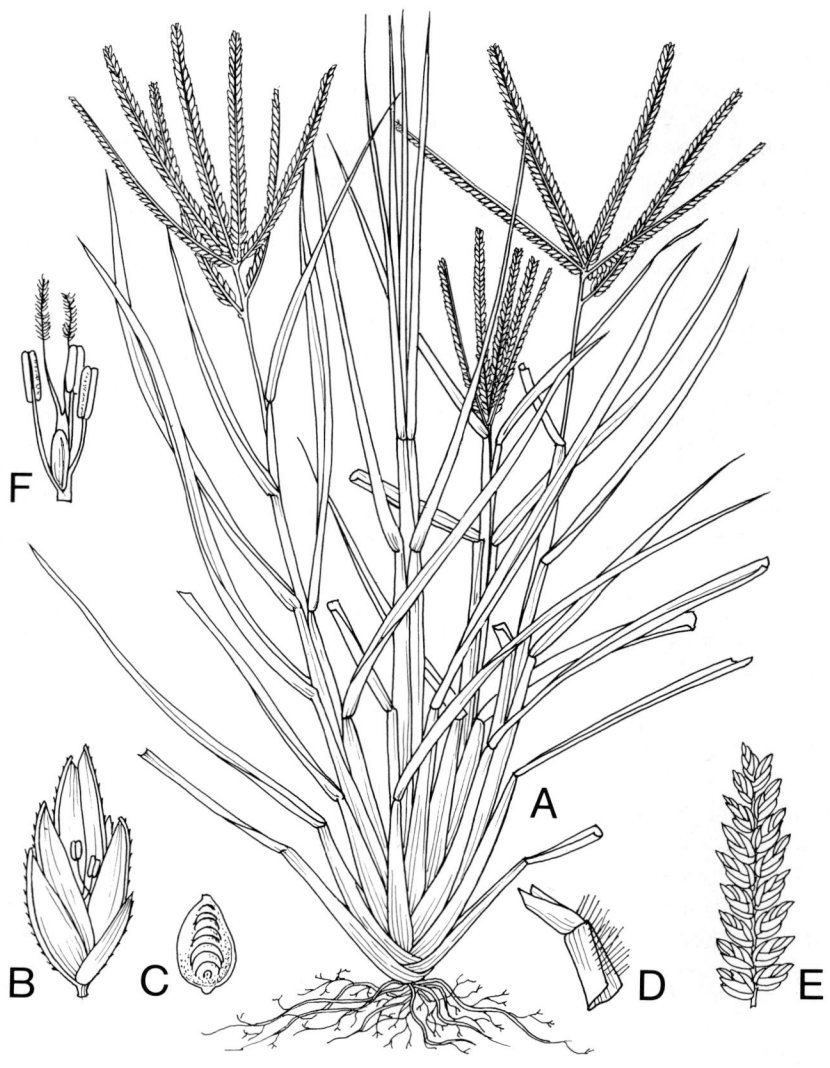

Figure 15.62. Eleusine indica. **A**-Habit **B**-Spikelet **C**-Grain **D**-Junction of leaf sheath and lamina **E**-Portion of spike **F**-Flower

when used as a pre-emergent spray. Use a minimum of 340 litres of water per hectare. Seedlings up to the four-leaf stage can be controlled by paraquat at 570 ml of a 200 g AI/litre product (e.g. Gramoxone) per 200 litres of water plus a surfactant at 250 ml/200 litres water. Spray until spray material runs off leaves (Tilley, 1977). In lawns in the United States potassium thiocyanate and disodium methylarsonate are recommended as effective control measures (Ivens, 1967). *E. indica* was killed in *P. maximum* and *B. decumbens* pastures on the Atherton Tableland, Queensland by atrazine above a strength of 0.9 kg AI/ha (Hawton, 1976). *Setaria sphacelata* seed crops can be selectively rid of *E. indica* by pre-emergence treatment with methabenzthiazuron at 1-2 kg/ha. *S. sphacelata* is susceptible to atrazine, simazine and terbutryne (Hawton, personal communication), Jones and Aliyu (1976) had some success with pre-emergent spraying of trifluralin at rates of 0.56 kg/ha AI in *Leucaena leucocephala* but yield of the legume was affected. Activated charcoal and Dacthal gave some control but more research is needed to clarify the effect.

Genetics and reproduction. 2n=18, 36 (Fedorov, 1974).
Suitability for hay and silage. It can be made into coarse hay and silage.
Chemical analysis and digestibility. See Table 15.34.
Palatability. It is eaten when young; when older the foliage is very tough.
Toxicity. It often contains prussic acid (*Cyanogenetic glucoside*), the main concentration being in the seeds varying from 0.015 to 0.019 percent, just below the theoretical potential danger level (Everist, 1974).
Cultivars. There are no cultivars registered. Bogdan (1977) records a sub-

TABLE 15.34 **Eleusine indica**

	DM	As % of dry matter				
		CP	CF	Ash	EE	NFE
Fresh, 4 weeks, Ghana	19.3	12.8	25.1	12.2		
Fresh, 8 weeks, Ghana	26.2	8.2	23.7	9.1		
Fresh, 12 weeks, Ghana	35.9	8.4	30.2	9.2		
Fresh, 16 weeks, Ghana	38.4	6.0	28.6	10.1		
Fresh, mature, Thailand	21.2	13.3	28.5	12.0	2.8	43.4
Hay, United States	92.1	3.5	35.8	9.9	1.1	49.7

	Animal	Digestibility (%)				
		CP	CF	EE	NFE	ME
Hay, United States	Sheep	0.0	44.0	10.0	34.0	1.19

Source: Göhl, 1975

species *africana* (Kennedy and O'Byrne), Phillips, a more robust form, a tetraploid with 2n=36, from East Africa occurring at higher altitudes than ssp. *indica*.

Economics. A worldwide weed of the tropics. It is one of the worst weeds of maize in Zimbabwe and South Africa (Ivens, 1967). It is the major weed problem in swards of *Panicum maximum, Setaria sphacelata* and *Brachiaria decumbens* grown for seed production on the Atherton Tableland, Queensland (Hawton 1976, 1978) and in the establishment of the browse legume *Leucaena leucocephala* at Samford, Queensland (Jones & Aliju, 1976). In India the seeds of *E. indica* are eaten by humans in times of drought.

Animal production. No figures have been cited.

Main attributes. Its aggressiveness and its easy establishment for stabilizing sandy soils.

Main deficiencies. Its problem as a weed, its occasional toxicity.

Eleusine jaegeri **Pilg.**

Common names. Manyatta grass, mafutiana (Kenya), akirma (Ethiopia).
Natural habitat. Tussock grassland or open forest at altitudes above 2 300 m in East Africa.
Distribution. Throughout tropical and southern Africa, in highlands of Kenya, Uganda, Ethiopia.
Description. Densely tufted perennial 120-150 cm high with compressed branching culms. Inflorescence of up to ten dark grey spikes on a short axis about 6 cm long. The leaves are stiff, up to 60 cm long with sharp edges. The inflorescence consists of three to seven stiff, one-sided, greyish-green spikes, up to 15 cm long, arising from the top 5-8 cm of the stem. The spikelets are numerous and densely crowded along the underside of the spike. They are 6 mm long and consist of four to five florets (Ivens, 1967) (see Fig. 15.63).
Altitude range. 2 000-3 000 m. It occurs at 2 600 m on the rim of Empakai Crater in the Ngorongoro Highlands and on the south rim at an elevation of 2 970 m in Tanzania.
Rainfall requirements. The 19-year average at Ngorongoro Crater is 900 mm per year. It requires a rainfall in excess of 625 mm.
Tolerance to herbicides. Dalapon at 5 and 10 kg AI/ha and glyphosate (Roundup) at 4 kg AI/ha applied in water at a volume rate of 300 litres per hectare with an Oxford Precision Sprayer operating at a pressure of 2.1 kg/cm^2 with size "O" fanjet nozzles, at midday in warm, sunny, dry, windless conditions reduced *Eleusine* populations within 24 hours. Dalapon at 5 kg AI/ha following cutting and removing the cuttings was the cheapest and most effective. This also encourages forage species. Apply only to *Pennisetum clandestinum* grassland.
Eradication. Destumping by cutting out individual plants and removing them is the most effective mechanical control with lesser beneficial results from cutting and removing and cutting and burning the *Eleusine*.
Chemical analysis. Dougall and Bogdan (1960) recorded 10.5 percent crude protein, 34.8 percent crude fibre, 6 percent ash, 2.4 percent ether extract and 46.3 percent nitrogen-free extract from fresh material at early bloom stage in Kenya — on a dry-matter basis.
Palatability. It is avoided by stock.
Economics. Widely used for making baskets in Ethiopia (Westphal, 1975). It is a frequent invader of such highland pastures in East Africa as Kikuyu (see Plate 36) or *Themeda triandra* where grazing is too intensive to permit occasional burning (Ivens, 1967). Common in old cattle bomas.
Further reading. Ivens, 1967.

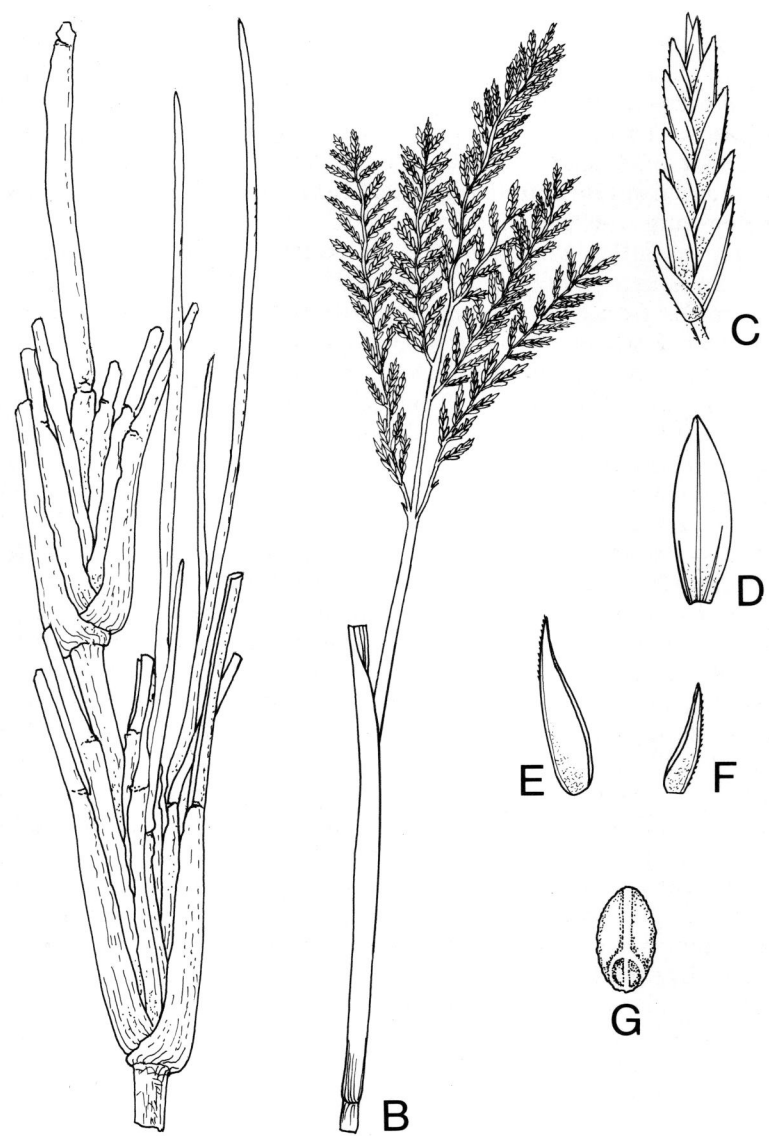

Figure 15.63. Eleusine jaegeri. **A**-Portion of culm **B**-Inflorescence **C**-Spikelet **D**-Lemma **E**-Upper glume **F**-Lower glume **G**-Grain

Enteropogon macrostachyus (Hochst. ex A. Rich.) Monro ex Benth.

Synonym. E. simplex (Schumach. ex Thorn.).
Common names. Bush rye (Kenya), mopane grass (Zimbabwe).
Natural habitat. Grassland and rocky outcrops in semi-arid climates.
Distribution. Throughout tropical Africa. Abundant between Sultan Hamud and Voi, Kenya (Bogdan & Pratt, 1967) and on Kongwa ranch, Tanzania (van Rensburg, 1969).
Description. Tufted annual or perennial about 90 cm high. Spikes solitary, about 15 cm long with glabrous spikelets 8-10 mm long; lemmas 2-3, awned. Leaves are scattered along the culm and the leaf-blades fold readily when dry, and are very finely pointed. They have a tuft of fine white hairs in the axil (see Fig. 15.64).
Season of growth. Summer.
Altitude range. Sea-level to 1 800 m.
Rainfall requirements. It occurs in an area receiving around 575 mm per annum (Bogdan & Pratt, 1967).
Drought tolerance. Good.
Soil requirements. It prefers loose sandy loams and loams, but will grow on alluvial silts and rocky soils (Bogdan & Pratt, 1967).
Sowing time and rate. Sow in the wet season at 7 kg/ha.
Number of seeds per kg. 176 000 spikelets with one seed each (Bogdan & Pratt, 1967).
Vigour of growth. It germinates readily and grows vigorously. In the Sahel it flowers in August and remains as standing hay through to February in Mauritania (Boudet & Duverger, 1961).
Palatability. It is palatable.
Seed production and harvesting. It is a very good seeder and seed can be collected rapidly by cutting the seed-heads or stripping the heads by hand (Bogdan & Pratt, 1967). It should lend itself easily to mechanical harvesting.
Economics. It has proved an excellent grass for reseeding the rangelands of Kenya under moderately dry conditions, as has been demonstrated in Baringo (Bogdan & Pratt, 1967).
Further reading. Bogdan & Pratt, 1967.

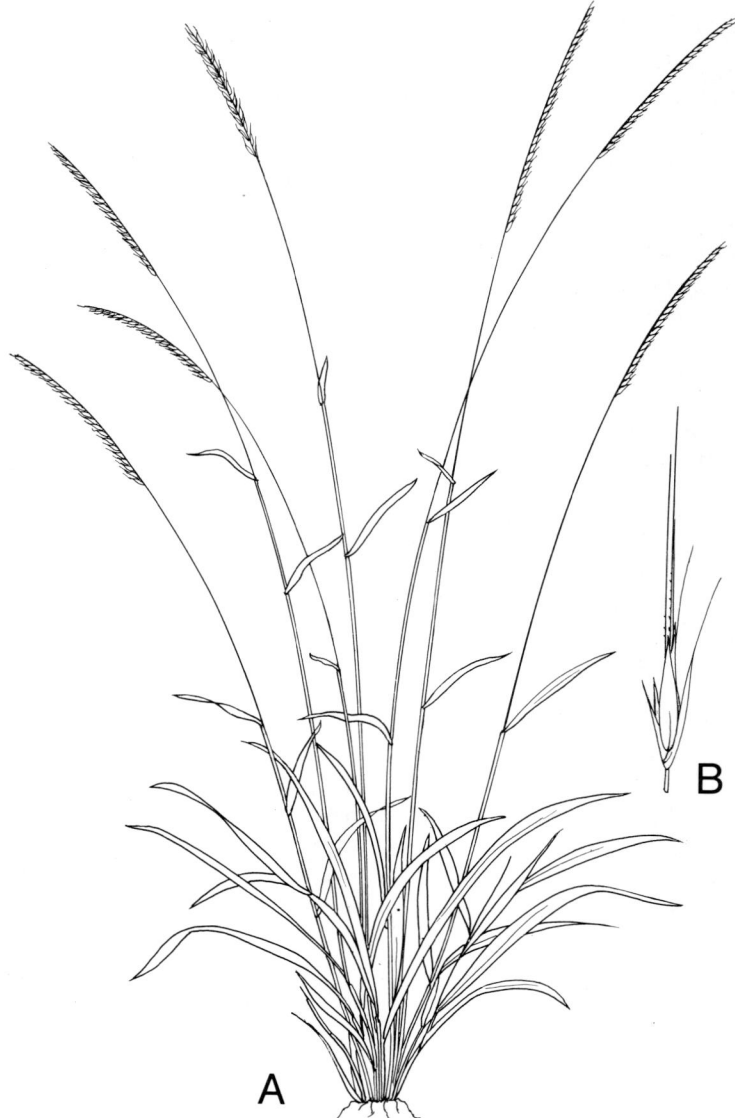

Figure 15.64. Enteropogon macrostachyus. **A**-Plant **B**-Spikelet

Enteropogon somalensis Chiov.

Natural habitat. Seasonally waterlogged sites with heavy soils (Bogdan & Pratt, 1967).

Distribution. It occurs most extensively in northern Kenya around Arba Johan, and in some areas of the southeast (Bogdan & Pratt, 1967).

Description. It resembles *E. macrostachyus* but it has a lower, more spreading habit, somewhat smaller spikelets and shorter awns. It is a tufted perennial (Bogdan & Pratt, 1967).

Economics. It is recommended for trial in range reseeding in Kenya on heavy soils and waterlogged sites.

Further reading. Bogdan & Pratt, 1967.

Entolasia imbricata Stapf

Common name. Bungoma grass (Kenya).
Natural habitat. It occurs naturally on vlei soils in swampy river valleys in East Africa (Mwakha, 1971).
Distribution. Tanzania, Kenya, Uganda and southern tropical Africa in sandy and swampy places by rivers and in miombo.
Description. A rhizomatous perennial, 60-120 cm high, glabrous except for the usually bearded nodes, leaf-blades flat, 10 mm wide. Inflorescence over 15 cm long with solitary, dense racemes 2-3 cm long appressed to the axis, having glabrous, green spikelets 4.5-5 mm long (Napper, 1965).
Season of growth. Summer.
Altitude range. 1 000-2 000 m in Tanzania.
Rainfall requirements. A wet-land species.
Tolerance to flooding. It will withstand flooding.
Number of seeds per kg. 300 000.
Response to defoliation. Cutting at intervals of one or two weeks affected the crop but at four- to eight-week intervals dry-matter production was increased with successive harvests (Mwakha, 1970).
Chemical analysis and digestibility. Herbage contained 19.5 percent dry matter, with 26.5 percent crude protein and 25.1 percent crude fibre (Mwakha, 1970).
Cultivars. A composite named Nzoia K6819 was selected at Kitale, Kenya.
Further reading. Mwakha, 1970.

Eragrostis caespitosa Chiov.

Synonym. E. basiilepis Pilg.
Common name. Cushion love grass (Kenya).
Natural habitat. Dry sandy places, grassland, semi-desert and abandoned cultivation.
Distribution. Northeast tropical Africa. Plentiful on Yatta Plateau, between Embu and Kitui, Kenya (Bogdan & Pratt, 1967).
Description. Erect tufted perennial up to 60 cm high, the basal shoots having hard yellow scales, culms rather wiry (Napper, 1965). There is no basal foliage, but the stems are densely set with short leaves (Bogdan & Pratt, 1967). It has numerous small purple panicles (see Fig. 15.65).
Altitude range. Sea-level to 1 800 m in Kenya (Bogdan & Pratt, 1967).
Rainfall requirements. It grows in a rainfall regime of 375-875 mm.
Chemical analysis and digestibility. Of moderate nutritive value only, containing about 8-9 percent crude protein in the dry matter at the flowering stage (Bogdan & Pratt, 1967).
Palatability. It is well grazed (Bogdan & Pratt, 1967).
Seed production and harvesting. It seeds well, but the seed is sometimes destroyed by smut. Seed can be collected by cutting the stems, though it sheds easily when ripe. The seed is in the form of naked caryopses and threshing and cleaning are easy (Bogdan & Pratt, 1967).
Animal production. It has never been used for reseeding rangeland, but Bogdan and Pratt (1967) recommend its trial in mixtures with other grasses on poor, sandy soils in eastern Kenya.
Further reading. Bogdan & Pratt, 1967.

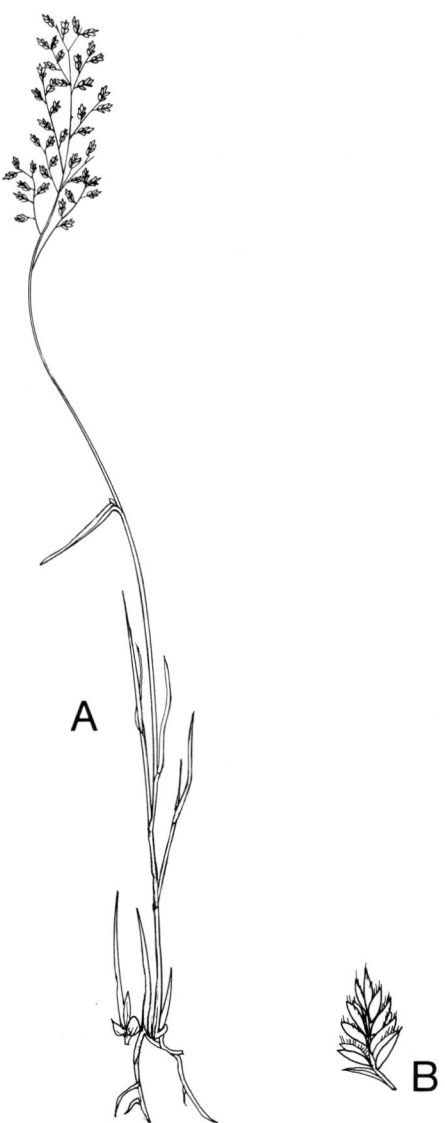

Figure 15.65. *Eragrostis caespitosa.* **A**-Plant **B**-Spikelet

Eragrostis chloromelas Steud.

Common name. Boer love grass (United States).
Distribution. Africa. Introduced to the United States.
Description. Plants with short, dense, basal tufts of filiform, curling leaves; inflorescence open and lax, branches filiform, flexible, purplish; spikelets spreading, not appressed to the branches. It is very similar to *E. curvula* in that both the culms are not branched (Chippendall, 1955). It is a tufted perennial. The basal leaves provide most of the forage. The seed-heads are distinctly diamond shaped (see Fig. 15.66).
Season of growth. Spring through summer to autumn.
Frost tolerance. It is easily killed by frost (Humphrey, 1960a).
Rainfall requirements. Around 625 mm (Bogdan & Pratt, 1967).
Drought tolerance. Adapted to semi-desert conditions and very drought resistant.
Soil requirements. Loose sandy loams and loams (Bogdan & Pratt, 1967).
Tolerance to salinity. It is not well adapted to alkaline soils.
Ability to spread naturally. It spreads well by seed.
Land preparation for establishment. A rough seed-bed prepared with a disc harrow is needed.
Sowing methods. Broadcast.
Sowing depth and cover. Surface sow and cover lightly.
Sowing time and rate. Sow in summer at about 200 g/ha.
Number of seeds per kg. 6.6 million (naked caryopses) (Bogdan & Pratt, 1967).
Genetics and reproduction. 2n=40, 60, 63 (Fedorov, 1974).
Chemical analysis and digestibility. No figures have been cited.
Palatability. Its palatability is rather low (Bogdan & Pratt, 1967) in Kenya but Humphrey (1960a) states that cattle make good use of Boer love grass in Arizona, especially in autumn when other grasses have dried off.
Seed production and harvesting. The seed is in the form of naked caryopses of good quality that are easy to handle (Bogdan & Pratt, 1967).
Economics. It is a valuable grass in semi-desert tropical grassland areas.
Animal production. It is hardy and persistent under moderately dry conditions and has been used with fair success in reseeding rangeland at West Pokot and around Baringo in Kenya (Bogdan & Pratt, 1967).
Main attributes. Its ability to produce green feed in the spring and continue into summer and autumn.
Further reading. Humphrey, 1960a.

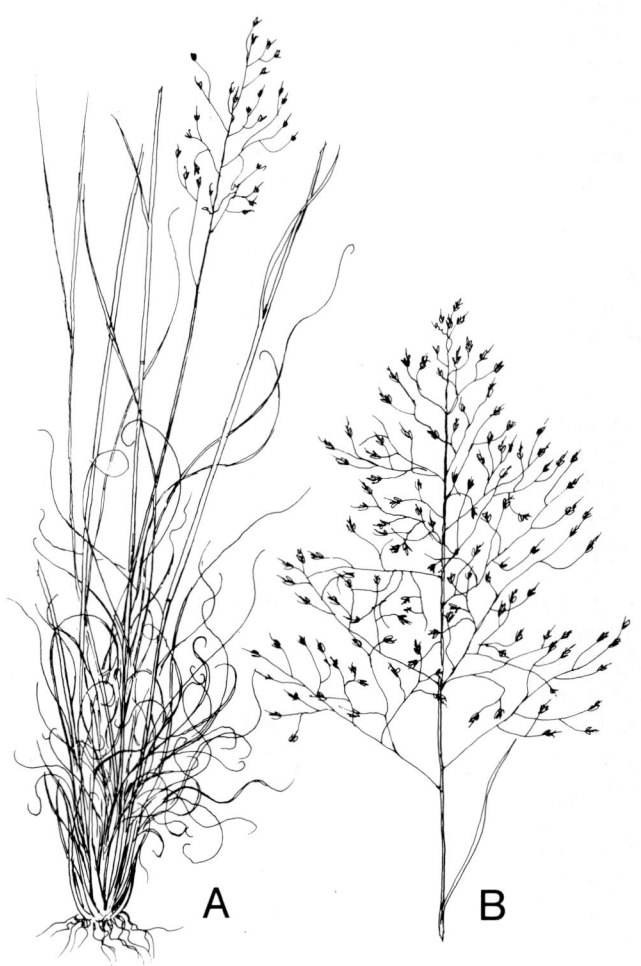

Figure 15.66. Eragrostis chloromelas. **A**-Plant **B**-Inflorescence

Eragrostis cilianensis (All.) Lutati

Synonym. E. major (L.) Host.
Common names. Stink grass (Australia), grey love grass (Kenya), black grass (New South Wales).
Natural habitat. Widespread as a weed, especially on poor soils.
Distribution. Native to the Mediterranean region, now widely distributed throughout the tropics, mainly as a weed. It is a natural dominant in the annual grasslands of northern Kenya (Bogdan & Pratt, 1967).
Description. Erect annual up to 90 cm high with geniculate or erect culms. Inflorescence a fairly open panicle. The leaf margins, nerves, panicle branches and lemmas are nearly always dotted with small, dark glands, some of which are raised, some depressed. Grains are almost spherical (Chippendall & Crook, 1976) (see Fig. 15.67).
Season of growth. Summer.
Altitude range. Sea-level to 2 250 m in Kenya.
Rainfall requirements. It occurs in rainfall regions down to 250 mm annually in Kenya but is more common in the 600-700 mm region in Australia.
Drought tolerance. As a free-seeding annual it escapes drought in a moderately dry year.
Number of seeds per kg. 4.4 million.
Vigour of growth and growth rhythm. It grows quickly and lasts only about three months.
Chemical analysis and digestibility. Even rather stemmy herbage can contain 15 percent crude protein in the dry matter (Bogdan & Pratt, 1967). Dougall and Bogdan (1960) recorded 15.3 percent crude protein, 29 percent crude fibre, 10.6 percent ash, 2.4 percent ether extract and 42.7 percent nitrogen-free extract from fresh material in late bloom in Kenya on a dry-matter basis.
Palatability. It is not very palatable but is eaten when young by cattle, horses and sheep. It has a disagreeable odour when fresh. The culms have a ring of glands below the nodes.
Seed production and harvesting. It is an efficient seeder, though the seed is tedious to collect in large quantities (Bogdan & Pratt, 1967).
Genetics and reproduction. 2n=20, 40 (Fedorov, 1974).
Economics. Usually this aromatic grass is regarded as a weed and unpalatable to stock, but it gives early feed. In Lesotho the grains are used in time of famine for human food (Smith, 1966). The seed is in the form of small, naked caryopses. One of the pioneer species to appear on denuded land in semi-arid areas. Bogdan and Pratt (1967) recommended it for seeding the annual grass zone in rangeland, particularly alluvial soils in Kenya. There are, however, more palatable grasses, for example *Dactyloctenium* spp. As *E. cilianensis* is a worldwide weed, it is best omitted from seeding programmes.

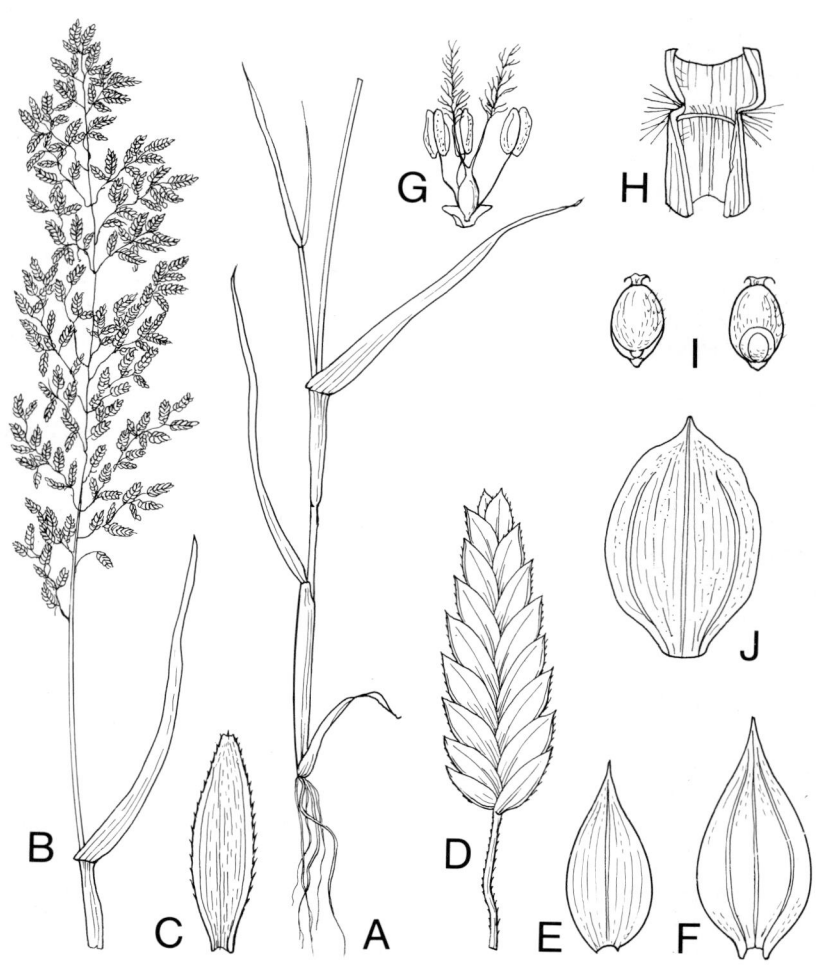

***Figure** 15.67. Eragrostis cilianensis.* **A**-Stem **B**-Inflorescence **C**-Palea **D**-Spikelet **E**-Lower glume **F**-Upper glume **G**-Flower **H**-Ligule **I**-Grain **J**-Lemma

Eragrostis curvula (Schrad.) Nees

Common names. Weeping love grass (South Africa, United States), African love grass (Australia), pasto llorón (Peru).
Natural habitat. Clearings in woodlands in trampled disturbed land, moist sandy soil.
Distribution. Native of Tanzania, now throughout South Africa, and introduced in several warm countries.
Description. Densely tufted perennial 90-120 cm high, with rigid narrow leafblades with inrolling margins. Panicle narrow with green or dark grey spikelets up to 2 mm wide having 6 florets. It is very similar to *E. chloromelas* but differs in having rigid and non-filiform branches and flat leaves (Chippendall, 1955). The foliage is exceedingly tough (see Fig. 15.68).
Season of growth. Spring and summer.
Optimum temperature for growth. It endures heat.
Minimum temperature for growth. Just above freezing.
Frost tolerance. It has survived temperatures as low as freezing in the southern Great Plains of the United States. At Samford, Queensland, *E. curvula* (CPI143218) produced dry matter at the rate of 52 kg/ha per day between March and July, during which 29 frosts were recorded (Strickland, 1973).
Altitude range. Sea-level to 3 500 m (originated near the equator in Tanzania at 1 000-1 600 m).
Rainfall requirements. 500-1 000 mm in the tropics and subtropics generally. It will grow in rainfall as low as 300 mm if sown in basins or contour furrows and mulched (Miller & Hafenrichter, 1958).
Drought tolerance. It is quite drought tolerant.
Tolerance to flooding. Not good. It will not grow on wet, seepy soils and will not tolerate standing water.
Soil requirements. It prefers sandy loams but will grow in a wide range of soils. It prefers a pH of 7.0-8.5 (Miller & Hafenrichter, 1958).
Tolerance to salinity. It is very tolerant of salinity and seed germinates well under high levels of soil sodium (Ryan, Miyamoto & Stroehlein, 1975). In Western Australia, *E. curvula* (CPI14369) was moderately tolerant (Rogers & Bailey, 1963).
Fertilizer requirements. It will grow on poor soils, but for high production it needs extra nitrogen. With no nitrogen in Oklahoma, United States, forage yield was 2 178 kg/ha, with 112 kg N/ha it yielded 8 309 kg/ha, and with 224 kg N/ha, 11 374 kg/ha. It also has a high potassium requirement and removed 3.8, 4.0 and 4.7 kg of potassium per 454 kg of forage at low, medium and high nitrogen rates (Altom, 1978). Botha and Hamburger (1953) got significant increases in response to nitrogen with the Ermelo strain but not to phosphorus. A positive nitrogen/phosphorus interaction only occurred with applications of nitrogen in excess of about 300 kg N/ha.

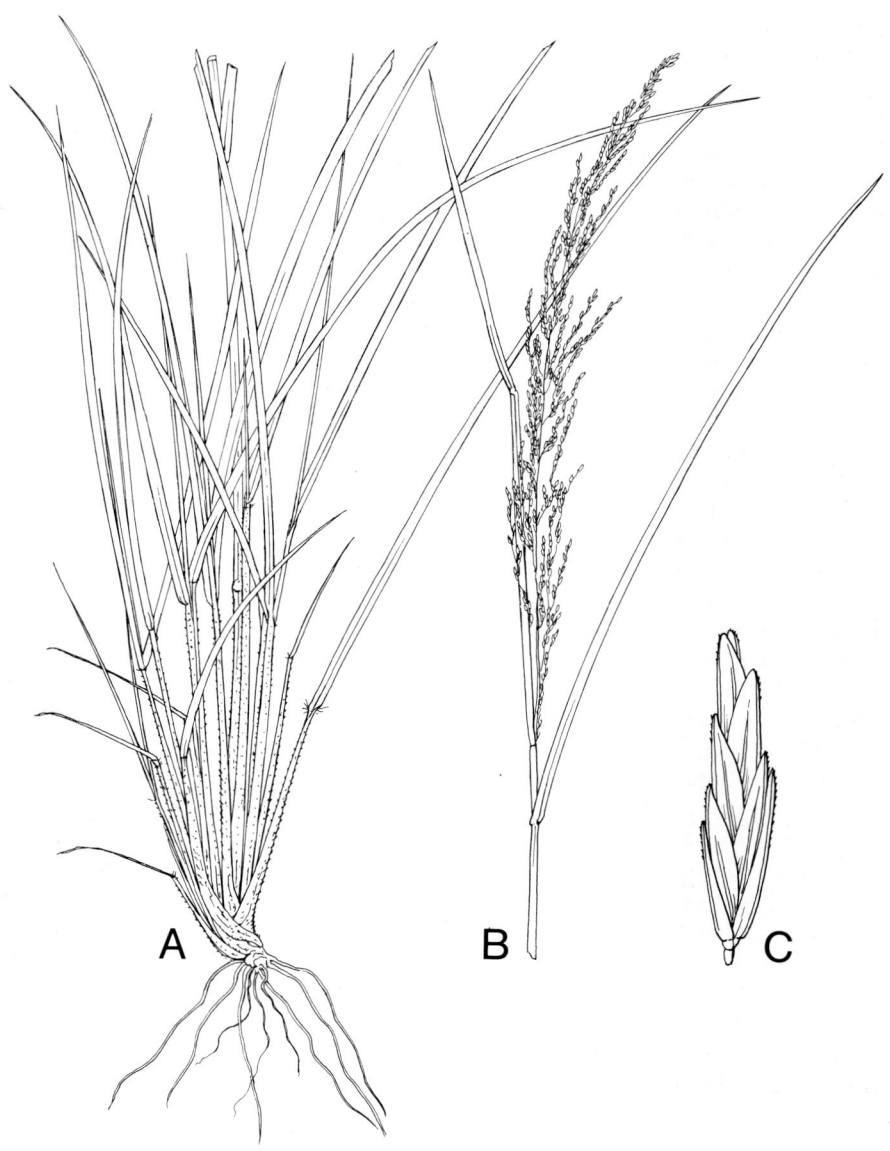

Figure 15.68. Eragrostis curvula. **A**-Habit, showing tufted root-stock **B**-Inflorescence **C**-Spikelet

Land preparation for establishment. A good seed-bed is preferred.
Sowing methods. Broadcast or drilled.
Sowing depth and cover. Do not cover over 0.5-1 cm.
Sowing time and rate. Sow late spring to late summer at 1 kg/ha broadcast or 0.25 kg in 1-m rows.
Number of seeds per kg. Approximately 3 850 000 or 110 000 flat spikelets.
Seedling vigour. Excellent.
Vigour of growth and growth rhythm. It starts growing early in the spring and continues until well into the autumn.
Response to photoperiod. It is indifferent to day length for flowering (Evans, Wardlaw & Williams, 1964), i.e. day neutral.
Compatibility with other grasses and legumes. In the United States it is sown with Korean lespedeza.
Response to defoliation. It is best subjected to rotational grazing to maintain the stand at moderate grazing pressure.
Grazing management. If it is sown in rows, an inter-row cultivation during the first year will help it compete with weeds, which it will do in succeeding years. Periodic mowing will be beneficial if stock cannot keep it eaten close to the ground. Davidson (1964) developed a system of management based on a heavy initial dressing of nitrogen and then annual maintenance dressings based on nutrient removal in milk, working on 80 percent return of nitrogen by the grazing animal and 50 percent nitrogen recovery in shoots.
Response to fire. It tolerates fire.
Genetics and reproduction. 2n=20, 40, 50, 80 (Fedorov, 1974). It is an obligate apomict (Brown & Emery, 1958).
Dry- and green-matter yields. At Stillwater, Oklahoma, United States, Pumphrey (1978) over four years obtained an average production of dry matter over the summer period from 1 July to 22 November of 3 178 kg/ha unfertilized, and 8 502 kg/ha fertilized with 224 kg N and 45 kg P_2O_5 per hectare. When fertilized with 450 kg N, 38 kg P and 58 kg K per hectare, the mean annual yield of *E. curvula* at Henderson Research Station, Zimbabwe over three years was 5 930 kg DM/ha (Rodel, 1970). At Samford, Queensland, dry-matter yields ranged from 13 000-27 000 kg/ha per year with eight-week cutting intervals (Strickland, 1973). Nitrogen was applied at 45 kg/ha at eight-week intervals. The grass was not irrigated. The mean yields were approximately double those for the four-week cutting interval. Under irrigation and with fertilization it yielded 28 000-32 000 kg DM/ha in southwest Australia (Roberts & Carbon, 1969).
Suitability for hay and silage. It makes good hay if cut before it becomes too tough, and combines well with lucerne in southern Africa.
Value as standover or deferred feed. It is grown for winter pasture in Florida.
Chemical analysis and digestibility. Digestibility results at Samford showed a range from 65 percent in spring to 49 percent in midsummer and 50 percent

in midwinter, with crude protein from 17.5 percent in spring to 6.25 percent in midsummer and 9.4 percent in midwinter (Strickland, 1973).

Palatability. The robusta types are well grazed by stock when young. Leigh (1961b) grouped *E. curvula* types into groups — 'curvula', 'robusta green', 'robusta intermediate', 'robusta blue' and 'chloromelas'. The three 'robusta' types were the most palatable, 'chloromelas' varieties and *E. plana* being intermediate and the 'curvula' varieties the least palatable.

Seed production and harvesting. This grass seeds heavily. It is harvested in early summer and again later in summer with a header-harvester or a hand sickle when one-third of the head has turned brown. Try to prevent scattering.

Seed yield. 30-225 kg/ha under good conditions. Larger seed is obtained from rows.

Cultivars. The 'robusta' types from Argentina yielded the highest in trials at Samford, Queensland, and the 'South African Robusta Blue' (CPI30380) was the highest individual yielder (Strickland, 1973). 'Witbank', 'Ermelo', 'Kromarrai' and 'American Leafy' are cultivars; 'Morpa' has been released in Oklahoma because it has better palatability and gives better animal production (12 percent) than common weeping grass (Shoop, McIlvain & Voight, 1976). 'Renner' was released in Texas because of better palatability than 'Ermelo'. It remains green during drought and heat, autumn and winter and into maturity (Dalrymple, 1978).

Value for erosion control. It is widely used in Kenya, Sri Lanka and the United States for stabilization of terraces, water discharge areas and banks of earth tanks. In Japan it has helped stabilize mountain slopes for at least three years (Endo, 1978).

Economics. **Eragrostis curvula** has been used successfully for oversowing the broad intermontane plains or *altiplanos* of the arid to semi-arid Puna proper in the province of Juyjuy in northern Argentina at 3 000 m elevation. The seed is sown in listed furrows at 2 kg/ha and covered with sheep manure. It takes 20-25 days to germinate. The seed is Tanganyika-type *E. curvula* grown locally in Buenos Aires (Tothill, 1978). *E. curvula* is one of the highest producing grasses in summer rainfall areas of temperate and cool subtropical areas of South Africa (Strickland, 1973).

Animal production. At Henderson Research Station, Zimbabwe, when fertilized at 270 kg N/ha and 35 kg P/ha per year and grazed over two summers by heifers at the rate of 12.4 per hectare, the mean maximum live-weight gain from *E. curvula* was 550 kg/ha (Rodel, 1970). At Deniliquin, New South Wales, irrigated *E. curvula* yielded 4 321 kg/ha unfertilized and 12 985 kg/ha per year fertilized with 480 kg/ha N (Squires & Myers, 1970). Stocked at 53 sheep per hectare over 130 days it gave a live-weight gain of almost 3 kg/ha; but at 70 sheep per hectare a live-weight loss of almost 4 kg per animal occurred.

In Oklahoma, Morpa weeping love grass showed steer gains of 1 kg per day during May and June, and 0.71 kg per day during July and August at a stocking rate of 1.5 steers per hectare during a 278-day grazing year (Pumphrey, 1978). Over a three-year period, Hereford steers showed 13 percent more live-weight gain (per animal) than those grazing the least palatable selections, and 12 percent more than those grazing common love grass, also of low palatability (Voight *et al.*, 1970).

Main attributes. Establishes easily, persists well under grazing. A tough grass with good cold tolerance, responds well to nitrogen, valuable in erosion control. Good palatability.

Further reading. Altom, 1978.

Eragrostis lehmanniana Nees

Common name. Lehmann love grass (United States).
Natural habitat. In cultivation.
Distribution. Native to South Africa, introduced to East Africa and India.
Description. A tufted perennial; culms 60-90 cm high, branched, branches repeatedly geniculated; leaf-blades narrow, eglandular, 1-3 mm wide with inrolled margins. Panicle 10-20 cm long, lax and open. It is distinguished from *E. curvula* and *E. chloromelas* in having papery lower leaf-sheaths with rounded nerves not very closely arranged. There are two varieties — var. *lehmanniana* and var. *chaunantha* (Pilg.). Strains introduced to the United States have prostrate stems rooting at the nodes (see Fig. 15.69).
Season of growth. Spring, summer and autumn.
Frost tolerance. Basal leaves remain green throughout the winter in southern California and stems stay green after autumn frosts, but temperatures below zero may kill established plants.
Altitude range. Below 1 700 m in southern California, but best at 1 000-1 500 m.
Rainfall requirements. Adapted to semi-arid tropical and subtropical summer rainfall areas. In California it grows in a rainfall regime of 250-375 mm.
Drought tolerance. It is quite tolerant of drought. Var. *chaunantha* flourishes in areas of low rainfall of 300-500 mm (Bor, 1960).
Soil requirements. It prefers light to medium soils of pH 7.0-8.5.
Tolerance to salinity. It tolerates high pH caused by calcium and magnesium rather than by sodium (Ryan, Miyamoto & Stroehlein, 1975).
Ability to spread naturally. It volunteers well in semi-desert grassland.
Land preparation for establishment. A well-prepared seed-bed is preferred, but for oversowing, rangelands are generally disc-harrowed.
Sowing time and rate. Sow in summer to early autumn at 250-500 g/ha.
Number of seeds per kg. 15.5 million.
Response to photoperiod. It is indifferent to day length for flowering (Evans, Wardlaw & Williams, 1964), i.e. day neutral.
Response to defoliation. It should not be too closely grazed.
Grazing management. It should become well established before being grazed. Only half the annual growth should be grazed off, but it can be continuously grazed for maximum production, though a late summer rest improved the total available carbohydrates, crude protein and phosphorus contents (Roberts & Opperman, 1966), and allows the grass to seed.
Response to fire. Warm-season fires may have an adverse effect, but burning four days after light rain in February, 1969, in Arizona (late winter) had little adverse effect on the grass (Pace, 1971).
Genetics and reproduction. 2n=40, 60 (Fedorov, 1974).

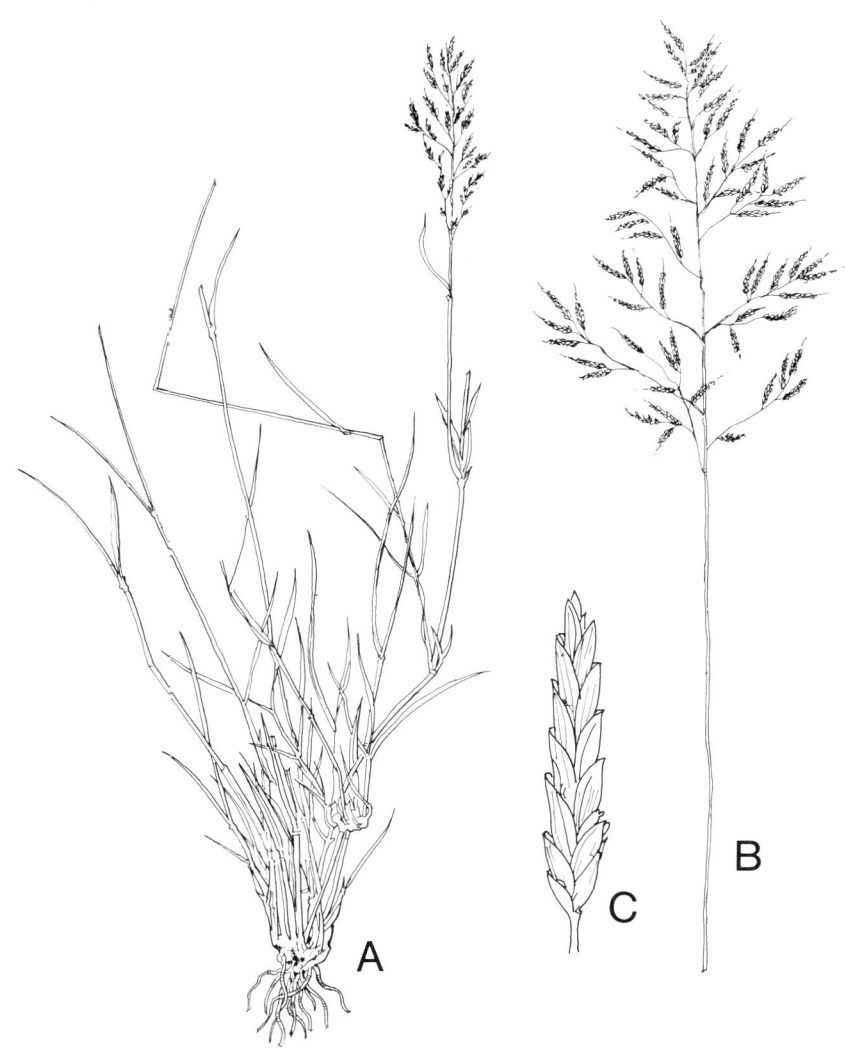

Figure 15.69. Eragrostis lehmanniana. **A**-Habit **B**-Inflorescence **C**-Spikelet

TABLE 15.35 *Eragrostis lehmanniana*

	DM	As % of dry matter				
		CP	CF	Ash	EE	NFE
Hay, late vegetative, South Africa		10.4	31.9	8.7	1.7	47.3
Hay, mature, South Africa		6.7	31.7	9.6	1.8	50.2
	Animal	Digestibility (%)				
		CP	CF	EE	NFE	ME
Hay, late vegetative, South Africa	Sheep	68.3	70.3	40.4	61.1	2.21
Hay, mature, South Africa	Sheep	58.6	63.0	51.5	59.4	2.04

Source: Göhl, 1975

Suitability for hay and silage. It is cultivated for hay in South Africa.
Chemical analysis and digestibility. See Table 15.35.
Palatability. It is palatable when green but of low palatability when mature.
Seed production and harvesting. It is a good seed producer and could be harvested by combine.
Value for erosion control. It is successful for reseeding rangeland in the southwestern United States and gives a rapid soil cover.
Economics. It is an imporant species in the sweet veld areas of South Africa and one of the best grasses for reseeding Arizona ranges (Humphrey, 1960a).
Further reading. Humphrey, 1960a.

Eragrostis superba Peyr.

Common names. Masai love grass (eastern Africa), heart-seed love grass (Zimbabwe), flat-seed love grass (southern Africa), Wilman love grass (United States).

Natural habitat. Open thicket and grassland on poor sandy soils, often as a weed.

Distribution. Native to southern and tropical Africa and introduced to India, Australia, the United States and elsewhere.

Description. Densely tufted perennial 30-90 cm high. Spikelets large, pale straw-coloured or slight purple-tinted, 5-9 mm wide and falling entire. The leaf-sheaths are smooth, keeled and persistent with a collar distinct, leaf-blades usually rolled. The individual spikelets are numerous and flattened somewhat, resembling rattle-snake rattles. The large seed-heads are rather ornamental (see Fig. 15.70).

Season of growth. It grows well in the spring in Arizona.

Minimum temperature for growth. It does not grow below −11°C (Humphrey, 1960a).

Frost tolerance. It was little affected by frost at Samford, Queensland (Strickland, 1973).

Altitude range. Sea-level to 2 000 m.

Rainfall requirements. It grows in a rainfall range of 500-875 mm.

Drought tolerance. It has good drought tolerance and is oversown into semi-arid land in Kenya.

Soil requirements. Prefers sandy soils but occurs also on clay loams and clays. A medium-textured deep soil neither strongly acid nor strongly alkaline is preferred.

Tolerance to salinity. It has a high tolerance to salinity and alkalinity and seed will germinate well (Ryan, Miyamoto & Stroehlein, 1975).

Ability to spread naturally. Excellent by seed (Millington & Winkworth, 1970).

Sowing methods. The mature spikelet holding two or three caryopses is sown sometimes under a nurse crop such as teff (*Eragrostis tef*) (Chippendall & Crook, 1976).

Sowing time and rate. Sow in the wet season at 55 kg spikelets per hectare.

Grazing management. Stands should not be grazed until the second summer after seeding. Sow enough to produce a good bulk of the grass so management can be applied according to its needs. In a mixed pasture it may be grazed out.

Genetics and reproduction. 2n=40 (Fedorov, 1974).

Dry- and green-matter yields. A dry matter of more than 24 000 kg/ha per year under an eight-week cutting interval was obtained by Strickland (1973) at

Samford, Queensland. Under a four-week cutting regime the yield was approximately one-half. In Gujarat, India, 3 104 kg green matter per hectare was recorded (Srinivasan, Bonde & Tejwani, 1962). At Himachal Pradesh University, Singh and Katoch (1975) obtained 114.3 kg DM/ha per day compared with 113.0 from *S. sphacelata,* 64.8 from *Heteropogon contortus* and 64.7 from *Bothriochloa bladhii.*

Figure 15.70. Eragrostis superba

Chemical analysis and digestibility. It has about 12 percent crude protein in the dry matter at an early-flowering stage with 30-35 percent crude fibre (Bogdan & Pratt, 1967).

Palatability. It is very palatable.

Seed production and harvesting. Seed can be collected easily from open grassland or at roadsides by stripping the ripe panicles. Mature spikelets, each with numerous florets, detach easily with the caryopses enclosed.

Animal production. E. superba, along with *Cenchrus ciliaris,* has been the basis of the seed mixtures used for large-scale reseeding in Kitui, Machakos and Baringo in Kenya (Bogdan & Pratt, 1967). It is used in moderately dry areas. Its cool season production was higher than *Paspalum* spp., pangola grass and *Setaria sphacelata* cv. Nandi (Strickland, 1973).

Main attributes. It is quick growing, shows green vegetative growth throughout the year and is very valuable in spring.

Main deficiencies. It gets stemmy and unpalatable near maturity and its nutritive value drops.

Further reading. Strickland, 1973.

Eragrostis tef (Zucc.) Trotter

Synonym. E. abyssinica (Jacq.) Link.
Common names. Teff, t'ef (Ethiopia).
Natural habitat. Usually in cultivation.
Distribution. Native of Ethiopia, introduced into other tropical countries.
Description. An annual forming scanty tufts; culms up to 120 cm high in selected cultivated plants, but often only 20 cm when growing as a weed, glabrous, finely striate. Leaf-blades narrow, folded. Panicle narrow, 18-20 cm long with adpressed branches at the base; spikelets grey or golden, 8 mm long with up to ten florets and rather large seeds (Napper, 1965). In Ethiopia two types are grown, one with white seeds (preferred) and one with brown seeds (see Fig. 15.71).
Season of growth. Summer.
Optimum temperature for growth. Maximum temperature is 25-28°C at 2 000 m.
Frost tolerance. It is susceptible to frost.
Altitude range. Sea-level to 1 800 m in Kenya. In Ethiopia, 1 800-2 400 m, at which height white teff disappears. Above 2 400 m brown teff is grown.
Rainfall requirements. In Ethiopia, it grows on an average rainfall of 950-1 500 mm. The maximum rainfall is 2 500 mm.
Tolerance to flooding. It can tolerate waterlogging (Westphal, 1975).
Soil requirements. Mainly sandy loams, but can grow on black soils (Westphal, 1975). A surface crust will kill off delicate young plants.
Fertilizer requirements. It is usually fertilized with farmyard manure in Ethiopia and is used in a rotation containing beans as a leguminous crop.
Land preparation for establishment. A very fine seed-bed is needed.
Sowing methods. It can be planted, broadcast or sown in rows and weeded.
Sowing depth and cover. Sow on the surface or no deeper than 1 cm (Bogdan, 1964). Cover by rolling or driving sheep across the area.
Sowing time and rate. July or August in Ethiopia at 15-20 kg/ha, or up to 40 kg/ha as a cover crop for moisture conservation in Kenya.
Number of seeds per kg. 2.5-3 million.
Dormancy. There is no dormancy.
Seedling vigour. The seedlings are small and delicate and should be carefully weeded. The crop may need thinning.
Vigour of growth and growth rhythm. It matures in ten to 12 weeks.
Genetics and reproduction. 2n=40 (Fedorov, 1974). An apomict.
Suitability for hay and silage. It is widely grown for hay in Transvaal and Orange Free State and in the United States. It is one of the faster growing hay crops known.

Figure 15.71. Eragrostis tef. **A**-Habit **B**-Spikelet

TABLE 15.36 *Eragrostis tef*

	DM	As % of dry matter				
		CP	CF	Ash	EE	NFE
Hay, late vegetative, South Africa	93.2	10.5	34.2	5.3	1.1	48.9
Hay, mature, South Africa	91.8	8.8	33.1	6.9	1.1	50.1
	Animal	Digestibility (%)				
		CP	CF	EE	NFE	ME
Hay, late vegetative, South Africa	Sheep	61.2	67.1	22.9	57.8	2.15
Hay, mature, South Africa	Sheep	57.2	74.5	43.2	60.5	2.24

Source: Göhl, 1975

Chemical analysis and digestibility. See Table 15.36 for results from South Africa.

Palatability. Very well grazed. The seed is eaten by wildlife and cattle, contributing significantly to their diet at certain times of the year.

Seed production and harvesting. It is a good producer of seed, which shatters easily. The heads are cut with a sickle when the panicles become greyish, cured in heaps in the field and then threshed by flailing or trampling with oxen.

Seed yield. 270-800 kg/ha.

Cultivars. No cultivars have been released, but there are wide ecotypic differences both in morphology and agronomic response. Very productive types can be selected.

Value for erosion control. Good (Narayanan & Dabadghao, 1972).

Diseases. A rust, *Uromyces eragrostides,* sometimes attacks it.

Economics. In Ethiopia the grain is used as human food, accounting for more than half the country's grain production. In east Welega (Ethiopia), crops of teff, barley and sorghum are sown in June and July and harvested in December. After harvest, the farmers enclose a plot of land to be used for next season and cattle use the pasture for ten to 15 nights to manure the field and are then moved to another area — the "shifting stable" system. Usually one year of teff is followed by beans, then barley and sorghum. In the Yerer-Kereyu Highlands of Shoa, east of Addis Ababa, teff is planted in well-prepared black cracking clays (Westphal, 1975). It is a good nurse crop for *Eragrostis curvula* pastures in South Africa (Chippendall & Crook, 1976).

Main attributes. Highly adapted to marginal rainfall areas and valuable for range reseeding.

Further reading. Westphal, 1975.

Eragrostis tremula (Lam.) Steud.

Common names. Bano, bannu (the Sudan), lehmleiche (Mauritania).
Natural habitat. Sandy soils, common in abandoned cultivation.
Distribution. Throughout tropical Africa, India and Burma in low rainfall areas.
Description. A short-lived grass up to 75 cm high. Panicle widely spreading, up to 15 cm long, with long-pedicelled yellow-green or purplish spikelets 10-30 cm long (Napper, 1965) (see Fig. 15.72).
Season of growth. Summer.
Altitude range. 1 000-1 750 m.
Soil requirements. It is a sand-loving species, being found in 29 out of 35 sandy sites in Mauritania (Boudet & Duverger, 1961).
Vigour of growth and growth rhythm. In the Sahel it begins growth in August and is standing hay through to June (Boudet & Duverger, 1961). It can make a second crop in the same season under favourable conditions.
Genetics and reproduction. 2n=20 (Fedorov, 1974).
Chemical analysis and digestibility. Analysis of material from the Sahel-Sudan zone in North Africa is set out in Table 15.37.
Palatability. It is quite palatable.
Economics. This annual grass grows abundantly in old cultivations in the lighter soils in Kordofan Province, the Sudan and is cut and carried into the villages to feed village livestock such as cattle, donkeys, goats and sheep (see Plate 37).
Further reading. Boudet & Duverger, 1961.

Figure 15.72. *Eragrostis tremula.* **A**-Habit **B**-Inflorescence **C**-Spikelet

TABLE 15.37 *Eragrostis tremula*

	DM	As % of dry matter		
		CP	CF	Ash
Flowering, September, the Sahel	60.0	6.9	34.5	4.7
Dry stems, December, the Sahel	97.0	3.4	38.7	5.8

Source: Boudet, 1975

Eriochloa fatmensis (Hochst. et Steud.) W.D. Clayton

Synonym. E. nubica (Steud.) Thell.
Natural habitat. Damp and swampy places in grassland and on lake shores, on heavy soils.
Distribution. Throughout tropical Africa. Introduced to India and Australia.
Description. An annual, 15-100 cm high with linear leaf-blades. Inflorescence narrow, of up to ten racemes 2-4 cm long, with biseriate aristulate silky spikelets, having the basal beadlike internode pale or purplish. Spikelets obtuse without a bristle (Napper, 1965). Differs from *Brachiaria* in having a tiny beadlike swelling below each spikelet (Rose-Innes, 1977) (see Fig. 15.73).
Altitude range. 500-1 750 m.
Rainfall requirements. About 250 mm per year (Bogdan & Pratt, 1967).
Drought tolerance. Being an annual, it has little drought tolerance, but escapes it by early seeding.
Tolerance to flooding. It tolerates seasonal flooding.
Soil requirements. It prefers the heavier alluvial silts and black cotton soils but will adapt to loams (Bogdan & Pratt, 1967).
Tolerance to salinity. Tolerant to slightly saline soils on lagoon flats (Rose-Innes, 1977).
Sowing time and rate. Early wet season at about 1.1 kg/ha (Bogdan & Pratt, 1967).
Number of seeds per kg. 1.1 million spikelets with one seed (Bogdan & Pratt, 1967).
Genetics and reproduction. 2n=36.
Chemical analysis and digestibility. It maintains its nutritive value well, and at full flowering may still contain 10 percent crude protein in the dry matter (Bogdan & Pratt, 1967).
Palatability. It is quite palatable.
Seed production and harvesting. It seeds well, with each spikelet containing a single caryopsis, but the seed sheds easily and is difficult to harvest (Bogdan & Pratt, 1967).
Economics. Adapts well to dry or wet conditions, being found in the Sudan in dry arid localities and also in inundated fields in Central Africa. It is eaten by all stock (Bor, 1960).
Animal production. Bogdan and Pratt (1967) recommend this grass for reseeding denuded alluvial flats in Kenya's arid zone where seasonal flooding occurs.
Further reading. Bogdan & Pratt, 1967.

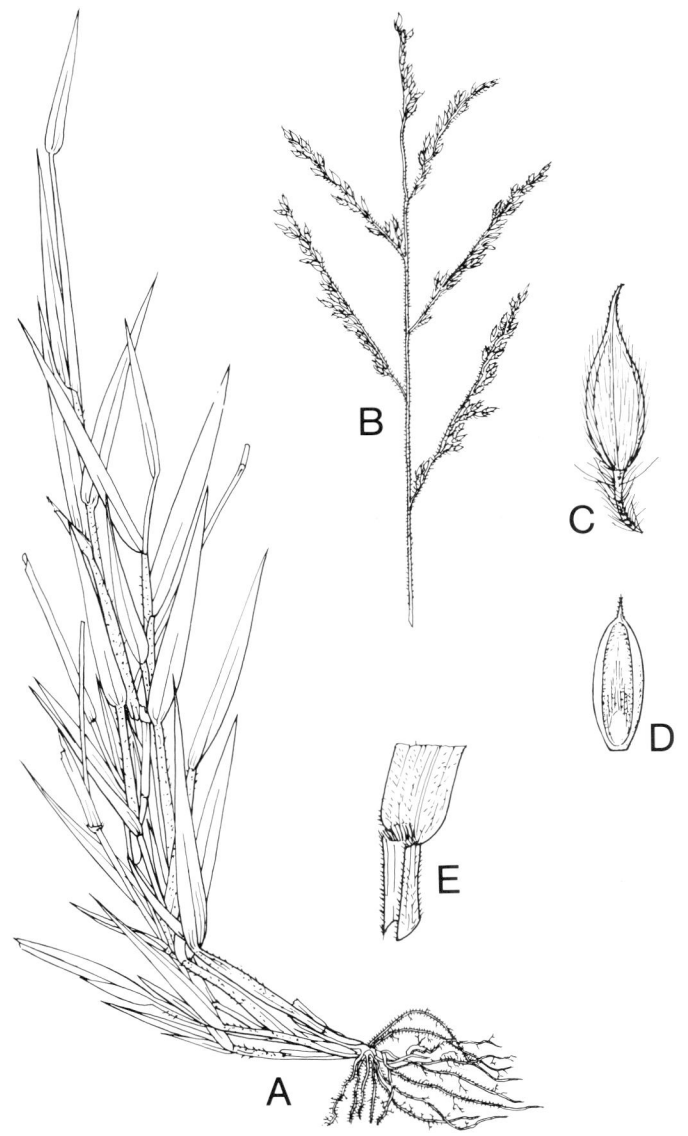

Figure 15.73. Eriochloa fatmensis. **A**-Base of plant **B**-Inflorescence **C**-Spikelet **D**-Upper fertile flower **E**-Ligule

Eriochloa punctata (L.) Desv.

Synonym. Eriochloa polystachya Kunth.
Common names. Janeiro (Costa Rica), carib grass (United States), malojilla (Fiji), lierba del Caribe (Cuba).
Natural habitat. Moist places.
Distribution. Native of the Caribbean region from the West Indies to Brazil.
Description. A glabrous, branching perennial, ascending from a decumbent base, commonly 1 m or more tall, with flat blades 10-15 mm wide and several to many narrowly ascending racemes (Hitchcock, 1927). Spikelets silvery, 3-4 mm long. Grain oblong, free within the hardened glume and palea (Cooke, 1958). It differs from Para grass in many ways: carib grass blooms throughout the year in the southern United States, whereas Para grass blooms from September to January; secondary racemes of Para are abundant, those of carib are sparse and the spikelets nearly sessile; carib grass has darker green, more glabrous, shorter and narrower leaves than those of Para; the flower stalks of carib grass are 15-30 cm shorter than those of Para, with stolons approximately 1 m long; carib grass is densely covered with hairs on the nodes and sparsely on the leaf-sheaths and on the nodes; carib grass is more palatable than Para grass (Judd, 1979) (see Fig. 15.74).
Latitudinal limits. About 24°N to 20°S.
Altitude range. Sea-level to 1 800 m. It grows best at 1 500 m in Costa Rica; at higher elevations production declines.
Rainfall requirements. Like Para grass, it is adapted to a hot, humid climate (Gonzalez & Pacheco, 1970). In Puerto Rico it grows in a rainfall range of 1 500-2 000 mm of evenly distributed rainfall.
Drought tolerance. It has little drought tolerance.
Tolerance to flooding. Selection 6017 withstands flooding in Colombia (CIAT, 1978).
Soil requirements. It adapts to a wide variety of soils but prefers fertile, moist sandy loams. In poor, dry soils it is short-lived, produces little and is invaded by weeds. Selection 6017 tolerates acid soils, high in iron and aluminium, in Colombia (CIAT, 1978) and Puerto Rico (Vicente-Chandler *et al.,* 1974).
Fertilizer requirements. It stands heavy, complete fertilizer application. Liming is needed with the acid ultisol soils and in Puerto Rico from 2 500-4 500 kg/ha are applied to bring the soil up to 70 percent base saturation. Two weeks after planting some 325 kg/ha of a 15:5:10 fertilizer mixture is applied and again three months later for intensive grazing. Yields increase linearly with nitrogen applications up to 440 kg/ha and more slowly to 880 kg/ha, and 1 760 kg/ha.
Land preparation for establishment. A good seed-bed is required.
Sowing methods. Propagated vegetatively by stem cuttings or division of root-stocks, in furrows 1 m apart.

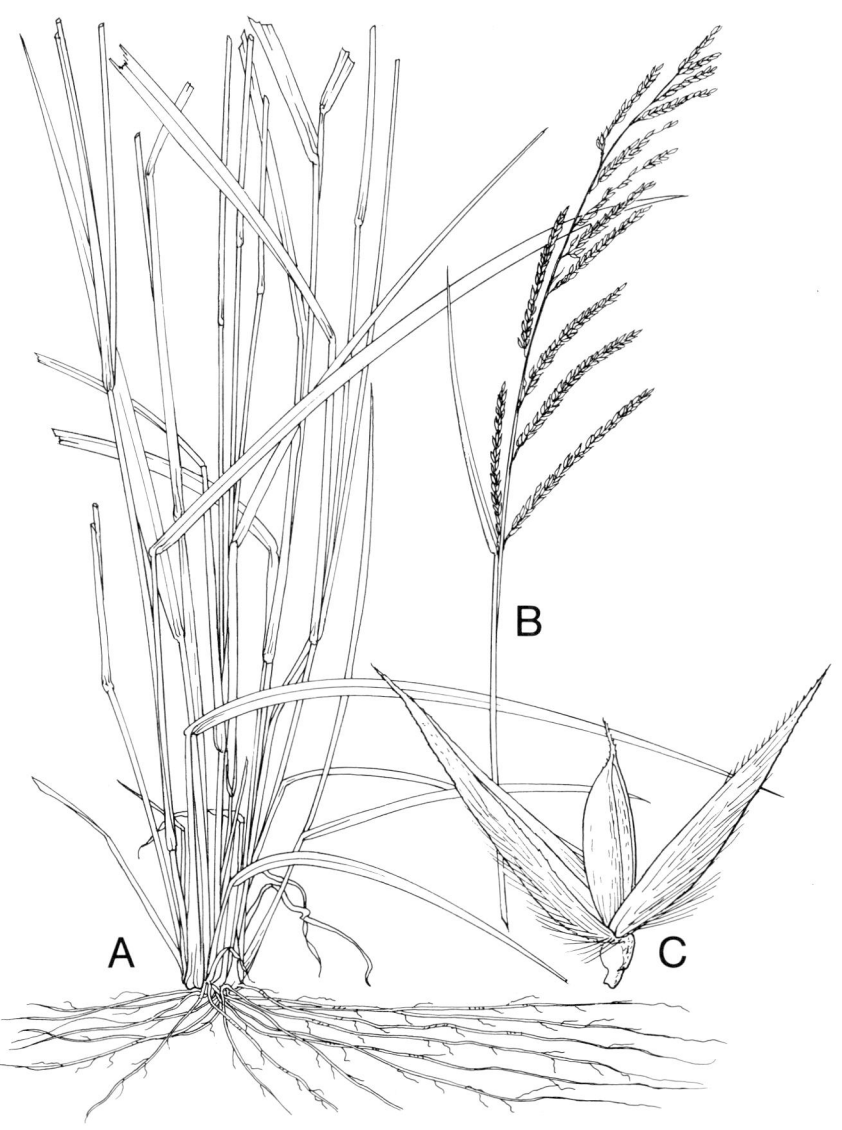

Figure 15.74. Eriochloa punctata. **A**-Habit **B**-Inflorescence **C**-Spikelet

Sowing depth and cover. Sow the cuttings 15-20 cm deep and cover with 5-7.5 cm of soil.
Sowing time and rate. Use 1 500 kg of mature cuttings per hectare in summer.
Ability to compete with weeds. Its vigorous growth suppresses weeds.
Tolerance to herbicides. Carib grass can be eradicated by applying 6.5 kg of dalapon (2,2 dichloropropionic acid) in 1 100 litres of water per hectare and repeating the application three weeks later if necessary (Vicente-Chandler *et al.*, 1974).
Response to defoliation. Heavily fertilized carib grass in Puerto Rico gave highest yields when cut at 90-day intervals producing 117 264 kg green forage per hectare per year, and 113 120 kg cut at 60-day intervals.
Grazing management. The grass should be renovated when needed and irrigated during the dry season. Put the cattle in to graze when the grass reaches 0.5-0.6 m in height and graze only once. It is usually cut for green chop.
Genetics and reproduction. 2n=36 (Fedorov, 1974).
Dry- and green-matter yields. It yields 175 t/ha of green material in Costa Rica of 19 percent dry matter and 1.6 percent crude protein (Gonzalez & Pacheco, 1970).
Suitability for hay and silage. It makes palatable hay. Medling (1972) made good silage in plastic bags at Gualaca, Panama when 10 percent molasses was added.
Chemical analysis and digestibility. In Costa Rica analysis of material at floral initiation revealed 6.08 percent crude protein, 29.66 percent crude fibre, 42.82 percent nitrogen-free extract, 1.35 percent ether extract and 10.09 percent ash on a 10 percent moisture basis (Gonzalez & Pacheco, 1970). It has high protein in Suriname (Dirven, 1963b) with a range of 5.6-10.3 percent, an average of 7.5 percent of the dry matter. Protein content increased from 6.4 percent when no nitrogen was applied to 10.2 percent with 880 kg N/ha in Puerto Rico (Vicente-Chandler *et al.*, 1974).
Palatability. It is quite a palatable grass.
Seed production and harvesting. The seed is rarely viable and the grass is propagated vegetatively.
Cultivars. Burkart (1969) records three forms — forma *intermedia*, var. *montevidensis* and var. *parodi*.
Diseases. Carib grass is attacked by a rust, *Uromyces leptodermus*, which causes minor defoliation during dry periods, but is of little importance. It is unaffected by pangola rust (*Puccinia oahuensis*) or grey leaf spot (*Piricularia grisea*).
Animal production. Figures for Carib grass production in Puerto Rico show that with a rainfall of 1 500-2 000 mm per year, evenly distributed, Carib grass fertilized with 5 t/ha annually of 15:5:10 fertilizer and cut every 40-60 days can support 7.75 steers of 270 kg each per hectare per year (Vicente-Chandler *et al.*, 1974).
Further reading. Gonzalez & Pacheco, 1970; Vicente-Chandler *et al.*, 1974.

Eustachys paspaloides (Vahl) Lanza and Mattei

Common name. Brown Rhodes grass (Zimbabwe).
Natural habitat. Dry grassland, open woodland, by roadsides and as a weed.
Distribution. Southern, Central, East and northeast Africa, naturalized in Florida.
Description. A tufted glaucous, shortly stoloniferous perennial; culms usually 30-60 cm high, sheaths strongly compressed, the blades folded, blunt at the apex; spikes 4-7 mm, brown; spikelets 7-22 mm long; two-flowered, the upper floret usually male; upper glume short-awned (Chippendall, 1955) (see Fig. 15.75).
Season of growth. Summer.
Altitude range. 1 000-2 000 m.
Rainfall requirements. About 550 mm (Bogdan & Pratt, 1967).
Tolerance to flooding. It withstands seasonal waterlogging.
Soil requirements. It is adapted to loose sandy loams, loams and black cracking clays (Bogdan & Pratt, 1967), also red loams and red earths.
Land preparation for establishment. A well-prepared seed-bed is preferable, but in reseeding programmes a rough disc-harrowing will suffice.
Sowing methods. Usually broadcast.
Sowing depth and cover. Sow on the surface and roll or cover with passage of bushes.
Sowing time and rate. In the wet season, at about 175 g per hectare (Bogdan & Pratt, 1967).
Number of seeds per kg. 770 000 to 880 000 spikelets with one seed each (Bogdan & Pratt, 1967).
Response to defoliation. It can stand heavy grazing (Roberts, 1970a & b).
Genetics and reproduction. 2n=36 (Fedorov, 1974).
Chemical analysis and digestibility. The herbage is leafy and contains about 10 percent crude protein in the dry matter at early flowering (Bogdan & Pratt, 1967).
Palatability. It is well grazed. In Bankerveld, South Africa, *E. paspaloides* fertilized with a complete fertilizer was the most sought-after species, with 70 percent utilization against 70 percent for *Themeda triandra* and 30 percent for *Heteropogon contortus* (Kruger & Edwards, 1972).
Toxicity. D.C. Steyn has recorded the presence of prussic acid (Chippendall & Crook, 1976).
Seed production and harvesting. Seed formation is erratic, and considerable effort may be required to collect large quantities of seed. The seed is small.
Economics. A valuable, nutritious grass in the natural veld in southern Africa, Kenya and Rwanda (Bouxin, 1975; Kruger & Edwards, 1972; Bogdan & Pratt, 1967).

Animal production. It has never been used for reseeding, but Bogdan and Pratt (1967) recommend its trial in Kenya's medium-rainfall eastern areas.

Figure 15.75. *Eustachys paspaloides.* **A**-Habit **B**-Inflorescence

Exotheca abyssinica (A. Rich.) Anderss.

Natural habitat. Montane grassland.
Distribution. Throughout tropical Africa, mainly in Central Africa.
Description. Densely tufted perennial up to 90 cm high. Inflorescence usually a single raceme-pair with green or purplish racemes, the upper with a 14-16 mm glabrous raceme base; spikelets 14-16 mm long with awns 70-100 mm long (Napper, 1965) (see Fig. 15.76).
Minimum temperature for growth. It grows where the minimum temperature is under 18°C.
Frost tolerance. Good.
Latitudinal limits. Mainly 0-18°S.
Altitude range. 1 000-3 000 m.
Rainfall requirements. It grows in a rainfall regime of 750-1 250 mm in Tanzania.
Tolerance to flooding. It grows on seasonally wet ground.
Soil requirements. It grows on poorly drained laterized soils on granite in the southern highlands near Njombe, Tanzania. The soils are poor in minerals, and mineralized licks are needed for cattle grazing the pastures. The pH is 5.0-6.0.
Response to fire. Burning every two years in October or November after the first rains maintained the *Exotheca abyssinica/Hyparrhenia bracteata* pasture (Compère, 1968).
Genetics and reproduction. 2n=20 (Fedorov, 1974).
Chemical analysis and digestibility. Dougall and Bogdan (1960) recorded 13.6 percent crude protein, 33.1 percent crude fibre, 5.8 percent ash, 2.4 percent ether extract and 45.1 percent nitrogen-free extract in the dry matter at the fresh, early bloom stage in Kenya.
Palatability. It has tough leaves and is grazed only when young.
Economics. It is a high-altitude species which tolerates low soil fertility and is not highly regarded as a pasture.
Animal production. *Exotheca* grasslands have a low carrying capacity, typically one animal per five to ten hectares and the nutritive value of the herbage is low throughout the year (Dougall, 1960). They are suitable for game and forest reserves. Under uncontrolled grazing the fibrous species increase and they are best improved by introduced species. In Rwanda and Burundi, Compère (1968) recorded a live-weight gain of 70-90 kg/ha on an *Exotheca abyssinica/Eragrostis boehmii* pasture.
Further reading. Vesey-Fitzgerald, 1963.

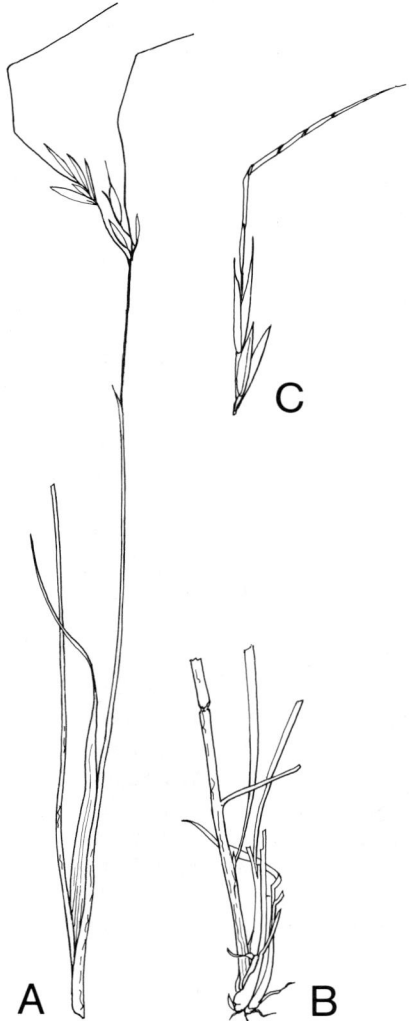

Figure 15.76. *Exotheca abyssinica.* **A**-Inflorescence **B**-Culms **C**-Pair of racemes

Hemarthria altissima (Poir.) Stapf and C.E. Hubbard

Common names. Red vlei grass, rooikweek (southern Africa), swamp couch (Zimbabwe), limpo grass (Florida), halt grass (Panama).
Natural habitat. Flooded areas, swamps and lakes, vleis.
Distribution. Tropical Africa, India, Burma. Introduced to the United States.
Description. A perennial with a creeping, branched rhizome; culms 30-100 cm high, compressed, usually decumbent and rooting from the lower nodes, generally branched; leaves smooth, glabrous, the blades up to 6 mm wide, usually folded; inflorescence a solitary raceme terminating the culm and its branches, these often in clusters, so there are several racemes from each node; racemes 5-12 cm long, 2-3 mm wide, spikelike, tapering toward the apex. Sessile spikelets 5-7 mm long, the pedicelled ones 6-9 mm long (Bor, 1960). During the greater part of the year the whole plant has a distinctive rust-red colour and it has a mat-forming habit (Chippendall & Crook, 1976). The leaf-blades, when dry, twist in corkscrew fashion (see Fig. 15.77).
Optimum temperature for growth. 31-35°C (Boyd & Perry, 1972). It was seriously affected by temperatures above 38°C.
Altitude range. 1 500-2 000 m, 1 200 m in Panama (Rattray, 1973).
Rainfall requirements. It requires a high rainfall.
Drought tolerance. It does not tolerate long droughts but can withstand short, seasonal droughts. In the dry months the plant assumes a distinct rust-red colour, hence the name of "rooikweek".
Tolerance to flooding. It tolerates flooding well.
Soil requirements. It tolerates acid soils (CIAT, 1978) and prefers moist, humid soils (Rattray, 1973).
Fertilizer requirements. It gave high yields in Venezuela when cut at 5 cm every 20 days and at 20 cm every 80 days, when fertilized with 840 kg N + 200 kg P + 100-200 kg K + 3 000-6 000 kg lime + Zn, Cu, B and Mo (Parra & Bryan, 1974).
Ability to spread naturally. It spreads rapidly by creeping rhizomes and culms rooting at the lower nodes.
Sowing methods. It can be propagated by cuttings placed in wet soil.
Response to photoperiod. Grown at 30-25°C in growth chambers at Florida University. *H. altissima* did not grow at nine hours; yielded slightly more dry matter at 15 hours than at 9.5 hours and did not respond to eight hours (Gaskins & Sleper, 1974).
Response to defoliation. At CIAT, Quilichao, Colombia, it was found to be most susceptible to clipping to ground level or 5 cm (CIAT, 1978).
Tolerance to fire. It will not tolerate burning.
Genetics and reproduction. The basic chromosome number of *H. altissima* is

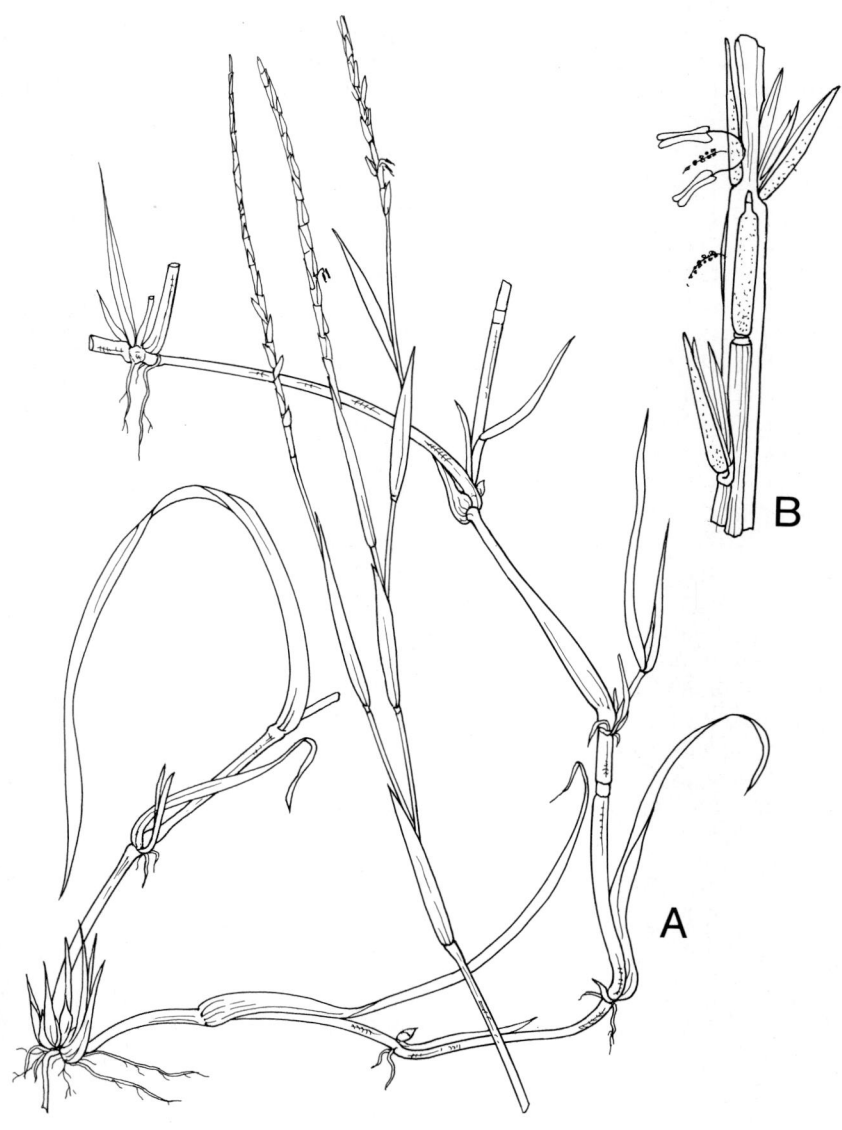

Figure 15.77. Hemarthria altissima. **A**-Habit   **B**-Portion of spike

TABLE 15.38 *Hemarthria altissima*

	As % of dry matter				
	CP	CF	EE	Ash	NFE
Flowering stage	2.50	31.98	1.39	5.31	48.82

Source: Verboom & Brunt, 1970

nine. Of 11 introductions, nine were diploid and two tetraploid (Schank, 1972) 2n=18, 20 (Fedorov, 1974) 36, 40.

Suitability for hay and silage. Medling (1972) in Panama made satisfactory silage in plastic bags, adding 10 percent molasses to the material.

Chemical analysis and digestibility. It is of good nutritional value. See Table 15.38. The OM digestibility of six pure and hybrid lines of *H. altissima* ranged from 38.5 percent in mature plants of a diploid line to 68.4 percent in five-week-old regrowth of a tetraploid line. Digestibility was inversely related to the percentage of cross-sectional area of stems occupied by vascular bundles.

Palatability. It is highly palatable and is valued as a fodder grass.

Toxicity. In Zambia, scouring occurs when cattle move from the fibrous forest grazing to the rich plains grasses consisting of *Echinochloa pyramidalis*, *E. scabra*, *Acroceras macrum*, *Hemarthria altissima*, *Leersia hexandra* and *Vossia cuspidata* and it may be two to four months before they regain condition (Veerboom & Brunt, 1970).

Seed production and harvesting. It is not a good seed producer.

Cultivars. No cultivars have been released. CIAT in Colombia has a selection, 663, under test.

Diseases. It has good disease resistance.

Pests. Sting nematodes (*Belonolaimus longicaudatus*) affected growth of *H. altissima* at soil temperatures of 18°C (Boyd, Schroder and Perry, 1972). The yellow sugar-cane aphid (*Sipha flava*) attacks some accessions, but other accessions exhibit a degree of resistance (Oakes, 1978).

Economics. *Hemarthria* germ plasm is undergoing agronomic, entomological and pathological evaluation by government and private research groups, and certain accessions have been distributed to Central and South America, West Indies and Hawaii and are undergoing field trials. Pasture is the primary use but it has been used successfully for hay and silage. In 1979, production in Florida from an area exceeding 6 000 ha was valued at over $US 1 million. In Lesotho, children eat the raw rhizomes.

Further reading. CIAT, 1978.

Heteropogon contortus (L.) Beauv. ex Roem. and Schult.

Synonym. Andropogon contortus L.
Common names. Black or bunch spear grass (Australia), tangle head (United States), pili grass (Hawaii), assegai grass (Zimbabwe).
Natural habitat. Open forest and woodland, grassland.
Distribution. Throughout the tropics and subtropics.
Description. A caespitose perennial, the culms erect to 75 cm, branching above; leaf-sheaths keeled, glabrous. Raceme solitary, 3.5-15 cm long with up to ten pairs of awnless spikelets at the base and an equal number of pairs above; the fertile sessile spikelets having awns 5-10 cm long (see Fig. 15.78, Plate 38).
Season of growth. In Queensland, 60 percent of the yield of dry matter is produced in summer between January and April (Shaw & Bisset, 1955).
Optimum temperature for growth. In Queensland the summer temperature ranges from 30-33.5°C.
Minimum temperature for growth. At lat. 23°30'S growth ceases at 21°C (Miles, 1949). It makes practically no growth in winter, irrespective of rainfall, and yields less than 50 kg DM/ha (Shaw & Bisset, 1955).
Frost tolerance. It tolerates frost well, but does not grow during the winter, regardless of frost.
Latitudinal range. In Queensland it occurs mainly between latitudes 19 and 27°S.
Altitude range. Sea-level to 2 000 m in the Himalayas. Less than 300 m in Hawaii.
Rainfall requirements. It occurs naturally in the 500-1 500 mm rainfall regime with a summer maximum, with 20-30 percent variability (Isbell, 1969).
Drought tolerance. It is fairly tolerant of short-term droughts but does not persist in semi-arid areas. It yields little in New Zealand during dry spells in January and February.
Tolerance to flooding. It does not tolerate flooding.
Soil requirements. H. contortus generally thrives best on sandy loams with a pH in the range of 5.0-6.0. It establishes with difficulty in heavy clay soils.
Tolerance to salinity. It cannot tolerate high levels of salinity (Isbell, 1969).
Fertilizer requirements. It is not usually fertilized. Weier (1977) showed that H. contortus under natural conditions has high nitrogenase activity associated with its roots and fixes some of its own nitrogen. In India, the application of 20 kg N/ha raised production from 3 340 kg to 4 330 kg/ha, while 40 kg N/ha raised it to 5 560 kg/ha (Dabadghao & Shankarnarayan, 1970). Responses up to 1 000 kg N/ha per year were recorded at Marandellas, Zimbabwe. At Rodd's Bay, Queensland, 't Mannetje (1972) obtained a linear response to increasing nitrogen, the level of response being linearly related to rainfall in both dry matter and nitrogen recovery.

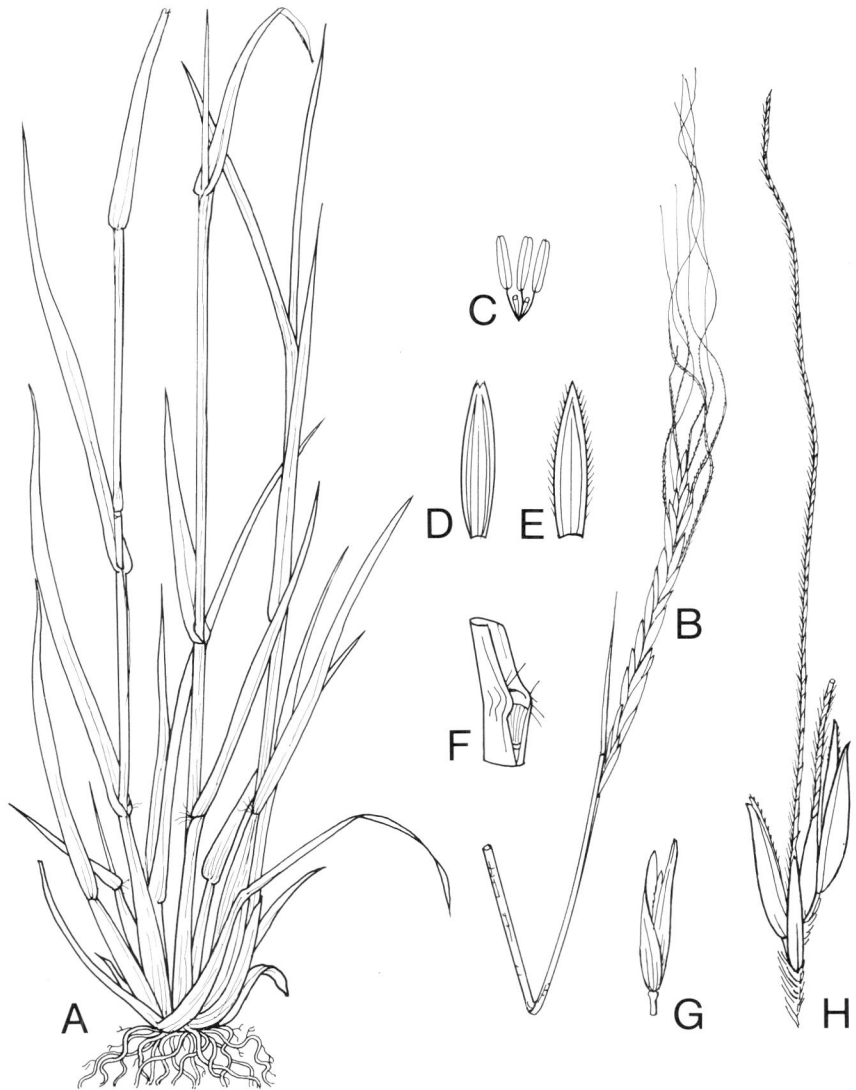

Figure 15.78. Heteropogon contortus. **A**-Habit **B**-Inflorescence **C**-Flower of male pedicelled spikelet **D**-Lower glume **E**-Upper glume **F**-Ligule **G**-Pedicelled spikelet of homogamous pair **H**-Two pairs of heterogamous spikelets

Ability to spread naturally. Excellent, especially if the country is burnt regularly. The seeds survive the burning by burying themselves. The awns twist and untwist as moisture changes, enabling the needle-like seeds to penetrate the soil surface (see Plate 39).

Land preparation for establishment. It is rarely sown. The existing natural pastures are utilized.

Vigour of growth and growth rhythm. It produces 90 percent of its growth during the warmer months between mid-October and mid-April at Rodd's Bay. Its winter contribution is small.

Genetics and reproduction. $2n = 20, 40, 44, 50, 60, 80$ (Fedorov, 1974). It is an obligate aposporous apomict. A higher proportion of flowers are male in inflorescences developed in long days than those developed in short days, and the most rapidly maturing types at any day length show a higher proportion of female inflorescences.

Suitability for hay and silage. Both have been made in India. Hay cut at late vegetative stage contained 5.9 percent crude protein in the dry matter, and at dough stage 3.5 percent. Immature material made into silage contained 6.6 percent crude protein in the dry matter (Göhl, 1975).

Chemical analysis and digestibility. Dabadghao and Shankarnarayan (1970) in India found the crude protein content of a *Heteropogon* community was 5 percent untreated, and 5.8 percent when treated with nitrogen. In Queensland the crude protein content, even when very young, does not rise above 10 percent and for the greater part of the growing season it is between 4 and 6 percent, dropping in winter to 2-3 percent. The digestibility of this protein is also low. Phosphorus figures as percentage of the dry matter ranged from 0.09-0.15, and calcium 0.23-0.30, indicating that on this soil the phosphorus figures were too low for an adequate diet for beef cattle (0.15-0.28 percent) and the calcium figures barely adequate (requirement: 0.15-0.37 percent) (Shaw & Bisset, 1955). Göhl (1975) lists analyses from Zimbabwe, Ghana and India. The fresh material from Ghana showed 9.4 percent crude protein in the dry matter at four weeks, 6.3 percent at eight weeks, 7.0 percent at 16 weeks and 2.5 percent at 36 weeks. It has low sodium levels (Playne, 1970a).

Palatability. It is palatable in the early vegetative stage, but unattractive as it matures.

Cultivars. There are no recognized cultivars, but two varieties are recognized in India (Dabadghao & Shankarnarayan, 1973).

Value for erosion control. It has proved useful in soil erosion control on 20° slopes in India (Misra, Ambasht & Singh, 1977).

Economics. This grass is highly esteemed as a summer fodder grass in India and it can be made into hay (Bor, 1960). The presence of the awned seed in wool causes "vegetable fault" and increases processing costs. In addition, the seed pierces the skin and penetrates the flesh of sheep, resulting in irritation and loss of wool production and downgrading of carcasses. This grass dominates the beef-raising areas in central coastal Queensland, from which a large proportion of fat cattle are supplied to the meatworks in autumn. From autumn to early summer it is unproductive.

Animal production. Generally in Queensland, Australia, the carrying capacity of native spear grass pastures is one beast to 3.5-4 ha in the southern area and 8-10 ha in the north of the region. Cattle lose weight in the winter and spring and take four to five years to reach market weight. On the granitic soils at Narayen, Queensland (lat. 25°60'S, 710 mm rain), native *Heteropogon contortus* pastures normally carry 0.27 steers per hectare and produce about 30 kg/ha per year live-weight gain. The addition of a legume such as *Stylosanthes guianensis* (Oxley fine-stem stylo) or *S. humilis* (Townsville stylo) plus superphosphate at 125 kg/ha per year improved the carrying capacity to 0.8 steers per hectare and live-weight gain was increased to 100 kg/ha per year. A fully sown pasture of Biloela buffel grass (*Cenchrus ciliaris*) and siratro (*Macroptilium atropurpureum*) grown in this soil and fertilized with 125 kg/ha per year superphosphate carried one beast per hectare and gave live-weight gains of 160 kg/ha per year ('t Mannetje, 1976). At Rodd's Bay, Queensland (lat. 23°50'S, 813 mm rain), *H. contortus* dominant pasture stocked at 0.27 beasts per hectare, which is the normal carrying capacity, gave an average live-weight gain of 84 kg/ha and 10 kg/ha per year over a seven-year period in which five of the seven years' rainfalls were below average. At a higher stocking rate of 0.62 beasts/ha, the live-weight gains were 47.3 kg per head and 12 kg/ha. When fertilized with 405 kg superphosphate and 58 kg potassium chloride and the trace element molybdenum, the live-weight gains at 0.62 beasts per hectare were 100 kg per head and 25 kg/ha. When *Stylosanthes humilis* was included with *H. contortus* without fertilizer, the legume improved the figures to a stocking rate of 0.77 beasts per hectare with live-weight gains of 121.4 kg per head and 60 kg/ha and in two of the years the carrying capacity reached 0.8 beasts per hectare. Moreover, half the steers in the *H. contortus/S. humilis* unfertilized pasture, and practically all the steers in the full treatment were marketed one year earlier (Shaw & 't Mannetje, 1970).

Main attributes. Its hardiness, perenniality, tolerance of fire and its early palatability. Its ability to grow on poor soils.

Main deficiencies. Its dominance in burnt areas. Its production of numerous robust awns which shed easily and cause damage to animals' skin and reduce wool values.

Further reading. Shaw & Bisset, 1955; Shaw and 't Mannetje, 1970; Tothill, 1970.

Hymenachne acutigluma (Steud.) Gilliland

Common name. Hymenachne (North Australia).
Natural habitat. In shallow water at the margins of swamps and slow rivers in the tropics of Australia and Papua New Guinea.
Distribution. Northern Australia, Papua New Guinea, Assam, Burma, Malaysia, Viet Nam and Polynesia.
Description. Tall, stoloniferous perennial, culms to 2 m; panicles narrow, 15 cm long (Henty, 1969) (see Fig. 15.79).
Season of growth. Perennial in the tropics.
Rainfall requirements. It is a swamp grass, more or less independent of rainfall.
Drought tolerance. It generally escapes drought because of the high soil moisture in its usual habitat, unless the drought is very prolonged.
Tolerance to flooding. It survives floods well and is aquatic in nature.
Soil requirements. It generally grows on heavy clays.
Tolerance to salinity. It grows in fresh water swamps.
Sowing methods. It is propagated by stolons.
Response to defoliation. It stands grazing well, but very heavy grazing by feral pigs and buffaloes in northern Australia leads to a reduction in density (Sturtz, Harrison & Falvey, 1975).
Palatability. It is very palatable.
Economics. It is an important grazing plant for swamp buffaloes in the Northern Territory, Australia. The buffaloes will submerge and graze it from below.
Animal production. The *H. acutigluma* plains are ideal for the swamp buffalo and live-weight gains of 0.27-0.31 kg per day have been recorded. Reproductive performance of buffalo is superior to that of cattle, with calving rates of 85 percent, compared with 50 percent for Brahman cross cattle (Graham, personal communication).

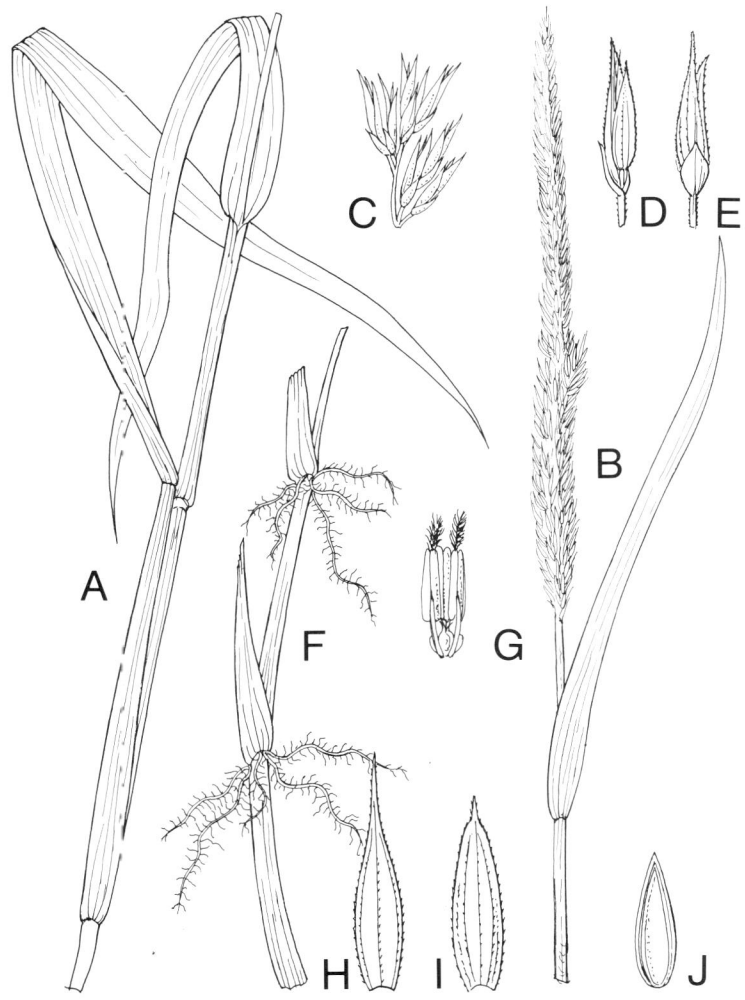

Figure 15.79. Hymenachne acutigluma. **A**-Culm **B**-Inflorescence **C**-Portion of inflorescence **D,E**-Spikelets **F**-Rooting nodes **G**-Flower **H,I**-Glumes **J**-Palea

Hymenachne amplexicaulis (Rudge) Nees

Synonym. H. *pseudointerrupta* C. Muell.

Common names. Canutillo, dal or bamboo grass (India), carrizo chico, cañuela blanca (Bolivia), bamboegras (Suriname).

Natural habitat. In shallow water at the margins of swamps and low rivers at low altitudes.

Distribution. South America.

Description. Tall, stoloniferous perennial, culms to 15 m; panicle narrow, 15 cm long. Semi-aquatic, rooting at the lower nodes with rather stout culms.

Chemical analysis and digestibility. Analyses from Suriname and India are shown in Table 15.39. *H. amplexicaulis* has a very high chlorine content (1.53 percent of the dry matter) and the sulphur content (0.43 percent of the dry matter) is higher than that of *Leersia hexandra* in Suriname. It also has a very high potassium content (3.39 percent of the dry matter) but low calcium values (0.02 percent of the dry matter) (Dirven, 1963a).

Palatability. It is eaten readily by cattle and buffaloes.

Economics. *H. amplexicaulis* is one of three dominant grasses in the Venezuelan llanos and, with *Leersia hexandra* and *Panicum laxum,* makes up 68 percent of the llanos flooded to a maximum depth of 50 cm during June or July to November. It is also a most important aquatic grass in Suriname and provides grazing in abandoned rice paddies (Dirven, 1963a) (see Fig. 15.80).

Animal production. In the Llanos Inundables of Venezuela and Colombia, *H. amplexicaulis* is the most commonly found grass in the diet of the capybara (*Hydrochoerus capybara*) comprising 34.96, 22.78 and 18.59 percent of the total grass eaten during the rainy season, at the end of the rains, and at the end of the dry season (Escobar & Gonzalez, 1976).

Further reading. Dirven, 1963a.

TABLE 15.39 *Hymenachne amplexicaulis*

		As % of dry matter				
		CP	CF	Ash	EE	NFE
Fresh, whole aerial part, Suriname		15.8	34.6	9.4	1.9	38.3
Fresh, stems only, Suriname		8.9	36.7	11.5	1.0	41.9
Fresh, leaves only, Suriname		22.6	32.4	7.2	2.8	35.0
Fresh, mid-bloom, India		9.4	22.1	12.2	2.3	54.0
Fresh, mid-bloom, India		7.5	29.2	12.9	1.4	49.0
Silage, mid-bloom, India		6.9	27.8	17.9	1.9	45.5
	Animal	Digestibility (%)				ME
		CP	CF	EE	NFE	
Fresh, mid-bloom, India	Oxen	61.5	60.5	37.9	67.0	2.11
Hay, mid-bloom, India	Oxen	42.4	70.7	39.1	60.6	2.00
Silage, mid-bloom, India	Oxen	43.9	69.3	40.9	60.3	1.88

Source: Göhl, 1975

Figure 15.80. Hymenachne amplexicaulis being grazed in an abandoned rice paddy, Suriname (**Source:** J.P. Dirven)

Hyparrhenia filipendula (Hochst.) Stapf

Common names. Tambookie grass (Australia), fine thatching grass (South Africa), fine hood grass (Kenya).

Natural habitat. Well-drained grassland and open woodland. Common in the African miombo.

Distribution. Africa, Sri Lanka, Burma, Australia.

Description. Slender perennial up to 150 cm high. Panicle narrow, loose, often over 30 cm long with greenish or purplish racemes up to 15 mm long. Panicle branches subtended by spathes, raceme pairs by spatheoles. At maturity the raceme forms a distinctive L-shaped unit. In var. *filipendula* the two racemes form an L-shaped unit, hence the common name "three o'clock thatching grass". The lower raceme is nearly sessile; the upper has a filiform base. Each raceme has a twisted awn that is hairy. The spikelets are glabrous. Var. *pilosa* has hairy spikelets and usually three to four twisted hairy awns to each pair of racemes (Chippendall & Crook, 1976) (see Fig. 15.81).

Altitude range. Sea-level to 2 250 m. In Zaire it succeeds *H. diplandra* at elevations below 1 700 m when *H. diplandra* formation is subjected to repeated crop growing (Risopoulos, 1966).

Rainfall requirement. It requires a rainfall in excess of 625 mm.

Fertilizer requirement. *Hyparrhenia* spp. grassland yield of dry matter increased from 3 671 kg/ha unfertilized to 7 594 kg/ha fertilized with 300 kg N/ha and in association with 105 kg/ha of P_2O_5 from 4 477 kg/ha to 9 883 kg/ha. It was concluded, however, that it was uneconomical to fertilize pure natural grassland and that legumes, e.g. *Desmodium* spp., should be introduced to make better use of fertilizer (Keya, 1973).

Response to defoliation. At Mt Makulu Research Station, Zambia, cutting *H. filipendula* pastures twice or four times a year (according to the frequency with which it reached a height of 30 cm) over a period of nine years changed the botanical composition. The dominance of shorter grasses, such as *Cynodon dactylon, Digitaria setivalva, Heteropogon contortus* and *Microchloa caffra,* offered an improvement in nutritive value (van Rensburg, 1968). In Uganda heavy stocking led to its replacement by *Brachiaria decumbens*, which is a much more nutritious grass (Harrington & Pratchett, 1974b).

Grazing management. It generally requires periodic burning late in the dry season. Biennial fires encouraged *H. filipendula* (Harrington & Pratchett, 1974b) and *Themeda triandra* in Uganda.

Response to fire. Burning late in the dry season in Zambia controls encroachment of *Acacia* spp., but if done too often it leads to decreased plant cover — after three years ground cover by *H. filipendula* was halved (Brockington, 1961).

Genetics and reproduction. 2n=40 (Fedorov, 1974).

Dry- and green-matter yields. Keya (1973) obtained 3 671 kg DM/ha over 14 months from unfertilized grass with linear increases up to 9 883 kg with 300 kg N/ha plus 105 kg P/ha from a *Hyparrhenia* grassland of *H. filipendula, H. cymbaria* and *Hyperthelia dissoluta* in Africa. In Zaire it produced 28 798 kg/ha and 25 627 kg/ha of green matter in 1958 and 1959, respectively (Risopoulos, 1966).

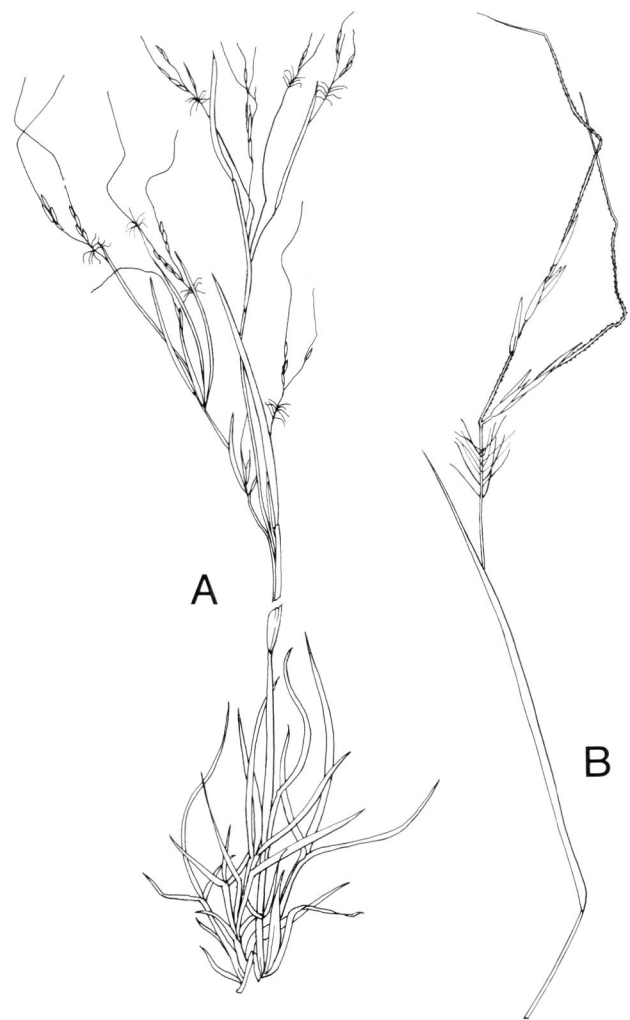

Figure 15.81. Hyparrhenia filipendula. **A**-Habit **B**-Pair of racemes

Chemical analysis and digestibility. Dougall and Bogdan (1958) recorded 6.6 percent crude protein, 36.3 percent crude fibre, 5.7 percent ash, 1.8 percent ether extract and 49.5 percent nitrogen-free extract in the dry matter at the fresh early bloom stage in Kenya. It is not regarded as a desirable grazing grass in Uganda (Harrington & Pratchett, 1972).
Palatability. Valuable grazing early in the rainy season. Less woody than other *Hyparrhenia* spp., but palatability falls toward maturity.
Toxicity. Ndyanabo (1974) recorded 0.46 percent total oxalic acid in the dry matter, but no toxicity.
Economics. It is commonly used for thatching.
Animal production. No specific quantitative figures have been found (see *H. hirta*).
Further reading. Risopoulos, 1966.

Hyparrhenia hirta (L.) Stapf

Synonym. Andropogon hirtus L.
Common names. Tambookie grass (Australia), coolatai grass (New South Wales), South African bluestem (United States), common thatching grass (southern Africa).
Natural habitat. Grassland, rocky places and open woodland.
Distribution. Mediterranean region, Near East, Iran, Iraq to northwest India, tropical eastern and southern Africa.
Description. Tufted perennial up to 90 cm high. Panicle loose, rather scanty; spatheoles 5 cm long with terminally exserted white or grey villous racemes, upper raceme base glabrous, fine, 4 mm long; pedicelled hairy spikelets 5-6 cm long. There are ten to 14 awns per raceme pair compared with two to six for *H. filipendula* (Napper, 1965) (see Fig. 15.82).
Season of growth. Summer-growing, with good autumn growth and some winter greenness.
Optimum temperature for growth. The seed germinates well over a range of 10-40°C.
Minimum temperature for growth. Germination rates of seed are low below 25°C (McWilliam, Clements & Dowling, 1970).
Frost tolerance. It is sensitive to frost and is killed in a hard winter in the United States (Robinson & Potts, 1950) but gives some winter growth in South and Western Australia (Greenwood, 1966).
Altitude range. 1 200-2 500 m.
Rainfall requirements. It grows satisfactorily with a rainfall of 500 mm or more, with a general range of 750-1 000 mm in Africa (Robinson & Potts, 1950).
Drought tolerance. Extremely drought tolerant and persistent.
Soil requirements. It has wide soil tolerance, including dry, hard, rocky soils and deep dry sands (Barnard, 1969).
Tolerance to salinity. In southwestern Australia, Rogers and Bailey (1963) found clones of CPI.5786 moderately tolerant.
Fertilizer requirements. It responds to a spring application of nitrogen. Two strains introduced to Australia, CPI.5786 and N.72, had low nitrogen requirements when heavily defoliated (Greenwood, 1966), but responded to summer applications and were affected by autumn applications in southwestern Australia.
Ability to spread naturally. It does not spread well by seed (Robinson & Potts, 1950).
Number of seeds per kg. 1 320 000.
Response to photoperiod. Flowering is accelerated by short days (Evans, Wardlaw & Williams, 1964).

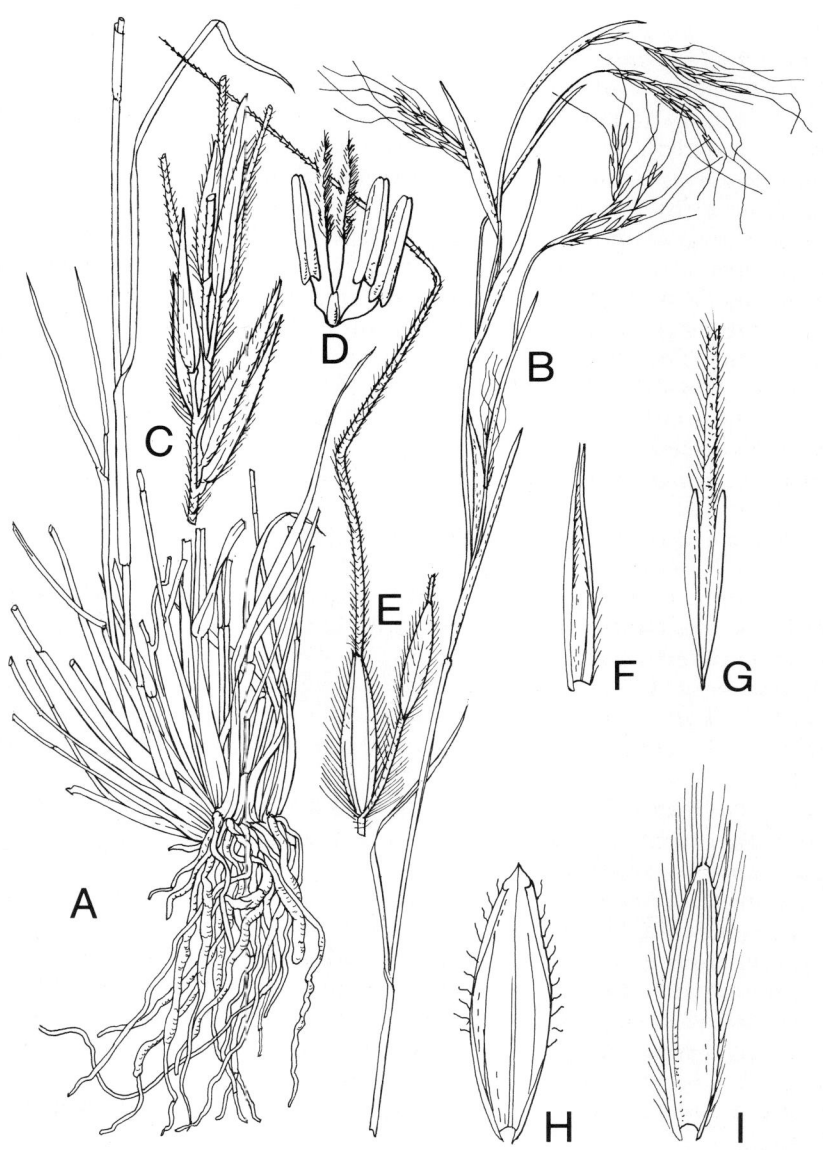

Figure 15.82. Hyparrhenia hirta. **A**-Habit **B**-Inflorescence **C**-Portion of inflorescence **D**-Flower **E**-Spikelet **F**-Lower lemma **G**-Upper lemma **H**-Upper glume **I**-Lower glume

Vigour of growth and growth rhythm. A vigorous grass, making good summer growth.
Response to defoliation. It stands heavy grazing and in fact requires it to prevent its running to seed and becoming inedible.
Grazing management. Graze heavily to prevent seeding and top-dress with 50 kg N/ha in midsummer.
Response to fire. It is usually burnt to destroy old growth and makes a good recovery after a burn.
Genetics and reproduction. The chromosome numbers are 2n=30, 40, 44, 60 (Fedorov, 1974). It is apomictic.
Dry- and green-matter yields. Under irrigation in southwest Australia it produced from 24 000-30 000 kg DM/ha with six fertilizer dressings totalling 168, 66 and 180 kg/ha of nitrogen, phosphorus and potassium respectively, plus lime and the trace elements calcium and zinc (Roberts & Carbon, 1969).
Suitability for hay and silage. It provides only fair hay and silage.
Value as standover or deferred feed. It becomes coarse and inedible if stood over.
Chemical analysis and digestibility. Karue (1974) records 3.2 percent crude protein, 38.1 percent crude fibre, 45.6 percent nitrogen-free extract, 1.8 percent ether extract and 49.6 percent total digestible nutrients in the dry matter.
Palatability. It is not very palatable, except for the young growth after burning.
Seed production and harvesting. Seed set is quite variable between plants and flowering is stimulated by decreasing temperature and rising humidity, or by increasing temperature and decreasing humidity (Robinson & Potts, 1950). It flowers over an extended period but is a poor and erratic seeder and sheds its seed readily (Barnard, 1969).
Cultivars. No cultivars have yet been established, but Humphries (1959) recorded 150 distinct forms in the introductions into Australia.
Value for erosion control. In southern Africa and the United States it is recognized as a useful conservation grass on hard stony soils and as a pioneer in vegetating eroded areas.
Economics. Besides being used for grazing, it is a useful thatching grass (Chippendall, 1955).
Animal production. A valuable fodder grass when young. Used for thatching, mat weaving and baskets. In the *Hyparrhenia* veld in Africa, cattle increase in weight and milk production from late November to March when pasture leaf protein is high (21 percent in late November, falling to 6 percent in early April). Thereafter, live weight and milk production decline (Smith, 1961).
Main attributes. Its ability to establish on hard stony soils and eroded land. Its drought tolerance and persistence.
Main deficiencies. Its poor and variable seed production and its variety of ecotypes.
Further reading. Greenwood, 1966; Robinson & Potts, 1950.

Hyparrhenia rufa (Nees) Stapf

Common names. Jaragua, faragua or yaragua grass, puntero (South America), veyale (Mali), senbelet (Ethiopia), yellow spike thatching grass (southern Africa).
Natural habitat. Seasonally flooded grassland and open woodland.
Distribution. Throughout tropical Africa, but widespread in Central and South America.
Description. A very variable perennial from 60-240 cm high. Panicle loose and narrow up to 50 cm long, with slightly spreading or contiguous racemes with shortly hairy or nearly glabrous spikelets 3.5-5 mm long. The rusty brown hairs on the spikelets and the racemes terminally exserted from the spatheoles distinguish it from *H. filipendula* and *H. hirta* (Napper, 1965). The flowering stems have little leaf. The sheaths of the leaves enclose about half the length of each internode, giving the culm and banded appearance (see Fig. 15.83).
Frost tolerance. It is susceptible to frosts.
Altitude range. Sea-level to 2 000 m in Colombia.
Rainfall requirements. 600-1 400 mm.
Drought tolerance. Good — on retentive soils withstands a dry season of six months in the llanos of Colombia and in Bolivia.
Tolerance to flooding. It stands waterlogging and temporary flooding, but not permanent flooding.
Soil requirements. Spain and Andrew (1977) found that the order of sensitivity to high aluminium soils was *Cenchrus ciliaris* cv. Biloela, very sensitive, *H. rufa, Panicum maximum* and *Melinis minutiflora,* somewhat sensitive, and *Paspalum plicatulum* and *Brachiaria decumbens*, relatively insensitive. Prefers black clays and latosols.
Fertilizer requirements. It gives a positive interaction with nitrogen and phosphorus, with 112 kg nitrogen and 56 kg superphosphate per hectare the most efficient application (Ortega, personal communication). However, it is one of the better grasses under low nitrogen and low phosphorus conditions (CIAT, 1978). It will not tolerate more than 250 kg nitrogen per hectare during the growing period.
Land preparation for establishment. A fully prepared seed-bed gives best results, but it will establish in a rough seed-bed or after a burn in natural grassland.
Sowing methods. The seed is broadcast or sown in 25-40 cm rows on a prepared and fertilized seed-bed. Root-stocks can also be planted, and it can be undersown in maize.
Sowing depth and cover. Broadcast seed and give it a light harrowing. The long, twisted awns on the 'seed' make it difficult to drill.

Sowing time and rate. Sow in Honduras early January to May at 15-20 kg/ha.
Seedling vigour. Selection 601 at CIAT has good seedling vigour.
Vigour of growth and growth rhythm. Growth is retarded when day-length is less than 12 hours, 15 minutes, during the growing season from October to April. Rattray (1973) recorded dry-matter production over a three-year period in Panama (see Table 15.40).

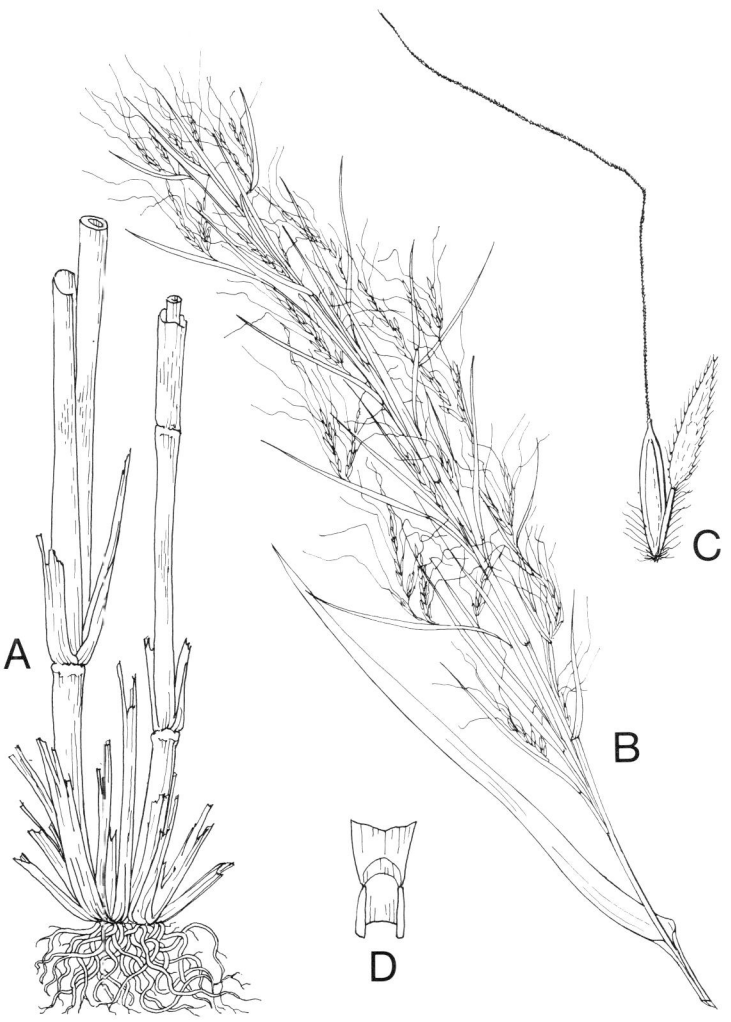

Figure 15.83. Hyparrhenia rufa. **A**-Habit **B**-Inflorescence **C**-Pair of pedicelled spikelets **D**-Ligule

TABLE 15.40 *Hyparrhenia rufa,* dry-matter production

Year	Percentage production			Total DM (tonnes/ha)
	Jan. - Apr.	May - Sept.	Oct. - Dec.	
1969	7	46	46	10.3
1970	26	61	13	8.0
1971	11	65	24	3.4

Source: Rattray, 1973

Response to photoperiod. It is a short-day plant.

Compatibility with other grasses and legumes. It combines well with legumes.

Ability to compete with weeds. It competes successfully with weeds and smothers them.

Response to defoliation. It stands close grazing (Semple, 1970) if applied rotationally and not continuously.

Grazing management. Jaragua grass must be grazed or mown so that the growth at no time reaches a height of more than 15 cm. This is attained in Costa Rica and Honduras by grazing at approximately one beast to 0.8 hectares throughout the year (Hogaboom, 1952). It takes about two years to establish a good stand by broadcasting seed, or four to five months if sown on a well-prepared seed-bed. Do not graze or cut within the first six months. Flowering stands should be mown or burnt. Undergrazing and no burning favour tussock formation and bare ground. A five-paddock system for beef breeding stock, grazing seven days on and 28 days off from June to November, and 14 days on and 56 days off from December to May, is recommended.

Response to fire. It tolerates seasonal burning.

Genetics and reproduction. 2n=20, 30, 36, 40 (Fedorov, 1974).

Dry- and green-matter yields. At the Naitama Station, Colombia, under a six-week cutting interval and 50 kg N/ha after each cut, leafy herbage accumulated to 75 cm or more, and contained 12-15 percent protein on a dry-matter basis. Dry-matter production averaged 4 500 kg/ha per year (Crowder, Chaverra & Lotero, 1970). In Honduras, 18 704 kg DM/ha was obtained at seven-week cuts when fertilized with 555 kg/ha of 10:15:15 fertilizer (Kemp, Mackenzie & Romney, 1971).

Suitability for hay and silage. It should be cut for hay and silage before flowering at a height of 60-70 cm and four to five cuts are obtained during a growing season. It has been used for silage in Minas Gerais (Brazil) (Alves & Silva, 1936). It makes good silage, but the fermentation is very slow.

Chemical analysis and digestibility. In Costa Rica, *H. rufa* at floral initiation

TABLE 15.41 *Hyparrhenia rufa*

	DM	As % of dry matter				
		CP	CF	Ash	EE	NFE
Fresh, vegetative, Brazil	29.7	9.2	28.9	14.9	2.6	44.4
Fresh, full bloom, Brazil	34.3	3.5	31.4	13.6	1.9	49.6
Fresh, milk stage, Brazil	35.5	2.8	33.7	11.5	1.5	50.5
Fresh, mature, Nigeria	24.5	4.4	32.3	19.5	1.8	42.0
30-day regrowth, January, N.W. Africa	31.7	8.1	31.4	14.1		
Hay, late vegetative, Brazil	86.3	6.5	35.0	17.9	2.3	38.3
Silage, late vegetative, Brazil	32.2	4.3	43.1	9.1	2.5	41.0

	Animal	Digestibility (%)				ME
		CP	CF	EE	NFE	
Fresh, vegetative, Brazil	Sheep	60.4	61.9	56.2	63.0	2.01
Fresh, full bloom, Brazil	Sheep	25.2	54.5	43.8	52.7	1.67
Fresh, milk stage, Brazil	Sheep	16.5	47.3	42.3	50.2	1.56
Fresh, mature, Nigeria	Cattle	18.2	66.6	11.1	40.7	1.44
Hay, late vegetative, Brazil	Sheep	55.7	53.5	51.7	63.3	1.80
Silage, late vegetative, Brazil	Sheep	44.4	55.2	44.6	47.2	1.73

Source: Göhl, 1975

contained 3.65 percent crude protein, 33 percent crude fibre, 33.55 percent nitrogen-free extract, 1.63 percent ether extract and 16.5 percent ash on a 10 percent moisture basis (Gonzalez & Pacheco, 1970). In Panama, cut at six-week intervals, it averaged 6-8 percent crude protein in the wet season and 4-6 percent in the dry season. Göhl (1975) has listed several analyses from Brazil and Nigeria in Table 15.41.

Palatability. It is not very palatable, especially as it approaches maturity (Semple, 1970).

Toxicity. Ndyanabo (1974) recorded 0.85 percent total oxalic acid in the dry matter, but no toxicity.

Seed production and harvesting. It produces abundant viable seed from which it is easily established. The stalks are cut off by hand in December (Honduras) in lengths of about 60 cm, dried in the field and then shaken to dislodge the seed, which must then be well dried. Seed germination is about 25 percent, decreasing to practically nil in ten months (Kemp, Mackenzie & Romney, 1971).

Diseases. It has good disease resistance.

TABLE 15.42 **Live-weight gain in the dry and wet seasons from** *Hyparrhenia rufa* **and** *Hyparrhenia rufa/Centrosema pubescens* **swards**

	Total live-weight gain (kg)		
	Hyparrhenia	*H. rufa* + centro	Advantage of legume/ month
Dry season (5 months)	131	182	10.2
Wet season (16 months)	727	829	6.4
Both seasons (21 months)	858	1011	7.3

Source: Stobbs & Joblin, 1966
NOTE: A mineral mixture of P, Ca, Co, I, Na, Mn and Fe plus cotton seed meal and molasses was fed as a supplement.

Pests. It has no insect problem.

Economics. H. rufa is a common native pasture plant throughout East Africa and Latin America, used mainly for beef cattle production. It has similar characteristics to the *H. contortus* pasture in near-coastal Queensland, without the troublesome awn. It is used in Africa as a coarse thatching grass and as a general purpose straw, and produces a useful pulp for paper.

Animal production. At the El Nus Station, Colombia, on steep slopes, two-year-old steers grazed on unfertilized *H. rufa* at one or two animals per hectare. Animals gained 0.37 kg each per day at the lighter rate, but weeds encroached because of low grazing intensity. At the heavier rate the animals gained 0.28 kg but the sward was damaged by heavy trampling (Crowder, Chaverra & Lotero, 1970). In Panama, Rattray (1973) cited Ortega's recorded live-weight gain of 0.30 kg per day from unfertilized grass and 0.45 kg from grass fertilized with 90 kg each of phosphorus and nitrogen per hectare, but it was uneconomic. A stocking rate of two 205-kg animals per hectare was optimum for unfertilized grass. The results obtained at Serere, Uganda, by Stobbs and Joblin (1966) with *H. rufa* alone and in combination with *Centrosema pubescens* are shown in Table 15.42.

Main attributes. Its ability to persist, and to produce high live-weight gains under heavy grazing demonstrates its value in African agriculture (Stobbs & Joblin, 1966). A *Panicum maximum/Stylosanthes guianensis* pasture at Serere gave 19 kg/ha more live-weight gain, but a *H. rufa*/centro pasture gave 73 more grazing days in the same experiment.

Main deficiencies. It is rather coarse.

Further reading. Tergas, Blue & Moore, 1971.

Hyperthelia dissoluta (Steud.) W.D. Clayton

Synonym. Hyparrhenia dissoluta (Steud.) Hutch.
Common names. Yellow thatching grass (Zimbabwe, South Africa), yellow hard grass (Kenya).
Natural habitat. Woodland and savannah on sandy soils.
Distribution. Throughout tropical Africa. Common in *Combretum* woodland.
Description. Tufted perennial up to 300 cm high. Panicle narrow and stiff; spikelets large, awns two per raceme pair, 50-88 mm long (Napper, 1965). Unlike the *Hyparrhenia* spp., the whole plant usually has a yellow and green appearance, apart from the spathes and dead leaves, which are often purple or red-brown. The culms are yellow, the sheaths that partly clasp them are green, and the stem is thus alternately yellow and green. The spikelets are mostly light green, the long, conspicuous arms yellow (Chippendall, 1955) (see Fig. 15.84).
Season of growth. Summer.
Altitude range. Sea-level to 3 000 m.
Rainfall requirements. It requires a rainfall in excess of 625 mm.
Soil requirements. It prefers sandy soils.
Response to fire. Late dry season burning of *Hyparrhenia* grassland in Zambia over three years reduced competition from shrubby *Acacia* spp., but halved the population of *H. filipendula*. However, *Hyperthelia dissoluta* withstood the fires and increased slightly in population (Brockington, 1961).
Genetics and reproduction. 2n=40 (Fedorov, 1974).
Dry- and green-matter yields At Gandajika, Zaire, it produced 25 895 kg/ha and 26 880 kg/ha of green matter in the years 1958 and 1959, respectively (Risopoulos, 1966).
Chemical analysis and digestibility. Analyses from Kenya, Niger, Ghana and Zimbabwe are shown in Table 15.43.
Palatability. Palatable when young, but too woody when mature. It is completely grazed in the early stage, but at later stages the stems are usually left ungrazed and only the leaves are eaten (Göhl, 1975).
Economics. Grazed early, but old material is used for thatching.
Animal production. No quantitative figures have been cited. It is an important grazing grass on the veld where it grows.
Further reading. Risopoulos, 1966.

Figure 15.84. Hyperthelia dissoluta. **A**-Portion of inflorescence **B**-Pair of racemes

TABLE 15.43 *Hyperthelia dissoluta*

	DM	As % of dry matter				
		CP	CF	Ash	EE	NFE
Fresh, early bloom, Kenya		12.9	33.7	8.8	3.0	41.6
Fresh, mid bloom, Niger		6.4	41.0	5.3	1.4	45.9
Fresh, 4 weeks, Ghana	25.6	16.7	21.6	10.8		
Fresh, 8 weeks, Ghana	21.9	10.6	25.2	7.8		
Fresh, 12 weeks, Ghana	31.8	8.7	31.8	4.9		
Fresh, 36 weeks, Ghana	38.8	5.2	28.0	6.5		
Hay, pre-bloom, Zimbabwe	92.6	6.8	36.9	5.8	1.5	49.0

	Animal	Digestibility (%)				ME
		CP	CF	EE	NFE	
Hay, pre-bloom, Zimbabwe	Cattle	62.1	71.0	40.2	61.9	2.27
Basal leaves, December, Sahelian Sudan	Cattle	4.3	35.0			0.41
Regrowth, June, Sahelian Sudan	Cattle	13.9	28.6			0.31

Source: Göhl, 1975

Imperata cylindrica (L.) Beauv.

Common names. Var. *major:* alang-alang or lalang (Malaysia), kunai (New Guinea), bladly grass (Australia), cotton wool grass, spear grass (Nigeria). Var. *africana:* silver spike (southern Africa), cogon grass (Philippines), cotranh (Viet Nam), illuk (Sri Lanka), yakha (Laos), gi (Fiji), sword grass (Zaire).
Natural habitat. Subhumid and humid grassland and open woodland.
Distribution. India, Australia, eastern and southern Africa and other warm temperate and tropical regions of the world.
Description. A perennial up to 120 cm high with narrow, rigid leaf-blades. Lower leaf-sheaths bearded at the mouth, upper usually glabrous; blades glabrous or hairy on the lower part, up to 100 cm long, often less, usually 3-10 mm wide, expanded; panicle 5-10 cm long; spikelets surrounded by hairs 10-15 mm long. *Imperata* roots penetrated to 58 cm in alluvial soil at Varanasi, India, with a production of 20 480 kg air-dried roots per hectare (Ramam, 1970). There are five varieties. *I. cylindrica* var. *africana* has the culm nodes usually glabrous and the spikelets 5 mm long, while in var. *major* the culm nodes are bearded and the spikelets to 3.5 mm long (Napper, 1965). Var. *europaea* occurs in Europe, var. *latifolia* in Tibet and var. *condensata* in Chile (Hubbard *et al.*, 1944) (see Fig. 15.85).
Optimum temperature for growth. 30°C, maximum 40°C.
Minimum temperature for growth. 20°C.
Altitude range. Sea-level to 2 000 m in the Himalayas.
Rainfall requirements. It grows over a wide rainfall range of 250-6 250 mm, with maximum performance over 1 500 mm.
Drought tolerance. It can survive quite long droughts because of its rhizomes.
Tolerance to flooding. It cannot stand continuous flooding and flooding is one method of control.
Soil requirements. It generally occurs on light-textured acid soils with a clay subsoil, but can tolerate a wide range of soils from strongly acidic to slightly alkaline, with a pH of 4.0-7.5, but germination is promoted by a pH of less than 5.0 (Sajise, 1973).
Fertilizer requirements. Chadokar (1977) found that yields of *Imperata* responded markedly to increased nitrogen rates, but nitrogen had less effect on improving the protein content.
Ability to spread naturally. It spreads readily by rhizomes and seed. If the rhizomes are cut by cultivation, propagation can take place from pieces with as few as two nodes.
Dormancy. There is no dormancy. Germination of the seed in the dark increased from 9 percent at 20°C, to 55 percent at 30°C, declined somewhat at 35°C, and was about 70 percent in the light with alternate 12-hour periods

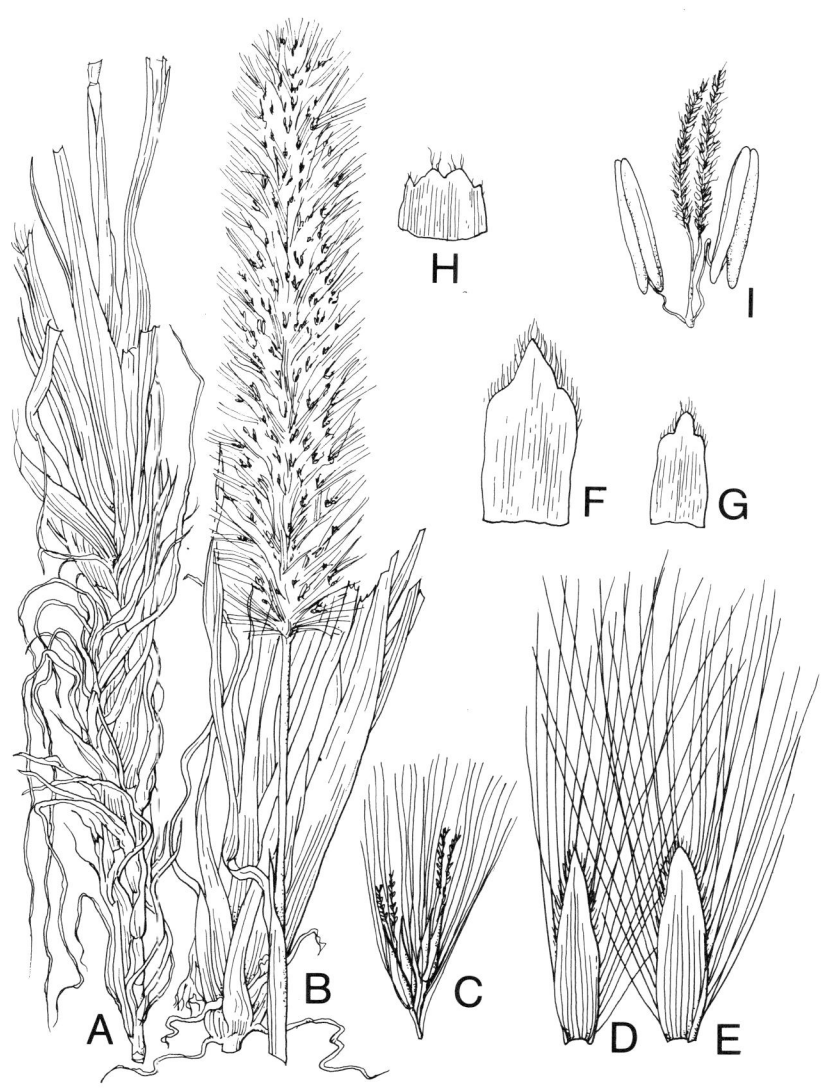

Figure 15.85. Imperata cylindrica. **A**-Habit **B**-Inflorescence **C**-Pair of spikelets **D**-Lower glume **E**-Upper glume **F**-Lower lemma **G**-Upper lemma **H**-Palea **I**-Flower

at 20 and 30°C. 0.2 percent KNO_3 solution increased germination in the dark, but not in the light. Germination declined gradually after 13 months' storage (Dickens & Moore, 1974).
Seedling vigour. It establishes well in a burn.
Vigour of growth and growth rhythm. More than 80 percent of shoots originate from the rhizomes less than 15 cm below the soil surface (Ivens, 1970).
Compatibility with other grasses and legumes. It usually forms a monospecific sward by repeated burning. Bor (1960) stated that in the Philippines the leguminous shrub *Leucaena leucocephala* can grow with it and upgrade its forage value.
Ability to compete with weeds. *Imperata* competes very successfully with weeds and suppresses them, but it is an important weed in its own right.
Response to defoliation. It cannot stand continuous heavy grazing and in Thailand it is superseded by weeds if grazed very heavily. Repeated cutting and rolling will weaken the stand and the rhizomes can be destroyed by systematic cultivation.
Grazing management. If the grass is not to be eradicated, it can be burnt periodically and grazed rotationally when 15-25 cm high.
Tolerance to fire. Frequent fires encourage the uniformity of an *Imperata* sward, and generally it occurs as a fire disclimax. The leaves burn readily, and regrowth from rhizomes is rapid (Chadokar, 1977).
Genetics and reproduction. 2n=20 (Fedorov, 1974).
Dry- and green-matter yields. In Indonesia it was found that the average number of shoots of *Imperata* at the places studied was 4.5 million per hectare, producing 11 500 kg of leaves and 7 000 kg of rhizomes (Soerjani, 1970). Chadokar (1977) followed the nutritive value of *Imperata cylindrica* at two-week cutting intervals after burning at Erap, Papua New Guinea. His results are given in Table 15.44.
Suitability for hay and silage. Most of the hay material is used for thatch and is not for fodder. It can be used as low-quality roughage in conjunction with concentrates (Soewardi *et al.*, 1974).
Chemical analysis and digestibility. Göhl (1975) lists analyses from Pakistan, India and Malaysia. See Table 15.45.
Palatability. It is eaten by livestock in the young stage, but avoided in the mature state. Elephants eat it in the Queen Elizabeth National Park, Uganda, during the dry season (Field, 1971). The extreme point and the margins of the leaves are sharp, causing irritation in the mouth, so cattle do not like it (Soerjani, 1970). The rhizomes are eaten by pigs.
Value for erosion control. It is effective in controlling erosion, but there are more useful fodder grasses which can also stabilize the soil. In eastern Nepal, of four grasses tested for soil binding — *Imperata cylindrica, Brachiaria mutica, Cynodon plectostachyus* and *Cymbopogon* spp. — *Imperata* produced the most "roots" (rhizomes and roots): 3 620 kg DM/ha in the top 7.5 cm of soil in its second year of growth and 4 574 kg/ha in its third year

TABLE 15.44 **Effect of cutting frequency on dry-matter yield and protein content of** *Imperata cylindrica*

Cutting interval (weeks)	Total DM yield (kg/ha)	Total period of growth (weeks)	No. of harvests	Yield of DM/week (kg/ha)	Crude protein (%)
2	2 191	20	10	110	9.81
4	3 059	12	3	255	7.00
6	4 833	18	3	269	6.00
8	4 960	16	2	310	5.00
10	3 452	20	2	173	4.19
12	2 078	12	1	173	4.19
14	2 265	14	1	162	3.75
16	2 255	16	1	141	3.75

Source: Chadokar, 1977

(Khybri & Mishra, 1967). It is used for stabilizing mine dumps in Zimbabwe (Hill, 1972).

Diseases. No major diseases affect it.

Pests. In Indonesia, a gall fly (*Urseoliella javanica*) attacks the apical meristem (Soerjani, 1970) but is itself heavily parasitized by a Chalcid wasp which destroys some 50 percent of the larvae. It exerts some biological control.

Economics. It covers more than 16 million hectares of waste land in Indonesia, with an annual increase of more than 150 000 hectares. Another 23 million hectares are still used for shifting cultivation, which serves as a source of area increase, and shifting cultivation increases by 100 000 hectares annually (see Plate 40). It is a noxious weed in rice, cotton, coffee, cinchona, tea, oil-palm, coconut, rubber and teak plantations. It is used for soil erosion control, mulch in coffee plantations, fodder, thatching, paper-making, packaging, fuel, ornamental purposes and for the sacrificial thread of the Hindu and as a bouquet material in the marriage ceremony in Java (Soerjani, 1970). The rhizomes and root extracts are used medicinally. In Lesotho rhizomes are eaten raw by herders and are used as a remedy for chest colds in children.

Animal production. The young shoots make good pasture (Henty, 1969). In Papua New Guinea, Holmes, Lemerle and Schottler (1976) recorded live-weight gains of 0.22, 0.25, 0.21 and 0.20 kg per day at stocking rates of 0.78, 0.94, 1.25 and 1.64 beasts per hectare for heifers grazing *Imperata* pasture, compared with the highest weight gain of 0.45 kg per day for heifers grazing a Hamil grass/legume pasture at a stocking rate of 1.69 and 2.17 beasts per hectare. In the Philippines, Magadan, Javier and Madamba (1974) recorded live-weight gains of cattle grazing *Imperata* at a stocking rate of one beast per

TABLE 15.45 *Imperata cylindrica*

	DM	As % of dry matter				
		CP	CF	Ash	EE	NFE
Fresh, early vegetative, Pakistan		6.6	34.6	7.9	3.3	47.6
Fresh, late vegetative, Pakistan		5.2	32.4	8.2	3.2	51.0
Fresh, late bloom, India		3.5	39.4	6.7	1.6	48.8
Fresh, 4 weeks, Malaysia	36.4	11.8	32.1	7.1	1.9	47.1
Hay, late vegetative, India		3.8	39.7	7.8	0.7	48.0
Regrowth, 15 days, wet season, Guinea Zone, Africa	28.7	8.7	39.2	7.6		
Regrowth, 6 days, dry season, Guinea Zone, Africa	24.4	11.2	39.1	8.4		
Regrowth, 12 days, dry season, Guinea Zone, Africa	27.4	8.5	40.6	5.6		
Regrowth, 18 days, dry season, Guinea Zone, Africa	29.2	8.7	40.3	7.3		
Aged regrowth		3.3	41.5	5.0		

	Animal	Digestibility (%)				
		CP	CF	EE	NFE	ME
Fresh, late bloom, India	Oxen	30.0	74.0	30.0	57.0	2.14
Hay, late vegetative, India	Oxen	34.0	59.0	40.0	55.0	1.88

Source: Göhl, 1975

NOTE: Nitrogen at 240 kg/ha lifted crude protein content from a range of 5.5-6.5% in the unfertilized grass to 7.18-7.50% in the fertilized material (Chadokar, 1977).

hectare as 0.27 kg per day or 100 kg per year compared with the gain on Para/centro pastures, which was more than three times this figure. In Florida, 52.4 kg beef per hectare were produced from grass alone. *Imperata* mixed with *Panicum repens,* unmanured, yielded 61.9 kg/ha. In the Thai highlands (Falvey & Andrews, 1979) the local cattle gain was about 16 kg live-weight gain per animal per year.

Further reading. Falvey, 1980; Falvey, Hengmichai & Hoare, 1979; Soerjani, 1970.

Ischaemum indicum (Houtt.) Merrill

Synonym. Ischaemum aristatum.
Common names. Batiki blue grass (Fiji), toto grass (Suriname), rattana (Panama).
Natural habitat. Seasonally wet or waterlogged areas, where it forms a dense mat.
Distribution. Peninsular India and Southeast Asia, widespread in Fiji.
Description. Culms are substoloniferous, rooting freely at the nodes, erect culms up to 60 cm tall. Leaf-sheaths tight, fringed toward the throat and hairy around the node, surfaces hairy, blade to 20 cm long and 9 mm wide. Inflorescence well exserted, of two closely apposed or somewhat divergent racemes 2-10 cm long, of pairs of spikelets, one sessile, one pedicelled, alternately, on one side of a triangular rachis. Perennial (see Fig. 15.86).
Season of growth. Summer.
Optimum temperature for growth. 32-35°C in Fiji.
Rainfall requirements. It is widespread throughout the wetter areas of Fiji and in India, over a rainfall range from 500-1 250 mm.
Drought tolerance. It has poor drought tolerance.
Tolerance to flooding. It tolerates temporary flooding.
Soil requirements. It is well adapted to clay soils of low fertility (Roberts, 1970a, b) but grows well on soils of high fertility (Soerjani, 1970). It is often confined to the hill country where there is good drainage. In India it is common on low-lying wet ground in black soils with pH from 7.0-8.5 (Dabadghao & Shankarnarayan, 1973).
Tolerance to salinity. It appears to have some tolerance to salinity.
Fertilizer requirements. It is rarely fertilized. Application of urea stimulates the growth of leaves as well as rhizomes, and also increases the soluble carbohydrate contents (Soerjani, 1970).
Ability to spread naturally. It can spread fairly rapidly by means of stolons and a little by seed on unploughable hill country in Fiji.
Land preparation for establishment. The land can be broken up sufficiently to take stem cuttings or be more fully prepared for seeding.
Sowing methods. It is more commonly propagated by cuttings, as seed viability is unpredictable.
Dormancy. Seed germination improves after nine to ten months' storage (Parham, 1960).
Seedling vigour. It is quite vigorous, especially on low-fertility soils (Partridge, personal communication).
Vigour of growth and growth rhythm. It seeds in April and May andt grows well in summer but has extremely low production in winter (Roberts, 1970b).
Response to photoperiod. It is a short-day plant, flowering in June and July.

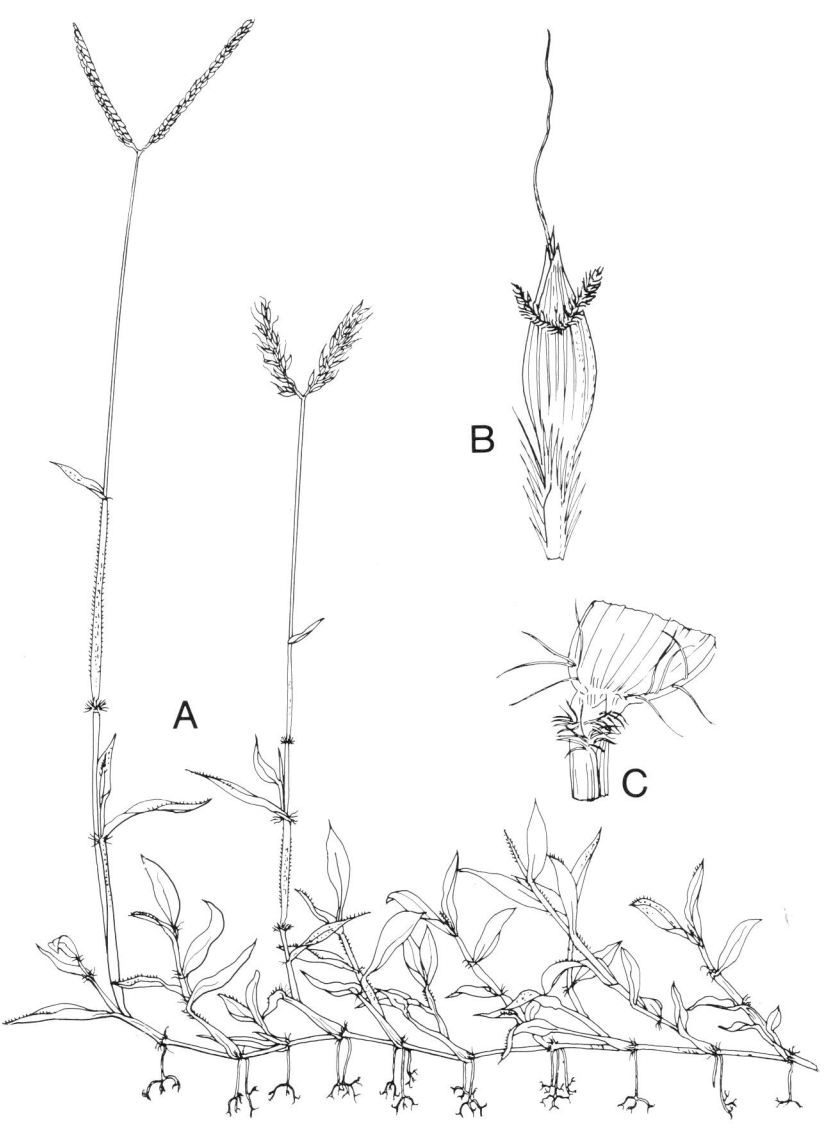

Figure 15.86. Ischaemum indicum. **A**-Habit **B**-Spikelet **C**-Ligule

Response to light. It grows very well in the shade of coconuts in the Solomon Islands (Gutteridge & Whiteman, 1978).
Compatibility with other grasses and legumes. Under coconuts in the Solomon Islands it combined very well with legumes in fertile alluvial soil. The legumes were *Centrosema pubescens, Pueraria phaseoloides* and *Desmodium heterophyllum*. In Fiji it also combines well with *D. heterophyllum*.
Ability to compete with weeds. Its stoloniferous habit is very effective in controlling weeds in Fiji (Roberts, 1970b).
Response to defoliation. It stands heavy grazing and recovers quickly. The sward opens up in winter at flowering.
Grazing management. Navua sedge is a problem in Fiji and farmers are adopting light grazing to combat the sedge along with rotational grazing (Partridge, personal communication).
Response to fire. There is little fire risk where this species grows.
Genetics and reproduction. 2n=36 (Fedorov, 1974).
Dry- and green-matter yields. In the Solomon Islands, Gutteridge and Whiteman (1978) obtained 10 800 kg DM/ha per year from a mixture of Batiki blue grass and puero. In Fiji an average yield of 9 107 kg DM/ha was obtained over a three-year period. The crude protein content was 8.1 percent (Roberts, 1970b). In Sarawak, dry-matter yields ranged from 3 141-9 299 kg/ha, with 192-940 kg crude protein per hectare with increasing nitrogen applications up to 896 kg/ha per year. Marked yield increases occurred only at 448 kg N/ha and its nitrogen recovery was poor. Yelf (1957) recorded a yield of 162 680 kg/ha green matter in Fiji. In Malaysia it yielded 212 800 kg/ha green matter.
Chemical analysis and digestibility. In the spring of 1967, the crude protein content of the dry matter was 10.3 percent for *Ischaemum indicum,* while the value for *Brachiaria humidicola* was 8.75 percent (Roberts, 1970b). In Suriname, the crude protein content varied from 14.2 percent of the dry matter at three weeks to 10.6 percent at six weeks and 8.2 percent at eight weeks (Appelman & Dirven, 1972).
Palatability. It retains its palatability if kept short. In Fiji it is eaten down before the cattle will graze *Axonopus affinis*.
Toxicity. It is reported to have a tendency to taint milk (Roberts, 1970b; Narayanan & Dabadghao, 1972) and has a detrimental effect on any tree crop.
Diseases. Often attacked by a smut in Malaysia, which destroys the inflorescence.
Pests. In 1967, a plague of army-worm (*Spodoptera mauritia*) caused damage (Roberts, 1970b).
Economics. With Para grass, it forms a large proportion of the dairy pastures in Fiji, but it is confined to the hill country (Roberts, 1970b). It is a useful pioneer grass and is vigorous and readily available.
Animal production. In the seasons 1966/67 and 1967/68, daily live-weight

TABLE 15.46 *Ischaemum indicum*

Stocking rate (beasts/ha)	LWG 1966/67		LWG 1967/68	
	(kg/head)	(kg/ha)	(kg/head)	(kg/ha)
6.25	0.29	4.15		
7.50	0.32	5.30	0.19	2.50
15.00			0.16	5.50

gains of young Friesian heifers grazed on Batiki blue grass fertilized with 250 kg N/ha and a grazing rotation of four days' grazing, 24 days' rest are shown in Table 15.46. These figures were lower than those obtained for Koronivia grass (*Brachiaria humidicola*) (Roberts, 1970b). In other experience in Fiji it has proved inferior to Pangola and Para grasses. At 3.75 cows per hectare, 345 kg milk fat per hectare in 365 days has been obtained, but this heavy grazing allowed the Navua sedge to build up to unacceptable levels.

Main attributes. Its ease and speed of establishment after clearing of rain forest, with its associated weed suppression.

Main deficiencies. The lack of commercial seed production.

Further reading. Roberts, 1970b.

Ischaemum magnum Rendle

Common name. Rumput melayu (Malaysia).
Natural habitat. In open spaces and in shade.
Distribution. Malay Peninsula, Burma and Borneo.
Description. A very robust species reaching a height of 2 m, the spikelets are unawned (Bor, 1960), 8-10 mm long, sessile, pedicel less than one-third the length of the sessile spikelet. It is a perennial, open-tussock grass with stout culms and a very strong root system.
Season of growth. Perennial.
Optimum temperature for growth. 30-35°C.
Minimum temperature for growth. No record is available. It grows at night temperatures of 22-24°C in its native habitat.
Frost tolerance. It is probably not tolerant of frost.
Latitudinal limits. It occurs mainly between 20°N and 5°S.
Rainfall requirements. It grows under high-rainfall conditions in Malaysia between 2 000 and 4 000 mm (Ng, personal communication).
Drought tolerance. It is moderately tolerant of drought during the dry season in Malaysia.
Tolerance to flooding. It is highly tolerant of flooding and waterlogged conditions.
Soil requirements. It has a wide soil tolerance — heavy clays, sands and marine peats.
Tolerance to salinity. Not known.
Fertilizer requirements. It responds well to fertilizers.
Ability to spread naturally. It spreads rapidly by shattered seed.
Land preparation for establishment. It will establish either by planting seed in a well-prepared seed-bed or by oversowing vegetative clumps into uncultivated fields (Ng, personal communication).
Sowing methods. It is sown by seed into a prepared seed-bed or by broadcasting root-stocks 1 m apart.
Sowing depth and cover. The seed should not be sown deeper than 4 cm. Germination is as good with surface sowing as with deeper planting in laboratory tests.
Sowing time and rate. In Malaysia it can be sown at any time when soil moisture is adequate.
Number of seeds per kg. About 420 000.
Dormancy. There is no post-harvest dormancy.
Seed treatment before planting. Dehulling the seed caused a rapid incease in germination in the laboratory, but may lead to infection by soil microorganisms and poor field germination (Ng & Wong, 1976).
Seedling vigour. Early growth is slow.
Vigour of growth and growth rhythm. It is not as vigorous as introduced tropical grasses.

Response to photoperiod. It appears to be day-neutral, flowering throughout the year in Malaysia with a day-length variation of only 30 minutes throughout the year. Peak flowering is in October-November.
Response to light. It tolerates shade well.
Compatibility with other grasses and legumes. It competes well with *Imperata cylindrica*, and the legumes *Centrosema pubescens* and *Stylosanthes guianensis* are compatible with it.
Ability to suppress weeds. Good. It may be a weed itself in certain areas.
Tolerance to herbicides. Highly tolerant.
Response to defoliation. It will stand heavy grazing and slashing, becoming more prostrate in habit.
Grazing management. This grass has not been subjected to grazing management trials. It is usually burnt yearly to obtain new growth.
Response to fire. It withstands fire very well.
Dry- and green-matter yields. In Malaysia a three-year mean yield of 13 740 kg DM/ha was achieved with an application of 224 kg N/ha per year. With the legumes calopo and tropical kudzu, the three-year mean was 10 812 kg DM/ha, and with siratro, 9 881 kg/ha per year when fully fertilized with basic fertilizer and trace elements. It was comparable with *Brachiaria decumbens* in yields alone, with nitrogen, and with legumes (Ng & Wong, 1976). Peak yield came from a ten-week cutting interval.
Chemical analysis and digestibility. *I. magnum* produced 846 kg/ha crude protein per year, about 100 kg/ha more than *B. decumbens* in Malaysia. Its crude protein content averaged 6 percent of the dry matter.
Palatability. It is quite palatable until maturity.
Seed production and harvesting. It produces abundant viable seeds, but ripening is uneven and the seed shatters readily. The best stage for harvesting is when the seed heads still retain some greenish tint, yet the spikelets are easily detached from the rachis by rubbing between the fingers.
Seed yield. 17 kg/ha have been harvested by hand from 40 days' regrowth, but loss of seed from bird attack was severe (Ng, personal comunication).
Minimum germination and quality required for commercial sale. No standards are available in Malaysia. Germination is usually 28-35 percent.
Value for erosion control. Because of its erect habit, it is not as useful as prostrate species.
Diseases. No serious diseases.
Pests. Mainly bird attack.
Economics. A productive indigenous grass in natural stands in Malaysia.
Animal performance. No research figures are available.
Main attributes. It is indigenous to Malaysia and seed is cheap. It stands heavy grazing. It tolerates a wide range of soils and is liked by cattle.
Main deficiencies. Its seed shattering and tendency to become a weed.
Further reading. Ng & Wong, 1976.

Iseilema laxum Hack.

Common names. Machuri (Uttar Pradesh), musel (Maharashtra), moshi (Gujarat) (India).

Native habitat. Favours low-lying situations where water stands for two to four months a year.

Description. Perennial, stems 15-50 cm long, ascending from a stout, hard, sometimes shortly creeping root-stock, very slender, simple or sparingly branched; root-fibres wiry. Leaves 7-15 cm long by 1-3 mm wide, linear; panicle occupying one-third to one-half of the stem, long, narrow, of distant axillary fascicles 6-12 mm long. Involucral spikelets truly whorled, 4 mm long. Pedicellate spikelets on long, ciliate pedicels. Bisexual spikelets narrowly lanceolate, 5 mm long (Cooke, 1958). Roots reach 105 cm, with most activity at 38 cm (Dabadghao & Shankarnarayan, 1973) (see Fig. 15.87).

Season of growth. Summer.

Altitude range. Sea-level to 760 m in Mysore, India.

Rainfall requirements. 500-1 375 mm in India.

Soil requirements. It favours medium- to fine-textured loamy soils with a pH of 6.1-7.4. Usually found in heavy, black, waterlogged soils (Whyte, 1964).

Tolerance to salinity. It is mildly tolerant, associated with *Sporobolus marginatus* (Whyte, 1964).

Land preparation. A good seed-bed is prepared.

Sowing methods. The seed is broadcast.

Sowing time and rate. Sow at 4.5-6.7 kg/ha.

Genetics and reproduction. 2n=8m24, 28, 36 (Fedorov, 1974). It is a variable species. Reproduction is sexual.

Dry- and green-matter yields. Fertilizer studies have shown that dry matter production can be boosted from 4 490 kg/ha to 6 370 kg/ha by applying 40 kg N/ha. There was no response to potassium. The forage yield varies little from year to year because of the wet habitat.

Suitability for hay and silage. In the *Dichanthium/Iseilema* zone in India it is used mainly for hay.

Chemical analysis and digestibility. See Table 15.47.

Palatability. One of the top-ranking fodder grasses, relished by cattle, and the main feed of the Ongole breed of cattle in Andhra Pradesh, India.

Seed yield. Seed loses viability quickly after 12 months.

Economics. It is so valued that it replaces paddy cultivation and brings a high price in the Bombay grass market (Dabadghao & Shankarnarayan, 1973). It can be grazed, but is usually cut for fodder.

Further reading. Whyte, 1964.

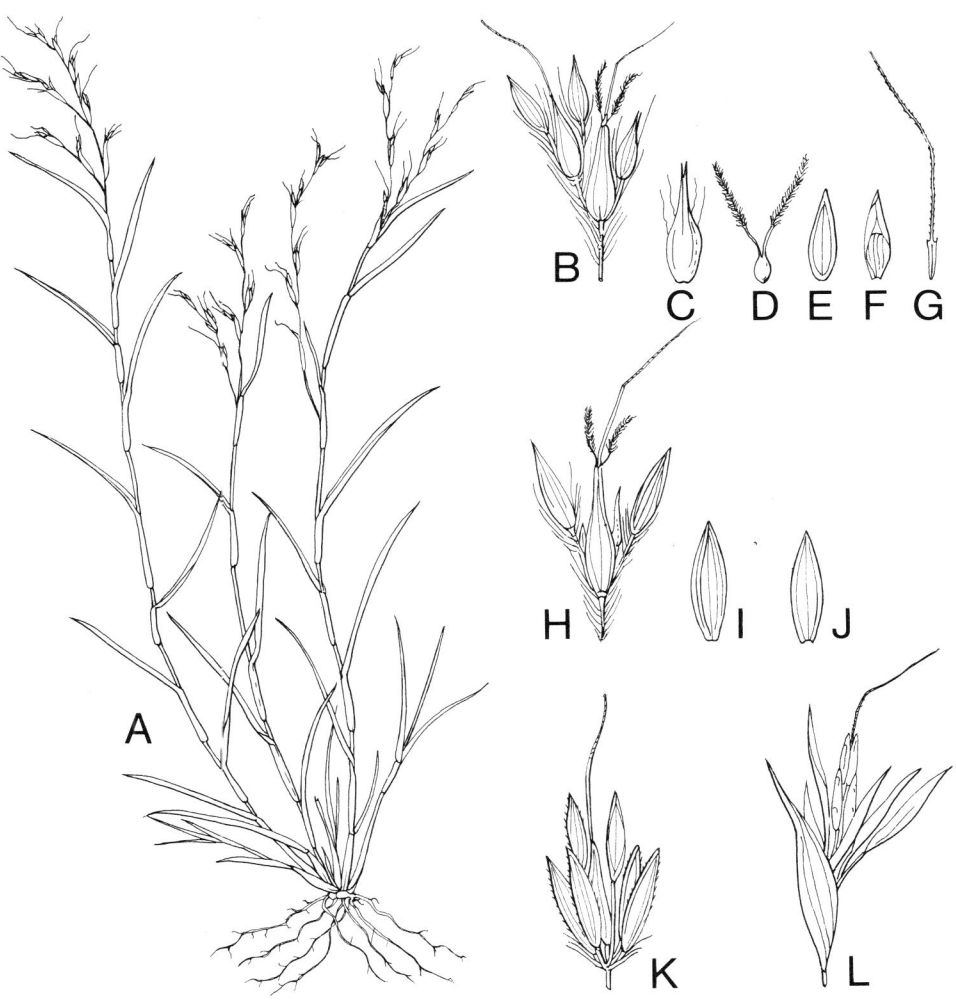

Figure 15.87 Iseilema laxum. **A**-Habit **B**-Two pedicelled male and two bisexual spikelets **C**-First glume of sessile spikelet **D**-Carpel **E,F,G**-First, second and third glumes of inner pedicelled spikelets **H**-Two pedicelled male and one sessile female or bisexual spikelet **I,J**-First and second glumes of involucral spikelet **K**-Involucral spikelets and three inner spikelets **L**-Cluster of spikelets with spathes

TABLE 15.47 *Iseilema laxum*

	As % of dry matter								Reference
	CP	CF	Ash	EE	NFE	CaO	P$_2$O$_5$	K$_2$O	
Before flowering, Bombay	5.06	34.24	11.64	1.36	47.70	0.49	0.17		Whyte, 1964
In flower, Bombay	3.69	38.79	9.85	0.99	46.68	0.52	0.16		Whyte, 1964
In seed, Bombay	2.79	34.52	11.80	1.10	49.79	0.59	0.29		Whyte, 1964
Pre-flowering, India	5.3								Dabadghao & Shankarnarayan, 1970
Dead, ripe, India	3.0								Dabadghao & Shankarnarayan, 1970
Flowering, India	3.7	38.8	9.8	1.00	46.70	0.36	0.07	0.7	Sen & Ray, 1964

Iseilema membranaceum (Lindl.) Domin

Synonym. *Anthistiria membranacea* Lindl; *Iseilema actinostachys* Domin.
Common names. Flinders grass, small Flinders grass (Australia).
Natural habitat. Open grassland.
Distribution. In Western Australia, South Australia, New South Wales and Queensland, Australia.
Description. A quick-growing, annual, glabrous grass, sometimes forming dense leafy tufts of 15 cm, the branching stems often elongated to 30-60 cm. Leaves flat. The grain is borne among the small leaves over almost all of the plant. Midrib conspicuous because of the folding of the leaf, ligule short, membranous, truncate. Distinguished by its very small racemes, very shortly bearded, and the scabrous involucral spikelets (see Fig. 15.88).
Season of growth. Early in the wet season in late spring to summer. The plant disintegrates at maturity in about six weeks to two months.
Optimum temperature for growth. Between 30 and 40°C.
Frost tolerance. It does not tolerate frosts.
Latitudinal limits. About 30°S to 18°N.
Altitude range. 160-300 m in Queensland.
Rainfall requirements. In its natural habitat it vegetates regions receiving 375-500 mm of annual rainfall with summer dominance.
Drought tolerance. It grows so quickly that it tends to escape droughts.
Tolerance to flooding. It will tolerate temporary flooding.
Soil requirements. It grows on cracking grey and brown clay soils of pH 7.0 or slightly higher, occasionally on other soil types.
Tolerance to salinity. It will grow on soils with a pH above 7.0 but it does not appear to tolerate salinity.
Fertilizer requirements. It is not fertilized, but it grows in fertile soil with adequate calcium and phosphorus and its growth is stimulated by the nitrogen resulting from the "birch" effect, releasing soil nitrogen immediately after rain.
Ability to spread naturally. It spreads rapidly by seed.
Compatibility with other grasses and legumes. It is associated with *Astrebla* spp. in the Mitchell grass grasslands in Australia, being an annual which vegetates the spaces between the *Astrebla* tussocks during the wet season in association with *Brachyachne convergens* and *Dactyloctenium radulans*. Few legumes other than sparse *Rhynchosia minima* are associated with it.
Response to defoliation. It does not stand heavy stocking and breaks up at maturity.
Response to fire. It is destroyed by fire.
Suitability for hay and silage. It makes an excellent, nutritious hay if harvested at flowering.

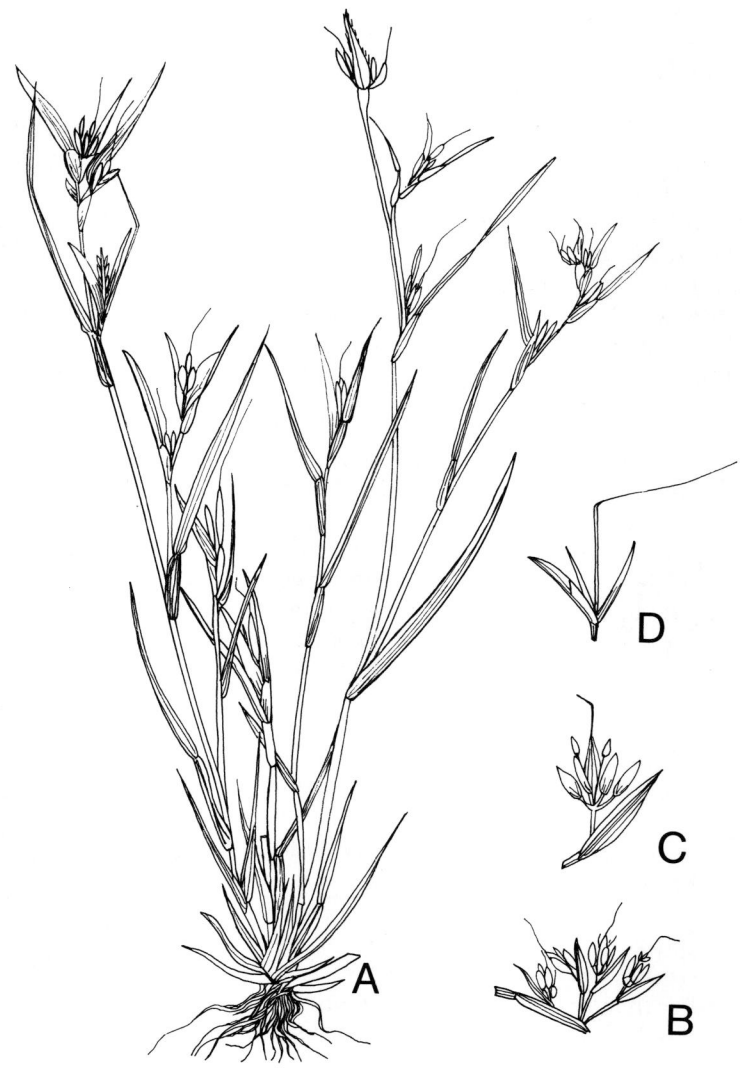

Figure 15.88. Iseilema membranaceum. **A**-Habit **B**-Spikelet cluster **C**-Male spikelet **D**-Fertile spikelet

Chemical analysis and digestibility. Analyses of hayed-off mature material in northwest Queensland show the following dry-matter content: crude protein, 2-3.9 percent; crude fibre, 36.3-42.4 percent; ash, 11.6-14.2 percent; ether extract, 1.3 percent; nitrogen-free extract, 43.4-46.9 percent; CaO, 0.28-0.43 percent and P_2O_5, 0.058-0.1 percent.
Palatability. Its palatability is questioned, but stock will retrieve broken pieces of the plant from the ground after maturity. Generally there is ample other feed available with it.
Seed production and harvesting. It seeds freely in November. Seed is rarely collected.
Minimum germination and quality required for commercial sale. 55 percent germinable seeds, 78 percent purity (Queensland).
Economics. Flinders grass is a common annual component in the Mitchell grass (*Astrebla*) association on the heavy cracking clays of western Queensland grasslands. It is very palatable when young but soon matures and disintegrates and much blows away. It does not provide a large part of the merino sheep's annual diet (Davidson, 1954; Lorimer, 1978).
Animal production. No figures have been cited. In areas where it occurs with *Astrebla* spp., the annual carrying capacity is rated at one sheep to two hectares, but because of its fast growth it is usually only available for three months and much of the feed is wasted.
Further reading. Davidson, 1954.

Iseilema vaginiflorum **Domin**

Common name. Red Flinders grass (Australia).
Natural habitat. Open grassland.
Distribution. Northern and central Queensland.
Description. A caespitose annual, up to 75 cm tall, with erect, slender culms, geniculate at the base. Stems often purple at the base, whole plant turning red at maturity. Similar to *I. membranaceum,* but larger and leafier (see Fig. 15.89).
Economics. It is very palatable and makes excellent hay. It occurs generally after heavy rains, associated with Mitchell grass, but on tighter soils and after cultivation it may occur as a pure stand. It seeds heavily and the seed is quickly shed. It could be a useful grass for short-term cover when normal rains arrive late.

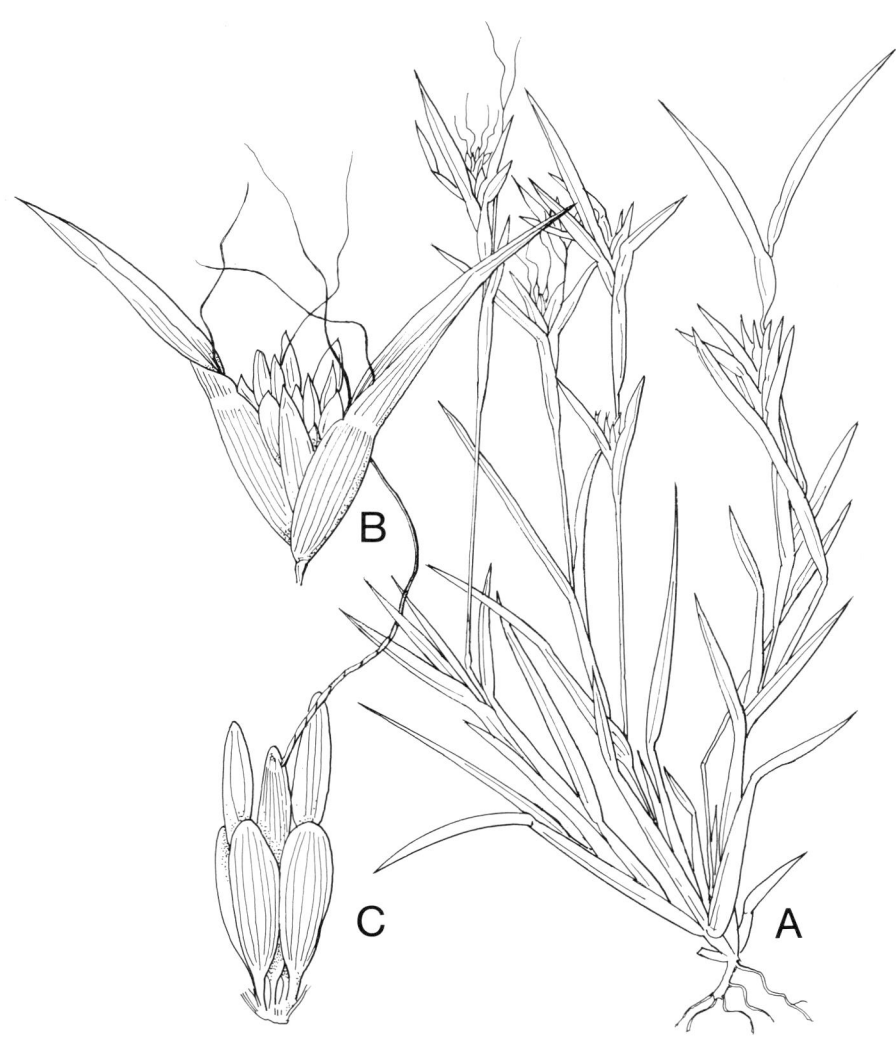

Figure 15.89. Iseilema vaginiflorum. **A**-Habit **B**-Raceme **C**-Cluster of spikelets

Ixophorus unisetus (Presl) Schlecht.

Synonym. Urochloa uniseta Presl; *Panicum unisetum* Trin.; *Setaria uniseta* Fourn.
Common names. Honduras grass (Costa Rica), Mexican grass (Hawaii).
Natural habitat. Low thickets, ditches and wet places, at low altitudes.
Distribution. Native to America between Mexico and Colombia.
Description. Erect or spreading, 50-150 cm tall, blades 15-30 cm or even as much as 60 cm long, as much as 4 cm wide, panicles 10-20 cm long, the racemes approximately 3-6 cm long; spikelets about 4 mm long, the bristles 3-10 mm long (Hitchcock, 1930). It forms closed clusters of succulent stems. The leaves wrap around the stems from base to top (see Fig. 15.90).
Season of growth. Summer.
Optimum temperature for growth. 20-24°C.
Frost tolerance. It does not tolerate frost.
Altitude range. Sea-level to 1 500 m; does best in warm coastal areas.
Rainfall requirements. It requires heavy rainfall.
Drought tolerance. Growth stops in dry weather.
Soil requirements. It requires fertile, moist soils, well supplied with organic matter.
Fertilizer requirements. It requires a fertile soil well supplied with organic matter and will require a complete fertilizer mixture if the soil is deficient.
Land preparation for establishment. Fully prepare and cultivate land or burn and sow in ashes.
Sowing methods. Drill into a prepared seed-bed or broadcast in ashes of burn. It can be propagated vegetatively, in rows 75-100 cm apart, cuttings 60 cm apart in row.
Sowing time and rate. Sow in early summer at 20 kg/ha.
Compatibility with other grasses and legumes. It associates well with forage legumes, especially tropical kudzu (*Pueraria phaseoloides*), which is sown at 3 kg/ha.
Ability to compete with weeds. It needs good weed control in the early phases of establishment, but afterwards it competes well with weeds.
Response to defoliation. It will not stand soil compaction by trampling, so heavy defoliation will affect the stand, and overgrazing should be avoided. It is better used as a soilage crop, cut and fed green.
Grazing management. Graze the grass when it reaches 70-100 cm in height, or when the first florets appear, and cease grazing when the pasture is reduced to a height of 30-40 cm (Gonzalez & Pacheco, 1970).
Dry- and green-matter yields. It yields 177 000 kg/ha per year in Costa Rica (Gonzalez & Pacheco, 1970) with 15 percent dry matter and 1.9 percent crude protein from four to five cuts per year. In Sri Lanka, it produced 19 749 kg DM/ha per year (Pathirana & Siriwardene, 1973).

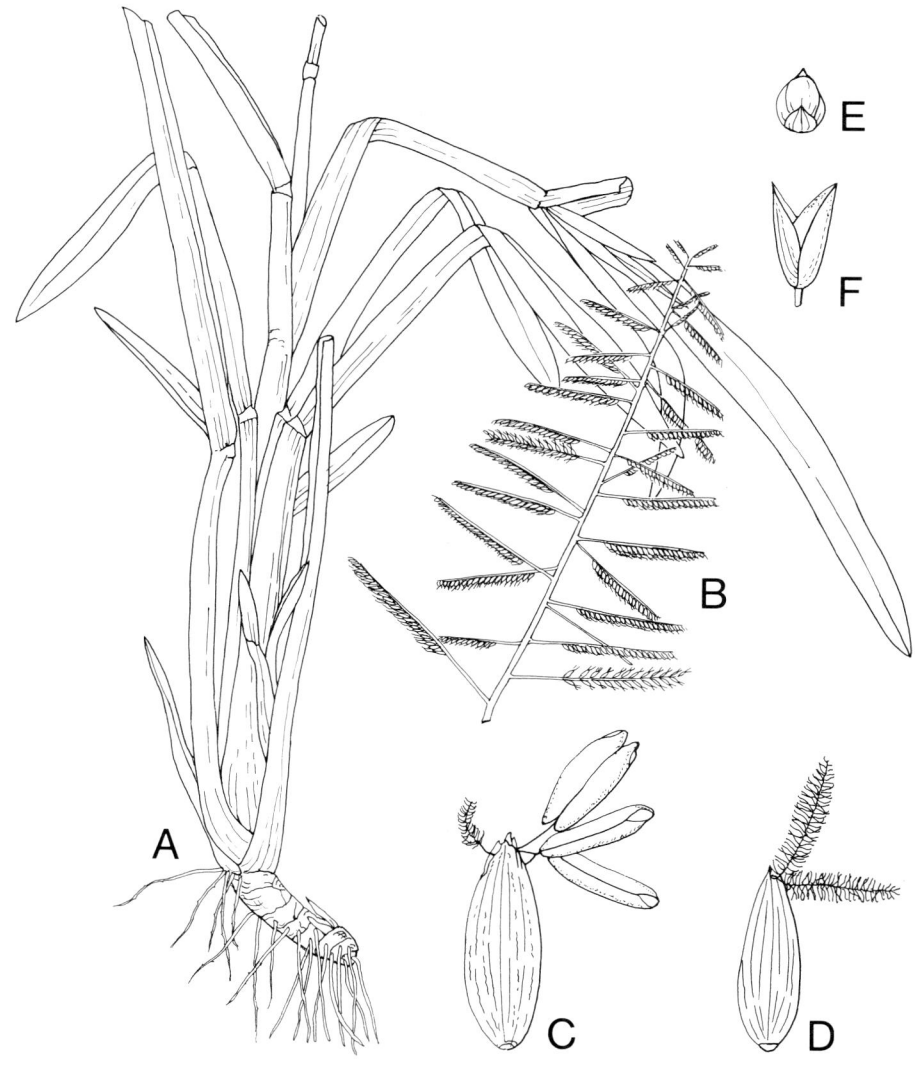

Figure 15.90. Ixophorus unisetus. **A**-Habit **B**-Inflorescence **C**-Bisexual floret **D**-Pistillate floret **E**-Grain **F**-Glumes

Suitability for hay and silage. It makes very good hay in Costa Rica.
Chemical analysis and digestibility. In Costa Rica the analysis at floral initiation was 12.88 percent crude protein, 27.01 percent crude fibre, 34.7 percent nitrogen-free extract, 3.13 percent ether extract and 12.88 percent ash (Gonzalez & Pacheco, 1970) on a 10 percent moisture basis. In Cuba, fresh mature growth contained 16.1 percent dry matter and the dry matter contained 7.9 percent crude protein, 30.3 percent crude fibre, 10 percent ash, 0.5 percent ether extract and 51.3 percent nitrogen-free extract (Calvino, 1952). In Sri Lanka, four weeks' growth had 14.08 percent dry matter and 14.52 percent crude protein, while at six weeks the figures were 15.76 percent dry matter and 13.04 percent crude protein when fully fertilized (Pathirana & Siriwardene, 1973).
Palatability. It is extremely palatable, with succulent leaves and stems throughout the plant. It does not persist when grazed (Göhl, 1975).
Seed production and harvesting. Honduras grass produces seeds which are abundant, but of low viability, and are shed as soon as they mature. However, this shed seed ensures the continuity of the pasture.
Animal production. The average carrying capacity of Honduras grass at the Los Diamantes Experiment Station in Costa Rica is two to three beasts per hectare per year.
Further reading. Gonzalez & Pacheco, 1970.

Lasiurus hirsutus **Boiss.**

Synonym. L. scindicus Henr.; *Elyonurus hirsutus* (Forsk.) Boiss.
Common names. Sewan grass (Rajasthan, India), karera (Pakistan).
Natural habitat. Large bushy thickets in sandy deserts.
Distribution. North Africa, Mali, Niger, Ethiopia, Iraq, southern Pakistan and northwest India.
Description. A perennial, almost sub-woody at the base, with wiry, glaucous stems, leaf-blade thin with a setaceous tip, racemes 10 cm long, densely white villous, spikelets often three at each node, two sessile and one pedicelled; sessile spikelets 6-9 mm long, the lower glume flat, hirsute (Hutchinson & Dalziel, 1954). It has the maximum quantity of root material in the top 0.03 m^3 of soil (see Fig. 15.91).
Latitudinal limits. 25-27°N.
Rainfall requirements. It occurs in rainfalls below 250 mm a year. One of the dominant species in arid zones with as low a rainfall as 12.5 mm in Rajasthan (Dabadghao & Shankarnarayan, 1973).
Drought tolerance. Excellent.
Soil requirements. It grows best on alluvial soils or light brown sandy soils with a pH of 8.5.
Tolerance to salinity. It has a moderate tolerance to salinity.
Fertilizer requirements. At Jodhpur, Rajasthan, application of 0-40 kg N + 20 kg PO_5 + 25 kg K_2O/ha gave a yield of 4 860 kg/ha. Increasing nitrogen from 0-20 kg/ha increased yield from 3 500 to 5 810 kg/ha, but yields declined with further increases in nitrogen rates (Bajpai & Jain, 1971).
Sowing methods. At Jodhpur, Rajasthan, it is best sown in rows 30-90 cm apart and weeded once (Chakravarty & Verma, 1972). In Thal, Pakistan, it is germinated in earthen tubes about 30 cm long and then transplanted in these tubes into the sand dunes at distances of 2 × 2, 3 × 2, or 3 × 3 m, with the top of the tube buried 10 cm in the sand. Protect the plants from wind, using reed windbreaks, till well established (Khan, 1968).
Sowing time and rate. Transplant the tubes during rain or within 24 hours after rain has fallen (Khan, 1968).
Dormancy. Germination increases with storage up to four months.
Seed treatment before planting. Chakravarty and Bhati (1969) made 5 mm diameter pellets containing one or two spikelets with a mixture of fine silt and cow dung. Germination was not affected and the best stand came from two spikelet pellets stored for four months.
Compatibility with other grasses and legumes. Sowing with *Vigna radiata, V. aconitifolia* and *Cyamopsis tetragonoloba* increased forage production by 17-36 percent when rainfall was adequate, in the first two years, but the legumes failed in the third year due to drought (Daulay, Chakravarty & Bhati, 1972).

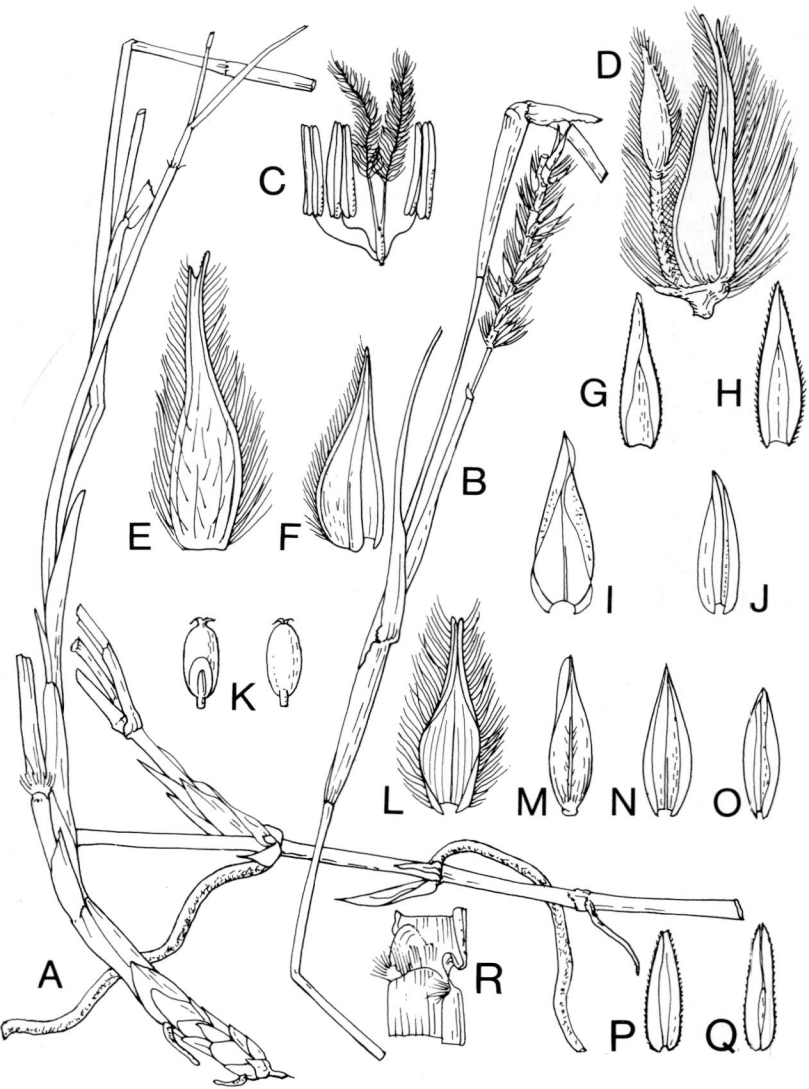

Figure 15.91. *Lasiurus hirsutus.* **A**-Habit **B**-Inflorescence **C**-Flower **D**-Pair of spikelets. *Sessile spikelet:* **E**-Lower glume **F**-Upper glume **G**-Upper palea **H**-Lower palea **I**-Lower lemma **J**-Upper lemma **K**-Grain. *Pedicelled spikelet:* **L**-Lower glume **M**-Upper glume **N**-Lower lemma **O**-Upper lemma **P**-Lower palea **Q**-Upper palea **R**-Ligule

Response to defoliation. At Jodhpur, Rajasthan, the highest dry-matter yields were obtained by cutting at 30-day intervals to a height of 15 cm (Dabadghao, Roy & Marwaha, 1973).

Genetics and reproduction. $2n=56$.

Dry- and green-matter yields. 2 700-10 500 kg fresh forage per hectare (Verma & Chakravarty, 1969) and dry-matter yields of 1 500 kg/ha (Chakravarty & Verma, 1972). Well-established swards in western Rajasthan yielded 3 400 kg DM/ha (Ahuja, 1972).

Chemical analysis and digestibility. No figures have been cited.

Palatability. Good, one of the first grasses to disappear under the impact of grazing.

Cultivars. Strain 318 is recommended by Prasad and Singh (1973) for cultivation under dryland conditions in western Rajasthan.

Value for erosion control. It is useful stabilizing sand dunes in Iraq (Dougrameji & Kaul, 1972).

Economics. One of the most important arid zone grasses in northwest India.

Animal production. At Jodhpur, Rajasthan, carrying capacity was 4.2, 0.29 and 7.1 cattle units per hectare in 1968/69, 1969/70 and 1970/71 respectively, with annual rainfalls of 178.8, 92.7 and 594.8 mm respectively (Gupta, Saxena & Sharma, 1972).

Main attributes. Its drought resistance, its persistence on sand dunes and its palatability.

Further reading. Chakravarty & Bhati, 1969; Chakravarty & Verma, 1972.

Leersia hexandra Sw.

Common names. Swamp rice grass, swamp cut grass (southern Africa), lambedora grass (Venezuela).
Natural habitat. Swamps and dams and ditches in standing water, or as floating grass islands as in Logtak Lake, Manipur, India.
Distribution. Throughout the tropics and subtropics.
Description. A scrambling, stoloniferous perennial with leaf-blades 5-13 mm wide, and growing to 40-60 cm high. Leaves bright green, very rough and unpleasant to handle. Spikelets like rice, but much smaller. They are scabrid and often strongly flushed with brick-red or orange, an unusual colour in grass spikelets. Panicle branches are nearly always zig-zag, at least after the spikelets have fallen (Chippendall & Crook, 1976) (see Fig. 15.92).
Season of growth. Perennial.
Latitudinal limits. About 30°N and S.
Altitude range. Sea-level to 2 200 m.
Rainfall requirements. It grows in a rainfall regime from 750-5 000 mm, in swamps.
Drought tolerance. It survives well into drought until the swamps dry out.
Tolerance to flooding. Excellent. It is an aquatic grass. In Suriname it occurs in creeks, ditches and swamps, in paddy fields, and on some drier clay soils (Dirven, 1963a).
Soil requirements. It generally grows on heavy-textured clay soils in swamps and valley bottoms.
Genetics and reproduction. 2n=48 (Fedorov, 1974).
Suitability for hay and silage. It makes quite good hay but is difficult to harvest from swamps and is usually cut when swamps dry out.
Chemical analysis and digestibility. Analyses from Tanzania and India are shown in Table 15.48.
Palatability. Palatable when young, but eaten when old if there is a scarcity of feed. Its sharp edges may make it less acceptable.
Toxicity. In Zambia, scouring occurs when cattle move from the fibrous forest grazing to the rich plains grasses consisting of *Echinochloa pyramidalis, E. scabra, Acroceras macrum, Hemarthria altissima, Leersia hexandra* and *Vossia cuspidata,* and it may be three to four months before they regain condition (Verboom & Brunt, 1970).
Economics. Leersia hexandra is pan-tropical in its occurrence and is one of the mostly highly regarded of the swamp grasses for grazing, particularly in the dry season. In the Llanos Inundables of Colombia and Venezuela it is the second most important food of the giant rodent, the capybara (*Hydrochoerus capybara*), and comprises 29.16 percent of the diet during the rainy season, 15.25 percent at the end of the rains, and 7.74 percent at the end of the dry season (Escobar & Gonzalez, 1976).

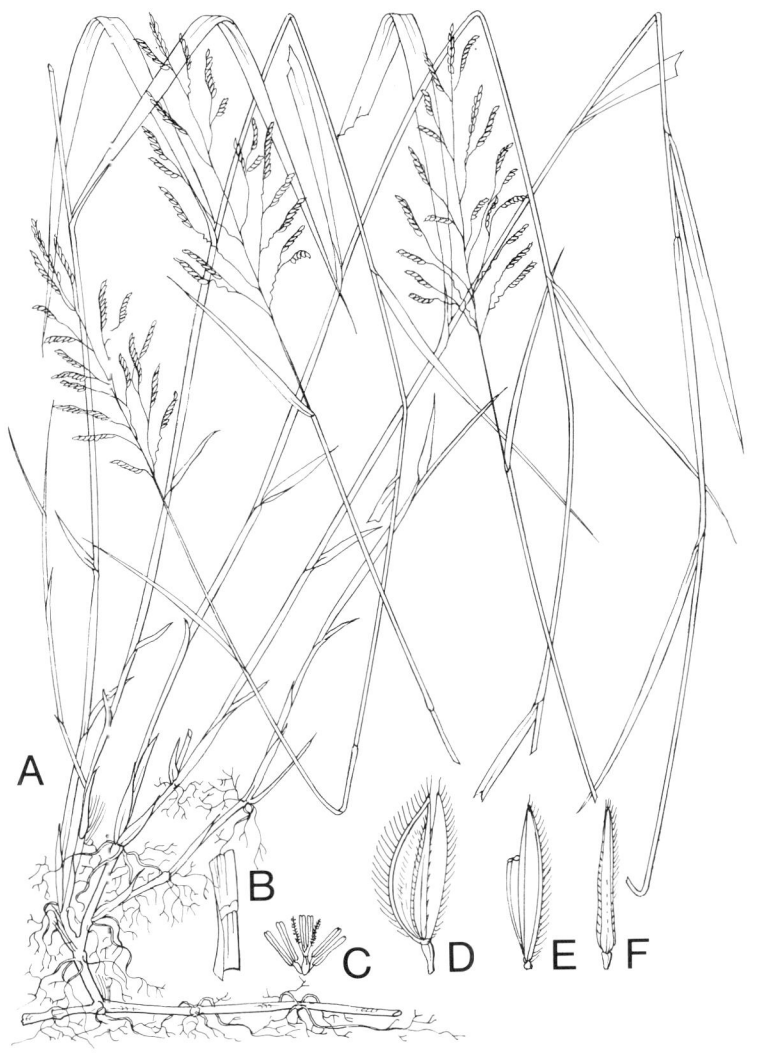

Figure 15.92. Leersia hexandra. **A**-Habit **B**-Ligule **C**-Flower **D**-Spikelet **E**-Palea, side **F**-Palea, back

TABLE 15.48 *Leersia hexandra*

	DM	As % of dry matter				
		CP	CF	Ash	EE	NFE
Fresh, vegetative, Tanzania	30.0	10.1	25.60	10.40	1.80	52.10
Fresh, early bloom, Tanzania		5.8	28.40	16.70	2.16	47.00
Flowering, Zambia[a]		7.5	30.34	9.42	2.98	49.76
Hay, India		6.3	31.40	14.90	1.50	45.90

	Animal	Digestibility (%)				
		CP	CF	EE	NFE	ME
Vegetative	Sheep	69.3	60.9	33.0	71.8	2.27
Early bloom	Oxen	40.0	63.0	23.0	54.0	1.70
Hay	Oxen	38.0	66.0	31.0	50.0	1.72

[a] Mineral content as percent of DM was Ca, 0.48; Na, 0.14; P, 0.11; K, 0.4; and Cl, 0.27 (Verboom & Brunt, 1970).
Source: Göhl, 1975

Animal production. In the Venezuelan llanos the pasture consists of *Panicum laxum*, *Leersia hexandra* and *Hymenachne amplexicaulis* (68 percent) with *Eleocharis mutata* and bushes as minor components. At the lowest stocking rate of 1.6 capybaras per hectare a decrease in *P. laxum* occurred. At three beasts per hectare a further decrease in *P. laxum* occurred with an increase of *Leersia hexandra* and weeds. At six beasts per hectare both *L. hexandra* and *P. laxum* declined and weeds and pioneer species increased (Ojasti, 1976). It is an important pasture grass for cattle, but no cattle performance data have been found.

Further reading. Dirven, 1963a; Talapatra, 1950.

Leptochloa obtusiflora Hochst.

Natural habitat. Widespread in open thicket and grassland on sandy and clay soils, often a weed.
Distribution. Central-southern and northeast Africa. Plentiful in the Coast and Eastern Provinces of Kenya (Bogdan & Pratt, 1967).
Description. Perennial, 90-120 cm high with branching culms. Panicle of five to many erect spikes on a rather short axis; glumes about half the length of the spikelets, six to nine florets (see Fig. 15.93).
Altitude range. Sea-level to 1 800 m.
Rainfall requirements. About 575 mm (Bogdan & Pratt, 1967).
Drought tolerance. Excellent.
Soil requirements. Adapted to loose sandy loams, loams and alluvial soils (Bogdan & Pratt, 1967), and red clays in Kenya.
Land preparation for establishment. Broadcast the seed on the surface and give a light cover if possible.
Sowing time and rate. Sow in summer at 350 g/ha (Bogdan & Pratt, 1967).
Number of seeds per kg. 3.3 million florets with one seed each (Bogdan & Pratt, 1967).
Genetics and reproduction. 2n=20 (Fedorov, 1974).
Chemical analysis and digestibility. In the leafier types, the content of crude protein at the early-flowering stage can be as high as 18 percent and the crude fibre less than 30 percent of the dry matter (Dougall & Bogdan, 1960). The ash level is 9.5 percent, ether extract 1.7 percent, nitrogen-free extract 42.4 percent, calcium 0.6 percent and phosphorus 0.19 percent.
Palatability. It is quite palatable.
Seed production and harvesting. It is a very good seeder and seed can be collected easily by hand stripping the panicles. Cutting with sickles may result in considerable losses, as the seed sheds easily when mature. Seventy to 100 plants may supply a kilogram of seed. The spikelets have several florets each, which at maturity break off to make the seed (Bogdan & Pratt, 1967).
Economics. It has not yet been tried in Kenya for reseeding the range, but Bogdan and Pratt (1967) recommend its trial throughout the semi-arid areas.
Main attributes. Its leafy early growth, drought resistance and high seed production.
Further reading. Bogdan & Pratt, 1967.

***Figure* 15.93.** *Leptochloa obtusiflora*

Leptothrium senegalense Kunth

Synonym. Latipes senegalensis Kunth.
Common names. Hook grass (Kenya), tougourit (Mauritania).
Natural habitat. Sandy, arid soils in open bush and grassland.
Distribution. North Africa extending to Kenya, common in arid Sudan.
Description. Tufted short-lived annual grass 15-60 cm high. Inflorescence spikelet up to 18 cm long with very characteristic warted spikelets 3-4 mm long. The raceme terminates in a curved point and has two spikelets, each with a single caryopsis (see Fig. 15.94).
Season of growth. Summer.
Frost tolerance. It will not tolerate frost.
Altitude range. Sea-level to 2 000 m.
Rainfall requirements. It is common in the 300-400 mm zone in Kordofan Province, the Sudan, and throughout the Sahel.
Drought tolerance. Excellent, but being an annual it can only hay off when it matures at the end of the rains. It seeds heavily and this replenishes the pasture.
Soil requirements. It is usually found on sandy desert soils.
Land preparation for establishment. It will establish on bare ground in sandy soils.
Sowing method. Broadcast the seed on sandy soil.
Sowing time and rate. Early wet season at 3.5 kg/ha.
Number of seeds per kg. 352 000 short racemes of two spikelets, each with one caryopsis (Bogdan & Pratt, 1967).
Vigour of growth and growth rhythm. In the Sahel it flowers from August to February and is in full vegetative stage in June (Boudet & Duverger, 1961).
Feeding value. It maintains its palatability right through the dry season in the Sahel (Boudet & Duverger, 1961).
Seed production and harvesting. It seeds well. The seed is in the form of a short raceme of the panicle, which at maturity can be easily stripped.
Value for erosion control. It is an efficient colonizer of bare ground and was used in reseeding the Mivea area of Embu and Baringo, Kenya (Bogdan & Pratt, 1967).
Economics. A valuable grass for arid climates.
Animal production. Its value is limited by its small size and fibrous, stemmy herbage of rather low protein content (Bogdan & Pratt, 1967) but it is palatable and a valuable food for camels in the arid areas.
Further reading. Bogdan & Pratt, 1967; Boudet & Duverger, 1961.

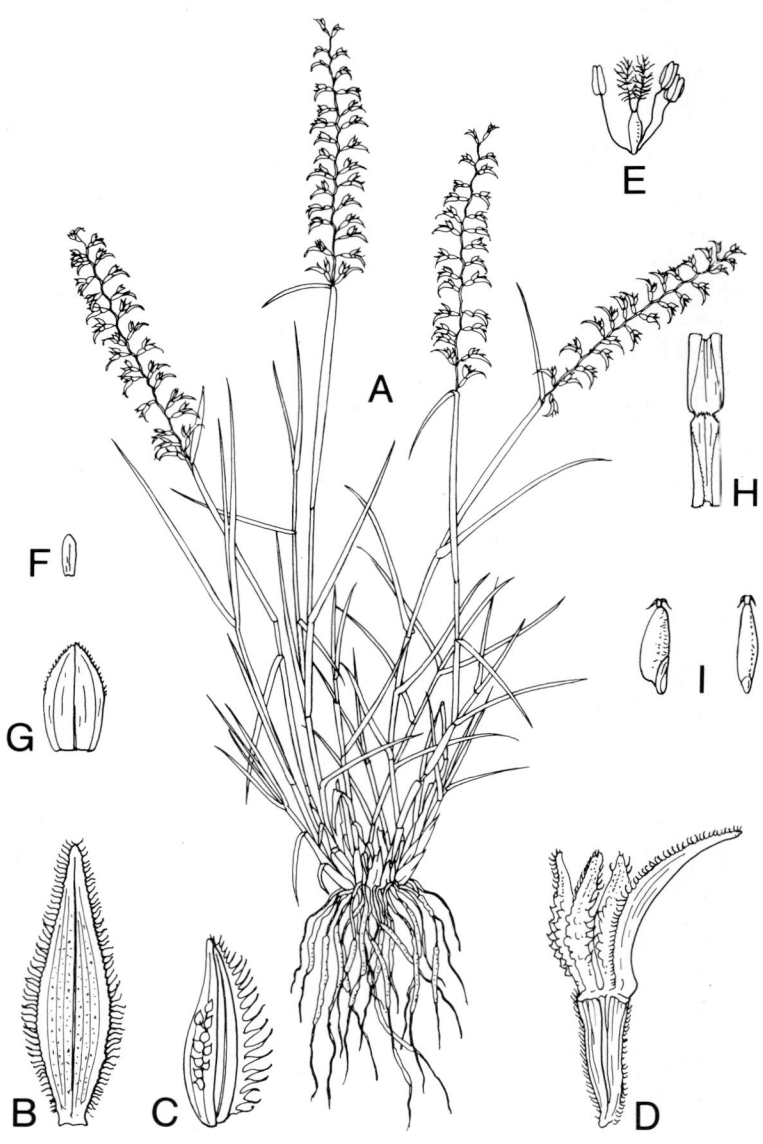

Figure 15.94. Leptothrium senegalense. **A**-Habit **B**-Lower glume **C**-Upper glume **D**-Spikelet **E**-Flower **F**-Palea **G**-Lemma **H**-Ligule **I**-Grain

Loudetia simplex (Nees) C.E. Hubbard

Common names. Common russet grass, besem grass (Africa).

Natural habitat. Common on open grassland and in poorly drained, high-rainfall sandy soils, but does not extend into the woodland.

Description. A densely tufted perennial with culms 20-90 cm high, simple, lowest leaf-sheaths densely hairy at the base, splitting into fibres, leaves otherwise hairy or glabrous. The blades exceedingly variable in length and width, usually 15-40 cm long and up to 7 mm wide. Panicle 40 cm long, open and loose. Spikelets 10-13 mm long, unequally and often long-pedicelled, solitary or in pairs, light, dark or dull brown, glumes truncate, obtuse, glabrous, callus of upper floret 0.5-1.3 mm long, two-toothed, bearded with white hairs, the lemma hairy (Chippendall, 1955) (see Fig. 15.95).

Altitude range. 300-2 750 m in Tanzania.

Rainfall requirements. It generally occurs in rainfalls varying from 750-1 000 mm.

Drought tolerance. It is not very tolerant of drought.

Tolerance to flooding. It is common on seasonally flooded valley grasslands or dambos in Central Africa (Vesey-Fitzgerald, 1963).

Soil requirements. In Tanzania it vegetates an infertile red earth (latosolic soil) derived from granitic rock and low in organic matter, lime, phosphorus and potash. In Zambia it is common on poor, sandy soils. (Verboom & Brunt, 1970). In Ghana it is common on rocky hillsides and shallow soils overlying impermeable ironstone hardpan or bedrock. Soil texture varies from sand to clay with a pH range from 5.0 to 6.0. It is also common on riverine plains, on a range of soils with a pH from 5.2 to 7.5, with mottled subsoils.

Vigour of growth and growth rhythm. The grasses vegetate and come into flower during the rains, but set seed and turn a reddish-yellow colour when their life cycle is completed, even if the soil has not dried out. *Loudetia* flowers early. After the rains, the whole herb mat dries off and usually burns. After the fires there is a rather sparse growth of green leaves from the fire-scorched perennial cushions and a little dry-season flowering by several grass species. The main regrowth from the perennial cushions, however, does not occur until after the rains have commenced (Vesey-Fitzgerald, 1963).

Response to fire. In Njombe, Tanzania, burning this pure vegetation every other year in October gave the best grass production. Burning annually in June, soon after the end of the rains, caused a vigorous growth of herbaceous plants. With neither burning nor grazing, the grass lost its vigour (van Rensburg, 1952).

Genetics and reproduction. 2n=24, 40, 60 (Fedorov, 1974).

Chemical analysis and digestibility. Dougall and Bogdan (1958) recorded 10.4 percent crude protein, 38.0 percent crude fibre, 5.6 percent ash, 1.8 per-

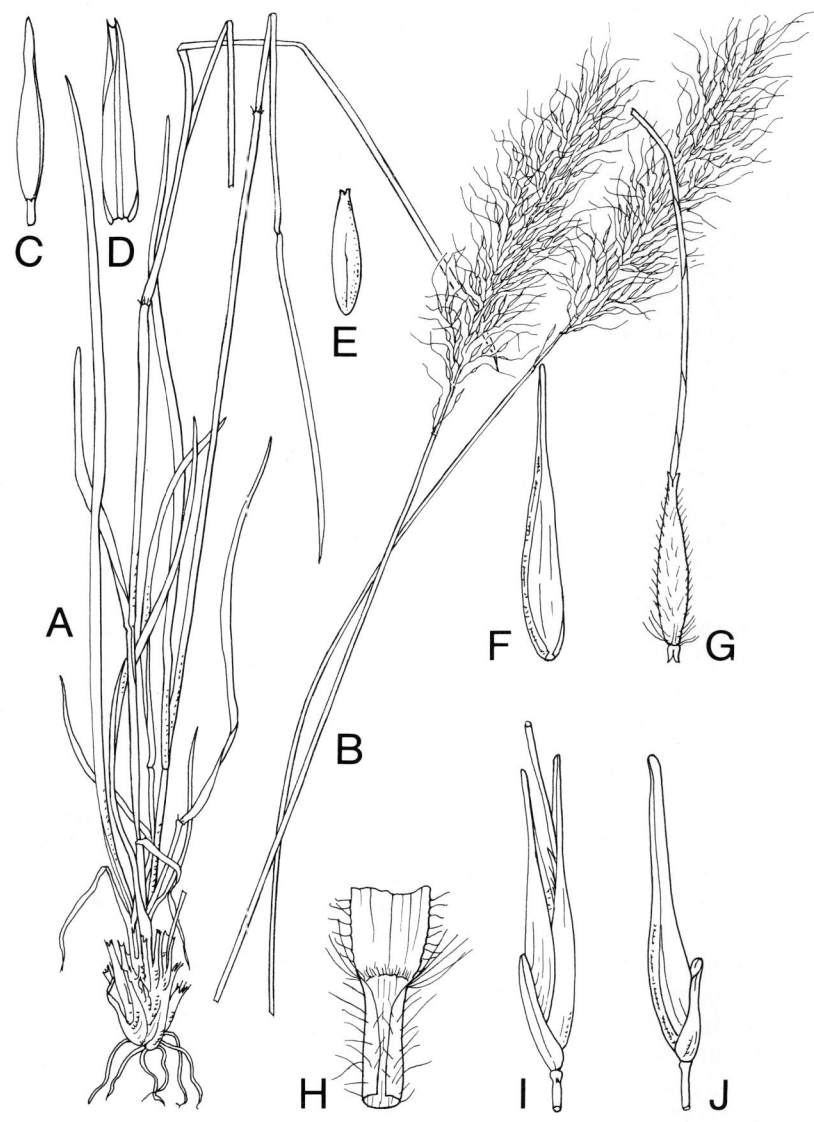

Figure 15.95. Loudetia simplex. **A**-Habit **B**-Inflorescence **C**-Lower palea **D**-Upper palea **E**-Caryopsis **F**-Lower lemma **G**-Upper lemma **H**-Ligule **I**-Spikelet **J**-Glumes

cent ether extract and 44.2 percent nitrogen-free extract in the dry matter of fresh material in early bloom in Kenya.
Palatability. It has low palatability (Verboom & Brunt, 1970).
Economics. Where the species is dominant, it is possibly an indication of veld mismanagement (Chippendall & Crook, 1976).
Further reading. Lamotte, 1979; Vesey-Fitzgerald, 1963.

Melinis minutiflora Beauv.

Common names. Molasses grass (Australia), gordura (South America), calinguero (Costa Rica), melado (Cuba), herbe à miel, Venezuela grass (India).

Natural habitat. Grassland, shady places and rocky slopes in subhumid and humid climates.

Distribution. Tropical and southern Africa and Brazil, introduced to many tropical countries as a fodder grass and now naturalized.

Description. Tufted perennial up to 150 cm high, often sticky, with a characteristic odour of molasses or cumin. Pubescent leaf-blades. Panicle 10-30 cm long with small glabrous spikelets 1.5 to nearly 2.5 mm long, awn 6-16 mm (Napper, 1965) (see Fig. 15.96).

Optimum temperature for growth. Ludlow and Wilson (1970b) found growth at 30°C was 1.36 times greater than at 20°C.

Minimum temperature for growth. Mean temperature of the coldest month ranges from 6.1-14.5°C (Russell & Webb, 1976).

Frost tolerance. It is sensitive to frost, and repeated heavy frost will kill it.

Latitudinal limits. 15.9-30.5° N and S (Russell & Webb, 1976).

Altitude range. 800-2 000 m.

Rainfall requirements. It needs moderate to high rainfall in excess of 750 mm. The normal range is 960 to 1 706 mm (Russell & Webb, 1976).

Drought tolerance. Relatively drought-hardy over a dry season of four to five months.

Tolerance to flooding. It does not tolerate flooding.

Soil requirements. It is tolerant to soils of fairly low fertility, high aluminium (Spain & Andrew, 1977) and light texture but will respond to more fertile soils. It does well in ashes left from a scrub burn, and on steep hillsides and road cuttings (see Plate 41). It needs good drainage.

Tolerance to salinity. It does not tolerate salinity.

Fertilizer requirements. As a pioneer species sown on the ashes of scrub burns, initial fertility may be high enough for establishment. The critical value of phosphorus as a percentage of the dry matter at the immediate pre-flowering stage is 0.18.

Ability to spread naturally. Molasses grass spreads quickly under favourable conditions.

Land preparation for establishment. It is usually established on burnt country to give a quick cover to suppress weeds. A rough cultivation will usually suffice if a burn is not obtainable.

Sowing methods. It is usually sown by seed, broadcast on a clean seed-bed and mixed with sawdust or rice hulls for even distribution. It can be undersown with cereal crops.

Sowing depth and cover. It is surface sown, with or without a light covering, and should be sown no deeper than 2.5 cm (Bogdan, 1964).
Sowing time and rate. It is best to sow just before the expected normal rainy season at 1.5 kg/ha or more.
Number of seeds per kg. Spikelets ("seed") 6-15 million.
Dormancy. There is no dormancy problem.

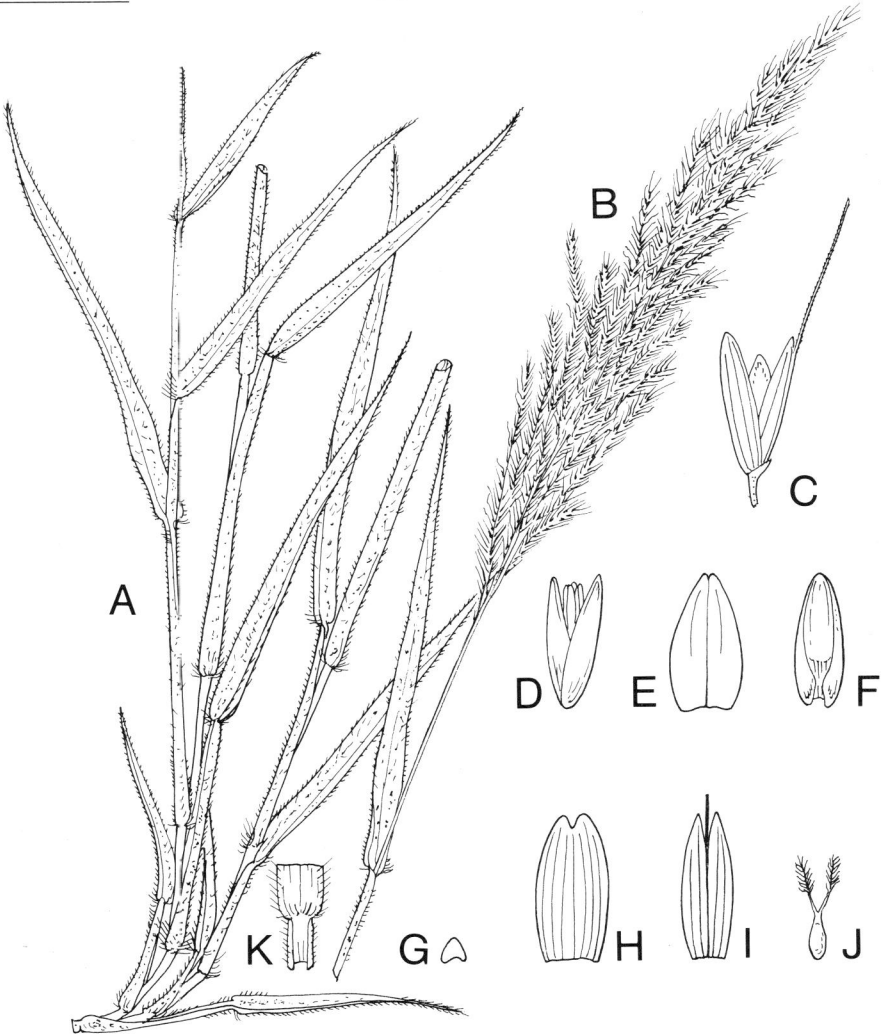

Figure 15.96. Melinis minutiflora. **A**-Habit **B**-Inflorescence **C**-Spikelet **D**-Upper floret **E**-Lower lemma **F**-Palea **G**-Lower glume **H**-Upper glume **I**-Upper lemma **J**-Carpel **K**-Ligule

Seed treatment before planting. The seed can be hammer-milled to improve germination and seed handling.

Seedling vigour. Excellent. It establishes quickly.

Vigour of growth and growth rhythm. It is a vigorous grass and is valuable as a pioneer species to suppress weeds and hold disturbed soil (ashes or a finely prepared cultivation) against erosion.

Response to photoperiod. It is a short-day plant.

Response to light. Molasses grass tolerates partial shade.

Compatibility with other grasses and legumes. Molasses grass usually dominates other grasses initially but it combines well with legumes, for example *Centrosema pubescens* in Brazil, *Neonotonia wightii* (glycine), *Macroptilium atropurpureum, Desmodium* spp., etc. An aqueous mixture of molasses grass, siratro seed and fertilizer is sprayed on newly established highway edges in Queensland, Australia, to effect quick stabilization. It is a transient grass and should not be the only species sown.

Ability to compete with weeds. Outstanding on newly burnt land in Laos (Thomas & Humphreys, 1970), and on roadsides in areas difficult to cultivate. In the Andes it is grown up to 2 000 m to suppress weed growth (Roseveare, 1948).

Tolerance to herbicides. It can be killed by spraying with 2.2-DPA at 2.3 kg of a 740 g AI/kg product (e.g. Dowpon) plus paraquat at 85 ml of a 200 g AI/litre product (e.g. Gramoxone) plus wetting agent at 250 ml per litre of water (Tilley, 1977).

Response to defoliation. It does not stand grazing below 15 cm because the crowns are well above the ground.

Grazing management. It should be well established before grazing and then grazed sparingly. Heavy stocking thins it out.

Response to fire. When mature it will burn so fiercely that its own seeds and roots are killed, leaving the land clear for future plantings such as Guinea grass and centro (Henty, 1969).

Genetics and reproduction. $2n=36$ (Fedorov, 1974). It is apomictic (Barnard, 1969).

Dry- and green-matter yields. In Colombia, dry-matter yields reach 6 000-8 000 kg/ha per year. This yield is doubled with 150 kg N/ha (Crowder, Chaverra & Lotero, 1970). In Fiji an average yield of 4 814 kg/ha of dry matter with a crude protein content of 6.8 percent was obtained over a three-year period (Roberts, 1970a, b). In Nigeria, annual dry-matter yield at Agege was 6 500 kg/ha (Adegbola, 1964).

Suitability for hay and silage. Medling (1972) made satisfactory silage in plastic bags when 10 percent molasses was added.

Value as standover or deferred feed. In São Paulo, Brazil, molasses grass spelled during a growing season could be cut in the winter without affecting root reserves or subsequent spring growth (da Rocha *et al.*, 1960).

Chemical analysis and digestibility. In Costa Rica analysis of material at floral

initiation revealed 8.97 percent crude protein, 25.20 percent crude fibre, 44.89 percent nitrogen-free extract, 3.61 percent ether extract and 7.33 percent ash in the dry matter on a 10 percent moisture basis (Gonzalez & Pacheco, 1970). Göhl (1975) lists analyses from Laos, Puerto Rico, India and Kenya.
Palatability. It is very palatable to stock.
Toxicity. No toxicity has been reported by Everist (1974).
Seed production and harvesting. It generally only produces seed in the lower latitudes.
Seed yield. It yields up to 280 kg/ha by hand harvesting. Jones (1973) records 134 kg/ha.
Minimum germination and quality required for commercial sale. 30 percent germinable seeds; 40 percent purity (Queensland). Germinate at 20-30°C moistened with water. Germination is increased by exposure to light.
Cultivars. No cultivars are registered in Australia. In Kenya there were differences between the ordinary cultivated form and local wild Kenya ecotypes which varied among themselves in many characters. Two of these ecotypes have been named 'Mbooni' and 'Chania River'. They form more even stands and are resistant to "small leaf" disease, but give lower seed yields (Bogdan, 1960). In Brazil four more-or-less distinct cultivated varieties are recognized: 'Roxo', 'Cabelo de Negro', 'Francana' and 'Branco'. 'Roxo' is the most widely used (Barnard, 1969).
Value for erosion control. Excellent in high-rainfall areas and as a temporary cover in subtropical areas of lower rainfall.
Diseases. A "small leaf" disease is present in Kenya.
Pests. There are no major pests.
Economics. It is an important pioneer grazing species to give cover on newly cleared land. In Zaire the indigenous people claim it has insect-repellent properties and use it as bedding for sitting fowls and bitches about to give birth. In Manipur, India, it is believed mosquitoes avoid it, possibly both the odour and viscid hairs being repellent (Bor, 1960).
Animal production. Using upgraded San Martinero cattle, daily gains of 0.48 percent per head were obtained in Colombia with a stocking rate of one animal per hectare (Crowder, Chaverra & Lotero, 1970).
Main attributes. Its quick establishment and ground cover which suppresses weeds, and, when used as a pioneer plant, its inflammability at maturity, paving a way for establishment of more productive pastures.
Main deficiencies. Its susceptibility to fire. It should not be sown as the sole grass species in an area, as it is transient.
Further reading. Bogdan, 1960; da Rocha *et al.,* 1960.

Oryza sativa L.

Common name. Rice.
Natural habitat. Swampy areas.
Distribution. Throughout the tropical world as a crop.
Description. An annual grass with erect culms 0.6-2 m tall usually with four to five tillers. Inflorescence a loose terminal panicle of perfect flowers; each panicle branch bearing a number of spikelets, each with a single floret. Each flower is surrounded by a lemma and palea at the base of which are two small glumes. The lemmas may be awnless or variously awned. The rice grain enclosed by the lemma and palea (hull) varies in size, texture and colour. Each panicle holds 100-150 seeds (see Fig. 15.97).
Season of growth. Summer.
Optimum temperature for growth. A mean temperature above 21°C.
Frost tolerance. It will not stand frosts.
Rainfall requirements. Rain-grown rice usually requires an annual rainfall in excess of 1 500 mm. Most of the world's rice is grown under irrigation and water is supplied as required. The amount may be in the vicinity of 7-9 million litres per hectare.
Drought tolerance. It has little drought tolerance.
Tolerance to flooding. The *japonica* or paddy varieties are aquatic and grow in water at various depths. Deep water varieties will grow in up to 6 m of water. Upland or *indica* varieties are not adapted to flooding.
Soil requirements. Upland varieties need fairly good drainage but *japonica* varieties need a heavy, relatively impervious soil. The pH is generally not important, as rice has the capacity to neutralize the soil on which it is growing.
Tolerance to salinity. Its extensive root system not only loosens the soil and renders it more permeable, to facilitate leaching, but also creates an acidic environment which reduces pH values in alkaline soils to neutrality. The improved cultivars 'IR-8' 'IR-8-68' and 'Jaya' exhibit high tolerance to salinity as do the tall *indica* cultivars 'Jhona 349', 'Damodor' and 'MCM' (Yadav, 1975).
Fertilizer requirements. In paddy rice, land preparation for planting usually involves some incorporation of organic matter, either from a previous grass/legume pasture, green manure crop or from plants cut and transported to the field. A basic phosphorus and potash dressing may be required but nitrogen fertilization is a main determinant of yield. Half the nitrogen may be applied at transplanting with the remainder at ear initiation, and the application may be 150-250 kg/ha.
Land preparation for planting. Upland rice and large-scale paddy cultivation where the seed is drilled prior to flooding need a well-prepared seed-bed, as for wheat. Paddy rice fields are usually ploughed and harrowed several times

in the wet state as a "mudding-up" operation which incorporates organic matter and levels the land.
Sowing methods. The seed may be drilled into dry land or sown in nurseries and the seedlings later transplanted into a wet paddy-field. Also, seed may be pre-germinated and broadcast into the mud in the paddy-field.
Number of seeds per kg. About 40 000.

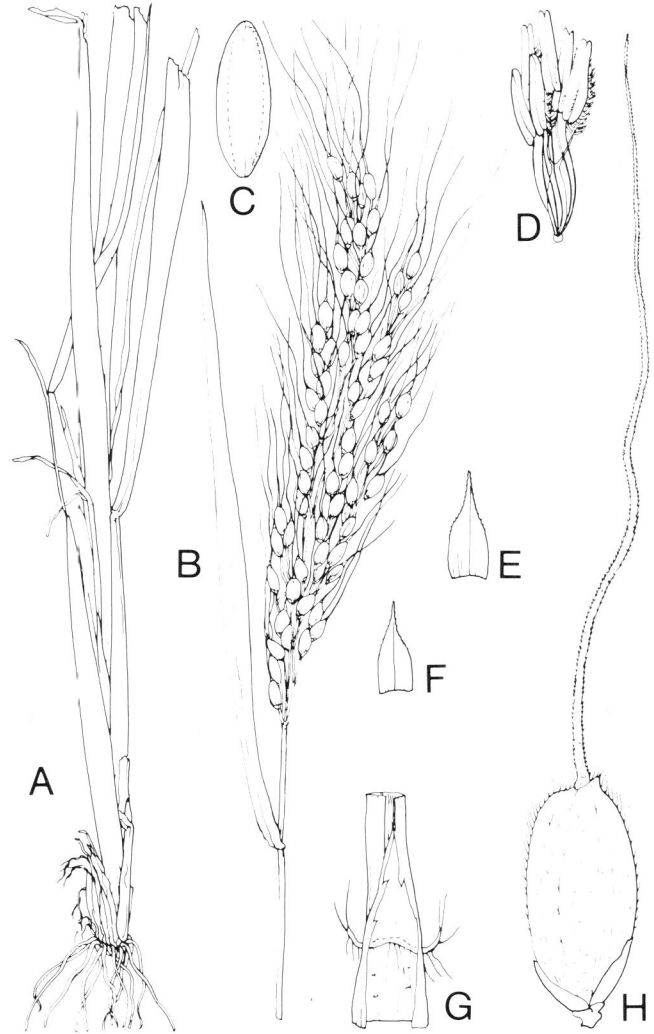

Figure 15.97. Oryza sativa. **A**-Habit **B**-Inflorescence **C**-Grain **D**-Flower **E,F**-Lemmas **G**-Ligule **H**-Spikelet

Dormancy. The seeds of *indica* (upland) varieties have post-harvest dormancy for up to three months, which can be broken by storage at 8°C under very moist conditions immediately after harvest. The seeds of *japonica* varieties have no dormancy period.

Response to photoperiod. It flowers over short and medium day lengths (Evans, Wardlaw & Williams, 1964).

Tolerance to herbicides. Rice tolerates the herbicides used to control its main grassy weed pest, *Echinochloa crus-galli* (barnyard or water grass). It will tolerate the use of 2,4-D amine at 1-1.5 kg AI/ha and 2,4,5-T amine at 1-1.5 kg AI/ha at the late tillering stage before panicle initiation for the control of broad-leaved weeds.

Genetics and reproduction. 2n=24 (Fedorov, 1974). Some tetraploid (2n=48) exist, but the diploid are more numerous.

Dry- and green-matter yields. Yields of paddy straw after harvest varies considerably. A yield of 5 000-15 000 kg/ha may be expected.

Suitability for hay and silage. Practically all the paddy straw from rice crops in the tropics is conserved as hay for animal feeding and is usually stacked around poles in the house compound. Medling, (1972) under high rainfall conditions (3 997 mm) at Gualaca, Panama, found that baling the straw with large roller balers was effective and that the straw bales would stay in the field

TABLE 15.49 *Oryza sativa*

	DM	As % of dry matter				
		CP	CF	Ash	EE	NFE
Fresh, vegetative, India		7.0	25.9	18.0	1.8	47.3
Fresh, dough stage, India		5.8	29.5	18.3	2.2	44.2
Fresh, regrowth after harvest, Trinidad	32.8	9.0	28.3	15.9	1.8	45.0
Hay, India	85.0	8.2	32.0	15.7	1.8	42.3
Straw, Philippines	80.8	3.9	33.5	21.4	2.1	39.1
Straw, India	93.8	2.4	36.6	16.5	0.9	43.7
Silage of straw, India		5.9	30.0	11.4	1.7	51.0

	Animal	Digestibility (%)				
		CP	CF	EE	NFE	ME
Fresh, regrowth	Sheep	58.0	63.0	35.3	69.4	2.05
Hay	Oxen	50.0	74.0	22.0	55.0	1.90
Straw	Cattle	9.8	60.6	24.8	39.1	1.34
Straw	Zebu	0.0	60.7	46.7	42.4	1.50

Source: Göhl, 1975

without major deterioration, while grass hay (*Hyparrhenia rufa* and other grasses) suffered serious deterioration.

Chemical analysis and digestibility. Göhl (1975) records analyses of vegetative material — see Table 15.49.

Palatability. The hay is fairly palatable, but not nutritious enough for maintenance.

Toxicity. Excess feeding of rice straw leads to toxicity because of high oxalates, which will bind the calcium in the diet. This effect can be reduced by soaking the straw in water or neutralizing it with a weak solution of calcium carbonate or calcium hydroxide (Göhl, 1975).

Seed production and harvesting. The rice is either hand harvested and later threshed or mechanically harvested with combines and hulled and cleaned.

Seed yield. Japonica varieties, 4 000-4 500 kg/ha; *indica,* 1 000-1 500 kg/ha of milled rice.

Minimum germination and quality required for commercial sale. 70 percent germinable seeds, 98.8 percent purity in Queensland.

Cultivars. There are numerous cultivars, each suited to a particular ecological niche.

Diseases. Numerous diseases occur, affecting the leaves for use as straw for feeding animals. The more important ones are blast, caused by *Piricularia oryzae,* leading to brown and shrivelled leaves; brown spot, caused by *Cochliobolus miyabeanus;* and narrow brown leaf spot, caused by *Cercospora oryzae.*

Pests. Army-worms and grasshoppers cause major leaf damage, and stem borers affect the stalk.

Economics. Rice is one of the world's two major human food crops, the other being wheat. Rice straw is retained for feeding draught animals in most rice-producing countries in the tropics. Rice bran is fed to domestic animals when not required for human consumption in dry years. Paddy straw provides 80 percent of the organized roughage for India and a large part of the roughage for animals in other rice-producing countries where draught animals are used.

Animal production. The inclusion of urea and molasses improved the voluntary intake of paddy straw from 1.5 to 2.1 kg/100 kg live weight and dry-matter intake was 1.9-1.98 kg/100 kg live weight. Molasses-urea also increased digestibility coefficients of crude fibre and crude protein. This supplement increased body weight from 3.5 to 5.5 kg per head in 100 days. Paddy straw plus 1 percent urea and 18-20 percent molasses on a dry-matter basis is maintenance (Singh & Kushwaha, 1974). Sunn hemp (*Crotalaria juncea*) hay, with 12.7 percent digestible crude protein and 65 percent total digestible nutrients, provided maintenance for bullocks as a supplement to paddy straw in the proportions of 1:3 and 1:2. The bullocks consumed 1.95-2.11 kg DM/50 kg live weight (Reddy & Murty, 1962).

Further reading. Chandraratna, 1963; Finlay, 1975.

Panicum antidotale Retz.

Common names. Blue panic (Australia, United States), giant panic (Australia), perennial Sudan grass (United States), bansi (India).
Natural habitat. Sand dunes and dry river beds in northwest Pakistan, Afghanistan and Iran.
Distribution. Native to India, now introduced to many countries as a pasture grass. Cultivated in some countries.
Description. Tufted perennial up to 150 cm high, glabrous but with woolly bud scales at the base. Panicle up to 30 cm long, dense, with 3 mm-long acute spikelets, the lower glume half as long as the spikelet. Glumes with broad, membranous margins (Napper, 1965). It has short, thick, bulbous rhizomes and deep roots and blue-green leaves (see Plate 42).
Season of growth. Summer.
Frost tolerance. Susceptible to frost damage, but will retain some greenness in mild winters.
Latitudinal limits. About 35°N.
Rainfall requirements. It is best adapted to areas of summer rainfall where annual precipitation is 500-750 mm, or irrigated land. It can grow in areas with less than 130 mm of rain in Rajasthan, India.
Drought tolerance. It has a high degree of drought tolerance, but will respond readily to summer storms. It has a very deep root rystem.
Tolerance to flooding. It will tolerate temporary flooding.
Soil requirements. Grows best on fertile soils and is more demanding than buffel grass. It prefers heavy loams or dark clay soils high in lime and does not do well on sandy soils that are acid or low in organic matter (Trew, 1954).
Tolerance to salinity. It has some tolerance to salinity but more to alkalinity caused by sodium and magnesium than to the chlorides (Ryan, Miyamoto & Stroehlein, 1975).
Fertilizer requirements. Responds markedly to nitrogen, but a basic NPK fertilizer may be needed according to soil composition. At Jodhpur, India, application of 30 kg N + 30 kg P_2O_5 + 20 kg K/ha increased yields by 273 percent (Singh & Chatterjee, 1968).
Ability to spread naturally. There is a very slow spread from fallen seed.
Land preparation for establishment. A fine seed-bed is preferable, either from mechanical preparation by ploughing or harrowing, or by scrub burning and sowing in the ashes.
Sowing methods. It is propagated by seed, either drilled in rows or broadcast. In India it is sown in rows 45 cm apart.
Sowing depth and cover. When drilled, cover no more than 1 cm; when broadcast, sow on surface and, if possible, give a light cover. In ashes the burial

from broadcasting is sufficient. In the Sudan it germinates from a depth of 5 cm (Abd-El-Rahman & El-Monayeri, 1967).
Sowing time and rate. Sow just before the expected rainy season at 6-7 kg/ha broadcast, or 1.25 kg/ha in rows 1 m apart.
Number of seeds per kg. 1 299 000 (Queensland); 1 445 000 (United States).
Dormancy. There is some post-harvest dormancy for a maximum of two years' germination of 80 percent at five to eight years, declining to 25 percent at 11 years and 3 percent at 13 years (Myers, 1940).
Seedling vigour. The seed germinates well, but plant development is slow for the first six to eight weeks (Barnard, 1969).
Vigour of growth and growth rhythm. It makes strong growth earlier in spring than buffel grass and continues strong through the summer, but becomes woody at maturity.
Response to light. It will grow in partial shade around buildings, but prefers full sunlight.
Tolerance to herbicides. It can be sprayed with 2,4-D to control broad-leaved weeds.
Response to defoliation. It cannot withstand heavy, close grazing. It needs to be utilized before running to seed, as the flowering stalks become hard and woody. Cutting at 20-day intervals to 10 cm in a wet year, and at 30-day intervals to a height of 15 cm in a normal year, gave highest yields in Rajasthan, India (Dabadghao, Roy & Marwaha, 1973).
Grazing management. Blue panic is ready to graze when well established, and needs heavy intermittent grazing to keep it at a nutritious stage. It requires 25-30 cm of stubble left after cutting or grazing. Stems rapidly become hard and woody and should be grazed or cut before flowering. The grass often grows too fast for the cattle; surplus should be made into hay. The stemmy material left at grazing should be removed by mowing or slashing to allow fresh growth to arise from the base. About 35 kg/ha of nitrogen can be applied after every grazing.
Response to fire. It is resistant to fire.
Genetics and reproduction. 2n=18, 36 (Fedorov, 1974). Reproduction is sexual.
Dry- and green-matter yields. In Gujarat, India, 4 733 kg green matter per hectare were harvested (Srinivasan, Bonde & Tejwani, 1962). From 2 500 to 6 000 kg/ha of hay can be anticipated. Under irrigation in Iowa, United States, it yielded 4 780 kg/ha per year over three years (Trew, 1954).
Suitability for hay and silage. It makes good hay and fair silage if cut at the flowering or milk stage (Trew, 1954).
Value as standover or deferred feed. The coarse woody stem is of little value, but a few leafy side shoots can provide a little grazing.
Chemical analysis and digestibility. Göhl (1975) records fresh early pasture with 18.8 percent crude protein in the dry matter, and fresh mature material with as low as 8.4 percent crude protein in Pakistan. In India figures of

7.3 percent crude protein, 40.5 percent crude fibre, 7.9 percent ash, 1.2 percent ether extract and 43.1 percent nitrogen-free extract were recorded (Sen & Ray, 1964).

Palatability. Young growth up to flowering is extremely palatable, but it should not be allowed to become too coarse.

Toxicity. Reported to have a high oxalate content (over 4 percent) in Queensland (Mathews & Sutherland, 1952).

Seed harvesting methods. The seed ripens unevenly and a lot of it shatters. To obtain more seed, it can be cut with a reaper and binder before shattering and subsequently threshed. Direct heading will give a low yield. It is hand harvested in India.

Seed yield. Rain grown yields reach 100-160 kg/ha, irrigated 250-600 kg/ha.

Minimum germination and quality required for commercial sale. 50 percent germinable seeds, 80 percent purity (Queensland). Germinate at 20-30°C in water.

Cultivars. There are several varieties recognized in the United States. Cultivar A-130 was derived from seed introduced from Australia. Strain 341 is recommended for use under arid conditions in western Rajasthan, India (Prasad & Singh, 1973).

Value for erosion control. It is used extensively for erosion control in the flood plains of the United States, mainly to protect against wind erosion, and is sown in rows at right angles to the prevailing wind. It is not very effective for control of water erosion (Srinivasan, Bonde & Tejwani, 1962).

Diseases. It has no major diseases.

Pests. There are no major pests.

Economics. It is sometimes used in native medicine, which probably prompted the specific name "antidotale". A useful summer forage.

Animal production. It has given good cattle production in burnt *Acacia cambagei* country along with *Cenchrus ciliaris* in Queensland in an annual rainfall regime of 425-500 mm with summer dominance.

Main attributes. Its palatability, deep-rootedness and drought tolerance.

Main deficiencies. The woody stems of mature plants and its uneven seed setting and seed shattering.

Further reading. Trew, 1954.

Panicum coloratum L.

Common names. Coloured Guinea grass (Kenya), small panicum (southern Africa), small buffalo grass (Zimbabwe), Klein grass (United States), keria grass (Kenya).
Natural habitat. Grassland and open woodland on heavy clay soils.
Distribution. Throughout tropical Africa, introduced to the United States and Australia.
Description. A very variable perennial, from 8-9 to over 100 cm high. Panicle 6-25 cm long with obtuse or subacute green and purple spikelets 2.5-3 mm long with a small, rounded or abruptly acuminate lower glume.
Season of growth. Summer.
Optimum temperature for growth. The mean annual temperature ranges from 17.7 to 21.7°C (Russell & Webb, 1976).
Minimum temperature for growth. The mean temperature for the coldest month ranges from 5.8-11.8°C (Russell & Webb, 1976). Cv. Kabulabula has good winter growth (Roe, 1972).
Frost tolerance. It is susceptible to frost but usually recovers.
Latitudinal limits. 13.5-30.3°N and S (Russell & Webb, 1976).
Altitude range. Sea-level to 1 000-2 000 m.
Rainfall requirements. It grows in areas with a rainfall in excess of 500 mm. The range is 650-1 700 mm (Russell & Webb, 1976).
Soil requirements. It occurs chiefly on red and black clay soils in Kenya.
Tolerance to salinity. An introduction from eastern Africa, CPI 14375, close to cv. Kabulabula, tolerated moderate salinity (Evans, 1967b).
Fertilizer requirements. *P. coloratum* was found to fix 23 kg N/ha in 100 days in southern Texas (Wright, Weaver & Holt, 1976). It responds well to nitrogen.
Land preparation for establishment. It requires a well prepared seed-bed.
Sowing methods. Drilling or broadcasting.
Sowing depth and cover. The best sowing depth is about 1 cm but *P. coloratum* germinated from a depth of 5 cm in the Sudan (Abd-El-Rahman & El-Monayeri, 1967).
Sowing time and rate. Sow in the wet season at 11-16 kg/ha (2-3 kg/ha in Texas).
Dormancy. In cv. Kabulabula, the tight envelopment of the caryopsis by the lemma and palea delays germination. This can be overcome by a light mechanical scarification (Strickland, 1978) but scarified seed will not remain viable after laboratory storage of three years whereas unscarified seed stored for three years remained viable and there was still some hardseededness (Roe & Williams, 1969).
Seedling vigour. In the Sudan, *P. coloratum* germinated after only 10 mm of rain, when planted at 1 cm (Abd-El-Rahman & El-Monayeri, 1967).

Genetics and reproduction. 2n=18, 32, 36, 44, 54 (Fedorov, 1974). It cross-pollinates, with some lines incompatible (Hutchinson & Bashaw, 1964).

Dry- and green-matter yields. In the wet zone of Vita Levu, Fiji, Roberts (1970a, b) recorded an average of 4 517 kg/ha of dry matter with 9 percent crude protein over three years.

Seed production and harvesting. Seed is matured over a long period, well in excess of 15 days with no peak maturation, and at the end of the period virtually all the seed has been shed. Direct heading yielded only 19 percent of the seed; cutting with a reaper and binder, drying under cover and subsequently threshing gave 42 percent; cutting with reaping hook, drying in the field and threshing gave 49 percent; and collecting seed several times by hand shaking gave 62 percent of possible yield (Roe, 1972).

Seed yield. About 400 kg/ha if most of the seed is collected.

Minimum germination and quality required for commercial sale. 20 percent germinable seed and 80 percent purity in Queensland.

Cultivars.
- 'Bushman Mine' — a tufted, erect perennial up to 1 m high that spreads by long creeping stems. It is deep rooting and very drought resistant, yet it will grow in heavy, seasonally waterlogged soils. It will thus tolerate a wide range of soil conditions and is suitable for use in low or high rainfall areas. It is palatable to cattle, makes good hay and responds well to nitrogen. It mixes well with legumes. Seed production is generally poor and establishment is usually by root-stock or stem cuttings. It was developed in Botswana from indigenous stock (Thorp, 1979).
- 'Kabulabula' — introduced as CPI 16796, has good winter growth and produced 5 810 kg DM/ha (1 852 kg leaf) in autumn (14 March - 16 July) in southeastern Queensland when fertilized with 30 kg P, 75 kg K and 300 kg N/ha. Spring growth was 2 181 kg/ha from 15 August to 30 October. Recovery of nitrogen was 53.9 percent (Ostrowski & Fay, 1979).
- 'Selection 75' — used in Texas.

Further reading. Roe, 1972; Roe & Williams, 1969.

Panicum coloratum L. var. *makarikariense* Goossens

Common names. Makarikari grass (Australia), Makarikari panicum (southern Africa).
Natural habitat. Occurs in warm, dry bushveld in Africa.
Distribution. Collected from the Makarikari pan in Botswana. Introduced widely.
Description. Cultivar Bambatsi is an erect, tussocky perennial, shortly rhizomatous, seldom stoloniferous. Culms robust, glaucous, branching and erect to a height of 1.5-1.8 m; nodes geniculate, slightly enlarged, leaf-blade 46 cm long and 13 mm wide in the prominent wide opaque midrib. Auricle absent. Inflorescence a large, open, nodding panicle 25-33 cm long, rachis grooved and partly flattened. Spikelets 2 mm long. Seed: the lemma and palea closely invest the caryopsis; the "seed" is ovoid, 2.25 mm long, smooth, shiny, grey-black. About 95 percent of the population have the erect habit, about 5 percent have the more spreading and stoloniferous habit described for cv. Pollock (Barnard, 1972) (see Fig. 15.98).
Season of growth. Summer.
Optimum temperature for growth. About 35°C.
Minimum temperature for growth. It makes no growth during the winter.
Frost tolerance. Recovers better than green panic after winter. Cultivar Pollock is quite frost tolerant, with 85 percent survival after the first winter on the Darling Downs (Jones, 1969), but cv. Kabulabula is very susceptible to frost.
Latitudinal limits. 13.5-30.3°N and S (Russell & Webb, 1976).
Altitude range. Sea-level to 2 000 m.
Rainfall requirements. It fits into a 500-1 000 mm rainfall belt with a dominant summer incidence.
Drought tolerance. A reasonable degree of drought tolerance (Bott, 1978). In its centre of origin it exists on flood plains receiving as little as 375 mm/year.
Tolerance to flooding. Stands waterlogged conditions extremely well (Bott, 1978) (see Fig. 15.99).
Soil requirements. Adapted to self-mulching, high-fertility, black clay soils where poor aeration conditions are common (Lloyd, 1970).
Tolerance to salinity. It is one of the better grasses to vegetate somewhat saline areas.
Fertilizer requirements. In addition to basic phosphorus and potash where required, it responds well to increasing nitrogen up to 900 kg/ha per year with a 20-30 percent recovery in the tops (Lloyd, 1970).
Ability to spread naturally. It will spread slowly from shattered seed and by stolons.
Land preparation for establishment. Prepare a good seed-bed with a 5 cm mulch if possible (Lloyd & Scateni, 1968).

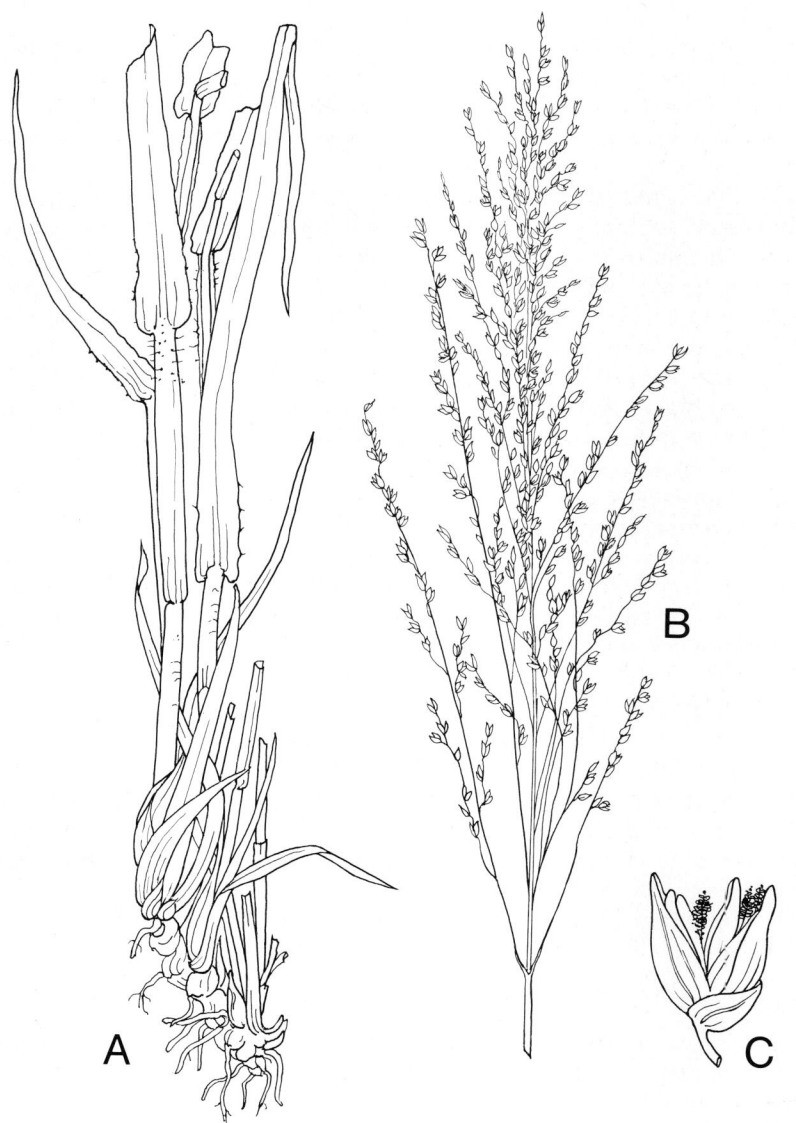

Figure 15.98. *Panicum coloratum* var. *makarikariense*.   **A**-Habit   **B**-Inflorescence
C-Spikelet

Figure 15.99. Panicum coloratum var. *makarikariense* cv. Bambatsi showing good growth one month after a flood (**Source:** E.R. Anderson, Queensland Department of Primary Industries)

Sowing methods. Drilling on the contours in small drill furrows at 1.5 cm with fluted roller-press wheels following gives excellent stands (Wilson, 1978). Seed may be broadcast and rolled in afterwards. Sow in rows 90 cm apart for inter-row cultivation or 30-45 cm for irrigation.

Sowing depth and cover. Surface, to no deeper than 2 cm (Bogdan, 1964; Lloyd & Scateni, 1968).

Sowing time and rate. Early or late wet season at 2-4 kg/ha. Midsummer seeding encourages too much weed competition.

Number of seeds per kg. 962 000 ('Bambatsi', Queensland).

Dormancy. The seed shows initial dormancy. The seed requires a ripening period of six months after harvest.

Seedling vigour. The seedlings have poor competitive ability (Bott, 1978) but improve later (Lloyd, 1970).

Vigour of growth and growth rhythm. It grows rapidly during late spring and summer but is dormant in winter (Lloyd, 1970).

Response to light. It prefers full sunlight.

Compatibility with other grasses and legumes. It will combine well with lucerne (*Medicago sativa*).

Tolerance to herbicides. *P. coloratum* showed good tolerance to atrazine when used as a pre-emergence and post-emergence spray on black clay soils on the Darling Downs, southeastern Queensland — up to 4 kg/ha (Scateni, 1978).

Ability to compete with weeds. Weed competition may be a problem early in its life because of its slow establishment (Lloyd & Scateni, 1968).
Response to defoliation. It withstands heavy grazing when established, but graze lightly for the first six months down to 7.5 cm when the first flower-head appears (Lloyd & Scateni, 1968) to encourage tiller development.
Grazing management. It should be grazed lightly in its first year, but when established can withstand heavy stocking. Spell during the summer and autumn if possible to preserve green leaf for the winter.
Response to fire. It will survive annual fires.
Genetics and reproduction. It is cross-pollinated, with some lines being incompatible (Hutchinson & Bashaw, 1964).
Dry- and green-matter yields. Under experimental conditions it produces over 20 000 kg/ha per year with a dressing of 650 kg N/ha per year (Lloyd, 1970).
Suitability for hay and silage. It makes useful hay in southern Africa with 9 percent crude protein (59 percent digestible) and 60 percent total digestible nutrients in the dry matter (Göhl, 1975). Medling (1972) made good silage in plastic bags in Panama when 10 percent molasses was added.
Value as standover or deferred feed. It is excellent, as it bears green leaf throughout the winter.
Chemical analysis and digestibility. Fresh, early bloom material contained 18.9 percent crude protein, 28.6 percent crude fibre, 11.0 percent ash, 2.6 percent ether extract and 38.9 percent nitrogen-free extract in the dry matter (Dougall & Bogdan, 1958).
Palatability. It is very palatable.
Toxicity. No toxicity has been reported by Everist (1974).
Seed production and harvesting. *P. coloratum* ripens over a long period, from the top to the bottom of the seed-head, hence seed harvesting should be by repeated beater or stripper harvesting rather than direct heading (Roe, 1972). It is desirable to have a mixture of lines to ensure good seed setting (Humphreys, 1975). Harvest when one-third of the seed has shattered.
Seed yield. Roe (1972) recorded 410 kg/ha by collecting shattered seed, 123 kg/ha by direct heading. Let the seed sweat in a 15-25 cm heap for two days to ripen more seed, then dry seed thoroughly.
Minimum germination and quality required for commercial sale. 80 percent purity and 20 percent germination in Queensland. Germinate at 20-35°C, moistened with water. Scarify seed.
Cultivars.
- 'Bambatsi' — collected at Bambatsi Lake on the Marczamnyana or Nata River in southern Zimbabwe. It is a dark-seeded, erect form that can be distinguished from most other such forms by its superior seed set. The seed ripens unevenly and shatters readily. Tolerates flooding. It is slow to establish. It is the most frost-tolerant cultivar.
- 'Pollock' — derived from seed from the Department of Agriculture in

South Africa. It differs from 'Bambatsi' in habit and growth. It is a leafy, ascending type, stoloniferous from the lowermost three to four nodes; strongly tussocky. Leaves smaller than 'Bambatsi', with leaf-blade 30-28 cm long, 9 mm wide. Inflorescence denser than 'Bambatsi', but still with lowermost branch solitary. In spaced swards it develops crowns 90-180 cm in diameter, useful where soil conservation is needed and waterlogging occurs. It is palatable, more frost-tolerant than 'Bambatsi' if grazed heavily in the autumn. It ripens unevenly, as does 'Bambatsi', and shatters heavily. Seed yield is only about half that of 'Bambatsi'.

- 'Burnett' — derived from seed from Botswana in 1954. A tall, tussocky, semi-erect type with greater ability to spread from the lower nodes than 'Bambatsi'. In this respect it is intermediate between 'Bambatsi' and Pollock, having about half the characteristics of each of the others. Leaves large like 'Bambatsi'. Seed dark like 'Bambatsi', has shown some frost tolerance, winter growth is slow but new tillers are produced then, if moisture is available. Its seed also ripens unevenly and shatters badly (Barnard, 1972).
- 'Bushman Mine' — selected at Henderson Research Station near Harare, Zimbabwe. Is drought tolerant (Chippendall & Crook, 1976).
- 'Prinshof 11/12' — like 'Bambatsi' but produces little seed.
- 'Thilo Creeping Panicum' — produces runners rooting at the nodes.
- 'Prinshof 14/12' — similar to 'Thilo'. These latter three produce little seed (Whyte, Moir & Cooper, 1959).
- *Panicum coloratum* var. *kabulabula* Codd — introduced as 'CPI 16797' to Australia. At Westwood in the Fitzroy Basin, central subcoastal Queensland, *P. coloratum* var. *kabulabula* established well on the alluvial soils and exhibited excellent seedling vigour. It also grew well with the legume *Macroptilium atropurpureum* on a red-brown prairie-like (clay 27-41 percent) ridge soil (Hall, 1970).

Value for erosion control. Cultivar Pollock, because of its large crown development, is useful in erosion control. Sown on terraces at Machakos, Kenya, *P. coloratum* var. *makarikariense* did not prevent erosion (Thomas, 1975).

Pests. It has no serious pests.

Animal production. It is used increasingly for leys in Africa and Australia. On the Darling Downs black clay soils of southeast Queensland, grazing *P. coloratum* var. *makarikariense* cultivars at 7.5 sheep per hectare produced more than 15.5 kg wool per hectare and live-weight gains averaged 29 percent.

Main attributes. The ability of the cultivars to grow on heavy, self-mulching, black clay soils.

Main deficiencies. Uneven seed set and seed shattering, and lack of winter production.

Further reading. Lloyd, 1970, 1971; Lloyd & Scateni, 1968.

Panicum maximum Jacq.

Common names. Guinea grass (Australia, United States), zaina, pasto Guinea (Peru), gramalote (Puerto Rico).

Natural habitat. Grassland and open woodland and shady places.

Distribution. From tropical Africa, but introduced in many countries.

Description. A tufted perennial, often with a shortly creeping rhizome, variable 60-200 cm high, leaf-blades up to 35 mm wide tapering to fine point; panicle 12-40 cm long, open spikelets 3-3.5 mm long, obtuse, mostly purple red, glumes unequal, the lower one being one-third to one-fourth as long as the spikelet, lower floret usually male (Chippendall, 1955). Upper floret (seed) distinctly transversely wrinkled (see Figs. 15.100, 15.101).

Season of growth. Summer.

Optimum temperature for growth. The mean range is 19.1-22.9°C (Russell & Webb, 1976).

Minimum temperature for growth. Mean temperature for the coldest month ranges from 5.4-14.2°C (Russell & Webb, 1976).

Frost tolerance. It will not tolerate heavy frosts, but recovers from light frosts with the return of warm weather.

Latitudinal limits. 16.3-28.7°N and S (Russell & Webb, 1976).

Altitude range. Sea-level to 2 500 m.

Rainfall requirements. It requires a rainfall usually in excess of 1 000 mm per year. With a summer dominance, cv. Gatton and creeping Guinea do not tolerate very wet conditions. Range 780-1 797 mm (Russell & Webb, 1976).

Drought tolerance. It does not tolerate severe drought. On an oxisol at Carimagua, Colombia, it dried the profile to a depth of 60 cm in the dry season, where *Andropogon gayanus* dried it to over 120 cm depth (CIAT, 1978).

Tolerance to flooding. It does not tolerate waterlogging.

Soil requirements. It will grow on a large range of soils, but produces poor stands on infertile types. It is well adapted to sloping, cleared land in rain forest areas where it will support heavy stocking. It will tolerate acid conditions if drainage is good. On an ultisol at Quilichaco, Colombia, *P. maximum* gave its maximum yield at 70 kg P_2O_5/ha per year and on an oxisol at Carimagua, Colombia, maximum yields were obtained at 100 kg P_2O_5/ha (CIAT, 1978).

Tolerance to salinity. It has little tolerance.

Fertilizer requirements. The optimum content of phosphorus in the dry matter was determined by Falade (1975) as 0.185 percent. Inoculation with *Spirillum lipoferum* increased yield by 480 kg DM/ha without nitrogen and 1 021, 1 690 and 1 930 kg/ha with 20, 40 and 80 kg N/ha, respectively (Quesenberry et al., 1976). Phosphorus at 24 kg/ha and nitrogen at 137 kg/ha are required in north Queensland, but soil fertilizer experiments are required to diagnose needs on

various soils. Hendrick concluded that at nitrogen levels above 45 kg/ha, phosphorus and potassium may become limiting to *P. maximum* in western Nigeria (Ademosun, 1973). It tolerates high aluminium (Spain, 1979).
Ability to spread naturally. It spreads slowly by seed, but needs fertile soil.
Land preparation for establishment. Full seed-bed preparation is generally required for Guinea grass establishment.

Figure 15.100. Panicum maximum. **A**-Habit **B**-Inflorescence **C**-Spikelet **D**-Lower glume **E**-Upper glume **F**-Upper lemma

Sowing methods. Drilling on the contour in small drill furrows and pressing in with press wheels (Wilson, 1978) gives an excellent stand. Sowing sods at intervals of 0.6 m in rows 1.25 m apart is successful but laborious. In Sri Lanka, it has been found that close planting of *P. maximum* cuttings (with a spacing of 15 × 45 cm) increases yield. Transplanting of *P. maximum* seedlings is more reliable than that of root cuttings, especially if they have recently

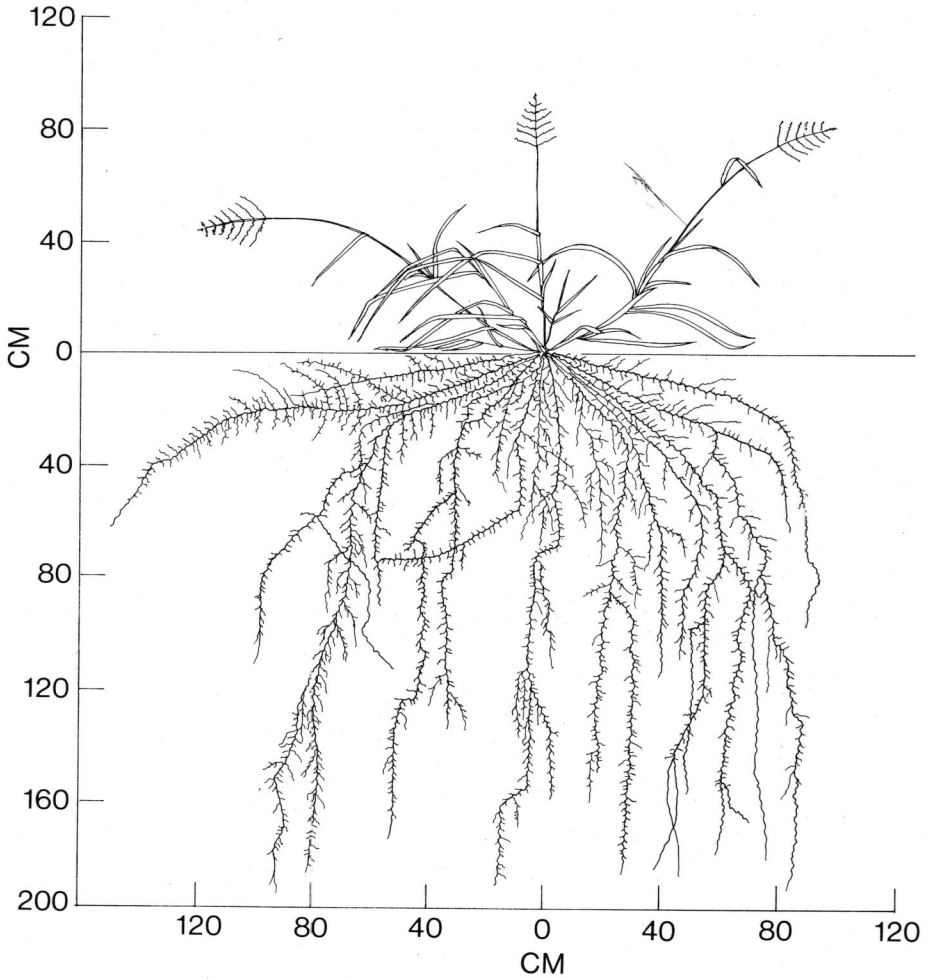

Figure 15.101. Vertical section showing the root system of *Panicum maximum* seven months after planting

started to show new growth after rain. In Puerto Rico it is also generally sown by clumps of roots (Vicente-Chandler *et al.*, 1953). One hectare will provide material for five hectares of planting.

Sowing depth and cover. Sowing depth should be no more than 1.5 cm. Rickert (1970) has shown better germination by using a straw mulch at 8 000-10 000 kg/ha to cover the surface-sown seed.

Sowing time and rate. Sow in spring or early summer, so the pasture is established before the extreme heat of summer, at 3-6 kg/ha (1-2 kg for 'Hamil', 3.5-4.5 for 'Common').

Number of seeds per kg. 1 750 000; 1 030 000 ('Hamil'); 2 200 000 (United States).

Dormancy. The quality of the seed improves for some months after harvest.

Seed treatment before planting. It does not require any special treatment except ageing.

Seedling vigour. It has good seedling vigour.

Response to photoperiod. It is a short-day plant (Wang, 1961).

Response to light. It is fairly tolerant of shading, and in its natural habitat inhabits woodlands throughout subhumid Africa.

Compatibility with other grasses and legumes. Guinea combines well with the legume centro (*Centrosema pubescens*) and this is a common pasture mixture for the wet tropics. In Brazil, 'Coloniao' Guinea, centro and siratro are used successfully. Guinea and *Stylosanthes guianensis* is a successful mixture (see Plate 43, Fig. 15.102). Puero (see Fig. 15.103) and glycine also combine well.

Ability to compete with weeds. In the wet tropics, weed competition is severe. However, a well-established Guinea grass pasture, well-fertilized, will suppress weeds.

Tolerance to herbicides. To control weeds in *Panicum maximum*, atrazine (2-chloro-6-ethylamino-4-isopropylamino-1,3,5-triazine) can be used. Gatton panic survived over 4.5 kg AI/ha on the Atherton Tableland, Queensland, whereas most of the associated weeds — *Nicandra physaloides, Raphanus raphanistrum, Argemone ochraleuca, Ageratum conyzoides, Sida cordifolia* and *Eleusine indica* were killed with the low concentration of 0.9 kg AI/ha (Hawton, 1976). *Panicum maximum* is a major weed itself in sugar-cane fields, due to its ability to grow under poor conditions. It can be killed by a pre-emergent spray of 2,4-D sodium salt at 4.5 kg/ha of an 840 g AI/kg product (e.g. Hormicide). No wetting agent is required when used as a pre-emergent spray. Use a minimum of 340 litres of water per hectare. For seedlings in the five-leaf stage, use Diuron at 2.5 kg/ha of an 800 g AI/kg product (Karmex, Diuron) applied in a minimum of 340 litres of water per hectare. For mature plants use 2,2-DPA at 2.3 kg of a 740 g AI/kg product (Shirpon, Dowpon) plus paraquat at 85 ml of a 200 g AI/litre product (e.g. Gramoxone) plus wetting agent at 250 ml per 200 litres of water. Spray to the point of runoff (Tilley, 1977).

Response to defoliation. Guinea grass stands a good deal of defoliation but

Figure 15.102. A nil-phosphorus plot in a sown pasture mixture of *Panicum maximum* cv. Hamil and *Stylosanthes guianensis,* Belyana, northern Queensland, Australia (**Source:** J.K. Teitzel, Queensland Department of Primary Industries)

should not be grazed or cut below about 30 cm for permanence (McLeod, 1972).

<u>Grazing management.</u> In the wet tropics it is necessary to let this pasture become well-established before grazing so that it can compete with weeds. Guinea usually seeds in autumn; do not graze a new pasture until after this seeding period. Guinea cannot be grazed below 35 cm, or it will recover slowly. Adjust the stocking rate to maintain this height. Rotational grazing will give better control of pasture growth. Mowing or slashing is useful to control excess growth and weeds, but do not mow below 35 cm, and not after mid-autumn, as it will give slow regrowth and encourage winter weeds. Do not graze under extremely wet conditions, as trampling damages pastures growing in boggy ground (see Fig. 15.104).

<u>Response to fire.</u> It is tolerant of fire.

<u>Genetics and reproduction.</u> The somatic chromosome numbers are $2n=18$, 32, 48 (Fedorov, 1974). It is a facultative apomict in which both apospory and pseudogamy occur (Warmke, 1954, quoted by Javier, 1970). The amount of sexual reproduction varies from 1-5 percent depending on the variety.

<u>Dry- and green-matter yields.</u> In the year 1973/74 at South Johnstone, Queensland, cv. Makueni produced more than 60 000 kg DM/ha when 300 kg/ha of nitrogen was applied (Middleton & McCosker, 1975). Vicente-Chandler, Silva and Figarella (1959) obtained 26 846 kg DM/ha with 440 kg N/ha, cut at 40-day intervals, in Puerto Rico.

Figure 15.103. Panicum maximum growing with Pueraria phaseoloides at Tully, Queensland, Australia (lat. 18°S, rainfall 4 320 mm) (**Source:** R. Bruce, Queensland Department of Primary Industries)

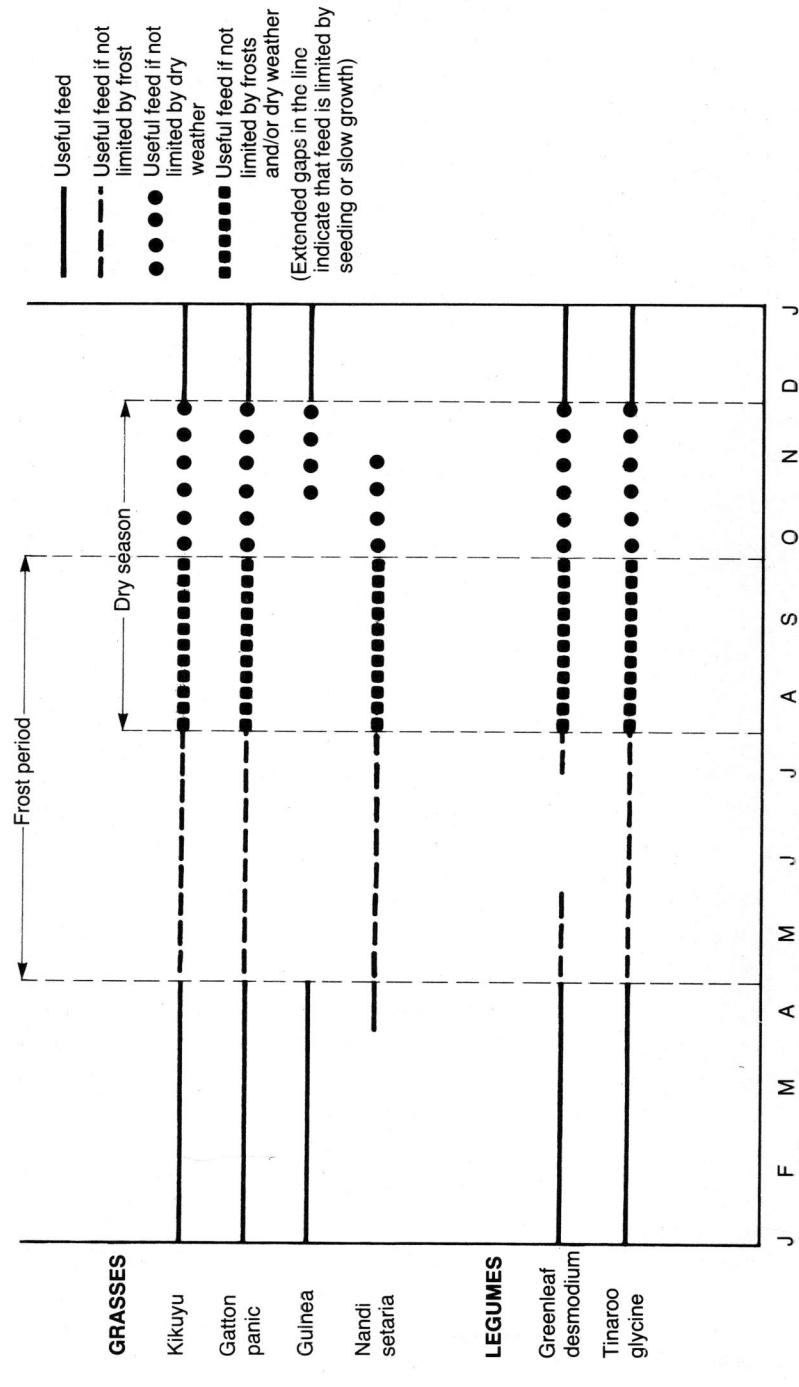

Figure 15.104. Feed supply patterns of some useful pasture species on the Atherton Tableland, Queensland, Australia (lat. 17°13'S, rainfall 715 mm, deep latosolic soil) **(Source:** Quinlan & Edgley, 1975)

Suitability for hay and silage. It has been used successfully for silage at Mpwapwa, Tanzania (Semple, 1970). Silage was also made in Brazil (Cezar *et al.*, 1976), Nigeria (Miller, Clifton & Cameron, 1963) and Australia (Teitzel, 1969). It also makes useful hay in Thailand (Göhl, 1975).

Value as standover or deferred feed. 'Hamil' and 'Coloniao' Guinea grasses are reasonably palatable when mature, and provide good roughage to use in conjunction with urea molasses licks.

Chemical analysis and digestibility. In Costa Rica, analysis of flowering material revealed 7.81 percent crude protein, 30.62 percent crude fibre, 40.88 percent nitrogen-free extract, 2.33 percent ether extract and 8.36 percent ash in the dry matter on a 10 percent moisture basis (Gonzalez & Pacheco, 1970). Göhl (1975) has records from Tanzania, Malaysia and Thailand with crude protein varying from 5.3 to 25 percent of the dry matter.

Palatability. It is very palatable.

Toxicity. No clear-cut evidence of toxicity with this grass is recorded by Everist (1974). Ndyanabo (1974) recorded 0.28 percent of total oxalic acid in the dry matter but no toxicity.

Seed production and harvesting. Seed ripens unevenly, and is shed as it matures. Javier (1970), in the Philippines, found the highest seed yield (19 percent recovery) was obtained when the panicle had shed 40-60 percent of its spikelets, which occurred about 12 to 14 days from panicle emergence. Harvesting is usually done by direct heading.

Seed yield. Javier (1970) recorded 48-156 kg/ha; Paretas *et al.* (1972), 395 kg/ha from three cuts in Cuba; Fernando (1958), 100 kg/ha in Sri Lanka.

Minimum germination and quality required for commercial sale. 25 percent germinable seed; 40 percent purity in Queensland. Germinate at 20-35°C, moistened in water. Germination is promoted by light (Ballard, 1964).

Cultivars.

- 'Hamil Panic' — seed obtained from Jack Hamil of Daintree, north Queensland, its source unknown. A tall, tufted perennial to 2.5-3 m, more robust and coarser than 'Common Guinea' and more like 'Coloniao'. Dense, stiff hairs on the basal leaf-sheath distinguish it from 'Coloniao'. Leaves blue-green, less hairy than 'Common Guinea'. Adapted to frost-free, warm conditions and fertile scrub soils. Grows vigorously during the wet season, when it is very palatable but is less so as it hays off. Seed set is poor. Probably apomictic.

- 'Coloniao' — a giant robust type with thick, fleshy stems growing to 3 m. The leaves are a distinct blue-green, 80-90 cm long and 25-30 mm broad, glabrous, the sheath is glabrous, except for short hairs on the sheath margin toward the junction of sheath and blade. The seed-head is 20-50 cm long, 15-40 cm wide, dark green, and the spikelet (seed) outer glume is glabrous (Middleton & McCosker, 1975).

- 'Embu' (creeping Guinea) — has a semi-erect, rambling habit, rooting freely from the nodes, producing aerial roots from the lower nodes. It grows to 1-1.5 m. The leaves are light green to green, the leaf-blades have a few short surface hairs, and are 20-30 cm long and 12-16 mm wide. There are occasional short hairs on the leaf surface, and sparse short hairs on the lower outside of the sheath near the node junction; occasional hairs on the lower stem internodes. The panicle is 15-20 cm long, 12-15 cm wide, green, and the spikelet outer glume is glabrous (Middleton & McCosker, 1975). Has good winter growth and is a leafy, light-seeded cultivar.

- 'Common Guinea' (common Guinea grass) — medium height 1.8-2 m, erect canopy, fine stems, green leaves 70-80 cm long, 15-18 mm wide, sparsely hairy upper surface, few on lower surface, sheaths moderately hairy on outside surface, density increasing toward node. Stems hairless, panicle 15-40 cm long, 12-30 cm wide, green. Spikelet outer glume hairless. It is the most widely used cultivar, but variable, and cv. Riversdale was selected to replace it.
- 'Coarse Guinea' — giant, robust, with thick woody stems, 2.5-3 m high, dark green leaves, 80-90 cm long, 25-30 mm wide, sparse to moderately dense, short hairs giving rough feel to the leaf. Moderately dense, long, stiff bristle hairs on outside surface, increasing in density toward the junction of blade and leaf. Sheath painful to handle. Sow 'Hamil' together with 'Coarse Guinea' to make it easier for stock to penetrate (Walsh, 1959).
- 'Gatton Panic' — derived from seed introduced from Zimbabwe. Less robust and not as coarse as cv. Hamil, but more robust than 'Petrie' (green panic, *P. maximum* var. *trichoglume*). It has broader and longer leaves, with a more prominent midrib and more scabrid margins than green panic, and its spikelets are glabrous. Adapted to 760-1 000 mm rainfall. A little more vigorous, drought resistant, and persistent and more palatable than green panic, flowers later and responds better to nitrogen. Apomictic. The seed takes on a silver sheen when mature. Harvest when 80 percent has silver sheen and 5-10 percent seed has shattered. 1 401 000 seeds per kilogram.
- 'Makueni' — introduced from Kenya in 1965, it has given better cool-season growth than other Guinea grasses. It is easily distinguished from other Guinea grasses because the whole plant is covered with dense, whitish, soft hairs, giving it a furry feel. The outer seed coat is also hairy and can be so distinguished under magnification. It is very like the seed of green panic and can only be identified after germination. 'Makueni' is erect, tufted, with the leaf canopy slightly drooping, and is not so tall as 'Hamil', 'Coloniao' and 'Coarse Guinea'. It is readily grazed and gives good live-weight gains (Middleton & McCosker, 1975). Not as productive as cv. Riversdale, and is difficult to establish in the presence of weed competition (Teitzel & Middleton, 1979). 1 143 000 seeds per kilogram.
- 'Likoni Guinea' — recommended for the high-rainfall areas (1 000-1 270 mm) on the Kenya coastal strip in association with *Macroptilium atropurpureum* and *Neonotonia wightii*.
- 'Ntchisi' — used in Zambia and propagated by vegetative planting mate-

rial (Thorp, 1979). Stems hairless, panicle 20-60 cm long, 15-40 cm wide, distinctive dark brown colour, spikelet outer glume hairless (Middleton & McCosker, 1975).
- 'Riversdale' — a selection from 'Common Guinea' to be continued as a certified seed line (Middleton, 1977).

In Puerto Rico five cultivars are distinguished:
- 'Local' or 'Common' — resistant to drought and heavy grazing. Suitable for drier areas.
- 'Gramalote' — a robust form in more humid areas, but always infested with leaf spot.
- 'Borinquen', 'Broad-leaf' and 'Fine-leaf' — under test.

In Jamaica two cultivars are sown:
- 'Silky Guinea' — a very leafy type for drier areas.
- 'St Mary's Cowgrass' — more robust and stemmy, grown in more humid parts. In Mauritius, cv. Sigor was highly productive, nutritious and drought resistant (Wright, 1961). In Brazil, 'Common' is the ordinary robust form; and 'Sempre-verde' (*P. maximum* var. *gongylodes*) is a fine-leaved, drought-resistant type with the base of the culms expanded.

Value for erosion control. Its great bulk aids in erosion control, but its generally tussocky growth (except for cv. Embu) makes it less valuable than other species.

Diseases. Bunt has interfered with Guinea grass seed production in Kenya in the Rift Valley (Semple, 1970). In Puerto Rico a leaf spot is caused by *Cercospora fusimaculosus*. In Colombia the inflorescence has been attacked by *Fusarium* spp. and a smut (*Ustilago* sp.) (CIAT, 1978).

Animal production. Cultivar Makueni and common Guinea grass are capable of giving live-weight gains of up to 0.8 kg per animal per day (Middleton & McCosker, 1975). The selection (cv. Riversdale) from common Guinea grass consistently gave in excess of 600 kg/ha annual live-weight gain in association with legumes in north Queensland (Middleton, 1977). In São Paulo State, south-central Brazil, with an annual rainfall of 1 154 mm, of which 80 percent occurs in summer, an annual live-weight gain from Nellore strain Zebu steers of 241 kg/ha was obtained from unfertilized 'Coloniao' Guinea grass, and a mean of 586 kg/ha from a similar pasture fertilized with 200 kg N/ha annually over a seven-year period. Basic phosphorus and sulphur were applied. Application of nitrogen during the cool season gave earlier marketing to obtain a 15-30 percent higher price, but time of application had no overall advantage in live-weight gain.

Richards (1965) in Jamaica recorded annual live-weight gains ranging from 816-1 262 kg/ha from irrigated Guinea grass fertilized with 158 kg N/ha per year. At Utchee Creek, near South Johnstone in north Queensland, at a stocking rate of 4.2 beasts per hectare, first year annual live-weight gain increased from 377.4 kg/ha on Guinea grass alone, to 464.4 kg/ha with the inclusion of the legume *Centrosema pubescens*. A further increase to 601.1

kg/ha was obtained by the addition of 169.5 kg/ha of nitrogen. In the second year the corresponding live-weight gains for pure Guinea grass, Guinea grass/centro and Guinea grass with nitrogen were 315, 481 and 731 kg/ha respectively. Each kilogram of applied nitrogen produced an average of 3.96 kg of live-weight gain (Grof & Harding, 1970). In a comparison of cultivars at South Johnstone, the average annual live-weight gains per hectare were 756 kg for 'Common Guinea', 698 kg for 'Hamil' and 633 kg for 'Coloniao' over three years. Cultivar Embu pastures did not persist (Mellor, Hibberd & Grof, 1973a). An irrigated Guinea grass/centro pasture at Ayr, north Queensland, under 41 weeks' grazing with shorthorn beef cattle gave a daily live-weight gain of 0.68 kg per head. From January to March, gains fell due to high day temperatures and high humidity (Allen & Cowdry, 1961). In the Solomon Islands, Gutteridge and Whiteman (1978) obtained a yield of 11 700 kg/ha per year from a mixture of *P. maximum* cv. Hamil and *Centrosema pubescens,* under coconuts.

Main attributes. Its wide adaptation, quick growth and palatability, ease of establishment from seed and good response to fertilizers (Harding, 1972).

Main deficiencies. 'Common Guinea' has two main weaknesses: the bulk of its growth occurs in summer and, in recent years, commercial Guinea grass seed has been contaminated with less desirable types, such as 'Coarse Guinea'. Rapid summer growth and quick subsequent deterioration result in management difficulties (Hartley, 1950).

Further reading. Grof & Harding, 1970; Middleton & McCosker, 1975; Motta, 1953.

Panicum maximum var. *trichoglume* (K. Schum.) C.E. Hibberd

Common names. Green panic (Australia), castilla (Peru), slender Guinea grass (Kenya).
Natural habitat. Forest fringes.
Distribution. Native to Africa, common in India and introduced to Australia.
Description. A tufted, tall, summer-growing perennial, differing from common Guinea grass in being smaller and less robust, in having finer stems and leaves, and in having the glumes of its spikelets covered with fine hairs. Its six- to eight-noded stems normally grow to 1 m, with crowns up to 15-30 cm in diameter. Compared with *P. maximum* cv. Gatton, the lower surface of the leaves and the leaf-sheaths are sparsely hirsute or villose rather than finely pubescent; the midrib of its leaf is less prominent and more hirsute; its leaf margins are less scabrid and its ligule is a ring of long downy hairs, rather than short straight bristles (Barnard, 1972).
Season of growth. Good growth in early spring, better than buffel and Rhodes, continuing through summer and autumn. Leaf production dwindles with the onset of flowering in early summer.
Optimum temperature for growth. Ludlow and Wilson (1970b) found dry-matter production was 3.47 times greater for cv. Embu, 21.4 times greater for cv. Hamil, 15 times greater for common Guinea grass and 9.76 times greater for cv. Gatton at 30°C than at 20°C.
Minimum temperature for growth. It responds quickly to mild weather in winter.
Frost tolerance. Only slightly tolerant — more susceptible than Rhodes grass.
Latitudinal limits. About 30°N and S.
Altitude range. Sea-level to 2 000 m.
Rainfall requirements. It has a wide range, between 650 and 1 780 mm in Queensland, but is not as well suited to high coastal rainfall as common Guinea; it does better a little inland from the coast. It does not thrive at over 2 000 mm annual rainfall.
Drought tolerance. Moderate (Tsiung, 1976), more tolerant than Rhodes grass, less than buffel.
Tolerance to flooding. It is killed by a few days of saturated soil and is unsuitable for intensive irrigation.
Soil requirements. It does best on deep scrub loams of high fertility, but performs well on basaltic uplands of prairie and black soils and sandy loams of reasonable fertility. Tolerates soil pH from 5.0-8.0. Deep sands are unsuitable.
Tolerance to salinity. It will tolerate soil pH to 8.0.
Fertilizer requirements. It responds fairly well to nitrogen on poor soils and will gradually disappear when fertility declines. Linear responses to nitrogen

up to 440 kg/ha per year were recorded by Gartner (1966) on the Atherton Tableland, Queensland, but the most efficient response by green panic was at 55 kg N/ha. Production of green panic, at 154 kg/ha, is of the order of that produced in combination with the legume *Neonotonia wightii* cv. Tinaroo. Phosphorus, nitrogen and sulphur are generally needed on the Darling Downs' black clay, Queensland (Swann, 1973). A pale leaf colour may be due to nitrogen or sulphur deficiency. Regular use of single superphosphate will supply the sulphur requirement (Delaney, 1975). The critical value of phosphorus as a percentage of the dry matter at the immediate pre-flowering stage is 0.19.

Ability to spread naturally. If allowed to seed it will gradually extend its population by new seedlings, especially on the edge of scrubs and near watercourses.

Land preparation for establishment. A fine seed-bed or ashes needed.

Sowing methods. Either drill into a prepared seed-bed or broadcast in ashes from land or air. Drilling in a contour furrow with press wheels gives an excellent stand (Wilson, 1978). Sowing at 1 cm depth in a companion crop of lucerne, oats or wheat gives good results (Bott, 1978).

Sowing depth and cover. Sow no deeper than 1 cm and cover lightly. A "Triad" seed planter applying a narrow band of gypsum over the row to prevent soil crusting is successful (Bott, 1978).

Sowing time and rate. When drilling or sowing from the air, sow during the rainy season and one week after a scrub burn at 4 kg/ha. Two plantings, one in spring and one in midsummer, may give a more reliable farm sequence.

Number of seeds per kg. 1 225 000 (Queensland).

Dormancy. It has a long period of dormancy, reaching maximum viability 18 months after harvest.

Seedling vigour. It has good seedling vigour.

Vigour of growth and growth rhythm. About 80 percent of the production occurs in the summer months, October to March, in north Queensland (Gartner, 1966).

Response to light. One of the outstanding features of green panic is its ability to grow in partial shade, which also protects it from frost. It grows right up to tree trunks.

Compatibility with other grasses and legumes. It combines very well with the tropical legumes suited to its environment, and with lucerne (*Medicago sativa*), especially on basaltic loamy soils and deep sandy loams in Queensland. It grows well with siratro (*Macroptilium atropurpureum*), *Neonotonia wightii* and several other legumes in Fiji and Queensland. It is compatible with buffel grass (*Cenchrus ciliaris*), and Nandi setaria but not with Rhodes grass. *Sorghum almum* can be a temporary companion grass in new sowings.

Ability to compete with weeds. In its normal environment it can successfully suppress weeds.

Tolerance to herbicides. Green and Gatton panics showed a high tolerance to post-emergence spray of atrazine at 1-2 kg/ha of 80 percent product, which killed seedlings of *Salvia reflexa* (Wilson, 1978). Green panic was unaffected by the application of 1.68-2.24 kg AI/ha of Fenoprop (2-(2,4,5-trichlor-phenoxy) propionic acid), as the propylene glycol ether ester (PGEE) with 0.1 percent nonionic wetter which exerted satisfactory control of chickweed (*Drymaria cordata*) in the *Panicum maximum* var. *trichoglume/Neonotonia wightii* pasture on the Atherton Tableland, Queensland (Hawton, Quinlan & Shaw, 1975).

Response to defoliation. Green panic will stand reasonable, but not heavy, defoliation.

Grazing management. Stock at reasonable intensity and then allow about six weeks for the pasture to recover before the next grazing. If in association with lucerne, stock fairly heavily for one week at each grazing and allow the lucerne to recover. Continuous grazing of the lucerne will cause it to die out. Allow the green panic to set seed at least every two years.

Response to fire. It does not stand hot burning, and burning should only occur where woody weeds need controlling (Skovlin, 1971).

Genetics and reproduction. P. maximum is a pseudogamous apomict.

Dry- and green-matter yields. At Koronivia, Fiji, a yield of 26 781 kg/ha was obtained over an 11-month period (Roberts, 1970a, b). Up to 10 000 kg

Figure 15.105. Harvesting *Panicum maximum* var. *trichoglume* on the Darling Downs, Queensland, Australia (**Source:** W. Bott, Queensland Department of Primary Industries)

DM/ha has been obtained in southeast Queensland during a growing season. Henzell (1976) increased dry matter yield from 7 250 kg/ha to 13 130 kg/ha with the application of 167 kg N/ha at Narayen, Queensland.

Suitability for hay and silage. It makes good hay and silage when cut at flowering stage. A green panic-siratro mixture is often used (Kelly, 1972).

Value as standover or deferred feed. It is a useful standover feed.

Chemical analysis and digestibility. It has a high nutritive value even when mature and frosted (Milford, 1960a, b).

Palatability. It is extremely palatable.

Toxicity. Rare cases of unconfirmed toxicity have been reported, but its long use for grazing makes the risk of toxicity very small (Everist, 1974).

Seed production and harvesting. It flowers from early summer to late autumn. Wait until the ripest seed has started to shed to ensure ripe seed. Seed does not ripen evenly and shatters badly; early harvesting may collect too much immature seed, and late harvests will miss seed which has shed. Harvest with an autoheader when 5-10 percent seed has shattered (see Fig. 15.105). Dry the seed thoroughly before storing or it may overheat. Seed may have to be stored for some time to reach satisfactory germination standards.

Seed yield. In Cuba, 327 kg/ha were harvested from two cuts, with 60 cm spacing and 300 kg NPK fertilizer (Paretas *et al.,* 1972).

Minimum germination and quality required for commercial sale. 20 percent germination and 70 percent purity are required in Queensland, Australia.

Cultivars. Only one cultivar is registered, cv. Petrie, the characteristics of which are being described. It was named after A.A. Petrie of 'Madoora', Gayndah, Queensland, who first grew the grass commercially.

Value for erosion control. Because of its tussocky nature it is not specifically suited to erosion control, but it exerts some influence.

Diseases. No major diseases are encountered.

Pests. No major pests occur.

Animal production. Given reasonable climatic conditions and management, a carrying capacity of better than one beast to two hectares can be anticipated for a green panic/legume pasture, or from a nitrogen-fertilized stand of pure green panic. With beef cattle, an annual live-weight gain of 140 kg per head could be expected with a pure green panic pasture, or 180 kg with a legume included with it (Delaney, 1975). At Narayen, Queensland, green panic pastures fertilized with 300 kg N/ha per year produced an average live-weight gain in steers of 180 kg per head per year of first-grade carcass, compared with 146 kg per head per year of second-grade carcass from unfertilized or superphosphate-fertilized pastures (Silvey, 1977a, b). Under irrigation, in conjunction with *Leucaena leucocephala,* on the island of Kauai, Hawaii, the pasture produced 9 770 kg milk and 400 kg beef per hectare per year (Plucknett, 1970).

At Kairi in the Atherton Tableland (lat. 17°14'S, 700 m altitude and 1 248 mm rainfall, of which 830 mm occurs from January to April) from green panic/Tinaroo glycine pastures, Jersey cows produced 2 480 kg milk and 114 kg butterfat per lactation and Friesian cows 4 100 kg milk and 137 kg butterfat (Cowan, Byford & Stobbs, 1975). These yields compare favourably with those recorded from other tropical dairying areas (Colman & Holder, 1968; Stobbs, 1971).

Main attributes. Its ability to grow in shade, its palatability, and its ability to combine with other grasses and legumes.

Main deficiencies. Its uneven seed setting and its lack of persistence in poor soils without adequate fertilizer.

Further reading. Delaney, 1975; Young, Fox & Burns, 1959.

Panicum miliaceum L.

Common names. White French millet, red French millet (Australia), proso, hog millet, brown corn millet, broom corn millet (United States), vari (India).

Natural habitat. Cultivation.

Distribution. Grown since prehistoric times as a grain crop — worldwide distribution in suitable climates.

Description. It has coarse, woody, hollow stems from 30-120 cm, but usually to 60 cm. Stems round or flattened, 6-8 mm thick at the base, covered with hairs. The stem and outer chaff are green or sometimes yellowish- or reddish-green when the seed is ripe. When threshed, most of the seed remains in the inner chaff or hull. The hulls are of various shades and colours including white, cream, yellow, red, brown, grey and black. The bran or seed coat is always creamy white (Martin & Leonard, 1959) (see Figs 15.106, 15.107).

Season of growth. Summer.

Frost tolerance. It is susceptible to frost.

Latitudinal limits. About 30°N and S.

Rainfall requirements. It generally grows in areas receiving a rainfall within the range of 500-750 mm with a summer dominance.

Drought tolerance. It survives hot weather better than other millets.

Soil requirements. It prefers sandy loams to clay loams, but has a wide soil range. Germination difficulties may be encountered in heavy, self-mulching clays.

Tolerance to salinity. In the Ukraine, Chapko (1977) found *P. miliaceum* to have high tolerance to Na_2CO_3 in the soil.

Fertilizer requirements. Soil tests show the need for a basic complete fertilizer.

Ability to spread naturally. It will grow readily from scattered seed, but is usually sown as a crop.

Land preparation for establishment. A good, fine, firm seed-bed is required for good germination.

Sowing methods. As a grain crop it is normally drilled in through a small seeds box.

Sowing depth and cover. Sow at about 2.5 cm, harrow and roll to compress the soil around the seed.

Sowing time and rate. Sow mid to late summer at 10-11 kg/ha.

Number of seeds per kg. 176 000.

Dormancy. There is no seed dormancy.

Seed treatment before planting. If seed-harvesting ants are a problem, dust with lindane. To control head smut, treat seed with a mercury or copper carbonate dust.

Seedling vigour. Excellent.

Vigour of growth and growth rhythm. The plants tiller freely and may thus compensate for a poor initial stand. It flowers in 68 days and matures in 90-100 days.

Figure 15.106. Panicum miliaceum. **A**-Plant **B**-Panicle **C,D**-Spikelet, front and back **E**-Grain (**Source:** USDA Farmers' Bulletin 1433)

Figure 15.107. Panicum miliaceum (**Source:** N.J. Douglas, Queensland Department of Primary Industries)

Response to photoperiod. It requires short day lengths for flowering (Evans, Wardlaw & Williams, 1964).

Response to light. It needs full sunlight for growth.

Compatibility with other grasses and legumes. It is usually sown as a pure crop for grain, but may be combined with cowpea for grazing. Because the stems and leaves are hairy and fibrous, it is not very attractive as a grazing or hay crop.

Ability to compete with weeds. It does not compete successfully with weeds.

Tolerance to herbicides. Weed competition may be suppressed by spraying with MCPA at a strength of not more than 0.25 kg acid equivalent per hectare. Apply in the tillering stage before seed-heads form in the sheath.

Response to defoliation. It recovers very poorly from defoliation.

Grazing management. It is not usually used for grazing in Australia. It is grown widely in the USSR as a fodder plant in association with vetches and also as a cover crop in establishing lucerne (Romanov, 1976).

Genetics and reproduction. 2n=36, 40, 49, 54, 72 (Fedorov, 1974).

Suitability for hay and silage. Because of its hairy nature it is not very suitable for these purposes and other millets are to be preferred. It has been made into silage in Romania. Hay made from flowering plants is poor.

Chemical analysis and digestibility. See Table 15.50.

TABLE 15.50 **Panicum miliaceum**

	DM	As % of dry matter				
		CP	CF	Ash	EE	NFE
Hay	86.6	12.5	33.9	6.6	2.5	44.5
Straw, India		4.8	35.5	8.9	1.2	49.6

	Animal	Digestibility (%)				ME
		CP	CF	EE	NFE	
Hay	Sheep	56.5	59.9	40.9	60.8	2.10

Source: Göhl, 1975

Palatability. It is palatable but hairy.

Toxicity. No toxicity has been noted.

Seed production and harvesting. It is harvested by combine or, where the seed shatters, it can be cut early with a reaper and binder, cured and subsequently threshed.

Seed yield. Yields of up to 1 500 kg/ha can be obtained.

Minimum germination and quality required for commercial sale. 75 percent germinable seed and 97.3 percent purity in Queensland.

Cultivars. Numerous varieties of *Panicum miliaceum* are used throughout the world. Queensland uses 'White French Millet', with a creamy yellow glistening grain, and 'Red French Millet', with a red glistening grain. The United States uses 'Yellow Manitoba', 'Turghai' and 'Early Fortune'. 'Turghai' gives the highest yields. A variety called 'Deerbrook' from Czechoslovakia is grown in Wisconsin, United States. It has grey-green stripes on the hulls. Another variety, 'Crown', grown in Canada, has greenish-grey hulls (Martin & Leonard, 1959).

Economics. It is an important food crop in the USSR, the Near East and India. It seems to have the lowest water requirement of any grain crop (308 litres of water per kilogram of dry matter produced).

Animal production. Usually cultivated for bird seed and poultry. A useful summer catch crop for emergency use if rain for planting is seasonally late.

Further reading. Douglas, 1970.

Panicum pilosum Dalz. and Gibs.

Common names. Bladhi (India).
Natural habitat. Moist ground and open woods.
Distribution. India, Mexico and the West Indies to Brazil and Ecuador.
Description. Spreading or ascending panicles 5-15 cm long, the numerous dense racemes 1-3 cm long, rather closely arranged along the main axis, the rachides stiffly ciliate; spikelets glabrous, 1.5 mm long (Hitchcock, 1927).
Economics. Bladhi is a rain-fed crop and is cultivated in India, on poorer soils with moderate rainfall. In the Deccan it is cultivated in hilly areas on light soils. It is sown in June. The field is prepared with a harrow and the seed is carefully sown, either by broadcasting or with a seed drill. About 6-7 kg of seed per hectare are evenly distributed over the land. Seedlings are delicate in the early stages; but once they begin to tiller, a dense growth soon covers the soil. The crop requires no further attention till harvest. No irrigation or manure is applied. The crop is ready for harvest by October.

Bladhi is much like *Setaria italica,* but larger. The ripe earhead is reddish-brown with bristles, while the ripe earhead of rala is smooth and of a pale yellow colour. The grain is husked by pounding and is a poor farmer's crop. It is boiled and eaten whole or sometimes ground into flour. The crop tillers profusely and provides good quality fodder. No disease or pest of any importance has been reported from this crop (Solomon, 1953).
Further reading. Solomon, 1953.

Panicum repens L.

Common names. Torpedo grass (United States), cheno (India), limanota (Zambia), creeping panicum (Zimbabwe), couch panicum (southern Africa), panic rampant, muran (Iraq).
Natural habitat. Lake shores, and seasonal and permanent swamps.
Distribution. Malaysia, Africa, Sri Lanka, India, Burma, Thailand, United States.
Description. A rhizomatous, creeping perennial, rooting at the base, 30-90 cm tall. Leaf-blades usually inrolled when dry, 5-15 cm long and 5-12 mm wide with scattered hairs on the upper surface. Inflorescence an open panicle 6-20 cm long, branches ascending, spikelets 3 mm long, acute and gaping at the tip. Fruit glossy white. Young shoots covered by leaf-sheaths (hence "torpedo grass") (see Fig. 15.108).
Season of growth. Makes its main growth in summer.
Frost tolerance. The leaves are easily killed by frost.
Altitude range. Sea-level to 1 800 m.
Rainfall requirements. Adapted to areas with a winter rainfall, it will not survive hot dry seasons.
Drought tolerance. It tolerates drought, as the rhizomes remain alive in long dry periods.
Tolerance to flooding. Panicum repens grows well even after several days in standing water. It is frequent on lake edges, edges of dams and in swamps throughout the tropics (Sayer & Lavieren, 1975).
Soil requirements. Generally found on sandy soils, but some strains grow on heavy clay. The soils are always wet and of alluvial origin. It is useful on copper-deficient soils.
Tolerance to salinity. Very good; it occurs on saline sands in western Zambia (Verboom & Brunt, 1970).
Ability to spread naturally. It is aggressive and can be spread by ploughing.
Sowing methods. It is usually propagated by rhizomes. In India, is sown by seed.
Sowing depth and cover. Sow on the surface and roll or cover lightly.
Sowing time and rate. It is sown in summer at 11 kg/ha in Gujarat, India.
Vigour of growth and growth rhythm. It is sown at the end of January in Gujarat, flowers in May, and is cut green or allowed to ripen for seed.
Response to light. It does not tolerate dense shade.
Ability to compete with weeds. It can invade other pastures and can become a weed along drainage ditches, where it becomes difficult to eradicate (Gilliland *et al.*, 1971).
Response to defoliation. Resistant to grazing and trampling, but rapidly becomes sod-bound.

Grazing management. It will stand heavy stocking, and can be renovated with a deep disc-harrowing when it becomes sod-bound.
Genetics and reproduction. 2n=36, 40, 45 (Fedorov, 1974).
Dry- and green-matter yields. It is usually not sown as a pasture, but one farmer in Taiwan claims it will produce 100 tonnes green matter per hectare per year under irrigation when top-dressed with 200 kg urea per hectare after

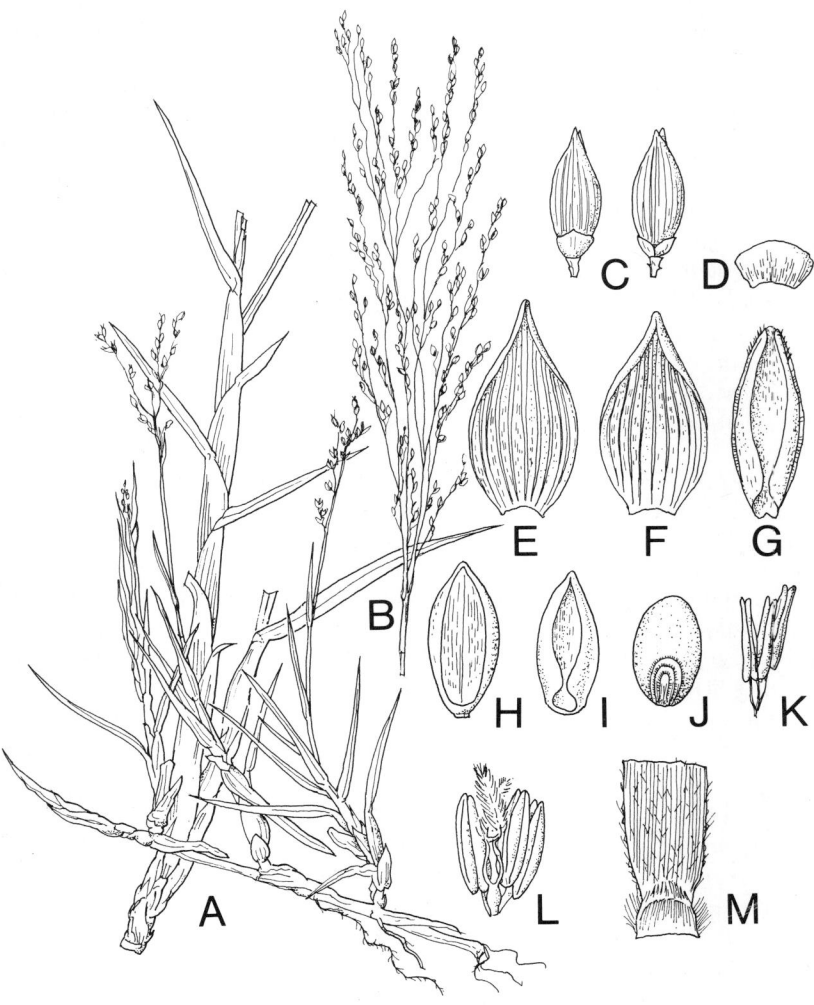

Figure 15.108. Panicum repens. **A**-Habit **B**-Inflorescence **C**-Spikelet **D**-Lower glume **E**-Upper glume **F**-Lower lemma **G**-Palea **H**-Upper floret **I**-Palea **J**-Grain **K**-Male flower **L**-Hermaphrodite flower **M**-Ligule

each of five cuts per year (Manidool, personal communication). Yields fall off after several years under unfavourable drought conditions (Thorp, 1979). At Gujarat, unirrigated, it yielded 2 096 kg green matter per hectare (Srinivasan, Bonde & Tejwani, 1962). At Laguna, Philippines, Furoc and Javier (1976) harvested 62 000 kg green matter per hectare from an irrigated, abandoned rice field.

Chemical analysis and digestibility. Göhl (1975) quotes analyses by Lim (1968); 28.3 percent dry matter, 24 percent crude protein, 22.6 percent crude fibre, 13.4 percent ash, 2.1 percent ether extract, 37.9 percent nitrogen-free extract in the dry matter of fresh material cut at four weeks in Malaysia.

Palatability. It is extremely palatable and nutritious over a long growing season, but at the mature stage the old leaves tend to become tough (Thorp, 1979) and are neglected by stock.

Seed production and harvesting. Seed production is poor. The ripe crop is cut into sheaves, dried, and the seed beaten out on boards in Gujarat.

Seed yield. 1 300 kg grain per hectare in India (Solomon, 1953).

Value for erosion control. It is a useful grass for binding coastal sands and lake shores. It is used to fix mine dumps in Zimbabwe (Chippendall & Crook, 1976). It was also used for stabilizing the steeper slopes of ponds (20-30°) in Zambia where cattle approach to drink water (Verboom & Brunt, 1970). In Gujarat, it proved to have the greatest root-binding capacity of several grasses, but gave poor above-ground yields (Srinivasan, Bonde & Tejwani, 1962).

Economics. It is a valuable pasture grass in a number of tropical countries and it provides feed in paddy areas, particularly for draught cattle and buffaloes. It is also cut by hand from roadsides and edges of paddy-fields to feed to dairy cattle. In Iraq it is an important grazing plant for swamp buffaloes.

Animal production. No figures have been cited.

Main attributes. Its adaptation to wet conditions; its production and palatability when young.

Main deficiencies. It can become a serious weed of arable land and is difficult to eradicate (Thorp, 1979). It is a poor seed producer.

Further reading. Furoc & Javier, 1976.

Panicum sumatrense Roth ex Roem. and Schult

Synonym. P. miliare Lamk.
Common name. Sava, kutki (India), little millet.
Distribution. Southeast Asia, Malaysia and northern India.
Description. An annual with erect or geniculate culms, 0.3-1 m long. Leaf-blades linear, sheath sometimes hairy. Panicle contracted, 4-15 cm long; spikelets persistent, 2-3.5 mm long; lower glume orbicular, apiculate (Bor, 1960); lower floret paleate. The caryopsis is glabrous, striated and brown. The grain is slightly larger than that of *P. psilopodium*.
Economics. Cultivated to some extent in the poorer parts of central India. Cattle are very fond of the straw, which, in southern India, is used largely as a fodder. The crop can be grown on very poor soil (Bor, 1960). Generally grown on light red soils and on hillsides as a rain-fed crop, never irrigated. Usually propagated by drilled seed, but can be transplanted; one hand-weeding is necessary when the plants are 30 cm tall. The yield of grain is about 750-850 kg/ha. The grain is cooked like rice or sometimes ground into flour and made into bread. The protein content of the grain is about 7.7 percent. The straw is used in making bricks and cement (Solomon, 1953). The green plant has potential as a quick-growing fodder which tolerates both drought and waterlogging. It grows in India up to 2 000 m elevation, where it matures in 14 weeks.
Genetics and reproduction. $2n = 36$ (Fedorov, 1974).
Further reading. Mann, 1946; Solomon, 1953.

Panicum trichocladum K. Schum.

Common names. Donkey grass, ikoka (Tanzania).
Natural habitat. Forest undergrowth and edges of bush forest.
Distribution. Throughout Central, East and southern Africa.
Description. A slender perennial with long trailing stems which may root at the nodes. Leaf-blades narrowly lanceolate. Panicle 5-16 cm long with very slender branches and pedicels; spikelets oblong, obtuse, 2.5-3 mm long (see Fig. 15.109).
Season of growth. At Mlingano, with 1 000 mm rainfall in north coastal Tanzania, it remains green throughout the year (van Voorthuizen, 1971).
Altitude range. Sea-level to 2 000 m.
Soil requirements. At Mlingano, it grows on deep red loam soils derived from gneiss (van Voorthuizen, 1971).
Ability to compete with weeds. It is an aggressive grass; it invaded pasture mixtures at Tanga, Tanzania (Hopkinson, 1970).
Tolerance to herbicides. In uncultivated land it can be controlled with TCA (trichloracetic acid), but in sisal, dalapon has been effective. Monuron at 9 kg/ha also exerts some control (Ivens, 1967).
Genetics and reproduction. 2n=32 (Fedorov, 1974).
Economics. A common weed of arable crops and waste land, particularly infesting coffee and sisal crops in East Africa; it is troublesome in sisal bulb nurseries (Ivens, 1967).
Animal production. From analyses it would appear to be adequate for beef cattle maintenance, but for finishing cattle for slaughter its crude protein content is too low during the dry months (van Voorthuizen, 1971).
Further reading. van Voorthuizen, 1971.

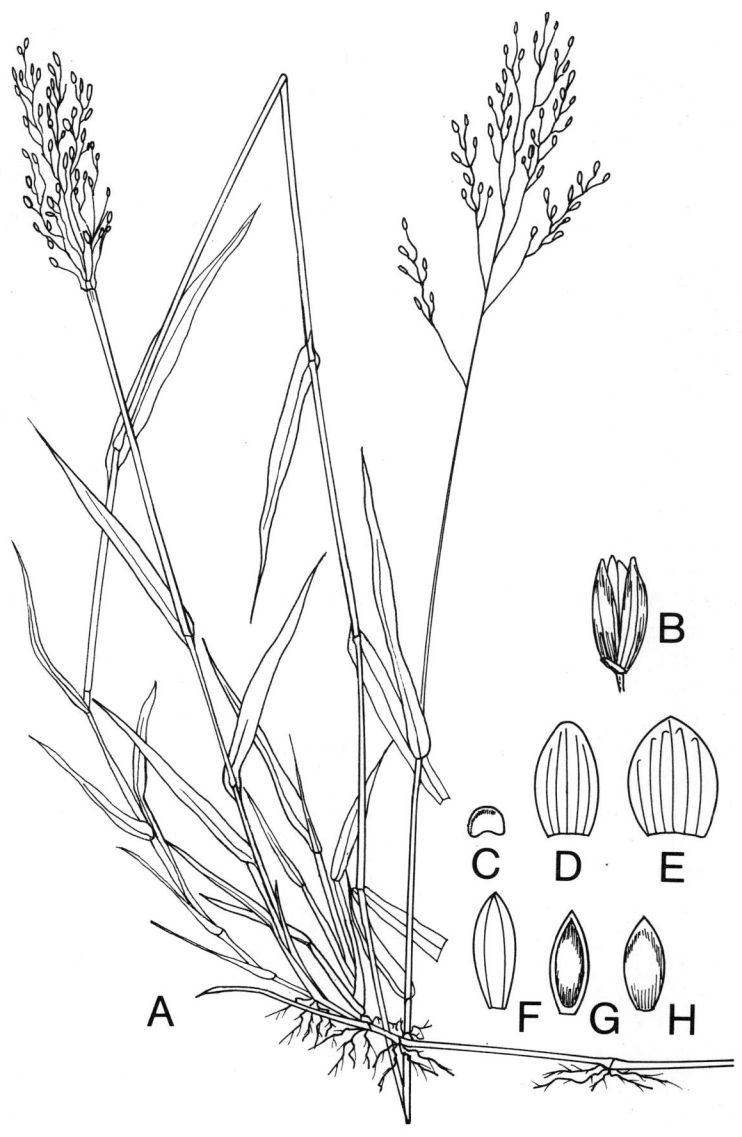

Figure 15.109. Panicum trichocladum. **A**-Habit and inflorescence **B**-Spikelet and pedicel **C**-Lower glume **D**-Upper glume **E**-Lemma of lower floret **F**-Palea of lower floret **G**-Upper floret **H**-Lemma of upper floret

Panicum turgidum Forsk.

Common names. Taman or tuman (Sudan), afezu (Nigerian Sahel), guinchi (eastern Sahara), thaman (Kuwait), markouba (Mauritania), du-ghasi (Somalia).
Natural habitat. Sand dunes on the edge of the Sahara, the arid Red Sea coast, and dunes in India.
Distribution. From Pakistan west through the Arabian peninsula to northern Africa (see Fig. 15.110).
Description. A perennial, growing as dense bushes up to 1 m tall. It bends over and roots at the nodes. Leaves few, stems hard, bamboo-like, solid, smooth and polished; 2.5-3 mm in diameter, emitting from the nodes panicles of branches in tufts from a swollen base. Panicle terminal, 3-10 cm long; spikelets 3-4 mm long, solitary (Cooke, 1958). The roots are remarkable for their clothing of root hairs to which fine sand adheres, giving them a felty appearance (Bor, 1960) (see Fig. 15.111).
Season of growth. Perennial.
Optimum temperature for growth. It is native to hot, dry, arid climates.
Latitudinal limits. 4-38°N, longitude 17°W-80°E.
Altitude range. From the Dead Sea Depression, at –380 m at Shor-es-Safiyeh, to 3 200 m in the Tibesti Mountains of the central Sahara.
Rainfall requirements. It occurs largely within the 250 mm isohyet.
Drought tolerance. Remarkable. In the open tussock communities in Mauritania and the western Sahara plants survive by dissociating themselves from one another rather than growing in association. The root-stock is stout and the root fibres strong and woody; the root hairs bind particles of fine sand by the extrusion of a glue which allows them to absorb more moisture from the soil (see *Brachiaria dura*).
Soil requirements. It is usually found on deep dune sand, but will grow in a well-drained latosol.
Fertilizer requirements. There is little response to nitrogen, but some to phosphorus and potash.
Ability to spread naturally. The plant usually spreads by the bending over of the stems until the nodes reach the ground, where they take root to form a new plant.
Land preparation for establishment. No preparation is necessary in the sandy environment in which it grows.
Dormancy. Grains will not germinate and establish unless 20-30 mm of rain, or its equivalent in irrigation water, is supplied, even though subsequent stages of growth are more or less tolerant to drought. Thus seedlings exist rarely, and reproduction is mainly vegetative (Williams & Farias, 1972).
Vigour of growth and growth rhythm. In the Sahel it begins flowering in

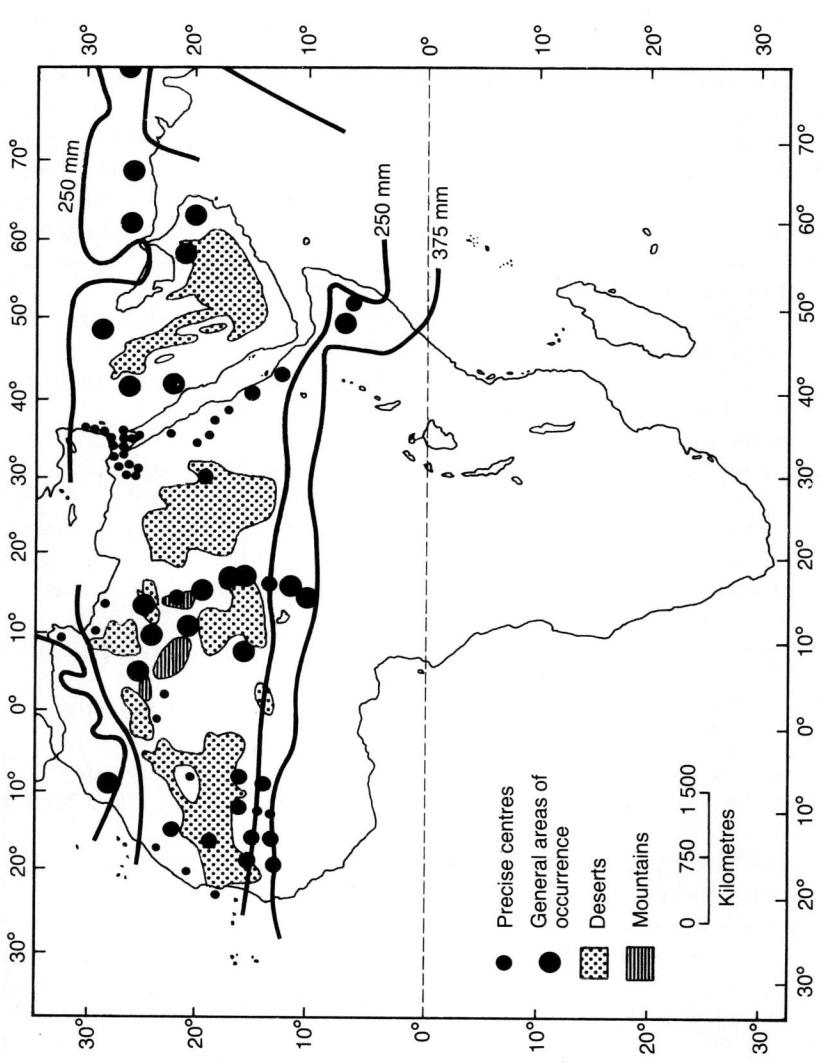

Figure 15.110. Distribution of *Panicum turgidum* (**Source:** Williams & Farias, 1972)

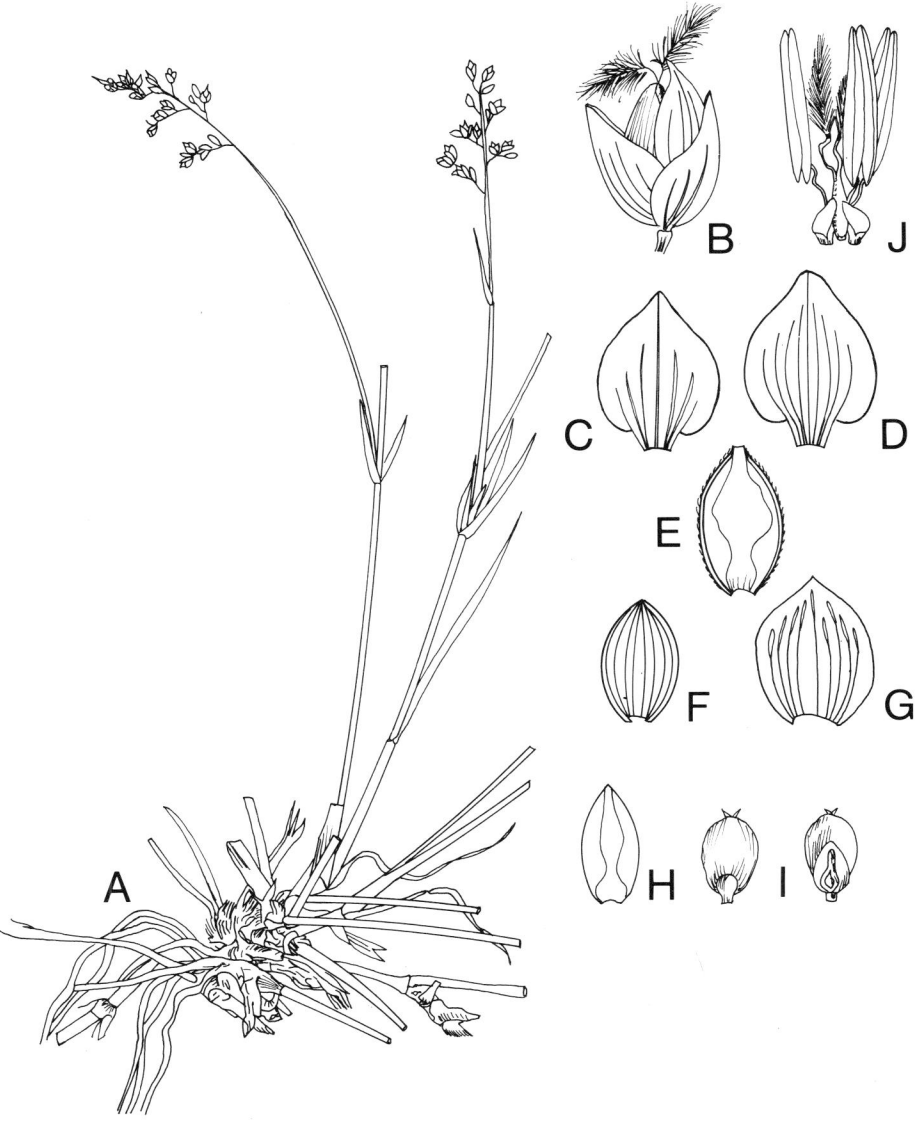

Figure 15.111. Panicum turgidum. **A**-Habit **B**-Spikelet **C**-Lower glume **D**-Upper glume **E**-Palea **F**-Upper lemma **G**-Lower lemma **H**-Palea **I**-Grain **J**-Flower

TABLE 15.51 *Panicum turgidum*

	DM	As % of dry matter		
		CP	CF	Ash
Flowering	59	4.9	36.4	8.7
Young culms, 20 cm	49	4.4	38.9	7.0
Stubble, 5 - 20 cm	49	3.6	40.2	5.6

Source: Boudet, 1975

August, continues flowering through to February and is mature in June (Boudet & Duverger, 1961). The tuft grows again each year.

Genetics and reproduction. 2n=18 (Fedorov, 1974).

Chemical analysis and digestibility. Figures from the Sahel are shown in Table 15.51.

Palatability. The young leaves and shoots are very palatable; even in the dry state it is still eaten by camels and donkeys.

Cultivars. There is a variation within the species, and there are forms with high grain yields. The vegetative yields of these forms in Near Eastern collections were up to twice those from Mauritania, especially at low levels of nutrients.

Value for erosion control. It is valuable for fixing dunes in the 100-400 mm rainfall areas. In the neighbourhood of the Red Sea, *P. turgidum* covers the whole of the coastal plain.

Economics. The Tuareg inhabitants of the Ahaggar Mountains in the central Sahara eat the grain (Bor, 1960); it is ground into a flour and made into porridge. It is also used for thatch, and mats (the Tuaregs use the stems with a weft of thin leather strips). The ashes are added to tobacco for chewing, and the powder from ground stems is used for healing wounds (Williams & Farias, 1972).

Main attributes. Its drought tolerance, sand-binding characteristics and grain production.

Main deficiencies. Its woodiness.

Further reading. Williams & Farias, 1972.

Panicum whitei J.M. Black

Common names. Pepper grass (Australia), sugar grass, pigeon grass.
Natural habitat. Lightly-flooded plains on the edges of the Channel Country in Queensland, Australia.
Distribution. All mainland Australian states except Victoria.
Description. A leafy annual growing rapidly after rain to 120 cm; culms geniculate, branched, erect, usually slender, shiny; nodes three to five, internodes often angular and shallowly grooved (see Fig. 15.112).
Season of growth. Early spring and summer rains germinate the seed.
Optimum temperature for growth. 25-35°C.
Latitudinal limits. 26-28°S.
Altitude range. 100-150 m.
Rainfall requirements. It occurs in the 300-500 mm annual rainfall belt, but grows only in the moister areas.
Drought tolerance. Its early seeding allows it to escape drought.
Tolerance to flooding. It tolerates shallow seasonal flooding.
Soil requirements. It occurs naturally on brown, black or grey clays, which have been seasonally inundated and are drying; also on red earth.
Fertilizer requirements. It occurs usually on soils reasonably high in phosphorus.
Ability to spread naturally. It spreads readily from annual seeding.
Suitability for hay and silage. Satisfactory hay and silage has been made from the grass.
Value as standover or deferred feed. It will stand as dry standing hay for over 12 months during the dry season, if rain does not fall to encourage mould growth.
Chemical analysis and digestibility. Contains almost 20 percent protein in seedling stage, 7.5 percent in the dry matter at maturity and 5 percent in the dry hay stage (Allen, 1949).
Palatability. Its palatability is variable; in general it is palatable and subject to preferential grazing. It seems to be more palatable when it is dry.
Toxicity. It has been mentioned as a cause of photosensitization in sheep.
Seed production and harvesting. It produces an abundance of seed and matures quickly. It could be harvested by direct heading.
Economics. It is an important component of the annual grasses which grow quickly after summer flooding of some 3 million hectares of Channel Country by the Georgina, Diamantina and Bulloo rivers and Cooper's Creek in Queensland.
Further reading. Whitehouse, Ogilvie & Skerman, 1947.

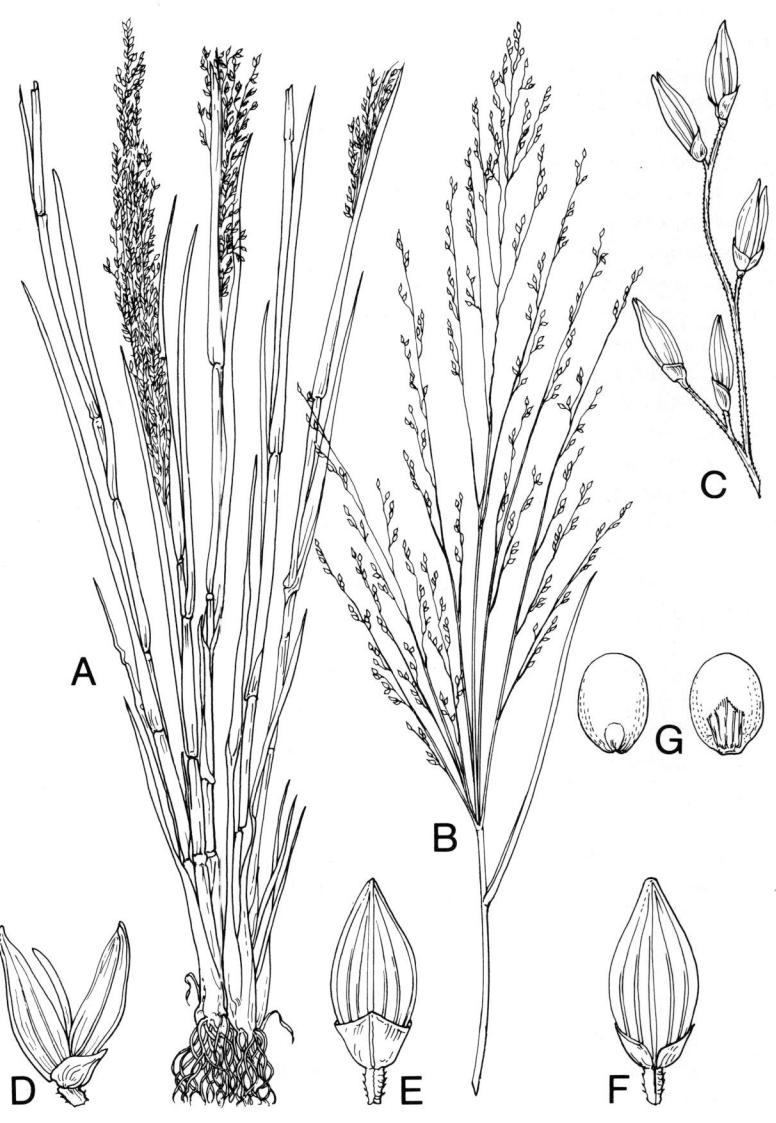

Figure 15.112. *Panicum whitei.* **A**-Habit **B**-Inflorescence **C**-Portion of panicle **D**-Opened spikelet **E,F**-Spikelet, front and back **G**-Grain

Paspalidium desertorum (A. Rich.) Stapf

Natural habitat. Annual grass zone of northern Kenya.
Description. A tufted perennial with numerous ascending stems with many nodes and long, narrow, somewhat succulent leaves (Bogdan & Pratt, 1967) (see Fig. 15.113).
Season of growth. Summer.
Rainfall requirements. About 375 mm (Bogdan & Pratt, 1967).
Drought tolerance. Excellent.
Soil requirements. It prefers loams and alluvial silts (Bogdan & Pratt, 1967).
Sowing methods. Broadcast on roughly disc-harrowed seed-bed.
Sowing time and rate. Early wet season, at 1.2 kg/ha (Bogdan & Pratt, 1967).
Number of seeds per kg. 1.2 million spikelets with one seed (Bogdan & Pratt, 1967).
Chemical analysis and digestibility. At an early flowering stage it showed a crude protein content of the dry matter of nearly 14 percent (Bogdan & Pratt, 1967).
Palatability. It is well grazed by cattle.
Economics. It is a useful pasture plant in the dry areas of the Sudan and Kenya.
Further reading. Bogdan & Pratt, 1967.

Figure 15.113. Paspalidium desertorum

Paspalum dilatatum Poir.

Common names. Paspalum (Australia), dallis grass (United States).
Natural habitat. Moist grassland.
Distribution. Native to the humid subtropics of southern Brazil, Argentina and Uruguay; now widely distributed.
Description. A leafy, tufted perennial with clustered stems arising from shortly creeping rhizomes; culms to 1 m; inflorescence of 3-5 racemes; spikelets ovate, about 3 mm long, fringed with silky hairs (Henty, 1969). The racemes have spikelets overlapping in rows along one side of a flattened axis (Chippendall & Crook, 1976) (see Fig. 15.114).
Season of growth. Spring and summer, declining at flowering in summer.
Optimum temperature for growth. Adapted to the humid subtropics. 30°C is optimal for leaf growth (Mitchell, 1956), 27°C for tillering and 22.5°C for flowering (Bennett, 1959).
Minimum temperature for growth. Seed production is inhibited at temperatures below 13°C (Knight, 1955). Mean temperature of coldest month, 2-10°C (Russell & Webb, 1976).
Frost tolerance. It is susceptible to frost but more tolerant than Rhodes grass. Its underground root-stock allows it to persist and recover from frost.
Latitudinal limits. About 28°N and 35°S (Russell & Webb, 1976).
Altitude range. Sea-level to 2 000 m.
Rainfall requirements. It requires a minimum of about 750 mm of annual rainfall; does best in a rainfall of about 1 250 mm, and in irrigated pastures. Maximum recorded, 1 650 mm (Russell & Webb, 1976).
Drought tolerance. The underground root-stock gives it considerable drought tolerance once it is established.
Tolerance to flooding. It is sensitive to flooding when actively growing, but is less so during its dormant period, when it tolerates inundation of up to one week's duration. Its density increased slightly under periodic 48-hour flooding (Squires & Myers, 1970).
Soil requirements. It grows best in heavy, moist, fertile, alluvial and basaltic clay soils.
Tolerance to salinity. It has little tolerance to salinity.
Fertilizer requirements. Paspalum needs high fertility and responds to a basic complete fertilizer mixture and subsequent dressings of nitrogen. Adequate nitrogenous fertilizer will stimulate the competitive ability of paspalum over associated *Axonopus* spp., whereas lack of fertilizer gradually allows *Axonopus* spp. to dominate. Linear nitrogen responses occurred up to applications of 135 kg N/ha, with recoveries of 72-80 percent (Colman & Lazenby, 1970). Cassidy (1971) obtained substantial growth increases when nitrogen at 224 kg/ha was applied during early summer (October-November), and a large growth rate increase from 7.8 to 56 kg DM/ha per day when applied at the height of its growing season (November to January). In another case, 89 kg

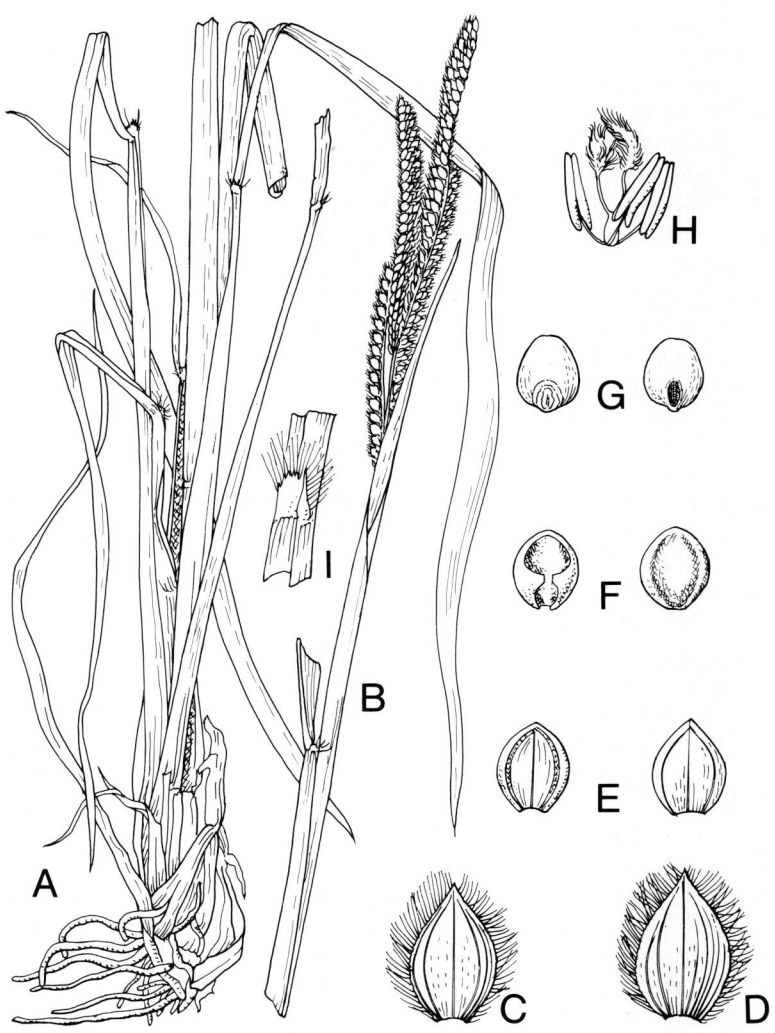

Figure 15.114. Paspalum dilatatum. **A**-Habit **B**-Inflorescence **C**-Lower lemma **D**-Upper glume **E**-Upper lemma, front and back **F**-Upper palea, front and back **G**-Grain **H**-Flower **I**-Ligule

N/ha were applied in April (early autumn) and increased dry-matter production from 39.2 to 54.9 kg/ha per day; in mid-May (late autumn) dry matter increased from 20.2 to 31.4 kg/ha per day. The use of strategic applications of nitrogen to paspalum can thus play an important role in extending the grazing season of this grass. The critical value for phosphorus expressed as a percentage of the dry matter at the immediate pre-flowering stage is 0.25.

Ability to spread naturally. It spreads readily by seed where conditions are suitable for germination.

Land preparation for establishment. A good, fine seed-bed prepared by ploughing, discing and harrowing gives best results. In favourable conditions a rough ploughing may suffice.

Sowing methods. It is generally drilled or broadcast as seed. It is often seeded into rice stubble in Texas and Louisiana (Bennett, 1973).

Sowing depth and cover. Surface sowing or drilling to a depth of 1-1.5 cm, and lightly covered is the usual practice.

Sowing time and rate. It is best sown just before the expected rainy season, at 9-14 kg/ha, but it can be sown at any time from spring to late summer.

Number of seeds per kg. 500 000 to 750 000.

Dormancy. There is some post-harvest dormancy (Whittet, 1965).

Seedling vigour. Good.

Vigour of growth and growth rhythm. It begins growth in the spring and grows vigorously in early summer. At Samford, Queensland (lat. 27°22′S) the growth rate declines rapidly in midsummer, is low in late autumn and dormant in winter (Shaw *et al.*, 1965). It flowers throughout the growing period (see Fig. 15.115).

Response to photoperiod. It is generally unresponsive to photoperiod, but a day length of 14-16 hours is best for seed production.

Response to light. It does not grow well in shade.

Compatibility with other grasses and legumes. As fertility declines paspalum pastures can be invaded by *Axonopus* spp. In planted pasture mixtures it is often sown with temperate grasses and clovers which make maximum growth when paspalum is comparatively dormant. It becomes sod-bound, and clovers can be sod-seeded into it during the dormant period with adequate fertilizer. It forms a very productive pasture with white clover (*Trifolium repens*).

Ability to compete with weeds. When fully established, paspalum competes well with broad-leaved weeds, but as fertility declines weedy grass species, e.g. *Axonopus* spp., invade.

Response to herbicides. To control paspalum in the young stage, use paraquat at 570 ml of a 200 g AI/litre product (e.g. Gramoxone) per 200 litres of water plus surfactant at 250 ml/200 litres water. Spray to the point of run-off. Mature plants can be sprayed with glyphosate at 2 litres of a 360 g AI/litre product (e.g. Round-up) per 200 litres of water in three applications, ten days apart (Tilley, 1977).

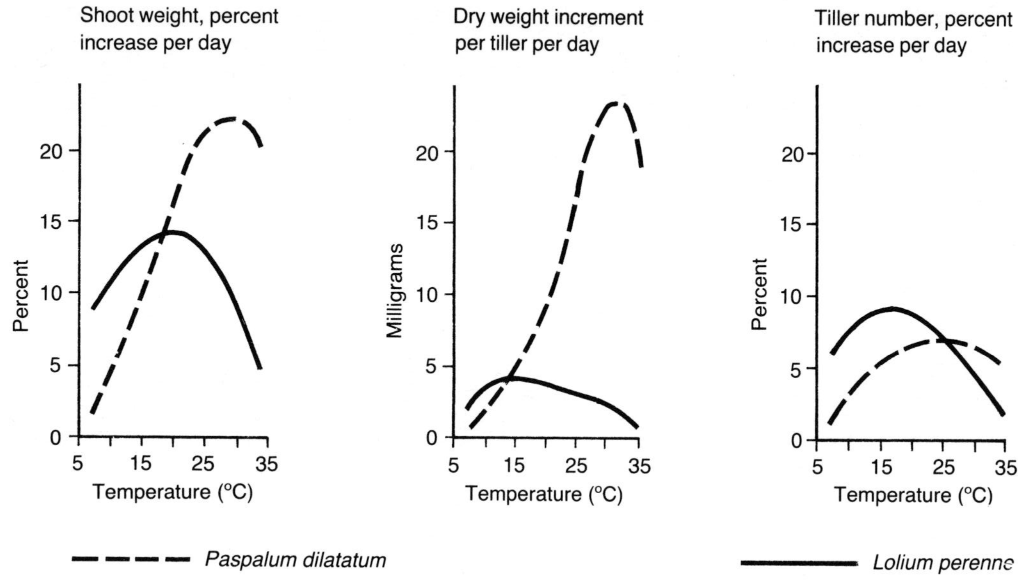

Figure 15.115. Growth of *Lolium perenne* and *Paspalum dilatatum* over a range of constant temperatures (**Source:** Mitchell, 1956)

———— *Lolium perenne*

– – – – *Paspalum dilatatum*

Response to defoliation. Paspalum will withstand heavy defoliation and, having an underground root-stock, it is protected from heavy grazing and trampling by livestock. Grazed no shorter than 5-7.6 cm it will produce up to three times the forage that it would if grazed lower (Bennett, 1973). Increasing frequency of defoliation reduces yields, but some recovery occurs with increasing application of nitrogen (Colman & Lazenby, 1970).

Grazing management. It should be kept grazed during the growing period to prevent if from flowering and becoming relatively unpalatable. This will also prevent ergot infection of the seed-head, which can cause poisoning. If cattle graze infected plants in the sphacelial stage, the sticky ergot clings to the face and legs of animals, assisting its spread and soiling the skin. As the paspalum sward ages it often becomes sod-bound and should be renovated periodically by ploughing, disc-harrowing or deep ripping (see Plate 44). In somé cases, a mole drainer will help to aerate and drain wet soils growing paspalum.

Response to fire. Paspalum pastures are seldom subject to fire, but, if burnt, they quickly recover from the root-stock when conditions are again favourable.

Genetics and reproduction. The chromosome number of the common type is $5x=50$. It is an obligate apomict by apospory and pseudogamy (Barnard, 1969). Bashaw and Forbes (1958) list it at $2n=40$ (sexual and apomictic) and $2n=50$ (apomictic).

Dry- and green-matter yields. At Samford, Queensland, an annual yield of 15 000 kg dry matter was recorded by Davies (1970). In Fiji an average yield of 5 311 kg DM/ha with a crude protein content of 9.9 percent was obtained over a three-year period (Roberts, 1970a, b). In the United States yields from 1 230-12 000 kg/ha are obtained (Bennett, 1973).

Suitability for hay and silage. Paspalum is suitable for both purposes. It should be cut before flowering to obtain the best quality hay. Paspalum which has gone to seed has a low feed value. Paspalum made good silage, but the pH was about 4.8 and the concentration of volatile acids below 5 percent of the dry matter, while NH_3-N accounted for nearly 20 percent of the nitrogen (Levitt *et al.*, 1962, 1964, 1965).

Value as standover or deferred feed. As the grass matures it declines markedly in feeding value, so it is best cured as hay or silage rather than left as standing material in the field.

Chemical analysis and digestibility. See Table 15.52.

Palatability. Paspalum in the pre-flowering stage is very palatable, but when infected at flowering by ergot its palatability declines rapidly.

Toxicity. The grass itself is not known to be toxic, but seed-heads parasitized by the ergot fungus *Claviceps paspali* can be toxic due to pyridine alkaloids in the sclerotia, which appear in late summer or autumn. Affected animals at first show excitement, distrust of people and a tendency to attack. Later they tremble, appear to lack muscular control, stagger and may fall. They recover

TABLE 15.52 *Paspalum dilatatum*

	DM	As % of dry matter				
		CP	CF	Ash	EE	NFE
Fresh, mid bloom, Trinidad	19.1	9.9	35.7	8.2	1.8	44.4
Fresh, 1st cut, early bloom, Tanzania	25.0	6.7	31.8	11.5	1.6	48.4
Fresh, 2nd cut, early bloom, Tanzania	30.0	6.2	31.9	10.3	1.5	50.1
United States		6.2-19.6	26-33		1.9-3.2	34-42
Fresh, leaves only, 50 cm, South Africa		13.3	36.2	6.9		
Fresh, stems only, 50 cm, South Africa		8.8	43.2	6.4		

	Animal	Digestibility (%)				
		CP	CF	EE	NFE	ME
Fresh, mid bloom, Trinidad	Sheep	54.1	69.4	31.7	58.2	2.10
Early bloom, 1st cut, Tanzania	Sheep	49.3	77.4	56.3	73.8	2.39
Early bloom, 2nd cut, Tanzania	Sheep	53.2	76.5	53.3	75.8	2.46

Source: Göhl, 1975

in a few days if removed from infected areas in the early stages of excitement (Everist, 1974).

Seed production and harvesting. Paspalum seeds freely, but the seed ripens from the tip of the racemes downwards and shatters as soon as it is ripe. It is thus hard to harvest, and viability is often low. Seed production is also affected by ergot infection. A day length of 14-16 hours and high temperatures are best for seed production, which is inhibited by temperatures below 13°C (Knight, 1955). Seed is rather slow to establish, but will remain dormant in the ground for months awaiting satisfactory germination conditions (Whittet, 1965). Harvesting should begin when 60-80 percent of the seed-heads are a light brown colour (Bennett, 1973). For storage, dry the seed at 60°C to a moisture content of 7-10 percent.

Seed yield. 90-500 kg/ha. The seed remains viable for two years (Jones, 1973).

Minimum germination and quality required for commercial sale. 60 percent germinable seed, 60 percent purity in Queensland. It is germinated at 20-35°C, moistened with KNO_3 solution. Germination is increased by exposure to light.

Cultivars. Bashaw and Forbes (1958) found three distinct cytological groups in the species: a yellow-anthered, erect type with pubescent spikelets and 40 chromosomes, in which the meiotic behaviour was regular and the mode of reproduction sexual; a semi-prostrate, purple-anthered strain with 40 chromosomes and extremely irregular meiotic behaviour, reproducing apomictically; and the common type with purple anthers and 50 chromosomes, reproducing apomictically. The second type has been recognized as a variety under the name of 'Prostrate' by the Georgia Coastal Main Experiment Station, United States. Two other varieties were released by the Louisiana Agricultural Experiment Station, United States, namely 'B-230' and 'B-430'. Both are alleged to have better seed production than the common type, and 'B-230' has a longer growing period. No varieties are registered in the United States nor on the OECD list for 1967 (Barnard, 1969).

Value for erosion control. Where paspalum is effectively established it exerts almost full erosion control. It is used to stabilize mine dumps in South Africa (Chippendall & Crook, 1976).

Diseases. The main disease of paspalum is ergot, caused by *Claviceps paspali*. The disease first appears in the form of a dark, sticky exudate from each spikelet or "seed". This sticky mass, produced during the "sphacelia" stage, contains many tiny spores which spread the disease to clean seed-heads. This stage gives rise to the "sclerotia", a kind of dormant spore that lodges in the infected spikelets, replacing the ovaries and grain. These are round, yellowish-grey bodies, 3 mm across, dry and firm. In autumn they ripen and fall to the ground, remaining dormant until the following spring. These are toxic. Preventing the paspalum from seeding helps to control the disease (Everist, 1974). Anthracnose (*Colletotrichum graminicola*) and leaf blight (*Helminthosporium microplus*) also attack paspalum.

Pests. On Queensland's Atherton Tableland it is attacked by root-destroying white grubs (*Lepidiota caudata* and *Rhopaea paspali*), which reduce pasture productivity (Quinlan & Edgley, 1975). The sugar cane borer (*Diatraea saccharalis*) sometimes attacks it (Bennett, 1973).

Economics. Paspalum is one of the most important summer forage grasses, and was one of the earliest species adopted for improved pastures.

Animal production. In the Murrumbidgee Irrigation Area in Australia, *P. dilatatum* pastures can carry 25 sheep per hectare during the growing season. At Badgery's Creek near Sydney (mean temperature of the coldest month 11.5°C, hottest 23.5°C) by planned irrigation, fertilizer supply and adding winter-growing species by sod-seeding, unsupplemented Friesian cows consistently produced more than 10 000 litres of milk per hectare per year. The paspalum content ranged from less than 5 percent in winter to 70 percent in spring and summer (September to February); the pastures supported all the nutritional needs of 2.5 cows per hectare in winter and 5 cows per hectare in spring and summer (Crofts & Pearson, 1977). Squires and Myers (1970) showed that paspalum was better than other warm-season grasses (*Cenchrus ciliaris, Eragrostis curvula, Panicum coloratum* and *Sorghum almum*) under irrigation as a pasture for sheep at Deniliquin, New South Wales, Australia (35°30'S).

Main attributes. Its palatability, productivity, ability to stand heavy grazing and trampling. Its compatibility with white clover.

Main deficiencies. Its heavy seeding and ergot susceptibility, low productivity and tendency to become sod-bound. Its short grazing season has led to its replacement by *Setaria* spp.

Further reading. Gardner, 1956; Whittet, 1965.

Paspalum distichum L.

Common names. Salt-water couch (eastern Australia), sea-shore paspalum (United States, Western Australia), grama bobo, grama salada (Peru), water couch grass (Malaysia), grama de mar (Cuba).

Natural habitat. A littoral species occurring in sands and muds near the sea-shore, and in saline soils and swamps (Barnard, 1969).

Distribution. Native to Africa and the Americas; now widely distributed throughout the tropics.

Description. A perennial with long creeping rhizomes and stolons; culms erect, from 15-60 cm. Leaves stiff, narrow, about 15 cm long; racemes usually two; spikelets elliptical, 3.5-4 mm long. It differs from *P. paspaloides* in that the upper glume is glabrous with the mid-nerve sometimes suppressed; the leaf-blades are usually narrower, up to 4 mm wide, often less, folded and with inrolled margins; racemes up to 4 cm long, often less, usually spreading horizontally or deflexed; lower glume absent (Chippendall, 1955) (see Fig. 15.116).

Season of growth. A summer-growing perennial.

Frost tolerance. Leaf-blades turn brown and deteriorate after the first frost, but stolons survive.

Latitudinal limits. About 30°N and S.

Altitude range. Just above sea-level.

Rainfall requirements. It occupies salt seepage areas in the 400-750 mm rainfall area of Western Australia. It must have moist areas in summer.

Drought tolerance. It needs good summer rain, but persists during the dry season.

Tolerance to flooding. It will tolerate waterlogged conditions and periodic flooding in salt swamps and by tidal waters (Colman & Wilson, 1960) (see Fig. 15.117).

Soil requirements. Adapted to marshy, brackish conditions and saline soils which are moist in summer.

Tolerance to salinity. Excellent. In Western Australia it has grown successfully in salt seepage patches where the ground water just below the surface contained 3 000 mg of sodium chloride per litre. When the salt content was as high as 12 000 mg per litre however, it did not grow satisfactorily. It grows along the sea front in Suriname and is frequently flooded with sea water (Dirven, 1963a, b). It can stand lawn irrigation with water containing up to 14 000 mg per litre total soluble salts (Malcolm & Laing, 1976). Sea water is generally too saline for *P. distichum*.

Fertilizer requirements. In non-salty soils it responds to phosphorus and nitrogenous fertilizers.

Ability to spread naturally. Excellent, spreading by rhizomes and stolons.

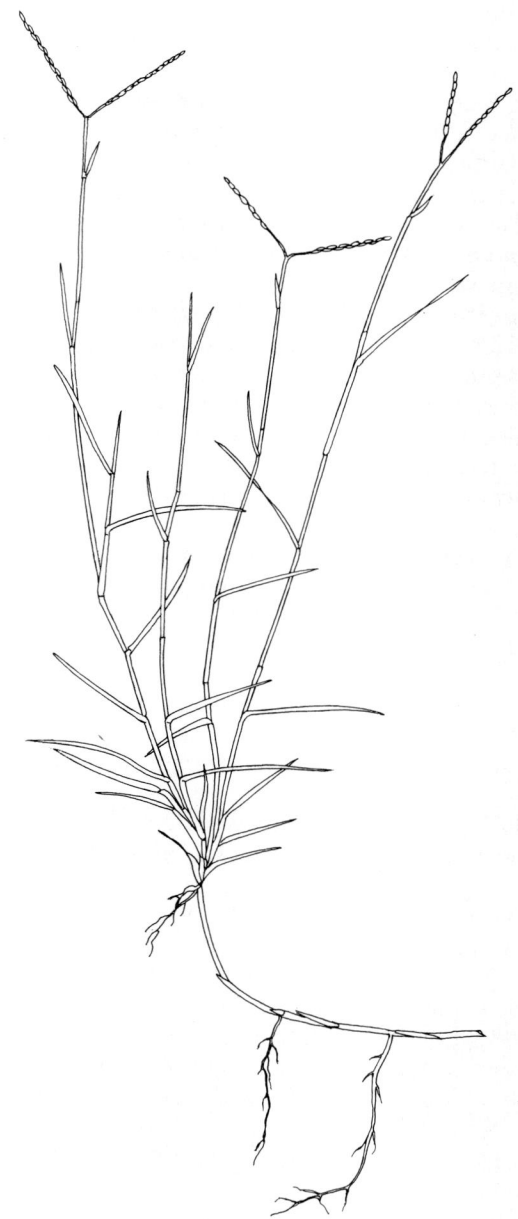

***Figure** 15.116.* Paspalum distichum

Land preparation for establishment. Minimum. Holes can be dug or the roots pushed into the moist soil.
Sowing methods. By pieces of rooted sod, about 6-8 cm square, at 1-m intervals.
Sowing depth and cover. The top of the sod should be planted at ground level.
Sowing time and rate. Sow in spring for a good strike in Western Australia (Burvill & Marshall, 1951).
Dormancy. Seed shows some dormancy which seems to require cold to break (Carpenter, 1958); it germinates best at 20-30°C.
Vigour of growth and growth rhythm. Stolons remain green all year, especially if growing in water.
Response to light. It grows as well as *Cynodon dactylon* and *Stenotaphrum secundatum* (buffalo grass) in shade, with better winter survival.
Ability to compete with weeds. It competes very successfully with weeds.
Response to defoliation. Once it is established it is virtually impossible to graze it out (Malcolm & Laing, 1976).
Grazing management. It is very productive if no more than half of the current season's growth (by weight) is grazed off. A 90-day grazing rest improves plant vigour and produces a forage reserve. Hard-surfaced soils can be cultivated to assist the runners' rooting. The plants should be well established before grazing is allowed.
Response to fire. Burning is not recommended as a management practice.

Figure 15.117. High-quality Angus cattle grazing *Paspalum distichum* on salt marsh rangeland in Louisiana, United States. Cattle are 1.6 kilometres from high ground, thanks to cattle walkways, without which they will normally enter the marsh to a distance of only 175 metres (**Source:** USDA Soil Conservation Service)

Genetics and reproduction. The somatic chromosome number is 2n=20, (sexual reproduction) (Bashaw, Hovin & Holt, 1970).

Dry- and green-matter yields. It is more productive than *Sporobolus virginicus* and common couch (Millington, Burvill & Marsh, 1951).

Feeding value. An important forage grass. In Suriname, Dirven (1963a, b) says the nutritional value of the grass is low and cattle grazing it are in poor condition.

Palatability. It is quite palatable.

Toxicity. No records of toxicity have been found.

Seed production and harvesting. *P. distichum* flowers freely in summer but some clones are markedly self-sterile so that little seed is produced. Some clones are reasonably self-fertile and cross-pollination between clones may result in satisfactory seed set (Carpenter, 1958).

Value for erosion control. It is useful in erosion control on salted lands and areas reclaimed from tidal influences.

Economics. Used by some Angolan farmers for composting the sandy dune soil of their vegetable farms (Rose-Innes, 1977). It is a very good lawn grass where only salty water is available, yet also does well with fresh water. It is good fodder grass, but may become a serious weed in irrigation channels. It can be a useful coastal sand binder in Australia.

Animal production. No figures are available. It is a useful fodder grass which stands heavy grazing.

Main attributes. Its adaptability to saline land, thus providing soil stabilization and beach protection, as well as light grazing.

Main deficiencies. Its low seed production.

Further reading. Burvill, 1956; Cameron, 1959; Carpenter, 1958; Logan, 1958; Malcolm & Laing, 1976.

Paspalum nicorae **Parodi**

Common name. Brunswick grass (United States).
Natural habitat. Moist sandy soils of the littoral.
Distribution. Southern Brazil, Argentina, Paraguay, Uruguay.
Description. A perennial with long, deep and vigorous rhizomes. Culms erect, generally less than 40 cm tall, with basal leaves and short internodes. Leaves erect, narrow, grey-green, 10-20 cm long and 2-3 cm broad with sparse hairs on the dorsal surface. Inflorescence grey-green, generally with two to five racemes, 2-5 cm long; spikelets oval to elliptical, 2.4-2.8 mm long and 1.5 mm broad; sterile lemma usually transversely wrinkled; glume with very short, fine hairs visible only under magnification. Seed dark hazel, glossy, and pronouncedly convex (Strickland, 1979, personnal communication). It is like Bahia grass (*P. notatum*), being perennial and sod-forming, but, unlike it, spreads by rhizomes rather than stolons (Beaty, Powell & Lawrence, 1970) (see Fig. 15.118).
Season of growth. Autumn.
Minimum temperature for growth. It has some cold tolerance, beginning growth in early spring (March) in Georgia, United States.
Fertilizer requirements. It responded to nitrogen up to 336 kg/ha in Georgia.
Ability to spread naturally. It spreads well by means of rhizomes (Beaty, Powell & Lawrence, 1970).
Vigour of growth and growth rhythm. Vegetative growth at Americus, Georgia, United States, begins in March (spring) and continues until November. Most forage is produced from 20 April to 20 September.
Response to light. It is shade tolerant; growth is significant under pine trees in the United States (Beaty, Powell & Lawrence, 1970).
Response to defoliation. It has been closely grazed for 20 years in the southeastern United States and Beaty, Powell and Lawrence (1970) cut it at intervals of one to six weeks to a height of 3 cm. The largest production with nitrogen fertilization occurred with cuts every three weeks. It regenerates well after grazing or cutting.
Genetics and reproduction. $2n=40$. It is apomictic and pseudogamous (Burson & Bennett, 1970; Bashaw, Hovin & Holt, 1970).
Palatability. It is well grazed.
Seed production and harvesting. Seed set was 18.6-48 percent for open pollination and 2.2-26.4 percent from self-pollination (Burson & Bennett, 1970).
Value for erosion control. It is excellent for the stabilization of waterways and surfaces of airfields.
Further reading. Beaty, Powell & Lawrence, 1970.

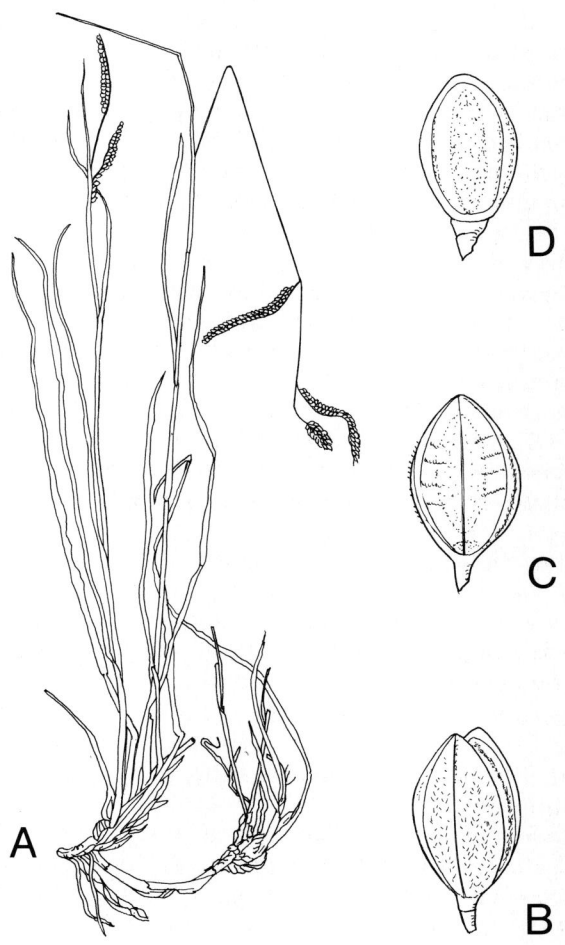

***Figure 15.118.** Paspalum nicorae.* **A**-Plant **B**-Spikelet with second glume **C**-Sterile spikelet and palea **D**-Floret

Paspalum notatum Flügge

Common names. Bahia grass (United States), jengi brillo (Costa Rica), batatais (Brazil), Paraguay paspalum (Zimbabwe), tejona (Cuba).

Distribution. It originated in South America and is widespread in the southern United States, Central and South America. Introduced into Africa and Australia.

Description. A low-growing perennial spreading by short, stout, woody runners and by seed. The runners have many large, fibrous roots which form dense, tough sods, even on drought-prone sandy soils. The leaf-blades are generally hairy on the margins and less than 1 cm wide. It seeds prolifically during the summer, the seed stalks 30-75 cm high, usually with two (sometimes three) racemes, each about 6 cm long. Seeds oval, yellowish-green, glossy and 3 mm in diameter (Wheeler, 1950) (see Fig. 15.119).

Season of growth. Spring, summer and autumn.

Optimum temperature for growth. 25-30°C. Mean 20.2°C ± 3.2 (Russell & Webb, 1976).

Minimum temperature for growth. Mean temperature of the coldest month 7.8°C ± 5.3 (Russell & Webb, 1976).

Frost tolerance. Good; 90 percent survival from the first winter in Queensland on the Darling Downs (Jones, 1969). It was eliminated by a temperature of –12°C in North Carolina, United States.

Latitudinal limits. About 25°N and 30°S (Russell & Webb, 1976).

Altitude range. Sea-level to 2 000 m (Anderson, 1969).

Rainfall requirements. It needs a moderate to high subtropical rainfall, with a minimum of about 750 mm per year. Mean 1 500 mm ± 586 (Russell & Webb, 1976).

Drought tolerance. Because of its deep root system it has good drought tolerance.

Tolerance to flooding. It is fairly tolerant of flooding. In New South Wales, Australia, it was submerged for 25 days, but it was non-productive during this time (Colman & Wilson, 1960).

Soil requirements. It can adapt to a wide range of soils, but is best suited to sandy soils.

Tolerance to salinity. No record has been found.

Fertilizer requirements. On Leon fine sand in Florida it responds well up to 224 kg N/ha per year (Beaty, Powell & Etheredge, 1963; Blue, 1970). On these poor sandy soils it fails to grow without 10-12 kg/ha of copper (as copper oxide or sulphate) because of soil deficiency (Hodges, Jones & Kirk, 1958). One or two early applications of 25-30 kg N/ha will hasten development. CSIRO workers in Queensland (Weier, 1976) have shown it has an active nitrogenase system and, over a growing season of 12 weeks, *P. notatum* fixed nitrogen at the rate of 4 kg N/ha per day. Schegel (1978) associates this with

Azotobacter paspali. Recovery rate of added fertilizer nitrogen increased yearly to the sixth year when it reached about 60 percent at 112 kg/ha and 70 percent at an application rate of 224 kg/ha (Blue, 1970).

<u>*Ability to spread naturally*</u>. It spreads rapidly by short stolons and seeds.

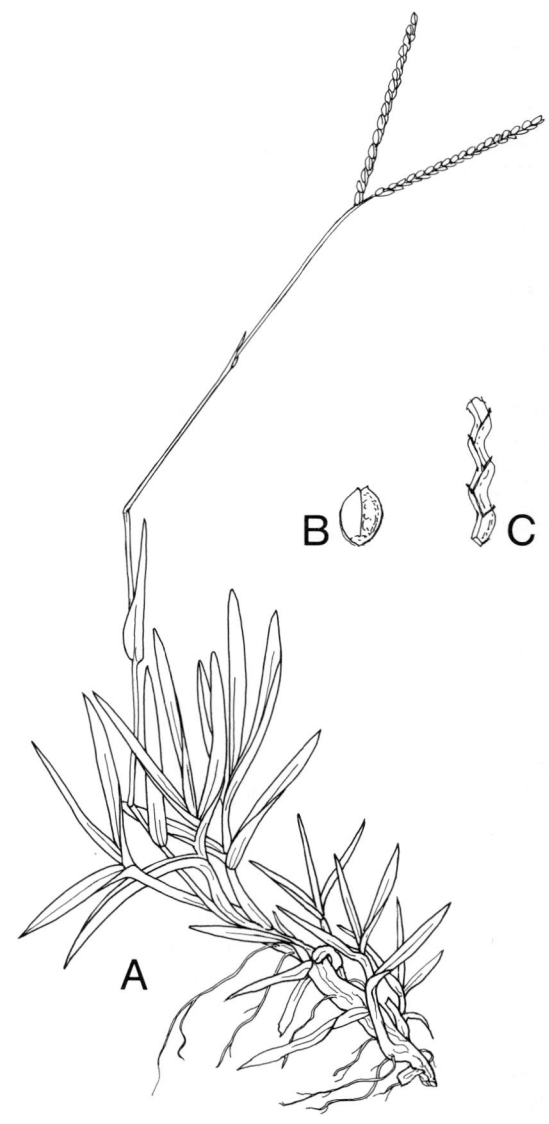

Figure 15.119. Paspalum notatum. **A**-Habit **B**-Spikelet **C**-Portion of raceme

Land preparation for establishment. A fully cultivated fine seed-bed is required.
Sowing methods. The seed is drilled into a fine seed-bed and rolled. Seed is also spread readily in animal dung and will germinate therefrom. Sods may also be laid.
Sowing time and rate. Drill seed in the spring, at 2-5 kg/ha.
Sowing depth and cover. Not deeper than 1 cm, rolled to cover.
Number of seeds per kg. 550 000 (Queensland), 330 000 (United States).
Dormancy. Germination is usually slow. It improves with up to three years' storage and then declines. Dormancy can be broken by treatment with sulphuric acid (Burton, 1945) and by hammer milling.
Seed treatment before planting. Hammer milling improves seed handling and germination; moistening with KNO_3 solution at 25-35°C also helps germination.
Seedling vigour. It is not vigorous and can be subject to weed competition, hence the need for a fine, clean seed-bed.
Vigour of growth and growth rhythm. After a slow start it grows rapidly when established and fertilized. It produces little feed during the winter.
Response to photoperiod. It is a long-day plant (Evans, Wardlaw & Williams, 1964).
Response to light. It does not grow well in shade.
Compatibility with other grasses and legumes. In the transition zone between temperate and subtropical species, Bahia grass tends to invade temperate species (Hoveland, Hasland & Rodriguez-Kabana, 1977). White clover and crimson clover can grow with it, if well fertilized. In Zimbabwe, Clatworthy (1970) grew it successfully in combination with *Trifolium semipilosum* and *Lotononis bainesii*.
Ability to compete with weeds. It can become weed infested early in its growth but when established it forms a dense sod, making it a useful lawn grass (Chippendall & Crook, 1976). It can itself become a weed.
Tolerance to herbicides. No records have been found.
Response to defoliation. Most of the forage of Bahia grass sods lies near the soil surface so it should be grazed closely and frequently to obtain the best quality material. Six-week cutting intervals were better than 2-, 3- or 4-week intervals (Beaty, Stanley & Powell, 1968).
Grazing management. Once established, Bahia grass can be grazed heavily to near soil level. It should be fertilized at up to 224 kg N/ha per year. Frequent defoliation over 24 months had little effect on the sward (Beaty, Brown & Morris, 1970). Cutting at six-week intervals gave the best yield and nutritive value in Brazil (Prates, 1977). In the early establishment period, mow the grass every three to four weeks to suppress weeds. Top-dress with nitrogen at 35 kg/ha during the first year, then increase as necessary. In Georgia, United States, 550 kg/ha of a 5:10:15 NPK mixture plus 110 kg N/ha is usually applied annually.

Response to fire. Winter burning is sometimes practised in the United States to get rid of dead litter. If the litter cover is light little harm is done, but if litter is abundant, such as after a seed harvest, damage can result (Stephens & Marchant, 1960).

Genetics and reproduction. 2n=20 (sexual), 2=40 (apomictic) (Bashaw, Hovin & Holt, 1970).

Dry- and green-matter yields. In Georgia, United States, yields of air-dried material varied from 5 947-7 240 kg/ha over a four-year period with five different cultivars (Stephens & Marchant, 1960). At Howard, southeastern Queensland, under a 1 075 mm summer-dominant rainfall, fully fertilized, it yielded a three-year mean of 6 859 kg DM/ha with CPI 9073 introduction (Evans, 1967a).

Suitability for hay and silage. It proved unsuitable for silage in Panama (Medling, 1972). It is not a very suitable hay plant because of its low yield, and it is hard to mow. If well fertilized and vigorous it can make useful hay.

Value as standover or deferred feed. It is not very useful because of its coarseness.

Chemical analysis and digestibility. In Costa Rica, analyses of material at floral initiation revealed 6.75 percent crude protein, 28.25 percent crude fibre, 43.32 percent nitrogen-free extract, 1.29 percent ether extract and 10.39 percent ash in the dry matter on a 10 percent moisture basis (Gonzalez & Pacheco, 1970). In Laos, Chavancy (1951) recorded 13 percent crude protein and 34.5 percent crude fibre in the dry matter of fresh material in late bloom.

Palatability. It is reasonably palatable in spring and summer, but is too coarse from midsummer to autumn (Burton, 1945). In the West Indies, Motta (1956) reported that it lacked palatability in trials held between 1948 and 1954. It is not as palatable as *Cynodon aethiopicus*.

Toxicity. No major toxicity has been reported. In some strains which are susceptible to paspalum ergot there may be slight toxicity.

Seed production and harvesting. Bahia grass ripens progressively over the summer in the United States and at no time is all the seed mature. Thus, a series of harvests with a beater or stripper gives the highest yields. Combine harvesting can be carried out, but yields are reduced. Seed should be dried thoroughly immediately after harvest.

Seed yield. 110-350 kg/ha. 112 kg N/ha early in the season will improve seed yields (Stephens & Marchant, 1960).

Minimum germination and quality required for commercial sale. 60 percent germinable seed, 60 percent purity in Queensland.

Cultivars.
- 'Common Bahia Grass' — described above.
- 'Paraguay' — 'Introduced to the United States from Paraguay. The leaf-blades are hairier and narrower than 'Common Bahia'. It seeds heavily and the seed is of good quality. It is very winter-hardy, but is too tough for grazing

from midsummer to autumn. It is palatable in spring and summer. It is a lawn grass.
- 'Pensacola' — narrow-leaved like 'Paraguay', but less hairy. The seeds are smaller than those of either 'Common' or 'Paraguay', and more seeds are produced per head. It seeds heavily, but the seed shatters badly. The seed germinates well and the grass establishes a sod quickly. It is fairly frost tolerant and growth starts early in the spring. It is fairly resistant to ergot.
- 'Argentine' — a medium broad-leaf type which makes a rapid and abundant growth and is more frost resistant than cv. Common and cv. Paraguay. It does not make early spring growth, but grows well in later summer and autumn.
- 'Wallace' and 'Tampa' — not used much. Cultivar Wallace gives low production and cv. Tampa is easily killed by frost.
- 'Tifhi-1' — hybrid which has outyielded cv. Pensacola at Tifton, Georgia, United States.
- 'Paraguay' — the main cultivar used at the Henderson Research Station, Zimbabwe, but several accessions promise to offer alternatives (Mills & Boultwood, 1978).

There is also a var. *saurae* Parodi.

Value for erosion control. Bahia grass is often used to stabilize terraces against erosion.

Diseases. Argentine Bahia grass is attacked by paspalum ergot (*Claviceps paspali*). *Helminthosporium micropus* sometimes causes damage.

Pests. Cultivar Paraguay 22 is resistant to the sting nematode (*Belonolaimus longicaudatus*) which affects cv. Pensacola and *Hemarthria altissima* (Boyd & Perry, 1972).

Economics. Bahia grass has resistance to internal root nematodes (root-knot and meadow nematodes) which are serious pests of tomatoes. In the United States, Bahia grass is often used as a ley in a four-year rotation to reduce the damage to tomatoes (Stephens & Marchant, 1960). The grass is easily ploughed out.

Animal production. In Georgia, cv. Argentine produced 405 kg beef per hectare, cv. Pensacola 439 kg/ha and 'Tifhi-1' 514 kg/ha per year (Johnson & Gurley, 1960). Stocking rates over a four-year period with five cultivars averaged about five beasts per hectare (Stephens & Marchant, 1960).

Main attributes. Bahia grass is a deep-rooted perennial which stands heavy grazing and forms a dense turf.

Main deficiencies. It is relatively unpalatable to cattle (but the least palatable cultivars give the best live-weight gains), it suffers in periods of drought and does not produce a large volume of herbage (Harker, 1962). As a lawn grass it is often difficult to mow.

Further reading. Burton, 1945; Stephens & Marchant, 1960.

Paspalum paspaloides (Michx.) Lams.- Scribn.

Common names. Water couch (Australia), gramilla blanca, pata de gallina, salaillo (Peru), knot grass (Hawaii), groffe doeba (Suriname), eternity grass (United States), gharib (Iraq).
Natural habitat. Freshwater swamps.
Distribution. Widely distributed over tropical regions.
Description. A creeping, perennial grass often growing in water. Leaves 4 cm long and 3 mm wide. The seed-head, which is carried on a stalk 5-30 cm long, consists of spikes with the seeds arranged in two rows along the axis. The "seeds" are elliptical, about 3 mm long and do not have hairs (see Fig. 15.120).
Season of growth. Summer.
Frost tolerance. It is affected by frost, but not usually killed.
Rainfall requirements. It grows in a rainfall regime of 500-1 500 mm, and in swamps where water accumulates.
Drought tolerance. It endures drought because of its moist environment.
Tolerance to flooding. It tolerates standing water and is a good soil binder but does not thrive in brackish water (Colman & Wilson, 1960).
Soil requirements. It usually grows in clay bottoms.
Tolerance to salinity. It tolerates moderate salinity (Leithead, Yarlett & Shiflet, 1971).
Fertilizer requirements. It is usually not fertilized, the accumulation of organic matter in its environment yielding sufficient sustenance.
Ability to spread naturally. It spreads quickly by rhizomes and stolons, and by seed under suitable conditions.
Sowing methods. It is propagated by roots and sods (Breakwell, 1923).
Tolerance to herbicides. To kill *P. paspaloides,* spray with diuron at 10 kg per hectare in 1 350 litres of water with 1 percent non-ionic wetting agent (Kleinschmidt & Johnson, 1977).
Response to defoliation. It stands heavy grazing.
Grazing management. For maximum production, no more than half of the current season's growth (by weight) should be grazed off. Grazing deferments of 60-90 days every two to three years during the growing season increase seed production and improve plant vigour (Leithead, Yarlett & Shiflet, 1971).
Response to fire. Controlled burning is not recommended. It withstands accidental burning if there is water above the soil level.
Genetics and reproduction. 2n=40, 48, 60 (Fedorov, 1974).
Suitability for hay and silage. It makes useful hay where soil conditions become dry enough for it to be harvested.
Palatability. It is readily grazed by cattle and horses in Australia.

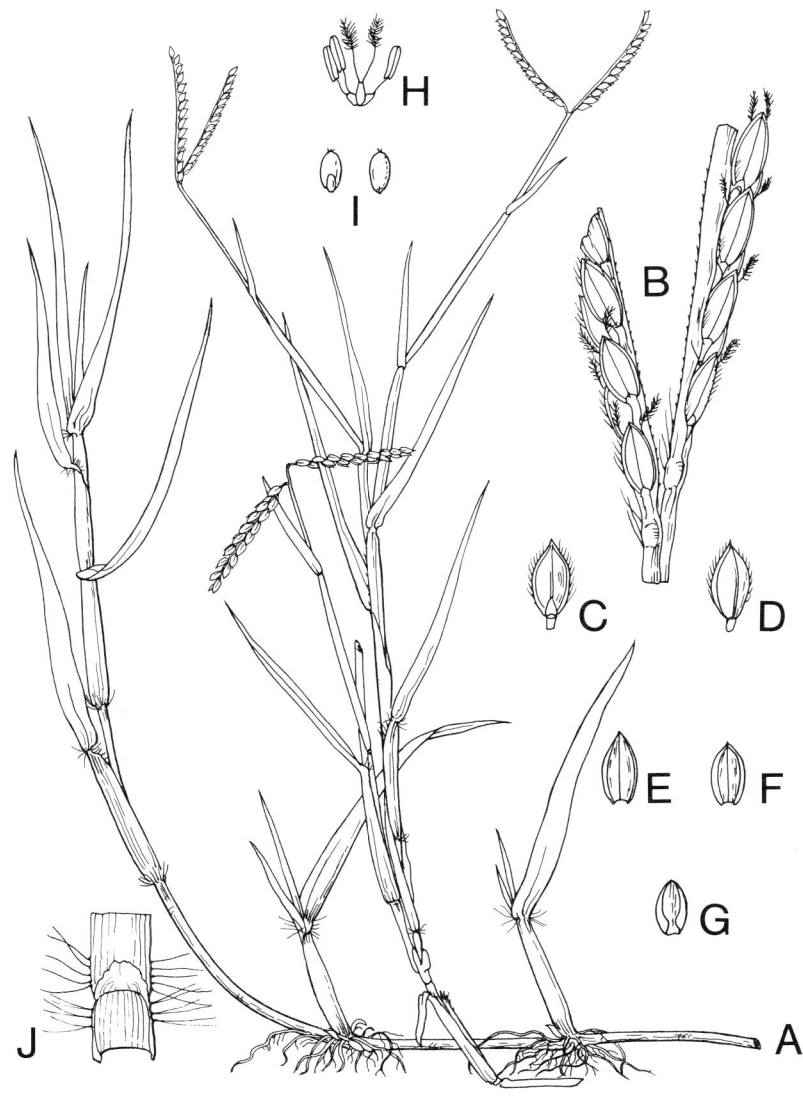

Figure 15.120. Paspalum paspaloides. **A**-Habit **B**-Portion of raceme **C,D**-Spikelets **E,F**-Lemmas **G**-Palea **H**-Flower **I**-Grain **J**-Ligule

Seed production and harvesting. It produces seed two to three times a year.
Cultivars. In Queensland there are two varieties: var. *normale,* a broad-leaved creeping grass which was introduced and has become naturalized; and var. *littorale,* with shorter-running underground stems, narrower leaves and erect stems, occurring only in coastal swamps.
Diseases. It is occasionally attacked by ergot of *P. dilatatum* and can then be toxic (Everist, 1974).
Economics. The seed is choice food for wild ducks on fresh water marshes. It is an important forage species for the swamp buffaloes of Iraq.
Animal production. No quantitative figures have been cited, but it is a valuable forage grass, especially as it is available when other grasses have dried off in the low-rainfall months.
Main attributes. Its palatability, its ability to grow in swampy places and tolerate flooding, its availability during the dry months.
Further reading. Leithead, Yarlett & Shiflet, 1971.

Paspalum plicatulum Michx.

Common names. Plicatulum (Australia), brown seed paspalum (United States).

Natural habitat. Adapted to marshy or aquatic habitats, usually growing in shallow water in the United States.

Distribution. Native to subtropical and tropical America. Widespread in Florida, Venezuela and Brazil, and introduced to Australia.

Description. Tufted perennial, with open, tussocky habit, up to 1.2 m high. Leaves usually about 40 cm long, 10 mm wide, folded at the base, pilose on the upper surface at base near margins, glabrous toward the top; leaf-sheaths glabrous, ligule 1.5 mm long. Inflorescence of 10-13 racemes, 2-6 cm long, 1.5-2 mm wide, usually one of a pair not developed at base of raceme; sterile lemma wrinkled just inside the margins, five-nerved, glabrous; glume pubescent, five-nerved. Seeds dark brown, shining (Barnard, 1972) (see Fig. 15.121).

Season of growth. Summer, early spring growth slower than *P. dilatatum* and *P. wettsteinii*. Reaches a marked summer peak and declines with the onset of flowering in late March.

Optimum temperature for growth. 18.9-23.3°C (Russell & Webb, 1976).

Minimum temperature for growth. Minimum temperature for coldest month, 6.1°C (Russell & Webb, 1976).

Frost tolerance. Susceptible to frost, but maintains nutritive value well after frosting. The crowns are not killed. Frosts seriously affect nutritive value of cv. Rodd's Bay.

Latitudinal limits. 17-28°S in Guatemala and 20°S to 31°N in the American tropics.

Rainfall requirements. At least 760 mm, preferably more than 1 000 mm up to 2 036 mm a year (Russell & Webb, 1976).

Drought tolerance. It has good drought tolerance, especially over short dry spells. Cultivar Rodd's Bay is the best for low-moisture conditions.

Tolerance to flooding. It is highly tolerant of waterlogging and flooding for short periods; flooding for 35 days at 7.5 cm did not affect it. Deeper water at 12.5 cm caused some wilting (Colman & Wilson, 1960).

Soil requirements. It is tolerant of a wide range of soils, including soils of low fertility which are too poor for *Paspalum dilatatum*. It is also fairly tolerant of high aluminium in soils (Spain & Andrew, 1977). It grows well on strongly acid to neutral, poorly drained clay loams and on excessively drained deep sandy soils (Leithead, Yarlett & Shiflet, 1971).

Tolerance to salinity. No record has been found.

Fertilizer requirements. A basic fertilizer of 250-650 kg/ha of superphosphate (with potassium if needed) with 100 kg N/ha at the beginning of the growing

season for seed production is used. Although it responds to nitrogen, it will produce better than other grasses under low soil nitrogen supply. Calcium deficiency shows when the distal ends of the young leaves turn brown, twist and die; seed setting may also be reduced (Humphreys, 1975).
Ability to spread naturally. It spreads quickly by seedlings.
Land preparation for establishment. Full land preparation is necessary to give a fine seed-bed, as the early vigour of the seedlings is not high.

Figure 15.121. Paspalum plicatulum. **A**-Plant **B,C**-Spikelet, front and back **D**-Fertile lemma

Sowing methods. Seed is drilled into a well-prepared seed-bed. *P. plicatulum* will also reproduce from short rhizomes.

Sowing depth and cover. 10-15 mm, rolled to cover.

Sowing time and rate. Early summer to flower in autumn at 2-3 kg/ha (Ostrowski, 1978).

Number of seeds per kg. 780 000 for cv. Rodd's Bay and 1 million for cv. Hartley.

Dormancy. There is no post-harvest dormancy, but seed remains viable for two years (Jones, 1973). Seeds germinate at 20-35°C, moistened with KNO_3 solution.

Seedling vigour. Not good.

Vigour of growth and growth rhythm. Plicatulum has a long growing period from October to April and a short flowering period, April-May (Davies, 1970) in Queensland, after which its growth declines.

Response to photoperiod. It has a short-day flowering response; floral initiation occurred at Brisbane (lat. 28°S) at a day length of 13.1 hours (Chadokar & Humphreys, 1974).

Response to light. It does not do well in shade.

Compatibility with other grasses and legumes. It combines well with a range of legumes to maintain a stable pasture. Silverleaf and greenleaf desmodium, siratro, lotononis, stylo and white clover have each formed a stable pasture with it (see Fig. 15.122). It does not become sod-bound. If sown with Nandi or Narok setarias, these latter will be selectively grazed but cv. Kazungula setaria will combine with it for some years before 'Kazungula' becomes dominant. Heavy grazing and slashing in the wet season will help to give plicatulum dominance (Bisset, 1975).

Ability to compete with weeds. Fair, due to its open, tussocky habit.

Tolerance to herbicides. It will not tolerate atrazine (Hawton, 1976).

Response to defoliation. *P. plicatulum* decreases in a pasture under continuous grazing; it responds to rests from grazing of about 30 days (Leithead, Yarlett & Shiflet, 1971).

Grazing management. Plicatulum is best grazed when it is leafy in spring and summer rather than saved for autumn grazing or haymaking. Mature growth is poorly accepted by cattle.

Response to fire. It will survive burning.

Genetics and reproduction. It is an aposporous apomict, 2n=40, or sexual 2n=20 (Bashaw, Hovin & Holt, 1970); 2n=20, 40, 60 (Fedorov, 1974).

Dry- and green-matter yields. Over a three-year period, the mean annual dry-matter yields per hectare were 11 340 kg/ha for cv. Bryan, 9 470 kg/ha for cv. Rodd's Bay and 10 560 kg/ha for cv. Hartley, when the yield for *P. dilatatum* over the same period was 5 670 kg/ha (Bisset, 1975).

Suitability for hay and silage. Plicatulum makes good silage but the fermentation is very slow. The pH level does not fall below 5.96, lactic acid concentration is less than 0.5 percent of the dry matter and NH_3 content is less than

Figure 15.122. Paspalum plicatulum cv. Bryan growing with Macroptilium atropurpureum at Bundaberg, Queensland, Australia (lat. 24°9′S, rainfall 1 695 mm) (**Source:** M. Hawley, Queensland Department of Primary Industries)

15 percent of the nitrogen (Catchpoole & Henzell, 1971).
Value as standover or deferred feed. Although the mature growth is unpalatable, it provides useful low-quality roughage to supplement with urea-molasses mixtures.
Chemical analysis and digestibility. 'Hartley' has a high nutritive value even when mature and frosted (Milford, 1960a & b). The daily intake of plicatulum was 359 g per sheep with 3.4 percent protein compared with buffel grass (cv. West Australian) at 740 g and 7.1 percent protein in Queensland, Australia.
Palatability. Cultivar Bryan is more palatable than cv. Rodd's Bay, but there are more palatable summer grasses.
Toxicity. No toxicity has been reported by Everist (1974), and in Puerto Rico a low figure of 0.02 percent of the dry weight for oxalates was recorded (García-Rivera & Morris, 1955). In Taiwan some phytotoxicity from roots has affected lettuce seedlings (Chou, 1977).
Seed production and harvesting. It flowers earlier, yields better and the seed viability is higher under high nitrogen; about 100 kg N/ha is the best application. The seed turns brown at maturity, harvest direct with a drum speed of 900-1 100 rpm (Humphreys, 1975) about 18-22 days after peak flowering (Kowithayakorn & Kannasoot, 1978).
Seed yield. 50-150 kg/ha according to variety and treatment; 920-2 620 kg/ha (Chadokar, 1978; Chadokar & Humphreys, 1970). Seed remains viable for two years (Jones, 1973).
Minimum germination and quality required for commercial sale. In Queensland seed must have 60 percent purity and give a germination of at least 40 percent.
Cultivars.
- 'Rodd's Bay' — introduced from Guatemala into Australia and developed at Rodd's Bay near Gladstone, Queensland. It grows faster than naturalized *Paspalum dilatatum* in autumn and summer and gives high yields compared with other types of paspalum at high levels of fertility and moisture. It flowers two to three weeks earlier than cv. Hartley, over a short period in late summer. Due to its erect habit, with the seed-heads well above the leaves, seed harvesting is easy. It differs from cv. Hartley in having slightly narrower leaves, and hairs on the leaf-blade when the plants are past the seedling stage (Barnard, 1972).
- 'Hartley' — (*Paspalum plicatulum* Mich. × var. *glabrum* Arech.). Introduced to Australia from Brazil by W. Hartley and J.R. Stephens in 1948. Tends to give slightly lower yields than cv. Rodd's Bay, but is of higher nutritive value. It grows on poorer soils than *P. dilatatum* and is tolerant of low, wet land. It flowers two to three weeks later than Rodd's Bay. The leaf-blade and sheath are both glabrous. Its seeding capacity is not as good as Rodd's Bay, and seed is not available now in Australia.
- 'Bryan' — has long hairs (to 5 mm) near the margins for the bottom 8 cm of the leaf-blade's adaxial surface, and a dense collar of hairs (about 2 mm

long) at the junction of sheath and blade; below this, there is a fairly dense zone of short hairs for 2 cm down the sheath (Loch, 1976). It is the most palatable of the three cultivars, but none are as palatable as most other sown grasses in similar situations.

Diseases. It is free from ergot and has no other disease problem.

Pests. It has no specific pests.

Animal production. In grazing experiments by CSIRO at Beerwah, Queensland, *P. plicatulum* cv. Rodd's Bay showed great persistence in combination with legumes under both rotational and continuous grazing at stocking rates of one beast to 0.4-0.6 hectares. However, the annual live-weight gain of 232 kg/ha was less than those from pastures based on *P. dilatatum* (272 kg) and *Digitaria decumbens* (290 kg) over an eight-year period. The plicatulum pasture was as good as the others from September to December (spring and summer) but inferior from January to August (late summer to late winter). Over a three-year period (from June 1972 to 1975), a pure cv. Bryan plicatulum pasture fertilized with 440 kg N/ha gave a mean annual live-weight production of 740 kg/ha when grazed at 5 steers per hectare (Bisset, 1975).

Main attributes. It produces fairly well in poor soils, does not become sod-bound and combines well with legumes. Suited to infertile coastal soils which are flooded in the wet season and then dry out rapidly (Harding, 1972).

Main deficiencies. Low nutritive value, nitrogen content and palatability compared with other tropical sown grasses such as Nandi setaria (*Setaria sphacelata* var. *sericea*) and green panic (*Panicum maximum* var. *trichoglume* cv. Petrie).

Further reading. Bisset, 1975; Bryan & Shaw, 1964; Humphreys, 1975; 't Mannetje, 1961; Milford, 1960a.

Paspalum scrobiculatum L.

Synonyms. P. *polystachyum* R. Br.; *P. commersonii* Lam.
Common names. Scrobic, scrobic paspalum (Australia), kodo millet.
Natural habitat. Semi-swamp forest, damp grassland and swamp.
Distribution. Widespread in Africa. Introduced to Australia from Zimbabwe in 1931.
Description. A loosely tufted, shallow rooting, short-lived perennial or annual with ascending, somewhat succulent branched stems up to 90 cm high, tufts up to 60 cm in diameter; culms with four to six nodes. Leaves up to 30 cm long and 12 mm wide, flat, soft, completely hairless on mature plants; ligule membranous, no auricles. Inflorescence of three to four racemes, 4-9 cm long, borne along a simple unbranched axis. Spikelets one-flowered. The one glume and sterile lemma are thin and papery at maturity and fall with the "seed". Seed ellipsoidal, flat on one side and markedly convex on the other, 2 mm long and 1.5 mm wide (Barnard, 1972) (see Fig. 15.123).
Season of growth. Summer, with a decline in autumn (April in Australia).
Optimum temperature for growth. The mean annual temperature ranges from 19.1-21.9°C (Russell & Webb, 1976).
Minimum temperature for growth. The mean minimum temperature of the coldest month ranges from 5.1-11.1°C (Russell & Webb, 1976).
Frost tolerance. Very frost sensitive, but partly frosted material retains a high degree of succulence until the spring, when new growth is made in southeast Queensland (Paltridge, 1955).
Latitudinal limits. 17.3-28.1°S (Russell & Webb, 1976).
Altitude range. Sea-level to 3 000 m.
Rainfall requirements. 890 mm or more, with reliable opening rains for establishment. A range of 806-1 638 mm (Russell & Webb, 1976).
Drought tolerance. It is not very drought resistant as it is relatively shallow-rooted.
Tolerance to flooding. It is very tolerant of flooding (Colman & Wilson, 1960).
Soil requirements. It has a wide soil range, from fertile clay loams to sandy loams.
Tolerance to salinity. No record has been found.
Fertilizer requirements. It requires nitrogen for its highest potential. Under low nitrogen and excess phosphorus it declines.
Ability to spread naturally. It readily spreads by self-sown seedlings.
Land preparation for establishment. It needs a very fine seed-bed prepared, as for wheat, by ploughing, discing and harrowing.
Sowing methods. It is best sown through a cereal drill in rows 1.3 m apart on a well-prepared seed-bed (Paltridge, 1955). In mixtures it can be broadcast or drilled.

Tropical Grasses

Sowing depth and cover. Not deeper than 1.5 cm, lightly covered and rolled.
Sowing time and rate. After good rains in late spring or early summer, at 3.5 kg/ha.
Number of seeds per kg. 660 000.
Dormancy. There does not appear to be post-harvest dormancy. Seed

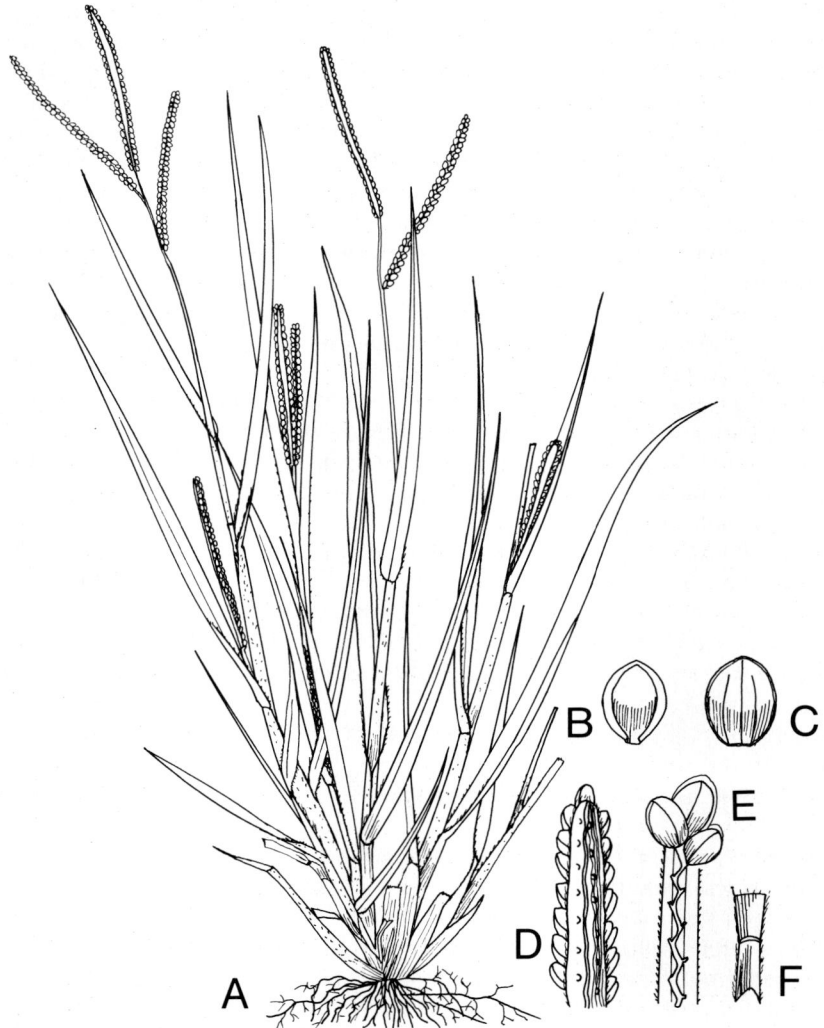

Figure 15.123. Paspalum scrobiculatum. **A**-Habit **B**-Upper lemma **C**-Lower lemma **D**-Raceme, back **E**-Raceme, front **F**-Ligule

remains viable for only one year. Germination will take place at 20-35°C, moistened with KNO_3 solution.
Seedling vigour. Good.
Vigour of growth and growth rhythm. It is a vigorous grass.
Response to light. It will not tolerate much shade.
Compatibility with other grasses and legumes. At Lawes, Queensland, Paltridge (1955) found that there was too much competition for moisture when scrobic and lucerne were sown together in deep, black clay soils; it was preferable to sow them in separate areas. In higher rainfall areas it is compatible with a range of tropical legumes. Bisset and Marlowe (1974) showed that *P. scrobiculatum* combined with siratro, but the mixture was invaded and dominated by *Heteropogon contortus* and *Cynodon dactylon* after a short period.
Ability to compete with weeds. Paltridge (1955) grew it in spaced rows, intercultivated when weed growth was eliminated.
Response to defoliation. It will not stand heavy grazing.
Grazing management. It requires an intermittent form of grazing to permit sufficient seeding for regeneration.
Genetics and reproduction. It is a tetraploid with somatic chromosome number $2n=40$. It is sexual in reproduction and lack of variation in the field possibly reflects a high degree of self-fertilization (Pritchard, 1970).
Dry- and green-matter yields. At Lawes, southeast Queensland, Paltridge (1955) obtained 2.9 t DM/ha per year; subsequently he harvested 2.3 t/ha unfertilized, 5.4 t/ha when fertilized with sulphate of ammonia, 5.4 t/ha in 1-m rows unfertilized, and 8.9 t/ha when fertilized. Under partial irrigation during dry weather, a seven-year stand of the grass grown at 1.3 m spacing yielded 10 t DM/ha. In southern Africa, a scrobic/lucerne pasture yielded 8.67 t DM/ha under irrigation, which was less than for *Chloris gayana*/lucerne and *Panicum maximum* var. *trichoglume*/lucerne (Muslera *et al.*, 1975).
Suitability for hay and silage. It made good silage in Panama when 10 percent molasses was added (Medling, 1972), and it makes excellent hay (Paltridge, 1955).
Value as standover or deferred feed. It makes good standover feed as it holds its nutritional value well into maturity.
Chemical analysis and digestibility. Sen and Mabey's (1965) results in Ghana are shown in Table 15.53. On unfertilized *P. scrobiculatum* pastures grown on black clay soils at Lawes, Queensland, the crude protein content of the dry matter grazed by sheep varied from 5.4-5.5 percent during the growing season to 2.6-3.8 percent during late winter and spring. The starch equivalent was 60 during the growing season (Paltridge, 1955). Andrew (1971) recorded 7.3 percent crude protein with 41 percent digestibility in Sri Lanka with 54 percent digestibility of the dry matter.
Palatability. Readily eaten and highly digestible up to flowering (Milford, 1960a), but low intake after frosting.
Toxicity. No toxicity has been recorded by Everist (1974). Ndyanabo (1974)

TABLE 15.53 *Paspalum scrobiculatum*

	DM	As % of dry matter		
		CP	CF	Ash
Fresh, 4 weeks	22.1	12.8	25.1	12.2
Fresh, 8 weeks	15.5	7.9	29.3	6.9
Fresh, 12 weeks	24.6	7.3	29.5	6.9
Fresh, 36 weeks	45.3	3.9	30.1	5.8

Source: Sen & Mabey, 1965

recorded 0.23 percent total oxalic acid in the dry matter, but no toxicity.

Seed production and harvesting. It flowers freely and sets seed over three to four months, beginning in January (Queensland). The seeds fall as they mature, which makes seed harvesting difficult. The seed can be picked up from the ground by a suction harvester (Paltridge & Coaldrake, 1943).

Seed yield. Paltridge (1955) recorded yields up to 2 500 kg/ha at Lawes, Queensland. Jones (1973) recorded 47 kg/ha.

Minimum germination and quality required for commercial sale. 40 percent germinable seed, 93 percent purity in Queensland.

Cultivars. The only cultivar recorded at present is cv. Paltridge and the foregoing characteristics refer to this plant.

Value for erosion control. Useful in its ecological niche, but there are better grasses available.

Diseases. It can be attacked by the paspalum ergot, *Claviceps paspali* (see *Paspalum dilatatum*) but is more resistant than paspalum.

Pests. It can be severely affected by felted grass coccid or mealy bug (*Antoninia* sp.), which reduces its persistence.

Animal production. It is capable of high levels of animal production. At Beerwah, southeast Queensland, over a seven-year period, the average annual live-weight gain per hectare was 297 kg, stocked at 1.6 and 2.5 beasts per hectare. This performance was better than for *Paspalum plicatulum, Digitaria decumbens* and *Paspalum dilatatum* (Bryan, 1968). At 1.6 beasts per hectare live-weight gain was 238 kg per hectare, and at 2.5 beasts per hectare the gain was 342 kg per hectare. At Lawes, Queensland, pure scrobic pastures carried 40.25 sheep per hectare over a four-month growing season compared with 28.25 sheep per hectare on *Urochloa panicoides*, 24.25 sheep on *Panicum maximum* and 22.75 sheep on *Chloris gayana* (Paltridge, 1955).

A grazing trial of siratro and scrobic pastures was conducted between 1966-71 in the Burnett district in the Queensland coastal subtropics, under an annual rainfall regime of 1 000-1 100 mm per year. Bisset and Marlow (1974) recorded that the scrobic population declined rapidly and the native *Heteropogon contortus* and *Cynodon dactylon* invaded and dominated the grass component, though siratro persisted. They concluded that the improved pasture mixtures should be used intensively in autumn-winter-spring, when their nutritive value is better than the native species.

Main attributes. Very palatable and highly digestible during summer; retains these characteristics later into maturity than other grasses.

Main deficiencies. Low crude protein and short life span. It is difficult to harvest its seed as it is shed very readily over a long period.

Further reading. Bryan, 1968; Paltridge, 1955.

Paspalum urvillei Steud.

Common names. Vasey grass, giant paspalum (Australia), regop paspalum, upright paspalum (southern Africa).

Distribution. A native of Argentina and Uruguay, it has spread to several tropical areas.

Description. An erect perennial, growing in tufts about 30 cm in diameter with many erect leaf-blades. The base of the stalks and leaf-sheaths is hairy and bluish in colour. The flower stalks are 60-200 cm tall, each flower cluster bearing six to 25 spikes. Flowering culms are produced over a long period (Wheeler, 1950). It differs from *P. dilatatum* in having erect culms, conspicuously hairy leaf-sheaths and ten to 30 or more racemes (see Fig. 15.124).

Season of growth. Summer.

Optimum temperature for growth. It flowers at 5.5°C sparingly, and improves up to 22.5°C.

Frost tolerance. It continues to grow in winter, except in very cold weather, and thus gives a long grazing period (Wheeler, 1950). It does not survive winter in Washington State, United States.

Latitudinal limits. About 32°N and S.

Altitude range. Sea-level to 1 000 m.

Rainfall requirements. It is a high-rainfall grass usually growing in regions of 1 000-1 500 mm, with summer dominance.

Drought tolerance. It can withstand severe drought.

Tolerance to flooding. It can grow on very wet land.

Soil requirements. It thrives best on heavy soils, but succeeds well on moist, sandy land. It has a wide soil range and does well in vleis in Zimbabwe.

Tolerance to salinity. No record has been found.

Fertilizer requirements. It will respond to basic complete fertilizer and top-dressings of nitrogen. There is some nitrogenase activity in association with its roots (Koch, 1977) in Hawaii.

Ability to spread naturally. With its heavy seed production it spreads fairly quickly under favourable moist soil conditions.

Land preparation for establishment. Full land preparation will give best populations.

Sowing methods. Sow with a drill, or broadcast. It is seldom planted, but is used where it is found in the United States (Bennett, 1973).

Sowing depth and cover. Not deeper than 1 cm; rolled to cover.

Sowing time and rate. Summer, at 22 kg/ha.

Number of seeds per kg. 970 000.

Dormancy. There is some post-harvest dormancy and the seed should be stored to improve germination (Wheeler, 1950).

Seedling vigour. Very vigorous.

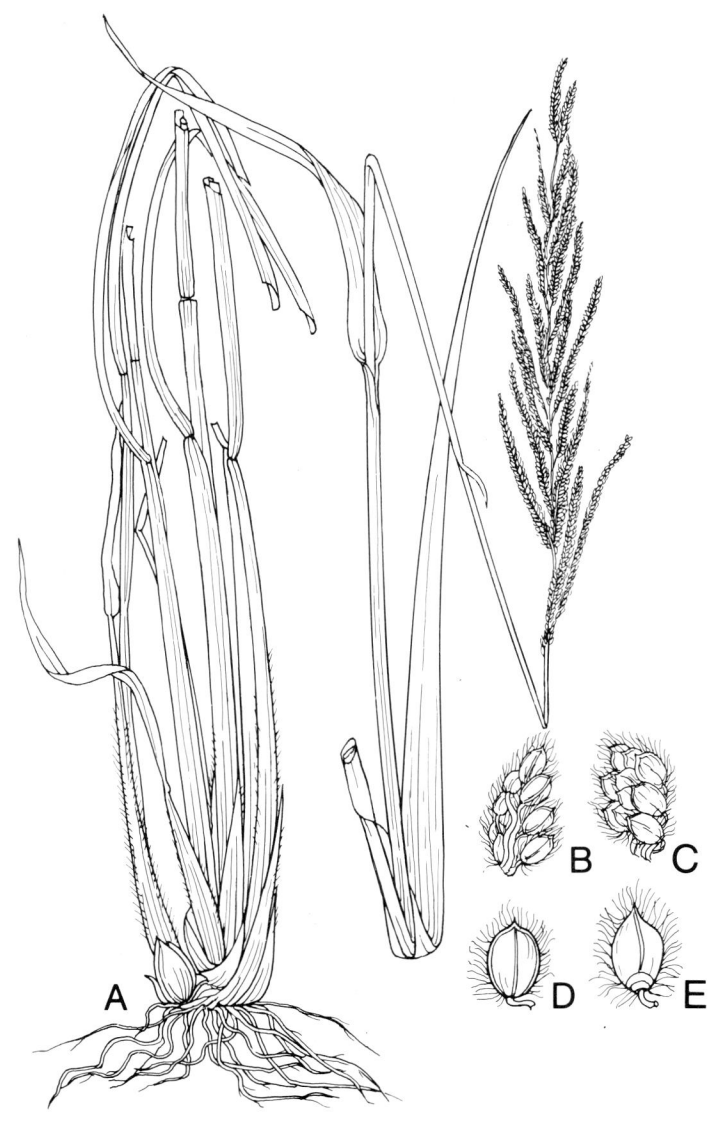

Figure 15.124. Paspalum urvillei. **A**-Habit **B,C**-Portions of inflorescence **D,E**-Spikelet, back and front

Response to photoperiod. Below 15.5°C, flowering is limited by light (Bennett, 1959).

Response to light. It prefers full sunlight and does not grow well in shade.

Compatibility with other grasses and legumes. It will combine well with most of the tropical legumes.

Ability to compete with weeds. Its vigorous, erect growth allows it to compete successfully with weeds. It can become a weed itself.

Tolerance to herbicides. To control this grass use 2,2-DPA at 2.3 kg of a 740 g AI/kg product (e.g. Shirpon, Dowpon, Ellapon) plus paraquat at 85 ml per 200 l of water. Spray to the point of run-off three times, at ten-day intervals. Alternatively a single application of glyphosate at 6 l/ha of a 360 g AI/l product (e.g. Round-up) can suffice (Tilley, 1977).

Response to defoliation. It is killed by heavy grazing. It is not as palatable as other species, quickly becoming coarse, and is thus avoided by stock.

Grazing management. It should be grazed sufficiently to prevent it flowering.

Response to fire. It will survive fire.

Genetics and reproduction. 2n=40, sexual (Bashaw, Hovin & Holt, 1970). 2n=40, 60 (Fedorov, 1974).

Suitability for hay and silage. It is cut for hay where it has become abundant in the United States. The hay is classed as good.

Value as standover or deferred feed. It becomes coarse and unpalatable as it matures.

Chemical analysis and digestibility. No figures have been cited.

Palatability. It loses its palatability and quality with age.

Toxicity. It is generally resistant to paspalum ergot, and no cases of poisoning have been reported (Everist, 1974).

Seed production and harvesting. Seed ripens very unevenly. It is fluffy and smaller than that of *P. dilatatum*. It is good practice to use the second crop for seed, the first crop being cut for hay.

Value for erosion control. It is rather too tussocky to control erosion.

Diseases. No major diseases occur.

Pests. No serious pests attack the plant.

Economics. It is used as a soilage crop in Sri Lanka (Andrew, 1971) and as a minor forage crop in the United States but it is generally regarded as a weed in subtropical coastal Australia, and stock do not eat it readily.

Main attributes. It will thrive on very wet lands, e.g. vleis in Zimbabwe.

Main deficiencies. Its tendency to become a weed, and its low palatability as it matures.

Further reading. Bennett, 1959.

Paspalum wettsteinii Hack.

Common name. Broad-leaf paspalum (Australia).
Distribution. Native to the humid subtropical areas of southern Brazil, Paraguay and northern Argentina; introduced to the United States and Australia.
Description. A semi-prostrate, tufted perennial with stoloniferous habit; grows to 90 cm high, tufts reach 100 cm in diameter. Culms erect, glabrous, unbranched, concealed by the leaf-sheaths; nodes pubescent. Ligules up to 2 cm long, membranous, surrounded by long hairs at the back. Leaf-blades rounded, up to 40 cm long and 3 mm broad, with wavy margins. Inflorescence a panicle with four to nine spike-like racemes. Spikelets are almost glabrous, and broadly oval; the upper glumes and lower sterile lemma are about equal, and equal the fertile floret; they are thinly membranous and three-nerved. Mature seeds are smaller and browner than those of *P. dilatatum;* its leaves are broader and the spikelets are less densely arranged (Barnard, 1969).
Season of growth. Early spring (September in Australia, ahead of *P. dilatatum*) to late summer seeding, after which little vegetative growth occurs.
Optimum temperature for growth. Growth commences at 13-15°C in spring. At 16-18°C, it produces 30 kg DM/ha per day (Kemp, 1975).
Minimum temperature for growth. Growth ceased in autumn at 12-14°C (Kemp, 1975).
Frost tolerance. Less frost tolerant than *P. dilatatum,* but it is not killed. The frosted material is readily eaten. It recovers earlier than most other grasses.
Rainfall requirements. Best adapted to a subtropical summer rainfall of 1 000-1 500 mm.
Drought tolerance. It has a high degree of drought resistance, comparable to Kikuyu grass (*Pennisetum clandestinum*).
Tolerance to flooding. Good; it survives moderate flooding and poor drainage.
Soil requirements. It is tolerant of a wide range of soils and, because of its relatively low nitrogen requirement, has the ability to compete with poorer grasses, such as mat grass.
Fertilizer requirements. It needs a basic treatment with nitrogen, phosphorus and potassium as indicated by soil tests. It responds well to nitrogen, increasing annual dry-matter yields from 639 kg/ha with no nitrogen to 2 299 kg/ha with 150 kg N/ha, and to 4 333 kg/ha with 300 kg N/ha during the autumn (Ostrowski & Fay, 1979).
Ability to spread naturally. Its spreading stems allow rooting from the nodes.
Land preparation for establishment. A good, fine, firm seed-bed is needed.
Sowing methods. The seed can be broadcast or drilled in, and rolled after

seeding. Apply 250 kg/ha superphosphate at sowing, N and K if needed.
Sowing depth and cover. Do not sow or cover too deeply.
Sowing time and rate. Summer, when soil moisture is good (usually from November to March in Australia), at 4.5 kg/ha broadcast or 2 kg/ha drilled.
Number of seeds per kg. 775 000 (Queensland).
Dormancy. There is some post-harvest seed dormancy which, for germination tests, is broken by treating the seed with 0.2 percent KNO_3 solution. Otherwise, the seed can be stored for a few months before planting.
Seedling vigour. Early growth is slow but, once established, it grows rapidly.
Vigour of growth and growth rhythm. It produces very early spring growth, better than other paspalums, Kikuyu and setaria. Seeding commences early in April (southeastern Queensland), after which little vegetative growth occurs (Cook, 1978) up to maturity in May-June (Kemp, 1975).
Compatibility with other grasses and legumes. It competes well with weedy grasses such as *Axonopus* spp. and is compatible with the subtropical legumes *Neonotonia wightii, Desmodium uncinatum, D. intortum, D. axillaris, Lotononis bainesii* and *Macroptilium atropurpureum* and with white clover (Barnard, 1969).
Response to defoliation. It stands grazing well.
Grazing management. As it matures later than *P. dilatatum,* it has a greater potential for summer production; it is easier to manage because it does not need frequent mowing. With careful selection of a companion legume it can give year-round grazing. At 8-10 weeks (when 22-30 cm high) graze quickly to 10 cm high, and then remove stock. If grazed intensely, pasture should be top-dressed with 65-90 kg N/ha. *P. wettsteinii* pasture can be top-dressed with 250 kg superphosphate/ha per year (and with 125 kg/ha of muriate of potash if potassium is deficient) and oversown with a temperate legume in autumn to give grazing in late winter and spring if the climate is suitable (Leggett, 1968).
Response to fire. It recovers well from fire.
Genetics and reproduction. It is believed to be apomictic.
Dry- and green-matter yields. Yields of green and dry matter are equal to or better than other useful grasses grown in a similar environment. At Mt Mee, southeastern Queensland, it yielded 4 333 kg DM/ha in autumn and 2 049 DM/ha in spring (Ostrowski & Fay, 1979).
Chemical analysis and digestibility. No figures have been cited.
Palatability. It is very palatable and selectively grazed in the presence of *P. dilatatum, Pennisetum clandestinum* and *Setaria sphacelata.*
Seed production and harvesting. It seeds abundantly, its seed-heads resembling those of *P. dilatatum,* but there is no seed production before April in Queensland.
Minimum germination and quality for commercial use. 40 percent germinable seed, 60 percent purity, in Queensland.
Cultivars. 'Warral' is described above.
Diseases. It is resistant to ergot. (Leggett, 1968).

Animal production. It has given good milk production in northern New South Wales and southeastern Queensland.
Main attributes. Its palatability and high production, especially in autumn and spring; its resistance to ergot.
Further reading. Leggett, 1968.

Pennisetum americanum (L.) Leeke

Synonym. P. *glaucum* (L.) R. Br.; *P. typhoides* (Burm.) Stapf and C.E. Hubb.
Common names. Bulrush millet, pearl millet, dukn (the Sudan), bajra (India), babala (Natal).
Natural habitat. Cultivation.
Distribution. Originated in central tropical Africa, but cultivated since 1200 BC in India. Now widely distributed in the drier tropics.
Description. A robust and free-tillering annual growing to a height of 3 m. Stems 10-20 mm thick; above each node is a shallow groove containing an axillary bud. Nodes slightly swollen; they bear a ring of adventitious root primordia at the basal end. Leaves flat, dark green and up to 8 cm wide. The inflorescence forms a compact, cylindrical, terminal, spike-like panicle. There are 870-3 000 spikelets on a panicle. Seeds small, 3-4 mm, wedge-shaped of various colours according to variety (see Fig. 15.125, Plate 45).
Season of growth. Summer.
Optimum temperature for growth. Summer temperatures should be high. Maximum germination occurs at a day/night temperature of 20/25°C (R.M. Hughes, 1979).
Minimum temperature for growth. 7.0°C ± 6.3. Low temperatures retard germination and at 10°C, photosynthesis is negligible (Russell & Webb, 1976).
Frost tolerance. Temperatures near 0°C are lethal.
Latitudinal limits. 14-32°N and S (Russell & Webb, 1976).
Altitude range. 800-1 800 m.
Rainfall requirements. It is grown in areas with an average annual rainfall of 125-900 mm, the lower rainfall areas using it as a grain crop where maize and sorghum fail. It is sown at low populations to allow each plant to find more soil moisture (see Plate 46). Where dry matter for forage is the consideration, a minimum rainfall of 500 mm is required. Late rainfall is important for grain development in weeks 5-12.
Drought tolerance. It is drought tolerant. Its roots may penetrate to 360 cm, although 80 percent of the root weight is in the top 10 cm.
Tolerance to flooding. It does not tolerate flooding, especially during the summer.
Soil requirements. Bulrush millet grows on a wide range of soils, from sands in the Sudan to clays. It is tolerant of very acid soils. It grows best in a well-drained fertile soil.
Tolerance to salinity. It is tolerant of salinity and was used for reclamation of salt lands in Sind because of its ability to take up salts (Tamhane & Mulwani, 1937; Ravikovitch & Porath, 1967). Soil salt concentrations of 1 400 to 2 600 ppm produced only slight tip burn (Smith & Clark, 1968).

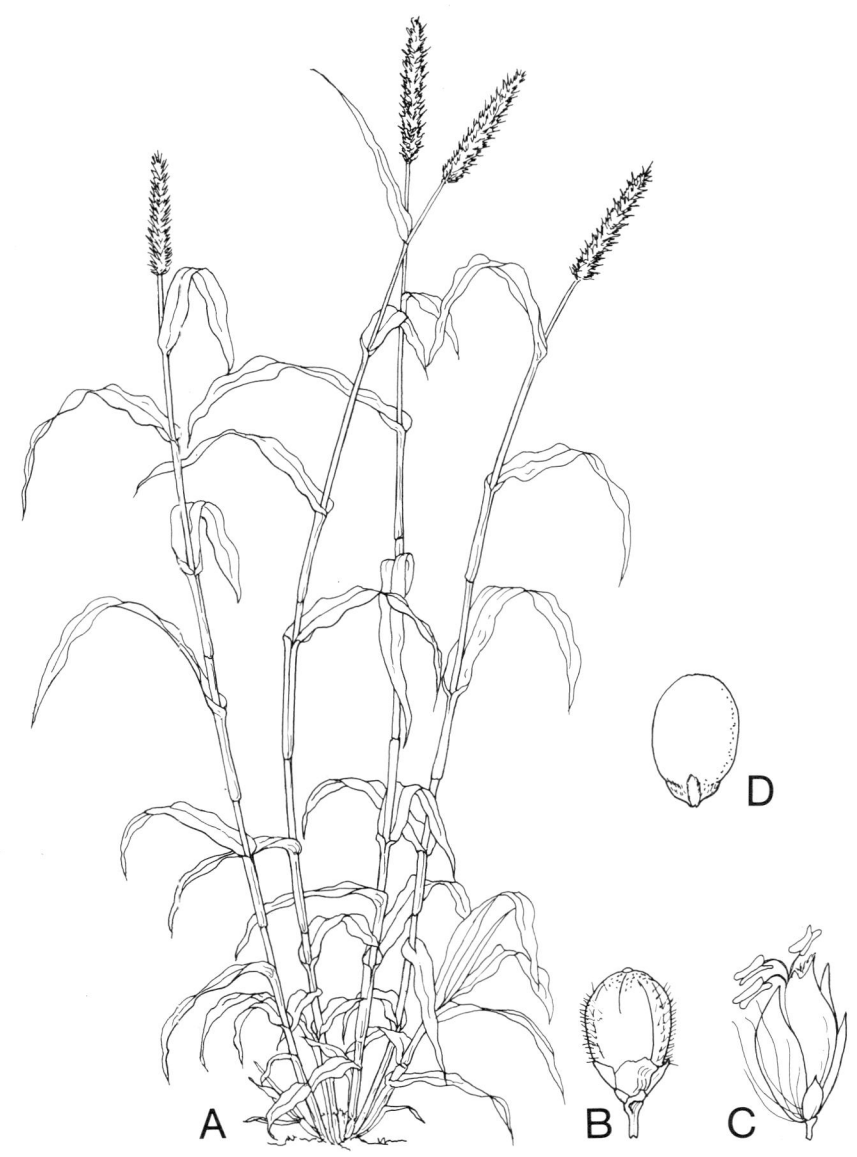

Figure 15.125. Pennisetum americanum. **A**-Plant **B,C**-Spikelets **D**-Grain

Fertilizer requirements. Bulrush millet is seldom manured by villagers in Africa; in India, farmyard manure may be used, and African nomads plant it on village cattle camps when the herds go on trek. The most common fertilizer element in use under cultivation is nitrogen at 60-100 kg N/ha, balanced with about half this level of P_2O_5, and potassium as needed. For use as fodder, higher nitrogen dressings may be used. Bulrush millet has an outstanding ability to recover deep accumulations of nitrate nitrogen from soils. In the United States, fodder dry matter responds to fertilizer nitrogen up to 400 kg/ha.

Ability to spread naturally. Practically nil.

Land preparation for establishment. For good crops it needs full seed-bed preparation as for cereals. In sandy soils in Africa, the ground is dug over with a hoe and weeded prior to planting.

Sowing methods. In peasant areas a few seeds are dropped in holes dug with a hoe, 45-90 cm apart according to rainfall, and covered. Mechanical drilling is common in developed countries.

Sowing depth and cover. Sowing depth varies from 13-50 mm, the optimum being 35-40 mm.

Sowing time and rate. Early summer, at 6-10 kg/ha is usual when drilled in rows 35-70 cm apart.

Number of seeds per kg. About 187 000.

Dormancy. Several reports state that the seed of pearl millet exhibits post-harvest dormancy of several weeks.

Seed treatment before planting. Where needed, it can be dusted with a combined insecticide-fungicide. A one-hour soak in 1 percent 2-chloroethanol plus 0.5 percent sodium hypochlorite solution was found to be effective in increasing germination rates.

Seedling vigour. Slow in the early stages of growth. It is good as temperature rises to 20-22°C.

Vigour of growth and growth rhythm. Norman (1962a) recognized three distinct development phases: an early tillering period, a period of rapid increase in dry weight and tiller height, and a period of head production. Full tiller production occurred in the fifth week with full light interception. Phillips and Norman (1967) recorded one of the highest growth rates recorded for any species when they measured the variety 'Ingrid Pearl', 14-16 weeks after sowing, as accumulating dry matter at the rate of 58 g/m^2 per day. Flowering occurs about the thirteenth week.

Response to photoperiod. Both day-neutral and short-day varieties exist. Burton and Powell (1968) suggested that short-day, photoperiod-sensitive, late-maturing millets should be superior to the other lines since they are leafier and have a better seasonal distribution of forage production. Grain production would best be improved by the use of photoperiod-insensitive types which mature early. The crop would thus escape drought and could be planted several times a year if conditions were favourable.

Compatibility with other grasses and legumes. It is usually grown as a pure stand. In India it has been grown with *Cajanus cajan*, the mixture providing a useful cover to reduce soil erosion.

Ability to compete with weeds. Most crops of pearl millet are sown in rows and cultivated between the rows. Where fodder crops are grown at high densities the crop canopy suppresses weed growth.

Tolerance to herbicides. Albert (1961) obtained effective control of the weedy *Digitaria ciliaris (sanguinalis)* and *Amaranthus* spp. by pre-emergence application of simazine at 1 kg, and atrazine and propazine at 1 and 2 kg/ha, without crop injury. 2,4-D at 0.5 kg/ha gave good weed control without crop injury if applied 21 days after sowing.

Response to defoliation. In the United States, three cuts of highly palatable green fodder are taken at six- to seven-week intervals. Late-maturing varieties are favoured for forage production. High regrowth yields after defoliation can best be obtained if the cutting height is above the apical meristem, and it is suggested that the crop be grazed rotationally when about 45 cm tall. Regrowth after later harvests declines rapidly (Begg, 1965).

Grazing management. Pearl millet should be subject to relatively frequent but lenient defoliation to maintain quality. The crop should not be allowed to grow above 1 m high before grazing starts. Forage intake varied from a high of 3.1 kg DM/100 kg body weight on immature forage to a low of 1.4 kg on mature forage over a five-year period (Ferraris, 1973). Density of tiller regrowth after cutting was reduced from 54 percent when cut at 4 weeks to about 3 percent when cut at 14-16 weeks.

Genetics and reproduction. 2n=14; the haploid chromosome number in pollen mother cells is thus 7. Burton and Powell (1968) consider millet to be an excellent plant for genetic and cytogenetic research, as the small number of large chromosomes and the clear meiotic stages allow detailed study. Interspecific hybridization of *P. americanum* has usually only been successful with *P. purpureum*. Bana grass is one such cross and is widely used in southeast Queensland as a wind-break on vegetable farms; it also provides useful fodder. A millet-breeding unit is centred on the EAAFRO, Serere Research Station in Uganda and at Coastal Plains Research Station, Tifton, Georgia, United States (see Plate 47).

Dry- and green-matter yields. A yield of 21 735 kg DM/ha was recorded at Katherine, Northern Territory, Australia. In Alabama, United States, forage dry-matter yields varied from 6 000-10 500 kg/ha with 40 kg N/ha applied at sowing and again for each cut.

In Queensland, Australia, Douglas (1974) recorded the following comparative dry-matter yields between Sudan grasses and pearl millet on a fertile, irrigated soil (see Table 15.54).

Suitability for hay and silage. Little hay has been made, and Norman and Stewart (1964) preferred a standing mature crop for dry-season grazing to conservation. However, the crop has been ensiled successfully in several

TABLE 15.54 **Comparative yields: Sudan grass and pearl millet**

Crop planted late Nov.	Harvested late Jan. (kg/ha)	Protein (%)	Regrowth early Mar. (kg/ha)	Protein (%)	Total yield (kg/ha)
Hybrid Sudan grass	9 524	16.2	3 696	14.4	13 220
Sudan grass	6 193	13.8	5 023	11.9	11 216
Hybrid pearl millet	8 528	16.2	1 073	16.4	9 601
Tamworth pearl millet	7 257	19.1	931	19.7	8 188
Katherine pearl millet	4 610	23.4	1 887	15.9	6 497
Ingrid pearl millet	3 889	23.5	2 268	13.0	6 157

Source: Douglas, 1974

countries (Ghana, Nigeria, the United States, Zimbabwe), and has proved the equal of maize silage when cut at eight to 12 weeks (full flowering). Chapman (1978), in Natal, found the best time to harvest was three weeks after flowering, when its dry-matter yield compared favourably with maize.

Value as standover or deferred feed. Norman and Stewart (1964) found the crop excellent for dry-season grazing by beef cattle, and live-weight gains averaged 296 kg/ha over 16 weeks at Katherine, Northern Territory, Australia, during a period when live weight on native pasture declined.

Chemical analysis and digestibility. The crude protein content depends on the age of the crop, young growth giving the highest proportion. Dry-matter digestibility ranges from 75.3 percent in young pearl millet leaves to 61.4 percent in old leaves. The lowest digestibility figure was 55 percent in mature, previously-grazed stands which were making slow recovery (see also *Digitaria ciliaris*).

Palatability. Young pearl millet is very palatable.

Toxicity. Grazing lactating cows on millet has led to marked butterfat depression, and it has been suggested (Schneider *et al.*, 1970) that high succinic and oxalic acids may be the cause. Under heavy nitrogen fertilization, high nitrate may be recorded. HCN contents are not sufficiently high to be hazardous to stock.

Seed production and harvesting. Seeds are ready to harvest three to four weeks after anthesis. They vary from 3 to 10 mg in weight. Uneven ripening of tillers necessitates multiple harvests where manual methods are used. The seed can be harvested directly by combines, but for tall varieties a roller attached in front of the comb will make the harvesting height easier to handle.

Seed yield. Millet hybrids have been known to yield up to 6 t grain per hectare, but yields in the Northern Territory of Australia have been nearer 600 kg/ha.

Minimum germination and quality required for commercial sale. 70 percent germinable seed, 97.3 percent purity (Queensland).

Cultivars.

- 'Katherine Pearl' — derived from seed introduced from Ghana and developed by CSIRO Australia, at Katherine, Northern Territory. It requires a growing season of three to four months from sowing to flowering, and a day length of 12-12.5 hours for flower formation. A high producer of dry matter and crude protein during the wet season, averaging almost 12 000 kg DM/ha per year over an 11-year period. Grain yields averaged 650 kg/ha. The crude protein content of the young plant reaches 28 percent but decreases to 8 percent at maturity. The seed is pearly white to grey.

- 'Ingrid Pearl' — introduced to Australia from West Africa. It has leaves which are less hairy, lighter green and wider than 'Katherine Pearl'. Seeds are smaller, yellow or greenish-grey, and very tightly packed in the seed-head. It flowers one to two weeks earlier than 'Katherine Pearl' and hence is more suited to a short wet season.

- 'Tamworth' — selected from crosses of cv. Gahi, bred in Georgia, United States. It is mainly used for late summer and autumn grazing in the coastal districts of New South Wales.

- 'MX 001' — the first hybrid *Pennisetum* millet produced commercially in Australia. It produces fine-leaved forage from vigorously tillering plants. It matures early- to mid-season, after making rapid initial growth (Douglas, 1974).

- 'Kawanda 4' — a high-yielding variety in Uganda, yielding 18 135 kg/ha of dry matter, with 10.2 percent protein.

- Hybrids P 99, P 97 and P 81 — resulting from crosses between *P. purpureum* and *P. americanum*. They have proved very productive in Uganda, P 99 being the best. Mugerwa and Ogwang (1979) suggest cutting at eight to ten weeks for direct feeding or conservation as silage. The yields of hybrids P 99, P 97 and P 81 were 20 726, 20 344 and 17 378 kg/ha of dry matter, and 9.8, 9.1 and 7.8 percent crude protein respectively.

- 'Starr' — a synthetic variety developed by pooling selfed seeds from a number of leafy, medium-tall, uniformly-maturing F_1 progenies from a wide cross (Burton & Powell, 1968).

- 'Tiflate' — a short-day, photoperiod-sensitive, late-maturing synthetic that remains vegetative throughout the long summer season in Georgia, United States. It is leafier, easier to manage, gives better seasonal distribution of forage, and lasts longer than 'Starr' (Burton & Powell, 1968).

- 'Gahi 1' — similar to 'Starr', but is capable of yielding 25-30 percent more forage. It is a first-generation chance hybrid (Burton & Powell, 1968).

- 'Tift 23A' — produces high forage yields in the United States and improves the quality (digestibility and disease resistance) of its hybrids (Burton, 1970).

- 'Millex 22' — a commercial variety of hybrid pearl millet in the United

States, produced by crossing selected males on Tift 23A pearl millet (Burton, 1970).
- 'Anand' — the most suitable fodder cultivar in Haryana, India (Singh et al., 1977).
- *Pennisetum americanum* × *P. purpureum* hybrids — hybrid pennisetums, such as Napier-bajra hybrid, elephant-bajra hybrid or hybrid Napier on cv. Gajraj and cv. Pusa Giant Napier, give very high fodder yields. The all-India trials yielded 200-400 tonnes green forage per hectare per year. It is palatable and readily eaten by cattle and sheep, and is a good standover forage for maintenance only. It is useful for silage. Seed-producing F_1 hybrids have not yet been obtained (Muldoon & Pearson, 1979).

Value for erosion control. In pure stands (for seed production) it affords little soil protection, but in dense stands (for forage production) or in conjunction with a legume, for example *Cajanus cajan* in India, it is useful.

Diseases. The main diseases, among many listed by Ferraris (1973), are smuts (caused by *Helminthosporium* spp.), downy mildew and top rot. In Queensland, a leaf spot is caused by a fungus, *Cercospora*.

Pests. In Africa one of the worst pests is the root parasite, *Striga hermonthica*, and less commonly *S. lutea*. The red-billed weaver bird, locusts and *Quelea quelea aethiopica* take heavy toll. *Heliothis armigera* attacks seed-heads, and the stem borer, *Coniesta ignefusalis,* is also damaging. Ferraris (1973) gives a full list of pests.

Economics. Pearl millet is an important grain crop in Africa where the rainfall is not secure enough for sorghum or maize. In the United States and Australia it is a useful, non-toxic forage to replace forage sorghum. The stalks are used in the dry tropics for home building.

Animal production. In southern Africa, pearl millet yielded an average of 25.2 tonnes of green matter (Haylett, 1961). Clark, Hemken and Vandersall (1965) found pearl millet equivalent to Sudan grass and a sorghum × Sudan grass hybrid for dry-matter yield, carrying capacity and milk yield for lactating cows. Carrying capacity varied from 4.7 to 6.7 cows per hectare per day with millet over a three-year period, and adjusted milk production averaged 19.8 kg per day. Body weight losses were least with millet. The grazing season averaged 121 days. At Katherine, in the Northern Territory, Australia, wet season grazing by beef cattle at a stocking rate of 2.5 beasts per hectare produced a live-weight gain of 102 kg per head in 20-24 weeks, an increase of 51 kg per head over native pasture (Norman, 1963b). Cattle grazing standing millet in the dry season made an average live-weight gain of 269 kg/ha over 16 weeks, during a period when animals grazing natural pasture lost weight (Norman & Stewart, 1964). Between January and March in the Macquarie Valley, New South Wales, irrigated cv. MX 001 yielded 18 950 kg DM/ha and 274 kg/ha of live-weight gain, 0.95 kg per day on a per caput basis (Upton, 1978). At Katherine, Northern Territory, Norman and Phillips (1968) con-

TABLE 15.55 **Data from grazing pearl millet at Katherine, Northern Territory, Australia**

	Grazing started on			
	28 Jan.	24 Mar.	29 Apr.	30 Mar.
Stocking rate (beasts/ha)	3.1	5.0	4.9	5.1
Date at end of gain	10 June	19 June	8 July	13 Aug.
Date at end of maintenance	27 June	3 Aug.	6 Aug.	2 Sept.
Period of gain (days)	133	87	70	75
Period of maintenance (days)	17	45	29	20
Total period (days)	150	132	99	95
Gain (kg/head)	86	45	33	24
Gain (kg//ha)	253	207	152	129
Grazing capacity, animal days/ha	465	660	489	485

Source: Norman & Phillips, 1968

ducted 18 grazing trials (from 1960 to 1967) with pearl millet (see Table 15.55).

Main attributes. It is the main cereal in semi-arid regions where sorghum cannot be profitable. It is a palatable, high-yielding summer forage, generally free from HCN; it can exploit soil nutrients to the full and tolerate water stress.

Main deficiencies. It is a little coarse for hay.

Further reading. Burton & Powell, 1968; Ferraris, 1973; Muldoon & Pearson, 1979; Vicente-Chandler, Silva & Figarella, 1959.

Pennisetum clandestinum Hochst. ex Chiov.

Common name. Kikuyu grass (after the Kikuyu people of Kenya, east of the Aberdare Mountains, where the grass thrives).

Natural habitat. Highland grassland on deep red, well-drained latosolic soils at the forest margins, and in grassy glades at an elevation of between 1 950 and 2 700 m in East and Central Africa (Ethiopia, Kenya, Tanzania, Uganda and Zaire).

Distribution. From Zaire and Kenya the grass has been introduced widely in tropical areas, especially Costa Rica, Colombia, Hawaii, Australia and southern Africa.

Description. A prostrate perennial which may form a loose sward up to 46 cm high when ungrazed, but under grazing or mowing assumes a dense turf. The grass spreads vigorously from rhizomes and stolons which root readily at the nodes, and are profusely branched. Short, leafy branches are produced from stolons, with leaf-blades strongly folded in the bud, later expanding to 44.5-114.3 mm long and 6 mm wide, tapering to subobtuse tips; leaf surface is sparsely and softly hairy. The ligule can be recognized by a ring of hairs, and the collar by its prominent pale yellow colour. The flower is small, consisting of a spike of two to four subsessile spikelets which are partly enclosed within the uppermost leaf-sheath. The spikelets are bisexual, or functionally unisexual. The florets are protogynous and the stamens are rapidly exserted on long filaments, usually in the early morning. The stigma is branched and feathery. The large seed (2 mm long) is dark brown, flat or ellipsoidal with a prominent style (Mears, 1970) (see Fig. 15.126).

Season of growth. Spring, summer and autumn.

Optimum temperature for growth. 16-21°C. It has a poor adaptation to high temperatures. Mean $18.8° \pm 2.8°$ (Russell & Webb, 1976).

Minimum temperature for growth. 2-8°C in Kenya (Mears, 1970). Mean $7.7° \pm 4°$ (Russell & Webb, 1976).

Frost tolerance. It tolerates an occasional frost but not sustained frosting.

Latitudinal limits. Mean 27°N and S (Russell & Webb, 1976).

Altitude range. Sea-level to 3 500 m.

Rainfall requirements. In its natural habitat, 1 000-1 600 mm/year either falling in one season or as a bi-modal rainfall (Mears, 1970). Mean $1\ 269 \pm 632$ mm (Russell & Webb, 1976).

Drought tolerance. Reasonably good because of its deep root system. It extends to 5.5 m, but only sparsely below 60 cm, with 90 percent of the total root weight found in the 0-60 cm layer. Added nitrogen improves the efficiency of water use.

Tolerance to flooding. It tolerates flooding well. It survived ten days' flooding at Quirindi, New South Wales (Dale & Read, 1975).

Soil requirements. Its natural occurrence is mainly on deep latosolic soils of good fertility, and it has quickly adapted to similar soils elsewhere. It also thrives on alluvial soils and on moist, sandy soils where the fertility has been raised by animal excreta or mineral fertilizer. It is an excellent colonizer and soil stabilizer in small paddocks around dairy bails, piggeries and feed-lots, and where non-toxic effluent is discharged from factories. It does require soils with good drainage.

Tolerance to salinity. Kikuyu lawns in western Queensland will tolerate saline soils if adequately watered to keep soil salts at depth (Everist, 1974). Russell (1976) also found that it had good salt tolerance.

Fertilizer requirements. Beyond basic nutrient requirements according to soil fertility, Kikuyu responds readily to nitrogen fertilizer which gives it a competitive advantage against *Axonopus* spp. and *Paspalum dilatatum* in Australia. Colman (quoted by Mears, 1970) in northern New South Wales obtained an efficiency response of 17-24 kg DM/ha/kg N applied. Responses in Colombia were recorded up to 150 kg N/ha. An effective association with the legume *Trifolium repens* (white clover), where the clover provides 25-60 percent of the pasture, reduced the need for nitrogen (Mears, 1970). Kikuyu does not give a good response to phosphorus except on markedly deficient soils, though phosphorus application increases the legume component. The

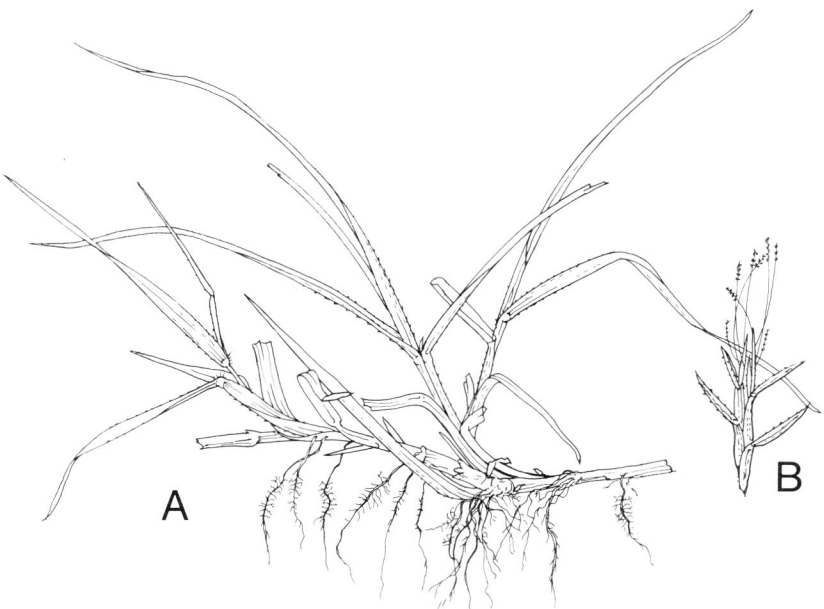

Figure 15.126. Pennisetum clandestinum. **A**-Habit **B**-Flower

critical level for phosphorus as a percentage of the dry matter at the immediate pre-flowering stage is 0.22. Potassium response is not likely unless intensive removal of the vegetative growth occurs. Symptoms of potassium deficiency appear as tip-burning and senescence of the lower leaves, and a reduced potassium content of the herbage (0.64-1 percent). Sulphur may also become deficient under heavy grazing or cutting. It is usually corrected in one normal superphosphate application (Mears, 1970).

Ability to spread naturally. Under favourable conditions of moisture and fertility, Kikuyu will spread rapidly from rhizomes and stolons, and from seeds germinating in dung pats (Wilson & Hennessy, 1977).

Land preparation for establishment. A properly prepared seed-bed is necessary for good establishment from seed. For stem and root cuttings a rougher seed-bed may suffice, as long as the vegetative material is adequately planted.

Sowing methods. Hand planting of vegetative stem and root cuttings has been traditional. Sprigs containing two or three nodes, planted on a 1-m grid is a usual plant spacing, but availability of sprigs and desired rapidity of establishment will decide procedure. For large areas, broadcasting sprigs (produced by putting plants through a chaff cutter) and then disc-harrowing them in will give adequate establishment if accompanied by a fertilizer mixture of nitrogen and phosphorus (Mears, 1970). Now that seed is available, well prepared seed-beds are essential as seed is costly. Pellet seed with activated charcoal at 1.3 kg a.c./ha and use atrazine at up to 4.5 a.c./ha. This reduces weed competition, especially from *Eleusine indica* and gives satisfactory stands (Cook & O'Grady, 1978). A drill with attached fluted furrow press wheels gives excellent results (Wilson, 1978).

Sowing depth and cover. Sow seed at approximately 5 mm depth, and roll to cover. Cultivar Whittet can germinate from a depth of 5.6 cm (Blair *et al.*, 1974).

Sowing time and rate. Plant in the spring, to compete against weeds at the rate of 2-4 kg/ha. The optimum temperature for germination is within the range of 19-29°C with the fastest rate at 29°C (Blair *et al.*, 1974). Sow after the topsoil reaches a temperature of 20°C.

Number of seeds per kg. About 40 000.

Dormancy. There is some initial hard-seededness, but it is modified by scarification during harvest or by passage through an animal (Mears, 1970).

Seedling vigour. The seedlings should be protected by slashing to reduce weed competition; vigour is enhanced by high nitrogen and phosphorus in the soil.

Vigour of growth and growth rhythm. On the Atherton Tableland, Queensland, it grows vigorously from December to April (lat. 17°13'S) but may be cut by frost from the first week in April to mid-July and be halted by dry weather from mid-July to late November (Quinlan & Edgley, 1975). Graphs of growth rhythm in southeastern Queensland and northern New South Wales are given in Figure 15.127.

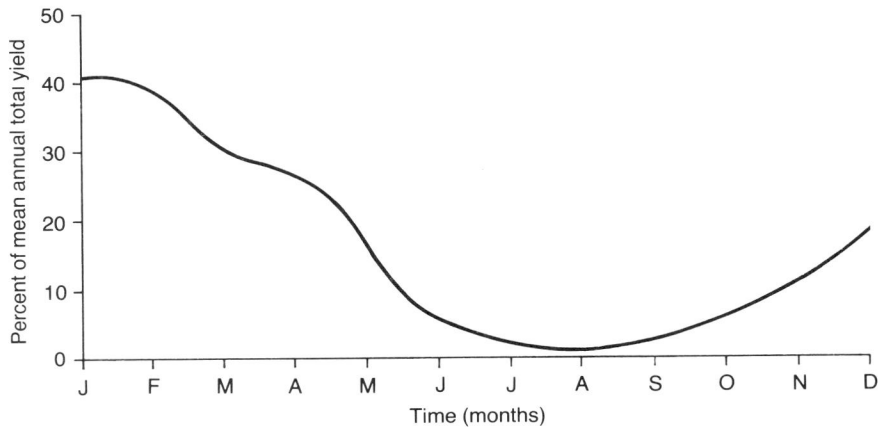

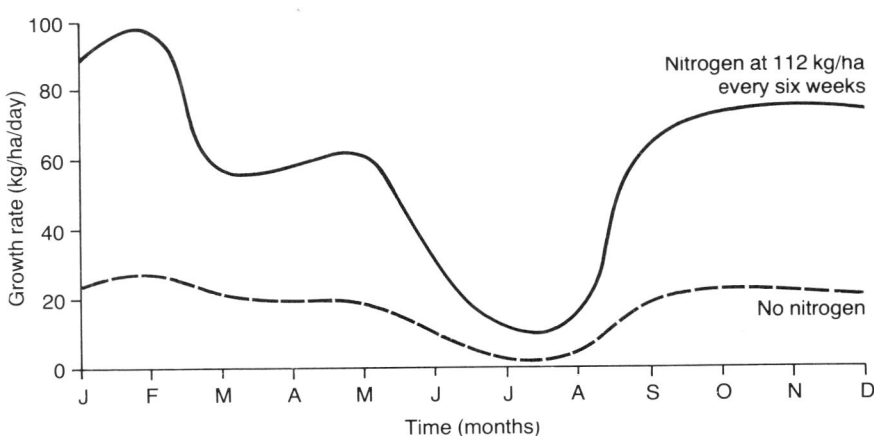

Figure 15.127. Growth rhythm of *Pennisetum clandestinum* at Crow's Nest, Queensland, Australia (lat. 27°3'S, rainfall 780 mm) and at Wollongbar, New South Wales, Australia (lat. 28°50'S, rainfall 1 700 mm) (**Source:** Jacobsen & Ivory)

Response to photoperiod. Flowering is not sensitive to changes in day length.
Response to light. Kikuyu does not grow well in shade.
Compatibility with other grasses and legumes. Under suitable conditions of soil and moisture, Kikuyu will dominate a pasture; most existing Kikuyu pastures are monospecific. With renovation and application of phosphorus-containing fertilizers, it can be combined with white and red clovers or *Desmodium uncinatum* and *D. intortum,* but management and fertilizer treatment must be good to maintain the mixture (see Plate 48). *Trifolium burchellianum* and *T. semipilosum* occur naturally with Kikuyu on the East African highlands, and some success has been achieved with the latter on the Atherton Tableland. Pure Kikuyu pastures, top-dressed with nitrogen, are usually more productive than grass/legume mixtures.
Ability to compete with weeds. With adequate moisture and fertility, Kikuyu will suppress weeds.
Tolerance to herbicides. Kikuyu itself may become a weed in cultivation. It can be killed by several herbicides including dalapon, but each successive seedling emergence must be treated. 2.2-DPA at 8 kg/ha will suppress it enough for sod-seeding another species into it. Atrazine at 1.0-1.5 kg/ha will eliminate weeds from a pure Kikuyu pasture.
Response to defoliation. Information has been obtained from cutting experiments. Colman (1966) at Wollongbar, northern New South Wales, found frequent cutting (every two weeks) reduced dry-matter yield by 54 to 25 percent, compared with a maximum yield at the 12-week cutting interval; this depression was greater in the presence of nitrogen fertilizer. Mean yield of nitrogen in the herbage fertilized with 224 kg N/ha and cut every two weeks was 176 kg/ha, compared with 131 kg/ha when cut at 12-week intervals. In Colombia, maximum production of green herbage and protein was obtained from a Kikuyu/clover sward cut to 5 cm every nine weeks. In north Queensland, a cycle of three to four weeks' grazing to 13 cm is preferred.
Grazing management. Close grazing or cutting designed to avoid the build-up of a dense mat of stolons is necessary to maintain temperate legumes with Kikuyu. Renovation of worn-out or degenerate pastures by mechanical ripping has no long-term effect unless accompanied by fertilization with inorganic nitrogen or the inclusion of a legume. Where *Neonotonia wightii* cv. Clarence was the associated legume, grazing every four weeks reduced the legume percentage, compared with the eigth- or 12-week grazing interval. The sward should be maintained with a dressing of at least 150 kg N/ha applied in split dressings in spring and autumn. If weeds are troublesome the pasture can be slashed. With a Kikuyu/tropical legume mixture, grazing to a height of 10-15 cm should take place every six to eight weeks. With strip grazing, 50 beasts per hectare per day can be grazed on a grass/legume mixture.
Response to fire. Where Kikuyu thrives there should be little fire risk, but in dry times the green top growth may hide a dry basal layer of dead leaves which can support a creeping fire. The plants soon recover.

Genetics and reproduction. The somatic chromosome number of Kikuyu is $2n=36$. Bisexual and male-sterile races exist. The Rongai strain is female-fertile. It has been suggested that apomictic reproduction occurs (Mears, 1970).

Dry- and green-matter yields. In northern New South Wales, a ceiling yield of 30 000 kg/ha of dry matter was obtained by applying 1 120 kg/ha of fertilizer nitrogen. On the Atherton Tableland, Queensland, a Kikuyu-dominant pasture produced 12 170 kg DM/ha per year.

Suitability for hay and silage. Although Kikuyu grass silage is palatable to dairy cattle, a considerable loss of dry matter occurs and digestibility of the silage is about 19.5 units lower than freshly-cut grass. Milling and pelleting the leaf for sheep resulted in a live-weight increase three times that of sheep fed the unmilled leaf ration (Hennessy & Williamson, 1976).

Chemical analysis and digestibility. Quality depends on frequency of defoliation and fertilizer applied. The high protein content of the leaves (rarely less than 12 percent) gives a high quality margin over the 8 percent crude protein required to obtain positive nitrogen balance, even with grass regrowth up to 100 days old. Digestibility of the dry matter is in the range of 60-70 percent. Kikuyu grass maintains high levels of digestible crude protein (Milford & Haydock, 1965) and of digestible organic matter (Holder, 1967). General phosphorus, potassium, calcium and magnesium levels in the herbage are adequate compared with other species, but in Hawaii some calcium deficiency in beef cattle has been recorded (Younge & Otagaki, 1958); in New South Wales, Australia, a mineral supplement of sodium, calcium and phosphorus increased calf live-weight gain by 27 percent (Kaiser, 1975).

Palatability. Very high.

Toxicity. It is rarely toxic. Lush grass growing on a heavily-manured, disused cow yard has caused nitrite poisoning (Everist, 1974; Quinlan & Edgley, 1975), and bloat (Said, 1971). In New Zealand, serious toxicity occurs spasmodically on Kikuyu pastures after rainfall in excess of 20 mm, grass temperatures above $14°C$ and invasion of pasture by army-worms. The toxin is unknown (Martinovich & Smith, 1973).

Seed production and harvesting. Seed production for commercial sale is relatively new. Repeated defoliation of the main shoots is essential to induce flowering from lateral shoots of Kikuyu (Evans, Wardlaw & Williams, 1964). Seed produced by fertile types is set so close to the ground it is difficult to harvest; hence, for seed-harvesting the Kikuyu pasture must be flat and even. Once established, the grass is mown to a height of about 25 mm and the cuttings removed. This stimulates flowering and seed production. The mower is then raised slightly so that at the next mowing, leaf growth is removed but the first crop of flowers is untouched. This promotes a second flush of flowers. Successive seed sets accumulate and the crop is mown at the end of the season, wind-rowed and threshed with a self-propelled combine (Quinlan, Shaw & Edgley, 1975).

Seed yield. 25 kg/ha from new stands, up to 500 kg/ha from established swards. Increasing the rate of nitrogen fertilizer from nil to 224 kg N/ha increased leaf production but decreased seed yield (Wilson & Rumble, 1975).

Minimum germination and quality required for commercial sale. In Queensland, 60 percent germinable seed of 93 percent purity.

Cultivars. Edwards (1937) recognized three ecotypes in Kenya.
- 'Rongai' — coarse, with broad leaves and thick stolons which develop rapidly after cutting; male sterile, anthers never exserted.
- 'Molo' — a finer plant with narrow leaves and more slender stolons which tend to throw up shoots from the centre crown after cutting; the stamens are never exserted (Parker, 1941), and the pollen is sterile.
- 'Kabete' — an intermediate form. The stamens are exserted, and functional pollen is produced.

Barnard (1972) registered two Australian cultivars.
- 'Whittet' — obtained from the Grassland Research Station, Kitale, Kenya, and developed at the Grafton Experiment Station, New South Wales. A taller, coarser, more broad-leaved and vigorous plant than the 'Kabete' ecotype above. It survives better than common Kikuyu under less fertile conditions. Seed is available from Grafton Experiment Station.
- 'Breakwell' — also developed at Grafton. It is more densely tillered than 'Whittet', more prostrate with narrower leaves, thinner stems and shorter internodes. The plants are female-fertile, but 15-20 percent are male-sterile. Seed is available at Grafton.

Value for erosion control. Kikuyu is excellent for erosion control, being used in a rainfall regime as low as 680 mm on black clays on the eastern Darling Downs, Queensland, although it prefers latosols in a higher-rainfall area. Its main function in the irrigation areas of New South Wales is to control erosion of irrigation channel banks, especially near regulators and water wheels (Read, 1975).

Diseases. Kikuyu yellows is common in northern New South Wales, especially in grass fertilized with nitrogen. To control it, return the area to cultivation of cash crops for a few years. A leaf-spot caused by *Pyricularia pennista* produces a spot surrounded by a yellow halo, and results in some leaf death, but is not of economic importance in a well-managed and fertilized pasture (Brands & Cook, 1976). A fungus disease caused by *Pyricularia grisea* causes high seedling mortality on the Atherton Tableland, Queensland, in wet seasons.

Pests. Larvae of the pasture scarab beetle (*Rhopea magnicornis*), Tarsonemus mites and soldier fly (*Atlermetapomia rubiceps*) have caused temporary damage to Kikuyu in Australia, but the effects are short-lived (Mears, 1970). In Hawaii, the hunting bull bug (*Sphenophorus vestitis*) and grass webworm (*Herpetogramma licarsicalis*) cause damage (Plucknett, 1970).

Economics. Kikuyu grass is essentially a high-quality grass for dairying and cattle finishing in high-altitude areas of the tropical and subtropical world; a useful lawn grass and soil stabilizer against erosion.

Animal production. Taylor (1941) recorded that from an area of 0.4 ha of fertilized Kikuyu grass in Natal, three Jersey cows grazed on a put-and-take system from October to May and fed a supplement of 0.45 kg maize meal at each milking time, produced a range of 8 260-15 550 kg milk per hectare (442-764 kg butterfat/ha) over a seven-year period. At Wollongbar in northern New South Wales, from a Kikuyu-based pasture fertilized with 336 kg N/ha, stocked at 4.94 cows/ha, 447 kg and 361 kg butterfat/ha were produced over two successive lactations (Kaiser & Colman, 1969). In Hawaii, beef production from fertilized *Desmodium canum*/grass mixtures was 587, 644, 706 and 806 kg/ha per year from native grass, Kikuyu, *Paspalum dilatatum* and pangola grass respectively (Younge, Plucknett & Rotar, 1964).

Main attributes. Kikuyu is a highly digestible, high protein, low fibre, palatable grass which responds readily to nitrogen, stands heavy grazing, holds soil against erosion and is an excellent lawn grass.

Main deficiencies. It does not easily lend itself to mixed grass/legume pastures, and may become a weed of cultivation.

Further reading. Mears, 1970; Quinlan, Shaw & Edgley, 1975.

Pennisetum pedicellatum Trin.

Common names. Annual kyasuwa grass (Nigeria), bara (Mauritania), deenanath grass (India).
Natural habitat. A secondary weedy invader of disturbed sites, road edges and fallows.
Distribution. Native of north tropical Africa and India.
Description. A tall, annual, bunch grass, up to 1 m high, branched from the base and above, leafy. Leaves 15-25 cm long by 4-10 mm wide, flat, glabrous. Racemes cylindrical, 5-12 cm long, dense-flowered; rachis glabrous, notched, outer bristles few, slender, short (about 3 mm long); inner bristles numerous (longest 9 mm) densely villous below the middle. Spikelets 4 mm long, usually solitary. It differs from *P. setosum* in having the inner bristles of the involucre densely villous while in *P. setosum* the inner bristles are laxly ciliate with long silky hairs (not villous) (Cooke, 1958) (see Fig. 15.128).
Season of growth. Summer.
Optimum temperature for growth. 30-35°C.
Frost tolerance. It has little frost tolerance.
Latitudinal limits. 20°N and S.
Rainfall requirements. In Bihar, India, it grows on a rainfall of 127 mm between June and September, from which it can grow and produce seeds. The usual rainfall range is 500-650 mm (Whyte, 1964).
Drought tolerance. It has good drought tolerance (Farinas, 1970). It persists well in northern Nigeria with a dry season of seven months (Foster & Mundy, 1961).
Soil requirements. It does best on fertile, loamy soils but, with manuring, can grow in sandy soils. It can tolerate both acidic and alkaline soils (Narayanan & Dabadghao, 1972).
Fertilizer requirements. It responds well to added nitrogen (Chatterjee, Roy & Bhattacharjee, 1974).
Ability to spread naturally. It spreads rapidly by self-sown seed (Whyte, 1964), regenerating each year.
Land preparation for establishment. It needs a well-prepared moist seed-bed.
Sowing methods. The seed is broadcast, or drilled in rows 45 cm apart in India.
Sowing depth and cover. It is either surface-sown or drilled at 1 cm.
Sowing time and rate. Just before the rainy season (May-July in India), at 1-2.2 kg/ha.
Vigour of growth and growth rhythm. In Bihar, it is the fastest-maturing grass. The number of days before flowering of four cultivars at Hissar, India, ranged from 105 for cv. P.p.3 to 124 for cv. P.p.H. It flowers in August in the Sahel and remains as standing hay through to June (Boudet & Duverger, 1961).

Compatibility with other grasses and legumes. It grows well in mixtures with *Phaseolus mungo* and *Melilotus alba* in India (Whyte, Moir & Cooper, 1959).
Response to defoliation. It can stand several cuts a year for green fodder.
Grazing management. It is generally used as a cut-and-carry green forage in India at ear emergence (80-90 days).

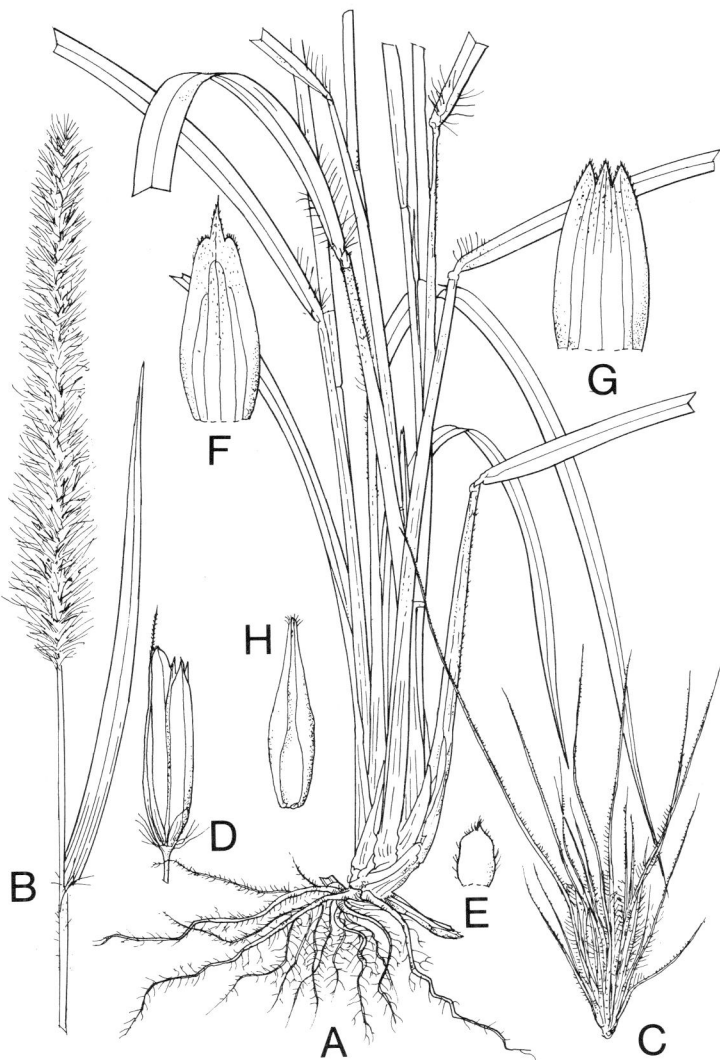

Figure 15.128. Pennisetum pedicellatum. **A**-Base of plant **B**-Inflorescence **C**-Involucre **D**-Spikelet and pedicel **E**-Lower glume **F**-Upper glume **G**-Lemma of lower flower **H**-Lemma of upper fertile flower

TABLE 15.56 *Pennisetum pedicellatum*

	DM	As % of dry matter					Reference
		CP	CF	Ash	EE	NFE	
Cultivar No. 3							
Hissar, India		11.35					
Hay, India		7.40	22.2	18.6	2.8	49.0	Sen & Ray, 1964
Bihar, India		5.63					Ghosh & Mukherjee, 1957
Fresh, bloom, Niger		12.50	35.2				Bartha, 1970
Growing, August							
Sahel-Sudan	20	7.70	35.0	13.0			
Flowering, Sept.							
Sahel-Sudan	30	7.80	38.2	15.3			
Dry stems, Oct.-Feb.							
Sahel-Sudan	95	2.80	44.0	7.8			

Genetics and reproduction. 2n=36, 48, 54 (Fedorov, 1974); 57 (Whyte, 1964). There is a wide range of growth forms. It is strongly apomictic (Whyte, 1964).

Dry- and green-matter yields. At the Punjab Agricultural University, Hissar, India, four cultivars of *P. pedicellatum* yielded from 96 207 to 109 875 kg green matter per hectare compared with 56 607 kg from sweet Sudan grass and 36 957 kg from sorghum (no fertilizer data given; Singh & Arora, 1970). It is cut two or three times a season, first 80 days after germination and subsequently at 60-day intervals. It has also yielded good hay in Nigeria and Sierra Leone (Whyte, 1964).

Suitability for hay and silage. It has been made into silage in Nigeria, Sierra Leone and India (Rains, 1963) and also into hay.

Chemical analysis and digestibility. Banerjee and Mandel (1974) recorded 55-77 percent total digestible nutrients for hay in India, with 3 percent crude protein. See also Table 15.56.

Palatability. It is very palatable to cattle in India (Banerjee & Mandel, 1974). It has a high leaf/stem ratio. It is not very palatable in the Sahel (Boudet & Duverger, 1961).

Toxicity. No toxicity has been recorded. The oxalic acid content of cultivar P.p.3 at Hissar was 1.69 percent, compared with 2.5 percent for *Pennisetum americanum* and 6.0 percent for *P. purpureum* (Singh & Arora, 1970).

Seed production and harvesting. It seeds abundantly and matures very quickly in India.

Seed yield. Up to 2 tonnes/ha (Whyte, 1964).

Cultivars. In India, the Punjab Agricultural University at Hissar has four cultivars whose characteristics are shown in Table 15.57. Variety G.73 is very good for overseeding overgrazed pastures (Whyte, 1964).

TABLE 15.57 *Pennisetum pedicellatum* cultivars

Cultivar	Average height of plant (cm)	Average no. tillers/plant	Days to 50% flowering
P.p.3	176.8	59	105
P.p.10	173.3	33	125
P.p.15	182.3	40	117
P.p.H	190.0	28	124

Value for erosion control. It is a valuable soil stabilizer in India.

Diseases. None observed at Hissar, India.

Pests. None observed at Hissar, India.

Economics. In India it is a valuable grazing grass for sheep, goats and cattle (Bor, 1960). It is also good as a short-term ley and soil stabilizer. In northern Australia it is a weed.

Animal production. At Hissar it was used in preference to Sudan grass and sorghum because it gave higher yields of disease- and pest-free green matter when irrigated (Banerjee & Mandel, 1974).

Main attributes. Its early flowering, high tiller number, high leaf/stem ratio, low oxalic acid content, and palatability (Singh & Arora, 1974).

Main deficiencies. Being an annual it provides only short-term grazing; can become a weed of cultivation.

Further reading. Chatterjee, Roy & Bhattacharjee, 1974; Mukerji & Chatterji, 1955.

Pennisetum polystachyon (L.) Schult.

Common names. Mission grass (Fiji), khachornchob (Thailand), thin Napier grass (India), nigolo (Mali).
Natural habitat. Grassland on sandy soils, and as a weed.
Distribution. Throughout the tropics, especially in Thailand and Fiji where it was introduced.
Description. An annual or perennial; culms simple or branched, the branches often flowering. Spikelets 3-5 mm; false spike 8-10 mm, rarely 6-15 mm wide, excluding the bristles; longest bristle 15-25 mm long, the others more than twice as long as the spikelet. When mature, the spikelets break off at the central axis together with the bristles (Chippendall & Crook, 1976). It produces few tillers per plant (Mishra & Chatterjee, 1968) (see Fig. 15.129).
Season of growth. Summer.
Optimum temperature for growth. It makes most growth at 32-35°C in Fiji.
Minimum temperature for growth. About 12°C in July in Fiji.
Latitudinal limits. It is common throughout Fiji from latitudes 17-18°S.
Altitude range. Sea-level to 1 500 m.
Rainfall requirements. It is a high-rainfall grass, but is also grown in semi-arid regions.
Drought tolerance. It is drought resistant and suitable for semi-arid areas in India (Narayanan & Dabadghao, 1972). In Fiji, growing leaf turns red if subjected to drought.
Tolerance to flooding. It tolerates flooding well and is good for waterlogged black clay soils.
Soil requirements. It vegetates the highly phosphate-deficient nigrescent soil at Sigatoka, Fiji. In Kenya, Thailand and India it is usually found on sandy soils. In India it tolerates both acid and alkaline soils.
Fertilizer requirements. It is usually not fertilized. With 448 kg/ha of superphosphate initially, followed by 228 kg/ha per year, *Desmodium heterophyllum* volunteers in the mission grass in Fiji. In India it is sown with farmyard manure and top-dressed annually with 158 kg ammonium sulphate.
Ability to spread naturally. It spreads readily by seed, which survives the annual burning.
Land preparation for establishment. It needs a well-prepared seed-bed.
Sowing methods. Seed is sown broadcast or in drills 15-22 cm apart in India, or planted out from nurseries as seedlings. In Fiji it has been planted vegetatively (Partridge, 1979a).
Sowing depth and cover. It should be surface-sown and rolled.
Sowing time and rate. At the beginning of the wet season, at 3.4-4.5 kg/ha in India.
Dormancy. The seed has no dormancy, and is viviparous in very wet weather.
Seedling vigour. Good, even in poor fertility conditions.

Vigour of growth and growth rhythm. It grows quickly, flowers in April and seeds in May/June in Fiji. After this, the flower stems lignify to a completely inedible straw, which, because of its bulk and height (2 m), prevents access to light and restricts the grazing animal to the lower green leaves (Partridge, 1975). It will remain green in the dry season in Fiji if grazed to prevent flowering. In India it is ready to cut three months from head emergence.

Response to photoperiod. It is a short-day plant, flowering mainly in May in Fiji.

Response to light. It is fairly shade tolerant, persisting under 80 percent shade under *Pinus caribaea*.

Compatibility with other grasses and legumes. Its clumpy growth habit allows legumes such as *Desmodium heterophyllum* and *Macroptilium atropur-*

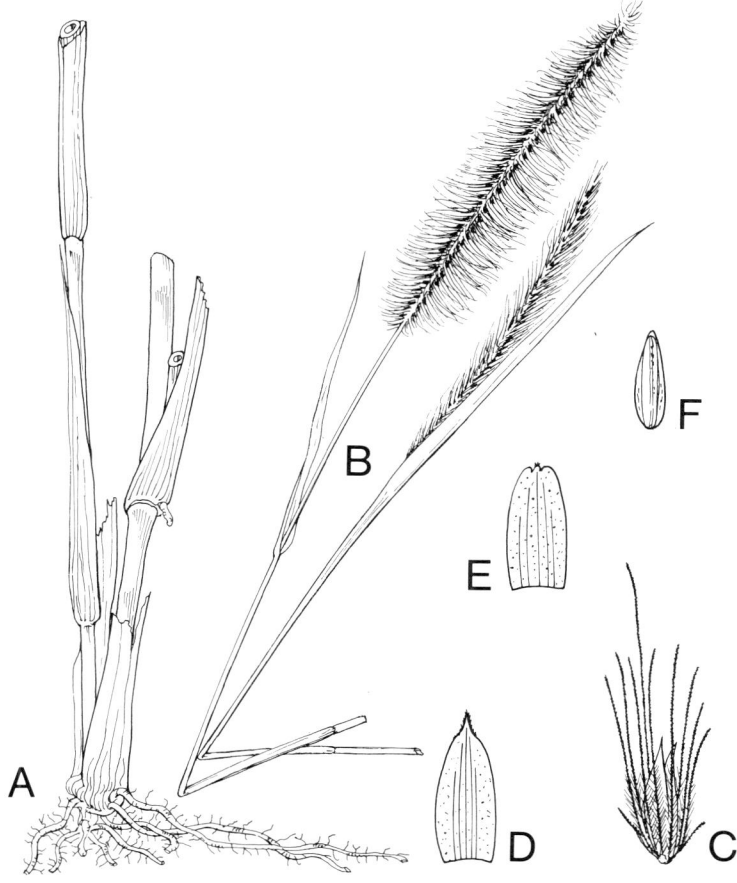

Figure 15.129. Pennisetum polystachyon. **A**-Plant **B**-Inflorescence **C**-Spikelet **D**-Upper glume **E**-Lower lemma **F**-Upper floret

pureum to grow with it. An initial application of 448 kg superphosphate per hectare, plus 224 kg/ha each year in Fiji causes *Desmodium heterophyllum* to appear spontaneously (Partridge, 1975), but the grass is not sufficiently productive to warrant fertilizing it alone. Growing siratro was very effective. In India it has combined successfully with *Atylosia scarabaeoides, Clitorea ternatea, Calopogonium mucunoides, Centrosema pubescens* and *Stylosanthes guianensis* (Singh & Chatterjee, 1968).

Ability to suppress weeds. It suppresses weeds well by its vigorous growth after burning.

Tolerance to herbicides. Fisher and Ive (1970) showed the density of *P. polystachyon* in an irrigated seed production plot of *Stylosanthes humilis* was greatly reduced by the use of chlorthal or of trifluralin (Treflan) at 1.1 kg AI/ha applied 14 days before irrigation.

Response to defoliation. *P. polystachyon* cannot stand heavy grazing; with fencing, it can be controlled by heavy stocking (Ellison & Henderson, 1973). In Fiji under heavy stocking it becomes invaded by *Desmodium heterophyllum* and other species (Partridge, personal communication).

Grazing management. The grass should be prevented from seeding to maintain its nutritive value. A six-week cutting interval gave better material than the 12-week interval. Complete burning, top-dressing with 450 kg/ha of single superphosphate and broadcasting siratro or *Stylosanthes guianensis* seed at 5 kg/ha each gave a good pasture mixture at Sigatoka, Fiji (Partridge, 1975; 1979a).

Response to fire. It tolerates annual fires; it constitutes a grassland representing a fire disclimax in northeast Thailand and Fiji. It will burn to ground level leaving a clean seed-bed suitable for easy legume establishment by oversowing (Partridge, 1975).

Genetics and reproduction. $2n=54$ (Fedorov, 1974).

Dry- and green-matter yields. In India it yields 3 360 kg of green fodder per hectare per year in three to four cuttings (Narayanan & Dabadghao, 1972). At Sigatoka, Fiji, Partridge (1975) obtained 1 500 kg DM/ha in January and March 1972, falling to near zero in July-August (see Fig. 15.130). Its yearly production of 1 390 kg/ha of green matter exceeded that of five other grasses (Partridge, 1979a).

Suitability for hay and silage. It makes useful hay if cut before maturity, but is usually cut and fed green to cattle in India (Narayanan & Dabadghao, 1972).

Chemical analysis and digestibility. See Table 15.58.

Palatability. The young grass is fairly palatable, but the mature material is ignored by stock.

Seed production and harvesting. Mishra and Chatterjee (1968) in India found cutting twice yearly and fertilizing with 38.9 kg N/ha and 22.2 kg P_2O_5/ha gave the highest "seed" yield. Caryopses constituted 30 percent of the total "seed" yield.

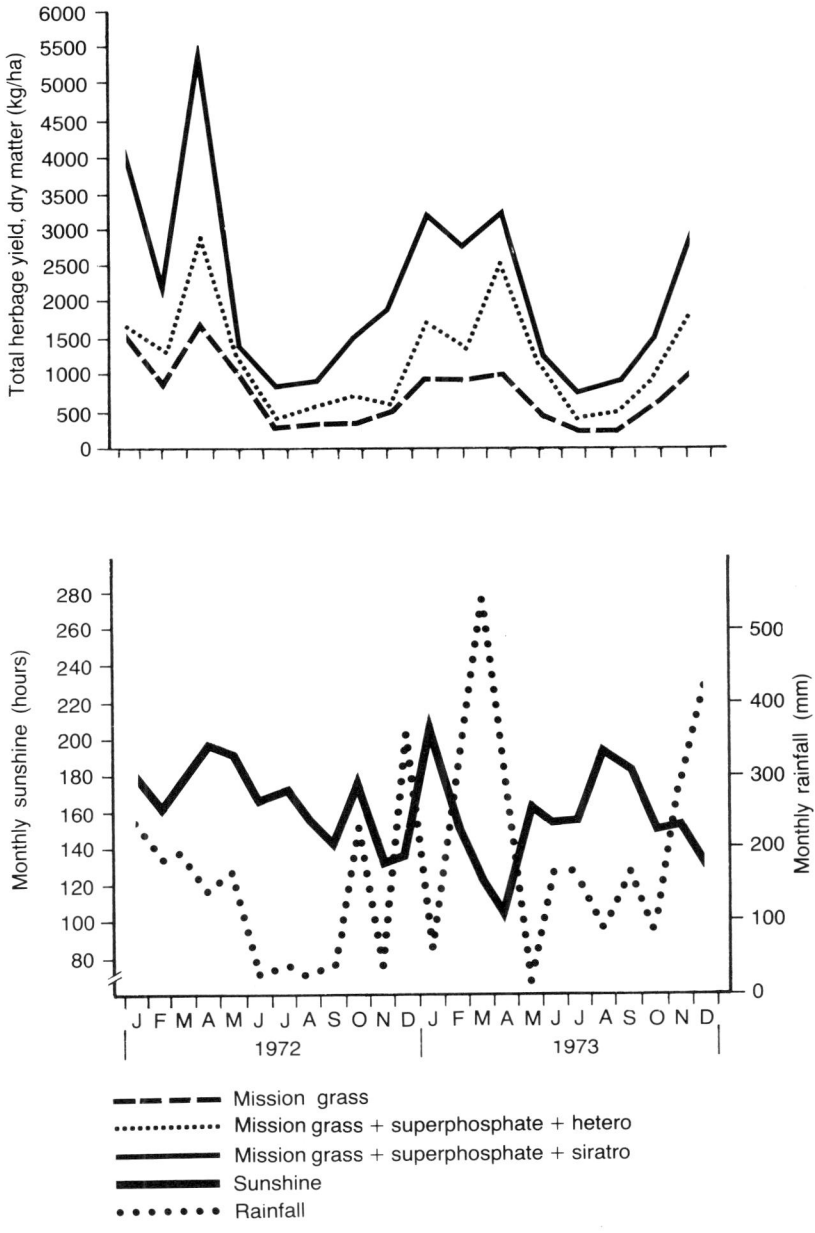

Figure 15.130. Seasonal dry-matter production of *Pennisetum polystachyon* in Fiji (**Source:** Partridge, 1975)

TABLE 15.58 *Pennisetum polystachyon*

	As % of dry matter				
	CP	CF	Ash	EE	NFE
Fresh, early bloom, Kenya	17.6	27.1	11.9	2.7	40.7
Fresh, early bloom, India	11.2	28.4	16.3	2.6	41.5
Fresh, 1st cut, India	17.4	23.0	16.0	1.4	42.2
Fresh, 2nd cut, India	12.3	31.4	10.4	1.7	44.2

Source: Göhl, 1975

Seed yield. From fertilized pasture, cut mid-January and mid-July in India, 420 kg/ha (Mishra & Chatterjee, 1968).
Cultivars. There are no registered cultivars.
Value for erosion control. It quickly covers the ashes of a fire and forms a dense tussock grassland, which prevents erosion.
Diseases. It does not suffer from any major diseases.
Pests. It has no serious pests.
Economics. Used as a "cut-and-carry" green fodder for cattle in Thailand and Fiji. As a fire disclimax, *P. polystachyon* grassland invades a good deal of the mountainous land in both these countries. It is generally regarded as a weed, but recently Partridge (1975, 1979a) has shown that it can be fertilized and combined with the legumes siratro, stylo and hetero to produce productive pastures.
Animal production. In Fiji, Partridge (personal communication) obtained the figures shown in Table 15.59.
Main attributes. Its ability to invade and dominate wet tropical areas after fire; producing a great bulk of fodder, preventing erosion and weed growth.
Main deficiencies. Its poor quality as it matures, and its susceptibility to overgrazing.
Further reading. Mishra & Chatterjee, 1968; Partridge, 1975; Roberts, 1970b.

TABLE 15.59 *Pennisetum polystachyon,* animal production, mean live-weight gain (kg/ha)

Superphosphate application	Steers/ha		
	1.5	2.5	3.5
Nil	155	220	265
220 kg/ha + stylo + hetero	205	275	300
440 kg/ha + stylo + hetero	375	375	375

Pennisetum purpureum Schumach.

Common names. Elephant or elefante grass, Napier grass, gigante (Costa Rica), mfufu (Africa).
Natural habitat. Damp grassland and forest edges, cultivation.
Distribution. Native to subtropical Africa (Zimbabwe) and now introduced into most tropical and subtropical countries.
Description. A robust perennial with a vigorous root system, sometimes stoloniferous with a creeping rhizome. Culms usually 180-360 cm high, branched upwards. Leaf-sheaths glabrous or with tubercle-based hairs; leaf-blades 20-40 mm wide, margins thickened and shiny. Inflorescence a bristly false spike up to 30 cm long, dense, usually yellow-brown in colour, more rarely purplish (Chippendall, 1955) (see Fig. 15.131, Plate 49).
Season of growth. Summer.
Optimum temperature for growth. Usually 25-40°C. Mean 21.1° ± 2.8°C (Russell & Webb, 1976).
Minimum temperature for growth. About 15°C. Mean minimum temperature of the coldest month 11.5° ± 5.4°C (Russell & Webb, 1976).
Frost tolerance. It is susceptible to frosts.
Latitudinal limits. Usually between 10°N and 20°S (Russell & Webb, 1976).
Altitude range. Sea-level to 2 000 m.
Rainfall requirements. Elephant grass grows best in high-rainfall areas (in excess of 1 500 mm per year), but its deep root system allows it to survive in dry times. Mean, 1 483 mm ± 620 (Russell & Webb, 1976).
Drought tolerance. It survives drought quite well when established because of its deep root system (see Plate 50).
Tolerance to flooding. It does not tolerate flooding.
Soil requirements. It grows best in deep, fertile soils through which its roots can forage. Deep, friable loams are preferable.
Tolerance to salinity. No record of salinity tolerance has been found.
Fertilizer requirements. A complete fertilizer mixture may be needed for establishment according to soil fertility. In Tobago, West Indies, a crop of elephant grass removed 463 kg nitrogen, 96 kg phosphorus and 594 kg potassium per hectare per year. The optimum phosphorus content of the dry matter for growth was determined as 0.248 percent for the purple type and 0.215 percent for the green variety (Falade, 1975). High rates of nitrogen generally give good responses (Walmsley, Sargeant & Dookeran, 1978) especially in the third and subsequent years when the native soil nitrogen has been exhausted (Vicente-Chandler *et al.*, 1953). The latter authors suggested that the highest yields could be expected from cutting at 12-week intervals and applying nitrogen after every cut.
Ability to spread naturally. It is usually planted, as it spreads slowly.

Land preparation for establishment. Full land preparation with ploughing and subsequent disc-harrowing and drilling will repay the cost of establishment of this perennial grass.

Sowing methods. Either root cuttings or stem pieces with at least three nodes are planted in the drills. When planting stem pieces, two nodes should be covered with soil, the third being exposed. One hectare of grass will provide

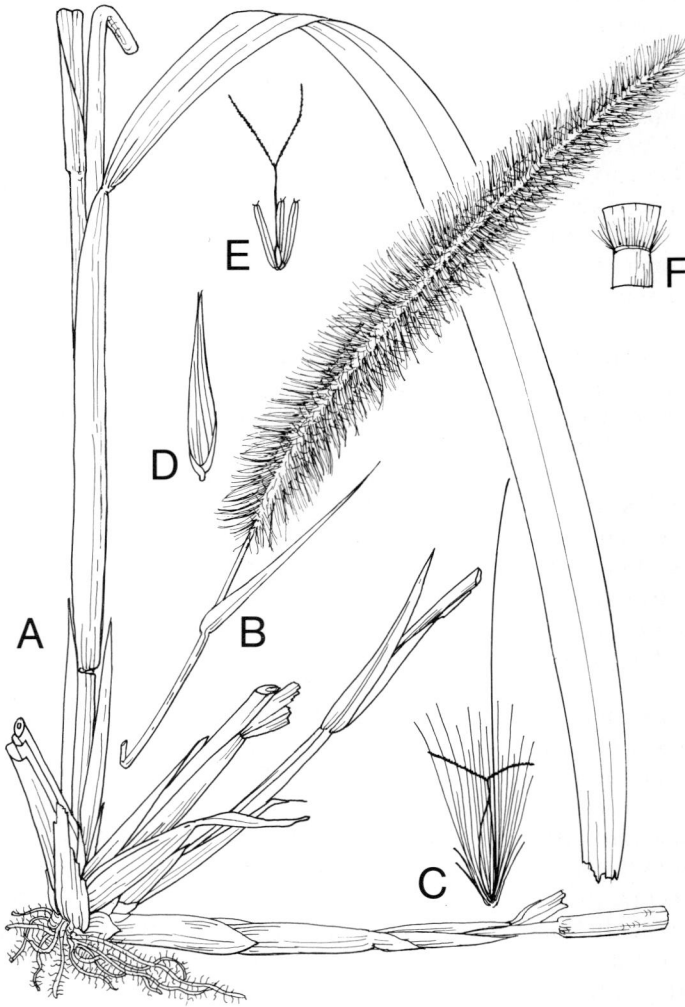

Figure 15.131. *Pennisetum purpureum.* **A**-Habit **B**-Inflorescence **C**-Spikelet with bristles **D**-Spikelet without bristles **E**-Flower **F**-Ligule

propagating material for 15-25 hectares. Planting rooted elephant grass pieces directly into an *Imperata* sward during the rainy season in the Philippines has had some success (Farinas, 1970).

Sowing depth and cover. Plant in furrows about 15 cm deep and cover with about 7.5 cm of soil initially, gradually filling as the plant grows.

Sowing time and rate. At the beginning of the wet season, at about 2 000 kg/ha of stem material.

Number of seeds per kg. 3 084 400 in the United States.

Vigour of growth and growth rhythm. It is a very vigorous grass.

Response to photoperiod. It is a short-day plant.

Response to light. It will grow in partial shade as a cut-and-carry fodder in tropical gardens, but produces better in full sunlight.

Compatibility with other grasses and legumes. It is generally grown as a pure pasture. However, it has been sown in alternate rows with such legumes as *Pueraria phaseoloides* in Puerto Rico, *Centrosema pubescens* (Venezuela) and *Neonotonia wightii* in Uganda. Cutting or grazing management will have to be adjusted to favour the legume to maintain a satisfactory mixed sward.

Ability to compete with weeds. When established, elephant grass will suppress weeds.

Tolerance to herbicides. To eradicate elephant grass, it should be burned off and any regrowth sprayed with 2,2-DPA at 4.5 kg of a 740 g AI/kg product (e.g. Shirpon, Dowpon) plus 250 ml wetting agent per 200 litres of water. Thoroughly wet the plants (Tilley, 1977).

Response to defoliation. Elephant grass will stand heavy grazing and provides a great bulk of feed (Harrison & Snook, 1971), especially if fertilized and irrigated. It is suited to rapid rotational grazing, which must not be severe enough to hinder regrowth (Ware-Austin, 1963). Only the leaves are eaten when the grass is near maturity. A height of 5 cm is best for cutting (Vicente-Chandler *et al.*, 1974).

Grazing management. Elephant grass is commonly used in a cut-and-carry system, feeding it in stalls, or it is made into silage. For grazing, it should be heavily stocked to maintain it in a lush vegetative form. The mature leaves are razor sharp and sometimes provide a problem for grazing cattle. The coarse stems produce new shoots and leaves called "lala" in Hawaii; the grass is best grazed when the new growth consists of five new leaves and associated stem growth. A stem plus "lala" takes a year to grow (Younge & Ripperton, 1960). Odhiambo (1974) showed no drop in nutritive value at Kitale, Kenya, in analyses taken at seven to 12 weeks. Grazing at six- to nine-week intervals at a height of about 90 cm gives good utilization. Nitrogen can be applied after each grazing or cutting in high-rainfall areas. Any coarse, leafless stems should be mowed.

Response to fire. Elephant grass will burn if dry enough, and produce new growth afterwards, but it is seldom dry enough to burn in its normal environment.

Genetics and reproduction. The somatic chromosome number is 2n=27, 28, 56 (Fedorov, 1974). It crosses readily with *Pennisetum americanum (P. typhoides)* to produce a rugged hybrid, bana grass, used for wind-breaks in vegetable areas in coastal Queensland.

Dry- and green-matter yields. Elephant grass gives heavy yields and Vicente-Chandler, Silva and Figarella (1959) established a world record production of 84 800 kg DM/year when it was fertilized with 897 kg N/ha per year and cut every 90 days under natural rainfall of some 2 000 mm per year. Other recorded yields are 35 500 kg DM/ha per year over three years in Tobago (Walmsley, Sargeant & Dookeran, 1978), 32 400 kg DM and 3 400 kg crude protein per hectare per year when cut every 56 days at CIAT, Colombia (Moore & Bushman, 1978), 20 800 kg DM/ha per year in Nigeria (Adegbola, 1964) and 40 000-50 000 kg green matter per hectare when cut each 35-40 days at the Tulio Ospina Station, Colombia (Crowder, Chaverra & Lotero, 1970).

Suitability for hay and silage. It makes good hay if cut when young but is too coarse if cut late in its annual growth cycle. It is more usually made into silage of high quality without additives. Silage losses have been 9 percent in India (Mahadevan & Venkatakrishnan, 1957) and 17 percent in Puerto Rico (Vicente-Chandler *et al.*, 1953). In Taiwan, elephant grass is widely used for the production of dehydrated grass pellets used as a supplementary stock feed (Manidool, personal communication).

Value as standover or deferred feed. If the grass is allowed to reach maturity before the last wet-season cut, it gives better dry-season use. On the Atherton Tableland, Queensland, it is used for dry-season feed by rolling at the end of winter, as it can make some winter growth during this period (Quinlan & Edgley, 1975).

Chemical analysis and digestibility. Göhl (1975) gives a list of chemical analyses and digestibilities from a wide range of conditions. Because of the importance of elephant grass this list is given in full in Table 15.60.

Palatability. It is highly palatable in the leafy stage.

Toxicity. García-Rivera and Morris (1955) recorded 2.48 percent of oxalates in the dry matter of elephant grass and 2.5 percent in the Merker variety but no toxicity was experienced. Ndyanabo (1974) recorded 3.1 percent total oxalates but again no toxicity.

Seed production and harvesting. Elephant grass does not produce much seed, and so is propagated vegetatively.

Cultivars.

● Var. *merkeri* (Merker grass) — similar to common elephant grass but has finer leaves and stems. It is cultivated widely in Puerto Rico and other West Indian areas. It is more drought resistant than common elephant grass but less productive and of lower feeding value (Whyte, Moir & Cooper, 1959). It is resistant to *Helminthosporium* sp. in Puerto Rico (Vicente-Chandler *et al.*, 1953).

TABLE 15.60 *Pennisetum purpureum*

	DM	As % of dry matter				
		CP	CF	Ash	EE	NFE
Fresh, vegetative, 40 cm, Tanzania	20.0	9.8	29.7	14.0	2.6	43.9
Fresh, vegetative, 80 cm, Tanzania	20.0	9.0	28.6	14.8	1.1	46.5
Fresh, early bloom, 240 cm, Tanzania	25.0	7.2	36.1	12.4	1.0	43.3
Fresh, tops only, 222 cm, Tanzania		13.2	32.9	10.3	2.4	41.2
Fresh, cut at 6-week intervals, Malaysia	19.0	10.0	31.6	15.3	2.1	41.0
Fresh, cut at 8-week intervals, Malaysia	19.5	9.7	33.3	16.4	1.5	39.1
Fresh, cut at 10-week intervals, Malaysia	21.0	7.6	35.2	14.8	1.4	41.0
Hay, vegetative, South Africa		15.1	34.9	12.1	2.4	35.5
Hay, mature, South Africa		7.5	40.3	11.7	1.4	39.1
Silage, 120 cm, Zimbabwe	23.5	6.8	35.8	13.7	0.9	42.8
Silage, 210 cm, Zimbabwe	21.4	4.2	35.3	15.2	1.2	44.1
Fresh, var. *merkeri*, late bloom, Puerto Rico	24.0	8.6	36.1	10.2	3.1	42.0
Fresh, var. *merkeri*, cut every 6 weeks, Malaysia	20.5	9.8	32.2	12.2	1.5	44.3
Fresh, var. *merkeri*, cut every 8 weeks, Malaysia	20.5	8.8	34.6	14.1	1.5	41.0
Fresh, var. *merkeri*, cut every 10 weeks, Malaysia	23.5	7.7	35.7	14.0	0.9	41.7
Fresh, Giant Napier, 4 weeks, 50 cm, Thailand	15.8	10.8	28.5	13.9	3.8	43.0
Fresh, Giant Napier, 6 weeks, 75 cm, Thailand	17.1	8.8	32.2	12.9	3.5	42.6
Fresh, Giant Napier, 8 weeks, 135 cm, Thailand	18.3	8.7	32.8	10.9	3.3	44.3
Fresh, Giant Napier, 10 weeks, 150 cm, Thailand	18.5	6.5	33.0	11.4	2.7	46.4
Fresh, Giant Napier, 12 weeks, 150 cm, Thailand	20.4	5.9	31.9	10.3	2.9	49.0

Continued

TABLE 15.60 *Pennisetum purpureum* (concluded)

	Animal	Digestibility (%)				
		CP	CF	EE	NFE	ME
Fresh, vegetative, 40 cm, Tanzania	Sheep	61.2	74.7	50.0	71.8	2.30
Fresh, vegetative, 80 cm, Tanzania	Sheep	54.4	60.5	45.0	62.2	1.92
Fresh, early bloom, 240 cm, Tanzania	Sheep	50.0	60.1	30.0	52.9	1.79
Fresh, tops only, Tanzania	Zebu	73.4	65.7	66.4	57.0	2.17
Hay, vegetative, South Africa	Oxen	73.2	77.2	66.5	67.8	2.44
Hay, mature, South Africa	Oxen	40.7	56.0	30.9	33.8	1.46
Merker grass, late bloom, Puerto Rico	Sheep	66.0	61.0	57.0	58.0	2.06
Giant Napier 75 cm, Thailand	Sheep	33.0	62.0	53.0	62.0	1.94
Giant Napier 135 cm, Thailand	Sheep	38.0	64.0	53.0	58.0	1.96
Giant Napier 10 weeks, 150 cm, Thailand	Sheep	50.0	60.0	54.0	54.0	1.87
Giant Napier 12 weeks, 150 cm, Thailand	Sheep	28.0	60.0	39.0	54.0	1.81

Source: Göhl, 1975

- 'Capricorn' — developed at Biloela Research Station, Queensland, for high rainfall areas receiving up to 2 500 mm/year. It is leafier, more palatable and later-flowering than the common type.
- 'Pusa Giant Napier' — performs well in Sri Lanka under good soil conditions, but is affected by *Helminthosporium* sp. (Pathirana & Siriwardene, 1973).

'Merkiron' and 'Costa Rica 532' are used in Colombia, and 'French Cameroons', 'Gold-Coast' and 'Cameroons' in Africa. 'Chadi' is recommended by Prasad and Singh (1973) for cultivation under arid conditions in West Rajasthan, India.

Value for erosion control. Elephant grass will give very effective control of erosion in its own ecological niche.

Diseases. The most common disease is blight caused by *Helminthosporium sacchari*. The best practice is to use a resistant variety.

Pests. No major pests have been recorded.

Economics. It is one of the most valuable forage, soilage and silage crops in the wet tropics.

Animal production. At the National Research Station, Kitale, Kenya, elephant grass was fertilized at the rate of 80-120 kg triple superphosphate per hectare and 120 kg sulphate of ammonia per hectare. It supplemented sown pastures during their decline in growth in April, May and June when the dairy cows were calving. The elephant grass yielded 11 480 kg DM/ha in the second season and 4 360 kg DM/ha in the third season, carrying 2.5 and 2.4 beasts per hectare, respectively (Ware-Austin, 1963). In Hawaii, elephant grass can produce as much as 336 000 kg of green forage per hectare per year (Takahashi, Moomaw and Ripperton, 1966) and live-weight gains as high as 549 kg/ha were obtained with beef cattle grazing mature elephant grass. In Colombia, 36 milking cows were maintained on forage from 2.5 hectares of elephant grass. They received a nutritional supplement concentrate ratio of 1 kg per 4 kg milk and averaged 15 litres of milk per day (Crowder, Chaverra & Lotero, 1970). At CIAT, Colombia, Moore and Bushman (1978) calculated that 1 hectare of high-quality elephant grass would provide enough forage to produce 3 tonnes live-weight gain in zebu-type cattle.

Main attributes. Its high dry-matter yield, especially with frequent cutting under fertilization and irrigation. Its suitability for silage and its deep and extensive root system which enables it to forage widely for moisture and nitrogen.

Main deficiencies. Its high fibre content at maturity, poor seed production, and susceptibility to frosts.

Further reading. Ware-Austin, 1963; Vicente-Chandler *et al.*, 1974.

The weedy *Pennisetum* spp.

Several of the genus *Pennisetum* are rugged tussock grasses which are avoided by cattle; they gradually invade improved pastures if they are not eradicated. *Pennisetum pedicellatum* and *P. polystachyon* are often regarded as weeds, but they have both been used in livestock production and can be useful when properly managed.

Pennisetum hohenackeri Steud. and C.E. Hubbard

Synonym. *P. catabasis* Stapf.
A perennial, 90-150 cm high with stout, compressed culms forming dense tussocks. The false spike is 10-25 cm long, dense with spikelets 6.5-9.0 mm long, surrounded by numerous bristles, the longest one up to 20 mm long. It occurs in swampy grassland or vleis in Uganda, Kenya and Tanzania (common on the Mara Plateau near Musoma, Tanzania), at elevations of 1 000-2 000 m. The chromosome number is 2n=18 (Fedorov, 1974). Dougall and Bogdan (1958) recorded 6.9 percent crude protein, 40.2 percent crude fibre, 5.8 percent ash, 2.0 percent ether extract and 45.1 percent nitrogen-free extract in the dry matter of fresh material in the early bloom stage in Kenya. It is of low palatability. It can be eradicated by ploughing or hoeing (see Fig. 15.132).

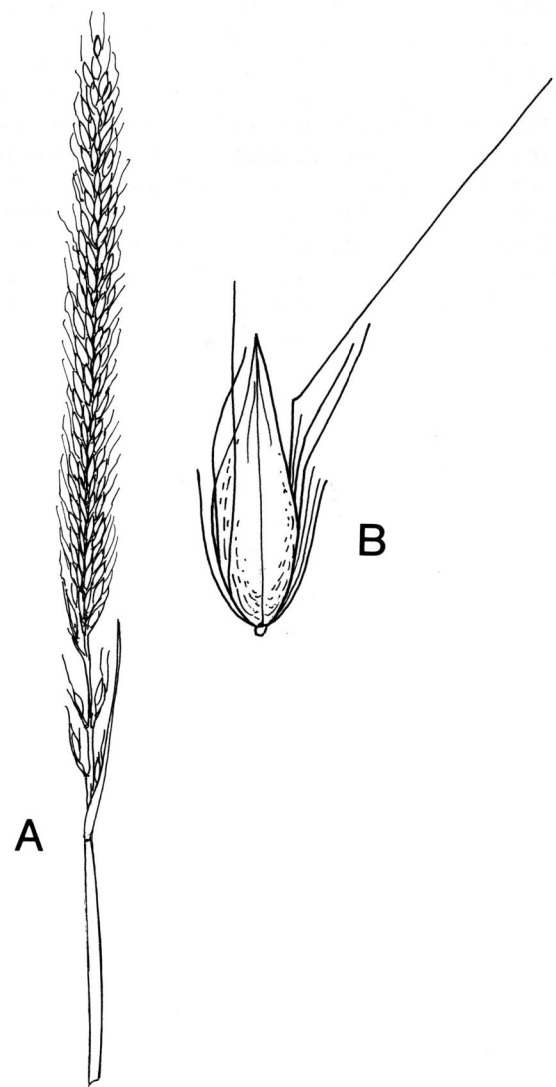

Figure 15.132. Pennisetum hohenackeri. **A**-Inflorescence **B**-Spikelet

Pennisetum schimperi A. Rich.

A densely tufted perennial up to 120 cm high, with culms usually hairy below the inflorescence. False spike dense, 4.5-9 cm long with 3.5-5 mm long spikelets surrounded by numerous bristles up to 12 mm long. It occurs in acacia tall-grass savannah throughout central and east tropical Africa and is common on the Ethiopian Highlands and in the highlands of Somalia, at elevations of 1 600-3 100 m in a rainfall regime of 500-750 mm per year. Dougall and Bogdan (1958) recorded 9 percent crude protein, 35.6 percent crude fibre, 7.3 percent ash, 2.9 percent ether extract and 45.2 percent nitrogen-free extract in the dry matter of fresh material in the early bloom stage in Kenya. However, it is of low palatability, and is generally avoided by grazing animals. It can be suppressed in red oat grass (*Themeda triandra*) grassland by burning during the dormant season after most of the grasses have ripened (Edwards, 1942). It can also be ploughed or hoed out if accessible.

Pennisetum setaceum (Forsk.) Chiov.

This grass is known as "alula" in Ethiopia and "fountain grass" in southern Africa. It is a densely tufted perennial, up to 100 cm high with long narrow leaf-blades. False spike dense, plumose, 10-23 cm long, often purplish, with one to three spikelets, 4.5-6.5 mm long, in a cluster surrounded by numerous ciliate bristles, up to 40 mm long (Napper, 1965). It resembles *P. villosum* but its spike is more cylindrical, 10-25 cm or more long and purple or rose-coloured (Chippendall, 1955). It occurs on rocky slopes in dry bush in Kenya, Tanzania, and western and northeast Africa at altitudes from sea-level to 2 000 m. Dougall and Bogdan (1958) recorded 15.9 percent crude protein, 31.3 percent crude fibre, 11 percent ash, 2.2 percent ether extract and 39.6 percent nitrogen-free extract in the dry matter of fresh material in the early bloom stage in Kenya. However, it has hard fibrous leaves, is not very palatable and little grazed. It is cultivated in Ethiopia and South Africa as an ornamental in parks and gardens (Chippendall, 1955).

Pennisetum spicatum (L.) Körn

This grass is called "bultuc" in Ethiopia. Its chromosome number is 2n=14 (Fedorov, 1974). It is cultivated in Ethiopia.

Pennisetum villosum (R. Br.) Fresen

This grass is known as "long-styled feather grass" in Australia ("foxtail" in Toowoomba, Queensland) and "feather-top" in the United States. It is a perennial tussock grass with a creeping rhizome; culms up to 90 cm high in cultivated plants, simple or branched from the lower nodes, with leaf-sheaths compressed and keeled, bearded at the mouth and usually hairy on the margins upwards; leaf-blades glabrous, 2-6 mm wide, expanded or folded. The inflorescence is a feathery spike 4-7 cm long, dense, light brown or green. It differs from *P. setaceum* in that its spike is usually ovoid and light brown. It is native to Africa but naturalized in the United States and Australia. The chromosome numbers are $2n=18, 27, 36, 45, 54$ (Fedorov, 1974). It is cultivated as an ornamental in Ethiopia, southern Africa and the United States. It has become a weed on the latosolic soil on basalt around Toowoomba, Queensland, at an altitude of 600-650 m, with rainfall of 750 mm per year.

Phragmites australis (Cav.) Trin. ex Steud.

Synonym. Phragmites communis Trin.
Common name. Common reed (Australia).
Natural habitat. Swamps, drains, moist headlands.
Distribution. Pan-tropical.
Description. A warm-season, rhizomatous, stoloniferous perennial growing 2-4 m high. Leaf-blades flat, smooth, 15-45 cm long, 1-5 cm wide. Seed-head an open panicle, purplish or tawny, flaglike appearance after seeds shatter. Spikelets open toward maturity to show a mass of dense soft hairs (Tothill & Hacker, 1973) (see Fig. 15.133).
Season of growth. Perennial.
Optimum temperature for growth. 30-35°C.
Rainfall requirements. It is a swamp grass and so requires high moisture conditions.
Drought tolerance. It survives droughts until the soil dries out.
Tolerance to flooding. It will tolerate considerable flooding but prefers very damp, rather than continually wet, conditions (Linedale, 1974). It does best where water level fluctuates from 15 cm below the soil surface to 15 cm above (Leithead, Yarlett & Shiflet, 1971).
Soil requirements. It grows best in firm mineral clays.
Tolerance to salinity. It tolerates moderate salinity, but grows mainly in brackish water.
Fertilizer requirements. It is not fertilized.
Ability to spread naturally. It spreads from an aggressive root system and stolons.
Land preparation for establishment. It is not planted; it occurs and spreads naturally.
Vigour of growth and growth rhythm. In the United States, growth starts in February (early spring) and foliage stays green until frost. New shoots grow from buds at nodes of old stems, stolons and rhizomes.
Response to light. It prefers full sunlight.
Compatibility with other grasses and legumes. It grows as a monospecific sward.
Ability to compete with weeds. It dominates other species and is usually a weed itself in drains and irrigation channels.
Tolerance to herbicides. It is controlled by a combination of mechanical and chemical means. Reeds are allowed to reach about 1 m in height before spraying, as lush growth is necessary for the herbicide to be effective. Dalapon is sprayed at 5.5 kg/ha, two or three times at ten-day intervals. One application of dalapon at 8.5 kg/ha plus 2.8 litres of commercial amitrole as a combination spray also gives good control. Amitrole alone at 12 litres/ha, sprayed first

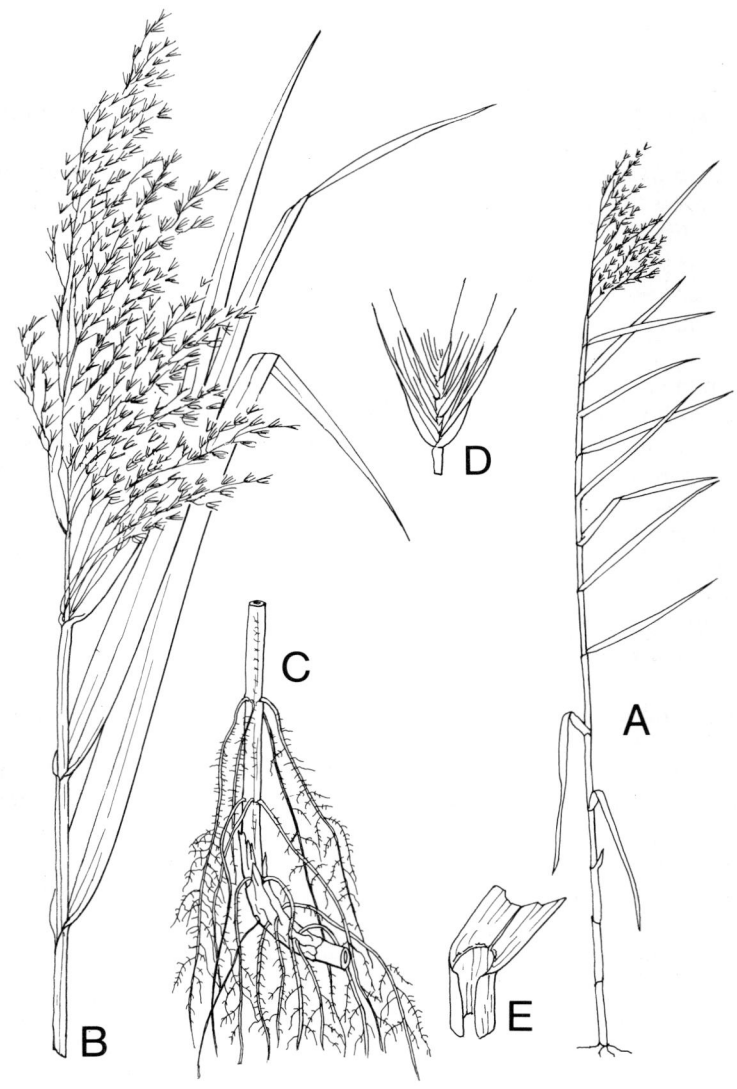

Figure 15.133. Phragmites australis. **A**-Habit **B**-Inflorescence **C**-Rooting culm **D**-Spikelet **E**-Ligule

before flowering (when the reed is commonly 2-3 m high), may be successful but gives erratic results. Patches of reed can be treated with heavy doses of dalapon or bromacil, at 17 kg/ha, to kill off a large proportion of the underground runners, but high cost and danger to (sugar cane) crops may preclude this heavy use. Most chemical control should take place during fallow periods when there is less danger to crops (Linedale, 1974).
Response to defoliation. It cannot stand prolonged heavy grazing. Its upright growth makes it easy for livestock to remove all the leaves.
Grazing management. For maximum production no more than half of the current year's growth (by weight) should be grazed off during the growing season. Grazing deferments of 60-90 days every two to three years during the growing season improve the plants' vigour. Water control that lowers the water level but does not drain the area increases production.
Response to fire. It tolerates burning if water is above the soil surface, but burning is not essential for management.
Genetics and reproduction. 2n=36, 48, 54, 96 (Fedorov, 1974).
Chemical analysis and digestibility. It provides high-quality warm-season forage but becomes tough and unpalatable after maturity.
Palatability. It is palatable only in the very young stage.
Economics. The common reed in southern Queensland coastal areas has spread out from its natural swampy areas into drains and sugar-cane farms, where it smothers young plants and ratoon crops, and offers strong competition to advanced cane. Its aggressive root system and good response to fertilizers are the main problems in its spread (Linedale, 1974). However, *Phragmites* stands are very important for wild life. The species is also widely used for thatching and matting and in some countries, e.g. Romania and Poland, it is harvested in large quantities as raw material for the paper and chemical industries (Cook, 1974). Common reed has been used in the southwestern United States for lattices in constructing adobe houses. Indians have used the stems for arrows and for weaving mats and nets (Leithead, Yarlett & Shiflet, 1971).
Animal production. No figures for animal production have been found.
Main attributes. It stabilizes banks and drains against erosion in non-agricultural areas. It is useful for temporary roofing, paper and arrows.
Main deficiencies. It is a problem in irrigation drains, roadside ditches and in some sugar-cane fields.
Further reading. Linedale, 1974.

Phragmites karka (Retz.) Steud.

Common names. Pit-pit (New Guinea), tropical reed (Australia).
Natural habitat. Along streams, in wet grassland and in swamps in Africa, India, New Guinea and northern Australia.
Description. A perennial reed with long rhizomes and robust, erect culms to 3 m. The leaves are 15-30 cm long and nearly 2.5 cm broad; inflorescence is a large plumelike panicle with capillary branches and small, slender spikelets. It is leafy up the panicle. *Phragmites* can be easily distinguished from *Arundo* and *Neyraudia* by the silky beards at the bases of the lowest panicle branches, which are absent in the other two (Dabadghao & Shankarnarayan, 1973) (see Fig. 15.134).

In New Guinea the reed occurs from near sea-level to at least 2 000 m. It thrives in a rainfall regime from 200 to 5 000 mm in swamps (India). It grows in standing water and is therefore tolerant of flooding. It usually grows in clay soils ranging from strongly acid (pH 4.5) to slightly alkaline (pH 7.5).
Genetics and reproduction. The chromosome number is 2n=36, 38, 48 (Fedorov, 1974).
Economics. It disappears quickly under the impact of cutting and burning. In New Guinea, *P. karka* swamps at 1 500 m elevation can be used for grazing, if drainage can be provided. The pit-pit is burnt in the dry season and the regrowth grazed at a high stocking rate (25-40 beasts/ha). The cattle graze shoulder to shoulder. A live-weight gain of almost 1 kg/head per day is possible. The regrowth can be grazed three to four times before the pit-pit is exhausted (Graham, personal communication).

It withstands heavy floods and is an excellent stabilizer of eroding river banks (Rose-Innes, 1977). *Saccharum robustum* Brandes and Jesueit ex Grassl. is also known as pit-pit in New Guinea.
Further reading. Rose-Innes, 1977.

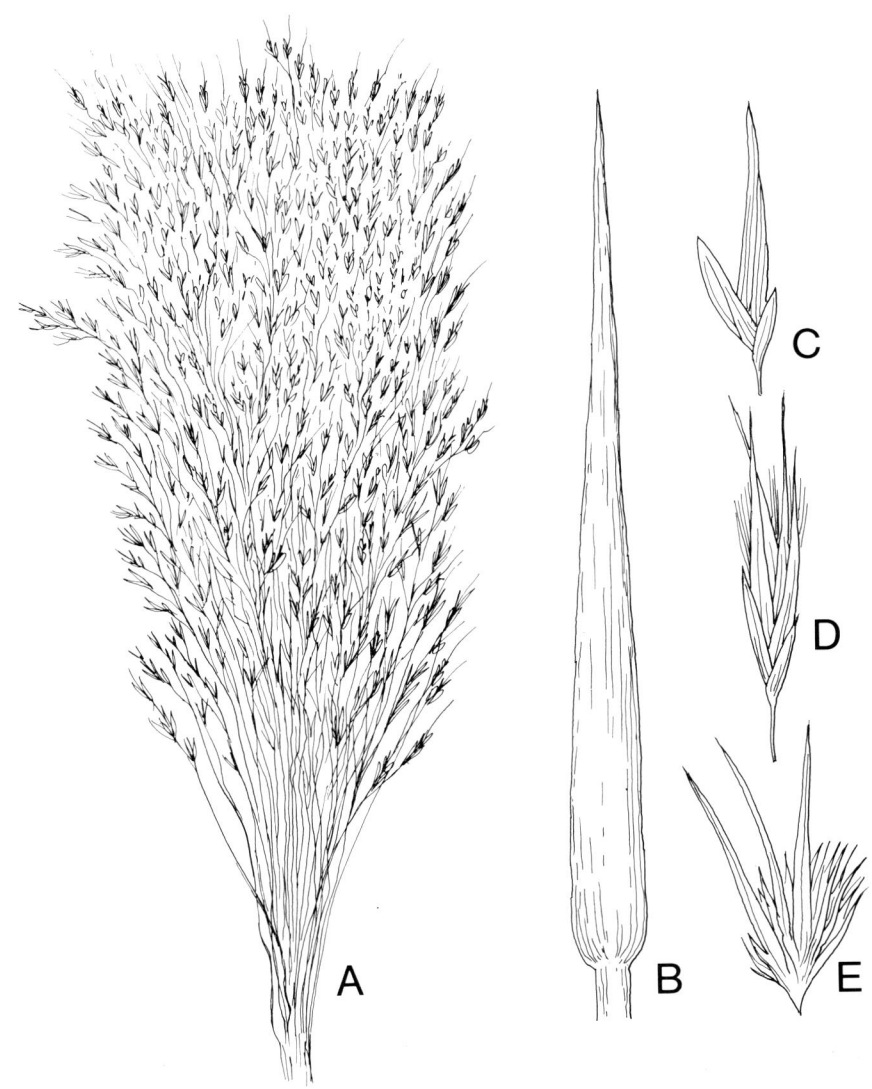

Figure 15.134. Phragmites karka. **A**-Inflorescence **B**-Leaf blade **C**-Glumes and lower lemma **D**-Upper lemmas **E**-Spikelet

Saccharum officinarum L.

Common names. Sugar cane (general), shunkora (Ethiopia).
Natural habitat. Tropical rain forest. Cultivated.
Distribution. First domesticated in India or Southeast Asia, now cultivated extensively in tropics and subtropics throughout the world.
Description. Cane to 5 m, leaves broad. Panicle large, plumelike, tapering from base to tip with silky spikelets. The sett, when planted, sends out roots to nourish the growing shoot from the node and, beneath the surface of the soil, the shoot forms a succession of very short joints; the buds of these germinate in turn to give rise to secondary shoots to form a "stool" below ground. These secondary shoots are fed by a further series of roots to produce a root mass, spreading to a depth of 30 cm or more and laterally for up to 1 m (see Fig. 15.135, Plate 51).
Optimum temperature for growth. Tillering increases with temperature up to 30°C (van Dillewijn, 1952).
Minimum temperature for growth. Stem elongation ceases at 18°C.
Frost tolerance. Sugar cane is susceptible to frost, the growing shoot and top "eyes" (buds) being the first to die, but the buds from the lower nodes may provide new growth, according to frost severity.
Latitudinal limits. 30°N and S.
Altitude range. 500-3 000 m.
Rainfall requirements. For economic sugar production an annual rainfall of 1 500 mm is regarded as the minimum; in lower rainfall areas, the "noble" varieties are irrigated.
Drought tolerance. It is fairly drought resistant, but production is low in drought periods.
Tolerance to flooding. Sugar cane will tolerate short floods, but, if approaching maturity, it will become lodged and the sugar content will decline.
Soil requirements. It has a wide range of soil tolerance, but drainage is essential. Heavy soils may be "bedded" to lift the soil level, and an open drainage furrow provided every five to ten rows.
Tolerance to salinity. Sugar cane gave maximum yields at EC_e 1.8 mmhos/cm, 50 percent of maximum at 10 mmhos/cm and nil at 18.7 mmhos/cm (Maas & Hoffman, 1976). In India, the varieties Co 75, Co 453, B 37172 and Co 1148 show good tolerance to salinity (Yadav, 1975).
Fertilizer requirements. Soil tests are usually conducted in commercial sugar-growing areas to determine these needs. There is usually a basic planting mixture of complete NPK fertilizer, followed by side-dressings of nitrogen during growth. In Hawaii "crop-logging", to decide fertilizer needs after planting, is carried out by tissue-testing.
Land preparation for establishment. As the crop may occupy the ground for

up to four years, thorough land preparation is required. Deep ploughing and deep ripping should be carried out and the final seed-bed prepared by disc cultivators.

Sowing methods. Sugar cane is propagated by burying whole stalks in furrows, then chopping the stalks into at least two-node lengths in the furrow. It can also be planted with a chopper-planter, cutting the stalk into two-node

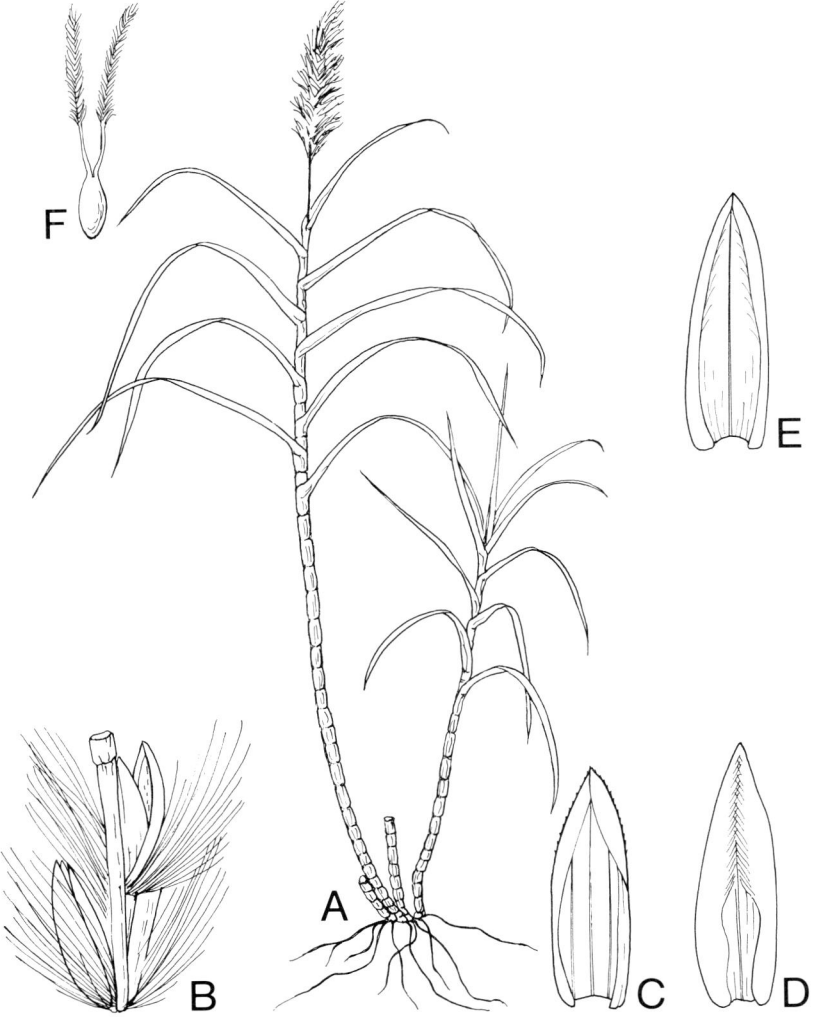

Figure 15.135. Saccharum officinarum. **A**-Habit **B**-Spikelets **C**-Upper glume **D**-Lower glume **E**-Lemma **F**-Carpel

lengths as it is fed into a planting chute. The setts are usually treated with a fungicide as they are planted.

Sowing depth and cover. The setts are planted in furrows 25 cm deep, placed 1.3-1.4 m apart, and covered lightly with soil until "tillering" (stooling) has progressed. Then the furrows are gradually filled by inter-row cultivation.

Sowing time and rate. Autumn and spring planting is common, at about 6 000-7 000 kg/ha.

Vigour of growth and growth rhythm. Sugar cane is a perennial. Growth is rapid in early summer, and sugar production increases in autumn, though it may decline if frosted. It matures in 12-14 months and is usually harvested then.

Response to photoperiod. It flowers in short and medium day lengths (Evans, Wardlaw & Williams, 1964).

Response to light. Sugar cane will grow in shade, but sugar production is aimed at the greatest use of incoming radiation to promote maximum photosynthesis.

Ability to compete with weeds. Sugar-cane land has to have thorough pre-planting preparation, inter-row tillage and herbicide treatment to suppress weeds until the cane is "out of hand", when the dense shade from the canopy will control weeds.

Tolerance to herbicides. Herbicides are used widely to suppress weeds in sugar cane, and are dealt with fully by Tilley (1977). See *Phragmites australis, Themeda quadrivalvis, Brachiaria subquadripara* and *Brachiaria mutica* for application information.

Response to defoliation. It is not usually grazed, the whole stalk being harvested at maturity. It will then grow again from the roots and produce a succession of ratoon crops, the number being dictated by the economics of retaining the crop. When the old "stool" is reduced by subsoil ploughing, it will give good regrowth after being shaved to ground level and fertilized.

Response to fire. Sugar cane is often burnt to ease harvesting. It is not killed, and will sucker from nodes or regrow from the "stool" afterwards.

Genetics and reproduction. 2n=60, 68, 80, 90 (Fedorov, 1974).

Dry- and green-matter yields. At Grafton, New South Wales, cv. Pindar yielded 149 000 kg green matter per hectare, and cv. 40 SN5819 produced 129 000 kg green matter per hectare (Mead & Norman, 1950). In Brazil, Zuniga, Sykes and Gomide (1967) recorded 69 900 kg and 66 200 kg DM/ha with two cultivars.

Suitability for hay and silage. Silage has been made from sugar-cane tops in Queensland (Skerman, 1941), Argentina (Bragadin & Diaz, 1957), Puerto Rico (Vicente-Chandler *et al.*, 1953) and Taiwan. The silage is very low in crude protein (1.4 percent of the green matter) and is fed to cattle, with concentrates, as low-quality, perennially-available roughage.

Value as standover or deferred feed. Sugar cane can stand in the field for several years and can be used in emergency as low-quality roughage.

Chemical analysis and digestibility. Where sugar cane is grown and harvested for sugar production, the tops are usually fed green, chopped for stall feeding ("chop-chop") or made into silage. Young sugar cane (two to three internodes) analysed in the Philippines showed 79.8 percent moisture, 1.8 percent crude protein, 9.6 percent crude fibre, 1.9 percent sucrose (Azman, 1951). More recently a product called "fith", consisting of de-rinded sugar-cane pith, has given excellent results in livestock feeding. Sugar cane itself provides negligible amounts of protein, and supplements are needed. Trials in the Caribbean (Donefer, James & Laurie, 1973) have shown that freshly harvested and processed sugar fith and cane tops (SF/CT) constituting 80 percent of the cattle's dry-matter ration resulted in weight gains averaging 0.9 kg per day during the traditional finishing period; additional energy supelementation from molasses or maize significantly increased gains. Dairy trials have indicated that SF/CT can constitute up to half of the total ration, replacing energy-rich feeds as well as supplying a source of succulent forage. Digestibility trials with sheep have indicated that a SF/protein supplement has a dry-matter digestibility averaging 70 percent. The authors conclude

> "Based on an average sugar-cane yield of 88 t/ha for Barbados, a projected annual live-weight gain per hectare would be 4 600 kg, with only protein/mineral supplement supplied in addition to SF/CT. An assumed world average sugar-cane yield of 50 t/ha could result in a 2 600 kg live-weight gain per hectare."

At Wollongbar, New South Wales, chopped ten-month-old sugar cane was fed as a supplement to cows grazing nitrogen-fertilized Kikuyu grass pasture (which is practically dormant during the winter). The sugar cane was very low in protein, containing only 0.62 percent in the dry matter, and *in vitro* digestibility was 52 percent. Milk production was 6.3 kg per cow per day of 4 percent fat-corrected milk, compared with 9.7 kg with oats and 9.6 kg with rye grass. Butterfat percentage was good at 4.9 percent in the milk produced; protein content was 3.58 percent.

Palatability. Sugar-cane stalks are quite palatable because of the sugar content, but the high fibre makes chewing a slow process.

Seed production and harvesting. Seed production is controlled by ecological factors. "Arrowing" (emergence of seed-heads) usually reduces the sugar yield, so the sugar cane is generally harvested before this would occur. Cane breeders encourage it artificially for cross-breeding purposes, and pollen can be deep frozen for future use.

Cultivars. Numerous cultivars are bred for sugar production, disease resistance, maturity, varying soils, dry conditions and flooding. They are available in sugar-producing countries.

Value for erosion control. Sugar cane can be used to hold soil and act as a wind-break, but retention of small areas for this purpose would endanger disease-quarantine efforts. For erosion or wind control, should be sown in rows on the contour.

TABLE 15.61 **Summary of growth results using sugar-cane fith, sugar-cane tops and pangola grass as forage, with a variety of energy and protein supplements**

Trial and phase	No. cattle	Type of forage	Energy supplement	Average daily gain (kg)
Cattle Trial I				
Phase 1	13	Pangola grass	None	0.53
	12	Sugar-cane fith	None	0.58
	13	Fith/sugar-cane tops	None	0.66
Phase 2	6	Pangola grass	None	0.81
	6	Pangola grass	Molasses	0.97
	6	Pangola grass	Maize	1.18
	6	Fith/tops	None	0.99
	6	Fith/tops	Molasses	1.08
	6	Fith/tops	Maize	1.26

Trial and phase	No. cattle	Type of forage	Energy supplement	Protein supplement	Average daily gain (kg)
Cattle Trial II					
Phase 1	8	Fith/tops	Molasses	Fish-meal	0.81
	8	Fith/tops	Molasses	Rape-seed meal	0.88
	7	Fith/tops	Molasses	Fish-meal/rape-seed meal	0.88
	8	Fith/tops	Maize	Fish-meal	0.98
	8	Fith/tops	Maize	Rape-seed meal	0.97
	8	Fith/tops	Maize	Fish-meal/rape-seed meal	0.93

Trial and phase	No. cattle	Type of forage	Energy supplement	Form of forage	Average daily gain (kg)
Cattle Trial II					
Phase 2	8	Fith/tops	None	Fresh	0.88
	8	Fith/tops	None	Ensiled	
	8	Fith/tops	Molasses	Fresh	1.00
	7	Fith/tops	Molasses	Ensiled	0.87
	8	Fith/tops	Maize	Fresh	1.09
	8	Fith/tops	Maize	Ensiled	0.94

Source: Donefer, James & Laurie, 1973

Economics. Sugar cane is one of the two main world sources of sugar for domestic and industrial use. Its products, such as molasses and sugar-cane tops, are available for livestock feeding and industrial use.

Animal production. Trials by Donefer, James and Laurie (1973) in the Carib-

TABLE 15.62 **Sugar-cane molasses**

Dry matter	76.40%
Total digestible nutrients	78.92%
Nitrogen	0.90% (equivalent to 5.6% crude protein)
Phosphorus	0.07%
Calcium	1.15%
Magnesium	0.61%
Sodium	0.10%
Potassium	5.19%
Chlorine	2.98%
Sulphur	0.73%
Copper	11.0 ppm
Zinc	11.6 ppm
Manganese	82.4 ppm
Iron	247.0 ppm

bean using sugar-cane fith and cane tops in comparison with chopped pangola grass with supplements are recorded in Table 15.61.

Molasses. The nutritive value of sugar-cane molasses based on all the sugar mills in Queensland, Australia, expressed as a percentage of the dry matter, is shown in Table 15.62. Table 15.63 shows the analyses in comparison with maize grain. Molasses is low in crude protein but supplies a lot of energy. It is also low in fibre and so has a laxative effect on cows if fed in large amounts. Protein, phosphorus, sodium and fibre should be added to molasses when feeding.

Molasses will give on average 0.7 kg of milk per kg of grain fed, and

TABLE 15.63 **Sugar-cane molasses versus maize grain: feeding value**

Component	Molasses	Maize grain
Water	20-25%	10%
Crude protein	0-6% on DM basis	10% of DM
Total digestible nutrients	81% of DM	83% of DM
Metabolizable energy	3.2 M cal/kg DM	3.3 M cal/kg DM
Crude fibre	—	2.2%
Phosphorus	0.5 - 0.1 %	0.3 - 0.35 %
Calcium	1.0 - 1.2 %	0.02 %
Total minerals (ash)	10.0 - 15.0 %	1.0 %

TABLE 15.64 Sugar-cane molasses: daily rations for cattle

Cattle type	Daily ration (kg)
Weaners and yearlings	2.0
Cows in good condition, early pregnancy	3.6
Cows in good condition, late pregnancy or early lactation	5.2
Cows in poor condition, late pregnancy or early lactation	6.4

maize grain will give 1 kg. The maximum intake of molasses per cow should be 3.6 kg per day. Milk production falls sharply when more than 25 percent of the dry-matter intake is molasses, that is, more than 4 kg per day.

Molasses may be used successfully for survival feeding of cattle when roughage supplies are limited. Daily amounts of molasses needed for survival are shown in Table 15.64.

It is advisable to add 30 g of urea for each kilogram of molasses. As a liquid, a suitable mixture is 80 percent molasses, 17 percent water and 3 percent urea. The urea can be dissolved first in the water, which makes the molasses easier to handle. Once the molasses has been mixed in, care must be taken to avoid fermentation. Cattle should be introduced slowly to this type of mixture; 1-2 kg per head for the first week, reaching full strength by the third week. In 3.5 kg of the mix are 130 g of urea, the required daily amount. A block lick can be made following the instructions in Table 15.65.

Further reading. Donefer, James & Laurie, 1973; King, Mungomery & Hughes, 1965.

TABLE 15.65 Molasses/urea block

Material	Parts
Crushed grain	40
Molasses	20
Coarse salt	20
Urea	10
Bone flour	7
Bone meal	7
Chrisphos	7
Meat meal	5

Source: Goodwin & Chamberlain, 1979
NOTE: Instructions for making molasses-urea blocks — Dissolve the urea in hot water, then add the molasses and then the other ingredients. Drop the mixture into a mould, such as half a 200-litre drum, and tamp down. Allow the block to harden for a few hours before feeding it.

Saccharum sinense L.

Common names. Uba cane, Japanese cane.
Distribution. Mainly grown in China and Japan, but was cultivated widely before the better "noble" canes were introduced for sugar production.
Description. A tall, hardy and vigorous cane with wide adaptability and early maturity. Stems slender with greenish-bronze, bobbin-shaped nodes, high fibre content and poor juice quality. Leaves up to 5 cm broad. Inflorescence: rachis with long hairs; glumes four; lodicules non-ciliate (Pursglove, 1976).
Season of growth. Perennial.
Frost tolerance. It cannot stand heavy frosts but will survive light frost. The growing point and upper buds (eyes) are killed but the lower buds survive and produce side shoots.
Latitudinal limits. About 30°N and S.
Altitude range. Sea-level to 300 m.
Rainfall requirements. It will grow in areas with a lower rainfall than the "noble" sugar canes, but generally used in the 750-1 000 mm rainfall regime.
Drought tolerance. Hardier than sugar cane. It will stand over well into the dry season.
Tolerance to flooding. It will not tolerate prolonged flooding.
Soil requirements. Adapted to poorer soils than the "noble" sugar canes (*S. officinarum*), but produces heavy crops in fertile soils under irrigation.
Fertilizer requirements. It is usually planted with a complete NPK fertilizer at 200-400 kg/ha, and subsequently side-dressed with 100-200 kg N/ha during growth. Soil or tissue tests will determine needs.
Land preparation for planting. A deep, well-prepared seed-bed is essential for useful yields.
Sowing methods. It is sown as stem cuttings (setts) in rows 1-1.5 m apart and 20-25 cm deep, and lightly covered until it tillers (stools). Later the drills are gradually filled by inter-row cultivation to reach ground level.
Sowing time and rate. Usually drilled in early spring if rainfall or irrigation is sufficient, but can be sown in autumn in frost-free areas. About 1 500-3 000 kg/ha of setts are planted.
Response to defoliation. It can be harvested frequently and the stools (lower stems and roots) will provide numerous ratoon crops, if watered and fertilized.
Management. It is cut for fodder every three to four months, or retained for cutting for drought fodder. It should be adequately fertilized especially with nitrogen and preferably irrigated.
Genetics and reproduction. 2n=116-120 (Fedorov, 1974). It is now thought to be a hybrid between *S. officinarum* and *S. spontaneum*. It is almost completely sterile.

TABLE 15.66 *Saccharum sinense*

	DM	As % of dry matter					Water
		CP	CF	Ash	EE	NFE	
Fresh, mature, Trinidad	23.4	10.3	32.1	6.3	2.5	48.8	
Fresh, whole plant, Suriname		8.3	34.1	9.2	2.4	46.0	
Fresh, leaves only, Suriname		9.5	33.4	8.8	2.8	45.5	
Fresh, stems only, Suriname		5.3	35.7	10.5	1.5	47.0	
	DM	As % of whole sample					Water
		CP	CF	Ash	EE	NFE	
Fresh, mature, Trinidad	32.4	2.96	9.8	1.75	0.49	17.4	67.6
Fresh, mature, Trinidad	24.7	1.44	7.5	2.40	0.56	12.8	75.3
Mature cane tops, Trinidad	27.6	1.32	9.28	1.67	0.48	14.85	72.4
Mature cane tops, Trinidad	25.1	1.26	8.1	1.5	0.64	13.6	74.9
Silage, Trinidad	19.1	5.3					

Source: Göhl, 1975; Harrison, 1942

Suitability for hay and silage. Paterson (1945) made good silage with four- to six-month-old Uba cane by chaffing and adding 9 litres of molasses per tonne. In Madagascar, it is also used for silage (Dufournet *et al.*, 1959).

Chemical analysis and digestibility. Harrison (1942) showed that uba cane (cut when mature) contained 23 percent dry matter, 1.4 percent digestible crude protein and 11.9 percent starch equivalent. Göhl (1975) records analyses in Table 15.66.

Cultivars. 'Uba' is an old variety with a high fibre content (17 percent), which will provide heavy yields and several ratoon crops, especially if fertilized and irrigated.

Economics. It was used widely for sugar production before the introduction of the low-fibre "noble" canes belonging to *S. officinarum*. It was also used for standover fodder for livestock during winter and for chewing. It is occasionally harvested for syrup.

Animal production. No figures have been cited.

Main attributes. Its ability to produce copious roughage at a time when summer pastures are low-yielding, its ability to produce several ratoon crops and its resistance to gumming disease.

Main deficiencies. Its relatively high fibre content.

Further reading. Pursglove, 1976.

Saccharum spontaneum L.

Common names. Wild cane, pit-pit (New Guinea).
Natural habitat. Common on river banks, alluvial plains, damp depressions and swamps as a fire disclimax grassland (Paijmans, 1976).
Distribution. Africa (Ghana), Asia to Melanesia.
Description. Rhizomatous erect grass 2-3.5 m, with stiff, rather slender culms, 0.6-1.25 cm in diameter. Leaves linear, lanceolate, channelled, about 1.25 cm wide. Panicle plumelike, rather narrow; spikelets 3-4 mm long, lodicules ciliate (Henty, 1969) (see Fig. 15.136).
Altitude range. Sea-level to 1 700 m in New Guinea.
Rainfall requirements. It prefers a high rainfall, usually in excess of 1 500 mm.
Drought tolerance. It has a good degree of drought tolerance.
Tolerance to flooding. It will tolerate some flooding.
Soil requirements. Adapted to a wide range of soils, generally of rather sandy types.
Sowing methods. Established by stem cuttings or division of rhizomes (see *S. sinense*).
Ability to compete with weeds. It develops an enormous root system and possesses the lightest of seeds. It flowers toward the end of the rains in India when the floods recede and expose bare mud flats, sand banks and islands, and eroded land. These areas are at once occupied by *S. spontaneum*, *Phragmites* and other grasses. If these stands are burnt in the earlier part of the year they will be ousted by *Imperata cylindrica* (Bor, 1960).
Tolerance to herbicides. Where *S. spontaneum* is a weed and needs control, it is best treated in hot weather by ploughing, followed by TCA sodium salt at 30 kg AI/ha. Any regenerating plants can be controlled by spraying with a mixture of dalapon and aminotriazole at 10 kg AI/ha (Singh, Pandey & Shankarnarayan, 1970).
Genetics and reproduction. 2n=40, 48, 54, 55, 56, 64, 80, 112, 120, 126, 128 (Fedorov, 1974).
Chemical analysis and digestibility. Kehar (1948) records the results shown in Table 15.67.
Palatability. It provides poor fodder, but is used to feed buffaloes in India.
Economics. The species flowers and fruits at the end of the rains in India and is therefore capable of colonizing areas such as soil and sand left bare by retreating floods. The root system is extremely extensive and the grass acts as an effective soil binder (Bor, 1960).

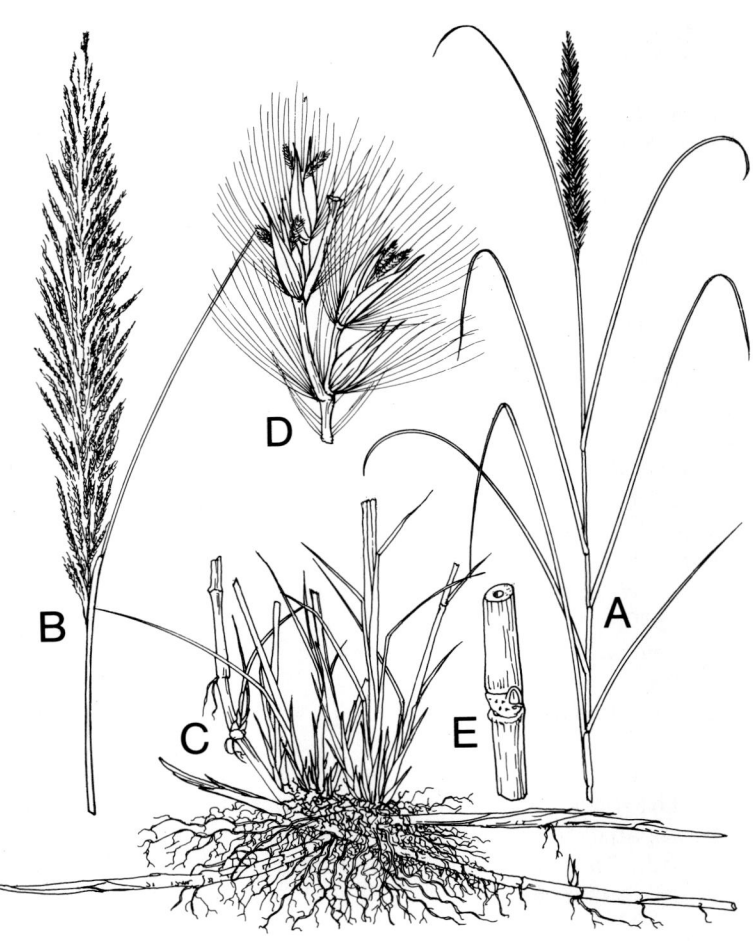

Figure 15.136. Saccharum spontaneum. **A**-Habit **B**-Inflorescence **C**-Base of plant **D**-Portion of raceme **E**-Culm node

TABLE 15.67 *Saccharum spontaneum*

		As % of dry matter				
		CP	CF	Ash	EE	NFE
Fresh, late vegetative, India		5.3	40.0	4.2	1.4	49.1
Fresh, stem cured, India		3.4	40.2	7.2	1.2	48.0
	Animal	Digestibility (%)				
		CP	CF	EE	NFE	ME
Fresh, late vegetative, India	Zebu	59.0	76.0	60.0	61.0	2.38
Stem cured, India	Zebu	11.0	66.0	30.0	35.0	1.61

Source: Göhl, 1975

Sehima nervosum (Willd.) Stapf

Common names. Rat's tail grass, white grass (Northern Territory, Australia).

Natural habitat. Rocky hills with grass, or open bush in partial shade.

Distribution. Central East Africa and the Sudan, Southeast Asia, Australia. The *Sehima/Dichanthium* association is very important in India (Whyte, 1957).

Description. Annual or perennial, culms densely tufted with leaf-blades up to 30 cm long. Racemes solitary, 7-12 cm long; sessile spikelets pale green, 8-10 mm long, with a long bristle from the upper glume, and an awn about 45 mm long from the lemma; pedicelled spikelets purplish (Napper, 1965). Maximum root activity occurs at 60 cm, with maximum depth of roots at 149 cm (see Fig. 15.137).

Altitude range. 100-2 750 m.

Rainfall requirements. 250-1 375 mm, with the optimum being up to 1 000 mm.

Drought tolerance. It survives the long dry season in northern Australia and dry seasons in India very well.

Soil requirements. It grows on lava and on black, seasonally waterlogged clays in Africa, and on lateritic red earths in northern Australia. It grows well on loamy sands with a pH of 6.5 in India, but grows best on black soils. There is a progressive increase in yield in India with increasing soil moisture up to field capacity in black soils (Dabadghao & Shankarnarayan, 1970).

Fertilizer requirements. It is normally adapted to its natural habitat without fertilizer. It has given response to nitrogen and phosphorus but not to potash in India (Dabadghao & Shankarnarayan, 1970) or in northern Australia (Arndt & Norman, 1959). The nitrogen response in India was in improved crude protein content but not in yield (Shankarnarayan *et al.*, 1977).

Land preparation. In India the land is well ploughed.

Sowing methods. Seed is broadcast at 11-13 kg/ha.

Vigour of growth and growth rhythm. In northern Australia, with the onset of effective rain in November/December, growth is rapid through the wet season, with flowering occurring in April/May (a little later than *Themeda australis*). It then remains dormant through the dry season from May to November. Virtually all tillers become reproductive in the season of their initiation (Arndt & Norman, 1959). In India it is somewhat difficult to establish in a new locality but, once established, it can stand cutting (Narayanan & Dabadghao, 1972).

Response to defoliation. Increasing the cutting interval from ten to 60 days in 1970-72 increased the yield of dry matter from 790 kg/ha to 4 100 kg/ha. Cutting at 15 cm was better than at 5 or 10 cm.

Tolerance to fire. It survives annual burning during the dry season in northern Australia.
Genetics and reproduction. 2n=20, 34, 40 (Fedorov, 1974).
Dry- and green-matter yields. In India, the dry-matter yield of a *Sehima* community can be raised from 4 126 kg/ha unfertilized to 7 561 kg/ha by applying 60 kg N/ha. Application of 40 kg P_2O_5/ha increased the yield from 5 824 kg/ha to 6 471 kg/ha. There was no response to potash (Dabadghao & Shankarnarayan, 1970).

Figure 15.137. Sehima nervosum. **A**-Habit **B**-Spikelets

TABLE 15.68 *Sehima nervosum*

	As % of dry matter					Reference
	CP	CF	Ash	EE	NFE	
Pre-flowering, India	5.4					Dabadghao & Shankarnarayan, 1970
Mature, India	2.3					Dabadghao & Shankarnarayan, 1970
Early bloom, Kenya	7.0	38.7	8.8	1.3	44.2	Dougall & Bogdan, 1958-60

Suitability for hay and silage. It is one of the most important grasses for hay in India.

Chemical analysis and digestibility. Various analyses as a percentage of the dry matter are listed in Table 15.68. Calculated on a whole-plant basis, Arndt and Norman (1959) recorded progressive monthly crude protein figures from 10.9 percent in December to 8.2, January; 5.9, February; 5.2, March; 4.4, April; 1.5, May; and 1.2 percent in June, which latter figure was maintained until October.

Palatability. Sehima nervosum was practically neglected by cattle during the growth period in northern Australia in favour of *Sorghum plumosum*, *Themeda australis* and *Chrysopogon latifolius*. In India it is regarded as one of the most palatable grasses and disappears quickly under grazing (Dabadghao & Shankarnarayan, 1973).

Economics. In India it is an excellent pasture grass. It is used for grazing, hay-making and cut forage. In northern Australia it constitutes about 3.2 percent of the tropical tall-grass pasture dominated by *Sorghum plumosum*, *Themeda australis* and *Chrysopogon fallax* on a lateritic red earth soil at Katherine, Northern Territory (lat. 14°3'S, altitude 100 m, annual rainfall 911 mm — 93 percent from October to March). It is neglected by cattle at most times (Arndt & Norman, 1959).

Further reading. Dabadghao & Shankarnarayan, 1970.

Setaria italica (L.) Beauv.

Synonyms. Panicum italicum L.; *Chaetochloa italica* (L.) Scribn.
Common names. Dwarf setaria, giant setaria, Hungarian millet, liberty millet, foxtail millet, red rala.
Distribution. Regarded as a native of China, it is one of the world's oldest cultivated crops. Cultivated extensively in the USSR, China and India but also widely elsewhere.
Description. An annual plant with stems that branch little, and with a well-developed, deep root system. The tubular stalk is filled with loose tissue. The leaf-blade is wide-lanceolate, long-acuminate, dense scabrous, and may have a brightly coloured midrib; leaf edges serrate. Leaf-sheaths longer than the nodes; collar indistinct, ligule small, short, thick. Inflorescence has main stalk with shortened branchings bearing spikes and bristles. Flowers two per spikelet, the upper bisexual. In cultivated varieties there are two to three bristles per spikelet. Fruit a caryopsis; grain of various colours; seeds enclosed in thin, papery hulls, largely removed by threshing, leaving free the small, convex seed, which is oval or elliptical (Malm & Rachie, 1971) (see Figs. 15.138, 15.139).
Season of growth. Summer.
Frost tolerance. It is intolerant of frost.
Latitudinal limits. 30°N and S.
Altitude range. Sea-level to 2 000 m.
Rainfall requirements. It is generally grown in the 500-700 mm rainfall areas with a summer maximum. Millets require less rainfall than sorghum and maize but success depends on strategic falls of rain.
Drought tolerance. It is fairly tolerant of drought; it can escape some droughts because of early maturity.
Tolerance to flooding. It cannot tolerate waterlogging.
Soil requirements. Preferably sandy loams to clay loams. Millets are difficult to germinate on heavy clay soils.
Fertilizer requirements. A complete fertilizer mixture, where soil tests show the need. For grazing, 55 kg N/ha is usually beneficial, but excess nitrogen causes lodging.
Ability to spread naturally. It will spread from scattered seed, but is usually planted.
Land preparation for establishment. A fine, firm seed-bed is needed for a good crop. In early land development an initial ploughing may be sufficient.
Sowing methods. The seed is usually drilled, but may be broadcast and harrowed in. Nitrogen fertilizer should not be put down the same chute as the seed.
Sowing depth and cover. Seed is best sown at 4-6 cm (deeper may result in

lower germination). Except in sandy soils, rolling after planting is desirable.
Sowing time and rate. Spring to late summer, depending on frost hazards, at 5-7 kg/ha. In light sandy soils a slightly lower sowing rate can be adopted.
Number of seeds per kg. 485 000 in the United States.
Dormancy. Seed dormancy is common in freshly harvested seed, but disappears by the following spring (Malm & Rachie, 1971).
Seed treatment before planting. In areas where seed-harvesting ants are troublesome, seed can be treated prior to planting with lindane.

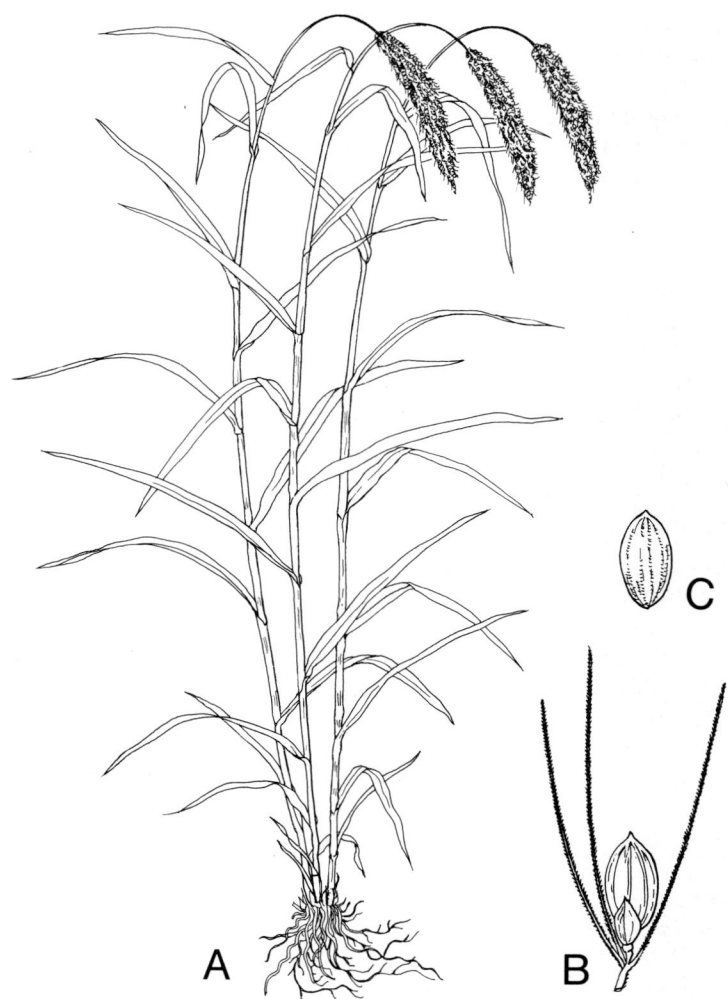

Figure 15.138. Setaria italica. **A**-Habit **B**-Spikelet **C**-Grain

Seedling vigour. Good.
Vigour of growth and growth rhythm. It grows quickly and flowers in about 56-62 days.
Response to photoperiod. Flowering is accelerated by short days (Evans, Wardlaw & Williams, 1964) and the flowers open both late at night and early in the morning. However, short-day, day-neutral and long-day varieties occur throughout the world (Malm & Rachie, 1971).
Compatibility with other grasses and legumes. It is usually sown as a pure crop but in India may be interplanted with finger millet (*Eleusine coracana*) or cotton (*Gossypium hirsutum*).
Response to defoliation. It is not often grazed, but can be used for this purpose with one or two grazings a season.
Response to fire. It is destroyed by fire.

Figure 15.139. Setaria italica cv. Panorama (**Source:** N.J. Douglas, Queensland Department of Primary Industries)

Genetics and reproduction. 2n=18 (Fedorov, 1974).
Dry- and green-matter yields. It yields about 15-20 tonnes of green matter per hectare, and 3.5 t/ha of hay.
Suitability for hay and silage. It makes good hay when cut at flowering and can also be ensiled (Malm & Rachie, 1971).
Chemical analysis and digestibility. The grain contains approximately 11.9 percent moisture, 9.7 percent protein, 1 percent fibre, 3.5 percent ether extract, 72.4 percent nitrogen-free extract and 1.5 percent ash. Göhl (1975) records other values in Table 15.69.
Palatability. It is extremely palatable.
Toxicity. Millet hay may be toxic to horses due to a glucoside setarian (Malm & Rachie, 1971).
Seed production and harvesting. Setaria millets seed heavily. Harvest with combines using a small seeds box. The grain must be dried thoroughly before storage or it may heat and spoil.
Seed yield. 800-900 kg/ha.
Minimum germination and quality required for commercial sale. 75 percent germinable seed, 97.5 percent purity (Queensland).
Cultivars. Two varieties are grown in Queensland.
- 'Giant Setaria' — a dual-purpose, tall-growing type producing a good body of leaf. It has a growing period of about 105 days. As a grazing crop, it should be subjected to heavy quick grazings. It gives a higher grain yield than dwarf setaria. The plant has no hair on the lower leaf-sheath, distinguishing it from dwarf setaria.
- 'Dwarf Setaria' — used only as a grain crop in Queensland, especially for the bird seed trade. It matures very quickly, in about 80 days. It is referred to as "panicum" in the Queensland trade circles. It produces less leaf than other

TABLE 15.69 *Setaria italica*

	DM	As % of dry matter				
		CP	CF	Ash	EE	NFE
Fresh, 8 weeks, Israel	35.6	9.0	33.7	10.1	2.2	45.0
Hay, southern Africa		7.6	45.1	9.7	1.7	35.9

	Animal	Digestibility (%)				ME
		CP	CF	EE	NFE	
Fresh, 8 weeks, Israel	Sheep	55.0	60.0	53.0	66.0	2.11
Hay, southern Africa	Sheep	57.2	65.5	47.3	58.1	2.07

Source: Göhl, 1975

millets and this, coupled with its quick maturing, reduces its moisture requirements (Douglas, 1970). Dwarf setaria bears a profusion of hairs on the lower leaf-sheath at 8-10 cm height.

● 'Panorama' — selected in Queensland; it yields more than the common variety and matures about a week later (Douglas, 1974).

Value for erosion control. It can be used as a quick-growing crop in contour strips in dense populations for erosion control.

Diseases. The crop is subject to leaf and head blast, caused by *Pyricularia grisea.* In India it is attacked by a smut, *Ustilago crameri,* and green ear caused by *Sclerospora graminicola.*

Pests. The millets are very susceptible to bird attack in the field, and mice and rat invasions.

Economics. One of the oldest cultivated crops. It was used in India, China and Egypt before there were written records. Millet is still used in eastern Europe for porridge and bread and for making alcoholic beverages. About 85 percent is used as foodgrain for humans and 6 percent for poultry. In the United States it is grown chiefly for hay.

Animal production. The setaria millets are usually not grazed. They may be fed off in dry times but are usually made into hay or harvested for grain, mainly for bird and poultry feed.

Main attributes. Its quick growth, which enables it to be grown as a short-term catch crop. Its adaptability to a wide range of elevations, soils and temperatures. Its heavy seeding, the grain being used for human consumption, and poultry and cage birds.

Main disadvantages. It is an annual and is not very suitable for continuous grazing.

Further reading. Douglas, 1974; Malm & Rachie, 1971.

Setaria porphyrantha Stapf

Common name. Purple pigeon grass.
Natural habitat. Common on cracking black earths.
Distribution. Native to Zimbabwe, where it is a minor forage species; introduced to Australia as CPI 124582.
Description. A tufted perennial on a short rhizome, 60-150 cm high. Culms geniculate, ascending; somewhat stout below; round, smooth and glabrous with a ring of short, silvery hairs at the insertion of the sheaths and the top of the peduncle. Leaf-sheaths light, striate and glabrous, ligule a narrow, densely and long ciliate rim; leaf-blade linear, up to 50 cm long and 49 mm wide. Inflorescence a dense, continuous false spike 6-18 cm long, 8 mm wide. Bristles six to nine to a cluster, tinged with purple, or dull purple all over. Glumes membranous, one-third to one-half the length of the spikelet, five- to seven-nerved. Lower floret male, upper floret perfect. Seed elliptical, 1 mm in diameter, 2 mm long, olive-green to yellow in colour (*Setaria, Setaria porphyrantha*, 1977).
Season of growth. Summer.
Optimum temperature for growth. It does better than most species in hot weather.
Frost tolerance. Very susceptible to frost but recovers in spring (Watt, 1976).
Rainfall requirements. In Zimbabwe, 500-700 mm.
Drought tolerance. Good, better than makarikari grass.
Soil requirements. It tolerates black soils.
Ability to spread naturally. It spreads readily from shattered seed.
Land preparation for establishment. A fine, well-prepared seed-bed is required.
Sowing methods. It is drilled into a good seed-bed.
Number of seeds per kg. 500 000 seeds with lemma and palea intact but glumes removed.
Seedling vigour. It establishes very quickly, much better than *Panicum coloratum* var. *makarikariense* cv. Bambatsi, *P. maximum* var. *trichoglume* and *Chloris gayana*.
Vigour of growth and growth rhythm. Midsummer growth is rapid, but growth in spring and autumn and overall production is slightly inferior to cv. Bambatsi. It flowers in six weeks (Truong, personal communication).
Response to photoperiod. It is day neutral, flowering through summer and autumn.
Ability to compete with weeds. It may invade other grasses itself.
Chemical analysis and digestibility. No figures have been cited.
Palatability. Palatable to cattle and sheep.
Seed production and harvesting. It seeds prolifically, but the seed shatters readily.

Seed yield. Up to 300 kg/ha has been recorded.
Value for erosion control. Its potential on black self-mulching clays is impressive, because of its ease of establishment. It is too tall for waterways, but good for strip cropping.
Main deficiencies. It becomes a weed.
Further reading. Setaria, Setaria porphyrantha, 1977; Stent, 1931; Watt, 1976.

Setaria sphacelata (Schumach.) M.B. Moss var. *sericea* (Stapf) W.D. Clayton

Synonym. Setaria anceps Stapf.
Common names. Setaria (Australia), golden timothy (Zimbabwe), golden bristle grass (southern Africa).
Natural habitat. Grassland, woodland and swampy places, usually on clay soils.
Distribution. Naturally confined to the African continent, now introduced into several tropical countries.
Description. Tufted perennial 45-180 cm high with the lower culm nodes compressed. Basal leaf-sheaths often nearly flabellate in arrangement. False spike dense with orange bristles and subacute spikelets, 2.5-3 mm long (Napper, 1966) (see Figs. 15.140, 15.141).
Season of growth. Early spring and summer.
Optimum temperature for growth. Mean 18-22°C (Russell & Webb, 1976).
Minimum temperature for growth. At temperatures below –4°C cv. Nandi dies. Mean temperature of the coldest month $8.6° \pm 3.4°$ (Russell & Webb, 1976).
Frost tolerance. Compared with other summer-growing grasses it is fairly frost tolerant. It will make some growth in winter if frosts are not too heavy. 'Nandi' is best adapted to cold (Hacker & Jones, 1969).
Latitudinal limits. 25.9°N and S $\pm$ 5.7° (Russell & Webb, 1976).
Altitude range. Sea-level to 3 300 m (Hacker & Jones, 1969), more common at 660-2 660 m.
Rainfall requirements. Mean 900-1 825 mm (Russell & Webb, 1976).
Drought tolerance. Only fairly drought tolerant. Cultivar Kazungula is the most tolerant of dry conditions.
Tolerance to flooding. S. sphacelata generally tolerates waterlogging over short periods, and cv. Kazungula is particularly tolerant (Hacker & Jones, 1969). Aquatic roots form at the submerged nodes, and leaves appear at these sites (Colman & Wilson, 1960).
Soil requirements. 'Kazungula' is the most tolerant of poor sandy and stony soils. 'Nandi' and 'Narok' prefer medium-textured, fertile soils. It is not common on alkaline or very acid soils, the majority of collections being made from soils in a pH range of 5.5-6.5.
Tolerance to salinity. Two hexaploid collections, CPI 32847 and CPI 32714 from near the Aberdare Mountains in Kenya showed some tolerance in southeastern Queensland and also tolerated cool conditions (Evans, 1967b), but generally it had low last tolerance (Russell, 1976).
Fertility requirements. A basal dressing of NPK is usually required. The critical level of phosphorus as a percentage of the dry matter at the immediate

pre-flowering stage is 0.21 (Andrew & Robins, 1971). The rate of potassium uptake is very high and the critical level for potassium in cv. Nandi is about 1 percent of the dry matter. Setaria responds markedly to nitrogen and in Queensland gave an average response over a four-year period of 30 kg dry matter and 3 kg protein for every kilogram of applied nitrogen (Hacker & Jones, 1969).

Ability to spread naturally. 'Kazungula', if allowed to seed, will spread well and form large crowns.

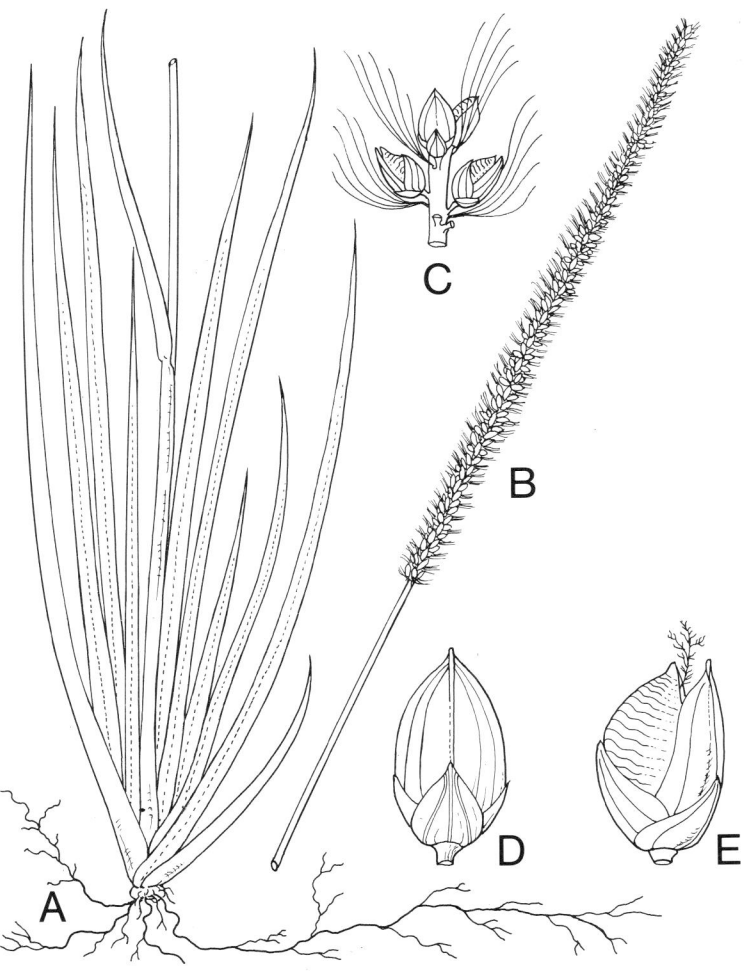

Figure 15.140. Setaria sphacelata. **A**-Habit **B**-Inflorescence **C**-Portion of raceme **D**-Spikelet, closed **E**-Spikelet, open

Figure 15.141. Setaria sphacelata var. *sericea* cv. Kazungula (**Source:** P. Mortiss, Queensland Department of Primary Industries)

Land preparation for establishment. A well-prepared seed-bed is preferred for establishment by seed.
Sowing methods. In the Philippines, propagation of new material by rooted cuttings or divided root-stocks has been successful, but drilling seed into a well-prepared seed-bed is better.
Sowing depth and cover. As the seed is small, it should be sown no deeper than 1.8 cm for cv. Kazungula and 2.5 cm for cv. Nandi, then lightly covered and rolled.
Sowing time and rate. Early to midsummer, at 1.5 kg/ha in conjunction with a legume such as *Desmodium intortum* at 1.15 kg/ha, or *Neonotonia wightii* at 4 kg/ha.
Number of seeds per kg. About 1.7 million for cv. Kazungula and 1.4 million for cv. Nandi.
Dormancy. Fresh seed has a germination inhibitor and should be stored for two months (Hacker & Jones, 1969). Germinate at 25-35°C, moistened with water. Exposure to light increases germination.
Seed treatment before planting. To control seed-harvesting ants, a seed-dressing of dieldrin at 10 g AI/kg seed can be given.
Seedling vigour. 'Nandi' is slow to establish but catches up later.

Vigour of growth and growth rhythm. Growth commences in early spring and continues at low autumn temperatures (Luck, 1979).

Response to light. 'Nandi' begins flowering in December (summer) in Australia, and flowers over a period of five months with a peak six weeks after head emergence. 'Kazungula' flowers one month later over a short period.

Compatibility with other grasses and legumes. The setarias compete successfully with Rhodes grass, green panic, paspalum and blue couch in coastal districts of Queensland, but generally should be the sole grass in mixtures of grass and legumes. This setaria combines well with white clover, *Neonotonia wightii, Desmodium intortum, D. uncinatum* and siratro. 'Kazungula' has little compatibility with legumes on the Atherton Tableland, Queensland (Quinlan & Edgley, 1975).

Ability to compete with weeds. When established, it suppresses most weeds. In the first season, cv. Nandi is troubled by weed competition but recovers well after the first grazing or mowing.

Tolerance to herbicides. For seed production on the Atherton Tableland, Queensland (lat. 17°13'S, altitude 715 m, rainfall 1 420 mm) the weeds *Nicandra physaloides* and *Eleusine indica* are troublesome. Competition with *E. indica* can be reduced by choosing a sowing time favourable to *Setaria* and unfavourable to *Eleusine,* that is, when the mean temperatures are less than 23°C (Hawton, personal communication). For cv. Narok planted at 5.5 kg/ha in rows 1 m apart, activated charcoal slurry sprayed at 675 litres/ha (168 kg charcoal) in a 2.5 cm band over the seed row, and diuron at 6.25-7.5 kg/ha sprayed at 350 litres/ha over the row eliminated both weeds (Hawton, 1979).

Response to defoliation. It withstands heavy grazing and forms a robust crown. The highest yield in a cutting trial at Redland Bay, Queensland, was obtained by cutting every three weeks at a height of 15 cm (Riveros & Wilson, 1970).

Grazing management. It should be lightly grazed until established, then heavily grazed to prevent it becoming stemmy. Early grazing may cause plants to be pulled up by the roots when the soil is moist. Undergrazing causes the plants to become coarse and to shade companion legumes, a real problem with cv. Kazungula. Heavy grazing in winter can clean up a pasture ready for early spring growth.

Response to fire. When established, setaria will survive an occasional fire quite well.

Genetics and reproduction. In 18, 36, 54 (Fedorov, 1974) cv. Nandi is a diploid 2n=18; cv. Kazungula and cv. Narok are tetraploids, 2n=36.

Dry- and green-matter yields. At Redland Bay, Queensland, Riveros and Wilson (1970) recorded dry-matter yields from 23 500-28 200 kg/ha over a six-month growing season. The grass was irrigated and supplied with 225 kg N/ha per year. The soil was a basaltic red loam (latosol).

Suitability for hay and silage. The setarias, especially cv. Kazungula, are rather tall and coarse for good quality hay at the flowering period, but cv. Nandi is satisfactory (Catchpoole, 1969). In association with *Desmodium intortum,* a very good acetic acid-type silage was made by Catchpoole (1970). The good quality was due to the marked ability of the two plants to resist degradation to volatile bases during ensilage. Cultivar Kazungula is used widely for silage in southern Africa.

Value as standover or deferred feed. Setaria is usually too coarse to be of much value as deferred feed, but it has a place as low-quality roughage, as a supplement to urea-molasses feeding. It is used for this purpose in Kenya and Uganda, but losses of crude protein and dry matter may reach 33 percent.

Chemical analysis and digestibility. Five-week-old *S. sphacelata* cv. Nandi contained 1.8 percent total nitrogen and 5.5 percent sugar on a dry-matter basis; at eight weeks, 1.9 percent nitrogen and 11.9 percent protein (Catchpoole, 1970). Dougall, in Kenya, showed a progressive fall in crude protein (from 15 to 5 percent), digestible crude protein (10-2 percent) total digestible nutrients (62-49 percent), while crude fibre increased from 23 to 42 percent over a four-month period (Hacker & Jones, 1969). Digestibility values of 70-72 percent of the dry matter have been recorded.

Palatability. The various cultivars are very palatable when young but less so as they approach maturity.

Toxicity. The setarias contain oxalates which can poison cattle. The amount of oxalate varies with the cultivar and stage of growth. Young plants contain more than older plants and strains highest in nitrogen are also highest in oxalate. Amounts of oxalates ranging from 3.7 percent in cv. Nandi to 7.8 percent in cv. Kazungula have been reported. Lactating cows and horses have been affected. Affected cattle have a staggering gait and diarrhoea, and then collapse and lie on their briskets. Rectal temperatures vary from 37.7 to 38.5°C. The muzzles are dry and rumination ceases. In eight days there is extensive subcutaneous oedema of the brisket and dewlap, and the animals die within three weeks of eating the grass. Death results from a build-up of calcium oxalate crystals in the kidney, which brings on acute hypocalcaemia. Poisoning rarely happens, but animals, especially lactating cows, should not be placed on young, luscious setaria pastures after a period of starvation (Everist, 1974). Horses, also, should be kept away from setaria pastures, as they can contract big-head disease (Cook, 1978). Feeding a calcium supplement, such as ground limestone or lucerne hay (containing calcium), can help control the disease.

Seed production and harvesting. Flowering occurs over a long period, and it is difficult to decide on a best harvesting date. Cutting and curing before threshing gives higher seed yields, but the crop is usually direct-headed. The pasture is well fertilized with 100-150 kg N/ha per harvest, and headed once in midsummer and again in autumn. Head when 10-15 percent of the seed has shattered at medium drum speed. Dry the seed immediately.

Seed yield. In Kenya, seed yields seldom exceed 330 kg/ha of 25 percent pure germinating seed (Hacker & Jones, 1969). Cultivar Nandi yields about 112 kg/ha.

Minimum germination and quality required for commercial sale. 20 percent germinable seed, 60 percent purity in Queensland.

Cultivars.
- 'Nandi' — originated in the highland Nandi district in Kenya, and introduced to Australia in 1961 and 1964 after selection at Kitale, Kenya. It is more sensitive to frost than cv. Kazungula, and top growth can be killed by heavy frosts, but its roots survive. It tolerates waterlogging; flowers earlier than cv. Kazungula (one to two months after commencement of rain in spring) and, if ungrazed, flowers through the season and becomes stemmy. It is cross-pollinated and seeds well. It establishes less readily than cv. Narok and cv. Kazungula. It has a high oxalate content, 3.22 percent (Ndyanabo, 1974).
- 'Kazungula' — native to Zambia, developed for grazing and hay. It is coarser and more robust than 'Nandi', and is tetraploid. The seed-heads are lighter than 'Nandi', and tend to bend. The coloration of the sheaths of the basal leaves is blue-green, and the stigmas are purple. Seed is slightly smaller than that of 'Nandi'. It flowers a month later than 'Nandi', in spring. It is hardier and more adaptable; more frost-tolerant under waterlogged conditions and a little more drought resistant, growing on as little as 575 mm annual rainfall. It has considerable tolerance to waterlogging and is suitable for areas frequently inundated with flood water, and for areas under irrigation. It has a high sodium content and far higher oxalate content than 'Nandi'. Cattle accept stubble grazing with 'Kazungula' more readily than with 'Nandi'. 'Kazungula' is fairly resistant to *Piricularia* leaf spot.
- 'Narok' — collected on the Aberdare Mountains, Kenya, and introduced to Australia in 1963. More robust than 'Nandi' but less so than 'Kazungula', it is greener in colour than both; some plants lack the red pigmentation at the base common to 'Nandi' and 'Kazungula'. Inflorescence rust-coloured, seed larger than 'Nandi'. Leaves broad, soft and hairless. It is a tetraploid. Its principal feature is its frost resistance. Negligible leaf damage occurs at temperatures of -3.3 to $-2.8°C$ but heavier frosting results in leaf kill. Grazing in winter should not be heavy. It is more nutritious than either 'Nandi' or 'Kazungula'. It is low in sodium and intermediate in oxalate content. Like 'Nandi', it is susceptible to leaf spot fungus (*Piricularia trisa*) under hot humid conditions (Barnard, 1972). 'Narok' gives low seed yields with consequent scarcity and high prices (Cook, 1978).

In southern Africa there are two other ecotypes:
- Gomoto-Mogolelo River ecotype — very tall, densely tufted with narrow, dark green leaves and fine stems; suitable for hay and silage.
- Du Toits or Middleveld ecotype — short, rhizomatous, with purplish leaves producing little seed; best suited to grazing (Whyte, Moir & Cooper, 1959).

Other ecotypes include:
- 'Bua River' — collected in Malawi, it is used for silage, hay or green chop.
- 'Du Toits Kraal' — from Zimbabwe, and not used outside South Africa, where it has been recommended for areas with 500-700 mm rainfall. It is drought resistant and retains some greenness and palatability into winter (Hacker & Jones, 1969).

In the Philippines several selections are under test (Farinas, 1970): 'Hairy K' and 'CRHS', selections from *S. sphacelata*; 'Decolores' and 'Mabolo', probably *S. sphacelata* var. *sericea* × *S. sphacelata* var. *splendida*; and 'Greencross A-X', probably a *S. sphacelata* cross.

Value for erosion control. 'Kazungula', with its large crowns, gives very good control of erosion when sown in contour strips.

Diseases. The leaf spot caused by *Piricularia trisa* attacks cv. Nandi and cv. Narok (but usually not cv. Kazungula) under hot, humid conditions. In Kenya a bunt disease caused by *Tilletia echinosperma* can devastate seed crops.

Pests. General pests such as army-worms and locusts can attack pastures.

Economics. The setarias are important pasture plants in Africa and have been introduced to other tropical areas. They do not contain prussic acid and so can replace *Sorghum* spp. They are nutritious and, though they contain oxalate, they usually give little trouble.

Animal production. In Kenya, live-weight gains from three pasture species over a three-year trial, without nitrogen fertilizer and without a legume, respectively, were 336 and 192 kg/ha from Nandi setaria, 369 and 220 kg/ha from Nzoia Rhodes grass, and 369 and 131 kg/ha from molasses grass. Hereford steers continuously grazing Nandi setaria and Samford Rhodes grass, fertilized with 330 kg N/ha each at Samford, Queensland, and stocked at 2.5 and 4 steers per hectare, gained a mean of 575 and 522 kg/ha per year on Nandi setaria and 535 kg/ha on the Samford Rhodes grass. In the first two years the animals on Nandi setaria gained significantly more weight at the higher stocking rate than did those on Rhodes grass (Hacker & Jones, 1969).

Main attributes. *Setaria sphacelata* var. *sericea* is palatable, establishes easily from seed, persists under grazing on a wide range of soils, gives high yields of digestible energy, has some cold tolerance, gives early spring growth, responds to fertilizer and will cross-pollinate.

Main deficiencies. Heavy summer seeding of cv. Nandi and cv. Kazungula is a disadvantage (Quinlan & Edgley, 1975); susceptibility to frost in low-lying areas, and its oxalate content.

Further reading. Hacker & Jones, 1969; Luck, 1979.

Setaria sphacelata (Schumach.) M.B. Moss var. *splendida* (Stapf) W.D. Clayton

This variety differs from *S. sphacelata* var. *sericea* in being larger and more robust, and in having the lower parts of the culms and the basal leaf-sheaths compressed and keeled (Chippendall, 1955).

It sets little viable seed and is propagated by division of root-stocks or from tillers. It flowers later than the other setarias in Brisbane, Australia. It is extremely palatable and is used for soilage and silage in Zaire. It can be cut monthly. Analyses and digestibilities are recorded in Table 15.70.

Toxicity. Middleton and Barry (1978) found that only young leaf material of *S. sphacelata* var. *splendida* was high in oxalate, ranging from 4.5 percent (26 February) to 6.7 percent (19 June) for three-week regrowth.

TABLE 15.70 *Setaria sphacelata,* var. *splendida*

	DM	As % of dry matter				
		CP	CF	Ash	EE	NFE
Fresh, 120 cm, Tanzania		11.3	39.2	15.8	3.5	30.2
Fresh, 25-day growth, Zaire		11.4	27.8	12.1	3.0	45.7

	Animal	Digestibility (%)				
		CP	CF	EE	NFE	ME
Fresh, regrowth	Sheep	65.2	75.2	56.7	76.5	2.47

Source: Göhl, 1975

Sorghum almum Parodi

Common names. Columbus grass (Australia), five-year sorghum, sorgo negro, Sudan negro (Argentina).

Distribution. It originated in Argentina as a probable hybrid between *Sorghum halepense* and a member of the series *Arundinacea*. It has now been introduced into several tropical countries.

Description. A more robust species than *S. halepense* (q.v.), sometimes reaching 4.5 m in height. It is a short-term perennial. The most satisfactory method of distinguishing between the two is by the articulation of the pedicelled spikelet. In *S. almum* the spikelet breaks off with the uppermost portion of the pedicel at maturity; in *S. halepense* there is a clean abscission at the base of the spikelet. *S. almum* usually produces short rhizomes, more or less pointing upwards, which are not as extensive or aggressive as those of *S. halepense,* but reach a depth of 50 cm (Chippendall, 1955; Pritchard, 1964) (see Fig. 15.142, Plate 52).

Season of growth. Spring to autumn.

Optimum temperature for growth. Mean 19.1 ± 3.3°C (Russell & Webb, 1976).

Minimum temperature for growth. About 15°C. Minimum temperature of the coldest month 7 ± 5.3°C (Russell & Webb, 1976).

Frost tolerance. It is susceptible to frost, and winter killing occurs on the high plains of the United States. Regrowth from rhizomes occurs after light frosts.

Latitudinal limits. About 25°N and 30°S. Mean 28.9 ± 6.3° (Russell & Webb, 1976).

Altitude range. Sea-level to 700 m.

Rainfall requirements. It is usually grown within the annual rainfall range of 460-760 mm, but may be grown under irrigation, or in areas with up to 1 900 mm annual rainfall (Russell & Webb, 1976).

Drought tolerance. It is more tolerant to drought than maize, Sudan grass and Johnson grass, and has survived in areas receiving 200 mm of annual rainfall.

Tolerance to flooding. It will not tolerate prolonged flooding.

Soil requirements. It prefers a soil of high fertility, from light loams to heavy clays, with a pH range from 5 to 8.5.

Tolerance to salinity. Russell (1976) tested several tropical grasses for salt tolerance. The most tolerant were *Chloris gayana, Panicum coloratum, Pennisetum clandestinum, Sorghum almum* and *Digitaria decumbens* in that order.

Fertilizer requirements. It requires a high level of nutrition for yields to be maintained, and does well as a pioneer species sown in the ashes of burnt brigalow (*Acacia harpophylla*) in Queensland where there is a high initial nitrogen status after years of growth of this leguminous tree. It also responds very well to applied nitrogen. It also requires adequate phosphorus and

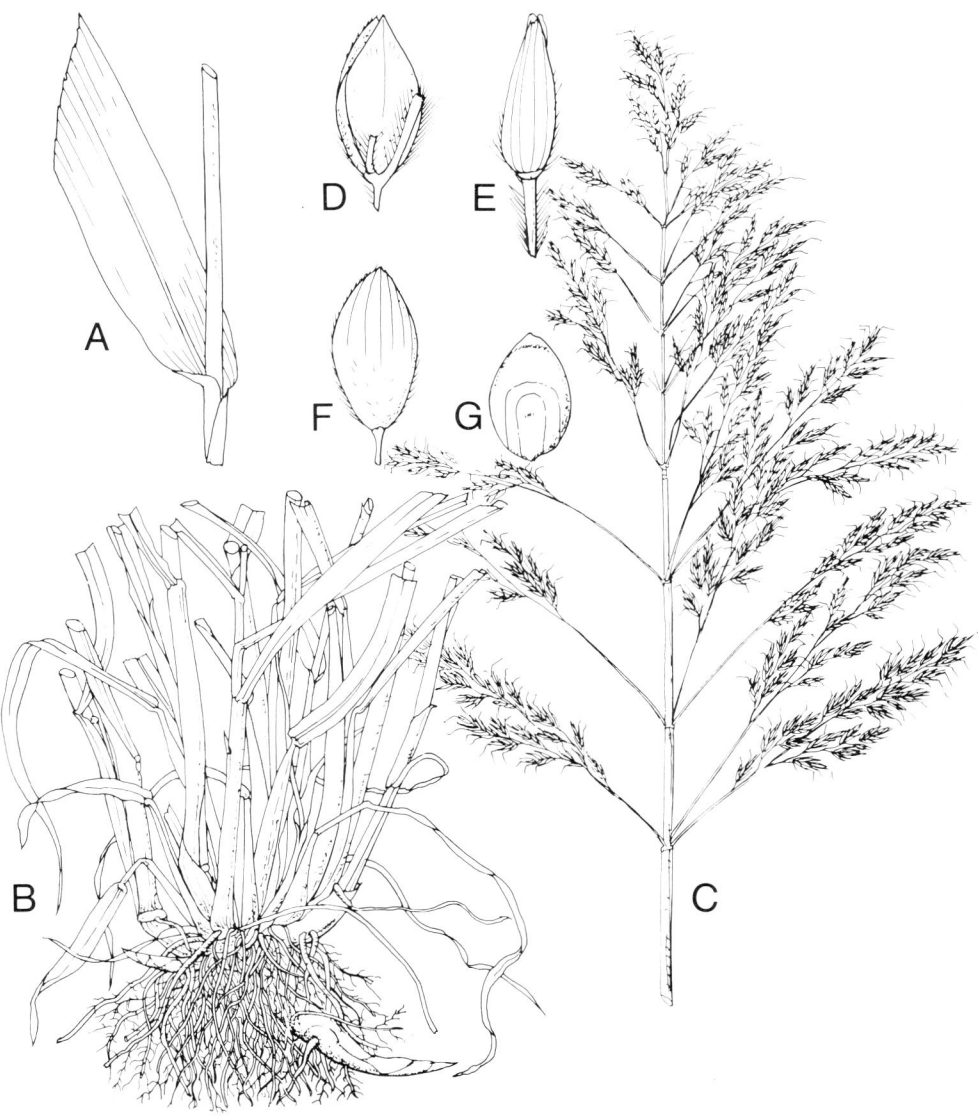

Figure 15.142. *Sorghum almum.*   **A**-Culm and leaf   **B**-Base of plant, showing rhizome
C-Inflorescence   **D**-Sessile spikelet   **E**-Male, pedicelled spikelet   **F**-Fertile spikelet   **G**-Grain

perhaps potash, as revealed by soil tests. The critical value for phosphorus in the dry matter at pre-flowering is 0.2 percent.

Ability to spread naturally. It does not spread quickly and is dependent on seeding into a cultivated seed-bed or ashes; shattered seed will germinate to fill gaps in a stand.

Land preparation for establishment. Use a fine, cultivated seed-bed, or the ashes from a recent scrub burn. This gives a seed-bed free from weed competition in the early seedling stage and aids even germination.

Sowing methods. It is sown through a seed drill adapted to small seeds. Wheat drills can be fitted with a smaller driving sprocket or the seed chutes reduced in diameter by longitudinal strips of leather nailed inside the seedbox and hanging down the chutes. Aerial seeding is used for the ashes of a scrub burn.

Sowing depth and cover. It is planted at 2-4 cm, depending on soil moisture, and lightly covered with harrows. Rolling following seeding will improve germination in drier soils.

Sowing time and rate. Usually spring to early summer when the soil temperature is above 15°C. For aerial seeding it must be sown at the beginning of the wet season. Rate, 1.25-2 kg/ha.

Number of seeds per kg. About 143 000 (Queensland).

Dormancy. S. almum seed has no dormancy, and will germinate immediately if it is sown into a moist seed-bed.

Seed treatment before planting. Dust with a combined fungicidal and insecticidal dust. Germinate pre-chilled seeds at 20-35°C, moistened with water (Prodonoff, 1966).

Seedling vigour. Excellent. It was ranked first of the *Sorghum* species in Texas, United States.

Vigour of growth and growth rhythm. It grows rapidly during spring and summer and begins flowering seven to eight weeks after planting. It persisted for three years at Samford, Queensland (Pritchard, 1964; Russell & Coaldrake, 1970).

Response to photoperiod. It is a short-day plant.

Response to light. It does not grow well in the shade.

Compatibility with other grasses and legumes. It is usually sown alone as a grazing crop, but in aerial seeding of newly burnt brigalow country in Queensland, Australia, a light seeding of *S. almum* is made with *Chloris gayana, Panicum maximum* var. *trichoglume* or *Cenchrus ciliaris*. The *S. almum* provides quick feed with little competition from the other grasses in the first year and helps to provide an early return in terms of beef production on the outlay for initial pasture establishment.

Ability to compete with weeds. S. almum cv. Crooble is able to suppress weeds (Pritchard, 1964).

Tolerance to herbicides. If Columbus grass does become a weed it can be eradicated by ploughing in most cases. If this is ineffective or not possible, control can be effected by post-emergence sprays as explained for Johnson grass (*Sorghum halepense*) by Marley (1978).

Response to defoliation. It stands heavy stocking and will give several grazings in a season, but it does not stand heavy trampling. Cutting at 5 cm every six to 12 weeks gave higher yields than cutting at 15 cm. Cutting every three weeks reduced yields (Santhirasegaram, Coaldrake & Salih, 1966).

Grazing management. S. *almum* should be grazed heavily once the crop is 50 cm high to prevent it from growing too coarse and from growing away from the grazing animals. Precautions must be taken to avoid prussic acid poisoning. It is advisable with any young *Sorghum* spp. crop to use a tester animal to graze the crop first for 20 minutes; if no toxicity is evident then the whole herd can be put on to the crop. It is best to give the herd only half an hour's grazing the first day, an hour the second, and then two to three hours the third, with a full day's grazing from then on. A rain-grown crop should provide two to three grazings per season. For maximum regrowth, stubble is left at 15 cm to renew growth.

Response to fire. It is rarely necessary to burn, but an established crop would survive a quick fire.

Genetics and reproduction. The somatic chromosome number is 2n=40 (Fedorov, 1974). It is predominantly cross-pollinated, but is also self-fertile.

Dry- and green-matter yields. Available dry-matter yields of 12 320, 10 640 and 5 040 kg/ha from nitrogen-fertilized S. *almum* were recorded when stocked at one beast to 0.6, 0.4 and 0.8 hectares, respectively (Yates *et al.*, 1964). At Taroom, Queensland (lat. 26°20'S, 550 mm rainfall) S. *almum* yielded 5 345 kg DM/ha (Russell & Coaldrake, 1970).

Suitability for hay and silage. It gives quite a good, though coarse, hay which is useful in the dry season. Mature crops (nine to 11 weeks old) make good silage in dry weather, and reasonable silage during the wet season if it is not wet by rain while ensiling it. Young crops up to seven or eight weeks old decompose badly during ensilage (Catchpoole, 1972).

Value as standover or deferred feed. It is not very useful for standover or deferred feed because the mature stem is not very palatable.

Chemical analysis and digestibility. See Table 15.71. In addition Minson and Milford (1966) and Minson (1972) carried out digestibility trials which included *Sorghum almum* at various stages of growth. From this, and some unpublished results, the figures in Table 15.72 were deduced.

Palatability. It is quite palatable, but not as readily eaten as annual sorghums (Pritchard, 1964).

Toxicity. In common with other *Sorghum* species, Columbus grass contains dhurrin, a cyanogenetic glucoside which can be toxic. The Queensland Department of Primary Industries recorded HCN equivalents of 0.06 and 0.081 percent in the plants. Danger is greatest in plants carrying young shoots, either from the base or from old stems. Hungry animals turned on to wet pasture are most susceptible. Affected animals breathe heavily, they

TABLE 15.71 *Sorghum almum*

	DM	As % of dry matter				
		CP	CF	Ash	EE	NFE
Fresh, wet season, 4 weeks, 85 cm	17.6	9.7	31.3	9.7	2.8	46.5
Fresh, wet season, 6 weeks, 140 cm	17.7	8.5	34.5	9.0	2.8	45.2
Fresh, wet season, 8 weeks, 160 cm	23.2	7.8	36.6	7.3	2.6	45.7
Fresh, dry season, 4 weeks, 85 cm	16.0	11.3	29.4	9.4	3.8	46.1
Fresh, dry season, 6 weeks, 140 cm	17.6	9.7	31.8	9.1	3.4	46.0
Fresh, dry season, 8 weeks, 165 cm	23.9	7.9	33.5	7.5	2.9	48.2

	Animal	Digestibility (%)				
		CP	CF	EE	NFE	ME
Fresh, wet season, 6 weeks	Sheep	39.0	54.0	53.0	49.0	1.73
Fresh, wet season, 8 weeks	Sheep	27.0	14.0	43.0	6.0	0.46

Source: Göhl, 1975

stagger about and display muscle tremor, become anchored and lie down. They die if not treated. Their mucous membranes remain red and do not become blue (Knott, personal communication).

Three treatments are effective for animals showing early signs of HCN poisoning:

1. Inject a mixture of sodium nitrate and sodium thiosulphate (photographic hypo) in water into a vein or under the skin. Recommended dose rates vary, but are approximately 3 g sodium nitrite and 15 g sodium thiosul-

TABLE 15.72 *Sorghum almum*

	Age of regrowth (months)			
	1	2	3	4
Dry-matter digestibility (%)	64.0	64.0	53.0	50.0
Intake (g/kg $W^{0.75}$/day)	53.0	55.0	46.0	48.0
Crude protein (%)	16.2	9.6	9.4	9.6
Crude protein digestibility (%)	75.0	65.0	58.0	56.0
Leaf (%)	49.0	41.0	29.0	24.0

phate in 20 ml water for cattle; l g sodium nitrite and 2 g sodium thiosulphate in 15 ml water for sheep. Excessive doses can cause nitrite poisoning.

2. Inject sulphuric ether under the skin; 10 ml for cattle, 5 ml for sheep.

3. Drench with sodium thiosulphate (photographic hypo) in water, 55 g in 550 ml for cattle; 10 g in 100 ml for sheep (Everist, 1974). This is the usual farm treatment.

Potentially toxic amounts of nitrite have been noted in some *S. almum* plants but no cases of nitrite poisoning have been reported.

Seed production and harvesting. It seeds heavily, but some seed may shatter. Harvesting is accomplished with a combine harvester using riddles suitable for the seed size and adjusting the blast to clean but not blow away the seed. Fields for seed production should be at least 1 km away from Johnson grass (*S. halepense*), with which it readily cross-pollinates.

Seed yield. 350-1 600 kg/ha.

Minimum germination and quality required for commercial sale. 70 percent germinable seed, 97.3 percent purity in Queensland (Prodonoff, 1966).

Cultivars.

- 'Crooble' — introduced to Australia from southern Africa and developed by CSIRO on O.C. Uebergang's property, "Crooble", in northwest New South Wales. An erect, robust perennial with numerous tillers and thick, short rhizomes which curve upwards to produce new shoots near the parental stool. Adapted to 460-760 mm rainfall with summer dominance, and soils of high fertility. Lasts up to ten years. It flowers in seven to eight weeks, and provides good hay, silage and grazing. Susceptible to leaf diseases, *Helminthosporium turcicum* (blight) and *Puccinia* spp. (rust). Easily eradicated.

- 'Nunbank' — not as palatable as cv. Crooble; it has disappeared from the market in Queensland.

Value for erosion control. It can be useful on eroded hillsides but needs nitrogen application to form an effective cover.

Diseases. Susceptible to leaf diseases (*Helminthosporium turcicum*, blight, and *Puccinia purpureum*, rust).

Pests. It can be attacked periodically by grasshoppers, army-worms and wild predators.

Economics. One of the most valuable summer forage and fodder crops in semi-arid to subhumid areas with rainfalls of 450-750 mm.

Animal production. In the Chaco Salteno in the province of Salta, northern Argentina, *S. almum* pastures have allowed Criollo × zebu cross animals to be grown to 400 kg live weight in two-and-a-half years, instead of four to five years under natural pastures. The pasture is grazed from November to April (Tothill, 1978). Stocked at 2.5 beasts per hectare, live-weight gain was 717 kg/ha over 23 months. When only green forage was considered, live-weight gain fell to zero when approximately 1 120 kg DM/ha still remained for the animals at the end of the growing season. This contained 83 percent *S. almum* stem and 15 percent other material, mainly *S. almum* leaf (Yates *et al.*, 1964).

Main attributes. A fast-growing, high-yielding, palatable, short-term summer perennial, suitable for giving quick grazing to help defray establishment costs; useful for silage. It has some drought and salinity tolerance.

Main deficiencies. Its short life and seed shattering. It may be difficult to eradicate in irrigated grain crops (Stevens, 1975).

Further reading. Pritchard, 1964; Yates *et al.*, 1964.

Sorghum bicolor (L.) Moench

Synonyms. *S. vulgare* Pers.; *Andropogon sorghum* (L.) Brot.

Common names. Sorghum (United States, Australia), durra (Africa), jowar (India), bachanta (Ethiopia).

Distribution. It probably originated in Ethiopia and has spread to other parts of Africa, India, Southeast Asia, Australia and the United States.

Description. Annual or short-term perennial, culms up to 4 m or more high, sweet except in grain types; panicle 8-40 cm long, loose or contracted; sessile spikelets 4-6 mm long. Var. *bicolor* (Pers.) Snowden: panicle loose and open, the upper branches slender and often drooping. Mature glumes of sessile spikelets either red or reddish brown, or straw coloured or yellowish, sometimes flushed with dark red or reddish brown; grain predominantly red or reddish brown.There are no creeping rhizomes and the mature sessile spikelets are persistent, less than twice as long as wide, not ridged in the middle, 3-4.5 mm wide (Chippendall, 1955). The *bicolor* sorghums are characterized by long, clasping glumes at least three-fourths as long as the broadly elliptical grain (de Wet, Harlan & Kurmarohita, 1972). (see Fig. 15.143).

Season of growth. Spring and summer through autumn until frosting.

Optimum temperature for growth. 30°C.

Frost tolerance. Sorghum is very susceptible to frost, but thick-stemmed, standing, sweet fodder sorghum will retain stem juiciness and sweetness for some time after the leaves are killed.

Latitudinal limits. 40°N and S.

Altitude range. Sea-level to 1 000 m.

Rainfall requirements. Mostly in an annual rainfall range of 400-750 mm. It is grown in areas which are too dry for maize.

Drought tolerance. The great advantage of sorghum is that it can become dormant under adverse conditions and can resume growth after relatively severe drought. Shoot removal lowers its capacity to withstand drought. Early drought stops growth before floral initiation and the plant remains vegetative; it will resume leaf production and flower when conditions again become favourable for growth. Late drought stops leaf development but not floral initiation (Wilson & Whiteman, 1965; Whiteman & Wilson, 1965).

Tolerance to flooding. Sorghum is intolerant of sustained flooding, but will survive temporary waterlogging.

Soil requirements. It has adapted to a wide range of soils, from the deep sands of the Goz to the heavy black cracking clays of the Gedaref, Sudan. Varieties to suit each have been selected. Good drainage, however, is necessary. Its deep rooting can extract water from low sources, though not as deep as *Pennisetum americanum* (pearl millet). Its soil pH range lies between pH 5 and 8.5.

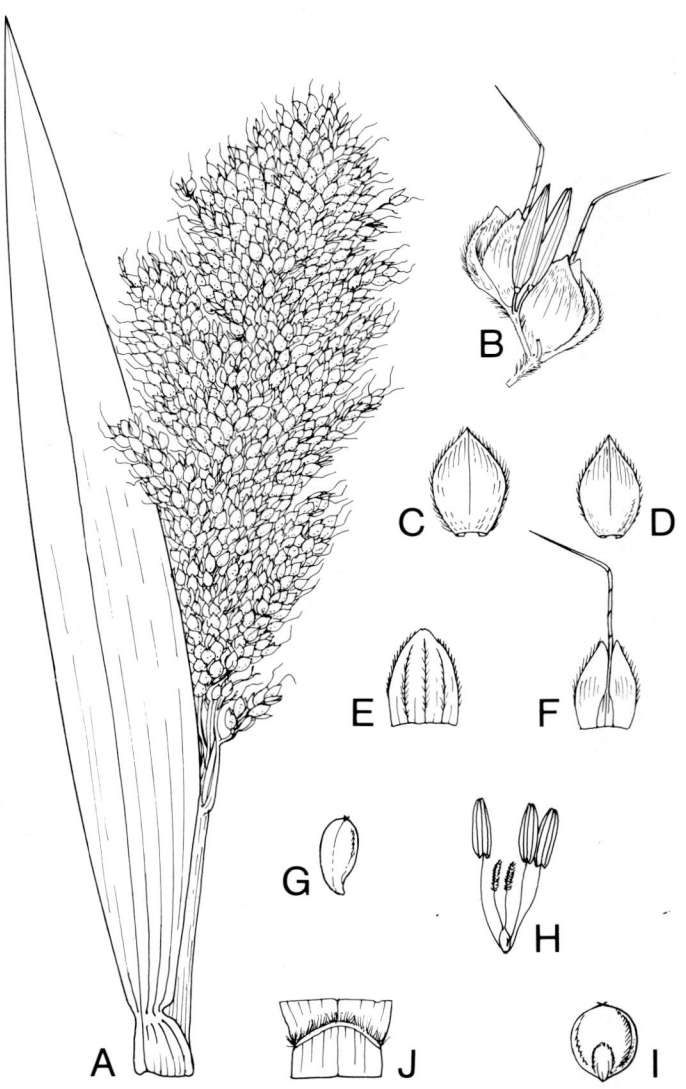

Figure 15.143. *Sorghum bicolor.* **A**-Leaf blade and inflorescence **B**-Pair of spikelets **C**-Lower glume **D**-Upper glume **E**-Lower lemma **F**-Upper lemma **G**-Palea **H**-Flower **I**-Grain **J**-Ligule

Tolerance to salinity. In some areas of California, accumulation of salt in the soil is sufficiently high to inhibit germination of seed, and young plants are injured. The tolerance of sweet sorghum to salt, however, appears to be relatively high after the plants become well established. The seed can be planted in beds 75-100 cm apart. In this single-row bed the seed is planted in a furrow made with a small plough mounted ahead of the planter shoe. The concentration of salt in the immediate vicinity of the seed is lower here than at the top of a high bed, and germination of seed is not inhibited (Price & Stokes, 1966). Sorghum has a high tolerance of sodium carbonate (Chapko, 1977).

Fertilizer requirements. These will be determined by soil type and rainfall. A basic dressing of NPK may be required, and the crop usually responds well to additional dressings of nitrogen during growth. A fallowed black clay may not need fertilizer. Rotation with a leguminous crop can give low-cost fertility build-up, for example, gum arabic (*Acacia senegal*) in the Sudan. Asher and Cowie (1974) showed that the effect of nitrogen deficiency on grain yield is greatest when the deficiency occurs early in the growing season. Low grain protein results when nitrogen deficiency occurs between anthesis and maturity.

Ability to spread naturally. Very low, except for loose seed.

Land preparation for establishment. Sorghum requires full seed-bed preparation for good performance.

Sowing methods. In developing countries the seed is often planted by hand hoe and covered, the spacing depending on expected rainfall. Small hand drills are available as a first step in mechanization; sophisticated grain and fertilizer drills for precision placement are used in advanced agriculture.

Sowing depth and cover. Sorghum seed is usually sown at 4-5 cm depth.

Sowing time and rate. Spring to summer after rain (soil temperature should be above 18.5°C) at 2-12 kg/ha, depending on soil-moisture expectancy and density of stand required, the heavier seeding being for forage production. Grain production in the 675-750 mm rainfall areas calls for 7-8 kg/ha of seed.

Number of seeds per kg. 28 600 to 61 000 in the United States.

Dormancy. Sorghum seed shows dormancy for the first month after harvest.

Seed treatment before planting. All *Sorghum* spp. seed should be dusted with a combined fungicidal/insecticidal dust before planting.

Seedling vigour. Good.

Vigour of growth and growth rhythm. Most sorghum plants take 90-120 days to mature. The boot stage is reached in 50-60 days, flowering in 60-70 days, with full grain maturity in 120 days. The daily water use is shown in Figure 15.144.

Response to photoperiod. Sorghum is a short-day plant. One variety began head differentiation in 23 days in a ten-hour photoperiod, compared with the norm of 39 days with a day length of 14 hours. Temperature also has an effect; flowering occurs earlier at 22-26°C than at 17-20°C.

Response to light. Sorghum requires full harnessing of incoming radiation for

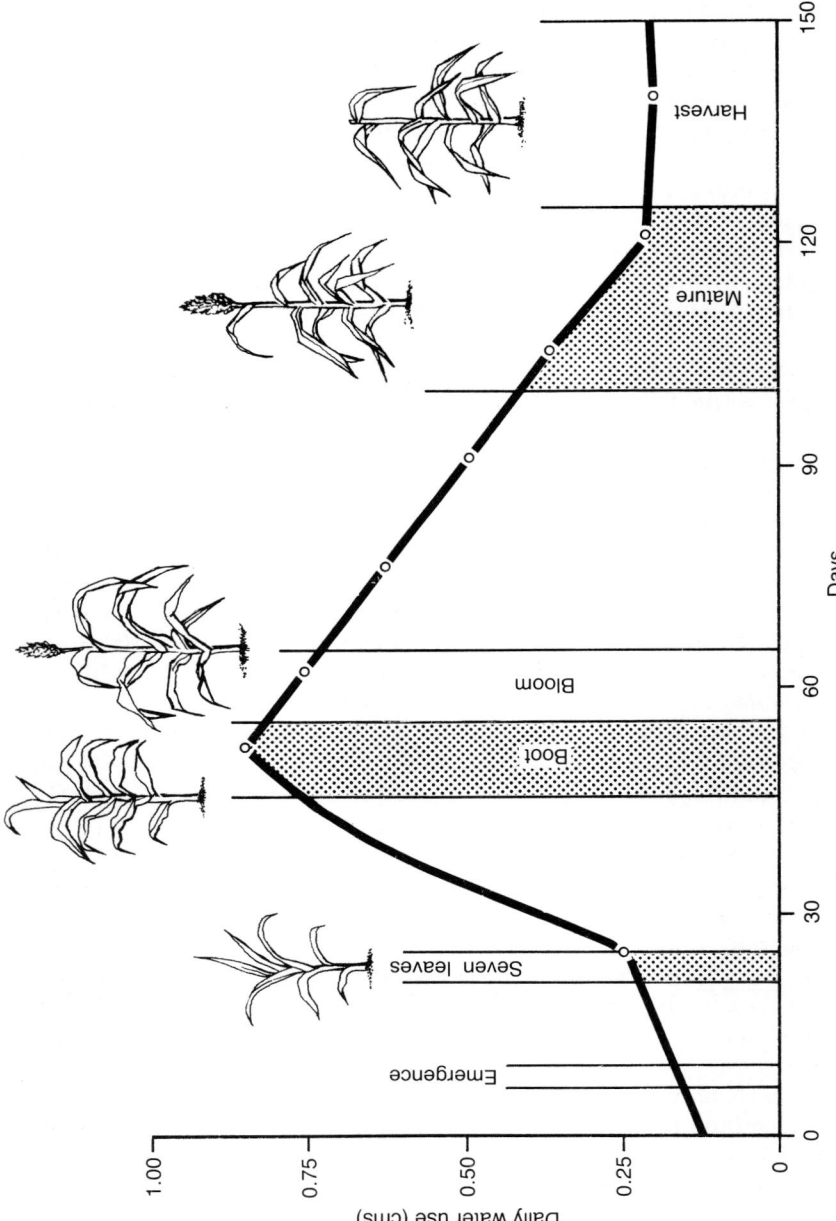

Figure 15.144. Water consumption of grain sorghum during the crop cycle (**Source:** Spears & Coffin, 1959)

high yields of grain and forage, and does not grow well in shade.

Compatibility with other grasses and legumes. It is generally grown as a pure crop in commercial grain production. In village crop areas it may be planted in rows alternating with other food or fibre crops in a rotation to spread labour and improve fertility. For forage and silage it is frequently grown with legumes, such as cowpea, to improve the nutritive value for grazing or stored fodder (see Plate 53).

Ability to compete with weeds. For grain production inter-row cultivation is frequently used. Where rows are close, weeds are suppressed by the shade of the crop canopy, but thorough seed-bed preparation is needed before planting to ensure a low weed population. Spraying with a pre-emergence weed-killer, e.g. atrazine at 1 kg AI/ha (Singh, Pandey & Shankarnarayan, 1970), completely controlled broad-leaved weeds such as *Portulaca* and *Amaranthus* spp. and the grassy weeds *Eleusine indica*, *Echinochloa colona*, *Brachiaria ramosa* and *Digitaria ciliaris*.

Response to defoliation. Sweet forage sorghum will stand a series of grazings where soil moisture and the temperature remain adequate, new branches and tillers being produced. Stalks may become thick and fibrous in the fodder types, and the forage or grass sorghums, such as Sudan grass, make better grazing.

Grazing management. In Queensland, Australia, the fodder sorghums are used mainly for silage production and the grain sorghums for grain. The fodder sorghums, however, are also used for autumn grazing by dairy and beef cattle to fill in a feed shortage between summer and winter grazing crops. The grain sorghums are valuable for grazing after the grain has been harvested and the crop residues (stubble, dropped seed-heads and regrowth, plus weeds) provide good autumn and winter roughage.

Response to fire. There is little trouble with fire, as the crop is generally fully utilized. The stubble can be burnt off to make way for cultivation for a new crop, but stubble retention can help prevent soil erosion and, if ploughed in, improves the organic status of the soil, especially if assisted by added nitrogen.

Genetics and reproduction. $2n=20$ (Fedorov, 1974). They cross-pollinate readily, thus seed-producing crops should be isolated by a distance of about 1 km from others.

Dry- and green-matter yields. Sweet sorghums yield 25 000-75 000 kg/ha green matter, according to soil fertility and rainfall. Grain sorghums yield 300-2 000 kg/ha of grain in India and Africa under rain-fed conditions, and irrigated hybrid sorghums in the United States produce 4 500-6 500 kg/ha of grain.

Suitability for hay and silage. Sorghum is made into a coarse hay in some countries and stored for later feeding. Frequently the grain is first harvested and the "stover" used for fodder, for example for work oxen in India, and for building, such as African rondels. Good hay was made in the Burdekin area of Queensland from 50-day growth of 'Zulu' forage sorghum which contained 16.2 percent crude protein in the dry matter (Thurbon, Byford & Winks, 1970). Sorghum is one of the best crops for silage because of its high yields, the sugar content and juiciness of its stalk and its adaptability to areas receiving too little rain to ensure crops of maize. The ensilage of sorghum also usually effectively stops stock losses from prussic acid poisoning.

Value as standover or deferred feed. It is quite a useful crop for standover and deferred feed, especially in frost-free areas, as it will continue to tiller and give a new green leaf for grazing while there is any moisture available in the soil.

Chemical analysis and digestibility. At harvest the grain sorghum stubble contains 50-70 percent moisture; it can be made into silage. The crude protein content will be about 6-7 percent (3-5 percent if irrigated) in the stubble; silage made from it will have 48-56 percent digestibility or, if dry, 48-50 percent. The material will generally need a mineral supplement of Ca:P of 1:1 for gestating and lactating cows, plus a vitamin A supplement (Corah, 1979). In Sri Lanka, Andrew (1971) recorded 20 percent dry matter, 3.9 percent digestible crude protein, 61.9 percent total digestible nutrients, 7.8 percent crude protein, 33.2 percent crude fibre, 1.5 percent ether extract, 51.2 percent nitrogen-free extract and 6.3 percent ash.

Palatability. The sorghums are all very palatable, especially in the young and flowering stages.

Toxicity. In common with other *Sorghum* spp., it can contain lethal amounts of prussic acid (see *S. almum* for details).

Seed production and harvesting. Sorghums usually give high yields of seed. The fodder sorghum grains contain higher tannin than the grain sorghum. This may affect egg production by poultry consuming a sorghum meal mixture; hence, grain sorghum is usually used.

Seed yield. The grain sorghums yield 300-2 000 kg grain per hectare in India and Africa under rain-fed conditions, and 4 500-6 500 kg/ha under irrigation for hybrid types in the United States and Australia.

Minimum germination and quality required for commercial sale. 70 percent germinable seed, 97.6 percent purity (Queensland).

Cultivars. There are numerous cultivars in use throughout the world and enquiries about the best cultivars and varieties for specific conditions should be made to agronomists within each country. Cultivars from the United States have been widely used in Australia, and Australian hybrids have also been produced to suit local conditions. Some Australian cultivars (Barnard, 1972) include the following:

- 'Saccaline' — probably introduced from the United States to Australia, but thought to have originated in Natal. A tall, erect annual with five to six culms per plant; pith solid, juicy and sweet. Seeds reddish-brown, elliptical, approximately 39 000 per kg. Foliage abundant but dries off rather early and the crop presents a somewhat stemmy appearance compared with cv. Sugar-

drip or cv. White African. Late maturing, flowering in 80 days in latitude 27-30°S and 65 days in latitude 23-24°S, in Australia. It produces heavy fodder yields (50 000-70 000 kg/ha green matter), palatable to maturity; it makes excellent silage.
- 'Sumac' — introduced to Australia from the United States, which in turn acquired it from Africa. Slightly shorter than 'Saccaline' with a short, compact panicle and plump, reddish-brown seeds. Its maturity date approximates that of cv. 'Saccaline'. Because of its lower height, it is a little easier to harvest for silage than 'Saccaline'.
- 'Sugardrip' — introduced to Australia from the United States. Similar to, but leafier than, 'Saccaline'. Stems are juicy and very sweet. Seeds are brown and small, rounded on top and pointed at the base (rather than elliptical, as in 'Saccaline'). It is late maturing, five to seven days behind 'Saccaline' in flowering. It stands well and does not lodge as readily as 'Saccaline' and 'Sumac'.
- 'White African' — introduced to Australia from the United States, but originally from Natal. Thicker, taller and with stronger stems than 'Saccaline', it tillers less, and its leaves are more widely spaced. Seeds white, approximately 40 000 per kilogram. Germination tends to be low. Late maturing, resists lodging, more resistant to leaf diseases than other cultivars. It carries well into the winter, but its juice is not so sweet. Very good for silage.
- 'Early Orange' — developed at Grafton Agricultural Research Station from a variety called 'Kansas Orange' in the United States. Similar to cv. Saccaline, with stalk a little stronger (but not as strong as 'White African' or 'Tracy'). Panicles medium compact; seeds orange-brown, ellipsoid, approximately 40 000 per kilogram. Early-maturing type, suitable for late sowing. Has a higher sugar content in its juice than 'Saccaline'.
- 'Tracy' — developed in Mississippi, United States for sugar production; introduced to New South Wales and Queensland. Similar in appearance to 'Saccaline', but leaf growth rather sparse in comparison to its height. Panicle small, erect, and cylindrical. Seed dark reddish brown and duller than those of 'Saccaline'; approximately 40 000 per kilogram. It is late-maturing, flowering two weeks after 'Saccaline' (about the same time as 'White African'). It resists lodging. The stems are palatable and juicy, with a high sugar content; they hold their juiciness into winter. Resistant to leaf diseases.

Value for erosion control. Row sorghum gives only reasonable protection from soil erosion but broadcast sorghum or sorghum sown in 20-cm rows gives good protection. Its quick growth is valuable in this regard.

Diseases. There are numerous diseases of sorghum. Leaf diseases are the most troublesome for forage producers. These are anthracnose caused by *Colletotrichum graminicola* (which can be overcome by using resistant varieties) and leaf blight caused by *Helminthosporium turcicum*. Charcoal rot (*Macrophomina phaseoli*) causes plants to lodge badly. Grain may be affected by covered smut (*Sphaceloztheca sorghi*) in which the seed is replaced by a sac of spores; fungicidal seed dressing before planting corrects this malady. The parasitic weed *Striga hermonthica* occurs in Africa (see colour Plate 54). Its seeds can only germinate when stimulated by a substance from the host root, and must be not more than 1 cm from it. The radicle attaches itself to the host root by a haustorium which penetrates the vascular system and parasitizes the sorghum below ground for three to six weeks; it then emerges and produces chlorophyll and photosynthates. Flowering commences ten to 20 days after emergence. It can be controlled by using a trap crop of Sudan grass, which is ploughed in after two months' growth.

Pests. From a forage point of view, grasshoppers would appear to be the worst pest, and feral pigs can cause havoc. Grain pests include the sorghum midge, *Contarinia sorghicola,* whose larvae feed on the developing seeds. Bird damage is also important and in Africa the weaver bird, *Quelea quelea,* causes major losses. Attempts to prevent damage by using awned varieties of sorghum give some hope of reducing losses. The high tannin content of the sweet sorghum seed is another deterrent, and early harvesting for silage avoids the main problem.

Economics. *Sorghum bicolor* is one of the major grain crops for human food throughout the drier areas of Africa and India. The fodder varieties are used widely for cut green fodder and silage, and for syrup production. The stalks are used for stover, roughage, thatch and fuel.

Animal production. The grain is used extensively for animal feeding in concentrate rations, with high-protein constituents. After the grain is harvested, the sorghum stubble gives quite useful grazing in winter when other feed may be scarce. Curran (1957) reported a grain yield of 1 609 kg/ha in central Queensland, and afterwards the stubble carried 5.5 sheep per hectare for a whole year. Grazing fodder sorghums in autumn to fill in a feed gap is also a sound practice. Onley and Sillar (1965) obtained an average live-weight gain of 0.86 kg per head per day over a seven-month period in northern Queensland from yearling steers grazing sweet sorghum at the rate of 1 beast/0.8 hectare.

Hintz and Gilliland (1976) showed that cv. Zulu, when sown in early October on the Darling Downs in Queensland, will yield 10 080 kg/ha green matter, of which about 5 000 kg/ha will be available from December to January. Two other grazings will be obtainable before winter, each of about 300 kg/ha. Nitrogen applied at 50 kg/ha with adequate subsequent rain after the first grazing will boost yields. If sown in late December to early January the crop yields about 5 000 kg/ha grazed in March-April with little regrowth. Cultivar Sugardrip, sown in October, will yield 8 400 kg/ha green matter, giving 3 900 kg/ha in December, plus two other grazings each of about 2 400 kg/ha. If sown in mid-January and fed late April to July it will yield about 5 600 kg/ha.

Main attributes. Sorghum has wide adaptability, is more drought-hardy than maize, makes excellent silage and has a wide range of uses.
Main deficiency. Its tendency to be toxic.
Further reading. Hintz & Gilliland, 1976; Martin & Leonard, 1959; Wyllie & Stirling, 1977.

Sorghum halepense (L.) Pers.

Common names. Johnson grass (United States, Australia, southern Africa), grama China, maicillo, sorguillo, sorgo de Alepo (Peru), Aleppo grass (southern Africa), Don Carlos (Cuba).

Natural habitat. In moist areas on river banks, in clay soils and wet sandy soils.

Distribution. Believed to be of Mediterranean origin (Meredith, 1955), but introduced very early to India (Bor, 1960); now widespread through the subtropics. Called "Johnson grass" after Colonel Johnson who first grew it in Alabama, United States.

Description. A strongly rhizomatous perennial, moderately stout, 50-200 cm tall. Culms arise from a stout, extensively creeping, scaly rhizome which is well rooted at the nodes; prop roots arise from the lower above-ground nodes (Tothill & Hacker, 1973). Leaf-blades have midrib pre-eminently white. Spikelets in pairs, one sessile, the other pedicellate, both dorsally flattened; at maturity, falling entire, together with the subtending joint and pedicellate spikelet. The sessile spikelet is 4.5-5.5 mm long and smaller than that of *S. sudanense* and *S. verticilliflorum* (see Fig. 15.145).

Season of growth. Spring to autumn.

Optimum temperature for growth. Tillering increases with temperature up to 27°C in a 12-hour day, and up to 32°C in a 16-hour day; at 32°C, growth is reduced (Ingle & Rogers, 1961).

Frost tolerance. It is susceptible to frosts but the rhizomes usually survive.

Rainfall requirements. It prefers semi-arid to subhumid areas, but does not do well in wet tropical places. A rainfall of 500-750 mm with a summer dominance is best.

Drought tolerance. It has good drought tolerance, the rhizomes surviving dry periods and shooting again after rain.

Soil requirements. It is essentially a crop for rich soils, and does not produce well on poor soils. It usually invades first-class alluvial soils and is enhanced by irrigation; it thrives in ditches.

Fertilizer requirements. Johnson grass gives linear yield increases up to 1 075 kg N/ha, but economical increases amounting to 9 tonnes per hectare are obtained with 538 kg N/ha (Bennett, 1973). A fertilizer mixture in the ratio of 6N:0.5P:0.8K is recommended. Nitrogen should be applied in split applications.

Ability to spread naturally. It spreads rapidly from seed and more gradually by rhizomes through the soil or from pieces distributed by machinery. Some is spread by livestock (see Plate 55).

Sowing methods. It is usually rolled into a well-prepared seed-bed, at 11-22 kg seed/ha. Rhizomes can be cut up by disc cultivation and spread by harrows.

Number of seeds per kg. 100 000-125 000 (Queensland); 286 000 (United States).
Dormancy. The seed requires after-ripening for a few months.
Compatibility with other grasses and legumes. Johnson grass will initially grow in association with grasses such as *Paspalum dilatatum, Chloris gayana* and *Cynodon dactylon,* but will gradually dominate them. It can become sod-bound, which reduces its productivity. It will grow in alfalfa (*Medicago sativa*) crops and can be controlled a little when the alfalfa is mown periodically for hay.
Ability to compete with weeds. It competes very successfully with weeds because of its shade effects and vigorous root-stock. It often becomes a weed in such crops as Sudan grass or even oats.

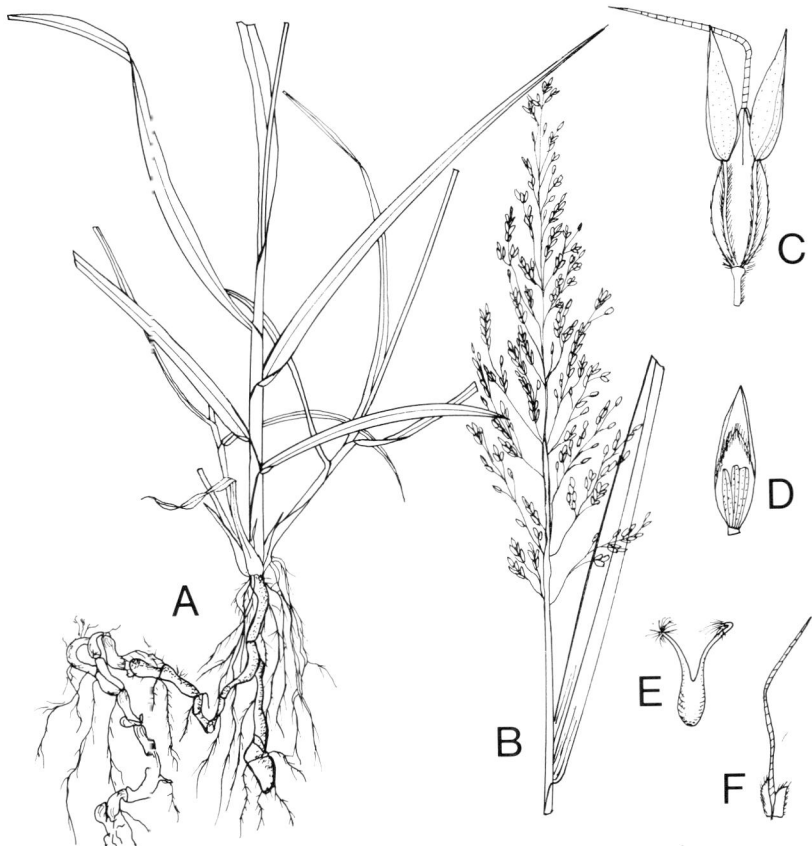

Figure 15.145. Sorghum halepense. **A**-Habit and rhizome **B**-Inflorescence **C**-Spikelet cluster **D**-Male floret **E**-Gynoecium **F**-Awned lemma

Seedling vigour. The seedling is very vigorous, germinating well and becoming aggressive.

Response to photoperiod. Flowering is accelerated by short days of 12 hours or less (Evans, Wardlaw & Williams, 1964). Plants at eight to 12 hours day length flowered at 27°C, but at 14 hours day length failed to flower (Ingle & Rogers, 1961).

Genetics and reproduction. The chromosome number is 2n=20, 40. It readily crosses with other sorghum species such as *S. sudanense*.

Dry- and green-matter yields. In Texas it yields 17-18 tonnes of hay per hectare under irrigation.

Suitability for hay and silage. In the southern United States it provides a good deal of hay. It is cut in the boot stage, and before the seed matures, two or three times a year. The coarse material dries slowly and thorough curing is necessary.

Value as standover or deferred feed. Frosted Johnson grass in which rough pea has been planted is used as winter feed in Alabama, Mississippi and Texas, United States. The peas provide green forage (Bennett, 1973).

Chemical analysis and digestibility. Table 15.73 shows analyses from India and the United States. Fertilized Johnson grass (cut at the boot stage) averaged 13.5 percent crude protein and 65.8 percent digestibility, compared to unfertilized grass at 7.2 percent crude protein and 47.5 percent digestibility (Bennett, 1973).

Palatability. It is very palatable in the early growing stage.

Toxicity. In common with other *Sorghum* species it is toxic at times due to its HCN content. A yield of 0.029 percent HCN was recorded at the Darling

TABLE 15.73 **Sorghum halepense**

	DM	As % of dry matter				
		CP	CF	Ash	EE	NFE
Fresh, 10 weeks, India	15.9	16.1	29.6	11.1	2.8	40.4
Fresh, 10 weeks, India	20.9	12.7	34.1	9.9	2.6	40.7
Fresh, 14 weeks, India	27.7	7.4	38.7	9.2	1.6	43.1
Fresh, 1st cut, India		10.3	35.9	8.2	2.3	43.3
Fresh, 2nd cut, India		5.1	36.4	9.4	1.5	47.6
Hay, United States	87.7	6.6	34.6	5.9	1.9	51.0

	Animal	Digestibility (%)				
		CP	CF	EE	NFE	ME
Hay, United States	Goats	45.0	58.0	40.0	54.0	1.91

Source: Göhl, 1975

Downs, Queensland, in September, 1958, and this would increase as growth accelerated in summer (Everist, 1974). For symptoms and treatment of the poisoning refer to *Sorghum almum.*

Seed production and harvesting. Johnson grass seeds heavily and shatters a good deal. It is harvested with combines if necessary.

Seed yield. Yields of 300 kg/ha are considered good.

Value for erosion control. Johnson grass will quickly vegetate suitable areas and act as an effective cover against wind and water erosion. However, its aggressive, weedy nature makes it less useful than other species which are available for this purpose.

Diseases. It is subject to leaf diseases similar to those that affect Sudan grass, i.e. leaf spot (*Cercospora sorghi*); zonate leaf spot (*Gloecercospora sorghi*) and anthracnose (*Colletotrichum graminicola*).

Pests. The sorghum midge (*Contarinia sorghicola*) and sorghum web-worm (*Celema sorghiella*) affect seed crops.

Economics. One of the worst weeds of cultivation in the subtropics throughout the world.

Animal production. Johnson grass is used for grazing where a decision is made to live with it rather than attempt eradication. It is cut for coarse hay over large areas of the southern United States.

Main attributes. Its perenniality for hay, its palatability.

Main deficiencies. Its problem as a weed overshadows its value for hay and pasture, and it is not now sown. Eradication is the best aim, if possible and economical.

Further reading. Marley, 1978; Martin & Leonard, 1959.

Sorghum sudanense (Piper) Stapf

Synonym. S. *vulgare* var. *sudanense* Hitchc.
Common names. Sudan grass (Australia, United States), garawi (Sudan).
Distribution. A native of tropical North Africa, now introduced widely through the tropics and subtropics.
Description. An annual with rather slender culms up to 300 cm high. Leaf-blades 8-15 mm wide. Panicle 15-30 cm long, open and loose when mature. Sessile spikelet 6-7 mm long, the glumes loosely hairy, at maturity glossy and often almost glabrous; upper lemma awned, awn up to 16 mm long. Pedicelled spikelet usually as long as the sessile but narrower. Caryopses are straw-coloured, brown, red-brown, buff or black. Unlike S. *halepense* and S. *verticilliflorum,* the panicles do not break up easily, and the pedicelled spikelets are persistent at maturity (Chippendall, 1955) (see Fig. 15.146).
Season of growth. Summer.
Optimum temperature for growth. Day/night temperatures 30/21°C (Sprague, 1943). Higher yields of top were obtained at 27.5-32.5°C than at 21°C (Sullivan, 1961).
Minimum temperature for growth. There was no growth at day/night temperatures of 12/4°C (Sprague, 1943); 15°C (Sullivan, 1961).
Frost tolerance. Sensitive to frost.
Latitudinal limits. About 30°N and S.
Altitude range. It generally grows from sea-level to 300 m in Australia.
Rainfall requirements. It is adapted to warm regions of low to medium rainfall and humidity, in the area of 600-900 mm annually, but is often grown under irrigation in low-rainfall areas. It gives heavy crops in areas of rainfall higher than 900 mm, but may be subject to more disease in these conditions.
Drought tolerance. It has reasonably good drought tolerance.
Tolerance to flooding. Little tolerance to wet conditions.
Soil requirements. It does best on the more fertile soils of medium to heavy texture, but will respond on light soils to irrigation and fertilizer.
Tolerance to salinity. Sudan grass gave its maximum yield at an electroconductivity of the soil extract of 3 mmhos/cm, 50 percent of maximum at 15 mmhos/cm, and nil at 26 mmhos/cm (Maas & Hoffman, 1977). In India, at Karnal it grew well on saline soil treated with gypsum (Yadav, 1975).
Fertilizer requirements. It requires a good level of phosphorus to help counteract cyanogenetic glucosides, and responds to medium nitrogen levels. High nitrogen may increase the glucosides. A soil test prior to planting will show the individual needs. Generally, 125-259 kg/N ha just before planting and 125-250 kg/ha superphosphate at seed planting are used.
Ability to spread naturally. Not very good. Unlike Johnson grass, it has no rhizomes.

Land preparation for establishment. A fine seed-bed is essential for a good plant population.
Sowing methods. It is best drilled into a good seed-bed, but some broadcasting into ashes is done.
Sowing depth and cover. The seed should be sown no deeper than 2.5 cm (preferably at about 1 cm) in moist soil, and rolled or press-wheeled.
Sowing time and rate. Spring to late summer (for late feed), at 8-12 kg/ha.
Number of seeds per kg. 95 000-120 000 (United States).

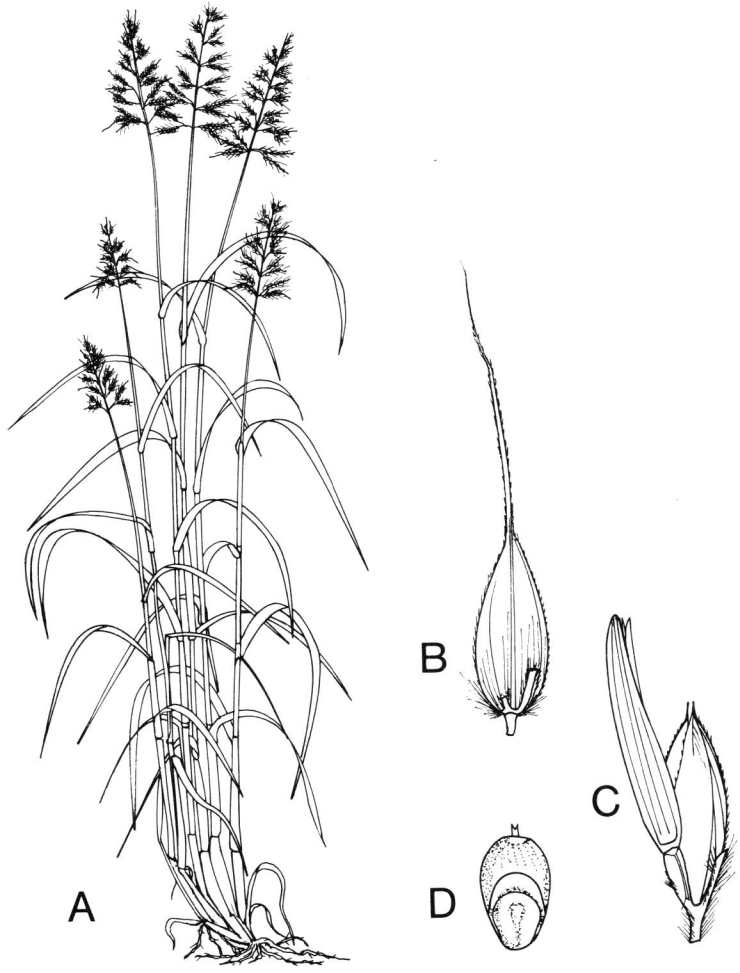

Figure 15.146. Sorghum sudanense. **A**-Habit **B**-Spikelet cluster **C**-Pedicelled, awned spikelet **D**-Grain

Dormancy. There is little post-harvest dormancy.
Seed treatment before planting. Seed must be treated with fungicidal dust. Germinate at 32°C in water.
Seedling vigour. Excellent.
Vigour of growth and growth rhythm. It grows very vigorously under good conditions and often outgrows the capacity of the grazing herd to maintain it at a nutritious stage before it begins to mature. It flowers in six to nine weeks.
Response to photoperiod. Flowering occurs at any time, accelerated by short days (Evans, Wardlaw & Williams, 1964).
Response to light. It grows little in the shade.
Compatibility with other grasses and legumes. Being an erect, tussock-like plant, it combines well with grasses and legumes. It is sometimes grown in conjunction with cowpea (*Vigna unguiculata*) for green fodder and silage.
Ability to compete with weeds. If established on a clean seed-bed it quickly smothers all weeds.
Tolerance to herbicides. See *Sorghum halepense*.
Response to defoliation. Sudan grass recovers quickly from grazing if the first grazing is taken at 30 cm high; it gives good grazing throughout the spring and summer if moisture is available.
Grazing management. A single animal should try grazing first. If there is no adverse effect (poisoning symptoms within half an hour) the remainder of the herd can be allowed to graze. It is best to give the whole herd only half an hour's grazing the first day, gradually increasing daily grazing time to a full day in four to five days. Graze down to about 20 cm, then shut up and graze again when the regrowth is 60 cm high.
Response to fire. Under annual cropping, fire is not a problem.
Genetics and reproduction. The somatic chromosome number is 2n=40 (Fedorov, 1974). Breeding is aimed at low HCN content, freedom from leaf diseases, juiciness and sweetness. It is cross-pollinated.
Dry- and green-matter yields. Under irrigation in the United States, it yields up to 25 tonnes of hay per hectare. For silage, yields of 12-25 tonnes per hectare have been recorded, depending on soil fertility and moisture.
Suitability for hay and silage. It makes reasonably good hay, although the stems may be a little coarse for good quality. It makes excellent silage when cut at flowering or a little later.
Value as standover or deferred feed. It is not usually allowed to remain in the field, and original first-season material will deteriorate in quality. However, if rainfall and fertility are good, a second and third crop of green material can provide valuable fodder during mild, frost-free winters and subsequent summers. Regrowth must be watched for prussic acid (cyanogenetic glucosides).
Chemical analysis and digestibility. Göhl (1975) records results from Chile and southern Africa, tracing the nutritive value throughout the growth, in Table 15.74.
Palatability. Sudan grass is extremely palatable.

TABLE 15.74 *Sorghum sudanense*

	DM	As % of dry matter				
		CP	CF	Ash	EE	NFE
Fresh, 20 cm, Chile	21.6	6.8	31.0	8.5	2.9	50.8
Fresh, 40 cm, Chile	21.6	6.9	28.5	9.4	2.5	52.7
Fresh, 80 cm, Chile	21.6	7.3	30.2	9.4	3.0	50.1
Fresh, 60 cm, southern Africa	19.3	15.4	23.4	9.9	3.7	47.6
Fresh, 65 cm, southern Africa	22.1	13.1	24.1	10.4	3.0	49.4
Fresh, 80 cm, southern Africa	20.0	11.1	26.4	9.5	3.1	49.9
Fresh, 90 cm, southern Africa	22.9	9.3	28.7	9.3	2.9	49.8
Fresh, 100 cm, southern Africa	22.0	9.3	29.1	9.5	1.8	50.3
Fresh, 125 cm, southern Africa	24.3	8.0	32.1	9.1	1.6	49.2
Hay, southern Africa		7.3	35.7	8.9	2.0	46.1

	Animal	Digestibility (%)				ME
		CP	CF	EE	NFE	
Fresh, 60 cm, southern Africa	Sheep	77.8	73.3	65.6	81.3	2.75
Fresh, 65 cm, southern Africa	Sheep	74.8	72.0	62.7	76.8	2.57
Fresh, 80 cm, southern Africa	Sheep	70.8	68.6	63.0	74.5	2.49
Fresh, 90 cm, southern Africa	Sheep	65.9	66.4	66.7	72.4	2.41
Fresh, 100 cm, southern Africa	Sheep	64.3	63.6	46.2	70.1	2.27
Fresh, 125 cm, southern Africa	Sheep	59.3	63.3	40.4	67.5	2.19
Hay, southern Africa	Sheep	47.7	62.9	54.1	62.0	2.08

Source: Göhl, 1975

Toxicity. Cyanogenetic glucosides, mainly dhurrin, occur in this species, as in other *Sorghum* spp. There is considerable variation in the amount of glucoside in different races and cultivars and under different soil and seasonal conditions (Everist, 1974). Poor sandy soils, young meristematic regrowth, low-phosphorus soils, cross-breeding, drought conditions and excess nitrogen all tend to increase the HCN content. Symptoms and treatment have been recorded under *Sorghum almum*.

Seed production and harvesting. The crop seeds heavily. It is harvested with conventional seed-harvesting combines, often fitted with rollers in front of the comb to bend the heads over to a reasonable harvesting height. Small areas can be hand harvested. Where humidity is high, cutting and drying prior to threshing may be necessary.

Seed yield. About 500 kg/ha.

Minimum germination and quality required for commercial sale. 70 percent

germinable seed, 97.6 percent purity (Queensland).
Cultivars. The first selection was made by Carl Wheeler of Kansas and sold as "Wheeler's Improved Sudan" by 1915. It was selected for seedling vigour, early maturity and uniformity of fine stems. 'California 23' was selected in 1930, and improved to eliminate black seed and other "off" types. The cultivars California 23, Greenleaf, Piper, Suhi-1 and Sweet-372 are listed under *Sorghum sudanense* in the list of cultivars eligible for certification under the OECD scheme in 1957. Cultivars listed in the United States are 'Georgia 337', 'Greenleaf', 'Lahoma', 'Suhi-1', 'Sweet-372' and 'Tift'. However, not all these cultivars are strictly *S. sudanense*. With the exception of 'California 23', they have been derived, directly or indirectly, from crosses between *S. sudanense* and *S. bicolor* cv. Leoti. In the Australian Herbage Plant Register such types as these, e.g. 'Lahoma' and 'SS6', have been designated *Sorghum* spp. hybrids with the common name of sweet Sudan grasses. The cultivars with *S. bicolor* strain in them usually have juicier and sweeter stems but may not be quite so well adapted to semi-arid conditions (Barnard, 1969).

- 'Piper' — listed in the United States under *S. sudanense,* but really a cross. Bred in Wisconsin, United States. A vigorous type, leaves blue-green with white midribs. Early maturing, dry stalked; 100 000 seeds/kg. Has strong seedling vigour, grows rapidly, recovers quickly from grazing and is low in prussic acid.

Value for erosion control. Sudan grass, unless sown thickly, is too tussocky for erosion control. It can give a quick cover if sown thickly but, being an annual, will give only temporary protection.
Diseases. Sudan grass is subject to several leaf diseases and head smut. Breeding work has helped provide some resistance to leaf diseases, and seed dusting before planting eliminates smut.
Pests. Army-worms and grasshoppers are periodic pests.
Economics. Sudan grass has been one of the most commonly used summer forage crops in subtropical areas of the world, especially for dairying.
Animal production. Milk production from cows fed solely on a diet of forage sorghum tends to be low, and a deficiency of sulphur in the diet was suspected. Large responses to sulphur with sheep grazing sorghums have been recorded. Trials with lactating cows at Samford, Queensland, where cows were fed 10 g per day of elemental sulphur resulted in a significant increase in milk production, in solids content and especially in protein, but butterfat content was lower. The increase in yield and protein content was probably due to increased intake of sorghum forage (Stobbs & Wheeler, 1978).
Main attributes. Extremely rapid growth of high palatability; nutritious green fodder supplying a number of grazings during the season.
Main deficiencies. It is often toxic due to cyanogenetic glucosides (prussic acid or HCN). It may grow coarse and be neglected; the hay from coarse material is of low nutritive value.
Further reading. Denman, 1958; Miles, L.G., 1949.

Sorghum spp. hybrids

Numerous hybrids between *Sorghum* spp. have been produced by plant breeders, and in recent years several fodder and forage types have been produced and released as cultivars. Some of these are listed here.

● 'SS6' (Sweet Sudan grass) — this is a selection from the American variety 'Sweet', obtained by crossing *S. sudanense* and *S. bicolor* cv. Leoti. It has been further selected for low prussic acid content and uniformity. It is an annual or short-lived perennial, shorter than Sudan grass; the leaves have cloudy instead of white midribs. The glumes are glassy and reddish brown or tan in colour, totally enclosing the seeds. Seeds smaller, longer and narrower than sweet sorghum, rich red or sienna in colour; 99 000 per kilogram. The stems are juicy and sweet. Adapted to rainfalls of 460-750 mm per year with summer incidence; rust and leaf spot diseases affect it at higher rainfalls. Relatively drought resistant. Flowers in 75 days at Tamworth, New South Wales (lat. 31.1°S). It is more vigorous and gives higher yields than Sudan grass, and retains its palatability longer after maturity. It makes good hay (Barnard, 1972).

● 'Lahoma' — originated at Texas Agricultural Experiment Station from a cross between *Sorghum sudanense* and *S. bicolor* cv. Leoti; introduced to New South Wales. It has large, wide, yellow-green leaves; it tillers freely. Seed colour ranges from apricot to sienna. It is relatively drought tolerant. Late-maturing, it produces high yields of palatable forage over a long period, and gives high seed yields. It recovers quickly after grazing. More resistant to leaf diseases than sweet Sudan.

● 'Sudax SX-11a' — an F_1 cross between *Sorghum sudanense* and *S. bicolor* in the United States. Requires annual purchase of seed, as the hybrid does not breed true. Annual, or short-lived perennial, 3-3.6 m high at maturity, usually 30-60 cm taller than 'Greenleaf' sweet Sudan and generally of a purple colour. Stem thicker than Sudan grass, juicy and reasonably sweet. Leaves eight to nine per stem, with cloudy midribs. The grain is red and ovoid, and threshes free of the glumes. Killed by frost, but stalks remain juicy for some time. It yields considerably higher than 'Lahoma' and sweet Sudan; not very suitable for hay but makes excellent silage. For summer grazing and autumn forage it is equal to 'Zulu'. In the Darwin area of Australia, its dry-matter production is almost as high as bulrush or pearl millet in the wet season, and its crude protein higher. It is superior to *S. almum* in yield and disease resistance, and to bulrush millet in its yield under irrigation.

● 'Zulu' — a cross between the male 'Redlan' grain sorghum and 'Greenleaf' Sudan grass, made by the Queensland Department of Primary Industries. The stem is erect, soft when young, sweet and juicy; taller than 'Greenleaf' Sudan. Grain is brown with a dark brown subcoat; black glumes enclose the grain. It is capable of rapid early growth, of producing a large bulk of green material quickly and of good recovery after mowing or grazing. It is one of the best sorghums for summer, autumn and winter grazing, and under irrigation.

- 'FS-22A' — an F_1 hybrid between grain sorghum and sweet sorghum, bred in Texas, United States. It grows to a height of 3-3.6 m at maturity with stout, juicy, sweet stems and many tillers. Leaves 12-14 per stem, with cloudy midribs. Stigmas yellow at flowering; glumes black; grain red, ovoid. This cultivar has an adaptation similar to cv. Sugardrip, and matures earlier than 'Tracy'. It does not lodge as early as 'Sugardrip'.
- 'Bantu' — A cross between the male-sterile 'Redlan' grain sorghum and 'Piper' Sudan grass by the Queensland Department of Primary Industries. Similar to 'Zulu' in appearance, but has a dry or pithy stem and white midrib. It is agronomically similar to 'Zulu', 'Sudan SX-11A', 'Bonanza' and 'Sordan'. It is resistant to the prevalent races of head smut (*Sphacelotheca reiliana*), whereas 'Zulu' is susceptible. This is the main reason for its development and release.
- 'Krish' — a hybrid between *Sorghum halepense* and *S. roxburghii*, developed by CSIRO, Australia. An erect perennial with numerous tillers and a solid, pithy stem about 3.8 cm thick; reaches 4 m in height. Rhizomes almost entirely absent. Glumes straw coloured, enclosing the seed. Seeds 160 000 per kilogram, caryopses 187 000 per kilogram. Requires a soil of high fertility. Growth from seed is slow and seedling vigour poor. The first defoliation should not occur before ten weeks; thereafter it will yield well till late in the season, but becomes coarse at maturity. It is more frost tolerant than *S. almum*. It matures late, but seed yield is low. It is resistant to common leaf diseases and sugar-cane mosaic (Pritchard, 1964).
- 'Silk' — a selection from a cross between cv. Krish and *S. arundinaceum*. Flowers later than *S. almum* and has a longer vegetative stage, shows better tolerance to frost and leaf diseases, and grows more vigorously (Silvey, 1977b).
- 'Sucro' — an F_2 selection from *S. almum* crossed with perennial sweet Sudan grass. It exhibits tolerance to frost and leaf diseases, has a higher soluble carbohydrate content in the cell sap of the leaves and stems, and has brown-coloured glumes making the "seed" easily distinguishable from Johnson grass (Silvey, 1977b).
- 'FS-26' — a hybrid between a grain sorghum and a sweet sorghum. Tall, free-tillering plant, reaching a height of around 3 m under favourable conditions. The stems are sweet and juicy and provide excellent standover feed. Sugar content reaches 20 percent. It is quick growing, has a large number of stems, excellent regrowth after the first cut, and high yields (Stevens, 1975).

Animal production. Irrigated 'Sudax' grazed rotationally at one-week intervals resulted in excess stem. Green lot feeding of 'Sudax' gave low gains of 3.1 kg per head per day for six-week-old fodder and 2.5 kg per head per day on eight-week old fodder in the Northern Territory, Australia (Blunt, 1969).

In the Macquarie Valley, New South Wales, Australia, 'Sudax SX11A' stocked at an average rate per hectare of 6.0, 6.0, 9.3 and 10.4 Angus yearlings (of approximately 200 kg/beast) gave dry-matter production of 18.95, 16.60, 19.41, and 31.10 t/ha respectively, and 274, 219, 249 and 334 kg beef per hectare. At no stage did HCN concentration of 'Sudax' reach lethal levels nor was the concentration correlated with animal performance.

Archer and Wheeler (1978) found that sodium and sulphur both help stock grazing sorghum. At Glen Innes, New South Wales, addition of NaCl to the diet of sheep or steers grazing *S. bicolor* × *S. sudanense* cv. Sudax SX6 increased their live-weight gain; the addition of sulphur produced a further small gain. For steers, average weight gains were 480 g per day with no dietary supplement, 610 g per day with NaCl, and 710 g per day with NaCl + S.

Further reading. Archer & Wheeler, 1978; Upton, 1978.

Spinifex hirsutus Labill.

Common names. Hairy spinifex, rolling spinifex, spring rolling grass, beach spinifex (Australia).

Natural habitat. Sea-shores.

Distribution. Australia, New Caledonia and New Zealand.

Description. A stout, dioecious, perennial grass, up to 30 cm tall, with strong, creeping stolons which root at the nodes. Plants are vegetatively similar, some male, some female or bisexual. The male inflorescence is a terminal cluster of stalked racemes, each cluster subtended by large, partly enclosing, silky-hairy bracts. The female or bisexual inflorescence is a large, globose, spiny head of numerous sessile racemes, each of which is reduced to a single spikelet, which is enclosed by a large, silky-hairy bract, the axis extending beyond the spikelet into a long, stout bristle, 10 cm or more in length (Tothill & Hacker, 1973) (see Fig. 15.147).

Season of growth. Perennial.

Optimum temperature for growth. Best temperatures for germination of the caryopses were 15-25°C and 20-35°C.

Soil requirements. It is essentially a plant for dune sands and sands of the sea-shores.

Tolerance to salinity. Excellent. It is continually exposed to salt-water spray.

Sowing methods. Seed will not germinate in the presence of light, and so must be buried in the sand by drilling. It can be propagated by cuttings.

Sowing depth and cover. Seed must be sown deeper than 1.25 cm to exclude light, but no deeper than 3.75 cm to ensure good germination. Deeper sowing leads to rotten seed in waterlogged conditions (Harty & McDonald, 1972). In the field, a depth of 3 cm has given satisfaction. Sown areas can be covered with brush wood.

Dormancy. Seed has no after-ripening requirements (Harty & McDonald, 1972).

Seed treatment before planting. Apart from removing the caryopses, soaking the seed in either fresh water or sea water does not improve germination (Harty & McDonald, 1972).

Grazing management. It is unpalatable to stock.

Genetics and reproduction. 2n=18 (Fedorov, 1974).

Seed production and harvesting. Inflorescences of beach spinifex have been harvested from open beach sites by hand and stored in hessian bags. Spikelets (the caryopsis enclosed in its lemma, palea and glumes) are usually attached to all or part of their associated spines. They were obtained from the inflorescences by several methods, the best of which was the use of a barley de-awning machine. This consisted of a cylinder, containing a centrally mounted spindle carrying long metal fingers, which rotates between fixed metal fingers

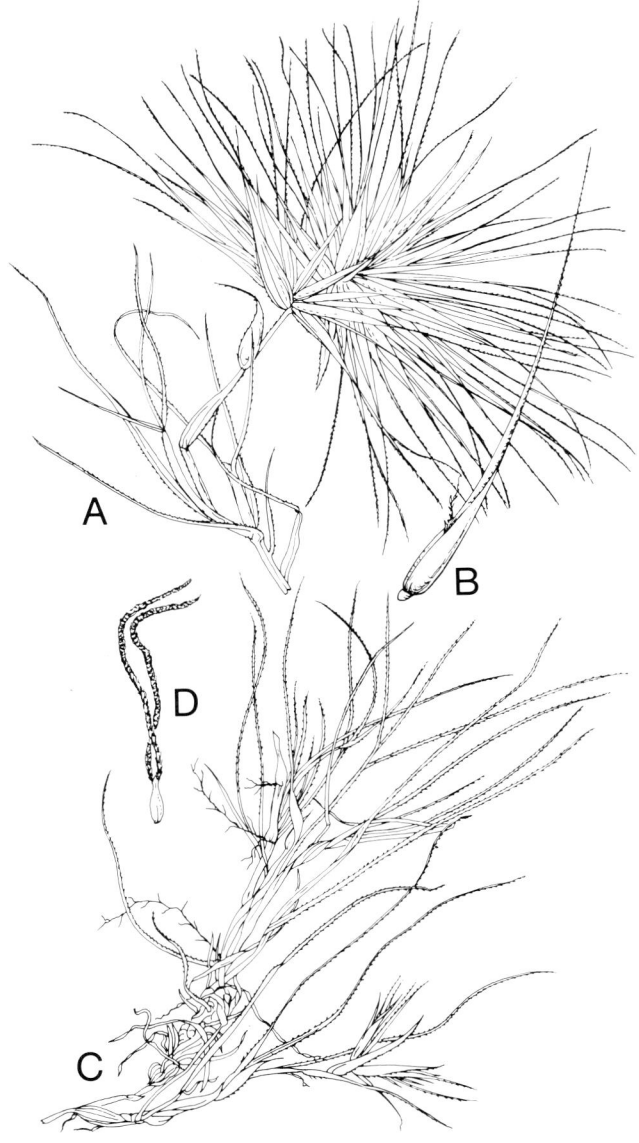

Figure 15.147. Spinifex hirsutus. **A**-Female inflorescence **B**-Female spikelet **C**-Male plant **D**-Male flower

attached to the inner surface of the cylinder. With this machine the spines were uniformly clipped short, and spikelets obtained in this way were free flowing and suitable for mechanical sowing. Hammer milling gave a high yield of caryopses, but was responsible for a great deal of breakage as the caryopses are soft. Partial threshing of the inflorescences, followed by winnowing, would help protect the caryopses from damage (Harty & McDonald, 1972).

Value for erosion control. It is excellent for the stabilization of sands on the sea-shore due to its vigorous stoloniferous habit. One plant can colonize a considerable area (Tothill & Hacker, 1973). It is used a good deal on the coasts of Australia, New Zealand and New Caledonia. It is also used in reclaiming coastal dunes after mining for zircon and rutile (Barr, 1965).

Further reading. Barr, 1965; Harty & McDonald, 1972; Maiden, 1894.

Sporobolus airoides Torr.

Common name. Alkali sacaton (New Mexico).
Natural habitat. On bottom lands and flats, sandy plateaux and washes.
Distribution. Alkaline soils in Mexico and the western United States.
Description. A coarse, tough perennial, 80-100 cm tall, growing in large dense clumps. The plant is pale green with a slightly greyish cast. The leaves are firm and fibrous, up to 50 cm long and 0.5 cm wide. The seed-heads are loose and open, with widely spreading branches, 30-45 cm long and 15-25 cm wide (Humphrey, 1960a, b) (see Fig. 15.148).
Altitude range. 800-2 000 m in Arizona.
Soil requirements. It occurs on fine-textured, often alkaline soils.
Tolerance to salinity. It can use drainage waters with salinities up to 10-15 mmhos conductivity (le Houérou, 1977a, b).
Land preparation for establishment. A good seed-bed is needed.
Sowing methods. Only large seeds more than one year old should be planted, when soil moisture is 14 percent or higher (one atmosphere tension or less), when probabilities of weekly precipitation are greatest and when soil temperatures are near 30°C. The planting site should be saturated with water just before planting. If storms do not yield at least 6 mm of rain within the first five days, the planting site must be rewatered to saturation.
Grazing management. Graze during the spring and summer only.
Suitability for hay and silage. It makes fair-quality hay when cut during the bloom stage.
Palatability. When growing vigorously it gives fair to good forage for horses and cattle, but not for sheep. When dry it is avoided by all stock (Humphrey, 1960a, b).
Economics. A major component of pastures in many valleys in New Mexico (Malcolm, 1971).
Further reading. Humphrey, 1960a.

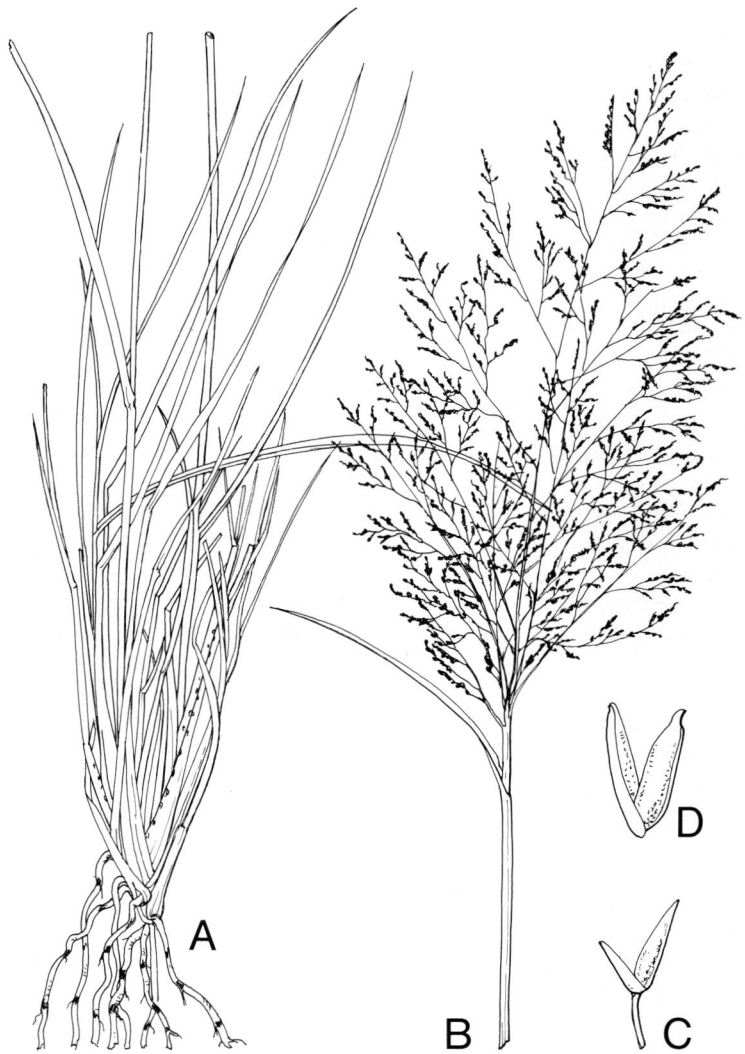

Figure 15.148. *Sporobolus airoides.* **A**-Habit **B**-Inflorescence **C**-Glumes **D**-Lemmas

Sporobolus helvolus (Trin.) Dur. and Schinz.

Synonym. S. *flagelliferus* Peter.
Common name. Okrich (Mauritania).
Natural habitat. Seasonally moist patches in dry grassland, mostly on dark clay or volcanic (often alkaline) soils, and in low-lying depressions carrying stagnant water (Gupta, Saxena & Sharma, 1972).
Distribution. Central, western and northeast Africa, northwest India.
Description. Tufted perennial with swollen culm bases, usually stoloniferous. The stolons grow vertically at first, and then bend over and root at the nodes. Panicle up to 12 cm long with erect branches, densely covered with short branchlets bearing few pale spikelets (Napper, 1965) (see Fig. 15.149).
Altitude range. 500-1 500 m.
Rainfall requirements. It grows in waterlogged soils (bottom lands) in arid and semi-arid areas.
Drought tolerance. Excellent. It is an important grass in the arid zone of Rajasthan.
Tolerance to flooding. It is adapted to moist soils.
Soil requirements. Common on alluvial silts and black clay soils in Kenya, also grows in volcanic ash (Göhl, 1975) and lacustrine deposits (Boudet & Duverger, 1961).
Tolerance to salinity. Excellent. It is used as grazing for the Chokla breed of sheep on saline rangeland in the arid zone of Rajasthan, India (Ahuja & Vishwanatham, 1976). It can tolerate drainage water with salinities up to 10-15 mmhos conductivity (le Houérou, 1977a, b).
Vigour of growth and growth rhythm. It flowers in August in the Sahel, matures in February and remains as standing hay till June (Boudet & Duverger, 1961). It revegetates each year.
Chemical analysis and digestibility. Göhl (1975) lists three analyses in Table 15.75.
Palatability. It is very palatable.
Seed production and harvesting. It does not normally produce good seed. If seeding types could be found, it would be a most useful grass for dry areas (Bogdan & Pratt, 1967).
Economics. A desert grass and fodder for camels. It is the dominant component of the sward developed on the sandy clays of the temporary ponds of M'Zerif, 30 km east of Timbédra, Mauritania, and is eaten well both at the beginning and end of the dry season (Boudet & Duverger, 1961). Saline rangeland in Rajasthan, India, dominated by salt-tolerant *S. helvolus*, is stocked at two- to four-month intervals by lambs (Ahuja & Vishwanatham, 1976).
Further reading. Boudet & Duverger, 1961.

Figure 15.149. *Sporobolus helvolus.* **A**-Habit **B**-Cluster of spikelets

TABLE 15.75 *Sporobolus helvolus*

	As % of dry matter				
	CP	CF	Ash	EE	NFE
Fresh, early bloom, Kenya	12.9	30.6	10.3	0.8	45.4
Fresh, vegetative, Pakistan	16.8	28.0	10.1	4.0	41.1
Fresh, mature, Pakistan	13.0	32.5	9.2	7.1	38.2

Source: Göhl, 1975

Sporobolus marginatus A. Rich.

Natural habitat. Dry grassland, often on alkaline soils.
Distribution. Northwest India, Baluchistan and tropical East Africa.
Description. A very variable tufted or stoloniferous perennial (rarely annual) 15-90 cm high, with well-grown stem leaves; basal leaf-sheaths glabrous, shining, often with ciliate margins. Spikelets pale- or greyish-green, but sometimes purplish, 1.5-2.5 mm long (Napper, 1965). The roots are very thick and are covered with a soft felt of root-hairs (Bor, 1960) (see Fig. 15.150).
Altitude range. 500-1 750 m.
Rainfall requirements. It is adapted to very dry conditions, around 300 mm per year (Bogdan & Pratt, 1967), 125-375 mm in India.
Drought tolerance. Excellent.
Soil requirements. It grows on a wide range of soils, from loose sandy loams to loams and alluvial silts. (Bogdan & Pratt, 1967).
Tolerance to salinity. Excellent, it inhabits saline soils (usar lands) in India and the Kafue flood plain in Zambia. It is seen on pure salt crust.
Ability to spread naturally. Covers saline soil well and helps minimize upward capillary movement of salts (Whyte, 1964).
Sowing time and rate. In summer, at 150-200 g/ha (Bogdan & Pratt, 1967).
Number of seeds per kg. 7.7 million (Bogdan & Pratt, 1967).
Chemical analysis and digestibility. Dougall and Bogdan (1960) in Kenya found 23.3 percent crude protein, 25.7 percent crude fibre, 9.3 percent ash, 3.1 percent ether extract and 38.6 percent nitrogen-free extract in the dry matter of fresh material in early bloom.
Palatability. It has low palatability (Verboom & Brunt, 1970).
Further reading. Bogdan & Pratt, 1967.

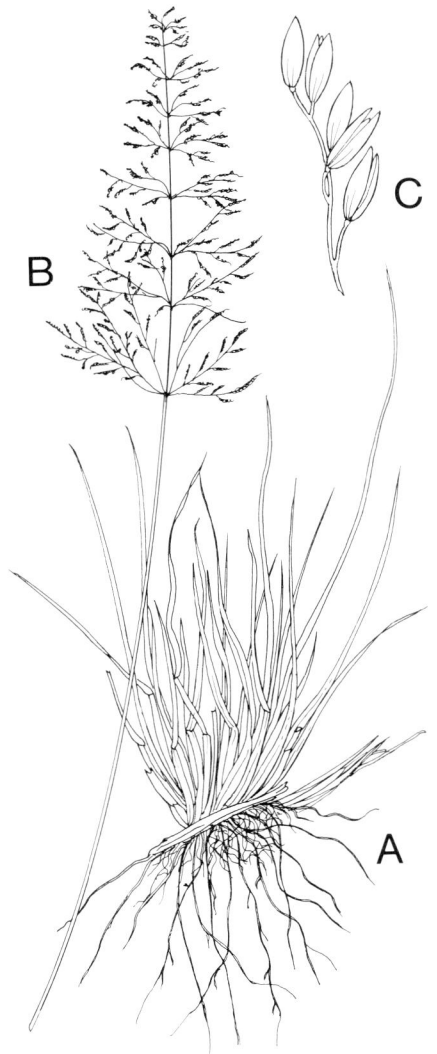

Figure 15.150. Sporobolus marginatus. **A**-Habit **B**-Inflorescence **C**-Spikelets

Sporobolus spicatus (Vahl) Kunth

Natural habitat. Alkaline sandy soils, often on lake shores, rarely on open grassland.
Distribution. Throughout tropical Africa, especially Kenya, Tanzania and Uganda.
Description. Stoloniferous perennial. Leaf-blades have in-rolled margins and a very hard, needle-like point, more rarely softer and flat. Spikes very dense, cylindrical and narrow, pale (Napper, 1965).
Altitude range. 300-2 000 m in Tanzania.
Soil requirements. Occurs in alkaline sandy soils, often on saline lake shores.
Palatability. High.
Economics. Dominates the grass cover of the salt pans of the Siloana Plains of Western Province, Zambia (Verboom & Brunt, 1970).

Sporobolus virginicus (L.) Kunth

Common names. Salt-water couch (Australia), beach drop-seed, sea-shore rush grass (Hawaii).
Natural habitat. Sand dunes just above high-water mark, and behind mangrove swamps.
Distribution. Along coasts in tropical Africa, western seaboard of India, Sri Lanka, Australia and the United States.
Description. Rhizomatous perennial with lanceolate, spine-tipped leaf-blades growing 15-40 cm tall, erect, from creeping, hard, scaly rhizomes. Inflorescence dense, spikelike, up to 15 mm wide with short appressed branches and pale spikelets (Napper, 1965). The panicle is shorter than many other *Sporobolus* spp., being not more than 7.5 cm long (Parham, 1955) (see Fig. 15.151).
Season of growth. Perennial.
Tolerance to flooding. It does best if the water level fluctuates from 5 cm above the soil surface to 15 cm below (Leithead, Yarlett & Shiflet, 1971). Prolonged inundation kills it.
Soil requirements. It has a wide range, from clays to sands.
Tolerance to salinity. It grows on highly saline marsh soils.
Vigour of growth and growth rhythm. It makes some growth all the year.
Genetics and reproduction. $2n = 18, 30$ (Fedorov, 1974).
Grazing management. For maximum production no more than half of current growth (by weight) should be removed in any season. Summer grazing deferments of at least 120 days are important to maintain vigour in the southern United States (Leithead, Yarlett & Shiflet, 1971).
Response to fire. Controlled burning results in lush, tender forage for winter grazing. Burning should be done not more often than every two years when water is above the soil surface. Allow 10 cm of regrowth after burning before grazing (Leithead, Yarlett & Shiflet, 1971).
Chemical analysis and digestibility. It is high in protein and minerals.
Palatability. Good.
Seed production and harvesting. It produces seed several times throughout the year.
Value for erosion control. A valuable stabilizer of wind-eroded shorelines (Rose-Innes, 1977).
Economics. It is an important grass for saline coastal or subcoastal areas throughout the tropical world.
Animal production. In northern Queensland it provides important grazing for beef cattle throughout the year, but especially during the dry season.
Main attributes. Its tolerance of salinity and its ability to stabilize sea-shores.
Further reading. Leithead, Yarlett & Shiflet, 1971.

Figure 15.151. Sporobolus virginicus. **A**-Habit **B**-Spikelets

Stenotaphrum secundatum (Walt.) Kuntze

Common names. Buffalo grass (Australia), St Augustine grass (Florida, United States).

Natural habitat. In moist swampy soil near the sea-shore in the United States and southern Africa.

Distribution. Native to North America, West Indies, Australia. Now widely distributed as a lawn grass.

Description. A hardy perennial, creeping extensively by means of branched rhizomes and many-noded stolons. Exceedingly variable in size, the culms rising above the ground for 6-40 cm or more, much branched from numerous nodes, the branches trailing, producing flowering stems or fin-shaped tufts of leaves. Leaf-sheaths strongly compressed and keeled; leaves nearly always glabrous except near the ligule, blades up to 12 mm wide, folded at first, then expanded, usually rounded or obtuse; ligule a fringe of short hairs. Inflorescence a false (or, rarely, a true) one-sided spike, 4-15 cm long, terminating the culm and each flowering branch; central axis thick, swollen, flat on one surface, deeply hollowed out on the other, each cavity containing a single spikelet or shoot spike of two to four spikelets borne alternately on either side of a wavy middle ridge. Spikelets 4.5-5 mm long, sessile, acute, awnless, glabrous, light green (Chippendall, 1955). St Augustine grass is more robust and taller than buffalo grass, which is used for lawns. *S. secundatum* var. *variegatum* is used as a decorative indoor plant (see Fig. 15.152).

Season of growth. Spring, summer and autumn.

Frost tolerance. It survives frosts.

Altitude range. Sea-level to 800 m.

Rainfall requirements. Grows in humid areas, along the coast.

Drought tolerance. Tolerant of short dry periods.

Tolerance to flooding. It will stand a good deal of flooding in Florida.

Soil requirement. It will grow on a wide range of soils, and is particularly adapted to the muck soils of the Florida Everglades coastal sands and alkaline soils. In Puerto Rico, cv. Roselawn does best on soils rich in lime, and on steep sandy soils (Vicente-Chandler *et al.*, 1953).

Tolerance to salinity. It is a sea-shore grass and will withstand salt spray (Wheeler, 1950).

Fertilizer requirement. It should be well fertilized, especially with nitrogen. In Florida, two applications of 125 kg/ha a year are recommended.

Ability to spread naturally. It spreads quickly by means of stolons. It does not produce seed.

Land preparation for establishment. Cuttings will establish in roughly prepared land, but good land preparation will usually pay.

Sowing methods. Vegetative material is used for new plantings. Rooted runners are dug or disc-harrowed into the soil, 30-40 cm apart in rows 60-80 cm

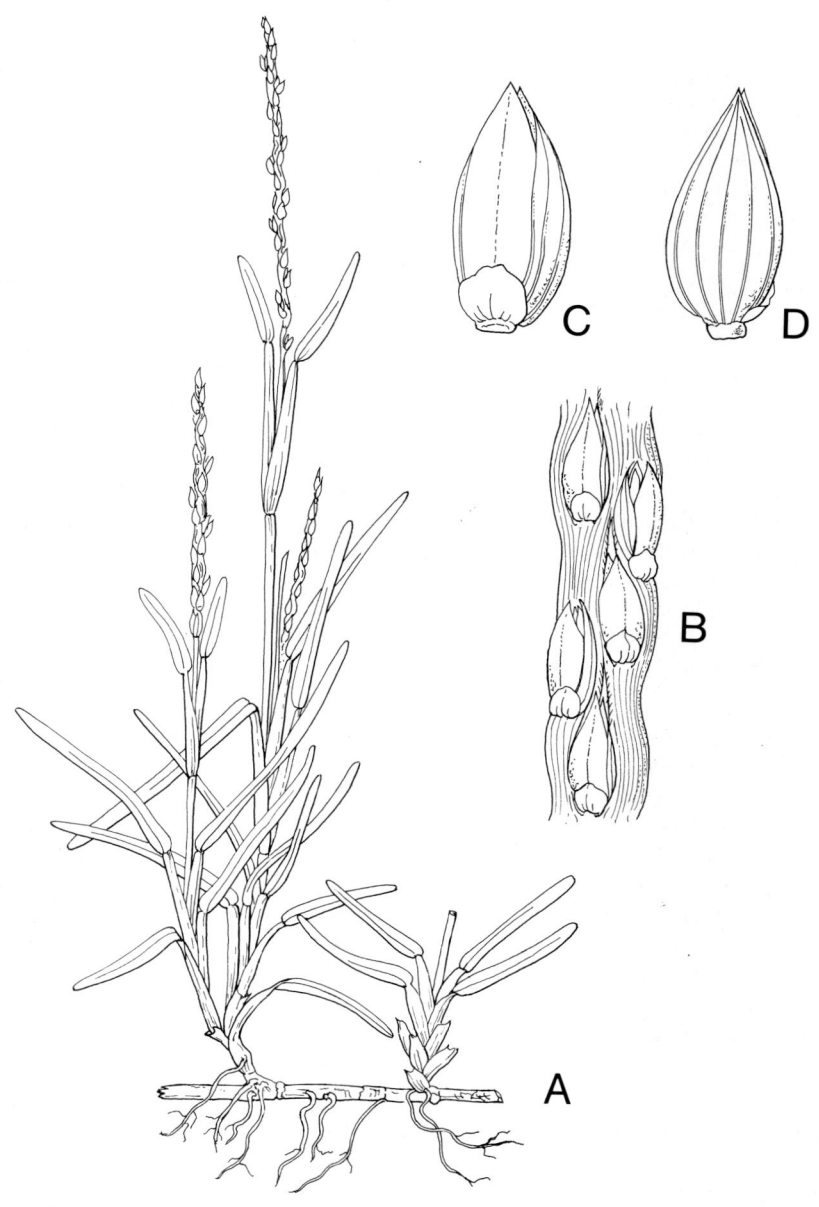

Figure 15.152. *Stenotaphrum secundatum.* **A**-Habit **B**-Portion of spike **C,D**-Spikelet, front and back

apart, and preferably rolled afterwards. A hectare of cuttings will plant about 10 ha of land.

Sowing time and rate. Early in the wet season.

Vigour of growth and growth rhythm. It is rather slow to cover the ground, but eventually provides a dense sward which crowds out weeds.

Response to light. It thrives in shaded areas and so is well adapted for lawns.

Response to defoliation. The creeping flat stems of St Augustine grass root to form dense sods which stand trampling and heavy grazing.

Grazing management. It should be grazed every second week down to 6 cm (Göhl, 1975), leaving sufficient leaf area for the plant to produce the carbohydrates it needs for growth without depleting its underground reserves (Vicente-Chandler *et al.*, 1953). It takes time to recover if grazed too closely. The herbage matures and becomes unpalatable very rapidly. An annual application of 350-500 kg/ha of 0:8:24 fertilizer is usually applied.

Genetics and reproduction. 2n=18, 20, 36, 54, 72 (Fedorov, 1974).

Suitability for hay and silage. It can be made into useful silage (Bennett, 1973).

Chemical analysis and digestibity. Göhl (1975) records only one analysis of hay. The hay contained 6.7 percent crude protein, 32.5 percent crude fibre, 3.7 percent ash, 2.7 percent ether extract and 54.4 percent nitrogen-free extract in the dry matter. The digestibility of the dry matter was 50.3 percent, of the crude protein 30.7 percent, and of the crude fibre 49.3 percent.

Palatability. Fairly palatable when young, but quickly loses palatability.

Toxicity. It contains about 1 percent of oxalates in the dry matter, but is not toxic (García-Rivera & Morris, 1955).

Cultivar. 'Roselawn' — used as a pasture forage (Bennett, 1973).

Value for erosion control. Excellent.

Diseases. It is subject to brownpatch in the United States.

Pests. Some damage is done by the cinch bug (*Blissus leucopterus*) in the United States.

Animal production. Kidder (1952) recorded a live-weight gain of 2 250 kg/ha in one year on a St Augustine grass pasture on organic soil in Florida. This result was never repeated. The experiment was conducted in an area with extremely favourable moisture and temperature conditions, while the animals were supplemented daily with 450 g of cotton-seed meal. In a ten-year grazing experiment on St Augustine grass cv. Roselawn in Florida, the average daily gain was 6.35 kg/ha from April to June, and 0.8 kg/ha during winter (Haines *et al.*, 1965). The pasture should carry seven yearlings per hectare all year, with surplus pasture made into silage (Bennett, 1973).

Main attributes. Its ability to form a dense sod, its suitability for lawns. Its ability to grow on the muck soils of Florida.

Main deficiencies. Its coarseness and general low productivity, its lack of seed production.

Further reading. Haines *et al.*, 1965.

Themeda australis (R. Br.) Stapf

Synonyms. *Anthistiria ciliata* Linn.; *Themeda triandra* Forsk.
Common name. Kangaroo grass (Australia).
Natural habitat. Grassland and open forest.
Distribution. Australia, and throughout New Guinea.
Description. A perennial, forming leafy tussocks, with culms to 1-1.75 m. Leaf-sheaths overlapping below, shorter than the internodes on the upper parts of the culms. Panicle narrow, may or may not nod, brownish; fertile spikelet, brown-hairy from the callus, the glumes brown to straw coloured, shining, short pubescent; a geniculate, twisted awn, 3-5 cm long from the fertile lemma (Henty, 1969). It is very similar to *T. triandra* (see Fig. 15.153).
Season of growth. Summer, starting growth in New South Wales in October-November, i.e. south of 32°S, but growing most of the year in the tropics.
Frost tolerance. It is tolerant of frost.
Latitudinal limits. South of 32°S it is a summer grass; further north it grows year round.
Altitude range. Near sea-level to 2 750 m.
Rainfall requirements. It grows over a wide rainfall range, about 450-1 250 mm.
Drought tolerance. Clumps survive long drought periods and produce rapid new growth.
Tolerance to flooding. It is intolerant of flooding.
Soil requirements. It grows over a wide range of soils. It does particularly well on upland basaltic red earths and prairie soils, but is found on sands and sandy loams.
Sowing methods. It is not usually sown as the seed is not very viable. Seed heads, spread to allow stock to trample them in, may provide some new plants. Seed does not readily pass through a drill, so it is broadcast. It can be propagated by root division.
Dormancy. *T. australis* has an after-ripening dormancy of six to ten months; this time differs with ecotypes. It can be broken by gibberellic acid, alternating temperatures, scarification or removal of glumes, or palea and lemma (Groves, 1976).
Response to photoperiod. Various strains flower in short, medium and long days (Evans, Wardlaw & Williams, 1964).
Response to light. It grows very well in lightly shaded woodland.
Response to defoliation. It will not stand heavy stocking and has disappeared from millions of hectares of Australian pasture land. It is prominent in enclosed and protected places such as railway lines. Defoliation during the growing period reduces its root growth.
Grazing management. Graze lightly during the growing season and more heavily during the dry season.

Figure 15.153. Themeda australis. **A**-Habit **B**-Inflorescence **C**-False involucre **D**-Spikelets

Response to fire. It survives fire very well.
Genetics and reproduction. The basic chromosome number is x = 10. About 95 percent of the population are tetraploids (2n=40), though diploid, triploid, tetraploid, pentaploid and hexaploid plants have been found (Hayman, 1960). The diploid plants are common in Tasmania and the central highlands of southern Australia. The tetraploid plants are inland types. Most show both sexual and aposporous behaviour (Evans & Knox, 1969).
Suitability for hay and silage. It makes reasonably good hay.
Chemical analysis and digestibility. Green flowering material showed only 0.95 percent nitrogen, 0.32 percent calcium and 0.17 percent phosphorus (Allen, 1949). In winter it is low in protein and digestible carbohydrates, and it is difficult to breed sheep on grazing lands on which this grass is dominant (Moore, 1970).
Palatability. The softer types are very palatable and hence have mainly disappeared. The coarser western forms are sparsely eaten.
Toxicity. No toxicity has been reported.
Seed production and harvesting. T. australis seeds in November and forms few perfect seeds. Many of the spikelets are male and barren, but there is usually one fertile one in the cluster. The grain often fails to mature. The seeds do not germinate readily. Hayman (1960) obtained best germination from domestic refrigeration for 48 hours, then germinating on filter paper moistened with 0.1 percent solution of potassium nitrate.
Value for erosion control. It has been proved useful for stabilizing black soil gullies at Surat, southwest Queensland.
Economics. Grassland where *T. australis* is dominant gives useful pasture after an initial burn, but under continuous grazing, productivity is likely to decline, with the entry of small, prostrate grasses and miscellaneous weeds (Henty, 1969). Conversely, the presence of almost pure stands of *T. australis* indicates a light grazing pressure or no grazing, e.g. inside fenced areas, along railway lines. Over 100 years' grazing, *Themeda australis* grass climax grassland has been transformed into a *Heteropogon contortus* subclimax; this change is non-reversible due to the loss of the *Themeda* seed source. Very few seedlings occur.
Animal production. In northern Australia, Norman (1970) studied the growth of beef shorthorn steers on a natural pasture dominated by *Themeda australis, Sorghum plumosum, Chrysopogon fallax* and *Sehima nervosum* growing on a lateritic red earth soil at Katherine. The average rainfall of 925 mm falls in the four months from December to March. He further studied the effect of adding Townsville stylo (*Stylosanthes humilis*) fertilized with 50-100 kg/ha superphosphate. The native pasture was stocked at 4-12 ha per head and the mixed grass/legume pasture at 1.2 ha per head. The steers on native pasture gained weight for only six months of the year, from the start of the wet season until one to two months after the rains ended. From late May to late November they lost about 20 percent of their peak May weight. Steers

on the grass/legume pasture gained weight for about 45 weeks of the year. Over a six- to eight-week period in October/November, from the time of the first early storms until heavy rains have initiated the main flush of pasture growth, cattle on all types of standing forage lost weight heavily.
Main attributes. Its wide adaptability, tolerance to fire, and fair palatability. Its frost tolerance and long grazing season (Hassall, 1976).
Main deficiency. Its poor seed production.
Further reading. Norman, 1970.

Themeda quadrivalvis (L.) Kuntze

Common names. Habana oat grass, grader grass (Australia).
Distribution. Widespread in India, introduced to Australia.
Description. An annual, growing to 1.4 m; it turns a distinctive orange-red colour as the seed-heads mature. The involucral spikelets are 6-7 mm long, obliquely lanceolate with stout, coarsely tuberculate hairs on the keels but glabrous elsewhere; fertile spikelet has a somewhat obtuse or acute (but not pungent) callus scarcely 1 mm long; spatheoles very acute (Blake, 1969) (see Fig. 15.154).
Season of growth. Summer.
Latitudinal limits. 17-27°S; in tropical India, to 28°N (Sillar, 1969).
Rainfall requirements. 500-1 250 mm in India; common in 1 000-1 500 mm in north Queensland.
Drought tolerance. It can withstand quite dry conditions in India.
Soil requirements. It is adapted to soils of moderate moisture content and declines on soils with a moisture content in excess of 19.3 percent. It requires moderate exchangeable calcium. Favours a sandy loam soil with a pH from 7.0-8.5, but grows on clay loams and well-drained lateritic soils.
Ability to spread naturally. It spreads rapidly by seed.
Number of seeds per kg. 633 000.
Vigour of growth and growth rhythm. It grows vigorously and flowers from October to January in Madhya Pradesh, India.
Ability to compete with weeds. It can compete successfully with broad-leaved weeds.
Tolerance to herbicides. It can be controlled by paraquat at 1.4 litres/ha of a 200g AI/l product (e.g. Gramoxone) plus surfactant at 250 ml per 200 litres of water, when the plant is at the young seedling stage of growth; plants must be thoroughly wetted. On well-grown grass, paraquat at 2.8 l/ha as above can be used, but increasing the rate to 400 litres of water per hectare; alternatively, 2,2-DPA at 2.3 kg of a 740g AI/kg product (e.g. Shirpon, Ellapon, Dowpon) plus TCA at 9 kg of a 940g AI/kg product (e.g. TCA grass-killer) per 200 l of water can also be used. A wetting agent at 250 ml per 200 litres of water must be added and the plants thoroughly wetted. For pre-emergence control of grass seedlings, trifluralin at 2.8 litres/ha of a 400g AI/l product (e.g. Treflan EC) can be used, but it must be well incorporated into the soil immediately following application. It will not control broad-leaved weeds (Tilley, 1977).
Response to defoliation. Cutting at full flowering gives some control.
Grazing management. It is not grazed very much. Allowing the grass to remain undisturbed suppresses seed germination by shading, and the shed seed soon deteriorates (Sillar, 1969).
Response to fire. Burning encourages germination.

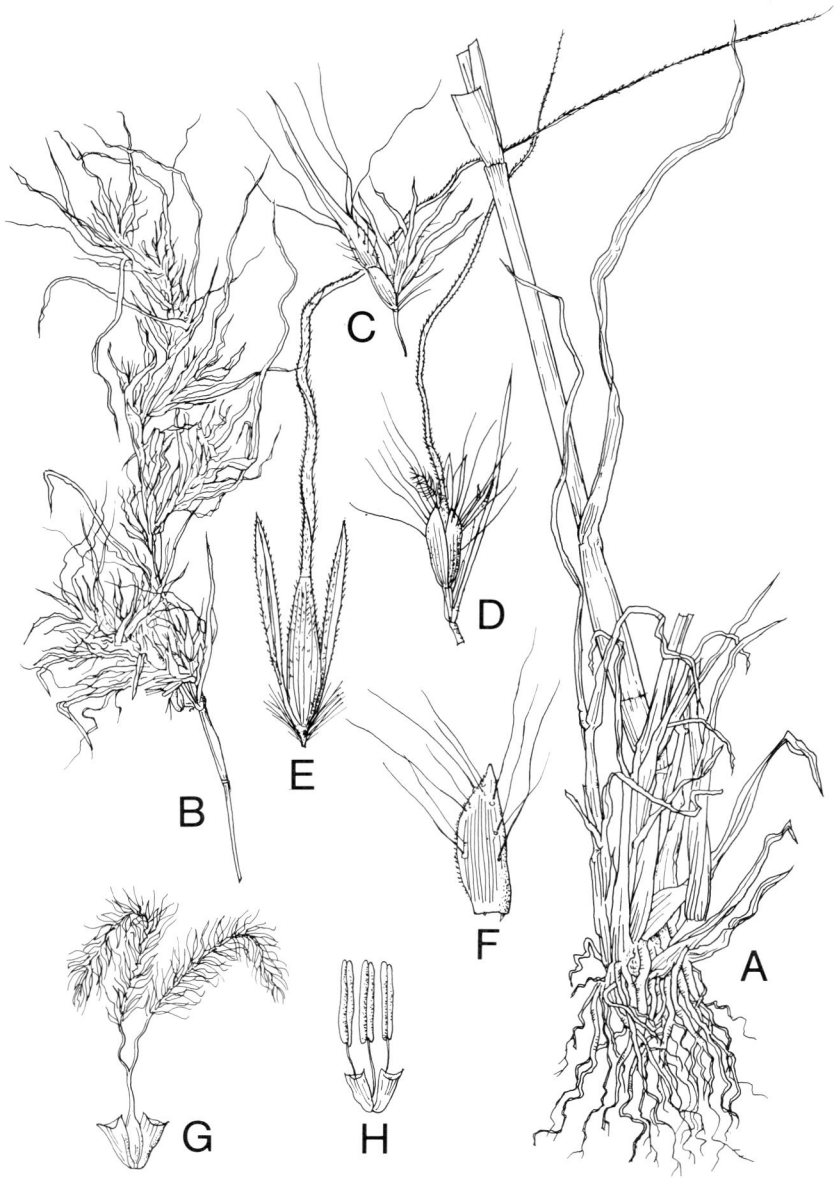

Figure 15.154. Themeda quadrivalvis. **A**-Habit **B**-Inflorescence **C,D,E**-Spikelet clusters **F**-Lower glume **G**-Female flower **H**-Male flower

Genetics and reproduction. 2n=40 (Fedorov, 1974).
Economics. It becomes a pest in Queensland's sugar-cane fields in poorly-grown sections of cane where cultivation has been neglected, but it will not grow in shade, such as under a good cane crop. It can be troublesome on headlands and roadways where it establishes very quickly. In Madhya Pradesh, India, it has some fodder value and is used for thatching.
Animal production. No animal production figures have been cited.
Main deficiencies. It has become an aggressive weed in sugar-cane fields in northern Queensland.
Further reading. Tilley, 1977.

Themeda triandra Forsk.

Synonym. Themeda australis (R. Br.) Stapf.
Common name. Red oat grass (Kenya).
Natural habitat. Widespread as grassland and in open woodlands on clay.
Distribution. All warm and tropical regions of the old world; abundant in East Africa where it constitutes 16 percent of the grasslands.
Description. Tufted perennial, 45-180 cm high. Panicle narrow, spatheate, up to 45 cm long; racemes reduced to a single awned fertile spikelet 5-6 mm long and two pairs of awnless spikelets (see Fig. 15.155).
    W.D. Clayton of Kew Gardens, London, is unable to separate this species from *T. australis* (R. Br.) Stapf. The following varieties are known: var. *burchellii* (Hach.) Domin, var. *trachyspathea* Goossens, var. *imberbis* (Retz.) A. Camus and var. *hispida* (Nees) Stapf.
Altitude range. Sea-level to 3 000 m in Africa, dominant at 1 300-3 000 m.
Rainfall requirements. It has a wide range. In places where it grows in areas with annual rainfall in excess of 760 mm it is not regarded as a good forage species. Where rainfall is less than 760 mm it is a major African forage species because of its abundance (Ndawula-Senyimba, 1972). Heady (1966) selected a 625-900 mm area to study botanical composition. In India it has a range from 1 000-6 250 mm.
Drought tolerance. It has some tolerance to drought.
Tolerance to flooding. It does not tolerate flooding; its proportion in a pasture increases with improving drainage (Ndawula-Senyimba, 1972).
Soil requirements. In Kenya and Tanzania, red oat grass forms almost pure stands on lateritic red earths (latosolic soils) of poor structure, low in lime, phosphorus and potash. It is also adapted to loose sandy soils, alluvial silts, and a wide range of other soils.
Sowing time and rate. Summer, at 20-30 kg/ha.
Dormancy. There is some after-ripening dormancy for approximately 12 months before a full germination potential is realized. Dormancy results from a combination of embryo dormancy and mechanically resistant glumes. Successful germination of spikelets entails the splitting of the tough upper glumes by radicles. Glume removal, plus treatment with gibberellic acid increases germination (Martin, 1975).
Response to defoliation. Ndawula-Senyimba (1972) showed that *T. triandra* persists best when cut at the end of the growing season. Frequent cutting shortens the life of the stand under semi-arid conditions. Under subhumid conditions, frequent cutting gives rise to a lawn.
Grazing management. At Rumuruti in Kenya it has been shown that red oat grass should be rotationally grazed in five blocks, with grazing during the most critical period of growth confined to only one year in every five. It can

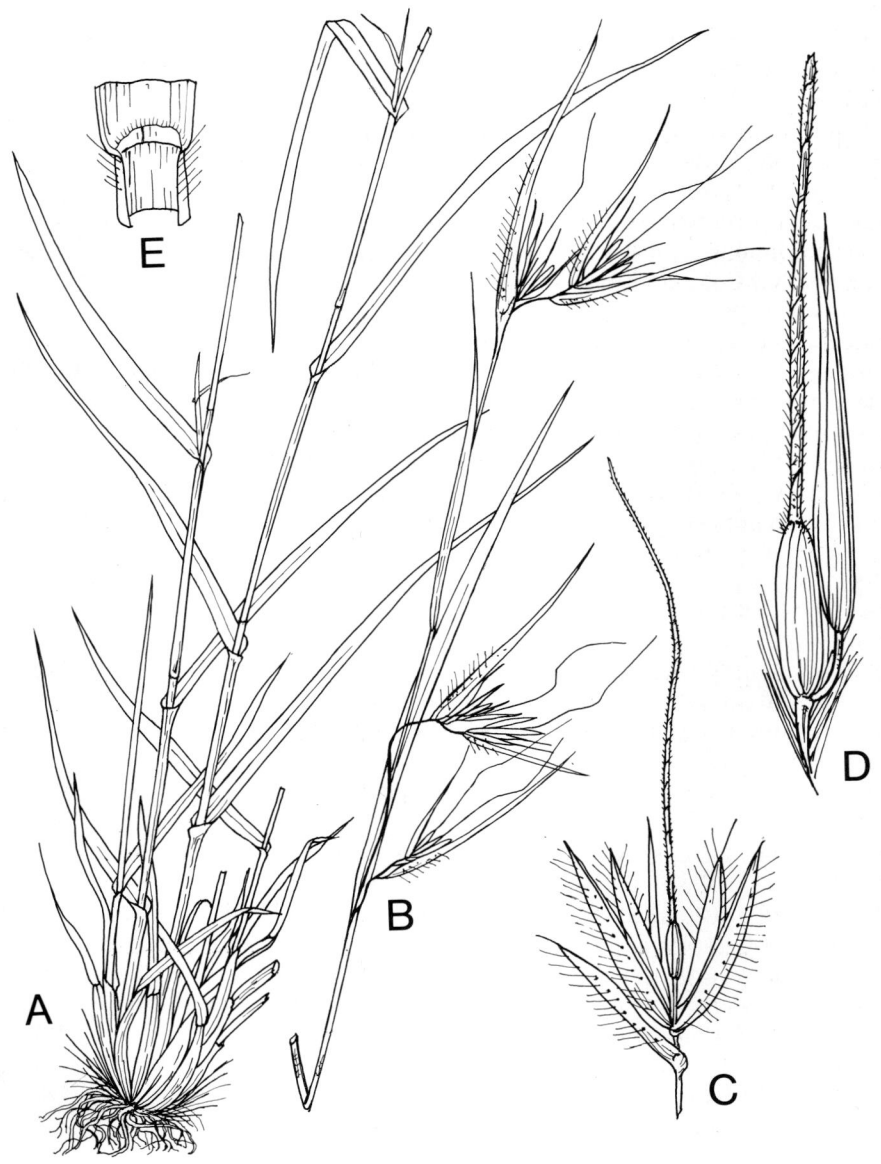

Figure 15.155. Themeda triandra. **A**-Habit **B**-Inflorescence **C**-Complex group of spikelets **D**-Awned hermaphrodite spikelet **E**-Ligule

thus be maintained well and kept highly productive at a stocking rate of one head of cattle to about five hectares (Henderson & Preston, 1959). The best time to graze *T. triandra* grassland is when 70 percent of the grass is green, that is, for a period of four weeks during the short rains (December-March), and six weeks during the long rains (May-August), both beginning about the sixth week of the grass's growth (Karue, 1975). Heady (1966) showed that grazing intensity was the main factor in determining the composition of *T. triandra* grassland. In southern Africa, Roberts and Opperman (1966) showed that an early summer (rather than late summer) rest period gave maximum production of dry matter, crude protein, roots growth reserves and flowering culms. Continuous grazing during the winter severely denuded *T. triandra* (Coetsee, 1975).

Response to fire. *T. triandra* is favoured by burning. It readily survives fires because the corkscrew-like awns, by alternate moistening and drying, drive the seeds about 2.5 cm into the soil. Some African studies show fires affect the soil only to a depth of 0.5 cm (Edwards, 1942). Burning followed by rain greatly increased germination of *T. triandra* in vacant areas (Ndawula-Senyimba, 1972). It is not found where protection from fire occurs (Göhl, 1975; Edwards, 1968).

Genetics and reproduction. 2n=20, 30, 40, 50, 60, 80 (Fedorov, 1974).

Suitability for hay and silage. It should be cut for hay at the stage of maximum dry-matter production — about eight weeks' growth during the long rains — but Marshall and Bredon (1967) say the hay is unlikely to be a satisfactory roughage.

Value as standover or deferred feed. The crude protein content of the hay is insufficient to meet the requirements of the grazing animal and would need a supplement to improve animal performance (Karue, 1975). Hay cut from a four-month-old stand had 3.4 percent crude protein in the dry matter. It is generally not highly regarded as a pasture (Harrington & Pratchett, 1972).

Chemical analysis and digestibility. A ten-hour intake trial (to simulate a ten-hour grazing day on the range) with Boran cattle at EAAFRO, Muguga, Kenya, gave a dry-matter intake of 70.87 ± 2.57 g/kg $W^{0.75}$ (not significantly different from Herefords). The dry matter of *T. triandra* hay contained 42.35 percent crude protein and 6.20 percent gross energy (Karue, 1975).

The chemical composition of the grass in dry and wet seasons is given by Karue (1974) as percentages of the dry matter in Table 15.76. Botha (1953) recorded 6.9 percent crude protein in the dry matter of fresh, vegetative material and only 2.7 percent in mature, fresh material. The digestibility of the crude protein with sheep was 51.9 percent for fresh, vegetative material and nil for the mature grass.

Palatability. Good when young, unpalatable when mature.

Seed production and harvesting. Seed is usually well formed, but harvesting is difficult as each plant produces a relatively small number of seeds which shed easily when ripe. The spikelets are awned and each contains a single

TABLE 15.76 **Themeda triandra**

	CP	CF	Ash	EE	NFE	DCP	TDN
Dry season	3.6	32.1	11.6	2.0	50.7	1.0	54.2
Wet season	4.6	27.8	17.4	2.4	47.8	1.6	58.0

Source: Karue, 1974

caryopsis. When threshed, the caryopses are mixed with a good deal of chaff and are not easy to separate.

Economics. Themeda triandra is an important grassland constituent of large areas of productive ranching land in the medium altitude-medium rainfall (around 1 000-2 000 m and 500-800 mm respectively) zones of eastern tropical and subtropical Africa.

Animal production. Weight gains of Boran steers were not significantly different at 1.76, 2.8 and 5.2 ha per head and averaged 0.29 kg per day, over one year. This varied seasonally from 0.68 kg per day to nil, with short periods of weight loss. Live-weight gain was less under a three-paddock/one-herd deferred rotation than it was with continuous grazing at 1.76 and 2.8 ha per head (McKay, 1971). In Uganda, Harrington (1973) recorded a live-weight gain of 0.3 kg per head per day for continuous grazing at 0.6 ha per head and 0.4 kg per head per day for continuous grazing at 2.4 ha per head. Karue (1975) estimated from dry matter and crude protein contents that the grass could carry a stocking rate of one 350-kg live-weight animal to 5 ha during the short rains on the Athi River ranch in Kenya, and, during the long rains, one 250-kg live-weight animal plus one 100-kg live-weight calf could be kept on one hectare. A year-long carrying capacity of one 250-kg live-weight animal to 5 ha is usually recommended. If seed were available in quantity, Bogdan and Pratt (1967) recommend its use in mixtures to reseed range at altitudes of about 1 800 m.

Main attributes. Its recovery after fire.

Main deficiencies. Early flowering, variation in palatability within swards, fire susceptibility.

Further reading. Heady, 1966; Marshall & Bredon, 1967; Ndawula-Senyimba, 1972.

Trachypogon spicatus (L.f.) Kuntze

Common names. Horo, danga (southern Africa), greybeard grass (Zimbabwe), arrow grass (Venezuela).
Natural habitat. Rocky grassland, vlei and seasonal swamps in East Africa, often associated with *Loudetia kagarensis*.
Distribution. Central and southeastern Africa; introduced to South America.
Description. A very variable perennial with slender or stout culms, 25-120 cm high. Leaves glabrous or hairy. Raceme solitary (rarely two to three) with pubescent spikelets, 7-8 mm long; the fertile spikelets with 35-70 mm long awns (Napper, 1965). There is a ring of hairs below the culm nodes which distinguishes *Trachypogon* from *Heteropogon contortus*. *Heteropogon* and *Trachypogon* both have velvety awns on the bisexual spikelets, but in *Heteropogon* the awns occur only in spikelets on the upper part of the raceme (Chippendall, 1955) (see Fig. 15.156).
Altitude range. 500-2 750 m; 1 650 m in the Transvaal high veld.
Rainfall requirements. It occurs naturally in the 750 mm annual rainfall area of Transvaal, with a high incidence of summer rainfall.
Genetics and reproduction. 2n=20 (Fedorov, 1974).
Chemical analysis and digestibility. Dougall and Bogdan (1958) recorded 5.7 percent crude protein, 40.2 percent crude fibre, 9.6 percent ash, 1.8 percent ether extract and 42.7 percent nitrogen-free extract in the dry matter of fresh material in early bloom in Kenya.
Palatability. It has low palatability (Verboom & Brunt, 1970).
Economics. A principal component of the drier, subhumid, undisturbed climax grassland of the Transvaal high veld with a summer dominant rainfall of 750 mm, with cold winters and frequent frosts.
Animal production. Live-weight gains occur over the early summer period, but the grasses mature early. In winter the feed value of the grassland is negligible and animals are stall-fed.

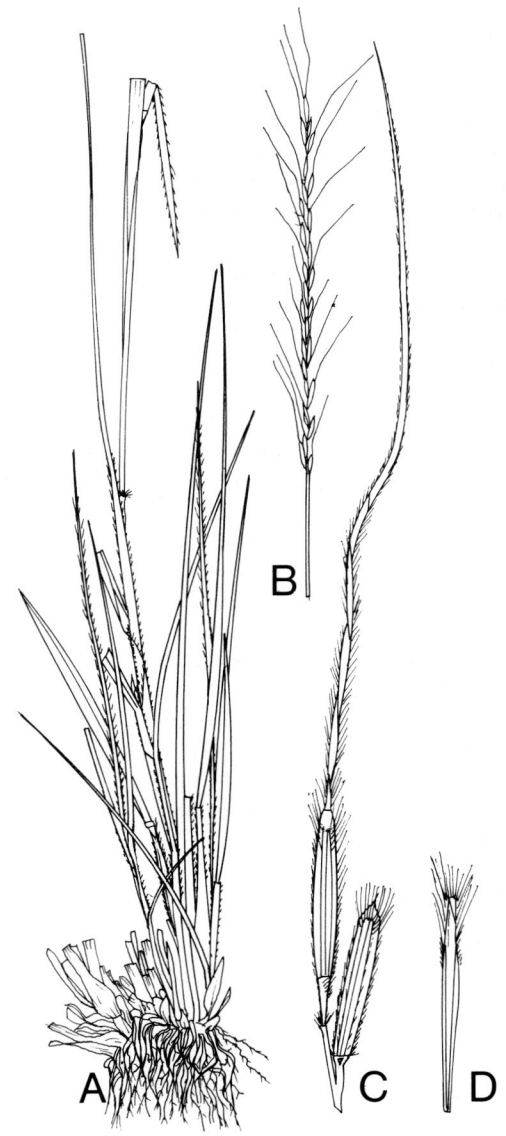

Figure 15.156. *Trachypogon spicatus.*   **A**-Base of plant   **B**-Inflorescence   **C**-Pair of spikelets   **D**-Lemma of fertile flower

Triodia pungens R. Br.

Common name. Soft spinifex (Australia).
Natural habitat. Sand dunes and sand plains.
Description. A somewhat glutinous tussock grass. Leaves rigid, spreading, sharp-pointed, the sheaths sometimes woolly. Panicle narrow, 8-15 cm long. Spikelets on slender pedicels, 8-12 mm long; outer glumes 6-8 mm long, glabrous, five- to seven-nerved; flowering glume purplish, cut halfway down into three broad, three-nerved lobes, silky-villous toward the base (see Fig. 15.157, Plate 56).
Response to fertilizers. T. pungens does not respond to fertilizers. At Yalleroi in central Queensland the dry-matter content of spinifex in the natural pasture fell from 78 percent unfertilized to 63 percent when fertilized with NPK, whereas that of other perennials and annual grasses increased (Edye *et al.*, 1964).
Grazing management. Rotational spelling of paddocks during the wet season every three to four years will allow the softer, edible, associated plants to build up a seed reserve to ensure their continuing presence. Burn every three to four years at the end of the dry season or after the first storms to remove old, dry, spiny material and promote soft growth for grazing (see Plate 57). Burning after the wet season destroys the softer edible plants. Uncontrolled grazing leads to complete removal of vegetation, increasing erosion and permanently reducing productivity (Bishop, 1973).
Response to fire. Annual burning followed by continuous heavy stocking increases the proportion of spinifex at the cost of associated softer edible plants.
Chemical analysis and digestibility. In the dry matter, Siebert, Newman and Nelson (1968) recorded a range of 2.8-4.3 percent crude protein, 27.6-35.8 percent crude fibre, 6.3-9.8 percent ash, 1-13.7 percent ether extract and 48.1-50.6 percent nitrogen-free extract, for two samples of dry material. One analysis of young regrowth (about 18 cm high after burning) yielded 7 percent crude protein in the dry matter.
Palatability. Not very palatable, but eaten in the absence of other forage.
Economics. Soft spinifex is the main constituent of the tussocky spinifex grasslands in arid Australia. It thrives on sand dunes, rocky slopes of laterized desert sandstone ranges and on solodic soils on the dry tropical plains. It is usually burnt by aboriginal tribes; the resin obtained from the burning material has been used to glue handles on stone axes and other implements. After burning, the young regrowth is grazed by cattle and sheep.
Further reading. Bishop, 1973.

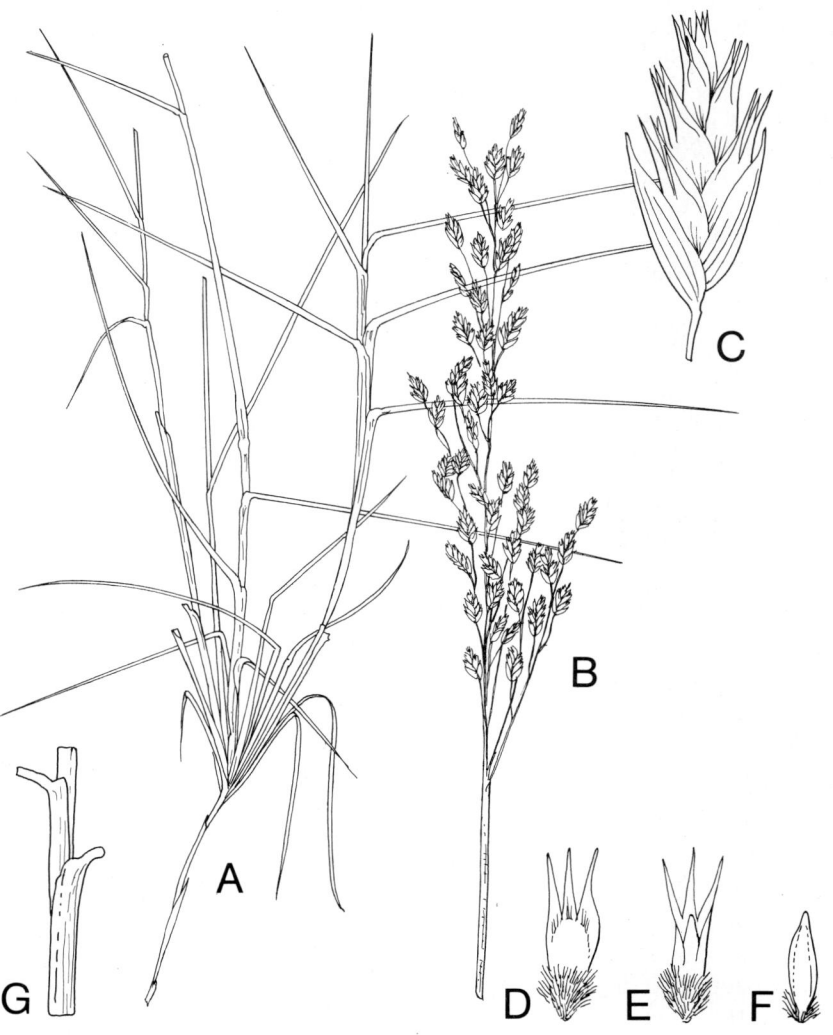

Figure 15.157. Triodia pungens. **A**-Habit **B**-Inflorescence **C**-Spikelet **D,E**-Lemma, back and front **F**-Palea **G**-Leaf sheaths

Tripsacum dactyloides (L.) L.

Common name. Eastern gama grass (United States).
Distribution. Western Hemisphere, United States to Brazil; Malaysia.
Description. Densely clumped grass with short, fibrous, woody rhizomes. Culms oval, stout, woody, solid, to 3-4 m tall, 3-5 cm thick at base, branching erect at centre of clump, geniculate peripherally, stilt-rooting from lower nodes, with a single ring of purple or mauve roots at the node, often growing through the persistent culm sheath; nodes glabrous, 5-14 cm long. Leaf-sheaths overlapping at base, clasping when young, lax and papery when old, often persistent, about 20 cm long; leaf-blades lanceolate-acuminate, to 1.5 m long and 10 cm wide, widest at about two-thirds of its length. Inflorescence to 30 cm long, terminal and axillary, of one to six racemes of unisexual spikelets, female basally for one-third to one-eighth of the length of the raceme, male distally (Gilliland *et al.*, 1971). It differs from *T. laxum* in that the inflorescence is stiff, and the male spikelets are longer (7-8 mm) (see Fig. 15.158).
Season of growth. Summer.
Frost tolerance. It is susceptible to frost.
Rainfall requirements. About 1 000-1 500 mm annually.
Tolerance to flooding. It does not tolerate standing water for long periods.
Soil requirements. It grows best on moist, well-drained, fertile soils.
Number of seeds per kg. 15 000 (United States).
Vigour of growth and growth rhythm. It makes major growth in early spring and stays green until frosts. It seeds from July to September in the United States.
Response to photoperiod. Flowering is accelerated by short days (Evans, Wardlaw & Williams, 1964).
Grazing management. It can be grazed during spring and summer, but deteriorates after frost and provides little winter grazing. Grazing is best if deferred at least 90 days every two to three years, to enable plants to produce seed.
Compatibility with other grasses and legumes. It is usually grown as a pure stand, and inclusion of legumes is difficult.
Response to defoliation. This grass should not be cut closer than 25 cm from the ground.
Grazing management. Cattle have difficulty in biting through the tough midribs of the leaves, and the shallow-rooted stools are easily uprooted. It makes very little growth in dry weather. It is persistent, and stands can be maintained almost indefinitely under sound management (Whyte, Moir & Cooper, 1959). Inter-row shallow cultivation helps control weeds but deep cultivation destroys the shallow roots. It is seldom grazed, but generally cut for soilage or silage at six- to ten-week intervals at a height of 25 cm; it is fertilized with nitrogen as necessary. Generally less productive than elephant grass (*Pennisetum purpureum*) and lower in nutritive value.

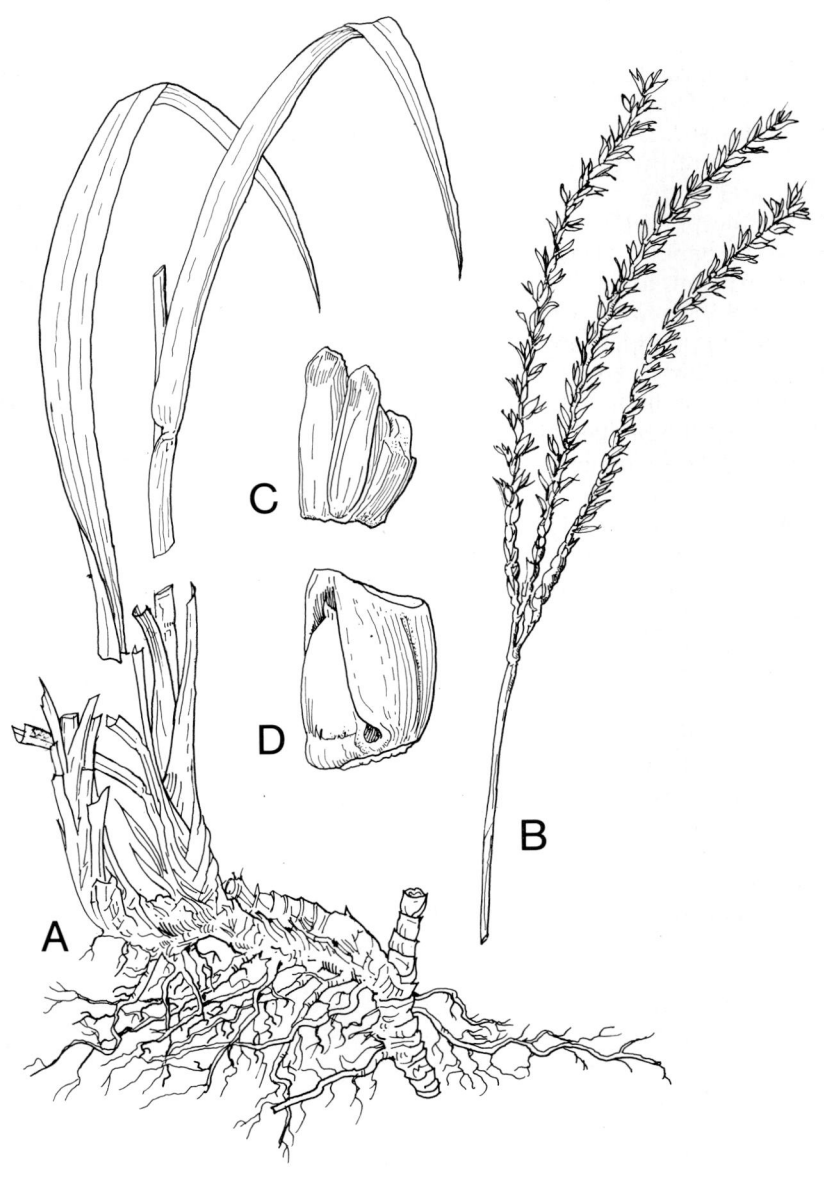

Figure 15.158. *Tripsacum dactyloides.* **A**-Culms and rhizome **B**-Inflorescence **C**-Female spikelets on rachis **D**-Male spikelet on rachis

Despite the above remarks, workers in Suriname found that, after the pasture was grazed for three years with rest periods of two months, a year without grazing put the pasture into excellent condition (Appelman & Dirven, 1972).

Genetics and reproduction. 2n=54, 70, 72 (Fedorov, 1974). It is a diplosporous apomict.

Dry- and green-matter yields. In Suriname it yielded 25 000 kg/ha of green matter in the first year without fertilizer, and 10 000 kg/ha in the second year (Appelman & Dirven, 1972).

Suitability for hay and silage. It is a choice hay plant and is usually managed for hay production in the United States, although no more than 50-60 percent of the current season's growth should be removed at any time during the growing season. For quality hay, it is cut at 15-20 cm when the seed-heads start appearing (Leithead, Yarlett & Shiflet, 1971).

Economics. The grass is being quite extensively planted on rubber estates in Malaysia as a soil conditioner in drained swamps, and for mulching. It provides good fodder (Gilliland *et al.*, 1971).

Further reading. Leithead, Yarlett & Shiflet, 1971.

Tripsacum laxum Scrib and Merr.

Synonym. Tripsacum fasciculatum Trin.
Common names. Guatemala grass, zacate prodigio (Latin America).
Distribution. Mexico and South America; now introduced to Sri Lanka and some other tropical countries, including Fiji.
Description. Culms stout, up to 3 m tall and 2.5 cm thick at base. Leaf-blades broad (to 9 cm wide), sheaths glabrous. Racemes slender, several in a terminal group; one male spikelet of a pair sessile, one pedicelled. It differs from *T. dactyloides* in having a slender inflorescence, and the male spikelets are 4 mm long (Gilliland *et al.,* 1971).
Season of growth. Summer.
Rainfall requirements. Humid areas with rich soils, moist, but well drained.
Drought tolerance. It has poor tolerance to drought.
Tolerance to flooding. It does not tolerate flooding.
Soil requirements. It needs rich soil, but will tolerate acidity and aluminium. In Suriname, it does best on podzolic soils.
Fertilizer requirements. Tripsacum removes 400 kg nitrogen, 80 kg phosphorus, 50 kg potassium, 50 kg calcium and 50 kg magnesium annually per hectare of soil, and so must be adequately fertilized (Risopoulos, 1966).
Sowing methods. Established by planting stem cuttings or rooted culms at the beginning of the rainy season, either in holes or in a plough furrow 50-65 cm apart in 1 m rows (about 10 000 per hectare; Risopoulos, 1966).
Vigour of growth and growth rhythm. Optimum production is reached six months after planting the cuttings, with four months between harvests.
Suitability for silage. Andrew (1971) states that it is capable of very high production. It makes useful silage (Boudet, 1975; Medling, 1972; Assis *et al.,* 1962). It lost 12 percent of its dry matter during ensilage (Paterson, 1945).
Chemical analysis and digestibility. This coarse tropical grass contains less than half as much digestible crude protein, and approximately three-quarters as much starch equivalent, as the fine grasses of the temperate zone. Harrison (1942) showed Guatemala grass cut at six weeks to contain 20 percent dry matter, 1.3 percent digestible crude protein and 7.9 percent starch equivalent. Göhl (1975) records numerous analyses and some digestibility figures from Suriname, the Philippines, Trinidad, Puerto Rico and Malaysia in Table 15.77.
Seed production and harvesting. It does not flower readily, and seed production is unusual except in its native habitat.
Economics. A good fodder plant, and much used in Sri Lanka as a soil binder and organic-matter builder in upland tea estates (Bor, 1960; Andrew, 1971). It is also used as a fodder grass in Fiji, Suriname, Malaysia and Puerto Rico.

It is more persistent than elephant grass (*Pennisetum purpureum*) but less productive and of lower nutritive value.
Animal production. It is used by dairy farmers in Fiji as green chop-chop for zero grazing (Roberts, 1970a, b). On the podzolic soils of the Lelydrop landscape in Suriname, planted cuttings of *T. laxum* after three years' grazing at intervals of two months gave a live-weight increase of 300 kg/ha, with a live-weight gain of 278 g per head per day over ten months (Appelman & Dirven, 1972).

In Brazil, a mixed silage of 50 percent *Tripsacum laxum,* 30 percent

TABLE 15.77 ***Tripsacum laxum***

	DM	As % of dry matter				
		CP	CF	Ash	EE	NFE
Fresh, 3 weeks, Suriname		15.9	31.4	9.6	2.8	40.3
Fresh, 4 weeks, Suriname		12.7	33.5	9.6	1.7	42.5
Fresh, 5 weeks, Suriname		10.9	33.2	8.8	1.4	45.7
Fresh, 6 weeks, Suriname		7.3	33.4	7.0	2.4	49.9
Fresh, 7 weeks, Suriname		7.1	35.9	6.5	2.4	48.1
Fresh, 8 weeks, Suriname		7.5	35.2	6.7	2.0	48.6
Fresh, 120 cm, Philippines	25.3	5.9	36.0	8.7	2.0	47.4
Fresh, mature, Trinidad	20.3	7.8	33.2	6.3	1.5	51.2
Fresh, 1st cut, fertilized, Puerto Rico	24.6	5.2	35.6	8.7	2.9	47.6
Fresh, 2nd cut, fertilized, Puerto Rico	30.4	4.6	31.2	8.2	2.7	53.3
Fresh, 8 weeks, Malaysia	20.0	12.0	35.0	14.0	1.5	37.5
Fresh, 10 weeks, Malaysia	19.0	8.4	34.7	15.8	1.0	40.1
Fresh, 12 weeks, Malaysia	19.5	5.1	35.9	16.4	1.5	41.1
Silage, chaffed, molasses, 9 litres/tonne	21.7	7.1				
Artificially dried, Suriname	88.7	9.3	37.5	5.3		
	Animal	Digestibility (%)				
		CP	CF	EE	NFE	ME
Fresh, 120 cm	Sheep	51.8	60.7	62.3	61.2	2.07
Fresh, mature	Sheep	50.5	69.7	46.7	63.9	2.24
Fresh, first cutting	Sheep	56.0	66.0	74.0	65.0	2.26
Fresh, second cutting	Sheep	58.0	60.0	74.0	72.0	2.33
Artificially dried, Suriname	Sheep	54.2	58.8			

Source: Göhl, 1975

Lablab niger and 20 percent *Saccharum officinarum* decreased milk yield by 10 percent, compared with maize silage; a *T. laxum* and *S. officinarum* silage reduced yield by 19 percent (Assis *et al.*, 1962).
Main attributes. Its high yield and persistence.
Main deficiency. Its poor seed production.
Further reading. Appelman & Dirven, 1972.

Urochloa mosambicensis (Hack.) Dandy

Synonym. Echinochloa notabile (Hook. f.) Rhind.
Common names. Sabi grass (Australia), gonya grass (Zimbabwe), common urochloa (southern Africa).
Natural habitat. In grassland, usually in sheltered places, or in disturbed areas.
Distribution. Southern Africa, East Africa, Burma.
Description. A perennial, variable in size and habit (Burt *et al.*, 1980) sometimes stoloniferous or with a creeping rhizome. Culms 120 cm or more high, sometimes rooting and branched from the lower nodes. Leaf-sheaths with a ring of soft hairs at the nodes; leaf-blades 18 mm wide, hairy. Inflorescence up to 15 cm long of four to 12 racemes, 2.5-9 cm long; spikelets 3-5 mm long, acuminate or shortly awned (Chippendall, 1955). It is distinguished from *U. panicoides* in having a tubercle-based bristle in the middle of the lower lemma in the fresh state (Whiteman & Gillard, 1971) (see Figs. 15.159, 15.160).
Season of growth. Summer.
Optimum temperature for growth. It can withstand high temperatures (Whyte, 1964).
Frost tolerance. It makes good winter growth, but is checked by frost. Light frosts did not affect it at Yarrowmere near Pentland in inland north Queensland (Burt, personal communication).
Latitudinal limits. Collected as far south as 25°S at 1 000 m elevation in Africa (Whiteman & Gillard, 1971).
Altitude range. 700-1 000 m, but more adapted to the lower end of this range.
Rainfall requirements. It needs a rainy season of 10 to 16 weeks in summer with an annual rainfall of 600-1 200 mm and a five- to nine-month dry season. It responds well to early wet season storms.
Drought tolerance. It is drought enduring.
Tolerance to flooding. It does not tolerate flooding (Anderson, 1970a, b).
Soil requirements. It will grow in a wide range of soils, from clay loams to sands, but appears to be more suitable for lighter soils with relatively high fertility. It can tolerate both acid and alkaline soils.
Tolerance to salinity. All species of *Urochloa* in India show high sodium content.
Fertilizer requirements. It may need a complete fertilizer for establishment, but it responds well to phosphorus and nitrogen. Weier (1977) has shown that it has high nitrogenase activity associated with its roots and can fix nitrogen.
Ability to spread naturally. It spreads well and becomes dominant in northern Australia after fires.
Land preparation for establishment. A well-prepared seed-bed is preferable.
Sowing methods. In India it is sown by seed or rooted slips. Seed is surface-

sown on to a fine seed-bed with 200 kg/ha superphosphate. Oversowing into natural pastures where the soil fertility has been improved gave *Urochloa* dominance after four years (Gillard, 1971).

Sowing depth and cover. It is surface-sown, and preferably rolled afterwards.

Sowing time and rate. Summer, at 1-6 kg/ha; or 2 kg/ha grass with 5-6 kg/ha *Stylosanthes humilis* (north Queensland).

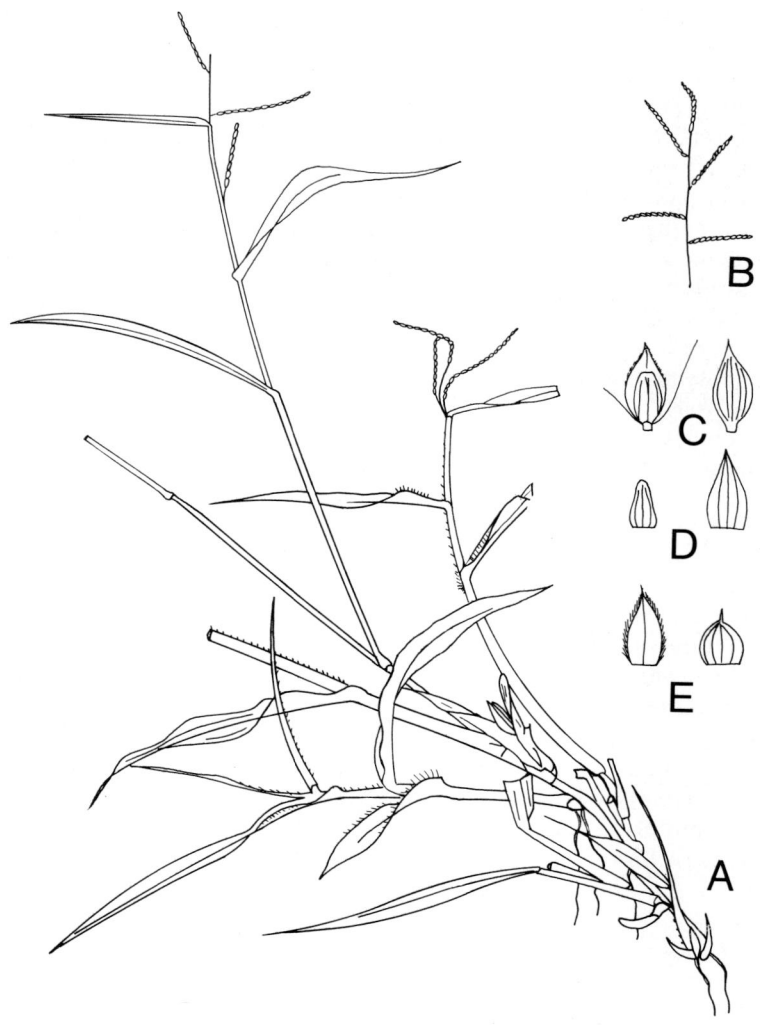

Figure 15.159. Urochloa mosambicensis.   **A**-Habit   **B**-Inflorescence   **C**-Spikelets **D**-Glumes   **E**-Lemmas

Figure 15.160. Urochloa mosambicensis (**Source:** D.R. Younger, Queensland Department of Primary Industries)

<u>Number of seeds per kg.</u> 60 600 (Queensland).
<u>Dormancy.</u> There is post-harvest dormancy, but after six to 12 months' storage germination is satisfactory (Harty, 1972), the delay being due to physical obstruction of the embryo by the enclosing lemma and palea. It germinates at 20-30°C, moistened with KNO_3 solution.
<u>Seedling vigour.</u> Good.
<u>Vigour of growth and growth rhythm.</u> Flowering usually occurs continuously from three to four weeks after the wet season begins, and the first seed is ripe in mid-December (Queensland). It can make good growth ahead of its companion legume.
<u>Response to photoperiod.</u> It is day neutral.
<u>Compatibility with other grasses and legumes.</u> It combines with *Stylosanthes humilis*, but must be carefully managed. Both *S. humilis* and *S. hamata* cv. Verano were successfully re-established in *Urochloa mosambicensis*-dominant pasture on Blain sandy loam by disc-harrowing and oversowing (Falvey, 1979).
<u>Ability to compete with weeds.</u> Good.
<u>Response to defoliation.</u> It stands quite heavy grazing.

Grazing management. At Katherine Research Station, northern Australia (lat. 14°35'S, rainfall 950 mm), heavy wet season grazing at the rate of 3.3 steers per hectare had no effect on the stand; continuous year-round grazing at 1.7 steers per hectare allowed the stand of *Urochloa* to increase slightly (Evans & Winter, 1976).

Response to fire. It recovers well from fire in northern Australia (Falvey, 1979).

Genetics and reproduction. The chromosome numbers are 28, 30, 42 (Fedorov, 1974). It is an aposporous apomict (Pritchard, 1970).

Dry- and green-matter yields. In India, Singh and Chatterjee (1968) recorded a dry-matter yield of 4 000 kg/ha when fertilized with 34 kg P and 44 kg N/ha (lat. 23°N, altitude 625 m, rainfall 1 320 mm). Good yields were also obtained in Zimbabwe under a 500 mm rainfall. In northern Australia in association with *Stylosanthes humilis* it yielded from 800 to 6 520 kg DM/ha (grass component), the higher figure being obtained with a fertilizer application of 25 kg P and 85 kg N/ha (lat. 19°S, altitude 50 m, rainfall 860 mm). The legume only contributed 25 kg/ha. Where both components yielded well (at 14°S, 200 m altitude, 800 mm rainfall) the grass yielded 4 117 kg DM/ha and the legume 3 438 kg/ha (Whiteman & Gillard, 1971). In Thailand, in association with *S. hamata* cv. Verano and native grasses, it produced a total of 11 245 kg/ha per year of which 8 208 kg/ha was stylo and 1 561 kg/ha sabi grass (Humphreys, 1978). In Fiji it yielded 73 tonnes DM/ha when fertilized with 450 kg/ha superphosphate; 23 percent of the yield was made in the dry season (Partridge, 1979a).

Suitability for hay and silage. It makes good hay.

Chemical analysis and digestibility. See Table 15.78.

Palatability. Very good, even when dry (Göhl, 1975). Voluntary intake was 50 percent better than that of *Heteropogon contortus* under similar conditions in Australia (Whiteman & Gillard, 1971).

TABLE 15.78 **Urochloa mosambicensis**

	DM	As % of dry matter				
		CP	CF	Ash	EE	NFE
Fresh, 1st cut, India		11.0	22.4	28.2	2.0	36.4
Hay, Zimbabwe	88.5	14.9	26.1	12.9	1.5	44.6

	Animal	Digestibility (%)				
		CP	CF	EE	NFE	ME
Hay, Zimbabwe	Cattle	45.6	57.1	62.9	65.1	1.96

Source: Göhl, 1975

Toxicity. No toxicity has been recorded by Everist (1974).

Seed production and harvesting. It is free-seeding, and three to five harvests per season may be obtained. It is harvested by direct heading.

Seed yield. Up to four direct-headed crops per year have been obtained in northern Australia. Seed yield is 100-130 kg/ha per year.

Minimum germination and quality required for commercial sale. 60 percent purity and 3 percent germination.

Cultivars. 'Nixon' — described above. Derived from CPI 6559 introduced from Harare, Zimbabwe, and developed by B. Nixon at Katherine, northern Australia.

Value for erosion control. It has been used successfully in India.

Diseases. It is not subject to major diseases.

Pests. It has no major pests.

Animal production. In the Tipperary region of northern Australia, under set stocking at one beast to 0.3-0.6 ha, cv. Nixon plus *S. humilis* has given up to 450 kg live-weight gain/ha per year (Austin, 1970). The pasture was sown with 200 kg/ha superphosphate and grazed at stocking rates of one beast to 0.4, 0.8 and 1.6 ha. When production had stabilized in the third year of the trial, respective live-weight gains were 360, 154 and 76 kg/ha in 344 days. The much better performance at the higher stocking rate was due to the much higher proportions of *S. humilis* maintained in the pasture. At the start of the experiment the pastures had a uniform 25 percent legume content, but two years later the legume proportions were 75, 35 and 8 percent for the 0.4, 0.8 and 1.6 ha per animal stocking rates respectively (Whiteman & Gillard, 1971).

Main attributes. Its quick growth and free-seeding habit, providing a quick early cover on overgrazed pastures, or when overseeding with cultivation.

Main deficiencies. It may be too palatable. It may not grow on heavy clays.

Further reading. Harty, 1972; *Urochloa, Urochloa mosambicensis*, 1974; Whiteman & Gillard, 1971.

Urochloa oligotricha (Fig. and De Not.) Henr.

Synonym. *U. bolbodes* (Steud.) Stapf.
Common name. Dubi grass.
Natural habitat. Open woodland, roadsides, and as a weed.
Distribution. Southern, Central and northeast tropical Africa.
Description. Tufted perennial, 30-60 cm high. Leaf blades 6-9 mm wide. Racemes dense, few to 20, spreading, on a finely-hairy axis; spikelets crowded, acuminate, 3-4 mm long, glabrous or with fringed margins (Napper, 1965). It is more densely tufted and robust than *U. mosambicensis*. It has a stout, short, creeping rhizome and the culms are sometimes bulbous at the base (Chippendall, 1955). One extreme form has been found to have a stolon of reasonable length (Burt, personal communication) (see Fig. 15.161).
Season of growth. Summer, but responds to out-of-season rains (Burt, personal communication).
Frost tolerance. Tolerates light frosts at Yarrowmere near Pentland, North Queensland.
Latitudinal limits. In Africa, Swaziland to Kenya; 14-20°S.
Altitude range. 1 250-1 800 m.
Rainfall requirements. It grows over a wide range, from 750 to 3 750 mm annually with various accessions.
Drought tolerance. It is a promising species for the low-rainfall tropics and has high drought tolerance (Burt, personal communication).
Tolerance to flooding. No record has been found.
Soil requirements. It seems to prefer lighter soils near Lake Victoria in Kenya, but appears to tolerate loams; it may be less successful on heavy clays.
Tolerance to salinity. It thrives on many alkaline soils in the Caribbean (Burt, personal communication).
Fertilizer requirements. According to soil test results.
Ability to spread naturally. It does not spread as readily as *U. mosambicensis* cv. Nixon, but better than *Cenchrus ciliaris*. It spreads by seed.
Land preparation for establishment. It does not need a fully prepared seedbed.
Sowing methods. It can be drilled or broadcast.
Sowing depth and cover. It gives satisfactory stands when broadcast, but it is preferable to roll afterwards.
Sowing time and rate. Just before the wet season, at 1.5-2 kg/ha.
Number of seeds per kg. 900 000 to 1.2 million (Queensland).
Dormancy. No dormancy problems have appeared in north Queensland.
Seed treatment before planting. Seed can be dusted with lindane to deter seed-harvesting ants.
Tolerance to light. It will grow under shade.

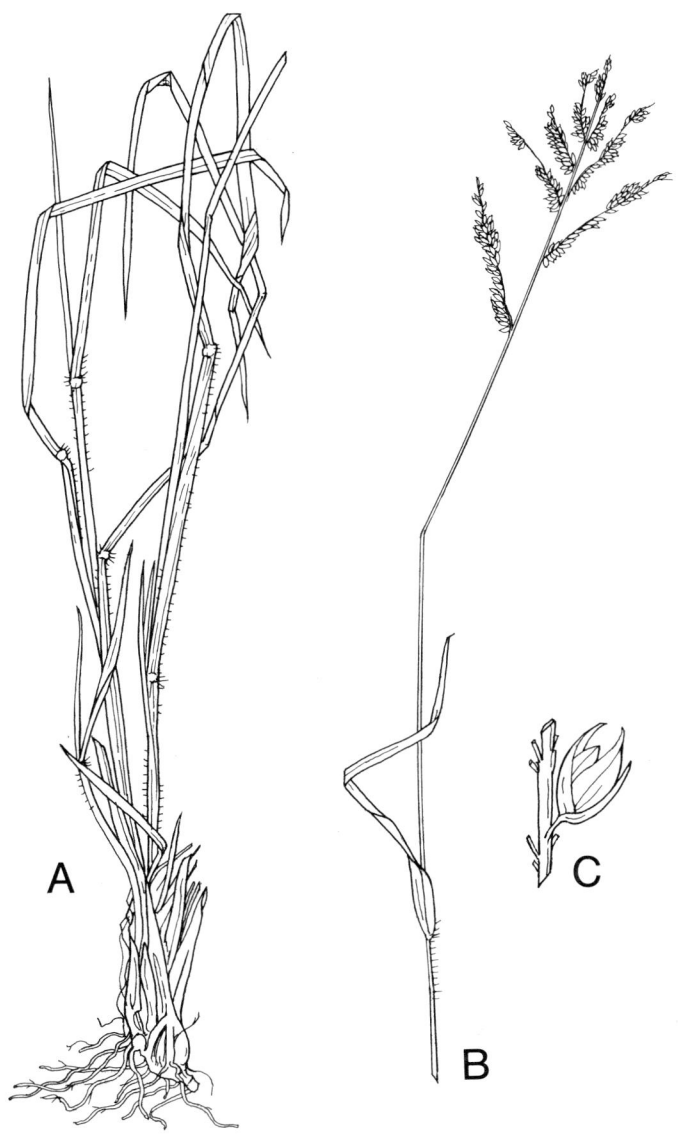

Figure 15.161. Urochloa oligotricha. **A**-Habit **B**-Inflorescence **C**-Spikelet

TABLE 15.79 *Urochloa oligotricha*

	DM	As % of dry matter				
		CP	CF	Ash	EE	NFE
Fresh, early bloom, Kenya		13.7	30.0	10.6	2.8	42.9
Hay, Zimbabwe	89.3	17.7	24.6	12.9	2.0	42.8

	Animal	Digestibility (%)				
		CP	CF	EE	NFE	ME
Hay, Zimbabwe	Cattle	37.6	53.0	66.2	65.6	1.88

Source: Göhl, 1975

Compatibility with legumes. Legumes will invade stands of *U. oligotricha*.
Genetics and reproduction. 2n=36 (Fedorov, 1974). It is apomictic (Brown & Emery, 1958).
Dry- and green-matter yields. In a preliminary trial with *U. oligotricha* CPI 45608 (in Cape York Peninsula, on a very poor yellow earth supporting low closed heath, with an average rainfall of 1 700 mm) Winter (1976) obtained a dry-matter yield of 5 780 kg/ha grass by the third year, compared with 15 965 kg/ha from 'Basilisk' signal grass.

U. oligotricha cut at 5 cm yielded 57 500 kg/ha of green matter when cut monthly, 67 000 kg/ha with bi-monthly cuts and 66 800 kg/ha when cut every three months (Semple, 1956).

Yields were poor at an altitude of 1 900 m with rainfall of 1 150 mm at latitude 1°N in Kenya, but good in Zimbabwe at 19°S, 200 m altitude and 500 mm rainfall (Whiteman & Gillard, 1971).

Schofield (1944), at South Johnstone, north Queensland, harvested 33 490 kg/ha of green matter per year when cut at two-month intervals and 33 600 kg/ha when cut at three-month intervals. The grass yielded 976.64 kg protein per hectare per year and 114.2 kg CaO and 99.9 kg P_2O_5 per hectare in the first 12 months.
Chemical analysis and digestibility. See Table 15.79.
Palatability. Very palatable.
Main attributes. Its good green leaf retention, rapid regrowth and high acceptability to animals; its wide ecological range and compatibility with legumes. Its ability to grow on alkaline soils and to spread by seeds.
Main deficiencies. May have seed production problems, and may not suit cracking clays.
Further reading. Whiteman & Gillard, 1971.

Urochloa panicoides Beauv.

Synonyms. U. *helopus* (Trin.) Stapf; *Panicum controversum* Steud.
Common names. Liverseed grass (Australia), Kuri millet (Zimbabwe), garden urochloa (South Africa).
Natural habitat. Sandy soils and loams, grassland and as a weed.
Distribution. Tropical Africa, Australia, Fiji and India.
Description. A tufted annual with culms 6-60 cm high, sometimes decumbent and rooting from the lower nodes, usually with flowering branches from several of them. Leaves usually loosely to densely hairy with tubercle-based hairs, blades up to 12 mm wide, expanded, light green, margins thickened and crinkled. Inflorescence up to 8 cm long of two to seven racemes up to 6 cm long; spikelets 4-5 mm long, solitary and almost sessile, forming two regular rows; glumes unequal, the lower one-quarter to one-third as long as the spikelet. This small lower glume readily distinguishes it from other South African species (Chippendall, 1955). Variety *panicoides* has glabrous spikelets, var. *pubescens* has pubescent spikelets (Simon, 1980) (see Fig. 15.162).
Season of growth. Summer.
Optimum temperature for growth. 25-40°C.
Minimum temperature for growth. About 15°C.
Frost tolerance. It is susceptible to frost; the leaves become brittle.
Altitude range. Sea-level to 1 750 m.
Rainfall requirements. It is best suited to a moderate annual rainfall of 675-800 mm.
Drought tolerance. Not very good; it grows vigorously in wet summers, seeds and dies out.
Tolerance to flooding. It will not survive the wet season in Fiji.
Soil requirements. It occurs mainly on black cracking clays in Queensland's Darling Downs and northern New South Wales.
Tolerance to salinity. No record has been found. It grows on soils of pH 7.0-7.5 on the Darling Downs, Queensland.
Fertilizer requirements. It is usually not fertilized on the Darling Downs, Queensland, which is inherently quite a fertile area.
Ability to spread naturally. In Queensland it seeds heavily and spreads rapidly by seed; especially vegetating overgrazed or bare areas on black clay downs.
Land preparation for planting. It needs some soil disturbance and a loose mulch in which the seed can germinate. Full seed-bed preparation is preferred.
Sowing methods. Drilled with suitable seed drill, or broadcast.
Sowing depth and cover. Surface sown or to depths of 1-1.5 cm and rolled after planting; a press wheel behind the seed chute can also be used.
Number of seeds per kg. 497 000.

Tropical Grasses

Dormancy. The seed requires some post-harvest ripening and should be held for 13 weeks at 34.5-44.9°C (Harty, 1972). Seed germinates at 20°C, moistened with KNO_3 solution.
Seedling vigour. Very vigorous.
Vigour of growth and growth rhythm. It germinates in spring and early summer and grows during summer. It seeds in late summer and dies out in autumn.
Response to light. In its native habitat it occurs in damp places or partial shade.

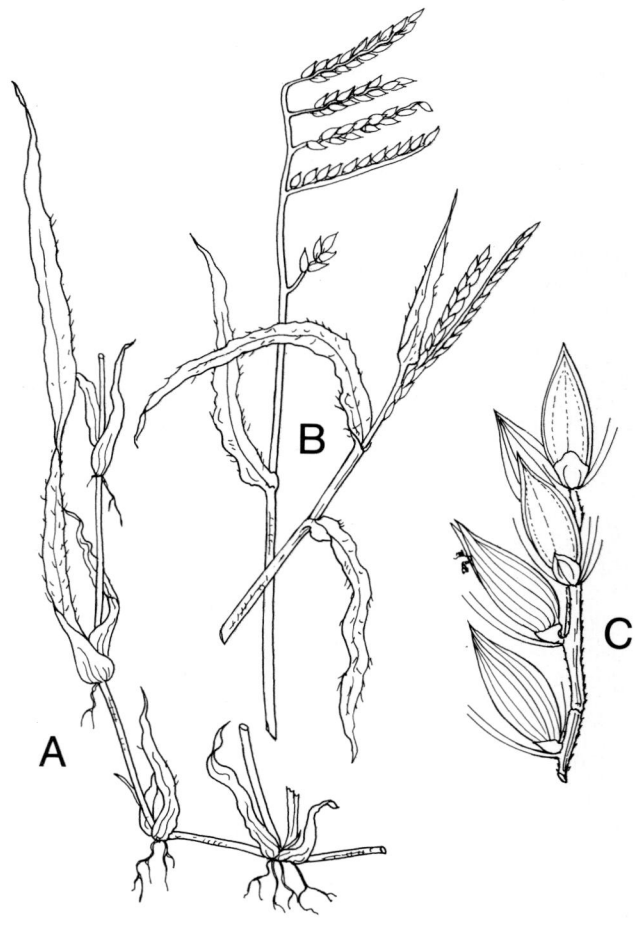

Figure 15.162. Urochloa panicoides. **A**-Habit **B**-Inflorescences **C**-Cluster of spikelets

Compatibility with other grasses and legumes. It generally dominates the areas it occupies during its life cycle.

Ability to compete with weeds. It can suppress annual weeds. It was used on the Darling Downs, Queensland, to vegetate overgrazed and bare areas to exclude *Salvia reflexa* and *Xanthium spinosum*.

Tolerance to herbicides. It can itself become a weed; pasture with *U. panicoides* often requires an additional cultivation of land being fallowed for winter cereals. One litre of 50-percent 2,4-D amine per hectare will destroy seedlings.

Response to defoliation. It gives only limited summer feed and does not stand constant grazing.

Grazing management. It is usually grazed only when necessary.

Response to fire. When it is dry enough to burn it has ceased its vegetative growth for the season.

Genetics and reproduction. The somatic chromosome number is 2n=30, 36 (Fedorov, 1974).

Dry- and green-matter yields. Good yields have been reported, without supporting figures, from India (665 mm rainfall) and Hawaii (1 500 mm rainfall); moderate to poor yields have been harvested in Fiji, southern Africa and Kenya (Whiteman & Gillard, 1971).

Suitability for hay and silage. In India, hay made from *U. panicoides* was able to maintain experimental sheep in positive nitrogen balance, and provided maintenance.

Value as standover or deferred feed. It is quite useful if allowed to stand over, and it is not adversely affected by wet weather. Being an annual it does not last long in the mature state.

Chemical analysis and digestibility. In India, Sen and Ray (1964) recorded 5.8 percent crude protein, 33.3 percent crude fibre, 13.2 percent ash, 1.4 percent ether extract, 46.3 percent nitrogen-free extract, 0.51 percent calcium and 0.31 percent phosphorus in the dry matter.

Dougall and Bogdan (1960) in Kenya recorded 14.7 crude protein, 29.9 percent crude fibre, 14.9 percent ash, 1.7 percent ether extract and 38.8 percent nitrogen-free extract in the dry matter of fresh material in late bloom. Milford (1960a, b) showed it to contain 14 percent crude protein in the young growth, dropping to 5 percent in dry and frosted material during the winter, with digestibility ranging from 57 to 34 percent.

Palatability. It is very palatable.

Toxicity. It has been reported as occasionally toxic; nitrate content of 0.88-4.9 percent potassium nitrate-equivalent in the dry matter has been recorded. Grass with levels of nitrate exceeding about 1.5 percent potassium nitrate-equivalent is potentially toxic (Everist, 1974).

Seed production and harvesting. Liverseed grass seeds heavily and could be harvested by pick-up harvester.

Minimum germination and quality required for commercial sale. 70 percent pure seed, and 20 percent minimum germination with a maximum of 29.5 percent inert material in Queensland.

Cultivars. There are no registered cultivars.

Value for erosion control. It was considered valuable for controlling wind and water erosion from the black clays of the Darling Downs, Queensland, but it has since been replaced by perennial grasses such as *Cenchrus ciliaris,* and it has tended to become a weed in land bare-fallowed for wheat. It is used in southern Africa to give a quick cover on denuded grassland.

Animal production. During four summer months *U. panicoides* maintained 27 sheep per hectare. This was substantially more than *Chloris gayana* but less than *Paspalum scrobiculatum* (Paltridge, 1955).

Main attributes. Its colonization of overgrazed and bare black clays to control *Salvia reflexa* and *Xanthium spinosum* in Queensland; its feed value (Milford, 1960a, b).

Main deficiencies. Its annual nature and its heavy seeding; its good germination of seedlings means that pastures may require an extra cultivation to eliminate *U. panicoides* from summer grain crops.

Further reading. Smith, 1940; Whiteman & Gillard, 1971; Whittet, 1965.

Vetiveria zizanioides (L.) Nash

Synonym. Andropogon zizanioides.
Common names. Khas-khas grass (Africa), vetiver (Europe), lacate violeta.
Distribution. Throughout Africa, India, Burma, Sri Lanka, Southeast Asia.
Description. Tall, stout perennial with an oblong panicle over 30 cm long which has whorled branches bearing spikelets 5-6 mm long, with a few tubercle-based short bristles (Napper, 1965). Under cultivation, the species does not flower (Chippendall, 1955) (see Fig. 15.163).
Altitude range. 300-1 250 m.
Rainfall requirements. 500-5 000 mm in India.
Tolerance to flooding. Good; it occurs on poorly-drained lands.
Soil requirements. It will grow on sandy loams to clay soils, on strongly acid to slightly alkaline soils with a pH range from 4-7.5, but prefers neutral to slightly alkaline soils.
Tolerance to herbicides. In India it can be controlled with dalapon at 11-17 kg/ha, or bromacil at 17-33 kg/ha (Ray, Agarawala & Fridrickson, 1975).
Response to defoliation. It stands very heavy grazing, especially in semi-arid areas of India.
Grazing management. It is usually burnt, and the tender regrowth grazed. The older leaves are too harsh for fodder (Gilliland et al., 1971).
Response to fire. Good.
Genetics and reproduction. 2n=20 (Darlington & Janaki, 1945).
Value for erosion control. It has proved useful for erosion control on 20° slopes in India (Misra, Ambasht & Singh, 1977) and is one of the most important soil-binding grasses in Fiji (Parham, 1955).
Economics. The aromatic roots are a source of vetiver oil, used chiefly in perfumery. In some Asian countries the roots are woven into coarse mats and hung in front of doors; they are moistened to cool and scent the air blowing through them (Chippendall, 1955). In India they have been used for matting to give fragrance to a room; they also yield a heavy essential oil, khas-khas or cuscus for perfumery. The roots are also used to provide an important ingredient in curry, khasu-khasu.

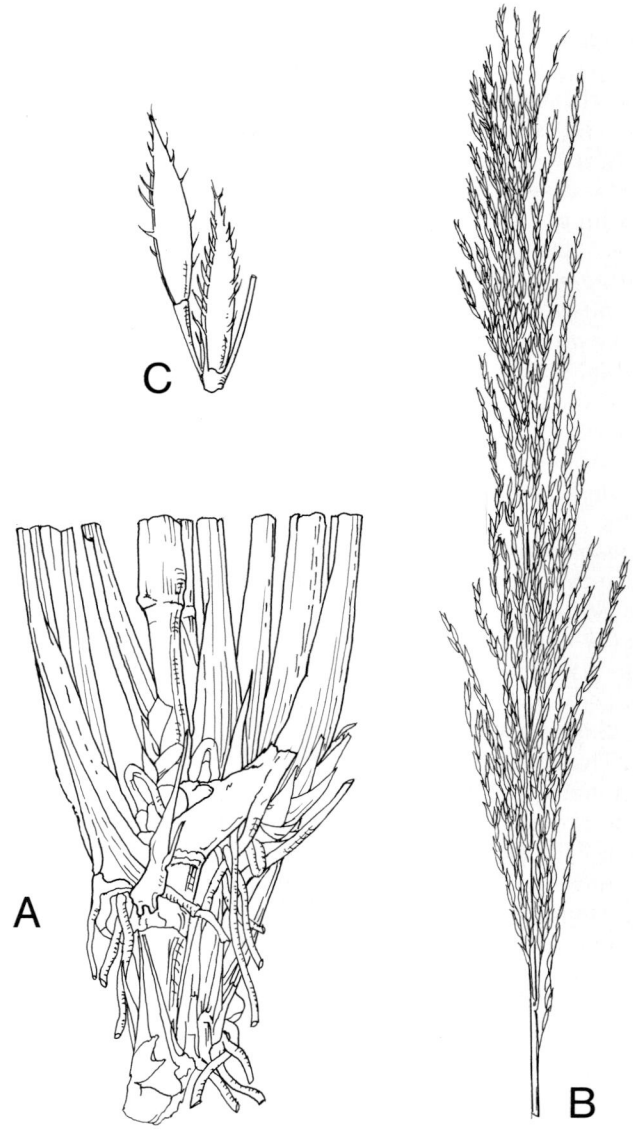

Figure 15.163. Vetiveria zizanioides. **A**-Base of plant **B**-Inflorescence **C**-Portion of raceme with one sessile and one pedicelled spikelet

Vossia cuspidata (Roxb.) Griff.

Natural habitat. Rooting on river banks and lake shores, and growing out over the water; it forms dense mats.
Distribution. Throughout tropical Africa and Southeast Asia.
Description. Perennial with submerged or floating glabrous culms. Racemes rarely seen, two to six on a short axis, or solitary 15-22 cm long; sessile spikelets up to 10 mm long, lower glume of both spikelets with a winged tail 5-30 mm long, rarely shorter (Napper, 1965). It usually grows in water up to 1 m deep at the margins of water holes, but is sometimes a floating grass; it then develops large numbers of roots and rootlets from the nodes of the spongy stem (Bor, 1960) (see Fig. 15.164).
Altitude range. 800-1 250 m.
Chemical analysis and digestibility. Verboom and Brunt (1970) recorded values as shown in Table 15.80.
Toxicity. In Zambia, scouring occurs when cattle move from the fibrous forest grazing to the rich plains grasses consisting of *Echinochloa pyramidalis, E. scabra, Acroceras macrum, Hemarthria altissima, Leersia hexandra* and *Vossia cuspidata,* and it may be three to four months before they regain condition (Verboom & Brunt, 1970).
Economics. Vossia cuspidata is the characteristic grass of the sump areas of the African flood plains which become more deeply flooded, and where water lies later into the dry season. This species forms dense, semi-floating beds, but it may not flower unless sufficient depth and duration of flooding prevails. Established *Vossia* stands persist for an indefinite period under less than optimal conditions. The foliage remains green long after the rains are over and the mat may not get burnt every year. It provides a favourite pasture for the heavier herbivorous animals which trample down the dense growth. Fresh shoots subsequently grow up from ground level and these form an excellent dry-season pasture. *Vossia* fringes the open water of deep pools, and when these dry up the exposed bare mud is colonized during the dry season by annual rosette herbs (Vesey-Fitzgerald, 1963).
Animal production. No quantitative figures have been cited.
Further reading. Vesey-Fitzgerald, 1963.

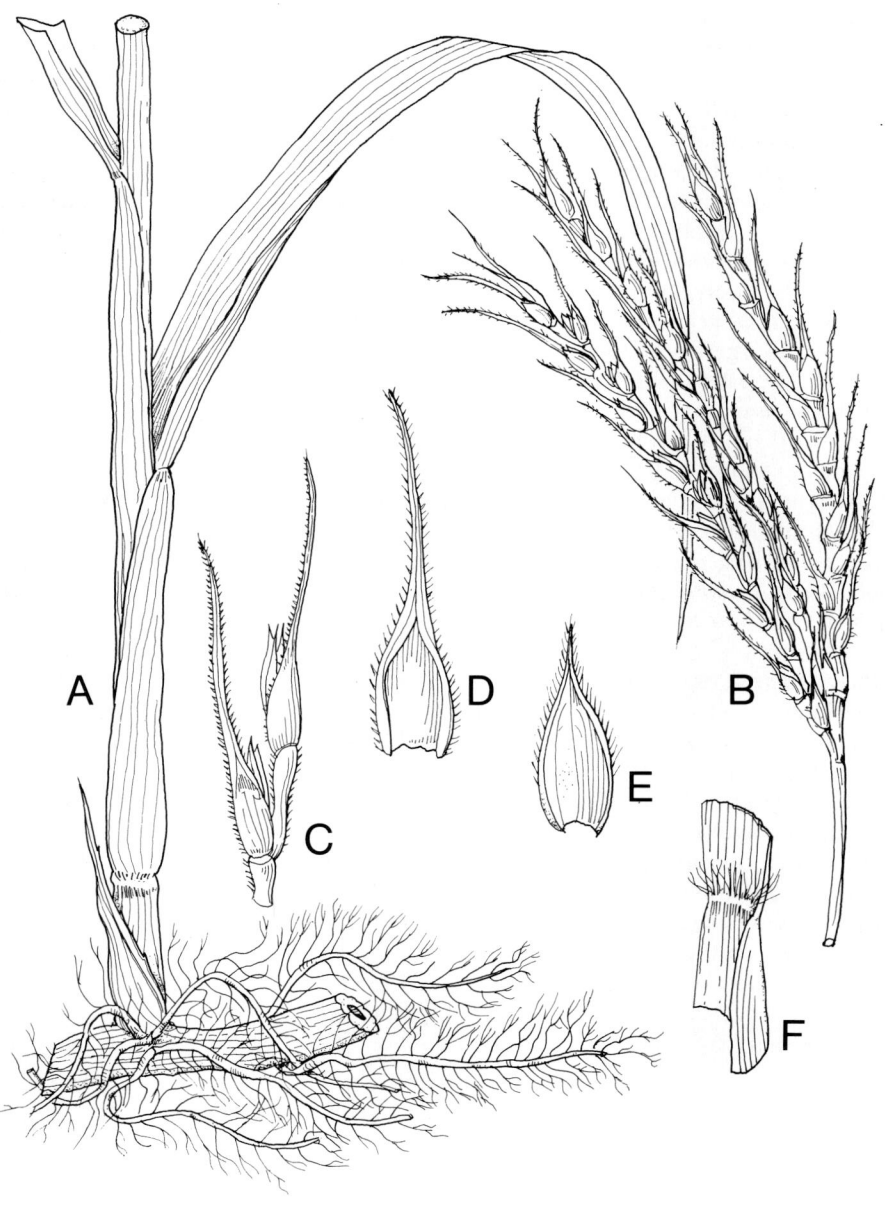

Figure 15.164. Vossia cuspidata. **A**-Base of plant **B**-Inflorescence **C**-Spikelet cluster **D**-Lower glume **E**-Upper glume **F**-Ligule

TABLE 15.80 *Vossia cuspidata*

	As % of dry matter										
	CP	CF	Ash	EE	NFE	Ca	Mg	Na	P	K	Cl
Young shoots	14.6	33.5	7.8	1.8	50.2						
Flowering	3.57	35.08	5.94	1.46	53.95	0.27	0.15	0.04	0.07	0.45	0.36

Source: Verboom & Brunt, 1970

Zea mays L.

Common names. Maize (Australia, United Kingdom), corn, sweet corn (United States), bok'olo (Ethiopia).
Distribution. Originated in Mexico or Central America, now pan-tropical; also grown as a summer crop in temperate Europe.
Description. A coarse annual, culms 60-80 cm high, straight, internodes cylindrical in the upper part, alternately grooved on the lower part with a bud in the groove. The stem is filled with pith. Leaf-blades broad. Has separate staminate (male) and pistillate (female) inflorescences. The staminate inflorescence is a tassel borne at the apex, the pistillate flowers occur as spikes (cobs) rising from axils of the lower leaves. The ovary develops a long style or silk which extends from the cob and receives the pollen from the tassel (see Fig. 15.165).
Season of growth. Summer.
Optimum temperature for growth. Peak germination was at 20-30°C and growth at 18-21°C (Hughes, 1979).
Minimum temperature for growth. 8.7°C for Kitale hybrids in Kenya.
Frost tolerance. It is very susceptible to frosts.
Latitudinal limits. It has a wide range, from 58°N in Canada to 40°S.
Altitude range. The period of flowering and to maturity varies greatly in East Africa. Allan (1973) divides Kenya into four zones:
- Zone A — below 200 m. Lowland tropics with high maximum and minimum temperatures. Quick-maturing varieties flower in two months and mature in four.
- Zone B — 200-1 200 m. Most of these areas have low rainfalls and little maize is grown.
- Zone C — 1 200-2 100 m. Contains over 90 percent of the maize grown in Kenya. The highest yields are regularly obtained in this area. The Kitale hybrids (prefixed by the number 6) flower in about three months and mature in six at 1 500 m. At 1 800 m, flowering is at 3.5 months and maturity at 7, while at 2 100 m, the figures are 4 and 8, respectively.
- Zone D — 2 100-3 200 m. Little maize is grown above 2 000 m, as only long-term varieties can survive in such high altitudes. At Ol Joro Orok (2 400 m) the maize takes 6.5 months to flower and more than a year to mature.

In Australia most of the maize is grown from sea-level to 500 m.
Rainfall requirements. An annual rainfall of more than 500 mm is needed, with best yields usually in the 1 200-1 500 mm area; it is often an irrigated crop. Kitale experiments show that the more rainfall after five weeks' growth, the higher the yield.
Drought tolerance. It is fairly drought tolerant up to five weeks, but there-

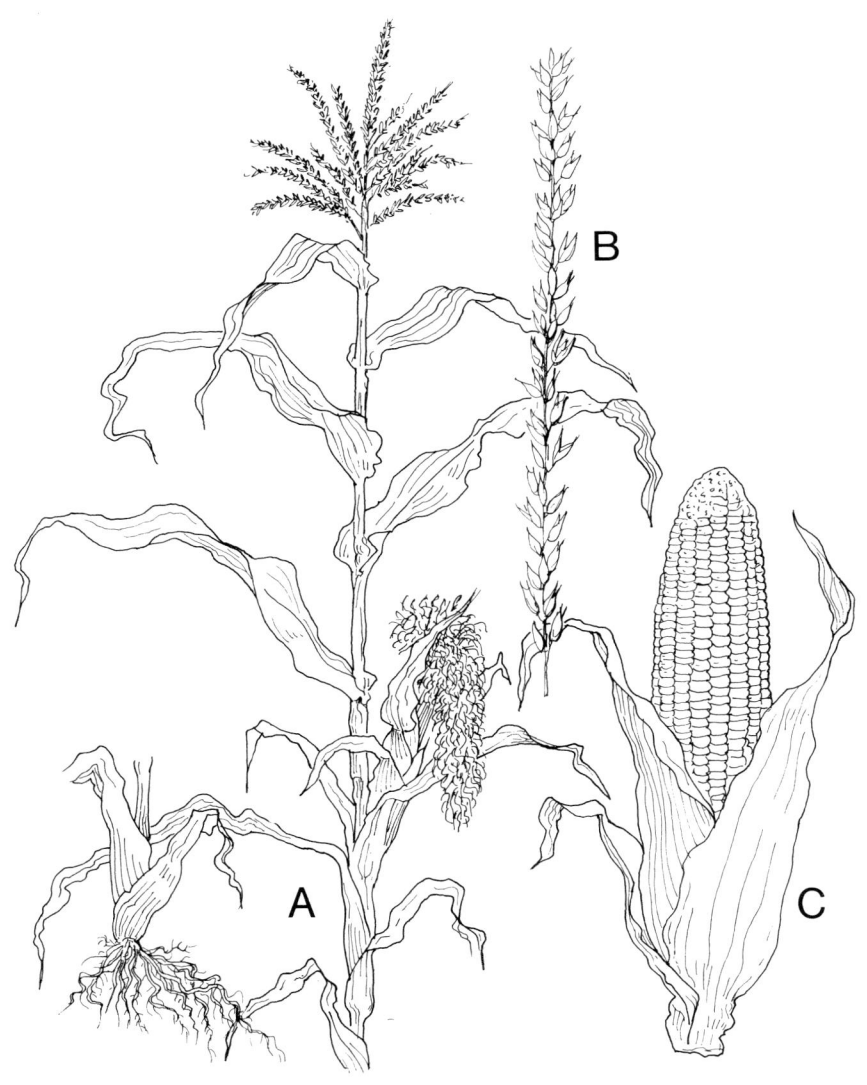

Figure 15.165. Zea mays. **A**-Habit with terminal male inflorescence and lateral female inflorescence **B**-Portion of male inflorescence **C**-Fruiting female inflorescence with enclosing bracts at base

after is very susceptible. Dry weather at pollination time seriously affects pollination and hence yields.

Tolerance to flooding. Maize has no tolerance to flooding.

Soil requirements. It requires a well-drained, fertile soil. Alluvial loams, deep latosols and clay loams are preferred.

Tolerance to salinity. Maize gave maximum yields at EC_e of 2 mmhos/cm, 50 percent at EC_e 9 mmhos/cm and nil at 15.3 mmhos/cm (Maas & Hoffman, 1977). Further studies showed that in water cultures, or on mineral soils with surface irrigation and continuous leaching, the maximum salt concentration in the soil saturation extract that does not reduce maize yields is about 1 100 mg/l total dissolved salts (EC_e - 1.7 dS/m). The maximum permissible salt concentration of irrigation water to sustain maize production is about 300 mg/l, an EC_w of 0.45 dS/m (Hoffman *et al.*, 1979).

Fertilizer requirements. The needs for maize are best determined by soil tests. It generally requires a complete fertilizer, with heavy demands from about 40 days until maturity (see Fig. 15.166). Zinc deficiency causes leaf chlorosis and can easily be overcome by the use of zinc sulphate.

Land preparation for establishment. A deep (20 cm) friable seed-bed should

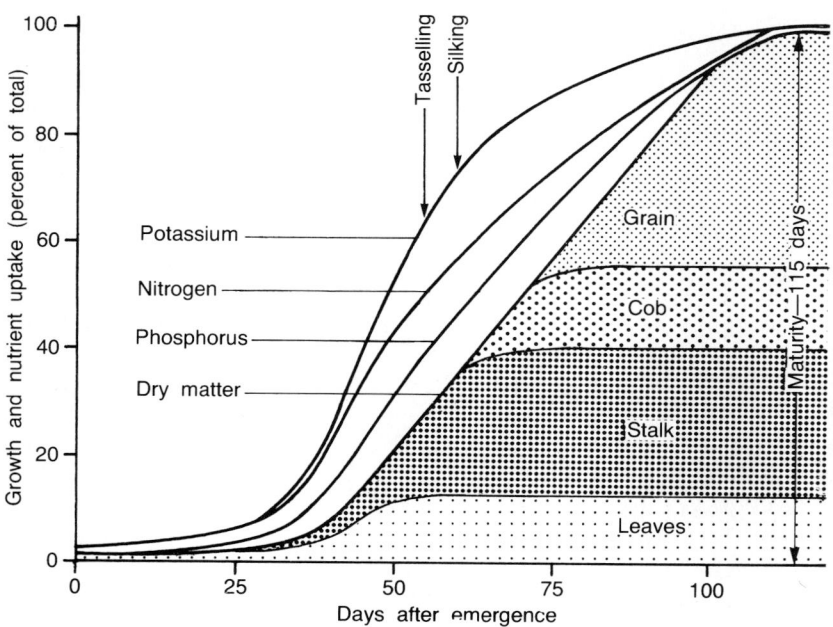

Figure 15.166. Uptake of nutrients in relation to dry weight of *Zea mays* (**Source:** American Potash Institute)

be prepared, as maize is comparatively shallow rooted and needs loose soil in which the roots can forage.

Sowing methods. It is usually drilled in rows for grain and fodder, though it can be broadcast thickly for turning in as a green manure.

Sowing depth and cover. Sow at 7.5-10 cm depth and cover using a tined instrument, then compact with a following press wheel.

Sowing time and rate. Late spring to midsummer, at row spacings of 105-135 cm, using 9-11 kg seed per hectare with populations from 25 000 to 70 000 plants per hectare.

Dormancy. There is little dormancy in maize seed.

Response to photoperiod. Some varieties are indifferent to day length, others require short days for flowering (Evans, Wardlaw & Williams, 1964).

Compatibility with other grasses and legumes. To improve the nutritive value of forage, legumes such as cowpea can be planted between rows of maize. *Lablab purpureus* is used in Brazil. Grass seed can be undersown in maize as a means of establishment.

Ability to compete with weeds. It has poor competitive ability until the crop canopy has closed.

Tolerance to herbicides. Pre-emergence treatment with atrazine or simazine at 1.5-2.5 kg/ha controls most weeds, while 2,4-D amine at 0.45-0.55 kg/ha used on crops 25 cm high is a useful post-emergence weed-killer.

Genetics and reproduction. 2n=20, 21, 22, 24 (Fedorov, 1974).

Dry- and green-matter yields. Yields of 10-50 tonnes of green matter per hectare are obtained for silage.

Suitability for hay or silage. Maize is conserved as stover in the United States and southern Africa, but the main use of the whole plant is as silage. It makes probably the best silage of the grass family, with heavy yields and high acceptability and without the need for additives. It is cut when the grain is full and glazed, in the medium dough stage.

Value as standover or deferred feed. After the grain is harvested the residue usually contains 3.5-4.0 percent crude protein; in a drought year it can be as high as 9 percent. The protein level usually does not decline as it stands in the field. Moisture content at harvest is 40-50 percent in the United States, too low to make silage. The digestibility is 40-50 percent at harvest, but falls to 36-38 percent after 40-60 days in the field. In the United States, good practice is to graze 25-35 percent immediately after harvest with one beast to 0.8-0.9 ha, and then harvest the rest for winter storage (Corah, 1979). In Kenya, it is usually dried in the field for several weeks after maturity (Acland, 1971).

Chemical analysis and digestibility. Göhl (1975) lists numerous analyses and digestibility figures for several countries in Table 15.81.

Palatability. Excellent for all green matter.

Seed production and harvesting. Maize matures in 80-120 days according to variety, and is either hand-picked and later threshed or is harvested with a combine.

TABLE 15.81 *Zea mays*

	DM	As % of dry matter				
		CP	CF	Ash	EE	NFE
Fresh, 8 weeks, Israel	15.7	8.9	31.2	10.2	1.9	47.8
Fresh, 10 weeks, Israel	21.9	10.0	31.5	8.7	1.4	48.4
Fresh, 7 weeks, irrigated, Israel	12.6	10.3	28.6	10.3	2.4	48.4
Fresh, 10 weeks, irrigated, Israel	18.1	8.8	30.9	10.5	2.2	47.6
Fresh, mid-bloom, fertilized, Puerto Rico	23.8	9.5	30.9	6.0	4.3	49.3
Fresh, milk stage, 200 cm, Tanzania	17.0	8.8	28.1	7.4	0.9	54.8
Fresh, whole plant, milk stage, Malaysia	16.0	11.3	29.4	8.1	1.9	49.3
Fresh, stems only, milk stage, Malaysia	13.0	7.7	46.2	8.5	0.8	36.8
Fresh, leaves and cob, milk stage, Malaysia	20.0	15.0	12.5	8.5	3.0	61.0
Dried stalks, Egypt	84.1	5.9	38.5	9.8	1.8	44.0
Dried stalks, southern Africa		6.3	35.0	7.4	1.3	50.0
Hay, southern Africa		7.0	27.0	6.9	1.4	57.7
Silage (clamp), mid-bloom, Tanzania		7.4	31.4	6.2	1.9	53.1
Silage (pit), milk stage, Tanzania		6.5	31.9	5.0	3.3	53.3
Silage (clamp), milk stage, Tanzania		6.2	31.9	6.8	2.4	52.7

	Animal	Digestibility (%)				ME
		CP	CF	EE	NFE	
Fresh, 7 weeks, Israel	Sheep	59.0	62.0	73.0	76.0	2.37
Fresh, 10 weeks, Israel	Sheep	55.0	60.0	69.0	71.0	2.22
Fresh, mid-bloom, Puerto Rico	Sheep	67.0	72.0	82.0	54.0	2.31
Fresh, milk stage, Tanzania	Sheep	56.8	66.5	22.2	72.8	2.35
Dried stalks, Egypt	Sheep	36.0	67.0	59.0	60.0	2.06
Dried stalks, southern Africa	Sheep	39.7	62.5	51.4	60.1	2.04
Hay, southern Africa	Sheep	37.1	64.1	50.7	66.3	2.18
Silage (clamp), mid-bloom, Tanzania	Sheep	35.0	71.7	69.4	74.7	2.46
Silage (pit), milk stage, Tanzania	Sheep	36.1	63.2	78.5	61.8	2.22
Silage (clamp), milk stage, Tanzania	Sheep	18.7	43.2	63.0	53.6	1.69

Source: Göhl, 1975

Seed yield. 1-4 tonnes of grain per hectare, which should be stored at 14 percent moisture or less.

Minimum germination and quality required for commercial sale. 80 percent germinable seed and 98.6 percent purity in Queensland.

Cultivars. Numerous cultivars are available throughout the world and contact should be made with local extension officers to ascertain what is the current preference to specific conditions. Most countries have bred their own cultivars to suit their varying conditions.

Dent maize (*Zea mays indentata*) is the main variety grown commercially for grain and fodder, but there are other types such as pod corn (*Zea mays tunicata*), a curiosity; flour corn (*Zea mays amylacea*) for human consumption; flint corn (*Zea mays indurata*), preferred for the European market, with horny endosperm; sweet corn (*Zea mays saccharata*), used as a vegetable; pop corn (*Zea mays everta*), used as a snack food. High-lysine corn has been improved for human nutrition. Open pollinated varieties have been used for a long time but now most of the commercial dent maize is either a single cross or a double cross hybrid bred for special areas, soils and climatic conditions.

Value for erosion control. As maize is usually a row crop and has a poorly developed root system, crops are very susceptible to erosion.

Diseases. Maize is subject to many diseases, chief of which are maize smut (*Ustilago maydis*), head smut (*Sphacelotheca reiliana*), and various stalk and ear rots such as *Gibberella* and *Diplodia*. They have been overcome by chemical seed treatment or by breeding resistant varieties.

Pests. A variety of pests are encountered. Chinch bug (*Blissus leucopterus*) is a major pest in the United States. The corn ear worm (*Heliothis armigera*) is a problem in Australia.

Economics. Maize is a major human food grain in Africa and in eastern Indonesia. Some 85 percent of the maize crop in the United States is fed to livestock as grain and silage.

Further reading. Martin & Leonard, 1959.

Zea mexicana (Schrad.) Reeves and Mangelsd.

Synonym. Euchlaena mexicana Schrad.
Common name. Teosinte (United States, Australia), malchari (India), maizillo (South America).
Distribution. Originated in Mexico, but introduced to many parts of the world.
Description. Culms tufted, 1 to 3 m tall; leaf-blades similar to maize (*Zea mays*); seeds or hardened joints of the pistillate rachis triangular or trapezoidal, smooth and bony, whitish (Hitchcock, 1930) (see Fig. 15.167).
Season of growth. Summer.
Frost tolerance. It cannot endure frosts.
Rainfall requirements. It requires a good rainfall, in the annual range of 750-1 250 mm.
Drought tolerance. It will not tolerate droughts.
Soil requirements. It requires a fertile soil, but has a wide tolerance of soil texture.
Fertilizer requirements. Needs a complete fertilizer.
Ability to spread naturally. It does not spread readily.
Land preparation for establishment. A well-prepared seed-bed is preferred.
Sowing methods. Seed is drilled or hoed into rows 90-100 cm apart.
Sowing depth and cover. 7-10 cm depth, well covered with soil.
Sowing time and rate. Early summer, at 10 to 60 kg/ha.
Number of seeds per kg. 15 400 (United States).
Response to photoperiod. It flowers in short day lengths (Evans, Wardlaw & Williams, 1964).
Compatibility with other grasses and legumes. It is often sown with legumes such as velvet bean and soybean to enhance its feed value.
Grazing management. It is commonly cut for green fodder, usually with two cuts a season, occasionally more.
Genetics and reproduction. 2n=20 (Fedorov, 1974). It hybridizes readily with maize and is used in breeding work to provide multiple cob production and increased tillers (see Plate 58).
Dry- and green-matter yields. Up to 70 tonnes green fodder per hectare from four to five cuts. In the Dominican Republic it has yielded 14 600 kg/ha green fodder and in the Philippines 29 900 kg/ha (Panikkar, 1951).
Suitability for hay and silage. It is commonly cut and fed green, but makes quite good silage.
Chemical analysis and digestibility. Panikkar (1951) records 4.46 percent crude protein, 32.2 percent crude fibre, 10.8 percent ash, 1.2 percent ether extract and 51.34 percent nitrogen-free extract, with 1.07 percent CaO and 0.36 percent P_2O_5 in the dry matter of a sample from Bihar, India. An Austra-

lian analysis revealed 7.27 percent crude protein, 27.67 percent crude fibre, 7.03 percent ash, 1.39 percent ether extract and 53.75 percent nitrogen-free extract.

Minimum germination and quality required for commercial sale. 50 percent germinable seed and 98.5 percent purity in Queensland.

Further reading. Panikkar, 1951.

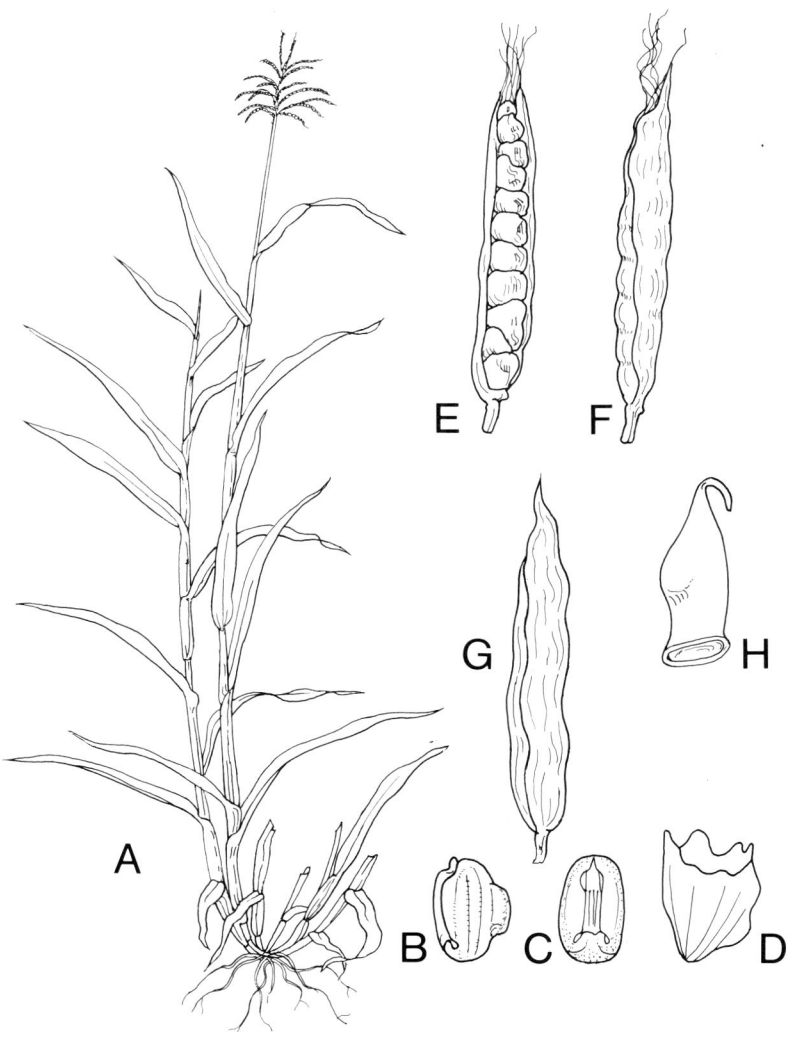

Figure 15.167. Zea mexicana. **A**-Habit **B,C**-Grain **D**-Lodicule **E,F**-Inflorescence (ear) **G**-Involucre **H**-Pistil

Bibliography

ABD-EL-RAHMAN, A.A. & EL-MONAYERI, M. Germination of some desert range plants
1967 under different conditions. *Flora* (Jena), 157: 229-238. *Herb. Abstr.* 39: No. 87.
ABU-EL-GASIM, E.H. & KAMBAL, A.E. Variability and inter-relations among characters in
1975 the indigenous grain sorghums of the Sudan. *E. Afr. Agric. For. J.*, 41: 125-133.
ACKERSON, R.C. & YOUNGNER, V.B. Responses of Bermuda grass to salinity. *Agron. J.*,
1975 67: 678-681.
ACLAND, J.D. *East African crops.* Rome, FAO.
1971
ADAMS, W.E., PEARSON, R.W., JACKSON, W.A. & McCREERY, R.A. Influence of lime-
1967 stone and nitrogen on soil pH and coastal Bermuda grass yield. *Agron. J.*, 59: 450-453.
ADDISON, K.B. Management systems on spear grass country. *Proc. 11th Int. Grassl. Congr.*,
1970 Surfers Paradise, Australia, 789-793.
ADEGBOLA, A.A. Forage crop research and development in Nigeria. *Nigerian Agric. J.*, 1:
1964 34-39.
ADEGBOLA, A.A. & ONAYINKA, E.A.O. A review of range management problems in the
1976 southern Guinea and derived savanna zones of Nigeria. *Trop. Grassl.*, 10: 41-51.
ADEGBOLA, A.A., ONAYINKA, E.A.O. & EWEJE, J.K. The management and improvement
1968 of natural grassland in Nigeria. *Nigerian Agric. J.*, 5: 4-6.
ADEMOSUN, A.A. A review of research on the evaluation of herbage crops and natural
1973 grasslands in Nigeria. *Trop. Grassl.*, 7:285-296.
ADEMOSUN, A.A. & KOLADE, J.O.Y. Nutritive evaluation of Nigerian forages. A compari-
1973 son of the chemical composition and nutritive value of two varieties of *Cynodon*. *Nigerian Agric. J.*, 10: 160-169.
AGGARWAL, R.K. & LAHINI, A.N. Influence of vegetation on the status of organic carbon
1977 and nitrogen of the desert soils. *Sci. Cult.*, 43: 533-535 [*Herb. Abstr.*, 49: No. 2394].
AGREDA, O. & CUANY, R.L. Efectos periódicos y fecha de floración en Yaragua (*Hypar-
1962 rhenia rufa*). *Turrialba*, 12: 146.
AGRICULTURAL RESEARCH COUNCIL. *The nutrient requirements of farm livestock.* Technical
1965 reviews and summaries. London, Agric. Res. Counc.
AHLGREN, G.H., ADEGBOLA, A.A., EWEJE, J.K. & SALAMI, A. The development of grass-
1959 land in the western region of Nigeria. *MANR-ICA Project Final Rept*, 61-13-050.
AHRING, R.M., TALIAFERRO, C.M. & RUSSELL, C.C. Establishment and management of
1978 Old World bluestem grasses for seed. *Univ. of Oklahoma Agric. Exp. Sta. Bull. (Tech.)*, T-149.
AHUJA, L.D. Range management in arid regions. *Bull. Indian Nat. Sci. Acad.*, 44: 95-102
1972 [*Herb. Abstr.*, 44: No. 2425].
AHUJA, L.D. & VISHWANATHAM, M.K. Growth of sheep of Chokla breed under different
1976 systems of grazing on a saline rangeland in the arid zone of Rajasthan. *Ann. Arid Zone*, 15: 102-105.

AJIJ SINGH SIDHU, 'T MANNETJE, L. et al. MARDI/CSIRO Pasture Project Serdang,
1977   Malaysia, Melbourne. *CSIRO Aust., Divn Trop. Crops and Pastures, Rept 1976*, 77: 144.
ALBA, J. DE, BASADRE, J.C. & MASON, D.D. Yield of imperial grass (*Axonopus scoparius*
1956   (Flugge) Hitch.) with chemical and organic manuring. *Turrialba*, 6:(4) 89-95.
ALBERT, W.B. The tolerance of sorghum, Sudan grass, browntop and Gahi millets for vari-
1961   ous herbicides. *Proc. 14th South. Weeds Conf.*, St Petersburg (Fl), USA, 77-85.
ALBERTS, H.W. The forage resources of Peru. *CAB Bull.*, 37.
1947
ALDON, E.F. Establishing alkali sacaton on harsh sites in the southwest. *J. Range Manage.*,
1975   28: 129-132.
ALEXANDER, R.A., HENTGES, J.F., MCCALL, J.T., LUNDY, H.W., GAMMON, N. & BLUE,
1961   W.G. The nutritive value of fall-harvested coastal Bermuda grass hay as affected by rate of nitrogen fertilization and stage of maturity. *J. Anim. Sci.*, 20: 93-98.
ALLAN, A.Y. Some effects of climatic factors on maize growing in Kenya. *E. Afr. Agric.*
1973   *For. J.*, 39: 10.
ALLEN, G.H. Notes on plants from southwest Queensland. Canberra, CSIRO. (Mimeo)
1949
ALLEN, G.H. & COWDRY, W.A.R. Yields from irrigated pastures in the Burdekin. *Queensl.*
1961   *Agric. J.*, 87: 207-213.
ALLEN, R.N., WONG, P., WALKER, J., GOODYEAR, G.J., WRIGHT, W.E. & RAND, J.R.
1975   Diseases and pests of kikuyu grass. In *Kikuyu — a research report*. Sydney, New South Wales Dept Agric.
ALTOM, W. Management of summer grasses: Fertilization, rotation and hay production.
1978   *Proc. Summer Grass Conf.*, Ardmore, Oklahoma, 48-69.
ALVES, L. & SILVA, M.T. DA [The conservation of forage: silos, silage and ensilage.] Rio de
1936   Janeiro, Min. da Agric. (In Portuguese)
ANDERSON, D., HAMILTON, L.P., REYNOLDS, H.G. & HUMPHREY, R.R. Reseeding desert
1957   grassland ranges in Southern Arizona. *Univ. Arizona Agric. Exp. Sta. Bull.*, 249.
ANDERSON, E.R. Pastures for flooded brigalow country. *Queensl. Agric. J.*, 96: 224.
1970a
ANDERSON, E.R. Effect of flooding on tropical pastures. *Proc. 11th Int. Grassl. Congr.*, Surf-
1970b   ers Paradise, Australia, 591-594.
ANDERSON, E.R. Flooding tolerance of *Panicum coloratum*. *Queensl. J. Agric. Anim. Sci.*,
1972   29: 173.
ANDERSON, E.R. The reaction of seven *Cenchrus ciliaris* L. cultivars to flooding. *Trop.*
1974   *Grassl.*, 8: 33-40.
ANDERSON, G.D. Promising pasture plants for northern Tanzania. *E. Afr. Agric. For. J.*,
1968   34: 106.
ANDERSON, R.C. Economic scaling of micro-habitats for some Costa Rican grasses. *Tur-*
1969   *rialba*, 19: 57-65.
ANDREW, C.S. Problems in the use of chemical analyses for diagnosis of plant nutrient
1968   deficiencies. *J. Aust. Inst. Agric. Sci.*, 34: 154-162.
ANDREW, C.S. & ROBINS, M.F. Effects of phosphorus on the growth, chemical composition
1971   and critical phosphorus percentages of some tropical pasture grasses. *Aust. J. Agric. Res.*, 22: 693-706.
ANDREW, W.D. Grassland improvement in the montane zone. Rome, FAO Report No. TA
1971   3040.
ANDREWS, A.C. & CROFTS, F.D. Hybrid Bermuda grass compared with kikuyu and com-
1979   mon couch in coastal New South Wales. *Aust. J. Exp. Agric. Anim. Husb.*, 19: 437-443.
ANDREWS, F.W. *The flowering plants of the Anglo-Egyptian Sudan. Vol. III, Compositae-*
1956   *Gramineae.* Arbroath, Scotland, Bungel and Co.

APPADURAI, R.R. Pasture development and management on marginal plantations in cen-
1975   tral Sri Lanka. *Proc. 3rd World Conf. Anim. Prod.*, 333-338.
APPADURAI, R.R. & GOONAWARDENE, L. Performance of three fodder grasses under high
1973   nitrogen applications in the mid-country wet zone of Sri Lanka. *J. Nat. Agric. Soc.* (Sri Lanka), 8: 31-37.
APPELMAN, H. & DIRVEN, J.G.P. [Guatemala (*Tripsacum laxum*) grass planting.] *De*
1972   *Surinaamse Landbouw*, 20: 2. (In Dutch)
ARCHER, K.A. & WHEELER, J.L. Response to salt and sulphur by cattle grazing sorghum.
1978   *Proc. Aust. Soc. Anim. Prod.*, 12: 172.
ARCHER, S.G. & BUNCH, C.E. *The American grass book*. Norman (Oklahoma), USA,
1953   Univ. of Oklahoma Press.
ARNDT, W. & NORMAN, M.J.T. *Native pasture on Tippera clay loam at Katherine, N.T.*
1959   CSIRO Aust., Divn Land Res. and Reg. Surv., Tech. Paper, 3.
ARONOVICH, S., SERPA, A. & RIBEIRO, H. Effect of nitrogen fertilizer and legumes upon
1970   beef production of Pangola grass pastures. *Proc. 11th Int. Grassl. Congr.*, Surfers Paradise, Australia, 796-800.
ARROYO, R.D. & TEUNISSEN, H. Comparative study of meat production from five tropical
1964   grasses. *Téc. pec-Méx.*, 3: 15-19.
ASARE, E.O. Effect of fertilization, height and frequency of cutting on herbage yield and
1970   nutritive value of *Cenchrus ciliaris* Linn. (buffel grass) in the forest region of Ghana. *Proc. 11th Int. Grassl. Congr.*, Surfers Paradise, Australia, 594-597.
ASHER, C.J. Mineral nutrition of tropical forage plants. Final Rept, UNDP/FAO project
1979   BRA.75/023.
ASHER, C.J. & COWIE, A.M. Grain sorghum: high yield, satisfactory protein content or
1974   both? *Proc. Agron. Soc. New Zealand*, 4: 79-82.
ASSIS, F. DE P., ROCHA, G.L. DA, MEDINA, P., GUARAGNA, R.N., BECKER, M. & KALIL,
1962   E.B. Value of simple and mixed silages in the diet of lactating cows. *Bol. Ind. Anim.* (São Paulo, Brazil), 20:25-33.
AULD, B.A. Chemical control of *Eupatorium adenophorum*, Crofton weed. *Trop. Grassl.*,
1972   6: 55-60.
AUSTIN, J.D.A. Looking for companion grasses. *Turnoff*, 2: 28-34.
1970
AZMAN, D.C. Palatability and chemical composition of soilage of sugar cane seedlings no.
1951   189 and Napier grass. *Philipp. Agric.*, 35: 333-337.
BAJPAI, M.R. & JAIN, K.C. Effect of levels of nitrogen on different grass species for arid and
1971   semi-arid regions of Rajasthan. *Indian J. Agric. Res.*, 5: 45-47.
BALLARD, L.A.T. Germination. *In* Barnard, C. ed. *Grasses and Grasslands*. London, Mac-
1964   millan.
BANERJEE, G.C. & MANDEL, L. Nutritive value of *Pennisetum pedicellatum* grass for adult
1974   sheep. *Indian Vet. J.*, 51: 620-625.
BANK OF NEW SOUTH WALES. *Pasture legumes and grasses.*
1975
BARNARD, C. *Grasses and grasslands.* London, Macmillan.
1964
BARNARD, C. *Herbage plant species.* Aust. Herbage Plant Registration Authority; Can-
1969   berra, CSIRO Aust., Divn of Plant Ind.
BARNARD, C. *Register of Australian herbage plant cultivars.* Canberra, CSIRO Aust., Divn
1972   of Plant Ind.
BARNARD, C. & FRANKEL, O.H. Grass, grazing animals and man in historic perspective. *In*
1964   Barnard, C. ed. *Grasses and grasslands.* London, Macmillan.
BARNES, D.L. Growth and management studies on Sabi panicum and star grass. *Rhod.*
1960   *Agric. J.*, 57: 399.
BARNES, D.L. A suggested broad approach to veld and pasture research in Rhodesia.
1972   (Mimeo)

BARR, D.A. Reclamation of coastal dunes after beach mining. *J. Soil Conserv. Serv.*
1965 (NSW), 21: 199-209.
BARRY, R.A. Genetic adaptation of grasses and legumes to tropical environments. *Trop.*
1975 *Grassl.*, 9: 109-116.
BARTHA, R. *Fodder plants in the Sahel zone of Africa.* Munich, Weltforum Verlag.
1970
BASHAW, E.C. Apomixis and sexuality in buffel grass. *Crop Sci.*, 2: 412-415.
1964
BASHAW, E.C. & FORBES, I., Jr. Chromosome number in microsporogenesis in dallis grass,
1958 *Paspalum dilatatum* Poir. *Agron. J.*, 50: 441-445.
BASHAW, E.C., HOVIN, A.W. & HOLT, E.C. Apomixis, its evolutionary significance and
1970 utilization in plant breeding. *Proc. 11th Int. Grassl. Congr.*, Surfers Paradise,
Australia, 245-248.
BATES, B.P. Adaptation and uses of summer grass varieties. *Proc. Summer Grass Conf.*,
1978 Ardmore, Oklahoma.
BEARDSLEY, D.W. Nutritive value of forage as affected by physical form. 2. Beef cattle and
1964 sheep studies. Symposium on forage utilization. *J. Anim. Sci.*, 23: 239-248.
BEATY, E.R, BROWN, R.H. & MORRIS, J.B. Response of Pensacola bahia grass to intense
1970 clipping. *Proc. 11th Int. Grassl. Congr.*, Surfers Paradise, Australia, 538-542.
BEATY, E.R., POWELL, J.D. & ETHEREDGE, W.J. Effect of nitrogen rate and clipping frequency on yield of Pensacola bahia grass. *Agron. J.*, 55: 3-4.
1963
BEATY, E.R., POWELL, J.D. & LAWRENCE, R.M. Response of Brunswick grass (*Paspalum*
1970 *nicorae,* Parodi) to N fertilization and intense clipping. *Agron. J.*, 62: 363-365.
BEATY, E.R., STANLEY, R.L. & POWELL, J.D. Effect of height of cut on yield of Pensacola
1968 bahia grass. *Agron. J.*, 60: 356-358.
BEGG, J.E. Seasonal changes in yield of dry matter, height of apical meristem, leaf area
1965 index and number of tillers per plant for *Pennisetum americanum. J. Agric. Sci.*,
65: 341.
BELL, R.H.V. A grazing system in the Serengeti. *Sci. Am.*, 225: 86-93.
1971
BENNETT, H.W. The effect of temperature upon flowering in *Paspalum. Agron. J.*, 51: 191-
1959 193.
BENNETT, H.W. Johnson grass, carpet grass and other grasses for the humid south. *In*
1962 Hughes, H.D., Heath, M.E. & Metcalfe, D.S., eds. *Forages.* 2nd ed. Ames
(Iowa), USA, Iowa State Univ. Press.
BENNETT, H.W. Johnson grass, dallis grass and other grasses for the humid south. *In* Heath,
1973 M.E., Metcalfe, D.S. & Barnes, D.L., eds. *Forages, the science of grassland
agriculture.* Ames (Iowa), USA, Iowa State Univ. Press.
BERNSTEIN, L. Salt-affected soils and plants. In *The problems of the arid zone.* Paris,
1962 Unesco.
BHIMAYA, C.P., REGE, N.D. & SRINIVASAN, V. Preliminary studies on the role of grasses
1958 in soil conservation in the Nilgiris. *J. Soil Water Conserv.* (India), 4: 113-117.
BIRCH, E.B. Nitrogen fertilization of weeping love grass (*E. curvula*) at Dohne. *Proc.*
1967 *Grassl. Soc. South. Afr.*, 2: 39-43.
BIRIE-HABAS, J. Trials with forage species at the agronomic station at Lake Alaotra. *Bull.*
1959 *Inst. Rech. Agron.* (Madagascar), 3: 68-74.
BISHOP, H.G. Gulf country pastures. *Queensl. Agric. J.*, 99: 257-262; 325-331; 363-368; 433-
1973 440.
BISSET, W.J. Plicatulum finds a place in coastal pastures. *Queensl. Agric. J.*, 101: 603-608.
1975
BISSET, W.J. The origin of *Bothriochloa insculpta* cv. Hatch. in Queensland. *Trop. Grassl.*,
1978 12: 208-209.
BISSET, W.J. & GRAHAM, T.G. Creeping blue grass finds favour. *Queensl. Agric. J.*, 104;
1978 245-253.

BISSET, W.J. & MARLOWE, G.W.C. Productivity and dynamics of two siratro-based pas-
1974　　tures in the Burnett coastal foothills of southeast Queensland. *Trop. Grassl.*, 8: 17-34.

BLAIR, G.J., HUGHES, R.M., LOVETT, J.V. & CAUSBY, M.G. Temperature, seeding depth
1974　　and fertilizer effects on germination and early seedling growth of kikuyu grass. *Trop. Grassl.*, 8: 163-179.

BLAIR, G.J., PUALILLIN, P. & SAMOSIR, S. Effect of fertilizers on the yield and botanical
1978　　composition of pastures in South Sulawesi, Indonesia. *Agron. J.*, 70: 559-562.

BLAKE, S.T. *Monographic studies in the Australian Andropogoneae, Pt.1.* Dept Bot., Univ.
1944　　of Queensland, 2: 1-62.

BLAKE, S.T. Taxonomic and nomenclatural studies on the Gramineae. No.1. *Proc. Royal*
1969　　*Soc. Queensl.*, 80: 55-84.

BLUE, W.G. Fertilizer uptake by Pensacola bahia grass (*Paspalum notatum*) from Leon fine
1970　　sand, a spodosol. *Proc. 11th Int. Grassl. Congr.*, Surfers Paradise, Australia, 389-392.

BLUE, W.G., GAMMON, N. JR., & LUNDY, H.W. Late summer fertilization for winter forage
1961　　in north Florida. *Proc. Soil Crop Sci. Soc. Florida*, 21: 56-62.

BLUNT, C.G. Irrigated forages. *CSIRO Aust., Divn Land Res. Ann. Rept, 1968-69.*
1969

BLUNT, C.G. Production from steers grazing nitrogen-fertilized irrigated pangola grass in
1978　　the Ord Valley. *Trop. Grassl.*, 12: 90-96.

BLYDENSTEIN, J., LOUIS, S., TOLEDO, J. & CAMARGO, A. Productivity of tropical pastures.
1969　　I. Pangola grass. *J. Brit. Grassl. Soc.*, 24: 71-75.

BOGDAN, A.V. A study of the depth of germination of tropical grasses: a new approach.
1959　　*J. Brit. Grassl. Soc.*, 19: 251-252.

BOGDAN, A.V. A molasses grass variety trial. *E. Afr. Agric. For. J.*, 26: 132-133.
1960

BOGDAN, A.V. The selection of tropical ley grasses in Kenya — general considerations and
1964　　methods. *E. Afr. Agric. J.*, 24: 206-217.

BOGDAN, A.V. Rhodes grass. *Herb. Abstr.*, 39: 1-13.
1969

BOGDAN, A.V. *Tropical pastures and fodder plants (grasses and legumes).* London, New
1977　　York, Longman.

BOGDAN, A.V. & KIDNER, E.M. Grazing natural grassland in western Kenya. *E. Afr.*
1967　　*Agric. For. J.*, 33: 31-34.

BOGDAN, A.V. & PRATT, D.J. *Reseeding denuded pastoral land in Kenya.* Nairobi, Repub-
1967　　lic of Kenya Min. Agric. Anim. Husb.

BOKYO, H. Salinity and aridity. *Biology (The Hague)*, 16.
1966

BOONMAN, J.G. Experimental studies on seed production of tropical grasses in Kenya.
1972　　4. The effect of fertilizer and planting density in *Chloris gayana. Neth. J. Agric. Sci.*, 20: 218-224.

BOOYSEN, P. DE V. & TAINTON, N.M. Grassland management: principles and practice in
1978　　South Africa. *Proc. 1st Int. Range. Congr.*, Denver (Colorado), USA, 551-557.

BOR, N.L. *The grasses of Burma, Ceylon, India and Pakistan.* London, Pergamon Press.
1960

BOR, N.L. *Flora of Iraq. Vol. 9. Gramineae.* Baghdad, Min. of Agric.
1968

BORGET, M. Yields and characteristics of five forage grasses on coastal sands in Cayenne,
1966　　French Guiana. *Agron. trop.*, (Paris), 21: 250-259.

BOSSER, J. *Graminées des pâturages et des cultures à Madagascar.* Tananarive, Madagascar,
1969　　ORSTOM.

BOTHA, J.P. Grass on old wattle lands in the Eastern Transvaal. *Farming S. Afr.*, 28: 270-
1953   272.
BOTHA, J.P. & HAMBURGER, H. Fertilization of *Eragrostis curvula* (Ermelo type). *Farming*
1953   *S. Afr.*, 28: 377-378.
BOTT, W. Pastures in the Dalby district. *Queensl. Agric. J.*, 104; 353-367.
1978
BOUDET, G. Management of savannah woodland in West Africa. *Proc. 11th Int. Grassl.*
1970   *Congr.*, Surfers Paradise, Australia, 1-3.
BOUDET, G. *Manuel sur les pâturages tropicaux et les cultures fourragères.* Paris, Alfort-
1975   Seine, Institut d'élevage et de médecine vétérinaire des pays tropicaux, Ministère
         de la coopération.
BOUDET, G. & DUVERGER, E. *Etude des pâturages naturels Sahéliens.* Paris, Alfort-Seine,
1961   Institut d'élevage et de médecine vétérinaire des pays tropicaux, Ministère de la
         coopération.
BOUXIN, G. Ordination and classification in the savanna vegetation of the Akagera Park
1975   (Rwanda). *Vegetatio*, 29: 155-167.
BOWDEN, B.N. An illustrated field key to the commoner Uganda grasses. *E. Afr. Agric.*
1962   *For. J.*, 27: 230.
BOWDEN, B.N. The root distribution of *Andropogon gayanus* var. *bi-squamulatus*. *E. Afr.*
1963a   *Agric. For. J.*, 29: 157-159.
BOWDEN, B.N. Studies on *Andropogon gayanus* Kunth. 1. The use of *Andropogon gayanus*
1963b   in agriculture. *Emp. J. Exp. Agric.*, 31: 267-273.
BOWDEN, B.N. Studies on *Andropogon gayanus* Kunth. 3. An outline of its biology.
1964   *J. Ecol.*, 52: 255-271.
BOYD, F.T., SCHROEDER, V.N. & PERRY, V.G. Sting nematodes attack forage grasses in
1972   Florida. *Sunshine State Agric. Res. Rept*, 14(6): 3-6.
BRAGADIN, E.A. & DIAZ, H.B. Silage made from tops and leaves of sugar cane. Its value
1957   for fodder. *Rev. Ind. Agric. Tucuman*, 41: 31-36.
BRANDS, L.E. & COOK, B.G. Whittet kikuyu in the Gympie district. *Queensl. Agric. J.*, 102:
1976   429-432.
BRAY, R.A. & PRITCHARD, A.J. Some aspects of selection for herbage quality in buffel
1976   grass (*Cenchrus ciliaris*). *Forage Res.*, 2: 1-7.
BREAKWELL, E.J. *The grasses and fodder plants of New South Wales.* Sydney, Dept
1923   Agric.
BREDON, R.M. & HORRELL, C.R. The chemical composition and nutritive value of some
1961-2   common grasses in Uganda. *Trop. Agric. (Trinidad)*, 38: 297-304; 39: 13-17.
BRITON, N.W. & PALTRIDGE, T.B. Preliminary note on photosensitization of sheep grazed
1941   on *Brachiaria brizantha*. *Proc. Royal Soc. Queensl.*, 52: 121-122.
BROADLEY, R.H. & ROGERS, D.J. Pests of pangola grass in north Queensland pastures.
1978   *Queensl. Agric. J.*, 104: 320-324.
BROCKINGTON, N.R. Studies on the growth of a *Hyparrhenia*-dominant grassland in North-
1961   ern Rhodesia. *J. Brit. Grassl. Soc.*, 16: 54-64.
BROWNE, D., WALSHE, M.J. & CONNIFFE, D. Irish research on problems in animal produc-
1970   tion experiments comparing legumes with fertilizer as the source of pasture nitro-
         gen. *Proc. 11th Int. Grassl. Congr.*, Surfers Paradise, Australia, 101-107.
BRYAN, W.W. *Lotononis bainesii* Baker, a legume for subtropical pastures. *Aust. J. Exp.*
1961   *Agric. Anim. Husb.*, 1: 4-10.
BRYAN, W.W. Grazing trials on the Wallum of southeastern Queensland. 1. A comparison
1968   of four pastures. *Aust. J. Exp. Agric. Anim. Husb.*, 8: 512-520.
BRYAN, W.W. Changes in botanical composition in some subtropical sown pastures. *Proc.*
1970   *11th Int. Grassl. Congr.*, Surfers Paradise, Australia, 636-639.
BRYAN, W.W. A review of research findings concerned with pastoral development on the
1973   Wallum of southeastern Queensland. *Trop. Grassl.*, 7: 175-194.

BRYAN, W.W. & EVANS, T.R. Coastal lowlands. Beef production from nitrogen-fertilized
1967   pangola grass. *CSIRO Aust., Divn Trop. Past. Ann. Rept, 1966-7.*

BRYAN, W.W. & EVANS, T.R. Soil characteristics after improved pasture at Beerwah,
1971   Queensland. *Aust. J. Exp. Agric. Anim. Husb.*, 11: 633-639.

BRYAN, W.W. & SHARPE, J.P. The effect of urea and cutting treatment on the production
1965   of pangola grass in southeastern Queensland. *Aust. J. Exp. Agric. Anim. Husb.*, 5: 433-441.

BRYAN, W.W. & SHAW, N.H. *Paspalum plicatulum* Michx., two useful varieties for pastures
1964   in regions of summer rainfall. *Aust. J. Exp. Agric. Anim. Husb.*, 4: 17-21.

BRYANT, W.G. Makarikari panic (*Panicum coloratum* (L.) var. *makarikariensis* Goossens)
1959   for erosion control. *J. Soil Conserv. Serv:* (NSW), 15: 146.

BRZOSTOWSKI, H.W. Influence of pH and superphosphate on establishment of *Cenchrus*
1962   *ciliaris* from seed. *Trop. Agric. (Trinidad)*, 39: 289-296.

BURBIDGE, N.T. *Australian grasses; Northern Tablelands of New South Wales. Vol. II.*
1968   Sydney, Angus and Robertson.

BURKART, A. *Flora Illustrada de Entre Rios (Argentina). Pt. II, Gramineas.* Buenos Aires,
1969   INTA.

BURROWS, W.H. New pasture plants in the mulga zone. *Queensl. Agric. J.*, 96: 321-324.
1970

BURSON, B.L. & BENNETT, H.W. Cytology, method of reproduction and fertility of
1970   Brunswick grass, *Paspalum nicorae* Parodi. *Crop Sci.*, 10: 184-187.

BURT, R.L. Growth and development of buffel grass (*Cenchrus ciliaris*). *Aust. J. Exp.*
1968   *Agric. Anim. Husb.*, 8: 712.

BURT, R.L., WILLIAMS, W.T., GILLARD, P. & PENGELLY, B.C. Variations within and be-
1980   tween some perennial *Urochloa* species. 1. Examination under spaced plant conditions. *Aust. J. Bot.*, 28(3): 343-356.

BURTON, G.W. Bahia grass types. *J. Amer. Soc. Agron.*, 38: 273-281.
1945

BURTON, G.W. Breeding of subtropical species for increased animal production. *Proc. 11th*
1970   *Int. Grassl. Congr.*, Surfers Paradise, Australia, A 56-A 63.

BURTON, G.W. & POWELL, J.B. Pearl millet breeding and cytogenetics. *Adv. Agron.*, 20:
1968   49-89.

BURVILL, G.H. *Paspalum vaginatum* for salted lands. *J. Agric., West. Aust.*, 5: 121-122.
1956

BURVILL, G.H. & MARSHALL, A.H. *Paspalum vaginatum* or sea-shore paspalum. *J. Dept.*
1951   *Agric., West. Aust.*, 28: 191.

BUTTERWORTH, M.H. Studies on pangola grass at ICTA, Trinidad. II. The digestibility of
1971   pangola grass at various stages of growth. *Trop. Agric. (Trinidad)*, 38: 189-193.

CABRERA, A.L. *Flora de la Provincia de Buenos Aires. Pt. II. Gramíneas.* Buenos Aires,
1970   INTA.

CALVINO, M. *Plantas forrajeras tropicales y subtropicales.* Mexico City, B. Trucco.
1952

CAMERON, D.G. Grasses tested for soil conservation. Results to April 1958. *J. Soil Conserv.*
1959   *Serv. (NSW)*, 15: 281-293.

CAMERON, D.G. & KELLY, T.K. Para grass for wetter country. *Queensl. Agric. J.*,
1970   96: 386-390.

CAMERON, D.G. & MULLALY, J.D. Studies with a range of grass cultivars in small plots at
1970   Biloela, central Queensland. *Queensl. J. Agric. Anim. Sci.*, 27: 55.

CAMPBELL, C.M., STANLEY, R.W., NOLAN, J.C. JR., HO-A, E.B., LENT, G., HAMADA, K.
1976   & SLATER, W. *Using sugar cane stripping silage for beef cows.* Hawaiian Agric. Exp. Sta. Res. Rept, 234.

Caro-Costas, R., Vicente-Chandler, J. & Burleigh, C. Beef production and carrying
1961   capacity of heavily fertilized, irrigated Guinea, Napier, and pangola grass pastures on the semi-arid coast of Puerto Rico. *J. Agric., Univ. Puerto Rico*, 45: 32-36.

Caro-Costas, R., Vicente-Chandler, J. & Figarella, J. Productivity of intensely-
1965   managed pastures of five grasses on steep slopes in the humid mountains of Puerto Rico. *J. Agric., Univ. Puerto Rico*, 49: 99-111.

Carpenter, J.A. Production and use of seed in sea-shore paspalum. *J. Aust. Inst. Agric.*
1958   *Sci.*, 24: 252-256.

Carver, L.A., Barth, K.M., McLaren, J.B., Fribourg, H.A., Connell, J.T. &
1975   Bryan, J.M. Nutritive value of Midland Bermuda pastures. *J. Anim. Sci.*, 40: 183.

Cassidy, G.J. Nitrogen-phosphate fertilizer for grass. *Queensl. Agric. J.*, 83: 235-238.
1957

Cassidy, G.J. Response of a mat-grass-paspalum sward to fertilizer application. *Trop.*
1971   *Grassl.*, 5: 11-22.

Catchpoole, V.R. The ensilage of sorghum at a range of crop maturities. *Aust. J. exp.*
1962   *Agric. Anim. Husb.*, 2: 101-105.

Catchpoole, V.R. Laboratory ensilage of *Setaria sphacelata* (Nandi) and *Chloris gayana*
1965   (CPI 16144). *Aust. J. Agric. Res.*, 16: 391-402.

Catchpoole, V.R. Preliminary studies on curing and storing Nandi setaria hay. *Trop.*
1969   *Grassl.*, 3: 65-74.

Catchpoole, V.R. The silage fermentation of some tropical pasture plants. *Proc. 11th Int.*
1970   *Grassl. Congr.*, Surfers Paradise, Australia, 891-894.

Catchpoole, V.R. Laboratory ensilage of *Sorghum almum* cv. Crooble. *Trop. Grassl.*, 6:
1972   171-176.

Catchpoole, V.R. & Henzell, E.F. Silage and silage-making from tropical herbage
1971   species. *Herb. Abstr.*, 41: 213-221.

Cezar, S.M., Barbosa, C., Mattos, J.C.A. de & Campos, B. do E.S. de. [Effect of poul-
1976   try waste, chopped maize with straw and ear, and Guinea grass silage on liveweight gain of beef cattle in feedlots.] *Bol. Ind. Anim.* (São Paulo, Brazil) 33(1): 1-7 (In Portuguese) [*Herb. Abstr.*, 49: No. 1064].

Chacón, E., Rodriguez-Carrásquez, S. & Chico, C.F. Efectos de la fertilización con
1971   nitrógeno sobre el valor nutritivo del pasto colorado (*Panicum coloratum*). *Agron. Trop.* (Maracay, Venez.) 21: 496-502.

Chacón, E. & Stobbs, T.H. Influence of progressive defoliation of a grass sward on the eat-
1976   ing behaviour of cattle. *Aust. J. Agric. Res.*, 16: 709-727.

Chadokar, P.A. Establishment of stylo (*Stylosanthes guianensis*) in Kunai (*Imperata cylin-*
1977   *drica*) pastures, and its effect on dry matter yield and animal production in the Markham Valley, Papua New Guinea. *Trop. Grassl.*, 11: 263-272.

Chadokar, P.A. Effect of rate and frequency of nitrogen application on dry matter yield
1978   and nitrogen content of para grass (*Brachiaria mutica*). *Trop. Grassl.*, 12: 127-132.

Chadokar, P.A. & Humphreys, L.R. Effects of time and nitrogen deficiency on seed pro-
1970   duction of *Paspalum plicatulum* Michx. *Proc. 11th Int. Grassl. Congr.*, Surfers Paradise, Australia, 315-319.

Chadokar, P.A. & Humphreys, L.R. Short day and plant age effects on flowering of *Pas-*
1974   *palum plicatulum*. *J. Aust. Inst. Agric. Sci.*, 40: 75.

Chakravarty, A.K. & Bhati, G.N. Study on pasture establishment technique. 2. Effect
1969   of pelleting on germination of *Lasiurus sindicus* seeds. *Ann. Arid Zone*, 8: 58-60.

Chakravarty, A.K. & Kackar, N.L. Selection of grasses and legumes for the arid and
1971   semi-arid zones. *J. Indian Bot. Soc.*, 50: 265-272.

CHAKRAVARTY, A.K. & VERMA, C.M. Studies on the pasture establishment technique.
1972   4. Effects of different spacings and weedings on establishment and forage production of *Cenchrus ciliaris* Linn., *Lasiurus sindicus* Henry and *Panicum antidotale* Retz. under arid conditions. *Ann. Arid Zone*, 11: 60-66.
CHANDRARATNA, M.F. *Genetics and breeding of rice.* London, Longman.
1963
CHAPKO, P.M. [Selection of fodder crops for cultivation on alkali soils.] *Visn. Sil's' kogos-*
1977   *pod. Nauki,* 1977(7): 50-55 (In Ukrainian) [*Herb. Abstr.*, 48: No. 3706].
CHAPMAN, H.L. & KRETSCHMER, A.E. Effect of nitrogen fertilizer on digestibility and feed-
1964   ing value of Pangola grass hay. *Proc. Soil and Crop Sci. Soc., Florida,* 24: 176-183.
CHAPMAN, J. Evaluation of maize and babala for silage production on an Avalon soil in
1978   northern Natal. *Agroplantae* 10(2): 45-47 [*Herb. Abstr.*, 49: No. 1850].
CHAPMAN, V.J. *Salt marshes and salt deserts of the world.* Braunschweig, Fed. Rep. of Ger-
1974   many, J. Cramer Verlag.
CHATTERJEE, B.N. ROY, B. & BHATTACHARJEE, K.K. *Pennisetum pedicellatum* as a short
1974   rotation forage crop for the eastern region of India. *J. Soil Wat. Conserv.* (India), 23: 47-53.
CHAUDRI, I.I., SHEIKH, M.Y. & ALAM, M.M. Halophytic flora of saline and waterlogged
1969   areas of West Pakistan plains. *Agric. Pak.*, 20: 404-414 [*Herb. Abstr.*, 41: 3007].
CHAVANCY, A. Compte-rendu des travaux du Centre d'expérimentation agronomique du
1951   Laos en 1947, 1948 et 1949. *Arch. des Recherches Agronomiques au Cambodge, au Laos et au Vietnam,* 10.
CHHEDA, H.R. & MOHAMED SALEEM, M.A. Effects of heights of cutting after grazing on
1972   yield, quality and utilization of *Cynodon* I.B.8 pasture in southern Nigeria. *Trop. Agric. (Trinidad),* 50: 113-119.
CHILD, R. The coconut industry in Portuguese East Africa. *World Crops,* 7: 488-492.
1955
CHILD, R. *Coconuts.* London, Longman Green.
1964
CHINNAMANI, S. Grasslands in Bellary black cotton soils. *Indian For.,* 94: 225-229.
1968
CHIPPENDALL, L.K.A. A guide to the identification of grasses in South Africa. *In* Meredith,
1955   D., ed. *The grasses and pastures of South Africa.* Parov (Cape Province), South Africa, Central Newsagency.
CHIPPENDALL, L.K.A. & CROOK, A.O. *240 grasses of southern Africa.* Salisbury, Rhodesia,
1976   M.O. Collins (Pvt) Ltd, Irwin Press (Pvt) Ltd.
CHOU, C.H. Phytotoxic substances in twelve tropical grasses. *Bot. Bull. Acad. Sinica*
1977   (Taibei), 18: 131-141.
CHRISTIAN, C.S. & SHAW, N.H. A study of two strains of Rhodes grass (*Chloris gayana*
1952   Kunth.) and lucerne (*Medicago sativa* L.) as components of a mixed pasture at Lawes in southeast Queensland. *Aust. J. Agric. Res.,* 3: 277-299.
CHRISTIE, E.K. A study of phosphorus nutrition and water supply on the early growth and
1975a   survival of buffel grass grown on a sandy red earth from southwest Queensland. *Aust. J. Exp. Agric. Anim. Husb.,* 15: 239.
CHRISTIE, E.K. A note on the significance of *Eucalyptus populnea* for buffel grass produc-
1975b   tion in infertile and semi-arid rangelands. *Trop. Grassl.,* 9: 243-246.
CIAT. Beef Program. 1978 Rept Cali, Colombia, Centro Internacional de Agricultura
1978   Tropical.
CLARK, N.A., HEMKEN, R.W. & VANDERSALL, J.H. A comparison of pearl millet, Sudan
1965   grass and sorghum-Sudan grass hybrid as pasture for lactating dairy cows. *Agron. J.,* 57: 266-269.

CLATWORTHY, J.N. A comparison of legume and fertilizer nitrogen in Rhodesia. *Proc. 11th*
1970   *Int. Grassl. Congr.,* Surfers Paradise, Australia, 408-411.
CLAYTON, W.D. A revision of the genus *Hyparrhenia. Kew Bull.,* additional series 11.
1969
CLAYTON, W.D. & HARLAN, J.R. The genus *Cynodon* L.C.Rich in tropical Africa. *Kew*
1970   *Bull.,* 24: 185-189.
CLAYTON, W.D., PHILLIPS, S.M. & RENVOIZE, S.A. *Flora of Tropical East Africa. Part 2.*
1974   *Gramineae.* Pothill, R.M., ed. Crown Agents.
COALDRAKE, J.E. Ecosystems of the coastal lowlands, southern Queensland. *CSIRO Aust.,*
1961   *Bull.,* 283.
COCONUT RESEARCH INSTITUTE. *Pasture under coconuts.* Luniwila, Sri Lanka. Adv. Leaflet
1966   No. 45.
COETSEE, G. Grazing of *Cymbopogon-Themeda* veld in the dormant period. *Proc. 10th*
1975   *Congr. Grassl. Soc. of Southern Africa,* Pietermaritzburg, South Africa, 10: 147-150.
COLMAN, R.L. *Growth curve studies on kikuyu grass.* Wollongbar, New South Wales,
1966   Agric. Res. Sta. Ann. Rept, 1965-66.
COLMAN, R.L. Kikuyu and animal production. Dairy cattle. In *Kikuyu — a research report.*
1975   N. S. W. Dept. Agric., 86(5).
COLMAN, R.L. & HOLDER, J.M. Effect of stocking rate on butterfat production of dairy
1968   cows grazing kikuyu grass pastures fertilized with nitrogen. *Proc. Aust. Soc. Anim. Prod.,* 7: 129-132.
COLMAN, R.L. & LAZENBY, A. Factors affecting the response of some tropical and temper-
1970   ate grasses to fertilizer nitrogen. *Proc. 11th Int. Grassl. Congr.,* Surfers Paradise, Australia, 392-397.
COLMAN, R.L. & WILSON, G.P.M. The effects of floods on pasture plants. *Agric. Gaz.,*
1960   *N.S.W.,* 71: 337-347.
COMPÈRE, R. Rational exploitation of high-altitude permanent *Exotheca abyssinica*
1968   Anders, and *Eragrostis boehmii* Hack. pastures in Rwanda and Burundi. *Bull. Rech. Agron. Gembloux,* 3: 583-604.
CONNIFFE, D., BROWNE, D. & WALSHE, M.J. Experimental design for grazing trials. *J.*
1970   *Agric. Sci.,* 74: 339-342.
CONNOLLY, T. Some comments on the shape of the gain-stocking rate curve. *J. Agric. Sci.,*
1976   86: 103-109.
CONWAY, A.G. A production function for grazing cattle. 3. An estimated relationship be-
1974   tween rate of liveweight gain and stocking rate for grazing steers. *Irish J. Agric. Econ. and Rur. Soc.,* 5: 43-45.
COOK, B.G. Pastures for the Gympie district. *Queensl. Agric. J.,* 104: 226-231.
1978
COOK, B.G. & O'GRADY, R. Atrazine in kikuyu grass establishment. A preliminary study.
1978   *Trop. Grassl.,* 12: 184-187.
COOK, C.D.K. *Water plants of the world.* The Hague, Dr W. Junk Publishers.
1974
COOK, S.J. Some ecological aspects of the establishment of aerial seeded pastures. *CSIRO*
1977   *Aust., Divn Trop. Crops and Pastures Rept 1976-77:* 102.
COOK, S.J. Establishment of pasture species oversown in spear grass pasture. *CSIRO Aust.,*
1978   *Divn Trop. Crops and Pastures. Ann. Rept 1977-78:* 10-11.
COOKE, T. *The flora of the Presidency of Bombay.* Calcutta, Sri Gouranga Press.
1958
CORAH, L. Utilizing crop residues. *Proc. Beef Cattle Conf. — Economics, Management and*
1979   *Alternative Feeding Systems,* Ardmore (Oklahoma), USA.
COWAN, R.T., BYFORD, I.J.R. & STOBBS, T.H. Effects of stocking rates and energy
1975   supplementation on milk production from tropical grass-legume pastures. *Aust. J. Exp. Agric. Anim. Husb.,* 15: 740-746.

COWARD-LORD, J., ARROYO-AGUILU, J.A., GARCÍA-MOLINARI, O. Proximate nutrient composition of ten tropical forage grasses. *J. Agric. Univ. Puerto Rico*, 58: 293-304; 305-311; 426-436.
1974

CREEK, M.J. & NESTEL, B.L. The effect of grazing cycle duration on liveweight output, with chemical composition of pangola grass, in Jamaica. *Proc. 11th Int. Grassl. Congr.*, Surfers Paradise, Australia, 1613-1616.
1965

CROFTS, C.F. & PEARSON, C.J. Irrigation, fertilizer and species management of a pasture for high milk production throughout the whole year, in a warm temperate climate. *Proc. Int. Meeting on Animal Production from Temperate Grassland*, Dublin.
1977

CROWDER, L.A.V., CHAVERRA, H. & LOTERO, J. Productive improved grasses in Colombia. *Proc. 11th Int. Grassl. Congr.*, Surfers Paradise, Australia, 147-149.
1970

CROWDER, L.A.V., MICHELIN, A. & BASTIDAS, A. The response of pangola grass (*Digitaria decumbens* Stent.) to rate and time of nitrogen application in Colombia. *Trop. Agric. (Trinidad)*, 41: 21-29.
1964

CSIRO. *Ann. Rept 1972-73.* Melbourne, Aust. Divn Trop. Agron., p. 20.
1973

CUEVAS, A.A. & GONZÁLEZ, M.E. Effect of burning on buffel grass (*Cenchrus ciliaris*) XV *Informe de investigación, 1975-76.* Monterrey (Nuevo León), Mexico, División de Ciencias Agropecuarias y Marítimas, Instituto Tecnológico de Monterrey (In Spanish) [*Herb. Abstr.*, 49: No. 1745].
1977

CULL, J.K. Establishing pasture on Eastern Downs. *Queensl. Agric. J.*, 100: 386-397.
1974

CUMMINS, D.G. Quality and yield of corn plants and component parts when harvested for silage at different maturity stages. *Agron. J.*, 62: 781-784.
1970

CUNARD, A.C. Maize agronomy. Part 1. Nutrition and silage. *World Crops*, 19: 20-27.
1967

CURRAN, L.T.F. There's more to grain sorghum than grain. *Queensl. Agric. J.*, 83: 173.
1957

CURRIE, J.A. Para grass—viewpoint from two sides of the fence. *Queensl. Cane Grow. Q. Bull.*, 39: 28-29.
1975

DABADGHAO, P.M. Types of grass cover in India and their management. *Proc. 8th Int. Grassl. Congr.*, Reading, UK, 227-230.
1960

DABADGHAO, P.M., ROY, R.D. & MARWAHA, S.P. The effect of interval and intensity of defoliation on the dry matter production of some important grass species of western Rajasthan. *Ann. Arid. Zone*, 12: 1-8.
1973

DABADGHAO, P.M. & SHANKARNARAYAN, K.A. Studies of *Iseilema*, *Sehima* and *Heteropogon* communities of the *Sehima-Dichanthium* zone. *Proc. 11th Int. Grassl. Congr.*, Surfers Paradise, Australia, 36-38.
1970

DABADGHAO, P.M. & SHANKARNARAYAN, K.A. *The grass cover of India.* New Delhi, Indian Council of Agric. Res.
1973

DAGG, M., HOSEGOOD, P.H. & MCQUEEN, M. Rooting habits of East African grasses. *EAAFRO Record of Research, Nairobi*, 18-24.
1967

DALE, A.B. & READ, J.W. Irrigation of kikuyu pastures. In *Kikuyu research report*, Mears A.D., ed., N.S.W. Dept Agric.
1975

DALRYMPLE, R.L. Kleingrass, old world bluestems, weeping love grass for quality summer forages. *Proc. Summer Grass Conf.*, Ardmore (Oklahoma), USA.
1978

DARLEY, E.C. Drought tests pastures in the Moree district. *Agric. Gaz. N.S.W.*, 78: 384-388.
1967

DARLINGTON, C.D. & JANAKI, A.E. *Chromosome atlas of cultivated plants.* London, George Allen and Unwin.
1945

DAS, R.B. Managing pastures in arid areas. *Indian Farming*, 23: 30-31.
1973

DAUBERMIRE, R. The ecology of fire in grasslands. *Adv. Ecol. Res.*, 5: 209-266.
1968
DAUBERMIRE, R. Ecology of *Hyparrhenia rufa* Nees. in derived savanna in northwestern
1972   Costa Rica. *J. Appl. Ecol.*, 9: 11.
DAULAY, H.S., CHAKRAVARTY, A.K. & BHATI, G.N. Intercropping of rainy season
1972   legumes in *Cenchrus ciliaris* L. and *Lasiurus sindicus* Henr. pastures in the arid zone of Rajasthan. *Indian J. Agric. Sci.*, 42: 148-151.
DAVIDSON, D.E. The Mitchell grass association of the Longreach district. *Univ. Queensl.*,
1954   *Dept Bot. Paper*, 3: 46-59.
DAVIDSON, D.E. Five pasture plants for Queensland. *Queensl. Agric. J.*, 92: 461-463.
1966
DAVIDSON, R.L. Theoretical aspects of nitrogen economy in grazing experiments. *J. Brit.*
1964   *Grassl. Soc.*, 19: 273.
DAVIES, J.G. Pasture development in the sub-tropics, with special reference to Taiwan.
1970   *Trop. Grassl.*, 4: 7-16.
DAVIES, J.G. & HUTTON, E.M. *Tropical and subtropical pasture species. In* Moore, R. Mil-
1970   ton, ed. *Australian grasslands*. Canberra, Aust. Nat. Univ. Press.
DAVIES, J.G., SCOTT, A.E. & KENNEDY, J.F. The yield and composition of a Mitchell grass
1938   pasture for a period of twelve months. *J. CSIRO Aust.*, 11: 127-139.
DAY, J.M., NEVES, M.C. & DÖBEREINER, J. Nitrogen fixation on the roots of tropical for-
1975   age grasses. *Soil Biol. Biochem.*, 7: 107.
DEANS, H.D., CHOPPING, G.D., SIBBICK, R., THURBON, P.N., COPEMAN, D.B. & STOKOE,
1976   J. Effect of stocking rate, breed, grain supplementation, nitrogen fertilizer level and anthelmintic treatment on growth rate of dairy weaners grazing irrigated pangola grass. *Proc. 11th Bienn. Conf. Aust. Soc. of Anim. Prod.*, Melbourne, 449-452.
DEGRAS, L., MATHURIN, P. & FÉLICITÉ, J. [Some facts about reproductive development in
1974   *Digitaria* spp. and *Panicum maximum*.] *Colloque sur l'intensification de la production fourragère en milieu tropical humide et son utilisation par les ruminants*, 24-29 May 1971, Paris, INRA (In French) [*Herb. Abstr.*, 46: No. 272].
DELANEY, N.E. Green panic is widely accepted. *Queensl. Agric. J.*, 101: 729-735.
1975
DELAWAULLE, J.C. The increasing unproductiveness of Africa south of the Sahara. *Bois*
1973   *For. Trop.*, 149: 3-20.
DENMAN, C.E. *Sudan grass,* Sorghum sudanense *(Piper) Stapf.* Stillwater (Oklahoma),
1958   USA. Agric. Res. Sta. For. Crop Leaflet No.15.
DEWALD, C.L. Grazing utility of summer annual forages. *Proc. Summer Grass Conf.*,
1978   Ardmore (Oklahoma), USA.
DEWEZ, J. The sowing of *B. ruziziensis* in the Ruzizi plain, Belgian Congo. *INEAC Inform.*
1959   *Bull.*, 8: 303-307.
DICKENS, R. & MOORE, G.M. Effects of light, temperature, KNO_3 and storage on germina-
1974   tion of cogon grass. *Agron. J.*, 66: 187-188.
DIENUM, B. & DIRVEN, J.G.P. Climate, nitrogen and grass. 5. Influence of age, light inten-
1972   sity and temperature on the production and chemical composition of Congo grass (*Brachiaria ruziziensis* Germain et Everard). *Neth. J. Agric. Sci.*, 20: 125-132.
DILLEWIJN, C. VAN. Botany of sugar cane. (Waltham, (Mass), USA). *Chron. Bot.*
1952
DIRVEN, J.G.P. The nutritive value of the indigenous grasses of Surinam. *Neth. J. Agric.*
1963a   *Sci.*, 11: 295-307.
DIRVEN, J.G.P. The protein content of Surinam roughages. *Surinam, Paramaribo, Agric.*
1963b   *Exp. Sta. Bull.*, 82.
DIRVEN, J.G.P. Yield increase of tropical grassland by fertilization. *Proc. 9th Congr. Int.*
1970   *Potash Inst.*, Antibes.

DIRVEN, J.G.P. Tropical roughage. *Thai J. Agric. Sci.*, 6: 323-334.
1973

DOAK, B.W. Some chemical changes in the nitrogenous constituents of urine when voided
1952   on pasture. *J. Agric. Sci.* (Cambridge), 42: 162-171.

DOGGETT, H. International aspects of *Sorghum* research. *E. Afr. Agric. For. J.*, 39.
1973

DONALD, C.M. The pastures of Southern Australia. *In* Leeper, G.W., ed. *The Australian*
1955   *environment.* Melbourne, CSIRO Australia, Melbourne Univ. Press. 68-82.

DONEFER, E., JAMES, L.A. & LAURIE, C.K. Use of a sugar cane-derived feedstuff for live-
1973   stock. *Proc. 3rd World Conf. Anim. Prod.*, Melbourne, May 1973, 563-566.

DOUGALL, H.W. Average nutritive values of Kenya feeding stuffs for ruminants. *E. Afr.*
1960   *Agric. For. J.*, 26: 119-128.

DOUGALL, H.W. & BOGDAN, A.V. The chemical composition of grasses of Kenya. *E. Afr.*
1958-60   *For. J.*, Part I, 24(1): 17-23; Part 2, 25(4): 241-244.

DOUGLAS, N.J. Tropical legumes on a blady grass burn. *Queensl. Agric. J.*, 91: 36-39.
1965

DOUGLAS, N.J. Grain and fodder millets. *Queensl. Agric. J.*, 96: 51-57.
1970

DOUGLAS, N.J. Millets for grain and grazing. *Queensl. Agric. J.*, 100: 469-476.
1974

DOUGRAMEJI, J. & KAUL, R.N. Sand dune reclamation in Iraq — present status and future
1972   projects. *Ann. Arid Zone,* 11: 133-134.

DOVRAT, A. & COHEN, Y. Regrowth potential of Rhodes grass (*Chloris gayana* Kunth) as
1970   affected by nitrogen and defoliation. *Proc. 11th Int. Grassl. Congr.*, Surfers
        Paradise, Australia, 552-554.

DOWLING, D.F. Comparative growth rate of different strains of cattle in a hot tropical envi-
1960   ronment. *Proc. Aust. Soc. Anim. Prod.*, 3: 184-191.

DOWNES, R.W. The potential of the genus *Dichanthium* in Queensland with particular
1969   reference to *Dichanthium sericeum*. Univ. of Queensland. (M. Agric. Sc. thesis)

DOYNE, H.C. Grass manuring in southern Nigeria. *Emp. J. Exp. Agric.*, 5: 248-253.
1937

DRADU, E.A.A. & HARRINGTON, G.N. Seasonal crude protein content of samples obtained
1972   from a tropical range pasture using oesophageal fistulated steers. *Trop. Agric.*
        (Trinidad), 49: 15-21.

DUFOURNET, R. ET AL. [Importance and future of forage species introduced into Madagas-
1959   car.] *Mém. Inst. Sci. Madagascar*, 9: 121-148. (In French)

DUNSMORE, J.R. & ONG, C.B. Preliminary work on pasture species and beef production in
1969   Sarawak, Malaysia. *Trop. Grassl.*, 3: 117-121.

DU PLOOY, J., LE BOUX, D.P. & COETZEE, P.J.S. The effects of leys on maize yields. *S. Afr.*
1965   *J. Agric. Sci.*, 8: 311-322.

EAVIS, B.W., CUMBERBATCH, E.R. ST J. & MEDFORD, D.L. Factors influencing regenera-
1974   tion of natural vegetation on reformed Scotland District soils of Barbados. *Trop.*
        Agric. (Trinidad), 51: 293-303.

EBERSOHN, J.P. & LEE, G.R. The impact of sown pastures on cattle numbers in Queens-
1972   land. *Aust. Vet. J.*, 48: 217-223.

EBERSOHN, J.P. & LUCAS, P. Trees and soil nutrients in southwestern Queensland.
1965   *Queensl. J. Agric. Anim. Sci.*, 22: 431.

EDEN, D.R.A. Pacific copra production near possible serious decline. *South Pac. Comm.*
1953   *Tech. Pap.*, 48: 1-32.

EDGLEY, W.H.R. & HARLE, J.G. Mat grass to milk. *Queensl. Agric. J.*, 100: 18-22.
1974

EDGLEY, W.H.R., TOW, P.G. & WALKER, R.W. Patterns of pasture and milk production
1968   from glycine-green panic pastures. *Aust. Grassl. Conf.*, Vol. 1, 26.

EDWARDS, D.C. Three ecotypes of *Pennisetum clandestinum* (Hochst.), kikuyu grass. *Emp.*
1937   *J. Exp. Agric.*, 5: 371.
EDWARDS, D.C. Grass burning. *Emp. J. Exp. Agric.*, 10: 219-231.
1942
EDWARDS, P.J. The long-term effects of burning and mowing on the basal cover of two veld
1968   types in Natal. *S. Afr. J. Agric. Sci.*, 11: 131-140.
EDYE, L.A., HUMPHREYS, L.R., HENZELL, E.F. & TEAKLE, L.J.H. Pasture investigations
1964   in the Yalleroi district of central Queensland. *Univ. Queensl. Dept Agric.*, 1(4): 133-172.
EDYE, L.A. & MILES, J.F. A comparison of sixty *Panicum* introductions in southeastern
1976   Queensland. *Trop. Grassl.*, 10: 70-88.
ELLISON, R.M. & HENDERSON, C.A. Beef commodity. Preliminary report to the Govern-
1973   ment of Fiji.
ENDO, J. [Studies on soil movement and its control on mountain slopes.] *Bull. Yamagata*
1978   *Univ.*, 8(1): 1-110 (In Japanese) [*Herb. Abstr.*, 49: No. 1968].
ENG, P.K., KERRIDGE, P.C. & 'T MANNETJE, L. Effects of phosphorus and stocking rate on
1978   pasture and animal production from a Guinea grass-legume pasture in Johore, Malaysia. *Trop. Grassl.*, 12; 188-197, 198-207.
ENGLER, A. & PRANTL, K. *Die natürlichen Pflanzenfamilien.* Leipzig, German Dem. Rep.,
1940   Wilhelm Engelman, Verlag.
ERASMUS, I.I. & SUD, A.D. Effect of commercial fertilizers on the natural grasslands on
1976   deep alluvial soils in the Ambala Siwaliks. *Indian For.*, 102: 90-95.
ESCOBAR, A. & GONZÁLEZ, J.E. [Regional variations on the relative frequency of plant
1976   species consumed by capybara (*Hydrochoerus hydrochaeris*)]. *Proc. 2nd Sem. Capybara and Cayman Crocodiles,* 1-4 December 1976, Maracay, Venezuela (In Spanish) [*Herb. Abstr.*, 49: No. 2755].
EUSSEN, J.H.H. & WIRJAHARDJA, S. Studies of an alang-alang (*Imperata cylindrica* L.
1973   Beauv.) vegetation. *Biotrop Bull.*, (Bogor, Indonesia), 6.
EVANS, L.T. & KNOX, R.B. Environmental control of reproduction in *Themeda australis.*
1969   *Aust. J. Bot.*, 17: 375-389.
EVANS, L.T., WARDLAW, I.F. & WILLIAMS, C.N. Environmental control of growth. *In* Bar-
1964   nard, C. ed. *Grasses and grasslands.* London, Macmillan.
EVANS, T.R. Preliminary evaluation of grasses and legumes for the northern Wallum of
1967a   southeast Queensland. *Trop. Grassl.*, 1: 143-152.
EVANS, T.R. A salinity problem in pasture establishment in the coastal lowlands of southern
1967b   Queensland. *J. Aust. Inst. Agric. Sci.*, 33: 216-218.
EVANS, T.R. Beef production from nitrogen-fertilized pangola grass (*Digitaria decumbens*)
1969   on the coastal lowlands of southern Queensland. *Aust. J. Exp. Agric. Anim. Husb.*, 9: 282-286.
EVANS, T.R. Influence of nitrogen fertilizer schedule on animal production. *CSIRO Aust.*,
1971a   *Divn Trop. Past., Ann. Rept 1970-71.*
EVANS, T.R. Symposium and field meeting with the South Coast Beef Producers' Associa-
1971b   tion on the ability of pastures in the coastal lowlands (Wallum) to meet year-round demands of beef cattle. *Trop. Grassl.*, 5: 43-53.
EVANS, T.R. Pangola rust. *CSIRO Aust., Divn Trop. Past., Ann. Rept 1971-72.*
1972
EVANS, T.R. & BIGGS, J. Breeding performance of Hereford cows grazing tropical pastures
1979   at Beerwah, southeast Queensland. *Trop. Grassl,* 13(3): 129-134.
EVANS, T.R. & BRYAN, W.W. Effects of soil, fertilizers and stocking rates on pasture and
1973   beef production on the Wallum, southeastern Queensland. 2. Animal response in terms of liveweight change and beef production. *Aust. J. Exp. Agric. Anim. Husb.*, 13: 530-536.
EVANS, T.R. & WINTER, W.H. Grazing systems on Townsville stylo pastures. *CSIRO Aust.*,
1976   *Divn Trop. Crops and Past., Ann. Rept 1975-6,* 32.

EVERIST, S.L. Response during 1934 season of Mitchell and other grasses in western and
1935 coastal Queensland. *Queensl. Agric. J.*, 43: 374-387.
EVERIST, S.L. Notes on some plants of western Queensland. *Queensl. Naturalist*, 14: 52-55.
1951
EVERIST, S.L. *Poisonous plants of Australia*. Sydney, Angus and Robertson.
1974
EVERIST, S.L. Under-exploited tropical plants with promising economic value. *US Nat.*
1975 *Acad. Sci.*, 9.
EVERS, G.W., HOLT, E.C. & BASHAW, E.C. Seed production characteristics and photo-
1969 period responses in buffel grass, *Cenchrus ciliaris* L. *Crop Sci.*, 9: 309-310.
FALADE, J.A. The effect of phosphorus on the growth and mineral composition of five trop-
1975 ical grasses. *E. Afr. Agric. For. J.*, 40: 342-350.
FALVEY, L. Sabi grass (*Urochloa mosambicensis*) as a component of Townsville stylo
1976 (*Stylosanthes humilis*) pasture. *Proc. Aust. Soc. Anim. Prod.*, 11: 337-340.
FALVEY, L. Review of existing knowledge of ruminants in the highlands of northern Thai-
1977 land. *Thai J. Agric. Sci.*, 10: 111-119.
FALVEY, L. Establishment of two *Stylosanthes* species in a *Urochloa mosambicensis*-domin-
1979 ant sward in the Daly River basin. *J. Aust. Inst. Agric. Sci.*, 45: 69-71.
FALVEY, L. *Imperata cylindrica* (L) Beauv. in animal production in South East Asia. *Trop.*
1980 *Grassl.*
FALVEY, L. & ANDREWS, A. Improved pastures in the Thai highlands. *Trop. Grassl.*, 13:
1979 154-156.
FALVEY, L., HENGMICHAI, P. & HOARE, P. Productivity of cattle grazing native highland
1979 pastures. *Thai J. Agric. Sci.*, 12: 61-69.
FAO. *China: recycling of organic wastes in agriculture*. Rome, Soils Bull. No. 40.
1977
FARINAS, E.C. Pasture legumes and grasses and other forage plants at the National Forage
1970 Park, Philippines (1958-1968). *Proc. 11th Int. Grassl. Congr.*, Surfers Paradise, Australia, 224-226.
FARNWORTH, J. *A trial of introduced and local forage species as pioneer crops for summer*
1974 *reclamation of sandy, saline, high carbonate soil at Hofuf, Saudi Arabia*. Univ. College North Wales, Gwynedd, Bangor, UK, Agric. Res. and Devel. Project, Joint publn Univ. College North Wales and Min. Agric. and Water, Saudi Arabia, No. 39.
FARNWORTH, J. *A comparison of twenty grasses for autumn reclamation of sandy, saline soil*
1977 *at Hofuf, Saudi Arabia*. Univ. College North Wales, Gwynedd, Bangor, UK, Agric. Res. and Devel. Project, Joint publn Univ. College North Wales and Min. Agric. and Water, Saudi Arabia, No. 76.
FARUQI, S.A. Range of morphological variation within the *Bothriochloa intermedia* com-
1969 plex. *Phyton* (Horn, Austria), 13: 285-303.
FEBLES-PEREZ, G., WHITEMAN, P.C. & HARTY, R.L. Seed production and germination in
1974 blue couch (*Digitaria didactyla*). *Aust. J. Exp. Agric. Anim. Husb.*, 14: 65-67.
FEDOROV, A. *Chromosome numbers of flowering plants*. Koenigstein, Fed. Rep. Germany,
1974 Otto Koeltz Science Publishers.
FERGUSON, J.E. Systems of pasture seed production in Latin America. *In* Sánchez, P.A. &
1979 Tergas, L.E., eds. *Pasture production in acid soils of the tropics*. Cali, Colombia, CIAT. 385-395. Series 03 EG-5.
FERNANDEZ, D.E.F. Intercropping with coconuts. *Ceylon Coconut Q.*, 23: 51-53.
1972
FERNANDEZ, D.E.F. Utilization of coconut lands for pasture development. *Ceylon Coconut*
1973 *Plant. Rev.*, 7: 14-19.
FERNANDO, G.W.E. Grassland farming in the dry zone. *Trop. Agric., (Sri Lanka)*, 114: 183-
1958 196.

FERNANDO, G.W.E. Preliminary studies on the association growth of grasses and legumes.
1961 *Trop. Agric.* (Ceylon), 117: 167-179.
FERRARIS, R. *Pearl millet* (Pennisetum typhoides). *Com. Agric. Bur. Pastures Field Crops*
1973 *Rev. Ser. 1.*, Berkshire, UK.
FIELD, C.R. Elephant ecology in the Queen Elizabeth National Park, Uganda. *E. Afr.*
1971 *Wildl. J.*, 9: 99-123.
FINLAY, M. Rice growing. *Queensl. Agric. J.*, 101: 227-233.
1975
FISHER, M.J. The recovery of leaf water potential following burning of two droughted trop-
1978 ical pasture species. *Aust. J. Exp. Agric. Anim. Husb.*, 18: 423-425.
FISHER, M.J. & IVE, J.R. The control of grass weeds in Townsville stylo (*Stylosanthes*
1970 *humilis*) experiments. *Aust. J. Exp. Agric. Anim. Husb.*, 10: 795.
FLOYD, A.G. Effect of fire upon weed seeds in the wet sclerophyllous forests of northern
1966 New South Wales. *Aust. J. Bot.*, 14: 243-256.
FOSTER, H.L. Crop yields after different elephant grass ley treatments at Kawanda
1971 Research Station, Uganda. *E. Afr. Agric. For. J.*, 37: 63-72.
FOSTER, W.H. & MUNDY, E.J. Forage species for northern Nigeria. *Trop. Agric.*
1961 (Trinidad), 38: 311-318.
FRENCH, M.H. Animal nutrition research. *Ann. Rept Dept Vet. Sci. Anim. Husb.*, Tan-
1932 ganyika, 34: 24-69.
FRENCH, M.H. *Emp. J. Exp. Agric.*, 9: 23.
1941
FRENCH, M.H. & LEDGER, H.P. Liveweight changes of cattle in East Africa. *Emp. J. Exp.*
1957 *Agric.*, 25: 10-18.
FRIBOURG, H.A., EDWARDS, N.C. & BARTH, K.M. *In vitro* dry matter digestibility of Mid-
1971 land Bermuda grass grown at several levels of N fertilization. *Agron. J.*, 63: 786-
788.
FUDECO (Fundación para el Desarrollo de la Región Centro Occidental de Venezuela).
Manual de Ganadería. 2nd ed.
FUROC, R.E. & JAVIER, E.Q. Integration of fodder production with intensive cropping
1976 involving rice. 1. Grass production from irrigated lowland rice field. 2. Herbage
weeds during juvenile stage of the rice crop. *Philipp. J. Crop Sci.*, 1: 146-148.
GANDARA, D., LOGGINS, P.E. & AMMERMAN, C.B. Characteristics of grass silages pre-
1962 served with zinc bacitracin and ground snap corn. *Dept Anim. Sci., Fla Agric.
Exp. Sta.*, Gainesville (Fla), USA. (mimeo)
GANDHI, R.T. Plants suitable for soil conservation. *Indian J. Agric. Sci.*, 27: 131-135.
1957
GARCÍA-RIVERA, J. & MORRIS, M.P. Oxalate content of tropical forage grasses. *Science*,
1955 122: 1089:1090.
GARDNER, C.A. *Flora of Western Australia.* Vol. 1, Part 1. *Gramineae.* Perth, Govt Printer.
1952
GARDNER, C.A. Paspalum grass (*Paspalum dilatatum* Poiret). *J. Dept Agric. West. Aust.*, 5
1956 (3rd series): 435-440.
GARTNER, J.A. The effects of different rates of fertilizer nitrogen on the growth, nitrogen
1966 uptake and botanical composition of tropical grass swards. *Proc. 10th Int. Grassl.
Congr.*, Helsinki, 223-227.
GARTNER, J.A. Effect of fertilizer nitrogen on a dense sward of kikuyu, paspalum and car-
1969 pet grass. 1. Botanical composition, growth and nitrogen uptake. *Queensl. J.
Agric. Anim. Sci.*, 26: 21-33.
GASKINS, M.H. & SLEPER, D.A. Photosensitivity of some tropical forage grasses in Florida.
1974 *Proc. Soil Crop Sci. Soc.*, 33; 20-21.
GATES, C.T., HAYDOCK, K.P., CLARINGBOLD, J.P. & ROBINS, M.F. Growth of sorghum
1966 varieties of three sorghum species at different levels of salinity. *Aust. J. Exp.
Agric. Anim. Husb.*, 6: 161-169.

GAUSMAN, H.W., COWLEY, W.R. & BARTON, J.H. Reaction of some grasses to artificial
1954   salinization. *Agron. J.*, 46: 412-414.
GIBSON, T.A. *Thai-Australia Highland Agricultural Project*. 1st Rept Canberra, Aust.
1976   Devel. Assistance Bureau.
GILL, R.S., RANA, N.D. & NEGI, S.S. Studies on the optimum stage of harvesting of Kharif
1970   grasses for hay making in the Kangra district. *J. Res. Punjab Agric. Univ.*, 7: 511-515.
GILLARD, P. The effect of stocking rate on botanical composition and soils in natural grass-
1969   land in South Africa. *J. Appl. Ecol.*, 6: 489-497.
GILLARD, P. *Urochloa mosambicensis*, an easily established perennial grass companion of
1971   Townsville stylo. *Trop. Grassl.*, 5: 131-135.
GILLET, H. Essai d'évaluation de la biomasse végétale en zone sahélienne (végétation
1967   annuelle). *J. d'agriculture tropicale et de botanique appliquées* (Paris), 14: 123-158.
GILLILAND, H.B., HOLTTUM, R.E., BOR, N.L. & BURKILL, H.M. *A revised Flora of*
1971   *Malaya*. 3. *Grasses of Malaya*. Singapore, Govt Printer.
GLEDHILL, D. Carpet grass (*Axonopus* spp.) in West Africa. *Proc. 9th Int. Grassl. Congr.*,
1980   São Paulo, Brazil, 1: 469-470.
GÖHL, B.O. *Tropical feeds. Feeds information, summaries, and nutritive value*. Rome,
1975   FAO.
GONZALEZ, E.R. & PACHECO, J.M. *Cultivo de pastos en Costa Rica*. San José, Min. de
1970   Agric. y Ganadería, Costa Rica. Boletín Técnico No. 51.
GOODWIN, P.J. & CHAMBERLAIN, A. Molasses in the Queensland dairying industry.
1979   *Queensl. Agric. J.*, 105: 450-463.
GORDON, C.J. Preserving grassland products for ruminant feeding. *In* Sprague, H.B., ed.
1974   *Grasslands of the United States*. Ames (Iowa), USA, Iowa Univ. Press.
GOULD, F.W. & BOX, T.W. *Grasses of the Texas Coastal Bend (Calhoun, Refugio, Aransas,*
1965   *San Patricio and northern Kleberg counties)*. College Station (Texas), USA, Texas A and M Univ. Press.
GOULD, F.W. & SHAW, R.B. *Grass systematics*. 2nd ed. College Station (Texas), USA,
1983   Texas A and M Univ. Press.
GRAHAM, N. McC. The net energy value of three subtropical forages. *Aust. J. Agric. Res.*,
1967   18: 137-147.
GRAHAM, T.W.G. & HUMPHREYS, L.R. Salinity response of cultivars of buffel grass. *Aust.*
1970   *J. Exp. Agric. Anim. Husb.*, 10: 725-728.
GRANIER, P. & LAHORE, J. [Report on irrigation trials at the Ecole d'Agriculture of Valabe,
1961   Provence, in 1959-60.] *Adv. agron.* (Paris) 1(1). (In French)
GREENWAY, H. Salinity, plant growth, and metabolism. *J. Aust. Inst. Agric. Sci.*, 39: 24-34.
1973
GREENWOOD, E.A.N. The response to nitrogen of two strains of *Hyparrhenia hirta* under
1966   irrigation and defoliation. *Field Sta. Record, Divn Pl. Ind., CSIRO Aust.*, 5(1): 29-36.
GRIEVE, C.M. & OSBOURN, D.F. The nutritive value of some tropical grasses. *J. Agric. Sci.*,
1965   65: 411-417.
GROF, B. Establishment of legumes in the humid tropics of northeastern Australia. *Proc.*
1965   *9th Int. Grassl. Congr.*, São Paulo, Brazil, 1137-1142.
GROF, B. Viability of seed of *Brachiaria decumbens*. *Queensl. J. Agric. Anim. Sci.*, 25: 149-152.
1968
GROF, B. Viability of para grass (*Brachiaria mutica*) seeds and the effect of fertilizer nitro-
1969   gen on seed yield. *Queensl. J. Agric. Anim. Sci.*, 26: 271-276.
GROF, B. & HARDING, W.A.T. Dry matter yields and animal production of Guinea grass
1970   (*Panicum maximum*) on the humid tropical coast of north Queensland. *Trop. Grassl.*, 4: 85-95.
GROVES, R.H. Growth and development of five populations of *Themeda australis* in
1976   response to temperature. *Aust. J. Bot.*, 23: 951-963.

GUILLOTEAU, J. The problem of bush fires and burns in land development and soil conserva-
1957   tion in Africa south of the Sahara. *Sols Afr.*, 4: 64-102.
GUMBS, F.A. & SHASTRY, M.V. Comparison of measured and calculated consumptive use
1978   of water by savanna grass (*Axonopus compressus* Swartz Beauv). *Trop. Agric. (Trinidad)*, 55: 33-38.
GUPTA, P.K. Apomixis in *Bothriochloa pertusa* (L) A. Camus. *Port. Acta. bio.* (A), 11(3-4):
1969-70   279-287.
GUPTA, P.K., SAXENA, S.K. & SHARMA, S.K. Above-ground productivity of three promis-
1972   ing desert grasses at Jodhpur under different rainfall conditions. *Symp. Ecophysiological Foundation of Ecosystems Productivity in Arid Zone.* 7-19 June, 1972, Leningrad, USSR. Jodhpur, Rajasthan, India, Central Arid Zone Research Institute [*Herb. Abstr.*, 45: 1061].
GUSMAN, H.W., COWLEY, W.R. & BARTON, J.H. Reaction of some grasses to artificial
1954   salinization. *Agron. J.*, 46: 412-414.
GUTTERIDGE, R.C. & ROBERTSON, A.D. Backyard forage, an important aspect of animal
1979   husbandry in Thailand. *Livestock Int.*, 32: 10-11.
GUTTERIDGE, R.C. & WHITEMAN, P.C. Pasture species evaluation in the Solomon Islands.
1978   *Trop. Grassl.*, 12: 113-126.
GUZMAN, M.R. DE. Pasture and fodder production under coconuts. *ASPAC Exten. Bull.*,
1974   45: 1-29.
GWYNNE, M.D. & NDAWAL-SENYIMBA, M.S. A punch-card method based on vegetative
1971   characters for identifying East African grasses. *E. Afr. Agric. For. J.*, 37: 334-352.
HACKER, J.B. Breeding improved *Setarias*. *Trop. Grassl.*, 3: 82-83.
1969
HACKER, J.B. Digestibility of leaves, leaf sheaths and stem of *Setaria*. *J. Aust. Inst. Agric.*
1971   *Sci.*, 37: 154-155.
HACKER, J.B. Seasonal yield distribution in *Setaria*. *Aust. J. Exp. Agric. Anim. Husb.*, 12:
1972   36.
HACKER, J.B. Variation in oxalate, major cations and dry matter digestibility of 47 introduc-
1974   tions of the tropical grass *Setaria*. *Trop. Grassl.*, 8: 145-154.
HACKER, J.B. *Digitaria smutzii* breeding programme. *CSIRO Aust., Divn Trop. Crops and*
1976   *Pastures, Rept. 1975-76:* 45.
HACKER, J.B. & JONES, R.J. The *Setaria sphacelata* complex — a review. *Trop. Grassl.*, 3:
1969   13-34.
HACKER, J.B. & JONES, R.J. The effect of nitrogen fertilizer and row spacing on seed pro-
1971   duction in *Setaria sphacelata*. *Trop. Grassl.*, 5: 61-73.
HAGGAR, R.J. The production of seed from *Andropogon gayanus*. *Rept 7th Int. Workshop*
1966   *Seed Pathology — Int. Seed Testing Assoc.*, Wageningen, the Netherlands, September 1964, 31: 251-259.
HAGGAR, R.J. Use of companion crops in grassland establishments in Nigeria. *Exp. Agric.*,
1969   5: 47-52.
HAGGAR, R.J. Season production of *Andropogon gayanus*. 1. Seasonal changes in yield
1970   components and chemical composition. *J. Agric. Sci.*, 74: 487-494.
HAGGAR, R.J. The effect of quantity, source and time of application of nitrogen fertilizers
1975   on the yield and quality of *Andropogon gayanus* at Shika, Nigeria. *J. Agric. Sci. Camb.*, 84: 529-535.
HAGGAR, R.J. & AHMED, M.B. Seasonal production of *Andropogon gayanus*. 3. Changes
1971   in crude protein content and *in vitro* dry matter digestibility of leaf and stem portions. *J. Agric. Sci.*, 75: 369-373.
HAGON, M.W. Germination and dormancy of *Themeda australis*, *Danthonia* spp., *Stipa*
1976   *bigeniculata* and *Bothriochloa macra*. *Aust. J. Bot.*, 24: 319-327.

HAINES, C.E., CHAPMAN, H.L., ALLEN, R.J. & KIDDER, R.W. Roselawn St Augustine
1965 grass as a perennial pasture for organic soils of south Florida. *Agric. Exp. Sta., Gainesville, Fla Bull.*, 689.

HALL, R.J. Pasture development in the spear grass region at Westwood in the Fitzroy Basin.
1970 *Trop. Grassl.*, 4: 77-84.

HALL, T.J. Cloncurry buffel grass (*Cenchrus pennisetiformis*) in northwestern Queensland.
1978 *Trop. Grassl.*, 12: 10-19.

HAMILTON, R.I., LAMBOURNE, L.J., ROE, R. & MINSON, D.J. Quality of tropical grasses for
1970 milk production. *Proc. 11th Int. Grassl. Congr.*, Surfers Paradise, Australia, 860-863.

HARDING, W.A.T. The contribution of plant introduction to pasture development in the
1972 wet tropics of Queensland. *Trop. Grassl.*, 6: 191-199.

HARKER, K.W. A fertilizer trial on *Paspalum notatum* pasture. 1. The effect on yields. *E.*
1962 *Afr. Agric. For. J.*, 27: 201-203.

HARKER, K.W. & NAPPER, D. *An illustrated guide to the grasses of Uganda.* Entebbe, Govt
1960 Printers.

HARLAN, J.R., WET, J.M.J. DE & RAWAL, K.M. Geographical distribution of the species
1970 of *Cynodon* L.C.Rich (Gramineae). *E. Afr. Agric. For. J.*, 36: 220-226.

HARLAN, J.R., WET, J.M. J. DE, RICHARDSON, W.L. & CHHEDA, H.R. *Studies on old-*
1961 *world bluestems. III.* Oklahoma State Univ. Tech. Bull. No. T-92.

HARPER, J.L. & BENTON, R.A. The behaviour of seed in soil. II. The germination of seeds
1966 on the surface of a water-supplying substrate. *J. Ecol.*, 54: 151.

HARRINGTON, G.N. *The fire ecology of the savannah grasslands of Ankole, Uganda.*
1973 *II. Shrubs.* Deniliquin, Australia, CSIRO Aust., Range. Res. Unit.

HARRINGTON, G.N. Fire effects on a Ugandan savanna grassland. *Trop. Grassl.*, 8(2): 87-
1974 101.

HARRINGTON, G.N. & PRATCHETT, D. Cattle diet on Ankole rangeland at different seasons.
1972 *Trop. Agric. (Trinidad)*, 50: 211-219.

HARRINGTON, G.N. & PRATCHETT, D. Stocking rate trials in Ankole, Uganda. I. Cattle
1974a weight gains at medium and heavy stocking rates under different managements. *J. Agric. Sci. Camb.*, 82: 497-506.

HARRINGTON, G.N. & PRATCHETT, D. Stocking rate trials in Ankole, Uganda. II. Botanical
1974b changes and the results of oesophageal sampling. *J. Agric. Sci. Camb.*, 82: 507-516.

HARRINGTON, G.N. & THORNTON, D.D. A comparison of controlled grazing and manual
1969 hoeing as a means of reducing the incidence of *Cymbopogon afronardus* Stapf. in Ankole pastures, Uganda. *E. Afr. Agric. For. J.*, 35: 154-159.

HARRIS, H.C. Adaptations of four grass species as factors of the environment. *J. Aust. Inst.*
1974 *Agric. Sci.*, 40: 49-50.

HARRISON, E. Digestibility trials on green fodder. *Trop. Agric. (Trinidad)*, 19: 147-150.
1942

HARRISON, R.E. & SNOOK, L.C. *The development of legume pastures on hill country in the*
1971 *Philippines.* Rome, FAO. Mission Rept Misc. No. 17.

HART, R.H. Forage yield, stocking rate, and beef gains on pasture. *Herb. Abstr.*, 42: 345-
1972 353.

HART, R.H. Stocking rate theory and its application to grazing on rangelands. *Proc. 1st Int.*
1978 *Rangeland Congr.*, Denver (Colorado), USA, 547-550.

HART, R.H., MARCHANT, W.H., BUTLER, J.L., HELLWIG, R.E., MCCORMICK, W.C.,
1976 SOUTHWELL, B.L. & BURTON, G.W. Steer gains under six systems of coastal Bermuda grass utilization. *J. Range. Manage.*, 29: 372-375.

HARTLEY, W. The global distribution of the tribes of the Gramineae. *Aust. J. Agric. Res.*,
1950 1: 355-373.

HARTLEY, W. The distribution of the grasses. *In* Barnard, C., ed. *Grasses and grasslands*.
1964   London, Macmillan.
HARTLEY, W. & WILLIAMS, R.J. Centres of distribution of cultivated pasture grasses and
1956   their significance for plant introduction. *Proc. 7th Int. Grassl. Congr.*, Palmerston North, NZ, 190-201.
HARTY, R.L. Germination requirements and dormancy effects in seed of *Urochloa mosam-*
1972   *bicensis. Trop. Grassl.*, 6: 17.
HARTY, R.L. & McDONALD, T.J. Germination behaviour in beach spinifex (*Spinifex hir-*
1972   *sutus* Labill.). *Aust. J. Bot.*, 20: 241-251.
HARVEY, J.M., BEAMES, R.M., HEGARTY, A. & O'BRYAN, M.S. Influence of grazing man-
1963   agement and copper supplementation on the growth rate of Hereford cattle in southeastern Queensland. *Queensl. Agric. Anim. Sci.*, 20: 137-159.
HASSALL, A.C. Native pasture management. *Trop. Grassl.*, 10: 59-60.
1976
HAVILAH, E.J. & MEARS, P.T. Plant nutrition research, beef cattle pastures of northern
1968   New South Wales. *Trop. Grassl.*, 2: 84-92.
HAWTON, D. Atrazine for weed control in the establishment of *Brachiaria decumbens* and
1976   *Panicum maximum. J. Aust. Inst. Agric. Sci.*, 42: 189-191.
HAWTON, D. Weed control in Narok *Setaria* using activated charcoal and diuron. *Queensl.*
1977   *Seed Producers Notes*. New Series. 10.
HAWTON, D. The effectiveness of some herbicides for weed control in *Panicum maximum*
1979   and *Brachiaria decumbens*, and some factors affecting the atrazine tolerance of these species. *Trop. Grassl.*, 14: 34-39.
HAWTON, D., QUINLAN, T.J. & SHAW, K.H. Control of chickweed (*Drymeria cordata*) in
1975   declining tropical pastures. *Trop. Grassl.*, 9: 229-233.
HAYMAN, D.L. The distribution and cytology of the chromosome races of *Themeda australis*
1960   in southern Australia. *Aust. J. Bot.*, 8: 58-68.
HAYLETT, D.G. Green manuring and soil fertility. *S. Afr. J. Agric. Sci.*, 4: 363-378.
1961
HEADY, H.F. Range management in the semi-arid tropics of East Africa according to prin-
1960a   ciples developed in temperate climates. *Proc. 8th Int. Grassl. Congr.*, Reading, UK, 223-225.
HEADY, H.F. *Range management in East Africa*. Nairobi, Govt Printer.
1960b
HEADY, H.F. Influence of grazing on the composition of *Themeda triandra* grassland. *E.*
1966   *Afr. J. Ecol.*, 54: 704-727.
HEGARTY, A. *Effect of nitrogen on subtropical pastures*. Aust. Agrost. Conf., Armidale.
1958   Paper 39. Melbourne, CSIRO.
HELLWIG, R.E. Effect of physical form on drying date of coastal Bermuda grass. *Trans.*
1965   *Amer. Soc. Agric. Eng.*, 8: 253-255.
HENDY, K. Review of natural pastures and their management problems on the north coast
1975   of Tanzania. *E. Afr. Agric. For. J.*, 41: 52-57.
HENDERSON, G.R. & PRESTON, P.T. *Fodder farming in Kenya*. Tech. Publn, Egerton, UK,
1959   Egerton Agric. College.
HENNESSY, D.W. & WILLIAMSON, P.J. The nutritive value of kikuyu grass (*Pennisetum clan-*
1976   *destinum*) leaf and the use of pelleted leaf in rations high or low in energy. *Aust. J. Exp. Agric. Anim. Husb.*, 16: 729-734.
HENRARD, J.T. *A monograph of the genus Aristida*. Leiden, the Netherlands, P.W.M. Trap.
1929
HENTY, E.E. *A manual of the grasses of New Guinea*. Lae, Papua New Guinea, Divn of
1969   Botany, Dept of Forests. Bull. No. 1.
HENZELL, E.F. The use of nitrogenous fertilizers on pastures in the subtropics and tropics.
1962   *Commonw. Bur. Pastures Field Crops Bull.*, 46: 161-172.

HENZELL, E.F. Nitrogen fertilizer responses of pasture grasses in south-eastern Queens-
1963   land. *Aust. J. Exp. Agric. Anim. Husb.*, 3: 290-299.
HENZELL, E.F. Sources of nitrogen for Queensland pastures. *Trop. Grassl.*, 2: 1-17.
1968
HENZELL, E.F. Use of nitrogenous fertilizers on subtropical pastures in Queensland. *J.*
1970   *Aust. Inst. Agric. Sci.*, 36: 206-213.
HENZELL, E.F. Growth response to nitrogen and water on a brigalow soil. *CSIRO Aust.*,
1976-77 *Divn Trop. Crops and Pastures, Rept 1975-76:* 55; *1976-77:* 95.
HENZELL, E.F., MARTIN, A.E. & ROSS, P.J. Recovery of fertilizer nitrogen by Rhodes
1970   grass. *Proc. 11th Int. Grassl. Congr.*, Surfers Paradise, Australia, 411-414.
HENZELL, E.F. & OXENHAM, D.J. Seasonal changes in nitrogen content of three warm-
1964   climate pasture grasses. *Aust. J. Exp. Agric. Anim. Husb.*, 4: 336-344.
HENZELL, E.F., PEAKE, D.C.I., 'T MANNETJE, L. & STIRK, G.B. Nitrogen response of pas-
1975   ture grasses on duplex soils formed from granite in southern Queensland. *Aust. J. Exp. Agric. Anim. Husb.*, 15: 498-507.
HILDER, E.J. & MOTTERSHEAD, B.E. The distribution of plant nutrients through free-
1963   grazing sheep. *Aust. J. Sci.*, 26: 88-89.
HILL, G.D. Grazing under coconuts in Morobe District. *Papua New Guinea Agric. J.*, 21:
1969   10-12.
HILL, J.R.C. The mine dump problem in Rhodesia. *Rhod. Agric. J.*, 69: 65-73.
1972
HINDMARSH, K. Studying the impact of cattle in central Australia. *Rural Res.*, 107: 4-9.
1980
HINTZ, C.W. & GILLILAND, R.T.J. Feed year programmes — what's involved. *Queensl.*
1976   *Agric. J.*, 102: 153-165.
HITCHCOCK, A.S. *The grasses of Ecuador, Peru, and Bolivia.* Vol. 24, part 8. Washington,
1927   Smithsonian Institution.
HITCHCOCK, A.S. *The grasses of Central America.* Vol. 24, part 9. Washington, Smith-
1930   sonian Institution.
HODGES, E.M., JONES, D.W. & KIRK, W.G. Grass pastures in central Florida. *Univ. Fla*
1958   *Agric. Exp. Sta. Bull.*, 484A.
HOFFMAN, G.R., MAAS, E.V., MEYER, J.L., PRITCHARD, T.L. & LANCASTER, D.R. Salt
1979   tolerance of corn in the Delta. *Calif. Agric.*, 33: 11-12.
HOGABOOM, H.G. Establishment and maintenance of pastures in the tropics. *Proc. 6th Int.*
1952   *Grassl. Congr.*, State College (Pennsylvania), USA, 2: 1479-1482.
HOLDER, J.M. Milk production from tropical pastures. *Trop. Grassl.*, 1: 135-141.
1967
HOLMES, T.H.G., LEMERLE, C. & SCHOTTLER, J.H. *Imperata cylindrica* for cattle produc-
1976   tion on Papua New Guinea. In *BIOTROP Workshop in Alang-Alang* (Imperata cylindrica), Bogor, Indonesia.
HOPKINS, B. The role of fire in promoting the sprouting of some savanna species. *J. West*
1963   *Afr. Sci. Assoc.*, 7: 154-162.
HOPKINS, B. Observations of savanna burning in the Olokemeji Forest Reserve, Nigeria. *J.*
1965   *App. Ecol.*, 2: 367-381.
HOPKINSON, D. The potential of some pasture plants in the sisal areas of the Tanga region
1970   of Tanzania. *E. Afr. Agric. For. J.*, 35: 299-310.
HORRELL, C.R. Herbage plants at Serere, Uganda, 1957-61. *E. Afr. Agric. For. J.*, 28: 174.
1963
HOSAKA, E.Y. *Pangola grass in Hawaii.* Univ. Hawaii Agric. Exp Sta. Ext. Circ., 342.
1954
HOSEGOOD, P.H. The root distribution of kikuyu grass and wattle trees. *E. Afr. Agric. For.*
1963   *J.*, 29: 60.

HOUÉROU, H.N. LE. Biological recovery versus desertization. *Econ. Geog.*, 52: 413-420.
1977a
HOUÉROU, H.N. LE. The grassland of Africa: classification, production, evolution and
1977b   development outlook. *Proc. 13th Int. Grassl. Congr.*, Leipzig, German Dem. Rep., 99-116.
HOVELAND, C.S., HASLAND, R.L. & RODRIGUEZ-KABANA, R. Improvement and adapta-
1977   tion of temperate grasses in a low latitude environment. *Proc. Int. Mtg Anim. Prod. Temperate Grassl.*, Dublin [*Herb. Abstr.*, 49: No. 1827].
HSU, CHIEN-CHANG. *Taiwan grasses.* Taipei, Taiwan Provincial Educ. Assoc.
1975
HUBBARD, C.E., WHYTE, R.O., BROWN, D. & GRAY, A.P. *Imperata cylindrica,* taxonomy,
1944   distribution, economic significance and control. *Imp. Agric. Bur. Jt Publ.*, 7.
HUDSON, W.H. *The purple land, an Uruguayan idyll.* London.
1895
HUGHES, K.K. *Assessment of dryland salinity in Queensland.* Queensl. Dept Primary Ind.,
1979   Divn Land Util., Rept, 79/7.
HUGHES, R.M. Effects of temperature and moisture stress on germination and seedling
1979   growth of four tropical species. *J. Aust. Inst. Agric. Sci.*, 45: 125.
HUMPHREY, R.R. *Arizona range grasses.* Univ. Ariz. Agric. Exp. Sta. Bull. No. 298.
1960a
HUMPHREY, R.R. *Forage production on Arizona ranges.* Univ. Ariz. Agric. Exp. Sta. Bull.
1960b   No. 302.
HUMPHREYS, L.R. Studies on the germination, early growth, drought survival and field
1959   establishment of buffel grass (*Cenchrus ciliaris*) and of Birdwood grass (*C. setigerus* Vahl) with particular reference to the Yalleroi district. *J. Aust. Inst. Agric. Sci.*, 25: 321.
HUMPHREYS, L.R. Pasture defoliation practice. *J. Aust. Inst. Agric. Sci.*, 32: 93.
1966
HUMPHREYS, L.R. Buffel grass (*Cenchrus ciliaris*) in Australia. *Trop. Grassl.*, 1: 123-134.
1967
HUMPHREYS, L.R. Agronomic techniques of pasture seed production. *Int. Training Course*
1973   *on Seed Improvement and Certification.* Canberra, Dept Foreign Affairs.
HUMPHREYS, L.R. *Tropical pasture seed production.* Rome, FAO.
1975
HUMPHREYS, L.R. *Tropical pastures and fodder crops.* London, Longman.
1978
HUMPHREYS, L.R. Dairy pasture development and seed production in Sri Lanka. Rome,
1979   FAO internal rept.
HUMPHREYS, L.R. & JONES, R.J. The value of ecological studies in establishment and man-
1975   agement of sown tropical pastures. *Trop. Grassl.*, 9: 125-131.
HUMPHRIES, A.W. *Hyparrhenia hirta* — a promising pasture species. *J. Aust. Inst. Agric.*
1959   *Sci.*, 25: 335-336.
HURST, E. *The poison plants of New South Wales.* Sydney, New South Wales Poison Plants
1942   Committee.
HUTCHINSON, D.J. & BASHAW, E.C. Cytology and reproduction of *Panicum coloratum* and
1964   related species. *Crop Sci.*, 4: 151.
HUTCHINSON, J. & DALZIEL, J.M. *Flora of west tropical Africa.* London, Crown Agents for
1954   Overseas Governments and Administrations.
HUTCHINSON, K.J. & KING, K.L. Sheep numbers and soil arthropods. *Search*, 1: 42.
1970
HUTCHINSON, R.J. A note on wool production responses to fodder conservation in pastoral
1966   systems. *J. Brit. Grassl. Soc.*, 21: 303-304.

HUTTON, E.M. Australian research in pasture introduction and breeding. *Proc. 11th Int.*
1970   *Grassl. Congr.*, Surfers Paradise, Australia, A1-A12.
HUTTON, J.B. Studies on the nutritive value of New Zealand dairy pastures. 1. Seasonal
1961   changes in some chemical components of pastures. *N.Z. J. Agric. Res.*, 4: 583-590.
IDNANI, M.A. Scope for utilization of human urine as manure. *Indian Farming*, 8: 31-32.
1947
INDIAN GRASSLAND AND FODDER RESEARCH INSTITUTE, JHANSI. *Ann. Rept, 1974.* New
1974   Delhi.
INGLE, M. & ROGERS, B.J. The growth of a western strain of *Sorghum halepense* under con-
1961   trolled conditions. *Amer. J. Bot.*, 48: 392-396.
INTERNATIONAL NETWORK OF FEED INFORMATION CENTRES. Data from International Net-
1978   work of Feed Information Centres. Rome, FAO.
ISBELL, R.F. The distribution of black spear grass (*Heteropogon contortus*) in tropical
1969   Queensland. *Trop. Grassl.*, 3: 35-41.
IVENS, G.W. *East African weeds and their control.* Nairobi, Oxford Univ. Press.
1967
IVENS, G.W. *Results of bush control experiments. 9. Effects of burning and goat browsing on
1970   bush growth at Katumani, Kenya.* Rept to Range Mgmt Divn, Min. of Agric. Anim. Husb., Nairobi.
IVORY, D.A. The effect of temperature on the growth of tropical pasture grasses. *J. Aust.*
1976   *Inst. Agric. Sci.*, 42: 113-114.
JACKOBS, J.A. *Improvement of grazing lands in Maharashtra and other Indian states.* Govt
1961   of India; US Int. Coop. Admin, Kansas State Univ. 11 CAKSU-3.
JAGOE, R.B. Beneficial effect of some leguminous shade trees on grassland in Malaya.
1949   *Malayan Agric. J.*, 32: 7-91.
JAVIER, E.Q. The flowering habits and mode of reproduction of Guinea grass (*Panicum
1970   maximum* Jacq.). *Proc. 11th Int. Grassl. Congr.*, Surfers Paradise, Australia, 284-289.
JENNINGS, D.H. The effects of sodium chloride on higher plants. *Biol. Rev.* (Dept of Bot.,
1976   Liverpool Univ., UK) 51: 453-486.
JIMÉNEZ, E.G. & PARRA, R. The capybara, a meat-producing animal of the flooded areas
1973   of the tropics. *Proc. 3rd World Conf. Anim. Prod.*, Melbourne, 81-86.
JOBLIN, A.D.H. The influence of night grazing on the growth rates of Zebu cattle in East
1960   Africa. *J. Brit. Grassl. Soc.*, 15: 212-215.
JOHNSON, J.R. & GURLEY, W.H. *Bahia grass.* Univ. of Georgia College of Agric., Agr. Ext.
1960   Serv. Circ. 464.
JOHNSON, J.R., McGILL, J.F. & GURLEY, W.H. *Coastal Bermuda for grazing, hay and si-
1960   lage.* Univ. of Georgia College of Agric. Agr., Ext. Serv. Circ. 355.
JOHNSON, R.W. Influence of age and environment on control of brigalow. *Proc. 2nd Aust.*
1965   *Weeds Conf.*, 1: 2-6.
JONES, D.I.H. Mineral content of some cultivated grasses grown in Northern Rhodesia.
1964   *Rhod. J. Agric. Res.*, 2: 57-59.
JONES, R.J. The effects of some grazed tropical grass-legume mixtures and nitrogen fer-
1967   tilized grass on total soil nitrogen, organic carbon, and subsequent yields of *Sorghum vulgaris. Aust. J. Exp. Agric. Anim. Husb.*, 7: 66-71.
JONES, R.J. The effect of nitrogen fertilizer applied in spring and autumn on the production
1970   and botanical composition of two subtropical grass/legume mixtures. *Trop. Grassl.*, 4: 97-109.
JONES, R.J. *Some seed problems associated with the use of tropical pasture species and
1973   methods of overcoming them.* Int. Training Course on Seed Improvement and Certification. Canberra, Dept Foreign Affairs.

JONES, R.J. Grass species, fodder conservation and stocking rate effects on nitrogen-fertilized subtropical pastures. *Proc. Aust. Soc. Anim. Prod.*, 11: 445-448.
1976

JONES, R.J. & ALIYU, A.S. The effect of *Eleusine indica*, herbicides and activated charcoal on the seedling growth of *Leucaena leucocephala* cv. Peru. *Trop. Grassl.*, 10: 195-203.
1976

JONES, R.J., DAVIES, J.G. & WAITE, R.B. The contribution of some tropical legumes to pasture yields of dry matter and nitrogen at Samford, southeastern Queensland. *Aust. J. Exp. Agric. Anim. Husb.*, 7: 57-65.
1967

JONES, R.J. & FORD, C.W. The soluble oxalate content of some tropical pasture grasses grown in southeast Queensland. *Trop. Grassl.*, 6: 201-203.
1972a

JONES, R.J. & FORD, C.W. Some factors affecting the oxalate content of the tropical grass *Setaria sphacelata*. *Aust. J. Exp. Agric. Anim. Husb.*, 12: 400-406.
1972b

JONES, R.J. & SANDLAND, R.L. The relation between animal gain and stocking rate. Derivation of the relations from the results of grazing trials. *J. Agric. Sci.*, 83: 335-342.
1974

JONES, R.J., SEAWRIGHT, A.A. & LITTLE, D.A. Oxalate poisoning in animals grazing the tropical grass *Setaria*. *J. Aust. Inst. Agric. Sci.*, 36: 41-43.
1970

JONES, R.M. Mortality of some tropical grasses and legumes following frosting in the first winter after sowing. *Trop. Grassl.*, 3: 57-63.
1969

JONES, T.N. & DUDLEY, B.F. Methods of field-curing hay. *J. Amer. Soc. Agric. Eng.*, 29: 159-161.
1948

JORDAN, S.M. Reclamation and pasture management in the semi-arid areas of Kitui District, Kenya. *E. Afr. Agric. J.*, 23: 84-88.
1957

JOZWIK, F.X. Some systematic aspects of Mitchell grasses (*Astrebla* F. Muell.). *Aust. J. Bot.*, 17: 359-374.
1969

JOZWIK, F.X. Response of Mitchell grasses (*Astrebla* F. Muell.) to photoperiod and temperature. *Aust. J. Agric. Res.*, 21: 395-405.
1970

JOZWIK, F.X., NICHOLLS, A.O. & PERRY, R.A. Studies on the Mitchell grasses (*Astrebla* F. Muell.). *Proc. 11th Int. Grassl. Congr.*, Surfers Paradise, Australia, 48-51.
1970

JUDD, I.B. *Handbook of tropical forage grasses*. New York, Garland STPM Press.
1979

KAISER, A.G. Response by calves grazing kikuyu pastures to grain and mineral supplements. *Trop. Grassl.*, 9: 191-198.
1975

KAISER, A.G. & COLMAN, R.L. Butterfat production from nitrogen-fertilized kikuyu. *Wollongbar Agric. Res. Sta. Ann. Rept, 1968-69.*
1969

KANNEGIETER, A. *Forage crop development*. Rome, FAO Rept TA. 2817. Govt of Ceylon.
1970

KARUE, C.N. The nutritive value of herbage in semi-arid lands of East Africa. 1. Chemical composition. *E. Afr. Agric. For. J.*, 40: 89-95.
1974

KARUE, C.N. Intake and digestibility of *Themeda triandra* hay by age-paired Boran and Hereford steers. *E. Afr. Agric. For. J.*, 41: 35-41.
1975

KASASIAN, L. *Weed control in the tropics*. London, Leonard Hill.
1971

KAYONGO-MALE, H. & THOMAS, J.W. Mineral composition of some tropical grasses and their relationships to the organic constituents and estimates of digestibility. *E. Afr. Agric. For. J.*, 40: 428-438.
1970

KELLY, T.K. Pasture hay on the central coast. *Queensl. Agric. J.*, 98: 384-386.
1972

KEMP, D.R. The seasonal growth of tropical pasture grasses on the mid-north coast of New South Wales. *J. Aust. Inst. Agric. Sci.*, 41: 50-51.
1975

KEMP, E.D.S, MACKENZIE, R.M. & ROMNEY, D.H. Productivity of pasture in British Honduras. III. Jaragua grass. *Trop. Agric. (Trinidad)*, 38: 161-171.
1971

KENDREW, J. *The climates of the continents*. Oxford, Clarendon Press.
1961

KERR, D.V. Potential of irrigated tropical pasture for the dairying industry. *Queensl. Agric.*
1979   *J.*, 105: 219-221.
KEYA, N.C.O. The effect of N and P fertilizers on the productivity of *Hyparrhenia* grass-
1973   land. *E. Afr. Agric. For. J.*, 39: 195-200.
KEYA, N.C.O., OLSEN, F.J. & HOLLIDAY, R. Comparison of seedbeds for oversowing a
1972   *Chloris gayana* (Kunth.) *Desmodium uncinatum* (Jacq.) mixture in *Hyparrhenia*
        grassland. *E. Afr. Agric. For. J.*, 37: 286-293.
KHAN, C.M.A. Sand dune rehabilitation in Thal, Pakistan. *J. Range Manage.*, 21: 316-321.
1968
KHYBRI, M.L. & MISHRA, D.D. Root studies on some selected grasses in eastern Nepal.
1967   *Indian For.*, 93: 400-406.
KIDDER, R.W. Ten of beef per acre of grass in 12 months. *Breeders Gaz.*, 117: 8.
1952
KIDDER, R.W., BEARDSLEY, D.W. & ERWIN, T.C. Photosensitization in cattle grazing
1961   frosted Bermuda grass. *Fla Agric. Exp. Sta. Bull.*, 630.
KING, N.J., MUNGOMERY, R.W. & HUGHES, C.G. *Manual of cane growing*. Sydney, Angus
1965   and Robertson.
KLEINSCHMIDT, H.E. & JOHNSON, R.W. *Weeds of Queensland*. Brisbane, Dept of Primary
1977   Industries.
KNIGHT, W.E. The influence of photoperiod and temperature on growth, flowering and
1955   seed production of dallis grass (*Paspalum dilatatum* Poir). *J. Am. Soc. Agron.*,
        47: 555-559.
KNIGHT, W.E. & BENNETT, H.W. Preliminary report of the effect of photoperiod and tem-
1953   perature on the flowering and growth of several southern grasses. *Agron. J.*, 45:
        268-269.
KNIPE, O.D. Influence of temperature on the germination of some range grasses. *J. Range*
1967   *Manage.*, 20: 298-299.
KNIPE, O.D. Light delays germination of alkali sacaton. *J. Range Manage.*, 24: 152-154.
1971
KNOX, R.B. Apomixis: seasonal and population differences in a grass. *Science*, 157: 325.
1967
KOCH, B.L. Association nitrogenase activity by some Hawaiian grass roots. *Plant Soil*, 47:
1977   703-706.
KOECHLIN, J. *Flore du Gabon. Gramineae*. Paris, Laboratoire de Phanérogamie.
1962
KOWITHAYAKORN, L. & KANNASOOT, T. Studies of flowering patterns and seed develop-
1978   ment in Guinea (*Panicum maximum* Jacq.) and plicatulum (*Paspalum
        plicatulum*) grasses. *Khon Kaen Univ. (Thailand) Pasture Improvement Project,
        Ann. Rept*, 1978: 104-108.
KRETSCHMER, A.E. *Stylosanthes humilis*, a summer-growing self-regenerating annual
1965   legume for use in Florida permanent pastures. *Proc. Soil Crop Sci. Soc. Fla*, 25:
        248-262.
KRETSCHMER, A.E. Four years' results with siratro (*Phaseolus atropurpureus* DC) in south
1966   Florida. *Proc. Soil Crop Sci. Soc. Fla*, 26: 238-245.
KRISHNA-MARAR, M.M. Inter-cultivation in coconut gardens, its importance. *Indian*
1953   *Coconut J.*, 4: 131-137.
KRISHNA-MARAR, M.M. Trial of inter-cultivation practices in coconut gardens. *Indian*
1961   *Coconut J.*, 14: 87-99.
KRUGER, J.A. & EDWARDS, P.J. Utilization and relative palatability of different grass
1972   species. *Proc. Grassl. Soc. South Afr.*, 7: 146-155.
KRUPKO, I. & DAVIDSON, R.L. An experimental study of *Stoebe vulgaris* in relation to graz-
1961   ing and burning. *Emp. J. Exp. Agric.*, 29: 175-180.

Kucera, C.L. *The grasses of Missouri.* Columbia, Univ. of Missouri Press.
1961

Lamotte, M. Structure and functioning of the savanna ecosystems of Lamto (Ivory Coast).
1979   In *Tropical grazing land ecosystems.* Paris, Unesco, 511-561.

Lamprey, H.F. Structure and functioning of the semi-arid grazing land ecosystem of the
1979   Serengeti region (Tanzania). In *Tropical grazing land ecosystems.* Paris, Unesco.

Laredo, M.A. & Minson, D.J. The voluntary intake, digestibility and retention time by
1973   sheep of leaf and stem fractions of five grasses. *Aust. J. Agric. Res.,* 24: 875-888.

Larkin, R.M. Treating timber and preparing for pastures. *Queensl. Agric. J.,* 91: 78-81.
1965

Lazarides, M. *The grasses of central Australia.* Canberra, Aust. Nat. Univ. Press.
1970

Lazarides, M. *Aristida L. (Poaceae, Aristideae) in Australia.* Canberra, CSIRO Aust.
1979   Herbarium.

Lebrun, J. [Botanical studies in the district of Ubangi.] *Bull. du Congo Belge,* 23: 135-146.
1932   (In French)

Leelavathy, K.M. Effect of rhizosphere fungi on seed germination. *Plant Soil,* 30: 473-
1969   476.

Leggett, E.K. Broadleaf paspalum. A new paspalum for the North Coast. *Agric. Gaz.*
1968   *NSW,* 78: 690-693.

Leigh, J.H. Behaviour of certain varieties of love grass when grown as spaced plants and in
1961a   broadcast stands. *Emp. J. Exp. Agric.,* 29: 265-268.

Leigh, J.H. The relative palatability of various varieties of weeping love grass (*Eragrostis*
1961b   *curvula* (Schrad.) Nees). *J. Brit. Grassl. Soc.,* 16: 135-140.

Leigh, J.H. & Davidson, R.L. *Eragrostis curvula* (Schrad.) Nees and some other African
1968   love grasses. *Plant Introd. Rev.* (Divn Plant Ind., CSIRO Aust.), 5: 21-40.

Leithead, H.L., Yarlett, L.L. & Shiflet, T.N. *100 native forage grasses in 11 southern*
1971   *states.* USDA Agric. Handbook No. 389.

Leslie, J.K. Average fate of seed after six weeks field exposure. *Queensl. J. Agric. Anim.*
1965   *Sci.,* 22: 17.

Leslie, J.K. Temperature and *Chloris gayana* germination. Univ. of Queensl. (Ph.D.
1970   thesis)

Levitt, M.S., Taylor, V.J. & Hegarty, A. Studies on grass silage from predominantly
1962   *Paspalum dilatatum* pastures in southeastern Queensland. 1. A comparative evaluation of the additives metabisulphite and molasses. *Queensl. J. Agric. Anim. Sci.,* 19: 153-175.

Levitt, M.S., Hegarty, A. & Radel, M.J. Studies on grass silage from predominantly
1964   *Paspalum dilatatum* pastures in southeastern Queensland. 2. Influence of length of cut on silages with and without molasses. *Queensl. J. Agric. Anim. Sci.,* 21: 181-192.

Levitt, M.S. & O'Bryan, M.S. Studies on grass silage from predominantly *Paspalum*
1965   *dilatatum* pastures in southeastern Queensland. 3. Influence of fertilization with nitrogen and method of harvesting on silages with and without the addition of molasses. *Queensl. J. Agric. Anim. Sci.,* 22: 109-123.

Lim, H.K. Animal feeding stuffs. Compositional data of feeds and concentrates. *Malay*
1968   *Agric. J.,* 46: 63-79.

Linedale, A.I. Fighting the common reed. *Queensl. Canegrowers' Quart. Bull.,* 38: 46-48.
1974

Lipschitz, N. & Waisel, Y. Existence of salt glands in various genera of the Gramineae.
1974   *New Phytol.,* 73: 507-513.

Lipschitz, N., Shomer-Ilan, A., Eshel, A. & Waisel, Y. Salt glands on leaves of Rhodes
1974   grass (*Chloris gayana,* Kunth.). *Ann. Botany,* 38: 459-462.

Little, D.A. Observations on the phosphorus requirement of cattle for growth. *Res. Vet.*
1980   *Sci.,* 28: 258-260.

LLOYD, D.L. Makarikari grass studies, Toowoomba, Queensland. *Proc. 11th Int. Grassl.*
1970   *Congr.*, Surfers Paradise, Australia, 230-235.

LLOYD, D.L. Growth and development of Makarikari grasses and three other sub-tropical
1971   species grown in the field at Toowoomba, Queensland. *Aust. J. Exp. Agric. Anim. Husb.*, 11: 525.

LLOYD, D.L. & SCATENI, W. Makarikari grasses for heavy soils. *Queensl. Agric. J.*, 94: 721-
1968   724.

LOCH, D.S. *Paspalum plicatulum* cultivars; a case of incorrect labelling. *Trop. Grassl.*, 10:
1976   219-220.

LOCH, D.S. *Brachiaria decumbens* (signal grass); a review with particular reference to
1977   Australia. *Trop. Grassl.*, 11: 141-157.

LOCH, D.S. Basilisk signal grass; a productive pasture grass for the humid tropics. *Queensl.*
1978   *Agric. J.* 104: 402-406.

LOCH, D.S. Coastal pasture seed production. *Trop. Grassl.*, 13: 183-185.
1979

LODGE, G.M. Effect of fertility level on the yield of some native perennial grasses in the
1979   northwest slopes, New South Wales. *Aust. Range. J.*, 1: 327-333.

LOGAN, J.M. Erosion problems on salt-affected areas. *J. Soil Conserv. Serv.* (NSW), 14:
1958   229-242.

LONG, M.I.E., THORNTON, D.D. & MARSHALL, B. Nutritive value of grasses in Ankole and
1969   Queen Elizabeth National Park, Uganda. II. Crude protein, crude fibre and soil nitrogen. *Trop. Agric. (Trinidad)*, 46: 43-46.

LORIMER, M.S. Forage selection by sheep grazing *Astrebla* spp. pastures in northwest
1978   Queensland. *Trop. Grassl.*, 12(2): 97-108.

LOVVORN, R.L. The effects of fertilization, species competition, and cutting treatments on
1944   the behaviour of dallis grass, *Paspalum dilatatum* Poir, and carpet grass, *Axonopus affinis* Chase. *J. Amer. Soc. Agron.*, 36: 590-600.

LOVVORN, R.L. The effect of defoliation, soil fertility, temperature, and length of day on
1945   the growth of some perennial grasses. *J. Amer. Soc. Agron.*, 37: 570-582.

LOWE, K.F. The value of a frost-tolerant *Setaria* component in mixed pastures for autumn
1976   saved feed in southeastern Queensland. *Trop. Grassl.*, 10: 89-97.

LOWE, K.F. & BOWDLER, T.M. Tropical grass and legume yield on a soloth soil in subcoastal
1977   southeastern Queensland. *Trop. Grassl.*, 11: 231-238.

LOWE, K.F., FILET, G.F., BURNS, M.A. & BOWDLER, T.M. Effect of sod-seeded siratro on
1977   beef production and botanical composition of native pasture in southeastern Queensland. *Trop. Grassl.*, 11: 223-230.

LUCK, P.E. *Setaria*, an important pasture grass. *Queensl. Agric. J.*, 105: 136-143.
1979

LUCK, P.E. & DOUGLAS, N.J. Dairy pasture research and development in the near north
1966   coast centred on Cooroy, Queensland. *Proc. Trop. Grassl. Soc. Aust.*, 6: 35-53.

LUDLOW, M.M. Physiology of growth and chemical composition. *In* Shaw, N.H. and Bryan,
1976   W.W. eds. *Tropical pasture research; principles and methods.* Com. Agric. Bur. Bull. 511.

LUDLOW, M.M. & WILSON, G.L. Studies on the productivity of tropical pasture plants.
1968   I. Growth analysis, photosynthesis and respiration of Hamil grass and siratro in a controlled environment. *Aust. J. Agric. Res.*, 19: 35-45.

LUDLOW, M.M. & WILSON, G.L. Studies on the productivity of tropical pasture plants.
1970a   II. Growth analysis, photosynthesis and respiration of 20 species of grass and legumes in a controlled environment. *Aust. J. Agric. Res.*, 21: 183-194.

LUDLOW, M.M. & WILSON, G.L. Growth of some tropical grasses and legumes at two tem-
1970b   peratures. *J. Aust. Inst. Agric. Sci.*, 36: 43-45.

LUDLOW, M.M. & WILSON, G.L. Net photosynthesis rates of tropical grass and legume
1970c   leaves. *Proc. 11th Int. Grassl. Congr.*, Surfers Paradise, Australia, 534-538.

MAAS, E.V. & HOFFMAN, G.J. Crop salt tolerance. Evaluation of existing data. *In* Dregne
1977 H.E., ed. *Managing saline water for irrigation. Proc. Int. Salinity Conf.*, Lubbock (Texas), USA, 187-198.
MAGADAN, P.B., JAVIER, E.Q. & MADAMBA, J.C. Beef production on native (*Imperata*
1974 *cylindrica* (L.) Beauv.) and para grass (*Brachiaria mutica* (Forsk.) Stapf) pastures in the Philippines. *Proc. 12th Int. Grassl. Congr.*, Vol. 3, part 1, 293-298.
MAGOON, M.L., & SHANKARNARAYAN, K.A. Forage on saline and alkaline land. *Indian*
1974 *Farming*, 24: 27-30.
MAHADEVAN, V. & VENKATAKRISHNAN, R. Losses in ensilage, and composition of Napier
1957 grass molassed salt silage. *Curr. Sci.*, (1): 16.
MAIDEN, J.H. Useful Australian plants. *Agric. Gaz. NSW*, 5: 833-835.
1894
MALCOLM, C.V. *Plant collection for pasture improvement in saline and arid environments.*
1971 West. Aust. Dept Agric. Tech. Bull. No. 6.
MALCOLM, C.V. *Production from salt-affected soils.* Report on visit to India and Pakistan.
1980 South Perth, West. Aust. Dept Agric.
MALCOLM, C.V. & LAING, I.A.F. Paspalum distichum *for salty seepages and lawns.* West.
1976 Aust. Dept Agric. Tech. Bull. No. 3696.
MALM, N.R. & RACHIE, K.O. *The setaria millets. A review of the world literature.* Lincoln
1971 (Nebraska), USA, Exp. Sta., Univ. of Nebraska College of Agric. No. S.B.513.
MANN, H.H. Millets in the Middle East. *Emp. J. Exp. Agric.*, 14: 208-216.
1946
'T MANNETJE, L. *A key based on vegetative characters of some introduced species of* Pas-
1961 palum *L.* CSIRO Aust., Divn Trop. Past. Tech. Paper No. 1.
'T MANNETJE, L. The effects of some management practices on pasture production. *Trop.*
1972 *Grassl.*, 6: 260-263.
'T MANNETJE, L. Relations between pasture attributes and live-weight gains on a sub-tropi-
1974 cal pasture. *Proc. 12th Int. Grassl. Congr.*, Moscow, 3: 299-304.
'T MANNETJE, L. Beef production from pastures on granitic soils. *CSIRO Aust., Divn Trop.*
1976 *Crops and Pastures Rept*, 1975-76: 10.
'T MANNETJE, L. Studies on buffel grass. *CSIRO Aust., Divn Trop. Crops and Pastures Rept,*
1977 1976-77: 28-29.
'T MANNETJE, L. The role of improved pastures for beef production in the tropics. *Trop.*
1978 *Grassl.*, 12: 1-9.
'T MANNETJE, L. & SHAW, N.H. Nitrogen fertilizer responses of a *Heteropogon contortus*
1972 and a *Paspalum plicatulum* pasture in relation to rainfall in central coastal Queensland. *Aust. J. Exp. Agric. Anim. Husb.*, 12: 28-35.
'T MANNETJE, L., SIDHU, A.S. & MURUGAIAH, M. Cobalt deficiency in cattle in Johore,
1976 liveweight changes and response to treatment. *MARDI Res. Bull.*, 4: 90.
MAPPLEDORAM, B. & THERON, E.P. Notes on the relative merits of four cultivars of *Era-*
1970 *grostis curvula* in Natal. *Proc. Grassl. Soc. South Afr.*, 5: 121-125.
MARLEY, J. Control of Johnson and Columbus grasses. *Queensl. Agric. J.*, 104: 95-97.
1978
MARRIOTT, S. & HARVEY, J.M. Bush hay conservation in northwestern Queensland.
1951 *Queensl. Agric. J.*, 73: 249-255.
MARSHALL, B. & BREDON, R.M. The nutritive value of *Themeda triandra. E. Afr. Agric.*
1967 *For. J.*, 32: 375-379.
MARTIN, C.C. The role of glumes and gibberellic acid in dormancy in *Themeda triandra*
1977 spikelets. *Physiol. Plant Pathol.*, 33: 171-176.
MARTIN, J.H. & LEONARD, W.H. *Principles of field crop production.* New York, Macmil-
1959 lan.
MARTIN, R.J. A review of carpet grass (*Axonopus affinis*) in relation to the improvement of
1975 carpet grass based pastures. *Trop. Grassl.*, 9: 9-20.

MARTOATMADJOUN, R.S. *Proc. BIOTROP Workshop on Alang-Alang* (Imperata cylin-
1976 drica), Bogor, Indonesia, 27-29 July.
MARTINOVICH, D. & SMITH, B. Kikuyu poisoning of cattle. 1. Clinical and pathological
1973 findings. *N.Z. Vet. J.*, 21: 55-63.
MATCHES, A.G. & MOTT, G.O. Estimating the parameters associated with grazing systems.
1975 *Proc. 3rd World Conf. Anim. Prod.*, Melbourne, May 1973, 203-208.
MATHEWS, R.H. & SUTHERLAND, A.K. The oxalate content of some Queensland pasture
1952 plants. *Queensl. J. Agric. Anim. Sci.*, 9: 317-334.
MAWSON, W.F. Brahman cattle grow faster than British in the north. *Queens. Agric. J.*, 82:
1956 173-179.
MCCOSKER, T.H. & TEITZEL, J.K. A review of Guinea grass (*Panicum maximum*) for the
1975 wet tropics of Australia. *Trop. Grassl.*, 9: 177-190.
MCILROY, R.J. *An introduction to tropical grassland husbandry*. Oxford, Oxford University
1964 Press.
MCKAY, A.D. Seasonal and management effects on the composition and availability of her-
1971 bage steer diet, and liveweight gains in a *Themeda triandra* grassland in Kenya.
 2. Results of herbage studies, diet selected, and liveweight gains. *J. Agric. Sci. Camb.*, 76: 9-25.
MCKENZIE, J.A. & CHHEDA, H.R. Comparative root growth studies of *Cynodon I.B.8*, an
1971 improved variety of *Cynodon* forage grass suitable for southern Nigeria and two
 other *Cynodon* varieties. *Nigerian Agric. J.*, 7: 91-97.
MCKENZIE, R.A. Induced calcium deficiency in horses grazing introduced pastures in
1978 Queensland. *Trop. Grassl.*, 12: 212.
MCLENNAN, L.W. Carpet grass. Has it a place in the north coast pastures? *Agric. Gaz.*
1936 *NSW*, 47: 555-558, 601-604.
MCLEOD, C.C. Borabu pasture and range development centre, northeast Thailand. Rept,
1972 New Zealand Colombo Plan.
MCMEEKAN, C.P. & WALSHE, M.J. The inter-relationship of grazing methods and stocking
1963 rate in the efficiency of pasture utilization by dairy cattle. *J. Agric. Sci. Camb.*,
 61: 147-163.
MCWILLIAM, J.E., CLEMENTS, R.J. & DOWLING, P.M. Some factors influencing the germi-
1970 nation and early seedling development of pasture plants. *Aust. J. Agric. Res.*, 21:
 19-32.
MCWILLIAM, J.E. & DOWLING, P.M. Factors influencing the germination and establishment
1970 of pasture seed on the soil surface. *Proc. 11th Int. Grassl. Congr.*, Surfers
 Paradise, Australia, 578.
MEAD, K.J. & NORMAN, J.B. The value of fodder cane. *Agric. Gaz. NSW*, 70: 296-299.
1950
MEARS, P.T. Kazungula setaria — tropical legume balance during establishment in pre-
1968 pared seedbeds. *Aust. Grassl. Conf.*, 1 Sect. 3a.
MEARS, P.T. Potassium and nitrogen requirement of kikuyu grass. *Wollongbar Agric. Res.*
1969 *Sta. Ann. Rept, 1968:* 9.
MEARS, P.T. Kikuyu (*Pennisetum clandestinum*) as a pasture grass: a review. *Trop. Grassl.*,
1970 4: 139-152.
MEARS, P.T. & HUMPHREYS, L.R. Nitrogen response and stocking rate of *Pennisetum clan-*
1974 *destinum* pastures. II. Cattle growth. *J. Agric. Sci. Camb.*, 83: 469-478.
MEDLING, P.C. *Mejora de pastos y cultivos forrajeros, Panamá. Forrajes, conservación y*
1972 *manejo de pastos*. Rome, FAO. AGP/PAN.10. Informe técnico 1.
MELLOR, W., HIBBERD, M.J. & GROF, B. Beef cattle liveweight gains from mixed pastures
1973a of some Guinea grasses and legumes on the wet tropical coast of Queensland.
 Queensl. J. Agric. Anim. Sci., 30: 259-266.
MELLOR, W., HIBBERD, M.J. & GROF, B. Performance of Kennedy ruzi grass on the wet
1973b tropical coast of Queensland. *Queensl. J. Agric. Anim. Sci.*, 30: 53-56.

MEREDITH, D. *The grasses and pastures of South Africa.* Johannesburg, Central News
1955   Agency Ltd.
MIDDLETON, C.H. Riversdale — a selected line of common Guinea grass. *Queensl. Agric.*
1977   *J.*, 103: 405-406.
MIDDLETON, C.H. & BARRY, G.A. A study of oxalate concentration in five grasses in the
1978   wet tropics of Queensland. *Trop. Grassl.*, 12: 28-35.
MIDDLETON, C.H. & MCCOSKER, T.H. Makueni, a new Guinea grass for north Queens-
1975   land. *Queensl. Agric. J.*, 101: 351-355.
MILES, J.F. *Plant introduction trials in central coastal Queensland.* CSIRO Aust., Divn Plant
1949   Ind., Rept No. 6.
MILES, L.G. Sorghum growing in Queensland. *Queensl. Agric. J.*, 69: 249-265.
1949
MILFORD, R. Nutritional values for 17 tropical grasses. *Aust. J. Agric. Res.*, 11: 139-148.
1960a
MILFORD, R. Nutritive value of subtropical pasture species under Australian conditions.
1960b   *Proc. 8th Int. Grassl. Congr.*, Reading, UK, 474-479.
MILFORD, R. & HAYDOCK, K.P. The nutritive value of protein in subtropical pasture species
1965   grown in southeast Queensland. *Aust. J. Exp. Agric. Anim. Husb.*, 5: 13-17.
MILFORD, R. & MINSON, D.J. The relation between the crude protein and digestible crude
1965   protein content of tropical pasture plants. *J. Brit. Grassl. Soc.*, 20: 177-179.
MILFORD, R. & MINSON, D.J. The digestibility and intake of six varieties of Rhodes grass
1968   (*Chloris gayana* Kunth.). *Aust. J. Exp. Agric. Anim. Husb.*, 8: 413-418.
MILLER, H.W. & HAFENRICHTER, A.L. New forage for the Western ranges. *West. Livestock*
1958   *J.*, January.
MILLER, T.B., RAINS, A.B. & THORPE, R.J. The nutritive value and agronomic aspects of
1964   some fodders in northern Africa. *J. Brit. Grassl. Soc.*, 19: 77-81.
MILLER, W.J., CLIFTON, C.M. & CAMERON, N.W. Ensiling characteristics of Coastal Ber-
1963   muda grass harvested at pre-head and full head stages of growth. *J. Dairy Sci.*, 46:
        727-732.
MILLINGTON, A.J., BURVILL, G.H. & MARSH, B. A'B. Salt tolerance, germination and
1951   growth tests under controlled salinity conditions. *J. Agric. West. Aust.*, 28: 198-
        211.
MILLINGTON, R.W. & WINKWORTH, R.E. Methods of screening introduced forage species
1970   for arid central Australia. *Proc. 11th Int. Grassl. Congr.*, Surfers Paradise,
        Australia, 235-239.
MILLS, P.F.L. & BOULTWOOD, J.N. A comparison of *Paspalum notatum* accessions for yield
1978   and palatability. *Rhod. Agric. J.*, 75: 71-74.
MINSON, D.J. The voluntary intake and digestibility in sheep of chopped and pelleted
1967   *Digitaria decumbens* (Pangola grass) following a late application of fertilizer ni-
        trogen. *Brit. J. Nutr.*, 21: 587-597.
MINSON, D.J. Animal nutrition. *Trop. Grassl.*, 2: 197-207.
1968
MINSON, D.J. The digestibility and voluntary intake of six *Panicum* varieties. *Aust. J. Exp.*
1971a   *Agric. Anim. Husb.*, 11: 19-25.
MINSON, D.J. The nutritive value of tropical pasture. *J. Aust. Inst. Agric. Sci.*, 37: 255-263.
1971b
MINSON, D.J. The digestibility and voluntary intake by sheep of six tropical grasses. *Aust.*
1972   *J. Exp. Agric. Anim. Husb.*, 12: 21-27.
MINSON, D.J. Effect of fertilizer nitrogen on digestibility and voluntary intake of *Chloris*
1973   *gayana*, *Digitaria decumbens* and *Pennisetum clandestinum*. *Aust. J. Exp. Agric.*
        Anim. Husb., 13: 153-157.
MINSON, D.J. Pasture management and animal nutrition. *Forage Res.*, 1: 1-10.
1975

MINSON, D.J. Chemical composition and nutritive value of tropical legumes. *In* Sherman,
1977   P.J. *Tropical forage legumes.* Rome, FAO.
MINSON, D.J. Relationships of conventional and preferred fractions to determined energy
1979   values. *Proc. Workshop on Analytical Methods,* Ottawa.
MINSON, D.J. & MCLEOD, M.N. The digestibility of temperate and tropical grasses. *Proc.*
1970   *11th Int. Grassl. Congr.,* Surfers Paradise, Australia, 719-722.
MINSON, D.J. & MILFORD, R. The energy values and nutritive value indices of *Digitaria*
1966   *decumbens, Sorghum almum* and *Phaseolus atropurpureus. Aust. J. Agric. Res.,*
       17: 411-423.
MINSON, D.J. & MILFORD, R. Intake and crude protein content of mature *Digitaria decum-*
1967   *bens* and *Medicago sativa. Aust. J. Exp. Agric. Anim. Husb.,* 7: 546.
MISHRA, M.L. & CHATTERJEE, B.N. Seed production in the forage grasses *Pennisetum*
1968   *polystachyon* and *Andropogon gayanus* in the Indian tropics. *Trop. Grassl.,* 2: 51-
       56.
MISRA, K.N., AMBASHT, R.S. & SINGH, A.K. Effect of some dominant grasses on the prop-
1977   erties of soil subjected to erosional stress. *Acta Bot. Indica,* 5 (2, supplement): 33-
       34 [*Herb. Abstr.,* 49: No. 1969].
MITCHELL, K.J. Growth of pasture species under controlled environment. I. Growth at vari-
1956   ous levels of constant temperature. *N.Z. J. Sci. Tech.,* 38A.
MOHAMED SALEEM, M.A., CHHEDA, H.R. & CROWDER, L.V. Effects of lime on herbage
1975   production and chemical composition of *Cynodon I.B.8* and on some chemical
       properties of the soil. *E. Afr. Agric. For. J.,* 40: 217-226.
MONNIER, F. & POIT, J. [Grazing problems in the Adamoua (Cameroon).] *Bois. for. trop.,*
1964   97: 3-16; 98: 13-25. (In French)
MOORE, A.W. The influence of fertilization and cutting on a tropical grass/legume pasture.
1965   *Exp. Agric.,* 1: 193-200.
MOORE, C.P. & BUSHMAN, D.H. Potential beef production on intensely managed elephant
1978   grass. *Trop. Agric.* 55: 335-341.
MOORE, C.W.E. Distribution of grasslands. *In* Barnard, C., ed. *Grasses and grasslands.*
1964   London, Macmillan.
MOORE, R.M. *Australian grasslands.* Canberra, Aust. Nat. Univ. Press.
1970
MOORE, R.M. & PERRY, R.A. Vegetation. *In* Moore, R.M., ed. *Australian grasslands.*
1970   Canberra, Aust. Nat. Univ. Press.
MOREL, G. & BOURLIÈRE, F. Relations écologiques des avifaunes sédentaires et migratrices
1962   dans une savane sahélienne du Bas-Sénégal. *Terre Vie,* 16: 371-393.
MORRIS, J.G. & GARTNER, R.J.W. The sodium requirements of growing steers given an all-
1971   sorghum grain ration. *Brit. J. Nutr.,* 25: 191-205.
MOTT, G.O. Grazing pressure and the measurement of pasture production. *Proc. 8th Int.*
1960   *Grassl. Congr.,* Reading, UK, 606-611.
MOTT, G.O., QUINN, L.R. & BISSCHOFF, W.V.A. The retention of nitrogen in a soil-plant-
1970   animal system in Guinea grass (*Panicum maximum*) pastures in Brazil. *Proc. 11th*
       Int. Grassl. Congr., Surfers Paradise, Australia, 414.
MOTTA, M.S. *Panicum maximum. Emp. J. Exp. Agric.,* 21: 33-41.
1953
MOTTA, M.S. Grazing management and production of some tropical grasses in British West
1956   Indies. *Proc. 7th Int. Grassl. Congr.,* Palmerston North, NZ, 539-546.
MOTTA, M.S. *Investigations 1958-9 — Jamaica.* Min. Agric. and Lands Bull. 58 (New
1968   Series).
MUGERWA, J.S. & OGWANG, B.H. Dry matter production and chemical composition of
1979   elephant grass hybrids. *E. Afr. Agric. For. J.,* 42: 60-65.
MUKERJI, S.K. & CHATTERJI, B.N. Culture of *Pennisetum pedicellatum* in Bihar for forage
1955   and soil conservation. *J. Soil Water Conserv. India,* 3: 161-163.

MULDOON, D.K. & PEARSON, C.J. The hybrid between *Pennisetum americanum* and *P.*
1979   *purpureum. Herb. Abstr.*, 49: 189-200.
MUNOZ, H., ROUX, H. & SEMPLE, A.T. Seasonal yields and palatability of elephant grass.
1961   *J. Anim. Sci.*, 20: 960.
MURATA, Y., IYAMA, J. & HONMA, T. Studies on the photosynthesis of forage crops.
1965   4. Influence of air temperature upon the photosynthesis and respiration of alfalfa and several southern-type forage crops. *Proc. Crop Sci. Soc.* (Japan), 34: 154-158.
MURRAY, R.M., TELENI, E. & PLAYNE, M.J. Sodium status of steers grazing tropical *Both-*
1976   *riochloa* pastures. *Proc. 11th Bienn. Conf. Aust. Soc. Anim. Prod.*, Adelaide, Australia.
MURTAGH, G.J. Chemical seed-bed preparation and the efficiency of nitrogen utilisation of
1971   sod-sown oats. I. Comparison of seed-beds. *Aust. J. Exp. Agric. Anim. Husb.*, 11: 299-305.
MURTAGH, G.J. The need for alternative techniques of productivity assessment in grazing
1975   experiments. *Trop. Grassl.*, 9: 151-158.
MURTAGH, G.J. Use of herbicide to reduce grass competition in a white clover sward. *Trop.*
1977   *Grassl.*, 11: 121-124.
MUSLERA, P.E., RATERA, G.C., AMBEL, E. & RUIZ, J.A. The potential and productivity
1975   of irrigated pastures in southern Spain. *In* Pasture and forage production in seasonally arid climates. *Proc. 6th General Meeting European Grassl. Federation*, Madrid, 177-182 [*Herb. Abstr.*, 49: 3: No.945].
MWAKHA, E. Observations on the growth of Bungoma grass (*Entolasia imbricata* Stapf.).
1970   *CSIRO Aust. Plant Introd. Rev.*, 7, (1).
MWAKHA, E. Germination depth of Bungoma grass, *Entolasia imbricata* Stapf. *E. Afr.*
1971   *Agric. For. J.*, 37: 26-28.
MYBURGH, S.J. The digestibility of African feeds. 1. The digestibility coefficients of some
1937   natural grasslands. *Onderstepoort J. Vet. Sci. Anim. Ind.*, 9: 165-184.
MYERS, A. Longevity of seeds of native grasses. *Agric. Gaz. NSW*, 51: 405.
1940
MYERS, A. Curly Mitchell grass. Reasons for poor field germination. *Agric. Gaz. NSW*, 53:
1942a   254.
MYERS, A. Germination of seed of curly Mitchell grass (*Astrebla lappacea* Domin.). *J. Aust.*
1942b   *Inst. Agric. Sci.*, 8: 31-32.
MYERS, R.J.K. *Environmental role of nitrogen-fixing blue-green algae and asymbiotic bac-*
1978   *teria*. Stockholm, Swedish Natural Sciences Research Council. Ecol. Bull. No. 26.
NAPPER, D.M. *Grasses of Tanganyika*. Dar-es-Salaam, Min. Agric. For. and Wildlife.
1965
NARAYANAN, T.R. & DABADGHAO, P.M. *Forage crops for India*. New Delhi, Indian Coun-
1972   cil of Agric. Res.
NATIONAL RESEARCH COUNCIL. *Nutrient requirements of beef cattle*. Washington, DC,
1968   NRC Publication No. 1137.
NAVEH, Z. & ANDERSON, G.D. Promising pasture plants for northern Tanzania. 4.
1967   Legumes, grasses and grass/legume mixtures. *E. Afr. Agric. For. J.*, 32: 282-304.
NDAWULA-SENYIMBA, M.S. Some aspects of the ecology of *Themeda triandra. E. Afr.*
1972   *Agric. For. J.*, 38: 83-93.
NDYANABO, W.K. Oxalate content of some commonly grazed pasture forages of Lango and
1974   Acholi districts of Uganda. *E. Afr. Agric. For. J.*, 39: 210-214.
NESTEL, B.L. & CREEK, M.J. Pangola grass. *Herb. Abstr.*, 32: 265-271.
1962
NG, T.T. & WONG, T.H. Comparative productivity of two tropical grasses as influenced by
1976   fertilizer nitrogen and pasture legumes. *Trop. Grassl.*, 10: 179-185.

NORMAN, M.J.T. *Performance of annual fodder crops under frequent defoliation at Katherine, N.T.* Melbourne, CSIRO Aust. Tech. Paper No. 19.
1962a

NORMAN, M.J.T. Performance of pasture grasses in mixtures with Townsville lucerne at Katherine, N.T. *Aust. J. Exp. Agric. Anim. Husb.*, 2(7): 221-227.
1962b

NORMAN, M.J.T. The short-term effects of time and frequency of burning on native pastures at Katherine, N.T. *Aust. J. Exp. Agric. Anim. Husb.*, 3: 26-29.
1963a

NORMAN, M.J.T. The pattern of dry matter and nutrient content changes in native pastures at Katherine, N.T. *Aust. J. Exp. Agric. Anim. Husb.*, 3: 119-124.
1963b

NORMAN, M.J.T. Seasonal performance of beef cattle on native pastures at Katherine, N.T., Australia. *Aust. J. Exp. Agric. Anim. Husb.*, 5: 227-231.
1965

NORMAN, M.J.T. Companion grasses for Townsville lucerne at Katherine, N.T. *J. Aust. Inst. Agric. Sci.*, 33: 14-22.
1967

NORMAN, M.J.T. Relationships between liveweight gain of grazing beef steers and availability of Townsville lucerne. *Proc. 11th Int. Grassl. Congr.*, Surfers Paradise, Australia, 829-832.
1970

NORMAN, M.J.T. & PHILLIPS, L.J. The effect of times of grazing on bulrush millet (*Pennisetum typhoides*) at Katherine, N.T. *Aust. J. Exp. Agric. Anim. Husb.*, 8: 288-293.
1968

NORMAN, M.J.T. & STEWART, G.A. Investigations on the feeding of beef cattle in the Katherine region, N.T. *J. Aust. Inst. Agric. Sci.*, 30: 39-46.
1964

NORMAN, M.J.T. & WETSELAAR, R. Losses of nitrogen on burning native pasture at Katherine, N.T. *J. Aust. Inst. Agric. Sci.*, 26: 272-273.
1960

NORTHERN TERRITORY RURAL NEWS MAGAZINE, 4:16

NORTON, B.W. & HALES, J.W. A response of sheep to cobalt supplementation on south-eastern Queensland pastures. *Proc. 11th Bienn. Conf. Aust. Soc. Anim. Prod.*, Adelaide, Australia, 393-396.
1976

NYASALAND. DEPARTMENT OF AGRICULTURE. *Ann. Rept, 1960-61.* Part 2. Zomba.
1962

OAKES, A.J. Pangola grass (*Digitaria decumbens* Stent.) of the Caribbean. *Proc. 8th Int. Grassl. Congr.*, Reading, UK, 386-389.
1960

OAKES, A.J. Replacing hurricane grass in pastures in the dry tropics. *Trop. Agric. (Trinidad)*, 45: 235-244.
1968

OAKES, A.J. Resistance in *Hemarthria* species to the yellow sugar cane aphid (*Sipha flava* Forbes). *Trop. Agric. (Trinidad)*, 55: 377-381.
1978

OAKLEY, K. The earliest fire-makers. *Antiquity,* 30: 102-107.
1956

OAKLEY, K. On man's use of fire, with comments on tool-making and hunting. *In* Washburn, S.L. ed. *Social life of early man.* London, Methuen.
1962

ODHIAMBO, J.F. The nutritive value of various growth stages of *Pennisetum purpureum. E. Afr. Agric. For. J.*, 39: 325-329.
1974

OJASTI, J. [Carrying capacity of pastures of the Llanos for capybara (*Hydrochoerus hydrochaeris*). 3. Preliminary data on floristic changes.] 2nd Seminar on Capybara and Cayman Crocodiles, Maracay, Venezuela, 1-4 December, 30-32 (In Spanish) [*Herb. Abstr.*, 49: No.2088].
1976

OKAMOTO, K., KAWATKE, M. & HORIUCHI, S. [Effect of osmotic suction on the germination of warm season grasses]. *J. Jpn. Soc. Grassl. Sci.*, 21: 16-20. (In Japanese)
1975

OKE, J.G. Some new improved strains of pasture grasses for Maharashtra State. *Indian For.*, 97: 654-657.
1971

OLIVEIRA, B.A.D. DE, FARIA, P.R. DE S., SOUTO, S.M., CARNEIRO, A.M., DOBEREINER, J. & ARONOVICH, S. Identification of tropical grasses with the C_4 pathway of photosynthesis from leaf anatomy. *Pasqui. Agropecu. Bras. Ser. Agron.*, 8(8): 267-271 (In Portuguese) [*Herb. Abstr.*, 45: No.2417].
1973

ONLEY, J.A. & SILLAR, D.I. Pasture-crop sequence for beef all the year round. *Queensl. Agric. J.*, 91: 470-473.
1965

O'Rourke, J.T., Terry, P.J. & Frame, G.W. Experimental results of *Eleusine jaegeri*
1976   Pilg. control in East African Highlands. *E. Afr. Agric. For. J.*, 41: 253-265.
Orr, D.M. A review of *Astrebla* (Mitchell grass) pastures in Australia. *Trop. Grassl.*, 9; 21-
1975   36.
Ostrowski, H. Pastures in coastal southeast Queensland. *Queensl. Agric. J.*, 104: 449-473.
1978
Ostrowski, H. & Fay, M.F. Cool season productivity of five tropical grasses in a high rain-
1979   fall area of southeast Queensland. *Trop. Grassl.*, 13: 149-153.
Ottosen, E.M., Brown, G.W. & Maraske, M.R. Strip grazing — advantage or disadvan-
1975   tage. *Queensl. Agric. J.*, 101: 569-570.
Owen, M.A. A note on the effect of a late application of nitrogenous fertilizer on hay qual-
1964   ity. *E. Afr. Agric. For. J.*, 29: 322-325.
Oyenuga, V.A. The composition and agricultural value of some grass species in Nigeria.
1957   *Emp. J. Exp Agric.*, 25: 237-255.
Oyenuga, V.A. The effect of frequency of cutting on the yield and composition of some
1959   fodder grasses in Nigeria. *J. Agric. Sci.*, 53: 27-28.
Oyenuga, V.A. Effect of frequency of cutting on the yield and composition of some fodder
1960   grasses in Nigeria (*Tripsacum laxum* Nash.). *W. Afr. J. Biol. Chem.*, 4: 46-62.
Pace, C.P. Effect of a February burn on Lehmann's love grass. *J. Range Manage.*, 24: 454-
1971   456.
Paijmans, K. *New Guinea vegetation.* Canberra, CSIRO Aust., Aust. Nat. Univ. Press.
1976
Paladines, O. & Leal, J.A. Pasture management and productivity on the Llanos Orien-
1979   tales of Colombia. In Sánchez, P.A. & Tergas, L.E., eds. *Pasture production in
       acid soils of the tropics.* Cali, Colombia, CIAT. 311-325.
Paltridge, T.B. *Studies in sown pastures for southeastern Queensland.* CSIRO Aust. Bull.
1955   274.
Paltridge, T.B. & Coaldrake, J.E. Technique for harvesting seed of *Paspalum*
1943   *scrobiculatum*. *J. Counc. Sci. Ind. Res.*, 16: 5-9.
Pandeya, S.C. & Pathak, S.J. Germination behaviour of some ecotypes of grass (*Cen-*
1978   *chrus ciliaris*) under dry storage and physical stress. *Proc. 1st Int. Rangeland
       Congr.*, Denver (Colorado), USA, 376-383.
Panikkar, M.R. Teosinte. *Indian J. Vet. Sci.*, 21: 201-204.
1951
Parbery, D.B. Multiple harvesting of irrigated sorghum in northern Australia. *J. Aust.*
1966   *Inst. Agric. Sci.*, 32: 44.
Paretas, J.J., Queseda, R.R., López, M. & Gómez, L. [Influence of fertilizer N and sow-
1972   ing distance on seed production in Guinea grass (*Panicum maximum* Jacq.) and
       green panic (*P. maximum* var. *trichoglume* Eyles).] *Estación Experimental de
       Pastos y Forrajes Indio Hatuey*, 1972: 41-52 (In Spanish) [*Herb. Abstr.*, 49: No.
       1810].
Parham, B.E.V. The blue grasses of Fiji. *Fiji Agric. J.*, 16: 104-107.
1946
Parham, J.W. *The grasses of Fiji.* Suva, Fiji, Dept of Agric. Bull. No. 30.
1955
Parham, J.W. The germination of Batiki blue grass seed. *Fiji Agric. J.*, 31: 71.
1960
Parker, C. Evaluation of new herbicides on some East African perennial weeds. *Proc. 4th
1970   E. Afr. Herbicide Conf.*, Arusha, United Rep. of Tanzania, 265-278.
Parker, D.L. Strain variation and seed production in kikuyu grass (*Pennisetum clandes-*
1941   *tinum* Hochst.). *J. Agric. South Aust.*, 45: 55.
Parra, C.O. & Bryan, W.B. *Preliminary study of pasture grasses in the Orinoco Delta.*
1974   Part 1. Management and fertilization. Paris, INRA, 135-143.

PARTRIDGE, I.J. The improvement of mission grass (*Pennisetum polystachyon*) in Fiji by
1975 top-dressing superphosphate and over-sowing a legume. *Trop. Grassl.* 9: 45-52.

PARTRIDGE, I.J. Evaluation of herbage species for hill land in the drier zones of Viti Levu,
1979a Fiji. *Trop. Grassl.*, 13: 135-148.

PARTRIDGE, I.J. Improvement of Nadi blue grass (*Dichanthium caricosum*) pastures on hill
1979b land in Fiji with superphosphate and siratro. Effects of stocking rates on beef production and botanical composition. *Trop. Grassl.*, 13: 157-164.

PARTRIDGE, I.J. & RANACOU, E. The effects of supplemental *Leucaena leucocephala*
1974 browse on steers grazing *Dichanthium caricosum* in Fiji. *Trop. Grassl.*, 8: 107-112.

PASK, C.W.W. Nitrogen on tropical pastures. *Trop. Grassl.*, 8: 123.
1974

Paspalum. *Queensl. Agric. J.*, 78: 249-57.
1954

PASSEY, H.B. & SMITH, H.N. *Grassland restoration. Part IV. Grassland management.* Tem-
1966 ple (Texas), USA, USDA Soil Conserv. Serv.

PATEL, B.M. *Animal nutrition in Western India.* Anand, Indian Council of Agricultural
1962 Research.

PATERSON, D.D. Silage investigations with tropical forages. *Trop. Agric. (Trinidad)*, 22: 43-
1945 48.

PATHIRANA, K.K. & SIRIWARDENE, J.A. DE. Studies on the yield and nutritive quality of
1973 herbage grasses in the mid-country of Sri Lanka. *Ceylon Vet. J.*, 21: 52-61.

PATIL, B.D. & GHOSH, R. Evaluation of grasses for soil conservation. *J. Soil Water Conserv.*
1963 *India*, 11: 125-129.

PAULL, C.J. Methods of sowing buffel grass. *Queensl. Agric. J.*, 99: 567-575.
1973

PAULL, C.J. & LEE, G.R. Buffel grasses in Queensland. *Queensl. Agric., J.*, 104: 57-75.
1978

PAYNE, W.J.A. The role of domestic livestock in the humid tropics. *Proc. Int. Meeting on*
1974 *Use of Ecological Guidelines for Development in the American Humid Tropics,* Caracas. Paper No.9.

PAYNE, W.J.A. Possibilities for the integration of tree crops and livestock production in the
1976 wet tropics. *J. Sci. Food Agric.*, 27: 888.

PAYNE, W.J.A., LAING, W.I., MILES, N.S. & MASON, R.R. Fodder and pasture investiga-
1955 tional work at Sigatoka, 1949-1953. *Fiji Agric. J.*, 55: 38-60.

PEAKE, D.C.I., STRICKLAND, R.W. & HACKER, J.B. Grazing evaluation of *Digitaria*.
1976 *CSIRO Aust. Divn Trop. Crops and Pastures Rept,* 1975-76: 37.

PEARSON, E., HILL, B. & ALLEN, K. Pasture species testing. *N.T. Rur. Mag.*, 4: 10-11.
1979

PEREIRA, H.C. *Land use and water resources.* Cambridge, UK, Cambridge Univ. Press.
1973

PEREIRA, H.C., WOOD, R.A., BRZOSTOWSKI, H.W. & HOSEGOOD, P.H. Water conserva-
1956 tion by fallowing in semi-arid tropical East Africa. *Emp. J. Exp. Agric.*, 29: 269-286.

PÉREZ INFANTE, F. Effect of cutting interval and N fertilizer on the productivity of eight
1970 grasses. *Rev. Cubana Cienc. Agríc.*, 4: 137-148.

PERRY, R.A. Pasture lands of the Northern Territory, Australia. *CSIRO Aust. Land Res.*
1960 *Series,* 5: 1-55.

PETERS, L.V. Millet breeding programme. *E. Afr. Agric. For. J.* Special issue 39.
1973

PETERSEN, R.G., LUCAS, H.L. & MOTT, G.O. Relationship between rate of stocking and
1965 per animal and per acre performance on pasture. *Agron. J.*, 57: 27-30.

795

PHILLIPS, L.J. & NORMAN, M.J.T. *A comparison of two varieties of bulrush millet* (Pennisetum typhoides S & H) *at Katherine, N.T.* Melbourne, CSIRO Aust., Divn Land Res. Tech. Memo No. 67/18.
1967

PITOT, A. & MASSON, H. Quelques données sur la température au cours des feux de brousse aux environs de Dakar. *Bull. inst. fr. Afr. noire*, 13: 711-732.
1951

PLAYNE, M.J. Buffering capacity of sweet sorghum: the effects of nitrogen content, growth stage and ensilage. *J. Sci. Food Agric.*, 14: 495.
1963

PLAYNE, M.J. The buffering constituents of herbage and of silage. *J. Sci. Food Agric.*, 17: 264.
1966

PLAYNE, M.J. The sodium content of some tropical pasture species. *Aust. J. Exp. Agric. Anim. Husb.*, 10: 32-35.
1970a

PLAYNE, M.J. Differences in nutritional value of three cuts of buffel grass for sheep and cattle. *Proc. Aust. Soc. Anim. Prod.*, 8: 511-516.
1970b

PLAYNE, M.J. Oxalate levels in pasture species. *CSIRO Aust., Divn Trop. Crops and Pastures Rept*, 1975-76, 77.
1976

PLUCKNETT, D.L. Productivity of tropical pastures in Hawaii. *Proc. 11th Int. Grassl. Congr.*, Surfers Paradise, Australia, A38-A49.
1970

POPPI, D.M., MINSON, D.J. & TERNOUTH, J.H. Studies of cattle and sheep eating leaf and stem fractions of grasses. 1. The voluntary intake, digestibility and retention time in the reticulo-rumen. *Aust. J. Agric. Res.*, 32(1): 99-108.
1981

POULTNEY, R.G. Preliminary investigations on the effect of fertilizers applied to natural grassland. *E. Afr. Agric. J.*, 25: 47-49.
1959

PRAJAPITA, M.C. Effect of different systems of grazing by cattle on *Lasiurus - Eleusine - Aristida* grassland in the arid region of Rajasthan. *Ann. Arid Zone*, 9: 114-124.
1970

PRASAD, M.V.R. & SINGH, R.P. Select your crop for Western Rajasthan. *Indian Farming*, 22: 13-15.
1973

PRATES, E.R. [Effect of rate of nitrogen and interval between cuts on yield and composition of two ecotypes of *Paspalum notatum* Flügge and of cultivar Pensacola *Paspalum notatum* Flügge var. saurae Parodi.] *Anu. Técnico do Instituto de Pesquisas Zootécnicas "Francisco Osorio"*, 4: 267-307 (In Portuguese) [*Herb. Abstr.*, 49: No.1742].
1977

PRATT, D.J. Re-seeding denuded land in Baringo District, Kenya. *E. Afr. Agric. For. J.*, 29: 78-91.
1963

PRATT, D.J. Re-seeding denuded land in Baringo District, Kenya. 2. Techniques for dry alluvial sites. *E. Afr. Agric. For. J.*, 29: 243-259.
1964

PRATT, D.J. & GWYNNE, M.D. *Rangeland management and ecology in East Africa*. London, Hodder and Stoughton.
1977

PRATT, D.J. & KNIGHT, J. Bush control studies in the drier areas of Kenya. V. Effects of controlled burning and grazing management on *Tarconanthus/Acacia* thicket. *J. Appl. Ecol.*, 8: 217-234.
1971

PRESTON, T.R. & LENG, R.A. Sugar cane as cattle feed. 1. Nutritional constraints and perspectives. *Wld Anim. Rev.*, 27: 7-12.
1978

PRICE, C. & STOKES, I.E. *Sweet sorghum growing in Southern California*. USDA Agric. Res. Serv. ARS 34-10.
1966

PRITCHARD, A.J. Comparative trials with *Sorghum almum* and other forage sorghums in south-east Queensland. *Aust. J. Exp. Agric. Anim. Husb.*, 4: 6-14.
1964

PRITCHARD, A.J. Meiosis and embryo sac development in *Urochloa mosambicensis* and three *Paspalum* species. *Aust. J. Agric. Res.*, 21: 649-652.
1970

PRODONOFF, E.T. The determination and maintenance of seed quality. *Trop. Grassl.*, 1: 91-98.
1966

PUMPHREY, J. A planned comparison of five warm season grasses. *Proc. Summer Grass Conf.*, Ardmore (Oklahoma), USA.
1978

PURCELL, D.L. & LEE, G.R. Effects of season and of burning plus planned stocking on
1970   Mitchell grass grasslands in central western Queensland. *Proc. 11th Int. Grassl. Congr.*, Surfers Paradise, Australia, 66-68.
PURSGLOVE, J.W. *Monocotyledons.* London, Longman.
1976
QUESENBERRY, K.H., SMITH, R.L., SCHANK, S.C. & BOUTON, J.H. *Nitrogen fixation by*
1976   *tropical grasses.* Gainesville, Florida, Univ. of Fla Dept Agron. FL.32611.
QUINLAN, T.J. & EDGLEY, W.H.R. Dairy pastures for the Atherton Tableland. *Queensl.*
1975   *Agric. J.,* 101: 28-36.
QUINLAN, T.J., SHAW, K.A. & EDGLEY, W.H.R. Kikuyu grass. *Queensl. Agric. J.,* 101:
1975   737-749.
QUINN, L.R., MOTT, G.O., BISSCHOFF, W.V.A. & DE FREITAS, L.M.M. Production of beef
1970   from winter vs. summer nitrogen-fertilized colonial Guinea grass (*Panicum maximum*) pastures in Brazil. *Proc. 11th Int. Grassl. Congr.*, Surfers Paradise, Australia, 832-835.
RAINS, A.B. *Grassland research in Northern Nigeria, 1952-1962.* Zaria, Nigeria, Ahmadu
1963   Bello Univ. Misc. Paper Samaru (Nigeria) No.1.
RAJARATNAM, D.T. & SANTHIRASEGARAM, K. Intensity of grazing trial. *Ceylon Coconut*
1963   *Q.,* 14: 38-39.
RAMAM, S.S. Root development in alluvial grasslands of Varanasi. *Indian For.,* 96: 100-110.
1970
RAMIA, M. *Las sabanas de Apure.* Caracas, Venezuela, Min. de Agric. y Cria.
1959
RAMIA, M. Tipos de sabanas en los Llanos de Venezuela. *Bol. Soc. Venez. Cienc. Nat.,* 27:
1967   264-288.
RAMIA, M. & FERNÁNDEZ, J.E. Pasture development in the humid tropics: its ecology and
1974   economy. *Proc. Int. Meeting on Use of Ecological Guidelines for Development in the American Humid Tropics,* Caracas. Paper No.10.
RAMSAY, J. & ROSE-INNES, R. Some quantitative observations on the effects of fire on the
1963   Guinea savannah vegetation of northern Ghana over a period of eleven years. *African Soils,* 8; 41-85.
RAO, R.S. Studies on the flora of Kutch, Gujarat State (India) and their utility in the
1970   economic development of the semi-arid region. *Ann. Arid Zone,* 9: 125-142.
RATTRAY, J.M. The grasses and grass associations of Southern Rhodesia. *Rhod. Agric. J.,*
1957   57: 197-234.
RATTRAY, J.M. *Matapos Research Station. Plant succession experiments.* Rept Sec. Fed.
1960a   Min. Agric. Rhod. and Nyasald.
RATTRAY, J.M. *Grass cover of Africa.* Rome, FAO. FAO Agric. Studies No.49.
1960b
RATTRAY, J.M. *Veld management.* Rept Sec. Fed. Min. Agric. Rhod. and Nyasald.
1962
RATTRAY, J.M. *Mejora de pastos y cultivos forrajeros. Panamá.* Rome, FAO. AGP:SF/
1973   PAN 10. Informe técnico 2.
RAVIKOVITCH, S. & PORATH, A. The effect of nutrients on the salt tolerance of crops. *Plant*
1967   *& Soil,* 26: 49-71.
RAY, B., AGARAWALA, S.B.D. & FRIDRICKSON, C.J. Control of perennial grasses in forest
1975   lands with application of herbicides. *Indian For.,* 101: 533-538.
RAYMOND, W.F. The efficient use of grass. *J. Brit. Grassl. Soc.,* 19: 81-89.
1963
READ, J.W. 1. [Kikuyu] in the Murrumbidgee Irrigation area. *In* Kikuyu — a research
1975   report. *Agric. Gaz. NSW,* 86: 2-24.
REDDY, M.R. & MURTY, V.N. Nutritive value of sunn hemp hay and its supplementation
1962   to paddy straw as a complete food for cattle. *Indian J. Anim. Sci.,* 42: 558-561.

REES, M.C. Winter and summer growth of pasture species in a high rainfall area of south-
1972 eastern Queensland. *Trop. Grassl.*, 6: 45-54.

REES, M.C., JONES, R.M. & ROE, R. Evaluation of pasture grasses and legumes grown in
1976 mixtures in southeast Queensland. *Trop. Grassl.*, 10: 65-78.

REES, M.C. & MINSON, D.J. Superphosphate effects on the nutritive value of pangola grass-
1976 sulphur response. *CSIRO Aust., Divn Trop. Crops and Pastures Rept, 1975-76:* 73-74.

REES, M.C. & MINSON, D.J. Fertilizer sulphur as a factor affecting voluntary intake, diges-
1978 tibility and retention by sheep. *Brit. J. Nutr.*, 39: 5-11.

REES, M.C., MINSON, D.J. & SMITH, F.W. The effect of supplementary and fertilizer sul-
1974 phur on voluntary intake, digestibility, retention time in the rumen, and site of digestion of pangola grass by sheep. *J. Agric. Sci.*, 82: 419-422.

REID, R. A numerical classification of sown tropical pasture regions based on the perfor-
1973 mance of sown pasture species. *Trop. Grassl.*, 7: 331-340.

RENSBURG, H.J. VAN. Grass burning experiments on the Msima River Livestock Farm,
1952 Southern Highlands, Tanganyika. *E. Afr. Agric. For. J.*, 17: 119-129.

RENSBURG, H.J. VAN. Growth and seasonal composition of natural grasslands in Zambia.
1968 *J. Brit. Grassl. Soc.*, 23: 51-52.

RENSBURG, H.J. VAN. *Management and utilization of pastures.* Rome, FAO. Pasture and
1969 Fodder Crop Studies No.3.

RENSBURG, H.J. VAN. Fire — its effect on grasslands including swamps in southern, central
1971 and eastern Africa. *Proc. 11th Ann. Tall Timbers Fire Ecol. Conf.*, Tallahassee (Fla), USA, 175-199.

RENSBURG, P.H.J.J. VAN. [Soil cultivation and the sowing of *Eragrostis curvula* on severely
1971 damaged *Themeda-Cymbopogon* veldt]. *Proc. Grassl. Soc. South. Africa*, 6: 93-100. (In Afrikaans)

REYNOLDS, S.G. Evaluation of pasture grasses under coconuts in Western Samoa. *Trop.*
1978 *Grassl.*, 12: 146-151.

RHIND, D. *The grasses of Burma.* Calcutta, Baptist Mission Press.
1945

RHOADES, E.D. Grass survival in flood pool areas. *J. Soil Water Conserv.*, 22: 19.
1967

RICHARDS, J.A. Productivity of Guinea grass (*Panicum maximum*). *Proc. 9th Int. Grassl.*
1965 *Congr.*, São Paulo, 1033-1035.

RICHARDSON, F.E. Chemical control of couch grass (*Digitaria scalarum*) by heavy applica-
1967 tions of sodium trichloracetate (Na-TCA). *Kenya Sisal Board Bull.*, 59: 29-33.

RICKERT, K.G. Some influences of straw mulch, nitrogen fertilizer and oat companion crops
1970 on establishment of Sabi panic. *Trop. Grassl.*, 4: 71-76.

RIDER, A.R. Hay and forage handling machinery and updated economics of various sys-
1979 tems. *Proc. Beef Cattle Conf. on Economics, Management and Alternative Feeding Systems,* Ardmore (Oklahoma), USA, 55-64.

RIPPERTON, J.C. & HOSAKA, E.Y. Vegetation zones of Hawaii. *Hawaiian Agric. Exp. Sta.*
1942 *Bull.*, 89.

RISOPOULOS, S.A. *Management and use of grasslands. Democratic Republic of the Congo.*
1966 Rome, FAO. Pasture and Fodder Crop Studies No.1.

RIVEROS, F. & WILSON, G.L. Responses of a *Setaria sphacelata-Desmodium intortum* mix-
1970 ture to height and frequency of cutting. *Proc. 11th Int. Grassl. Congr.*, Surfers Paradise, Australia, 666-668.

ROBERTS, B.R. & OPPERMANN, D.P.J. The influence of defoliation on carbohydrate status
1966 and nutritive value of perennial veldt grasses. *Proc. 10th Int. Grassl. Congr.*, Helsinki, 940-944.

ROBERTS, F.J. & CARBON, B.A. Growth of tropical and temperate grasses and legumes
1969 under irrigation in southwest Australia. *Trop. Grassl.*, 3: 109-116.

ROBERTS, O.T. Pasture improvement and research in Fiji. *South Pacific Bull.*, 20: 35-37.
1970a
ROBERTS, O.T. A review of pasture species in Fiji. I. Grasses. *Trop. Grassl.*, 4: 129-137.
1970b
ROBERTSON, A.D., HUMPHREYS, L.R. & EDWARDS, D.G. Influence of cutting frequency
1976 and phosphorus supply on the production of *Stylosanthes humilis* and *Arundinaria pusilla* at Khon Kaen, Northeast Thailand. *Trop. Grassl.*, 10: 33-39.
ROBINSON, B.P. & POTTS, R.C. The history of *Hyparrhenia hirta* and studies of its flowering
1950 habits and seed production. *Agron. J.*, 42: 395-397.
ROBINSON, C.S. Soil conservation in the Northern Territory. *J. Soil Conserv. Serv. (NSW)*,
1978 34: 101-105.
ROCHA, G.L. DA, MARTINELLI, D., CORREA, A.A., SILVEIRA, H. DA & CINTRA, B. Seasonal
1960 productivity of molasses grass in Brazil. *Proc. 8th Int. Grassl. Congr.*, Reading, UK, 378-381.
ROCHA, G.L. DA, MARTINELLI, D., CORREA, A.A., TUNDISI, A.G.A., LIMA, F.P. &
1962 KALIL, E.B. [Comparative value of grasses for meat production.] *Bol. Ind. Anim.*, 20: 289-296. (In Portuguese)
RODEL, M.G.W. Herbage yields of five grasses and their ability to withstand intensive grazing.
1970 *Proc. 11th Int. Grassl. Congr.*, Surfers Paradise, Australia, 618-621.
RODRIGO, E. Fodder grass experiment (Lunuwila). *Ann. Rept Coconut Res. Scheme,*
1945 *Ceylon.* 11.
ROE, R. Seed losses with different methods of harvesting *Panicum coloratum*. *Trop.*
1972 *Grassl.*, 6: 113-118.
ROE, R. & ALLEN, G.H. Studies on the Mitchell grass association in southwestern Queensland.
1945 2. The effect of grazing on the Mitchell grass pasture. *CSIRO Aust. Bull.*, 185.
ROE, R. & WILLIAMS, R.W. Viability of *Panicum coloratum* seed in storage. *Trop. Grassl.*,
1969 3: 141-142.
ROGERS, A.L. & BAILEY, E.T. Salt tolerance trials with forage plants in southwestern
1963 Australia. *Aust. J. Exp. Agric. Anim. Husb.*, 3: 125-130.
ROMANOV, V.A. [Sowing methods and rates and fertilizers for perennial herbage species for
1976 seed production.] *Agrotekhnika i biologiya sel'skokhozyaĭstvennykh kul'tur.* 1976: 59-65 (In Russian) [*Herb. Abstr.*, 49: No.909].
ROMNEY, D.H. Productivity of pasture in British Honduras. II. Pangola pasture. *Trop.*
1961 *Agric. (Trinidad)*, 38: 39-47.
ROSE-INNES, R. Fire in West African vegetation. *Proc. 11th Ann. Tall Timbers Fire Ecol.*
1971 *Conf.*, Tallahassee (Fla), USA, 147-173.
ROSE-INNES, R. *A manual of Ghana grasses.* Surbiton, Surrey, UK, Min. Overseas Devel.,
1977 Land Resources Divn.
ROSEVEARE, G.M. *The grasslands of Latin America.* Aberystwyth, Imp. Bur. Pastures
1948 Field Crops Bull. No.36.
ROTAR, P.P. *Grasses of Hawaii.* Honolulu, Univ. of Hawaii Press.
1968
ROTAR, P.P. & PLUCKNETT, D.L. Tropical and subtropical forages. *In* Heath, M.E., Met-
1973 calfe, D.S. & Barnes, R.E., eds. *Forages, the science of grassland agriculture.* 3rd edn. Ames (Iowa), USA, Iowa State Univ. Press.
ROYAL, A.J.E. & HUGHES, R.M. Winter-spring forage crops as supplements for dairy cows
1976 grazing kikuyu pastures. *Proc. 11th Bienn. Conf. Aust. Soc. Anim. Prod.*, Melbourne, 513-516.
RUSSELL, J.S. Comparative salt tolerance of some tropical and temperate legumes and tropical
1976 grasses. *Aust. J. Exp. Agric. Anim. Husb.*, 16: 103-109.
RUSSELL, J.S. & WEBB, H.R. Climatic range of grasses and legumes used in pastures. Result
1976 of a survey conducted at the 11th International Grassland Congress. *J. Aust. Inst. Agric. Sci.*, 42: 156-163.

RUSSELL, M.J. & COALDRAKE, J.E. Performance of eight tropical legumes and lucerne and
1970   of four tropical grasses on semi-arid brigalow lands in central Queensland. *Trop. Grassl.*, 4: 111-120.
RYAN, J., MIYAMOTO, S. & STROEHLEIN, J.L. Salt and specific ion effects on germination of
1975   four grasses. *J. Range Manage.*, 28: 61-64.
SAID, A.N. *In vivo* digestibility and nutritive value of kikuyu grass, *Pennisetum clandes-*
1971   *tinum*, with a tentative assessment of its yield of nutrients. *E. Afr. Agric. For. J.*, 37: 15-21.
SAJISE, P.E. Evaluation of cogon (*Imperata cylindrica* (L.) Beauv.) as a seral stage in Philip-
1973   pine vegetational succession. 1. The cogonal seral stage and plant succession. 2. Auto-ecological studies on cogon. *Cornell Univ. Dissertations Abst. Int.*, 33: 3040-3041.
SÁNCHEZ, P.A. & TERGAS, L.E. *Pasture production in acid soils of the tropics*. Colombia,
1980   CIAT.
SANDS, E.B., THOMAS, D.B. & PRATT, D.J. Preliminary selection of pasture plants for the
1970   semi-arid regions of Kenya. *E. Afr. Agric. For. J.*, 36: 49-57.
SANTHIRASEGARAM, K. The effect of pasture on yield of coconuts. *Ceylon Coconut Plant.*
1964   *Rev.*, 4: 43-46.
SANTHIRASEGARAM, K. The effect of monospecific grass swards on the yield of coconuts in
1966   the north-west province of Ceylon. *Ceylon Coconut Q.*, 17: 73-79.
SANTHIRASEGARAM, K., COALDRAKE, J.E. & SALIH, M.H.M. Yield of a mixed sub-tropical
1966   pasture in relation to frequency and height of cutting and leaf area index. *Proc. 10th Int. Grassl. Congr.*, Helsinki, 125-129.
SANTIAGO, A. Studies in auto-ecology of *Imperata cylindrica* (L.) Beauv. (1812). *Proc. 9th*
1965   *Int. Grassl. Congr.*, São Paulo, 1: 499-502.
SANTARI, A.M. Pengaruh vegetasi alang-alang dan belukar terhadrap beberapa sifat tanah.
1968   *Commun. Agric.*, 1: 10-15.
SAYER, J.A. & LAVIEREN, L.P. VAN. The ecology of the Kafue lechwe population of Zam-
1975   bia before the operation of hydro-electric dams on the Kafue river. *E. Afr. Wildl. J.*, 13: 9-37.
SCAILLET, M.M. High altitude fodder crops and fallows in Burundi. *Sols Afr.*, 10: 205-216.
1965
SCATENI, W.J. Pasture dry matter production and utilization. *Trop. Grassl.*, 1: 56-59.
1966
SCATENI, W.J. Atrazine tolerance in five tropical pasture grasses. *Trop. Grassl.*, 12: 35-39.
1978
SCAUT, A. *Détermination de la digestibilité des herbages frais*. Yangambi, Inst. natl l'étude
1959   agr. Congo Belge. INÉAC Sér. Sci. No. 81.
SCHAAFFHAUSEN, R. VON. Adlay or Job's tears. A cereal of potentially greater economic
1952   importance. *Econ. Bot.*, 6: 216-217.
SCHANK, S.C. Chromosome numbers in eleven new *Hemarthria* (limpo grass) introduc-
1972   tions. *Crop Sci.*, 12: 550-551.
SCHLEGEL, H.G. [Practical application of microbiology with changes of time.] *Angewandte*
1978   *Bot., Univ. Göttingen*, 52: 57-63 (In German) [*Herb. Abstr.*, 49: No.1200].
SCHNEIDER, B.A., CLARK, N.A., HEMKEN, R.W. & VANDERSALL, J.H. Relationship of
1970   pearl millet to milk fat depression in dairy cows. 2. Forage organic acids as influenced by soil nutrients. *J. Dairy Sci.*, 53: 305-310.
SCHOFIELD, J.L. The effects of season and frequency of cutting on the productivity of vari-
1944   ous grasses under coastal conditions in northern Queensland. *Queensl. J. Agric. Anim. Sci.*, 1: 1-58.
SCHOFIELD, J.L. Protein content and yield of grasses in the wet tropics as influenced by sea-
1945   sonal productivity, frequency of cutting and species. *Queensl. J. Agric. Anim. Sci.*, 2: 209-243.

SCHOTTLER, J.H., BOROMANA, A. & WILLIAMS, W.T. Comparative performance of cattle
1977 and buffalo on the Sepik Plains, Papua New Guinea. *Aust. J. Exp. Agric. Anim. Husb.*, 17: 550-554.
SEAMAN, D.E., MORSE, M.D., MILLER, M.D., HARVEY, W.A., BUSCHMANN, L.L., WICKS,
1968 C.M. & FISCHER, B.B. Controlling submerged weeds in rice. *Calif. Agric.*, 22: 11.
SEARCY, V.S. & PATTERSON, R.M. Renovating carpet grass pastures with herbicides and til-
1961 lage. *Proc. 14th South. Weed Conf.*
SEMPLE, A.T. *L'amélioration des herbages dans le monde.* Rome, FAO.
1956
SEMPLE, A.T. *Grassland improvement.* London, Leonard Hill.
1970
SEN, K.C. New Delhi, Indian Council of Agricultural Research. Bulletin No. 25.
1938
SEN, K.C. & RAY, S.N. *Nutrient values of Indian cattle feeds and the feeding of animals.* New
1964 Delhi, Indian Counc. Agric. Res. Bull. No.25.
SEN, K.M. & MABEY, G.L. The chemical composition of some indigenous grasses of coastal
1965 savanna of Ghana at different stages of growth. *Proc. 9th Int. Grassl. Congr.*, São Paulo, 763.
SERRÃO, A., FALESI, I.C., VIEGA, J.B. DE & NETO, J.F.T. Productivity of cultivated pas-
1979 tures on low fertility soils in the Amazon of Brazil. *In* Sánchez, P.A. & Tergas, L.E., eds. *Pasture production in acid soils of the tropics.* Cali, Colombia, CIAT. Series 03 EG-5.
Setaria, Setaria porphyrantha (purple pigeon grass). *J. Aust. Inst. Agric. Sci.*, 43: 180-82.
1977
SETH, S.K. A short note on *Sporoboletum* grasslands of Uttar Pradesh. *Indian For.*, 81: 185-
1955 190.
SHANKARNARAYAN, K.A. Systematic botany. *Central and Arid Zone Res. Inst. Sci. Prog.*
1962-63 *Rept,* 56-58.
SHANKARNARAYAN, K.A., DABADGHAO, P.M., KUMAR, R. & RAI, P. Effect of defoliation
1977 management and manuring on dry matter yields and quality in *Sehima nervosum, Cenchrus ciliaris* and *Cenchrus setigerus. Ann. Arid Zone,* 16: 441-454.
SHANTZ, H.L. The place of grasslands in the earth's cover of vegetation. *Ecology,* 35: 143-
1954 151.
SHARMA, S.K. & GUPTA, R.K. Effect of salts on seed germination of some desert grasses.
1971 *Ann. Arid Zone,* 10: 33-36.
SHAW, K.A. & QUINLAN, T.J. Dry matter production and chemical composition of Kenya
1978 white clover and some tropical legumes grown with *Pennisetum clandestinum* in cut swards on the Evelyn Tableland of north Queensland. *Trop. Grassl.,* 12: 49-57.
SHAW, N.H. Bunch spear grass dominance in burnt pastures in southeastern Queensland.
1957 *Aust. J. Agric. Res.,* 8: 325-334.
SHAW, N.H. & BISSET, W.J. Characteristics of a bunch spear grass (*Heteropogon contortus*
1955 (L.) Beauv.) pasture grazed by cattle in sub-tropical Queensland. *Aust. J. Agric. Res.,* 6: 539:552.
SHAW, N.H. & BRYAN, W.W. *Tropical pasture research: principles and methods.* Com.
1976 Agric. Bur. Bull. No. 511.
SHAW, N.H., ELICH, T.W., HAYDOCK, K.P. & WAITE, R.B. A comparison of seventeen
1965 introductions of *Paspalum* species and naturalized *Paspalum dilatatum* under cutting at Samford, southeastern Queensland. *Aust. J. Exp. Agric. Anim. Husb.,* 5: 423-432.
SHAW, N.H. & 'T MANNETJE, L. Studies on a spear grass pasture in central coastal Queens-
1970 land. The effect of fertilizer, stocking rate and oversowing with *Stylosanthes humilis* on beef production and botanical composition. *Trop. Grassl.,* 4: 43-56.

SHAW, N.H. & NORMAN, M.J.T. Tropical and subtropical woodlands and grassland. *In*
1970   Moore, R.M., ed. *Australian grasslands*. Canberra, Aust. Nat. Univ. Press.
SHELDRICK, R.D. The control of Siam weed in Nigeria. *J. Niger. Inst. Oil Palm Res.*, 5: 17-
1968   19.
SHELDRICK, R.D. Growing silage maize. *ADAS Q. Rev.*, 4: 177-186.
1972
SHELDRICK, R.D. Optimum cutting period for silage maize in western Kenya. *E. Afr. Agric.*
1975   *For. J.*, 40: 394-399.
SHETH, A.A., YU, L. & EDWARDSON, J. Sterility in pangola grass (*Digitaria decumbens*
1956   Stent.). *Agron. J.*, 48: 505-507.
SHOJI, K. & SUND, K.A. Drainage and salinity investigations at the Haft Tapeh Sugar Cane
1967   Project, Iran. *Proc. 12th Congr. of the Int. Soc. of Sugar Cane Tech.*, San Juan,
         Puerto Rico, 28 March-10 April 1965, 90-95.
SHOOP, M., MCILVAIN, E.H. & VOIGHT, P.W. Morpa weeping love grass produces more
1976   beef. *J. Range Manage.*, 29: 101-103.
SIEBERT, B.D., NEWMAN, D.M.R. & NELSON, D.J. The chemical composition of some arid
1968   zone pasture species. *Trop. Grassl.*, 2: 31-40.
SILCOCK, R.G. Drying temperature and its effect on viability of *Setaria sphacelata* seed.
1971   *Trop. Grassl.*, 5: 75-80.
SILCOCK, R.G. Factors influencing the establishment of permanent grasses on the lateritic
1976   red earth (mulga soils) of southwestern Queensland. *J. Aust. Inst. Agric. Sci.*, 42:
         111-112.
SILLAR, D.I. Control of grader grass (*Themeda quadrivalvis*). *Queensl. J. Agric. Anim. Sci.*,
1969   26: 581-586.
SILVEY, M.W. Perennial forage sorghum trials. *CSIRO Aust., Divn Trop. Crops and Pas-*
1977a   *tures Ann. Rept, 1976-77.*
SILVEY, M.W. Continuous pastures. *CSIRO Aust., Divn Trop Crops and Pastures Ann.*
1977b   *Rept, 1976-77.*
SIMON, B.K. *A key to Queensland grasses.* Brisbane, Dept Primary Ind. Bot. Branch Tech.
1980   Bull. No. 4.
SINGH, L.N. & KATOCH, D.C. Forage yield potential of some grasses. *Indian J. Agric. Res.*,
1975   9: 25-29.
SINGH, MANMOHAN & ARORA, W.D. Introduction and evaluation of *Pennisetum pedicel-*
1970   *latum* as summer fodder. *Proc. 11th Int. Grassl. Congr.*, Surfers Paradise,
         Australia, 621-624.
SINGH, MUKTER, PANDEY, R.K. & SHANKARNARAYAN, K.A. Problems of grassland weeds
1970   and their control in India. *Proc. 11th Int. Grassl. Congr.*, Surfers Paradise,
         Australia, 71-74.
SINGH, P. & KUSHWAHA, N.S. Utilization of paddy straw with urea and molasses as a feed
1974   for dairy cows. *Indian Vet. J.*, 51: 690-694.
SINGH, R., GUPTA, P.C., SAGAR, V. & PRADHAN, K. Note on variability in protein cell-wall
1977   constituents and *in vitro* dry matter digestibility in some important fodder strains
         of pearl millet. *Indian J. Agric. Sci.*, 47: 477-479.
SINGH, R.D. & CHATTERJEE, B.N. Growth analysis of perennial grasses in tropical India.
1968   1. Herbage growth in pure grass swards. 2. Herbage growth in mixed grass/
         legume swards. *Exp. Agric.*, 4: 117-125, 127-134.
SIVALINGAM, T. A study of the effect of N fertilization and frequency of defoliation on yield,
1964   chemical composition and nutritive value of three tropical grasses. *Trop. Agric.*
         (Sri Lanka), 120: 159.
SKERMAN, P.J. Silage from sugar cane tops. *Cane Grow. Q. Bull.*, 8: 124-125.
1941
SKERMAN, P.J. Vegetation in the Channel Country. In *The Channel Country of Southwest*
1947   *Queensland.* Brisbane, Bur. of Investigation.

SKERMAN, P.J. *Cropping for fodder conservation and pasture production in the wool-*
1958 *growing areas of western Queensland.* Brisbane, Univ. of Queensl. Dept Agric. Paper No. 3.
SKERMAN, P.J. *Pastures and livestock in the project area.* Athens, Doxiades Associates.
1966 Document DOX-SUD-A.47.
SKERMAN, P.J. *Tropical forage legumes.* Rome, FAO.
1977
SKERMAN, P.J. Types of utilization. In *Tropical grazing land ecosystems.* Unesco/UNEP/
1979 FAO. Natural Resources Res. Series No. 16.
SKOVLIN, J.M. The influence of fire on important range grasses in East Africa. *Proc. 11th*
1971 *Ann. Tall Timbers Fire Ecol. Conf.*, Tallahassee (Fla), USA, 201-217.
SMITH, C.A. The utilization of *Hyparrhenia* veld for the nutrition of cattle in the dry season.
1961 II. Veld hay compared with *in situ* grazing of the native forage, and the effects of feeding supplementary nitrogen. *J. Agric. Sci.*, 57: 311.
SMITH, C.A. Conservation and steer production at two stocking rates near Gatton, Queens-
1967 land. *CSIRO Aust., Divn Trop. Pastures Rept, 1966/67.*
SMITH, C.J. The effect of mulching on the establishment of pasture grasses. *Rhod. Zambia*
1966 *Malawi J. Agric. Res.*, 4: 129-132.
SMITH, D.T. & CLARK, N.A. Effect of soil nutrients and pH on nitrate nitrogen and growth
1968 of pearl millet and Sudan grass. *Agric. J.*, 60: 38-40.
SMITH, F.W. Potassium nutrition, ionic relations, and oxalic acid accumulation in three cul-
1972 tivars of *Setaria sphacelata. Aust. J. Agric. Res.* 23: 969.
SMITH, F.W. & DOLBY, G.R. Derivation of diagnostic indices for assessing the sulphur
1977 status of *Panicum maximum* var. *trichoglume. Commun. Soil Sci. Plant Anal.*, 8(3): 221-240.
SMITH, L.S. *Urochloa* grass. *Queensl. Agric. J.*, 53: 526-529.
1940
SOERJANI, M. *Alang-alang. Pattern of growth as related to the problem of control.* Bogor,
1970 Indonesia, SEAMEO Regional Center for Tropical Biology. Biotrop. Bull. No. 1.
SOEWARDI, B., SASTRADIPRADA, D., NASOETION, A.H. & HUTASOIT, J.H. *Studies on the*
1974 *alang-alang* (Imperata cylindrica *(L.) Beauv.) for cattle feeding. 1. The effects of carbohydrate sources in urea-containing concentrate on feed utilization.* Bogor, Indonesia, SEAMEO Regional Center for Tropical Biology. Biotrop. Bull. No. 8.
SOLOMON, S. *Crops of the Bombay State. Their cultivation and statistics.* Poona, India,
1953 Yeravda Prison Press. Bull. No. 186.
SPAIN, J.M. Pasture establishment and management in the Llanos Orientales of Colombia.
1979 *In* Sánchez, P.A. & Tergas, L.E., eds. *Pasture production in acid soils of the tropics.* Cali, Colombia, CIAT. Series 03 EG-5.
SPAIN, J.M. & ANDREW, C.S. Mineral characterization of species. Responses of tropical
1977 grasses to aluminium in water culture. *CSIRO Aust., Divn Trop. Pastures Rept, 1976-77.*
SPEARS, B.R. & COFFIN, L.C. *Growing grain sorghum.* Texas Agric. Ext. Serv. Bull. B210.
1959
SPRAGUE, M.A. & TAYLOR, B.B. Preservation of silage in plastic bags — a new method for
1965 research with the forages. *Proc. 9th Int. Grassl. Congr.*, São Paulo, 637-643.
SPRAGUE, V.G. The effects of temperature and day-length on seedling emergence and early
1943 growth of several pasture species. *Proc. Soil Sci. Soc. Amer.*, 8: 287-294.
SQUIRES, V.R. & MYERS, L.F. Performance of warm season perennial grasses for irrigated
1970 pastures at Deniliquin, southeastern Australia. *Trop. Grassl.*, 4: 153-161.
SRINIVASAN, V., BONDE, W.C. & TEJWANI, K.G. Studies on grasses and their suitability to
1962 stabilize and maintain bunds in the ravine lands of Gujarat. *J. Soil Water Conserv. India*, 10: 72-78.

STEEL, R.J.H. & HUMPHREYS, L.T. Growth and phosphorus response of some pasture
1974 legumes sown under coconuts in Bali. *Trop. Grassl.*, 8: 171-178.
STEINKE, T.D. The translocation of C_{14} assimilates in *Eragrostis curvula:* an autoradio-
1969 graphic survey. *Proc. Grassl. Soc. South. Afr.*, 4: 19.
STEINKE, T.D. & BOOYSEN, P. DE V. The regrowth and utilization of carbohydrate reserves
1968 of *Eragrostis curvula* after different frequencies of defoliation. *Proc. Grassl. Soc. South. Afr.*, 3: 105-110.
STEINKE, T.D. & NEL, L.O. The growth of veld in response to defoliation by various means
1967 in late winter and spring. *Proc. Grassl. Soc. South. Afr.*, 2: 113-117.
STENT, S.M. Preliminary list of the more common grasses of Southern Rhodesia. *Rhod.*
1931 *Agric. J.*, 28: 342-359.
STENT, S.M. & RATTRAY, J.M. The grasses of Southern Rhodesia. *Proc. Rhod. Sci. Assoc.*,
1933 32.
STEPHENS, J.L. & MARCHANT, W.H. *Bahia grass for pastures.* Georgia Agric. Exp. Sta.
1960 Bull. No. N.S.67.
STEVENS, G.R. Forage sorghums on the Darling Downs. *Queensl. Agric. J.*, 101: 721-728.
1975
STOBBS, T.H. The effect of grazing management upon pasture productivity in Uganda.
1969a I. Stocking rate. *Trop. Agric. (Trinidad)*, 46: 187-194.
STOBBS, T.H. The effect of grazing management upon pasture productivity in Uganda.
1969b II. Grazing frequency. *Trop. Agric. (Trinidad)*, 46: 195-200.
STOBBS, T.H. The effect of grazing management upon pasture productivity in Uganda.
1969c III. Rotational and continuous grazing. *Trop. Agric. (Trinidad)*, 46: 293-301.
STOBBS, T.H. Quality of pasture and forage crops for dairy production in the tropical re-
1971 gions of Australia. 1. Review of the literature. *Trop. Grassl.*, 5: 159-170.
STOBBS, T.H. The effect of plant structure on the intake of tropical pastures. 2. Differences
1973 in sward structure, nutritive value and bite size of animals grazing *Setaria anceps* and *Chloris gayana* at various stages of growth. *Aust. J. Agric. Res.*, 24: 821-829.
STOBBS, T.H. & JOBLIN, A.D.H. The use of liveweight gain trials for pasture evaluation in
1966 the tropics. II. Variable stocking rate designs. *J. Brit. Grassl. Soc.*, 21: 181-185.
STOBBS, T.H. & MINSON, D.J. Effect of plant structure on the intake of nutrients from trop-
1978 ical pastures. *CSIRO Aust., Divn Trop. Crops and Pastures, Ann. Rept, 1977/78.*
STOBBS, T.H. & WHEELER, J. Response by lactating cows grazing sorghum to sulphur
1978 supplementation. *Trop. Agric. (Trinidad)*, 54: 228-234.
STOCKER, G.C. & STURTZ, J.D. The use of fire to establish Townsville lucerne in the North-
1966 ern Territory. *Aust. J. Exp. Agric. Anim. Husb.*, 6: 277-279.
STRICKLAND, R.W. Dry matter production, digestibility and mineral content of *Eragrostis*
1973 *superba* Peyr. and *E. curvula* (Schrad.) Nees. at Samford, southeastern Queensland. *Trop. Grassl.*, 7: 233-241.
STRICKLAND, R.W. *Cynodon* spp. at Samford. *CSIRO Aust., Divn Trop. Crops and Pas-
1977 tures Rept, 1976/77.*
STRICKLAND, R.W. The cool season production of some introduced grasses in southeast
1978 Queensland. *Trop. Grassl.*, 12: 109-112.
STUBBS, W.C. & ARBUCKLE, J. These beef cattle had lucerne hay as winter feed supple-
1962 ment. *Queensl. Agric. J.*, 88: 449-454.
STURTZ, J.D., HARRISON, P.G. & FALVEY, L. Regional pasture development and
1975 associated problems. II. Northern Territory. *Trop. Grassl.*, 9: 83-91.
SUIJENDORP, H. Buffel and birdwood grasses. Two useful perennials. *J. Agric. West. Aust.*,
1953 2 (New Series): 492-493.
SULLIVAN, E.F. Effect of temperature and phosphorus fertilization on yield and composi-
1961 tion of Piper sudan grass. *Agron. J.*, 53: 357-358.
SUMAN, R.F., WOODS, E.G., PEELE, T.C. & GODBEY, E.G. Beef gains from differentially
1962 fertilized summer grasses in the Coastal Plain. *Agron. J.*, 54: 26-28.

SWANN, I.F. Green panic needs nitrogen and sulphur. *Queensl. Agric. J.*, 99: 273-275.
1973

SWEENEY, F.C. Changing blady grass plains at Tully into good pastures. *Queensl. Agric. J.*,
1961 87: 697-698.

SWEENEY, F.C. & HOPKINSON, J.M. Vegetative growth of nineteen tropical and subtropical
1975 pasture grasses and legumes in relation to temperature. *Trop. Grassl.*, 9: 209-217.

TAERUM, R. Comparative shoot and root growth studies on six grasses in Kenya. *E. Afr.*
1970a *Agric. For. J.*, 36: 94-113.

TAERUM, R. A study of root and shoot growth of three grass species in Kenya. *E. Afr. Agric.*
1970b *For. J.*, 36: 155-170.

TAKAHASHI, M., MOOMAW, J.C. & RIPPERTON, J.C. Studies on Napier grass. 3. Grazing
1966 management. *Hawaii Agric. Exp. Sta. Bull.*, 128.

TALAPATRA, S.K. The nutritive value of the indigenous grasses of Assam. II. The semi-
1950 aquatic grasses as cattle feeds. *Indian J. Vet. Sci.*, 20: 229-240.

TAMAYO, F. [Sand savannas.] *Bol. Soc. Venez. Cienc. Nat.*, 106: 18-19. (In Spanish)
1963

TAMHANE, V.A. & MULWANI, B.T. Removal of some of the injurious salts by ordinary farm
1937 crops. *Proc. 22nd Indian Sci. Congr.*, 1935: 363 [*Herb. Abstr.*, 1937: 73].

TAYLER, J.C. The relationship between growth and carcass quality in cattle and sheep. A
1964 review. *Emp. J. Exp. Agric.*, 32: 191-204.

TAYLER, J.C. Relationships between the herbage consumption or carcass energy increment
1966 of grazing beef cattle and the quantity of herbage on offer. *Proc. 10th Int. Grassl. Congr.*, Helsinki, 463-470.

TAYLER, J.C. Dried forages and beef production. *J. Brit. Grassl. Soc.*, 25: 180-190.
1970

TAYLOR, A.J. *Studies in pasture management. The composition of kikuyu grass under inten-*
1941 *sive grazing and fertilizing.* Union of S. Afr. Dept Agric. For. Bull. No. 203.

TEAKLE, L.J.H. Saline soils of Western Australia and their utilization. *J. Dept Agric. West.*
1937 *Aust., 14 (2nd Series):* 313-324.

TEITZEL, J.K. Pastures for the wet tropical coast. *Queensl. Agric. J.*, 95: 304-314; 380-385;
1969 464-471; 532-537.

TEITZEL, J.K., ABBOTT, R.A. & MELLOR, W. Beef cattle pastures in the wet tropics.
1974 *Queensl. Agric. J.*, 100: 98-105; 149-155; 185-189; 204-210.

TEITZEL, J.K., MCTAGGART, A.R. & HIBBERD, M.J. Pasture and cattle management in the
1971 wet tropics. *Queensl. Agric. J.*, 97: 25-30.

TEITZEL, J.K. & MIDDLETON, C.H. New pastures for the wet tropical coast. *Queensl. Agric.*
1979 *J.*, 105; 98-103.

TEITZEL, J.K., STANDLEY, J. & WILSON, R.J. Maintenance fertilizer strategies for wet trop-
1978 ical pastures. *Queensl. Agric. J.*, 104: 126-130.

TERGAS, L.E. & BLUE, W.G. Nitrogen and phosphorus in Jaragua grass (*Hyparrhenia rufa*
1971 (Nees.) Stapf.) during the dry season in a tropical savanna affected by nitrogen fertilization. *Agron. J.*, 63: 6-9.

TERGAS, L.E., BLUE, W.G. & MOORE, J.C. Nutritive value of *Hyparrhenia rufa* (Faragua)
1971 in the wet-dry Pacific region of Costa Rica. *Trop. Agric. (Trinidad)*, 48: 1-8.

TERNOUTH, J.H., POPPI, D.P. & MINSON, D.J. The voluntary food intake, ruminal reten-
1979 tion time and digestibility in cattle and sheep fed tropical grasses. *Proc. Nutr. Soc. Aust.*, 4: 152.

TERRY, P.J. Field evaluation of glyphosate, asulam and dalapon on African couch grass
1974 (*Digitaria scalarum*). *E. Afr. Agric. For. J.*, 39: 386-390.

THOMAS, D.B. *Perennial pasture plants for low rainfall areas of Ukambani.* Nairobi, Kenya,
1960 Min. of Agric.

THOMAS, D.B. Land use and soil erosion in parts of Kalama location, Machakos, Kenya.
1975 *Kijani*, 1: 16-17. Nairobi, Dept Agric. Mech. and Farm Planning.

Thomas, D.B. Pastures and livestock under tree crops in the humid zone. *Trop. Agric.*
1978 *(Trinidad),* 55: 39-43.
Thomas, D.B. & Pratt, D.J. Bush control studies in the drier areas of Kenya. IV. Effects
1967 of controlled burning on secondary thicket in upland Acacia woodland. *J. Appl. Ecol.,* 4: 325-335.
Thomas, R. & Humphreys, L.R. Pasture improvement at Na Pheng, central Laos. *Trop.*
1970 *Grassl.,* 4: 229-236.
Thorp, T.K. *A Zambian handbook of pasture and fodder crops.* Rome, FAO.
1979
Thurbon, P., Byford, I. & Winks, L. Evaluation of hays of *Dolichos lab-lab* cv. Rongai,
1970 a sorghum/Sudan grass cv. Zulu hybrid, and Townsville lucerne (*Stylosanthes humilis* HBK) on the basis of organic matter and crude protein digestibility. *Proc. 11th Int. Grassl. Congr.,* Surfers Paradise, Australia, 743-747.
Tilley, L.G.W. Chemical weed control. *Cane Grow. Q. Bull.,* 40: 68-116.
1977
Tiwari, S.D.N. The grasses of Madhya Pradesh. *Indian For.,* 81: 191-200.
1955
Todd, J.R. Seasonal variation in the composition of the grasses *Bothriochloa insculpta,*
1956 *Chloris gayana* and *Brachiaria dictyoneura* under rotational light grazing with a note on the persistence of the grasses. *J. Agric. Sci.,* 47: 29.
Tompsett, P.B. Factors affecting the flowering of *Andropogon gayanus* Kunth. *Ann.*
1976 *Botany,* 40: 695-705.
Tothill, J.C. Reproductive behaviour as it affects the distribution and patterns of
1970 responses of *Heteropogon contortus* in tropical and subtropical regions. *Proc. 11th Int. Grassl. Congr.,* Surfers Paradise, Australia, 30-32.
Tothill, J.C. A review of fire in the management of mature pasture with particular refer-
1971 ence to northeastern Australia. *Trop. Grassl.,* 5: 1-10.
Tothill, J.C. Flowering phenology of some native perennial tropical grasses from north-
1977 eastern Australia. *Aust. J. Ecol.,* 2: 199-205.
Tothill, J.C. *Research programmes for the development of natural pastures in northern*
1978 *Argentina.* Rept UNDP/FAO Project ARG/76/003.
Tothill, J.C. & Hacker, J.B. *The grasses of southeast Queensland.* St Lucia, Univ. of
1973 Queensl. Press.
Tothill, J.C. & Hacker, J.B. Polyploidy, flowering phenology and climatic adaptation in
1976 *Heteropogon contortus* (Gramineae). *Aust. J. Ecol.,* 1: 213-222.
Tow, P.G. Sowing rate, survival and productivity of green panic-glycine mixtures. *Queensl.*
1967 *J. Agric. Anim. Sci.,* 24: 141-148.
Trew, E.M. *Blue panic grass.* Texas Agric. Exp. Sta. Bull. B-245.
1954
Troll, C. Season climates of the earth. The seasonal course of natural phenomena in the
1966 different climatic zones of the earth. *In* Rodenwalt, E. and Juratz, H.J., eds. *World maps of climatology.* Berlin, Springer-Verlag.
Trollope, W.S.W. Fire, a rangeland tool in southern Africa. *Proc. 1st. Int. Range. Congr.,*
1978 Denver (Colorado), USA, 245-247.
Tsiung, N.T. Influence of drought stress on growth, photosynthesis, translocation and
1976 forage quality of *Panicum maximum* Jacq. var. *trichoglume. J. Aust. Inst. Agric. Sci.,* 42: 107-108.
Turner, F.W. *The forage plants of Australia.* Sydney, Govt Printers.
1891
Upton, W.H. Irrigated summer forage crops for beef cattle. *Agric. Gaz. NSW,* 89: 40-41.
1978
Urochloa, Urochloa mosambicensis. J. Aust. Inst. Agric. Sci., 40: 88-94.
1974

UTLEY, P.R., NEWTON, G.L., MONSON, W.G., HELLWIG, R.E. & MCCORMICK, W.C.
1978 Relationships among laboratory analyses of pelleted warm season grasses and animal performance. *J. Anim. Sci.,* 47: 276-282.

VALLIANT, A. [The improvement of pastures and cattle in the Chad regions south of the
1957 Sahara (Diamere, Logana-Cheri, Mandara)]. *J. agric. trop. et bot. appl.,* 4: 69-82. (In French)

VENDARGON, X.A. Observations on grass and legume mixtures at the Central Husbandry
1964 Station, Kluang. *J. Malay Vet. Med. Assoc.,* 3: 119-127.

VERBOOM, W.C. *Brachiaria dura,* a promising new forage grass. *J. Range Manage.,* 19: 91-
1966 93.

VERBOOM, W.C. & BRUNT, M.A. *An ecological survey of Western Province, Zambia, with*
1970 *special reference to the fodder resources. Vol. 2. The grasslands and their development.* Tolworth (Surrey), UK, Directorate of Overseas Surveys. Land Resources Divn Land Res. Study No. 8.

VERDCOURT, B. & TRUMP, E.C. *Common poisonous plants of East Africa.* London, Collins.
1969

VERMA, C.M. & CHAKRAVARTY, A.K. Study on the pasture establishment technique.
1969 4. Comparative efficiency of seedlings and rooted slips as a transplanting material in *Lasiurus sindicus* pasture. *Ann. Arid Zone,* 8: 52-57.

VESEY-FITZGERALD, D.F. Grazing succession among East African game animals. *J. Mammal.,*
1960 41: 161-172.

VESEY-FITZGERALD, D.F. Central African grasslands. *J. Ecol.,* 57: 243-274.
1963

VICENTE-CHANDLER, J., ABRUNA, F., CARO-COSTAS, R., FIGARELLA, J., SILVA, S. &
1974 PEARSON, R.W. *Intensive grassland management in the humid tropics of Puerto Rico.* Rio Pedras, Univ. Puerto Rico Agric. Exp. Sta. Bull. No. 233.

VICENTE-CHANDLER, J., CARO-COSTAS, R., PEARSON, R.W., ABRUNA, F., FIGARELLA, J.
1964 & SILVA, S. *The intensive management of tropical forages in Puerto Rico.* Rio Pedras, Univ. Puerto Rico Agric. Exp. Sta. Bull. No. 187.

VICENTE-CHANDLER, J., RIVERA-BRENES, L., CARO-COSTAS, R.R., RODRIGUEZ, J.P.,
1953 BONETA, E. & GARCÍA, W. *The management and utilization of the forage crops of Puerto Rico.* Rio Pedras, Univ. of Puerto Rico Agric. Exp. Sta. Bull. No. 116.

VICENTE-CHANDLER, J., SILVA, S. & FIGARELLA, J. The effect of nitrogen fertilization and
1959 frequency of cutting on the yield and composition of three tropical grasses. *Agron. J.,* 51: 202-206.

VINIJSANOND, J. A survey of diseases of pasture grasses and legumes, Khon Kaen University,
1978 Thailand. *Pasture Improvement Project, Ann. Rept,* 32-34.

VOGL, R.J. The role of fire in the evolution of Hawaiian flora and vegetation. *Proc. 9th Ann.*
1964 *Tall Timbers Fire Ecol. Conf.,* Tallahassee (Fla), USA.

VOIGHT, P.W., KNEEBONE, W.P., MCILVAIN, E.H., SHOOP, M.C. & WEBSTER, J.E. Palat-
1970 ability, chemical composition and animal gains from selections of weeping love grass, *Eragrostis curvula* (Schrad.) Nees. *Agron. J.,* 62: 673-676.

VOORTHUIZEN, E.G. VAN. A quality evaluation of four widely distributed native grasses in
1971 Tanzania. *E. Afr. Agric. For. J.,* 37: 384-391.

WALKER, B. Herbage plants for pasture improvement in western Tanzania. *E. Afr. Agric.*
1969 *For. J.,* 35: 1.

WALMSLEY, D., SARGEANT, V.A.L. & DOOKERAN, M. Effect of fertilizers on growth and
1978 composition of elephant grass (*Pennisetum purpureum*) in Tobago, West Indies. *Trop. Agric. (Trinidad),* 25: 329-334.

WALSH, S.R. Improved pastures will fatten cattle in Far North. *Queensl. Agric. J.,* 85: 576-
1959 592.

WALSHE, M.J. Grazing management and the productivity of grazing systems. *Proc. 3rd*
1975 *World Conf. Anim. Prod.,* Melbourne, 165-173.

WALTHALL, J.C. Bighead disease of horses at pasture. *Queensl. Agric. J.*, 103: 331-336.
1977
WANG, C.C. [Growth, flowering and forage production of some grasses and legumes in
1961 response to different photoperiods.] *J. Agric. Assoc. China*, 36: 27-52 (In Chinese) [*Herb. Abstr.*, 33: No. 864].
WARE-AUSTIN, W.D. Napier grass for milk production in the Trans Nzoia. *E. Afr. Agric.*
1963 *For. J.*, 28: 223-226.
WARMKE, H.E. Apomixis in *Panicum maximum*. *Amer. J. Bot.*, 41: 5-11.
1954
WATSON, E.R. & LAPINS, P. Losses of nitrogen from urine on soils from southwestern
1969 Australia. *Aust. J. Exp. Agric. Anim. Husb.*, 9: 85-91.
WATT, L.A. Evaluation of pasture species for soil conservation on cracking black clays,
1976 Gwydir District, New South Wales. *J. Soil Conserv. Serv.* (NSW), 32: 86-97.
WEBSTER, J.E., HOGAN, J.W. & ELDER, W.C. Effect of rate of ammonium nitrate fertiliza-
1965 tion and time of cutting upon selected chemical components and the *in vitro* rumen digestion of Bermuda grass forage. *Agron. J.*, 57: 323-325.
WEEKS, M.E. & YEGIAN, H.M. The place of silage in a forage utilization program.
1965 Researches on production problems and evaluation. *Proc. 9th Int. Grassl. Congr.*, São Paulo, 589-593.
WEIER, K.L. Nitrogen economy of pastures. *CSIRO Aust., Divn Trop. Crops and Pastures*
1976 *Rept, 1975-76:* 57.
WEIER, K.L. Nitrogen fixation by grass micro-organism associations. *CSIRO Aust., Divn*
1977 *Trop. Crops and Pastures Rept, 1976-77:* 97-98.
WEIER, K.L., MACRAE, I.C. & ALLEN, J. Nitrogen fixation associated with the root system
1978 of tropical grasses. *CSIRO Aust., Divn Trop. Crops and Pastures Ann. Rept, 1977-78:* 75-76.
WENDT, W.B. Responses of pasture species in eastern Uganda to phosphorus, sulphur and
1970 potassium. *E. Afr. Agric. For. J.*, 36: 211-219.
WESLEY-SMITH, R.N. Para grass in the Northern Territory; parentage and propagation.
1973 *Trop. Grassl.*, 7: 249-250.
WEST, O. *Fire in vegetation and its use in pasture management, with special reference to trop-*
1965 *ical and subtropical Africa.* Commonw. Bur. Pastures and Field Crops Rev. Ser.
WEST, S.H. Biochemical mechanism of photosynthesis and growth depression in *Digitaria*
1971 *decumbens* when exposed to low temperatures. *Proc. 11th Int. Grassl. Congr.*, Surfers Paradise, Australia, 514.
WESTPHAL, E. *Agricultural systems in Ethiopia.* Wageningen, the Netherlands, Centre for
1975 Agricultural Publishing and Documentation. Agric. Res. Rept No. 826.
WET, J.M.J. DE, HARLAN, J.R. & KURMAROHITA, B. Origin and evolution of Guinea sor-
1972 ghums. *E. Afr. Agric. For. J.*, 38: 114-119.
WHEELER, J.L. Experimentation in grazing management. *Herb. Abstr.*, 32: 1-7.
1962
WHEELER, W.A. *Forage and pasture crops.* New Jersey, van Nostrand.
1950
WHITE, C.T. *Inland pastures.* Brisbane, Queensl. Dept Agric. and Stock Bot. Rept No. 1.
1935
WHITEHOUSE, F.W., OGILVIE, C. & SKERMAN, P.J. *The Channel Country of southwest*
1947 *Queensland with special reference to Cooper's Creek.* Brisbane, Bureau of Inves- tigation, Lands Dept Tech. Bull. No. 1.
WHITEMAN, P.C. & GILLARD, P. Species of *Urochloa* as pasture plants. *Herb. Abstr.*, 41:
1971 352-357.
WHITEMAN, P.C. & WILSON, G.L. Effects of water stress on the reproductive development
1965 of *Sorghum vulgare* Pers. St Lucia. *Univ. Queensl. Dept Bot. Papers*, 4: 233-239.
WHITNEY, A.S. & GREEN, R.E. Legume contribution to yield and composition of
1969 *Desmodium* spp.-Pangola grass mixtures. *Agron. J.*, 61: 741-74.

WHITTET, J.N. *Pastures of New South Wales.* Sydney, Dept of Agric., NSW.
1965

WHYTE, R.O. Land utilization in the humid tropics. *Proc. 9th Pacific Sci. Congr.,*
1958   18 November-9 December 1957, Bangkok, Vol. 20, 143-148.

WHYTE, R.O. *The grassland and fodder resources of India.* New Delhi, Indian Counc. for
1964   Agric. Res.

WHYTE, R.O. *Grasslands of the monsoon.* London, Faber and Faber.
1968

WHYTE, R.O., MOIR, T.R.G. & COOPER, J.P. *Grasses in agriculture.* Rome, FAO. Agric.
1959   Group Studies No.42.

WIERINGA, G.W. The influence of nitrate on silage fermentation. *Proc. 10th Int. Grassl.*
1966   *Congr.,* Helsinki, 537-540.

WIGG, P.M. The role of sown pastures in semi-arid central Tanzania. *E. Afr. Agric. For. J.,*
1973   38: 375-382.

WIGG, P.M., OWEN, M.A. & MUKURASI, N.J. Influence of farmyard manure and nitrogen
1973   fertilization on sown pastures, seed yield and quality of *Cenchrus ciliaris* at Kongwa, Tanzania. *E. Afr. Agric. For. J.,* 38: 367-374.

WILLIAMS, C.H. & DONALD, C.M. Changes in organic matter and pH in a podzolic soil as
1957   influenced by subterranean clover and superphosphate. *Aust. J. Agric. Res.,* 8: 179-189.

WILLIAMS, J.T. & FARIAS, R.M. Utilization and taxonomy of the desert grass, *Panicum tur-*
1972   *gidum. Econ. Bot.,* 26: 13-20.

WILLMS, E.F., SCHLEIDER, H.E., WEIHING, R.M. & SORENSON, J.W. *Silage studies.* Rice-
1958   Pasture Exp. Sta. 1957-8. Texas Agric. Exp. Sta. Prog. Rept No. 2057.

WILLOUGHBY, W.M. Grassland management. *In* Moore, R.M., ed. *Australian grasslands.*
1970   Canberra, Aust. Nat. Univ. Press.

WILSON, G.L. & WHITEMAN, P.C. The influence of shoot removal on drought survival of
1965   sorghums. *Univ. Queensl. Dept Bot. Papers,* 4: 223-239.

WILSON, G.P.M. & HENNESSY, D.W. The germination of excreted kikuyu grass seed in
1977   cattle dung pats. *J. Agric. Sci.,* 88: 247-249.

WILSON, J.R. & 'T MANNETJE, L. Senescence, digestibility and carbohydrate content of buf-
1978   fel grass and green panic leaves in swards. *Aust. J. Agric. Res.,* 29: 503-516.

WILSON, G.P.M. & RUMBLE, C.J. The effect of seed rate and nitrogen fertilization on the
1975   yield of seed and by-product leaf of Whittet kikuyu grass at Grafton, New South Wales. *Trop. Grassl.,* 9: 53-58.

WILSON, J.R., TAYLOR, A.O. & DOLBY, G.R. Temperature and atmospheric humidity
1976   effects on cell wall content and dry matter digestibility of some tropical and temperate grasses. *N.Z. J. Agric. Res.,* 19: 41-46.

WILSON, R.G. Contour furrow and press wheel in pasture establishment. *Queensl. Agric. J.,*
1978   104: 315-319.

WINKS, L. Integration of native and sown pastures for increased animal production. *Trop.*
1975   *Grassl,* 9; 159-164.

WINTER, W.H. Preliminary evaluation of twelve tropical grasses with legumes in northern
1976   Cape York Peninsula. *Trop. Grassl.,* 10: 15-20.

WINTER, W.H., SIEBERT, B.D. & KUCHEL, R.E. Cobalt deficiency of cattle grazing
1977   improved pastures in northern Cape York Peninsula. *Aust. J. Exp. Agric. Anim. Husb.,* 17: 10-15.

WRIGHT, S.F., WEAVER, R.W. & HOLT, E.C. *Survey of grasses in Texas for acetylene*
1976   *returning activity.* College Sta. Texas, Dept Soil Crop Sci., TX77843.

WRIGHT, W.A. Results of work done on the production and conservation of fodder crops.
1961   *Rev. agric. sucr. Ile Maurice,* 40: 46-48.

WYCHERLEY, P.R. & YUSOF, A.A.B.M. *Grasses in Malayan plantations.* Kuala Lumpur.
1974   Malaysia, Rajvv Printers.

WYK, H.P.D. VAN, OOSTHUIZEN, S.A., MEYER, E.E., BREVIS, J.G. & GROBIER, J.H. The
1955   nutritive value of South African feeds. 3. Hay and pasture crops, silage, cereals, tubers and pods. *Union of S. Afr. Dept Agric. Sci.* Bull., No. 354.

WYLLIE, P.B. Silage in Queensland. *Queensl. Agric. J.*, 101: 709-781.
1975

WYLLIE, P.B. & STIRLING, G.D. Making grain sorghum pay in the near southwest. *Queensl.*
1977   *Agric. J.*, 103: 12-20.

YADAV, J.S.P. Improvement of saline soils through biological methods. *Indian For.*, 101:
1975   388-395.

YATES, J.J., EDYE, L.A., DAVIES, J.G. & HAYDOCK, K.P. Animal production from a *Sor-*
1964   *ghum almum* pasture in southeast Queensland. *Aust. J. Exp. Agric. Anim. Husb.*, 4: 326-335.

YATES, J.J., RUSSELL, M.J. & FERGUS, I.F. Effects and interactions of lucerne and subtrop-
1971   ical legumes in a *Sorghum almum* pasture. *Aust. J. Exp. Agric. Anim. Husb.*, 11: 651-671.

YELF, J.D. [*Effect of cutting times on pasture yields*]. Suva, Fiji, Dept Agric. Bull. No. 34.
1957   (In French)

YEPES, S. *Guía práctica de pastos y forrajes existentes en Cuba.* Habana, Cuba, Dpto Pastos
1975   y Forrajes, Dirección General Pecuaria.

YOUNG, N.D., FOX, N.F. & BURNS, M.A. A study of three important pasture mixtures in
1959   the Queensland tropics. *Queensl. J. Agric. Sci.*, 16: 199-215.

YOUNGE, O.R. & OTAGAKI, K.K. *The variation in protein and mineral composition of*
1958   *Hawaii range grasses and its potential effect on cattle nutrition.* Honolulu, Hawaii Agric. Exp. Sta. Bull. No. 119.

YOUNGE, O.R., PLUCKNETT, D.L. & ROTAR, P.P. *Culture and yield performance of*
1964   *Desmodium intortum and* D. canum *in Hawaii.* Honolulu, Hawaii Agric. Exp. Sta., Tech. Bull. No. 59.

YOUNGE, O.R. & RIPPERTON, J.C. *Nitrogen fertilization of pasture and forage grasses in*
1960   *Hawaii.* Honolulu, Hawaii Agric. Exp. Sta. Bull. No. 124.

YOUNGER, D.R. & GILMORE, J.M. Studies with pasture grasses on the black cracking clays
1978   of the Central Highlands. 1. Species evaluation. *Trop. Grassl.*, 12: 152-162.

YOUNGNER, V.B. Growth of U_3 Bermuda grass under various day and night temperatures
1959   and light intensities. *Agron. J.*, 51: 557-559.

YOUNGNER, V.B. & LUNT, O.R. Salinity effects on roots and tops of Bermuda grass. *J. Brit.*
1967   *Grassl. Soc.*, 22: 257.

ZERPA, H. & VILLALOBOS, H. [New forage grass for pasture]. *Agron. Trop. Maracay,*
1971   *Venez.*, 2: 117-121. (In Spanish)

ZUNIGA, M.P., SYKES, D.J. & GOMIDE, J.A. [Comparison of 13 forage grasses for cutting
1967   with or without fertilizers in Viscosa, Minas Gerais]. *Riv. Ceres*, 13: 324-343. (In Spanish)

Common names of tropical grasses

Compiled by Camille Trentacoste

abu shaar - *Cenchrus biflorus*
afezu - *Panicum turgidum*
African couch grass - *Digitaria abyssinica*
African foxtail - *Cenchrus ciliaris*
African love grass - *Eragrostis curvula*
African millet - *Eleusine coracana*
African star grass - *Cynodon plectostachyus*
akirma - *Eleusine jaegeri*
alabang X - *Imperata cylindrica*
alang-alang - *Imperata cylindrica*
alemán grass - *Echinochloa polystachya*
Aleppo grass - *Sorghum halepense*
alkali sacaton - *Sporobolus airoides*
alula - *Pennisetum setaceum*
Angleton grass - *Dichanthium aristatum*
anjan grass - *Cenchrus ciliaris*
antelope grass - *Echinochloa pyramidalis*
Antigua hay grass - *Dichanthium caricosum*
arrow grass - *Trachypogon spicatus*
assegai grass - *Heteropogon contortus*
Aucher's grass - *Chrysopogon aucheri*
Australian blue-stem - *Bothriochloa bladhii*
awnless barnyard grass - *Echinochloa colona*
azz - *Echinochloa colona*

Bahamas grass - *Cynodon dactylon*
Bahia grass - *Paspalum notatum*
bajra - *Pennisetum americanum*
bamboo grass - *Hymenachne amplexicaulis*
bana grass - *Pennisetum americanum* × *P. purpureum*
bannu - *Eragrostis tremula*
bano - *Eragrostis tremula*
bansi - *Panicum antidotale*
banti - *Echinochloa scabra*
bara - *Pennisetum pedicellatum*

Barbados sour grass - *Bothriochloa pertusa*
barley Mitchell grass - *Astrebla pectinata*
barnyard grass - *Echinochloa crus-galli*
barnyard millet - *Echinochloa crus-galli*
batiki bluegrass - *Ischaemum indicum*
beach drop-seed - *Sporobolus virginicus*
beach spinifex - *Spinifex hirsutus*
beach wire grass - *Dactyloctenium aegyptium*
bent grass - *Agrostis* spp.
Bermuda grass - *Cynodon dactylon*
Bermuda mejorado - *Cynodon plectostachyus*
bes-chaitgras - *Axonopus compressus*
besem grass - *Loudetia simplex*
big blue-stem - *Andropogon gerardii*
billion dollar grass - *Echinochloa crus-galli* var. *crus-galli*
birdwood grass - *Cenchrus setigerus*
black grass - *Eragrostis cilianensis*
black spear grass - *Heteropogon contortus*
bladhi - *Panicum pilosum*
blady grass - *Imperata cylindrica*
blue citronella grass - *Cymbopogon nardus*
blue panic - *Panicum antidotale*
blue-stems - *Bothriochloa* spp.
bluegrass - *Themeda triandra*
Boer love grass - *Eragrostis chloromelas*
Boer millet - *Setaria italica*
bongo grass - *Brachiaria ruziziensis*
bourgou - *Echinochloa scabra*
bread grass - *Brachiaria brizantha*
broad-leaf carpet grass - *Axonopus compressus*
broad-leaf paspalum - *Paspalum wettsteinii*
broom corn millet - *Panicum miliaceum*
brown beetle grass - *Diplachne fusca*

brown corn millet - *Panicum miliaceum*
brown Rhodes grass - *Eustachys paspaloides*
brown top grass - *Agrostis tenuis*
brown-flowered swamp grass - *Diplachne fusca*
brown-seed paspalum - *Paspalum plicatulum*
Brunswick grass - *Paspalum nicorae*
buffalo bean grass - *Rottboellia exaltata*
buffalo grass - *Stenotaphrum secundatum*
buffel grass - *Cenchrus ciliaris*
bull Mitchell grass - *Astrebla squarrosa*
bulrush millet - *Pennisetum americanum*
bunch spear grass - *Heteropogon contortus*
bungoma grass - *Entolasia imbricata*
bush rye - *Enteropogon macrostachyus*

cachi - *Axonopus scoparius*
calinguero - *Melinis minutiflora*
cama gueyana - *Bothriochloa pertusa*
carib grass - *Eriochloa punctata*
carpet grass - *Axonopus compressus, A. affinis*
castilla - *Panicum maximum* var. *trichoglume*
Caucasian blue-stem - *Bothriochloa caucasica*
Ceylon sheep grass - *Brachiaria brizantha*
Channel millet - *Echinochloa turneriana*
cheno - *Panicum repens*
citronella grass - *Cymbopogon nardus*
Cloncurry buffel grass - *Cenchrus pennisetiformis*
coarse Guinea - *Panicum maximum* cv. Coarse Guinea
coastal Bermuda grass - *Cynodon dactylon*
cogon grass - *Imperata cylindrica*
coloniao Guinea - *Panicum maximum* cv. Coloniao
coloured Guinea grass - *Panicum coloratum*
Columbus grass - *Sorghum almum*
common Guinea - *Panicum maximum* cv. Common Guinea
common needle grass - *Aristida adscensionis*
common reed - *Phragmites australis*
common russet grass - *Loudetia simplex*
common thatching grass - *Hyparrhenia hirta*
common urochloa - *Urochloa mosambicensis*
compressum - *Axonopus affinis*
Congo signal grass - *Brachiaria ruziziensis*
coolatai grass - *Hyparrhenia hirta*

cori grass - *Brachiaria subquadripara, B. miliiformis*
corn - *Zea mays*
cotranh - *Imperata cylindrica* var. *africana*
cotton wool grass - *Imperata cylindrica* var. *major*
couch finger grass - *Digitaria abyssinica*
couch grass - *Cynodon dactylon*
couch panicum - *Panicum repens*
crab grass - *Digitaria ciliaris*
creeping bluegrass - *Bothriochloa insculpta*
creeping Guinea grass - *Panicum maximum* cv. Embu
creeping panicum - *Panicum repens*
crow's foot grass - *Eleusine indica*
crowfoot grass - *Dactyloctenium aegyptium, Eleusine indica*
curly Mitchell grass - *Astrebla lappacea*
curly spear grass - *Aristida latifolia*
cushion love grass - *Eragrostis caespitosa*

dagusa - *Eleusine coracana*
dal grass - *Hymenachne amplexicaulis*
dallis grass - *Paspalum dilatatum*
danga grass - *Trachypogon spicatus*
dari - *Sorghum bicolor*
dark finger grass - *Eustachyus paspaloides*
deenanath grass - *Pennisetum pedicellatum*
Delhi grass - *Dichanthium annulatum*
devil's grass - *Cynodon dactylon*
dhaman grass - *Cenchrus ciliaris*
dhoub grass - *Cynodon dactylon*
dhurra grass - *Sorghum bicolor*
digit grass - *Digitaria decumbens*
donkey grass - *Panicum trichocladum*
du-ghasi - *Panicum turgidum*
dubi grass - *Urochloa oligotricha*
dukn - *Pennisetum americanum*
Dunn's finger grass - *Digitaria abyssinica*
durra grass - *Sorghum bicolor*
Durrington grass - *Axonopus affinis*
dwarf setaria - *Setaria italica*

eastern gama grass - *Tripsacum dactyloides*
elephant grass - *Pennisetum purpureum*
elkan blue-stem - *Bothriochloa ischaemum*
estrella - *Cynodon plectostachyus*
estrella de Africa - *Brachiaria brizantha*
eternity grass - *Paspalum paspaloides*

fairway wheatgrass - *Agropyron cristatum*
false citronella - *Cymbopogon nardus*
faragua grass - *Hyparrhenia rufa*

feather finger grass - *Hyparrhenia rufa*
feather-top - *Pennisetum villosum*
feather-top chloris - *Chloris virgata*
feather-top Rhodes grass - *Chloris virgata*
feather-top wire grass - *Aristida latifolia*
feterita - *Sorghum bicolor*
fine grass - *Bothriochloa radicans*
fine hood grass - *Hyparrhenia filipendula*
fine thatching grass - *Hyparrhenia filipendula*
finger millet - *Eleusine coracana*
five-year sorghum - *Sorghum almum*
flat joint grass - *Axonopus compressus*
flat-seed love grass - *Eragrostis superba*
Flinders grass - *Iseilema membranaceum*
forest blue grass - *Bothriochloa bladhii*
fountain grass - *Pennisetum setaceum*
fox-tail millet - *Setaria italica*
foxtail - *Pennisetum villosum*
French millet - *Panicum miliaceum*

gama grass - *Tripsacum dactyloides*
gamalote - *Paspalum fasciculatum*
gamarawal - *Echinochloa scabra*
gamba grass - *Andropogon gayanus*
garden urochloa - *Urochloa panicoides*
German grass - *Echinochloa polystachya*
gi - *Imperata cylindrica* var. *africana*
giant Bermuda grass - *Cynodon dactylon* var. *aridus*
giant button grass - *Dactyloctenium giganteum*
giant panic grass - *Panicum antidotale*
giant paspalum - *Paspalum urvillei*
giant reed - *Arduno donax*
giant setaria - *Setaria italica*
giant star grass - *Cynodon aethiopicus, C. nlemfuensis, C. plectostachyus*
golden bristle grass - *Setaria sphacelata* var. *sericea*
golden timothy - *Setaria sphacelata* var. *sericea*
gonya grass - *Urochloa mosambicensis*
goose grass - *Eleusine indica*
gordura - *Melinis minutiflora*
grader grass - *Themeda quadrivalvis*
grama de agua - *Echinochloa colona*
grama de caballo - *Eleusine indica*
gramalote - *Panicum maximum*
gramilla - *Digitaria ciliaris*
gramilla blanca - *Cynodon dactylon*
grass seed - *Chrysopogon aciculatus*
green couch - *Cynodon dactylon*

green panic - *Panicum maximum* var. *trichoglume*
green summer grass - *Brachiaria subquadripara*
grey beard grass - *Trachypogon spicatus*
grey love grass - *Eragrostis cilianensis*
Guatemala grass - *Tripsacum laxum*
guinchi - *Panicum turgidum*
Guinea corn - *Sorghum bicolor*
Guinea grass - *Panicum maximum*
guria grass - *Chrysopogon fulvus*

haakdoring - *Acacia litakunensis*
Habana oat grass - *Themeda quadrivalvis*
hairy crabgrass - *Digitaria ciliaris*
hairy herringbone grass - *Tetropogon spathaceus*
hairy spinifex - *Spinifex hirsutus*
halt grass - *Hemarthria altissima*
Harding grass - *Phalaris stenoptera*
haskanit - *Cenchrus biflorus*
Hawaiiano - *Cynodon plectostachyus*
heart-seed love grass - *Eragrostis superba*
hegari - *Sorghum bicolor*
herbe a miel - *Melinis minutiflora*
hierba-fina - *Cynodon dactylon*
hippo grass - *Echinochloa scabra*
hog millet - *Panicum miliaceum*
Honduras grass - *Ixophorus unisetus*
hood grass - *Hyparrhenia cymbaria*
hook grass - *Leptothrium senegalense*
hoop Mitchell grass - *Astrebla alymoides*
horo - *Trachypogon spicatus*
horsetail grass - *Chloris roxburghiana*
Hungarian millet - *Setaria italica*
hurricane grass - *Bothriochloa pertusa*
hybrid ryegrass - *Lolium perenne* × *L. multiflorum*
hymenachne - *Hymenachne acutigluma*

ikoka - *Panicum trichocladum*
illuk - *Imperata cylindrica*
imperial grass - *Axonopus scoparius*
Indian blue-stem - *Dichanthium caricosum*
Indian blue grass - *Bothriochloa pertusa*
Indian corn - *Zea mays*
Indian goose grass - *Eleusine indica*
Indian sandbur - *Cenchrus biflorus*
initi - *Cenchrus biflorus*
Italian ryegrass - *Lolium multiflorum*
itchgrass - *Rottboellia exaltata*

Janeiro - *Eriochloa punctata*
Japanese cane - *Saccharum sinense*

Tropical Grasses

Japanese millet - *Echinochloa utilis*
jaragua grass - *Hyparrhenia rufa*
jiribilla - *Dichanthium caricosum*
Job's tears - *Coix lacryma-jobi*
Johnson grass - *Sorghum halepense*
joint vetch - *Aeschynomene falcata*
jowar - *Sorghum bicolor*
jungle rice - *Echinochloa colona*

kabuta - *Cynodon dactylon*
kaffir - *Sorghum bicolor*
kangaroo grass - *Themeda australis*
karad - *Dichanthium annulatum*
karena - *Lasiurus hirsutus*
kase - *Chrysopogon aciculatus*
kavoronaisivi - *Eleusine indica*
kazungula grass - *Setaria sphacelata*
Kennedy ruzi grass - *Brachiaria ruziziensis*
Kenya sheep grass - *Brachiaria decumbens*
khachornchob - *Pennisetum polystachyon*
khas-khas grass - *Vetiveria zizanioides*
Kikuyu - *Pennisetum clandestinum*
King Ranch blue-stem - *Bothriochloa ischaemum*
kiri-hiri grass - *Cynodon dactylon*
Kleberg blue-stem - *Dichanthium annulatum*
Klein grass - *Panicum coloratum*
knot grass - *Paspalum paspaloides*
koda - *Paspalum coloratum*
kodo millet - *Paspalum scrobiculatum*
koluk katai - *Cenchrus ciliaris*
koracan millet - *Eleusine coracana*
korohiria grass - *Brachiaria humidicola*
Koronivia grass - *Brachiaria humidicola*
kra lekrab - *Dactyloctenium aegyptium*
kunai grass - *Imperata cylindrica*
kuri millet - *Urochloa panicoides*
kurrakan millet - *Eleusine coracana*
kutki - *Panicum sumatrense*
kweek grass - *Cynodon dactylon*
kyasuwa grass - *Pennisetum pedicellatum*

lalang - *Imperata cylindrica*
lambedora grass - *Leersia hexandra*
land grass - *Panicum laevifolium*
lautoka - *Bothriochloa bladhii*
Lehmann's love grass - *Eragrostis lehmanniana*
lehmleiche - *Eragostis tremula*
liberty millet - *Setaria italica*
lierba del Caribe - *Eriochloa punctata*
limanota - *Panicum repens*

limpo grass - *Hemarthria altissima*
lindi - *Dichanthium annulatum*
little millet - *Panicum sumatrense*
liverseed grass - *Urochloa panicoides*
long-awned water grass - *Echinochloa scabra*
long-styled feather grass - *Pennisetum villosum*
love grass (Malaysia) - *Chrysopogon aciculatus*
love grasses - *Eragrostis* spp.
lucuntu grass - *Ischaemum timorense*

ma yuen - *Coix lacryma-jobi*
machuri - *Iseilema laxum*
Mackie's pest - *Chrysopogon aciculatus*
maicillo - *Axonopus scoparius*
maize - *Zea mays*
makarikari grass - *Panicum coloratum* var. *makarikariense*
makarikari panicum - *Panicum coloratum* var. *makarikariense*
makarikariense grass - *Panicum coloratum* var. *makarikariense*
Malaysian love grass - *Chrysopogon aciculatus*
malojilla - *Brachiaria mutica*
malutania mopane grass - *Enteropogon macrostachyus*
mangrasi - *Eleusine indica*
maniene-ula - *Chrysopogon aciculatus*
manyatta grass - *Eleusine jaegeri*
markouba - *Panicum turgidum*
marvel grass - *Dichanthium annulatum*
Masai grass - *Pennisetum stramineum*
Masai love grass - *Eragrostis superba*
mat grass - *Axonopus affinis*
Mauritius grass - *Brachiaria mutica*
Mauritius signal grass - *Brachiaria mutica*
melado - *Melinis minutiflora*
Merker grass - *Pennisetum purpureum*
Mexican grass - *Ixophorus unisetus*
milanje finger grass - *Digitaria milanjiana*
Miles lotononis - *Lotononis bainesii*
milo - *Sorghum bicolor*
mission grass - *Pennisetum polystachyon*
Mitchell grass - *Astrebla* spp.
moda dhaman grass - *Cenchrus setigerus*
molasses grass - *Melinis minutiflora*
moshi - *Iseilema laxum*
mulga - *Acacia aneura*
muran - *Panicum repens*
musal grass - *Iseilema laxum*

nachni - *Eleusine coracana*
Nadi bluegrass - *Dichanthium caricosum*
naid grass - *Cymbopogon nardus*
Naivasha star grass - *Cynodon plectostachyus*
Nandi grass - *Setaria sphacelata*
Nandi setaria - *Setaria sphacelata* var. *sericea* cv. Nandi
Napier grass - *Pennisetum purpureum*
narrow-leaved carpet grass - *Axonopus affinis*
native millet - *Echinochloa turneriana*
native sorghum - *Echinochloa turneriana*
nawai grass - *Dichanthium caricosum*
never-fail grass - *Eragrostis setifolia*
Nigeria grass - *Pennisetum pedicellatum*
nigolo - *Pennisetum polystachyon*
Nile grass - *Acroceras macrum*
nudillo - *Axonopus compressus*
Nunbank setaria - *Setaria italica*

okrich - *Sporobolus helvolus*
onaga - *Andropogon gayanus*

palisade grass - *Brachiaria brizantha*
pangola digit grass - *Digitaria decumbens*
pangola grass - *Digitaria decumbens*
pangola river grass - *Digitaria pentzii*
panic rampant - *Panicum repens*
Para grass - *Brachiaria mutica*
Paraguay paspalum - *Paspalum notatum*
parana - *Brachiaria mutica*
pardegrao - *Echinochloa polystachya*
paspalum - *Paspalum dilatatum*
pearl millet - *Pennisetum americanum*
pepper grass - *Panicum whitei*
perennial ryegrass - *Lolium perenne*
phalaris - *Phalaris aquatica*
pigeon grass - *Heteropogon whitei*
pili grass - *Heteropogon contortus*
pinhole grass - *Bothriochloa insculpta*
pit-pit - *Phragmites karka, Saccharum spontaneum*
pitilla - *Dichanthium annulatum*
plains blue-stem - *Bothriochloa ischaemum*
plicatulum - *Paspalum plicatulum*
plume chloris - *Chloris roxburghiana*
plume grass - *Erianthus ravannae*
pongola grass - *Digitaria decumbens*
prasi-grasi - *Echinochloa polystachya*
proso - *Panicum miliaceum*
prostrate signal grass - *Brachiaria ruziziensis*
purple pigeon grass - *Setaria porphyrantha*

Queensland blue couch grass - *Digitaria didactyla*
Queensland bluegrass - *Dichanthium sericeum*
quick grass - *Cynodon dactylon*

ragi - *Eleusine coracana*
rapoka grass - *Eleusine indica*
rapoko - *Eleusine coracana*
rat's tail grass - *Sehima nervosum*
rattana - *Ischaemum indicum*
red Flinders grass - *Iseilema vaginiflorum*
red French millet - *Panicum miliaceum*
red-oat grass - *Themeda triandra*
red rala - *Setaria italica*
red vlei grass - *Hemarthria altissima*
reed - *Phragmites australis*
reflexed panic — *Paractaenum novae-hollandiae*
regop paspalum - *Paspalum urvillei*
Rhodes grass - *Chloris gayana*
Rhodesian andropogon - *Andropogon gayanus*
Rhodesian bluegrass - *Andropogon gayanus*
rice - *Oryza sativa*
robust star grass - *Cynodon nlemfuensis*
rolling spinifex - *Spinifex hirsutus*
rooikweek - *Hemarthria altissima*
rumput melayu - *Ischaemum magnum*
russet grass - *Loudetia simplex*
ruzi grass - *Brachiaria ruziziensis*

sabi grass - *Urochloa mosambicensis*
Sadabahar - *Andropogon gayanus*
salt-water couch - *Paspalum distichum, Sporobolus virginicus*
sava - *Panicum sumatrense*
savannah grass - *Axonopus compressus*
scrobic - *Paspalum scrobiculatum*
scrobic paspalum - *Paspalum scrobiculatum*
sea-shore paspalum - *Paspalum distichum*
sea-shore rush grass - *Sporobolus virginicus*
seed grass - *Chrysopogon aciculatus*
senbelet - *Hyparrhenia rufa*
seragoon grass - *Digitaria didactyla*
setaria - *Setaria sphacelata* var. *sericea*
sewan grass - *Lasiurus hirsutus*
Seymour grass - *Bothriochloa pertusa*
sheda grass - *Dichanthium annulatum*
shirohie millet - *Echinochloa utilis*
shunkora - *Saccharum officinarum*
Siberian millet - *Echinochloa frumentacea*

signal grass (Australia) - *Brachiaria decumbens*
signal grass (E. Africa) - *Brachiaria brizantha*
signal grasses - *Brachiaria* spp.
sila - *Coix lacryma-jobi*
silky blue-stem - *Dichanthium sericeum*
silver spike - *Imperata cylindrica* var. *africana*
six-weeks three-awn - *Aristida adscensionis*
slender buffel grass - *Cenchrus pennisetiformis*
slender Guinea grass - *Panicum maximum* var. *trichoglume*
small buffalo grass - *Panicum coloratum*
small Flinders grass - *Iseilema membranaceum*
small panicum - *Panicum coloratum*
soft spinifex - *Triodia pungens*
song-chang - *Echinochloa crus-galli*
sorghum - *Sorghum bicolor*
sorgo negro - *Sorghum almum*
South African blue-stem - *Hyparrhenia hirta*
spear grass - *Imperata cylindrica*
spring rolling grass - *Spinifex hirsutus*
St Augustine grass - *Stenotaphrum secundatum*
St Lucia grass - *Brachiaria brizantha*
star grass - *Cynodon nlemfuensis*
stink grass - *Eragrostis cilianensis*
stippel grass - *Bothriochloa insculpta*
Sudan grass - *Sorghum sudanense*
Sudan negro - *Sorghum almum*
sugar cane - *Saccharum officinarum*
sugar grass - *Panicum whitei*
summer grass - *Digitaria ciliaris*
Suriname grass - *Brachiaria decumbens*
swamp couch - *Hemarthria altissima*
swamp cut grass - *Leersia hexandra*
swamp millet - *Echinochloa turneriana*
swamp rice grass - *Leersia hexandra*
sweet corn - *Zea mays*
sweet pitted grass - *Bothriochloa insculpta*
sweet Sudan grass - *Sorghum sudanense* × *S. bicolor*
switch grass - *Panicum virgatum*
sword grass - *Imperata cylindrica* var. *africana*

taman - *Panicum turgidum*
tambookie grass - *Hyparrhenia filipendula, H. hirta*

tangle head - *Heteropogon contortus*
Tanner grass - *Brachiaria radicans*
tassel bluegrass - *Dichanthium tenuiculum*
t'ef - *Eragrostis tef*
teff - *Eragrostis tef*
teosinte - *Zea mexicana*
thangari - *Digitaria abyssinica*
thatch grass - *Hyparrhenia hirta*
thatching grass - *Hyparrhenia* spp.
thin Napier grass - *Pennisetum polystachyon*
Thurston grass - *Brachiaria subquadripara*
tigriston - *Cynodon dactylon*
torpedo grass - *Panicum repens*
toto grass - *Ischaemum indicum*
tougourit - *Leptothrium senegalense*
tropical reed - *Phragmites karka*
tuman - *Panicum turgidum*
Turkistan blue-stem - *Bothriochloa ischaemum*
two-finger grass - *Brachiaria subquadripara*
two-spiked panic - *Brachiaria subquadripara*

uba cane - *Saccharum sinense*
upright brachiaria - *Brachiaria brizantha*
upright paspalum - *Paspalum urvillei*

vari - *Panicum miliaceum*
vasey grass - *Paspalum urvillei*
Venezuela grass - *Melinis minutiflora*
veyale - *Hyparrhenia rufa*

water couch grass - *Paspalum distichum, P. paspaloides*
water grass - *Echinochloa crus-galli*
weeping love grass - *Eragrostis curvula*
weeping Mitchell grass - *Astrebla elymoides*
western millet - *Echinochloa turneriana*
wezzeg - *Cenchrus biflorus*
white buffel grass - *Cenchrus pennisetiformis*
white French millet - *Panicum miliaceum*
white grass - *Sehima nervosum*
white panicum - *Echinochloa frumentacea*
wild cane - *Saccharum spontaneum*
wild crab grass - *Digitaria ciliaris*
wild millet - *Echinochloa turneriana*
wild sorghum - *Echinochloa turneriana*
wildergrass - *Dichanthium aristatum*
Wilman love grass - *Eragrostis superba*
wool grass - *Anthephora pubescens*

woolly finger grass - *Digitaria smutzii*
woolly-butt - *Eragrostis eriopoda*
woolly-top Rhodes grass - *Chloris virgata*

yakha - *Imperata cylindrica* var. *africana*
yaragua grass - *Hyparrhenia rufa*
yellow blue-stem - *Bothriochloa ischaemum*

yellow hard grass - *Hyperthelia dissoluta*
yellow spike thatching grass - *Hyparrhenia rufa*
yellow thatching grass - *Hyperthelia dissoluta*

zaina - *Panicum maximum*

Common names of other plants

acha - *Digitaria exilis*
alfalfa - *Medicago sativa*
alsike clover - *Trifolium hybridum*
Alyce clover - *Alysicarpus vaginalis*
arb - *Arachis glabrata*
barley - *Hordeum vulgare*
barrel medick - *Medicago truncatula*
berseem - *Trifolium alexandrinum*
birdsfoot trefoil - *Lotus corniculatus*
black medick - *Medicago lupulina*
blue grama - *Bouteloua gracilis*
blue lupin - *Lupinus angustifolius*
brigalow - *Acacia harpophylla*
brome grass - *Bromus inermis*
burr medick - *Medicago polymorpha*
button medick - *Medicago orbicularis*
cashew nut - *Anacardium occidentale*
centro - *Centrosema pubescens*
Chewings fescue - *Festuca rubra* subsp. *comutata*
clustered clover - *Trifolium glomeratum*
cocksfoot grass - *Dactylis glomerata*
cowpea - *Vigna unguiculata*
crested wheatgrass - *Agropyron desertorum*
crimson clover - *Trifolium incarnatum*
fine-stem stylo - *Stylosanthes guianensis* var. *intermedia*
fodder beet - *Beta vulgaris*
gidgea - *Acacia cambagei*
glycine - *Neonotonia wightii*
hairy vetch - *Vicia villosa*
harbinger medick - *Medicago littoralis*
hetero - *Desmodium heterophyllum*
Hungarian vetch - *Vicia pannonica*
intermediate wheatgrass - *Agropyron intermedium*
kale - *Brassica oleracea* var. *viridis*
kapok - *Ceiba pentandra*
Kentucky bluegrass - *Poa pratensis*
ladino clover - *Trifolium repens*
lappa clover - *Trifolium lappaceum*
large hop clover - *Trifolium campestre*
leleswha bush - *Tarchonanthus camphoratus*
little blue-stem - *Schizachyrium scoparium*
lucerne - *Medicago sativa*
Malaysian love grass - *Chrysopogon aciculatus*
mangel - *Beta vulgaris*
marrow-stem kale - *Brassica oleracea* var. *viridis*
meadow foxtail - *Alopecurus pratensis*
mesquite - *Prosopis glandulosa*
monantha vetch - *Vicia angustifolia*
nal grass - *Arundo donax*
Natal grass - *Rhynchelytrum roseum*
Natal redtop - *Rhynchelytrum repens*
nut grass - *Cyperus rotundus*
oats - *Avena sativa*
orchard grass - *Dactylis glomerata*
pigeon pea - *Cajanus cajan*
puero - *Pueraria phaseoloides*
siratro - *Macroptilium atropurpureum*
stylo - *Stylosanthes guianensis*
tung oil - *Aleurites moluccana*

Index

Acacia spp., 23, 44, 96
  A. aneura, 23, 89
  A. brevispica, 41
  A. cambagei, 55
  A. cana, 23
  A. derpanolobium, 42
  A. flavescens, 41
  A. harpophylla, 21, 41, 52, 55
  A. senegal, 73, 679
  A. seyal, 41
  A. tortilis, 42
  A. xanthophloea, 42
Acroceras macrum, 129, *181-184*
adaptability to soils, 62
Aeluropodeae, 116, 117
Aeschynomene, 28
afezu - see *Panicum turgidum*
Africa, 11, 37, 40, 41, 85, 87, 88, 109
African couch grass - see *Digitaria abyssinica*
Ageratum conyzoides, 126
Agromyza sp., 74
Agrosteae, 7, 8, 10
aircraft seeding, 56
alang-alang - see *Imperata cylindrica*
Albizia falcata, 105
alemán grass - see *Echinochloa polystachya*
Aleurites moluccana (tung oil), 105
algae, 71
Alice Springs, 37
alkali sacaton - see *Sporobolus airoides*
alkali soils, 113
aluminium tolerance by grasses, 130
Amazon, 29
Amphilophis insculpta - see *Bothriochloa insculpta*
Amphilophis pertusa - see *Bothriochloa pertusa*
Anacardium occidentale (cashew nut), 105

Andes, 29
Andropogoneae, 7, 8, 9, 21, 26, 27
Andropogon spp., 11, 13, 15, 24, 26, 27, 29
  A. aciculatus - see *Chrysopogon aciculatus*
  A. amplectans, 41
  A. annulatus - see *Dicanthium annulatum*
  A. bicornis, 28
  A. callipes, 24
  A. caucasicus - see *Bothriochloa caucasica*
  A. condensatus, 28
  A. gayanus, 41, 42, 55, 58, 126, 129-131, *185-190*
  A. glomeratus, 27
  A. hirtiflorus, 28
  A. hirtus - see *Hyparrhenia hirta*
  A. incanus, 26
  A. intermedius - see *Bothriochloa bladhii*
  A. ischaemum - see *Bothriochloa ischaemum*
  A. nodosus - see *Dichanthium aristatum*
  A. pertusus - see *Bothriochloa pertusa*
  A. saccharoides, 27
  A. selloanus, 28
  A. sorghum - see *Sorghum bicolor*
  A. stolonifer, 24
Angleton grass - see *Dichanthium aristatum*
antelope grass - see *Echinochloa pyramidalis*
Anthephora pubescens, 88, *191-192*
Anthisteria ciliata - see *Themeda australis*
Anthisteria membranacea - see *Iseilema membranaceum*
aquatic grass, 129-130
Argentina, 25, 29
Aristida spp., 11, 15, 17, 23, 24, 26, 38
  A. adscensionis, 27, 88, *193-195*
  A. cognata, 28

A. hygrometrica, 21
A. ingrata, 21
A. latifolia, *196-198*
A. mutabilis, 74, 88
A. riparia, 28
A. sub-micronata - see *A. adscensionis*
Arizona, 24, 91
arrow grass - see *Trachypogon spicatus*
Artocarpus integrifolia (jack fruit), 105
Arundinaria pusilla, 20
Arundinella spp., 15, 16, 18, 29
　A. bengalensis, 18
　A. hispida, 28
ashes, establishment of grass in, 55
assegai grass - see *Heteropogon contortus*
Astrebla spp., 21, 36, 38, 40, 88, *199-205*
　A. elymoides, 23, 199, 202
　A. lappacea, 23, 72, 199, 204
　A. pectinata, 23, 199, 202, 204
　A. squarrosa, 23, 202, 204
Atalaya hemiglauca, 23
Axonopus spp., 93, 94, 95
　A. affinis, 24, 93, 94, 126, 129, *206-210*
　A. anceps, 28
　A. compressus, 24, 25, 125-131, *211-214*
　　A. purpusii, 28
　A. scoparius, 28, 129, *215-216*
Azolla, 71
Azospirillum brasiliense, 72
Azotobacter sp., 72
Azotobacter paspali, 72

Bahia grass - see *Paspalum notatum*
Balanites aegyptiaca, 40
bansi - see *Panicum antidotale*
Barbados, 117
barnyard grass - see *Echinochloa crus-galli*, *E. colona* (awnless)
batiki bluegrass - see *Ischaemum indicum*
bauhinia, 27
beach dropseed - see *Sporobolus virginicus*
beans, 60
belah, 21
Belmolaimus longicaudatus, 63
Bermuda grass - see *Cynodon dactylon*
Beyerinckia sp., 72
birdwood grass - see *Cenchrus setigerus*
bhadli - see *Panicum pilosum*
blady grass - see *Imperata cylindrica*
bluegrass, 21
blue-stems (see *Bothriochloa* spp.), 24
Boer love grass - see *Eragrostis chloromelas*
Bolivia, 29
Bothriochloa spp., 15, 18, 24, 41

B. bladhii, 18, 21, *217-220*
B. caucasica, *221-226*
B. ewartiana, 21
B. glabra, 126
B. insculpta, 15, 59, 72, 88, 126
B. intermedia - see *B. bladhii*
B. ischaemum, 122, 126, *227-229*
B. pertusa, 18, 24, 88, 126-131, *230-233*
Brachiaria spp., 15
　B. brizantha, 19, 20, 61, 107, 125-131, *234-237*
　B. decumbens, 19, 59, 60, 61, 67, 72, 97, 124-131, *238-242*
　B. dictyoneura - see *B. humidicola*
　B. dura, 62, *243-244*
　B. eminii - see *B. decumbens* and *B. ruziziensis*
　B. humidicola, 23, 125-131, *245-248*
　B. miliiformis - see *B. subquadripara*
　B. mutica, 19, 23, 28, 29, 59, 61, 62, 64, 70, 121-131, *249-253*
　B. obtusiflora, 74
　B. radicans, 254
　B. ruziziensis, 125, *255-259*
　B. subquadripara, 20, 125, *260-262*
Brachiachne convergens, 23
Brazil, 26, 27, 29
brigalow - see *Acacia* spp.
Briza, 26
broad-leaf paspalum - see *Paspalum wettsteinii*
bromacil, 110, 111
Bromus, 26
brown beetle grass - see *Diplachne fusca*
brown Rhodes grass - see *Eustachyus paspaloides*
brown seed paspalum - see *Paspalum plicatulum*
Brunswick grass - see *Paspalum nicorae*
buffalo grass - see *Stenotaphrum secundatum*
bulrush millet - see *Pennisetum americanum*
buffel grass - see *Cenchrus ciliaris*
bulldozers, 50, 53
bunch spear grass - see *Heteropogon contortus*, 21
bungoma grass - see *Entolasia imbricata*
burning, 38, 52, 106
burning frequency, 42
bush rye - see *Enteropogon macrostachyus*

caatinga, 27
Cahoon broadcaster, 56
California, 109

Calopogonium mucunoides, 105, 125
campo, 26
campo cerrado, 27
campo limpo, 27
carbohydrate root reserves, 35
carib grass - see *Eriochloa punctata*
Carimagua, 55
carpet grass - see *Axonopus compressus* and *A. affinis*
cassava, 60
cassia, 27, 28
Casuarina cristata, 21
Caucasian blue-stem - see *Bothriochloa caucasica*
Ceiba pentandra (kapok), 105
Cenchrus spp., 11, 15, 16, 17, 18, 24
    C. barbatus - see *C. biflorus*
    C. biflorus, 263-265
    C. ciliaris, 17, 18, 55, 58, 62, 72, 88, 89, 90, 115-131, *266-274*
    C. pauciflorus, 27
    C. pennisetiformis, 88, 121-131, *275-278*
    C. setigerus, 17, 18, 88, 117-122, *279-282*
centro - see *Centrosema pubescens*
Centrosema spp., 27
    C. pubescens, 61, 105, 107, 108, 125
cerrado, 27
Chaco, 25
Chaetochloa italica - see *Setaria italica*
channel country, 23
channel millet - see *Echinochloa turneriana*
Chile, 25
China, 71
Chlorideae, 7, 21, 117
Chloris spp., 21, 29
    C. barbata, 72, 117
    C. gayana, 55, 56, 59, 62, 74, 79, 88, 110, 114-131, *283-288*
    C. inflata, 24
    C. montana, 118
    C. mosambicensis, 88, *289-290*
    C. myriostachya - see *C. roxburghiana*
    C. roxburghiana, 41, 88, 127, *291-292*
    C. virgata, 24, 88, 127, *293-295*
Chrysopogon spp., 15, 18
    C. aciculatus, 20, 24, *296-297*
    C. aucheri, 15, 88, *298-299*
    C. fallax, 21, 31
    C. flulvus, 18
    C. zeylanicus, 19
chinch bug, 757
climatic zones, 2
Cloncurry buffel grass - see *Cenchrus pennisetiformis*

coastal Bermuda grass - see *Cynodon dactylon*
coconut, 123-125
cogon grass - see *Imperata cylindrica*
Coix lacryma-jobi, 130, *300-302*
Colombia, 25, 28, 59
Columbus grass - see *Sorghum almum*
compatibility with legumes, 61
compensatory gain in cattle, 31
Congo signal grass - see *Brachiaria ruziziensis*
continuous grazing, 36, 78
contract specifications - timber pulling, 51
Cordia cylindristachya, 104
corn - see *Zea mays*
corn ear worm - see *Heliothis armigera*
Costa Rica, 25, 28
couch grass - see *Cynodon dactylon*, *Digitaria didactyla*, *D. abyssinica*
cowpea - see *Vigna unguiculata*
crab grass - see *Digitaria ciliaris*
creeping bluegrass - see *Bothriochloa insculpta*
Crotalaria spp., 105
crowfoot grass - see *Dactyloctenium aegyptium*, *Eleusine indica*
Cuba, 29
cushion love grass - see *Eragrostis caespitosa*
Cymbopogon spp., 15, 29, 96, 97
    C. afronardus - see *C. nardus*
    C. confertiflorus, 19
    C. giganteus, 41
    C. nardus, 13, 96-97, *303-305*
    C. nervatus, 74
    C. rufus, 28
Cynodon spp., 15, 17, *306-307*
    C. aethiopicus, 127, *308-309*
    C. dactylon, 17, 18, 20, 24, 25, 26, 74, 75, 88, 115-122, 126-131, *310-315*
    C. nlemfuensis, 64, 70, 127, 130, *316-318*
    C. plectostachyus, 88, 127, *319-321*

Dactyloctenium spp., 23, 88
    D. aegyptium, *322-324*
    D. giganteum, 88, *325-327*
    D. radulans, 23, 88
    D. sindicum, 17, 18
dalapon, 110
dallis grass - see *Paspalum dilatatum*
dambo, 13
Danthonia sp., 23
deenanath grass - see *Pennisetum pedicellatum*
deferred grazing, 36

Desmanthus, 26
Desmodium spp.
   D. adscendens, 28
   D. barbatum, 28
   D. heterophyllum, 108, 109
   D. intortum, 105, 107
   D. microphyllum, 19
   D. ovalifolium, 60
   D. uncinatum, 665
dhoub grass - see *Cynodon dactylon*
dhurra or durra - see *Sorghum bicolor*
Dichanthium spp., 15, 16, 17, 18, 21, 29, 41
   D. annulatum, 17, 18, 127, *328-332*
   D. aristatum, 115-122, 127, 130, *333-335*
   D. caricosum, 23, 127, 130, *336-340*
      D. sericeum, 21, *341-343*
   D. superciliatum - see *D. tenuiculum*
   D. tenuiculum, 21, 88, *344*
digit grass - see *Digitaria decumbens*
Digitaria spp., 15, 23, 24, 63
   D. abyssinica, 13, 97-98, 127, 130, *345-347*
   D. adscendens - see *D. ciliaris*
   D. ciliaris, 26, 27, *348-350*
   D. decumbens, 59, 61, 63, 64, 70, 72, 74, 116-122, 127, 131, *351-356*
   D. didactyla, 127, *357-359*
   D. longiflora, 20
   D. marginatus - see *D. ciliaris*
   D. milanjiana, 41, *360-362*
   D. pentzii, 75, *363-365*
   D. sanguinalis - see *D. ciliaris*
   D. scalarum - see *D. abyssinica*
   D. smutzii, 127, *366-367*
   D. swynnertonii - see *D. milanjiana*
   D. vestita - see *D. abyssinica*
Diplachne fusca, 13, 118, 130, *368-369*
donkey grass - see *Panicum trichocladum*
drought tolerance, 62
dubi grass - see *Urochloa oligotricha*
dukn - see *Pennisetum americanum*

East African highlands, 13
Echinochloa spp., 13
   E. colona, 26, 130, *370-372*
   E. colonum - see *E. frumentacea*
   E. crus-galli, 24, 27, 129, 130, *373-375*
   E. frumentacea, *376-379*
   E. haploclada, 88, 130, *379-380*
   E. notabile - see *Urochloa mosambicensis*
   E. polystachya, 28, 130, *381-382*
   E. pyramidalis, 13, 118, 129, *383-386*
   E. scabra, 13, *387-390*

   E. spectabilis - see *E. polystachya*
   E. stagnina - see *E. scabra*
   E. turneriana, 23, 130, 391, 392
   E. utilis, 127, *393-396*
Ecuador, 25
elephant grass - see *Pennisetum purpureum*
Eleusine spp., 17, 26, 38
   E. coracana, *397-400*
   E. indica, 26, *401-404*
   E. jaegeri, *405-406*
Elyonurus spp., 29
   E. adjustus, 28
   E. hirsutus - see *Lasiurus hirsutus*
   E. latiflorus, 26
Enneapogon sp., 23
Enterobacter aerogenes, 72
Enteropogon spp.
   E. macrostachyus, 88, *407-408*
   E. simplex - see *E. macrostachyus*
   E. somalensis, 88, *409*
Entolasia imbricata, 130, *410*
Eragrostideae, 7, 8, 9, 10, 21
Eragrostis spp., 15, 17, 21, 23, 24, 26, 27, 29
   E. abyssinica - see *E. tef*
   E. basiilepsis - see *E. caespitosa*
   E. caespitosa, 88, *411-412*
   E. chloromelas, 88, *413-414*
   E. cilianensis, 88, *415-416*
   E. curvula, 75, 88, 115, 118, 127, *417-421*
   E. eriopoda, 88
   E. lehmanniana, 88, 115-121, 127, *422-424*
   E. major - see *E. cilianensis*
   E. maypurensis, 29
   E. setifolia, 88
   E. superba, 88, 115-122, *425-427*
   E. tef, 127, *428-430*
   E. tremula, *431-433*
Eremopogon foveolatus, 15
Eriachne sp., 23
Erianthus angustifolius, 26
Eriochloa spp., 21
   E. fatmensis, 88, 121, 130, *434-435*
   E. nubica - see *E. fatmensis*
   E. polystachya - see *E. punctata*
   E. punctata, 23, 64, 70, 130, *436-438*
erosion control - grasses, 126-129
espartillo grasses, 26
establishment of grasses, 55, 62
estrella - see *Cynodon plectostachyus*
Ethiopia, 15
Eucalyptus sp., 23, 40, 44, 52, 89
Eucalyptus grandis, 105

Euchlaena mexicana - see *Zea mexicana*
Eulalia fulva, 21
Eupatorium odoratum, 104, 126
Eustachys paspaloides, 88, *439-440*
Exotheca abyssinica, 15, *441-442*

feather-top Rhodes grass or feather-top chloris - see *Chloris virgata*
Fiji, 23, 40, 44, 93, 108, 125
fire, 40-44
fire temperatures, 42
fire tolerance, 62
Flinders grass - see *Iseilema membranaceum*
Flindersia maculosa, 23
flood tolerance, 62
Florida, 24, 72, 74, 93
fodder conservation, 138-161
forage utilization, 133-161
fox-tail millet - see *Setaria italica*
French millet - see *Panicum miliaceum*
frost, 62

gama grass - see *Tripsacum dactyloides*
gamba grass - see *Andropogon gayanus*
gavara, 19
Geijera parviflora, 21
German grass - see *Echinochloa polystachya*
germination, 59
giant button grass - see *Dactyloctenium giganteum*
giant paspalum - see *Paspalum urvillei*
gidgea - see *Acacia cambagei*
Gliricidium sepium, 105
Glycine wightii - see *Neonotonia wightii*
glyphosate, 103, 110
gonya grass - see *Urochloa mosambicensis*
gordura - see *Melinis minutiflora*
grader grass - see *Themeda quadrivalvis*
gramalote - see *Panicum maximum*
gramilla blanca - see *Cynodon dactylon*
Gran Chaco, 29
grass distribution, 7
grasses for special purposes, 113-131
grass/legume mixtures, 66, 67, 76, 79
grazing frequency, 79
grazing methods, 36, 77
green panic - see *Panicum maximum* var. *trichoglume*
green summer grass - see *Brachiaria subquadripara*
Guanacaste, 28
Guiana, 29

Guinea grass - see *Panicum maximum*
Guinea savannah, 42
Guyana, 93

Habana oat grass - see *Themeda quadrivalvis*
handling difficult grasses, 93-111
hariq cultivation, 73
haskanit - see *Cenchrus biflorus*
Hawaiian Islands, 23, 93
hay, 138-161
Heliothis armigera, 7, 57
Hemarthria altissima, 63, 130, 131, *443-445*
hetero - see *Desmodium heterophyllum*
Heterodendron oleifolium, 23
Heteropogon spp., 15, 29, 31
    H. contortus, 17, 18, 20, 21, 24, 31, 32, 41, 43, 45, 72, 88, 89, 128, *446-449*
    H. triticeus, 21
hippo grass - see *Echinochloa scabra*
Homolepis, 24
Honduras grass - see *Ixophorus unisetus*
hook grass - see *Leptothrium senegalense*
Hordeum, 26
horsetail grass - see *Chloris roxburghiana*
HPG/CSG grazing system, 38, 39
HUG/NSG grazing system, 52, 38, 39
Hungarian millet - see *Setaria italica*
hurricane grass - see *Bothriochloa pertusa*
Hymenachne spp.
    H. acutigluma, 129, *450-451*
    H. amplexicaulis, 60, 129, *452-453*
    H. pseudointerrupta - see *H. amplexicaulis*
Hyparrhenia spp., 11, 13, 15, 41, 134
    H. dissoluta - see *Hyperthelia dissoluta*
    H. filipendula, *454-456*
    H. hirta, 121, *457-459*
    H. lintonii, 41
    H. pseudocymbaria, 74
    H. rufa, 13, 25, 28, 56, 131, *460-464*
Hyperthelia dissoluta, *465-467*

illuk - see *Imperata cylindrica*
Imperata spp., 15, 18
    I. brasiliensis, 24
    I. cylindrica, 6, 18, 20, 21, 56, 100-112, 123, *468-473*
imperial grass - see *Axonopus scoparius*
improved pastures, 77
India, 15, 35, 38, 118
Indonesia, 102, 105
Ischaemum spp., 15

I. aristatum - see *I. indicum*
I. barbatum, 18
I. indicum, 23, 130, *474-477*
I. magnum, 130, *478-479*
Iseilema spp., 15, 21
    I. actinostachys - see *I. membranaceum*
    I. laxum, 122, 130, *480-482*
    I. membranaceum, 88, *483-485*
    I. vaginiflorum, 72, 88, *486-487*
Ixophorus unisetus, 130, *488-490*

Janeiro - see *Eriochloa punctata*
Japan, 119
Japanese cane - see *Saccharum sinense*
Japanese millet - see *Echinochloa utilis*
jaragua grass - see *Hyparrhenia rufa*
jiribilla - see *Dichanthium caricosum*
Job's tears - see *Coix lacryma-jobi*
Johnson grass - see *Sorghum halepense*
jowar - see *Sorghum bicolor*
jungle rice - see *Echinochloa colona*

kangaroo grass - see *Themeda australis*
Kennedy ruzi grass - see *Brachiaria ruziziensis*
Kenya, 15, 41
key species, 34
khachornchob - see *Pennisetum polystachyon*
khas-khas grass - see *Vetiveria zizanioides*
Kikuyu - see *Pennisetum clandestinum*
kind of grazing animal, 83
King Ranch, 53
King Ranch blue-stem - see *Bothriochloa ischaemum*
Klein grass - see *Panicum coloratum*
Koronivia grass - see *Brachiaria humidicola*
kunai grass - see *Imperata cylindrica*
kurrakan millet - see *Eleusine coracana*
kyasuwa grass - see *Pennisetum pedicellatum*

lalang - see *Imperata cylindrica*
land clearing, 49
Lasiurus sp., 15, 16, 17, 18, 38
    L. hirsutus, 17, 128, *491-493*
    L. sindicus - see *L. hirsutus*
Latin America, 25
Latipes senegalensis - see *Leptothrium senegalense*
leader and follower grazing system, 79
Leersia sp., 13, 28
    L. hexandra, 24, 129, *494-496*

Lehmann's love grass - see *Eragrostis lehmanniana*
Leptochloa fusca - see *Diplachne fusca*
Leptochloa obtusiflora, 88, *497-498*
Leptothrium senegalense, 88, *499-500*
Leucaena leucocephala, 71, 104, 105
leys, 73-76
liberty millet - see *Setaria italica*
limpo grass - see *Hemarthria altissima*
liverseed grass - see *Urochloa panicoides*
Live-weight, cattle, 32, 35, 63
llanos, 27, 28
Lolium, 26
Lotononis bainesii, 75
Loudetia spp., 13, 15, 96
    L. acuminata, 42, 188
    L. simplex, 13, *501-503*
love grass - see *Eragrostis* spp.

machuri - see *Iseilema laxum*
Mackie's pest - see *Chrysopogon aciculatus*
Macroptilium atropurpureum, 75, 78, 105, 106, 107, 109
maize - see *Zea mays*
Majestic plough, 102
makarikari grass - see *Panicum coloratum* var. *makarikariense*
Malaysia, 93, 102, 103, 125
Mallen Niche Seeder, 116
malojilla - see *Brachiaria mutica*
mana, 19
management of improved grassland, 77-85
management of natural pastures, 31
manyatta grass - see *Eleusine jaegeri*
maps - Africa - western and western equatorial regions, 12 - eastern and central region, 14
    Australia - pasture map, 22
    climatic zones of the world, 2
    grass tribes - distribution, 9
    grazing areas of the world, 4
    India - distribution of grass covers, 16
    Imperata cylindrica - distribution, 101
    soil groups of the world, 5
MARDI/CSIRO, 125
marvel grass - see *Dichanthium annulatum*
Masai love grass - see *Eragrostis superba*
mat grass - see *Axonopus affinis*
Mato Grosso, 29
Mauritia minor, 28
Mauritius signal grass - see *Brachiaria mutica*
mbuga, 13

Medicago sativa, 75
Meiboma, 27
Melastoma malabaricum, 126
Melinis minutiflora, 25, 28, 45, 56, 59, 64, 128, *504-507*
Merker grass - see *Pennisetum purpureum*
mesquite - see *Prosopis glandulosa*
Mesa de Guanipa, 28
Mexican grass - see *Ixophorus unisetus*
Mexico, 24, 27
millet - see *Echinochloa frumentacea, E. utilis, Pennisetum americanum, Eleusine coracana, Setaria italica, Panicum miliaceum*
Mimosa, 28
miombo, 15
mission grass - see *Pennisetum polystachyon*
Mitchell grass - see *Astrebla* spp.
Moghania macrophylla, 105
Morus australis (Mulberry), 105
Mojos, 29
molasses, 149
molasses grass - see *Melinis minutiflora*
montane grasslands, 19
MSMA, 110
mulch for establishment, 60
mulga, 21
multi-camp grazing system, 38

Nadi bluegrass - see *Dichanthium caricosum*
Naivasha star grass - see *Cynodon plectostachyus*
Napier grass - see *Pennisetum purpureum*
natural pastures, management, 31-44
nematodes, 74
Neonotonia wightii, 67
Neurachne sp., 23
never-fail grass - see *Eragrostis setifolia*
New Mexico, 24
Ngorongoro crater, 13
Nigeria, 74
night kraaling, 37, 85
Nile grass - see *Acroceras macrum*
nitrogen, 72, 74, 75
nitrogen fixation by grasses, 71-72
nurse crop, use of, 59
nutritive value, 61

oats, as nurse crop, 60
offset disc cultivator, 56
oil palm, 125-126
okrich - see *Sporobolus helvolus*

Ophiurus sp., 20
Orinoco, 27
Orscotiella javanica, 105
Oryza spp., 13
    O. sativa, 119, 130, *508-511*
Ottochloa nodosa, 125
overgrazing, 40
Owenia acidula, 23

Pacific Islands, 33
palisade grass - see *Brachiaria brizantha*
pampa, 25
Panama, 25, 140, 155
pangola grass - see *Digitaria decumbens*
Paniceae, 7, 8, 9, 10, 21, 27
Panicum spp., 15, 24, 26, 27, 28, 29, 114
    P. antidotale, 55, 88, 115-122, 128, 129, *512-514*
    P. colonum - see *Echinochloa colona*
    P. coloratum, 88, *515-516*
    P. coloratum var. *makarikariense*, 116-122, 125, 128, 130, *517-521*
    P. controversum - see *Urochloa panicoides*
    P. crus-galli - see *Echinochloa crus-galli*
    P. dichotomiflorum, 24, 27
    P. distachyum - see *Brachiaria subquadripara*
    P. elephantipes, 28
    P. italicum - see *Setaria italica*
    P. maximum, 19, 20, 28, 29, 41, 56, 59, 60, 70, 71, 72, 78, 79, 88, 125-131, *522-532*
    P. maximum var. *trichoglume*, 55, 80, *533-537*
    P. miliaceum, 119, *538-541*
    P. miliare - see *P. sumatrense*
    P. muticum - see *Brachiaria mutica*
    P. pilosum, 542
    P. porphyrrhizos, 119
    P. purpurescens - see *Brachiaria mutica*
    P. queenslandicum, 21
    P. repens, 13, 119, 128, 129, 130, *543-545*
    P. spectabile - see *Echinochloa polystachya*
    P. sumatrense, 546
    P. tennuissimum - see *Digitaria didactyla*
    P. trichocladum, *547-548*
    P. turgidum, 11, 15, 17, 128, *549-552*
    P. unisetum - see *Ixophorus unisetus*
    P. virgatum, 119
    P. whitei, 23, 130, *553-554*
Papua-New Guinea, 106, 125

Pará, 29
Paractaenum novae-hollandiae, 88
Para grass - see *Brachiaria mutica*
Paraguay, 29
Páramos, 25
Paraná, 26
Parochetus communis, 19
Paspalidium spp.
   P. caespitosum, 21
   P. desertorum, 88, *555-556*
   P. globoideum, 21
Paspalum spp., 18, 25, 26, 27, 28, 29
   P. commersonii - see *P. scrobiculatum*
   P. conjugatum, 24, 25, 110, *125-131*
   P. dilatatum, 93, 94, 128, *557-564*
   P. distichum, 119, 128, 130, *565-568*. See also *P. paspaloides*
   P. fasciculatum, 28
   P. gardnerianum, 28
   P. littorale - see *P. paspaloides*
   P. nicorae, 128, *569-570*
   P. notatum, 24, 25, 26, 28, 29, 72, 125, 128, *571-575*
   P. orbiculare, 24
   P. paspaloides, 24, 26, 27, 119, 130, *576-578*
   P. plicatulum, 29, 75, 125, 130, *579-584*
   P. polystachyon - see *P. scrobiculatum*
   P. scoparium - see *Axonopus scoparius*
   P. scrobiculatum, 125, 130, 136, *585-589*
   P. urvillei, 19, 130, *590-592*
   P. vaginatum - see *P. distichum*
   P. virgatum, 24
   P. wettsteinii, 130, *593-595*
pasture improvement, 47, 49
pastures under plantation crops, 174-179
patana, 19
pearl millet - see *Pennisetum americanum*
Pennisetum spp., 11, 41, 628
   P. americanum, 72, 75, 122, *596-603*
   P. catabasis - see *P. hohenackeri*
   P. cenchroides - see *Cenchrus ciliaris*
   P. ciliare - see *Cenchrus ciliaris*
   P. clandestinum, 13, 25, 61, 64, 93, 94, 98, 99, 116-122, 128, *604-611*
   P. glaucum - see *P. americanum*
   P. hohenackeri, *629-630*
   P. mezianum, 41
   P. mollissimum, 74
   P. pedicellatum, *612-615*
   P. polystachyon, 6, 20, 23, 40, 88, 108-109, 123, *616-620*
   P. purpureum, 13, 15, 19, 20, 45, 64, 70, 71, 74, 75, 125, 128, 129, *621-627*
   P. schimperi, *631*
   P. setaceum, *632*
   P. spicatum, *633*
   P. trisetum, 15
   P. typhoides - see *P. americanum*
   P. villosum, *634*
pepper grass - see *Panicum whitei*
persistence of grasses, 61
Peru, 25
Philippines, 125
Phragmites spp., 15, 16, 18, 20
   P. australis, 119-130, *635-637*
   P. communis - see *P. australis*
   P. karka, 18, 128, 129, *638-639*
pinhole grass - see *Bothriochloa insculpta*
pini baru tana, 19
Pinus elliotii, 105
Pinus merkusii, 105
pit-pit - see *Phragmites karka* and *Saccharum spontaneum*
plains blue-stem - see *Bothriochloa ischaemum*
plantation crop pastures, 122
Plectrachne sp., 21, 23
plume chloris - see *Chloris roxburghiana*
Poa, 26
proso millet - see *Panicum miliaceum*
Prosopis glandulosa, 27
prussic acid, 63
Puccinia oahuensis, 63
Pueraria phaseoloides, 45, 64, 105, 125
puero - see *Pueraria phaseoloides*
Puerto Rico, 59, 64
puntero - see *Hyparrhenia rufa*

Queensland bluegrass - see *Dichanthium sericeum*

Rajasthan, 118-122
rat's tail grass - see *Sehima nervosum*
red-oat grass - see *Themeda triandra*
red vlei grass - see *Hemarthria altissima*
reed - see *Phragmites australis*
reflexed panic - see *Paractaenum novae-hollandiae*
re-seeding the range, 87-91
resistance to diseases and pests, 63
Rhodes grass - see *Chloris gayana*
Rhynchelytrum repens, 72
Rhynchosia, 26
rice - see *Oryza sativa*
Rio Grande do Sul, 26

Rome plough, 102
rooikweek - see *Hemarthria altissima*
root plough, 55
root-raking, 53
root-ripping, 53
rotational grazing, 38, 77, 78
rotational resting, 36
Rottboellia exaltata, 74
rumput melayu - see *Ischaemum magnum*
russet grass - see *Loudetia simplex*

sabi grass - see *Urochloa mosambicensis*
Saccharum spp., 15, 16, 18
    S. arundinaceum, 18
    S. officinarum, 59, 119, *640-646*
    S. robustum, 20
    S. sinense, 59, *647-648*
    S. spontaneum, 18, 20, 129, *649-651*
Sacciolepis contracta, 24
salinity, grasses tolerant of, 113-122
saltwater couch - see *Paspalum distichum*, *Sporobolus virginicus*
Samoa, 125
scrobic - see *Paspalum scrobiculatum*
seashore paspalum - see *Paspalum distichum*
seed, 56, 58, 89
seed-harvesting ants, 58
Sehima, 15, 16
Sehima nervosum, 20, 21, 88, *652-654*
selection of pasture species, 61-72
seragoon grass - see *Digitaria didactyla*
Serengeti, 3, 42, 85
Sesbania grandiflora, 105
Setaria spp., 15
    S. anceps - see *S. sphacelata*
    S. geniculata, 24, 26
    S. italica, *655-659*
    S. porphyrantha, 128, *660-661*
    S. sphacelata, 13, 59, 62, 98, 116-122, 125-130, *662-668*
    S. sphacelata var. *splendida*, *669*
    S. verticillata, 24
sewan grass - see *Lasiurus hirsutus*
sheda grass - see *Dichanthium annulatum*
shifting cultivation, 40
shifting stable grazing, 79
signal grass - see *Brachiaria decumbens, B. brizantha*
siratro - see *Macroptilium atropurpureum*
Smithia blanda, 19
sod-seeding, 55
Somalia, 15

Sopubia ramosa, 106
Sorghum spp., 20, 63, 74, 90, 114
    S. almum, 55, 75, 116-122, *670-676*
    S. australiense, 21
    S. bicolor, 59, 75, 120, *677-685*
    S. halepense, 109-111, *686-689*
    S. plumosum, 21, 31, 72
    S. purpureosericeum, 74
    S. sudanense, 116-122, *690-694*
    S. vulgare - see *S. bicolor, S. sudanense*
    Sorghum spp. hybrids, *695-697*
Sorocaba, 26
Southeast Asia, 20
species for humid and subhumid areas, 65, 66
spinifex, 23
Spinifex hirsutus, 120, *698-700*
Spirillum lipoferum, 72
Sporoboleae, 7, 117
Sporobolus spp., 18, 24, 27, 29, 120
    S. airoides, 88, 114, 115, 120, *701-702*
    S. arabicus, 120
    S. asperifolius, 120
    S. capensis, 24, 28
    S. coromandeleanus, 120
    S. cubensis, 28
    S. flagelliferus - see *S. helvolus*
    S. helvolus, 88, 120, 130, *703-755*
    S. marginatus, 17, 18, 88, 121-122, *706-707*
    S. pallidus - see *S. arabicus*
    S. pyramidatus, 121
    S. spicatus, 13, 121, *708*
    S. virginicus, 21, 121, 130, *709-710*
Sri Lanka, 19, 125
star grass - see *Cynodon nlemfuensis*
St Augustine grass - see *Stenotaphrum secundatum*
Stenotaphrum secundatum, 24, 25, 121, 128, 130, *711-713*
stick-picking, 53
stink grass - see *Eragrostis cilianensis*
Stipa sp., 23, 26
stocking rate, 47, 61, 80-83
Stoebe vulgaris, 42
Striga hermonthica, 684
strip grazing, 80
stump-jump plough, 55, 56
stylo - see *Stylosanthes guianensis*
Stylosanthes spp., 26, 27, 105
    S. guianensis, 59, 60, 102, 109, 126-131
    S. hamata, 67
    S. humilis, 45, 46, 48, 67

Sudan, 41, 73, 74
Sudan grass - see *Sorghum sudanense*
sugar cane - see *Saccharum officinarum*
Suriname, 119
switch grass - see *Panicum virgatum*
Syzygium aromaticum (clove tree), 105

taman, tuman - see *Panicum turgidum*
tangle head - see *Heteropogon contortus*
Tanner grass - see *Brachiaria radicans*
Tanzania, 15, 42, 59, 74, 125
Tarchonanthus camphoratus, 41
teosinte - see *Zea mexicana*
Teramnus, 27
Tetrapogon mosambicensis - see *Chloris mosambicensis*
Tetrapogon villosus, 88
Texas, 24
Thailand, 108
thangari - see *Digitaria abyssinica*
thatching grass - see *Hyparrhenia filipendula* ("fine"), *H. hirta* ("common"), *H. rufa* ("yellow spike"), *Hyperthelia dissoluta* ("yellow")
Themeda spp., 15, 16, 18, 96
   T. anathera, 18
   T. arundinacea, 20
   T. australis, 20, 21, 31, *714-717*. See also *T. triandra*
   T. quadrivalvis, *718-720*
   T. tremula, 19
   T. triandra, 13, 15, 41, 88, 96, 98, *721-724*. See also *T. australis*
timber pulling, 50
torpedo grass - see *Panicum repens*
toxicity, 63
Trachypogon spp., 6, 28, 29
   T. montufari, 28
   T. plumosus, 28
   T. spicatus, *725-726*
Transvaal, 127
Tricholaena teneriffae, 88
Trifolium spp.
   T. repens, 93
   T. semipilosum, 67
Triodia spp., 21, 23
   T. ambigua - see *Diplachne fusca*
   T. pungens, *727-728*
Tripsacum spp.
   T. dactyloides, 27, 131, *729-731*
   T. fasciculatum - see *T. laxum*
   T. laxum, 19, *732-734*
Tristachya spp.
   T. chrysothrix, 29

T. leiostachya, 29
tropical tall grass, 21
Turkistan blue-stem - see *Bothriochloa ischaemum*

Uganda, 78, 79, 102, 140, 459
United States of America, 24
Uralepsis fusca - see *Diplachne fusca*
Urochloa spp.
   U. bolbodes - see *U. oligotricha*
   U. helopus - see *U. panicoides*
   U. mosambicensis, 72, 121, 128, *735-739*
   U. oligotricha, 72, 121, *740-742*
   U. panicoides, *743-746*
   U. uniseta - see *Ixophorus unisetus*
Uruguay, 26

vasey grass - see *Paspalum urvillei*
veld - sweet, 15
   - sour, 15
Venezuela, 25
Ventilago viminalis, 23
Vetiveria zizanioides, 130, *747-748*
Vicia, 26
Vigna hosei, 105
Vigna unguiculata, 692
villu, 19
vlei, 13
Vossia sp., 13
Vossia cuspidata, 13, 129, *749-751*

water couch - see *Paspalum paspaloides*
water grass - see *Echinochloa crus-galli*
weeping love grass - see *Eragrostis curvula*
West Indies, 24, 93
white grass - see *Sehima nervosum*
white panicum - see *Echinochloa frumentacea*
wildergrass - see *Dichanthium aristatum*
Wilman love grass - see *Eragrostis superba*
wind - effect on fire, 42
   effect on grazing, 38
wind-rowing pulled timber, 52
woolly-butt - see *Eragrostis eriopoda*
woolly finger grass - see *Digitaria smutzii*

Yacuma, 29

Zaire, 13
Zea mays, 122, *752-757*
Zea mexicana, 27, *758-759*
Zimbabwe, 74
Zornia, 26, 27

WHERE TO PURCHASE FAO PUBLICATIONS LOCALLY
POINTS DE VENTE DES PUBLICATIONS DE LA FAO
PUNTOS DE VENTA DE PUBLICACIONES DE LA FAO

- **ANGOLA**
Empresa Nacional do Disco e de Publicaçoes, ENDIPU-U.E.E.
Rua Cirilo de Conceiçao Silva, No. 7, C.P. No. 1314-C Luanda.

- **ARGENTINA**
Librería Agropecuaria S.A.
Pasteur 743, 1028 Buenos Aires.

- **AUSTRALIA**
Hunter Publications
58A Gipps Street,
Collingwood, Vic. 3066.

- **AUSTRIA**
Gerold & Co.
Graben 31, 1011 Vienna.

- **BAHRAIN**
United Schools International
PO Box 726, Manama.

- **BANGLADESH**
Association of Development Agencies in Bangladesh
1/3 Block F, Lalmatia, Dhaka 1209.

- **BELGIQUE**
M.J. De Lannoy
202, avenue du Roi,
1060 Bruxelles. CCP 000-0808993-13.

- **BOLIVIA**
Los Amigos del Libro
Perú 3712, Casilla 450, Cochabamba.
Mercado 1315, La Paz.

- **BOTSWANA**
Botsalo Books (Pty) Ltd
PO Box 1532, Gaborone.

- **BRAZIL**
Fundaçao Getulio Vargas
Praia de Botafogo 190,
C.P. 9052, Rio de Janeiro.
Libreria Nobel S.A.
Rua da Balsa 559
2910 São Paulo.

- **CHILE**
Librería - Oficina Regional FAO
Avda. Santa María 6700.
Casilla 10095, Santiago.
Teléfono 228-80-56.

- **CHINA**
China National Publications Import Corporation
PO Box 88, Beijing.

- **CONGO**
Office national des librairies populaires
P.B. 577, Brazzaville.

- **COSTA RICA**
Librería, Imprenta y Litografía Lehmann S.A.
Apartado 10011, San José.

- **CUBA**
Ediciones Cubanas, Empresa de Comercio Exterior de Publicaciones
Obispo 461, Apartado 605,
La Habana.

- **CYPRUS**
MAM
PO Box 1722, Nicosia.

- **CZECHOSLOVAKIA**
ARTIA
Ve Smeckach 30,
PO Box 790, 111 27 Prague 1.

- **DENMARK**
Munksgaand
Book and Subscription Service
P.O. Box 2146
Telephone: 4533128570
Telefax   : 4533129387.
DK 1016 Copenhagen K.

- **ECUADOR**
Libri Mundi, Librería Internacional
Juan León Mera 851,
Apartado Postal 3029, Quito.
Su Librería Cía. Ltda.
García Moreno 1172 y Mejía,
Apartado Postal 2556, Quito.

- **EL SALVADOR**
Librería Cultural Salvadoreña, S.A. de C.V.
7ª. Avenida Norte 121,
Apartado Postal 2296, San Salvador.

- **ESPAÑA**
Mundi-Prensa Libros S.A.
Castelló 37, 28001 Madrid.
Librería Agrícola
Fernando VI 2, 28004 Madrid.
Librería Internacional AEDOS
Consejo de Ciento, 391
08009 Barcelona

- **FINLAND**
Akateeminen Kirjakauppa
PO Box 128,
00101 Helsinki 10.

- **FRANCE**
Editions A. Pedone
13, rue Soufflot, 75005 Paris.

- **GERMANY, FED. REP.**
Alexander Horn
Internationale Buchhandlung
Kirchgasse 22,
Postfach 3340, 6200 Wiesbaden.
UNO Verlag
Poppelsdorfer Allee 55,
D-5300 Bonn 1.
S. Toeche-Mittler GmbH
Versandbuchhandlung
Hindenburgstrasse 33
6100 Darmstadt.

- **GHANA**
Ghana Publishing Corporation
PO Box 4348, Accra.

- **GREECE**
G.C. Eleftheroudakis S.A.
4 Nikis Street, Athens (T-126)
John Mihalopoulos & Son S.A.
75 Hermou Street,
PO Box 73, Thessaloniki.

- **GUYANA**
Guyana National Trading Corporation Ltd.
45-47 Water Street,
PO Box 308, Georgetown.

- **HAÏTI**
Librairie "A la Caravelle"
26, rue Bonne Foi,
B.P. 111, Port-au-Prince.

- **HONDURAS**
Escuela Agrícola Panamericana,
Librería RTAC
Zamorano, Apartado 93,
Tegucigalpa.
Oficina de la Escuela Agrícola Panamericana en Tegucigalpa
Blvd. Morazán, Apts. Glapson,
Apartado 93, Tegucigalpa.

- **HONG KONG**
Swindon Book Co.
13-15 Lock Road, Kowloon.

- **HUNGARY**
Kultura
PO Box 149, 1389 Budapest 62.

- **ICELAND**
Snaebjörn Jónsson and Co. h.f.
Hafnarstraeti 9, PO Box 1131,
101 Reykjavik.

- **INDIA**
Oxford Book and Stationery Co.
Scindia House, New Delhi 100 001;
17 Park Street, Calcutta 700 016.
Oxford Subscription Agency,
Institute for Development Education
1 Anasuya Ave, Kilpauk,
Madras 600010.

- **IRELAND**
Publications Section
Stationery Office,
Bishop Street,
Dublin 8.

- **ITALY**
FAO (see last column)
Libreria Scientifica
Dott. Lucio de Biasio "Aeiou"
Via Meravigli 16, 20123 Milano.
Libreria Concessionaria Sansoni S.p.A. "Licosa"
Via Lamarmora 45,
C.P. 552, 50121 Firenze.
Libreria internazionale Rizzoli
Galleria Colonna, Largo Chigi,
00187 Rome.

- **JAPAN**
Maruzen Company Ltd
PO Box 5050,
Tokyo International 100-31.

- **KENYA**
Text Book Centre, Ltd
Kijabe Street, PO Box 47540, Nairobi.

- **KOREA, REP. OF**
Eulyoo Publishing Co. Ltd
46-1 Susong-Dong, Jongro-Gu,
PO Box 362, Kwangwha-Mun,
Seoul 110.

- **KUWAIT**
The Kuwait Bookshops Co. Ltd
PO Box 2942, Safat.

- **LUXEMBOURG**
M.J. De Lannoy
202, avenue du Roi,
1060 Bruxelles (Belgique).

- **MAROC**
Librairie "Aux Belles Images"
281, avenue Mohammed V, Rabat.

WHERE TO PURCHASE FAO PUBLICATIONS LOCALLY
POINTS DE VENTE DES PUBLICATIONS DE LA FAO
PUNTOS DE VENTA DE PUBLICACIONES DE LA FAO

● **MEXICO**
Ediapsa
Librerías Cristal
Tehuantepec 170, Col. Roma Sur
06760 México, D.F.

● **NETHERLANDS**
Keesing b.v.
Hogeliweg 13, 1101 CB Amsterdam.
Postbus 1118, 1000 BC Amsterdam.

● **NEW ZEALAND**
Government Printing Office Bookshops
25 Rutland Street.
Mail orders: 85 Beach Road,
Private Bag, CPO, Auckland;
Ward Street, Hamilton;
Mulgrave Street (Head Office),
Cubacade World Trade Centre,
Wellington;
159 Hereford Street, Christchurch;
Princes Street, Dunedin.

● **NICARAGUA**
Librería Universitaria,
Universidad Centroamericana
Apartado 69, Managua.

● **NIGERIA**
University Bookshop (Nigeria) Limited
University of Ibadan, Ibadan.

● **NORTH AMERICA**
UNIPUB
4611/F, Assembly Drive,
Lanham, MD 20706-4391
Toll Free: 800 233-0504 (Canada).
         800 274-4888 (USA)
Fax 301-459-0056.

● **NORWAY**
Johan Grundt Tanum Bokhandel
Karl Johansgate 41-43,
PO Box 1177, Sentrum, Oslo 1.
Narvesen Info Center
Bertrand Narvesens vei 2
P.O. Box 6125 Etterstad
0602 Oslo 6.

● **PAKISTAN**
Mirza Book Agency
65 Shahrah-e-Quaid-e-Azam,
PO Box 729, Lahore 3.
Sasi Book Store
Zaibunnisa Street, Karachi.

● **PARAGUAY**
Agencia de Librerías Nizza S.A.
Casilla 2596, Eligio Ayala 1073,
Asunción.

● **PERU**
Librería FAO
Universidad Nacional Agraria
La Molina
Lima.
Librería Distribuidora "Santa Rosa"
Jirón Apurímac 375, Casilla 4937,
Lima 1.

● **PHILIPPINES**
International Book Center
5th Flr Filipinas Life Building
Ayala Avenue, Makati,
Metro Manila.

● **POLAND**
Ars Polona
Krakowskie Przedmiescie 7,
00-068 Warsaw.

● **PORTUGAL**
Livraria Portugal,
Dias y Andrade Ltda.
Rua do Carmo 70-74, Apartado 2681.
1117 Lisboa Codex.

● **REPUBLICA DOMINICANA**
Editora Taller, C. por A.
Isabel la Católica 309,
Santo Domingo, D.N.
Fundación Dominicana de Desarrollo
Casa de las Gárgolas,
Mercedes 4, Apartado 857,
Santo Domingo.

● **ROMANIA**
Ilexim
Calea Grivitei Nò 64066, Bucharest.

● **SAUDI ARABIA**
The Modern Commercial
University Bookshop
PO Box 394, Riyadh.

● **SINGAPORE**
Select Books Pte. Ltd
Tanglin Shopping Centre,
03-15 Tanglin Shopping Centre,
19 Tanglin Rd., Singapore 1024.

● **SOMALIA**
"**Samater's**"
PO Box 936, Mogadishu.

● **SRI LANKA**
M.D. Gunasena & Co. Ltd
217 Olcott Mawatha,
PO Box 246, Colombo 11.

● **SUISSE**
Librairie Payot S.A.
107 Freiestrasse, 4000 Basel 10.
6, rue Grenus, 1200 Genève.
Case Postale 3212, 1002 Lausanne.
Buchhandlung und Antiquariat Heinimann & Co.
Kirchgasse 17, 8001 Zurich.

● **SURINAME**
VACO n.v. in Suriname
Domineestraat 26,
PO Box 1841, Paramaribo.

● **SWEDEN**
Books and documents:
C.E. Fritzes Kungl. Hovbokhandel,
Regeringsgatan 12,
PO Box 16356, 103 27 Stockholm.
Subscriptions:
Vennergren-Williams AB
PO Box 30004, 104 25 Stockholm.

● **TANZANIA**
Dar-es-Salaam Bookshop
PO Box 9030, Dar-es-Salaam.
Bookshop, University of Dar-es-Salaam
PO Box 893, Morogoro.

● **THAILAND**
Suksapan Panit
Mansion 9, Rajadamnern Avenue.
Bangkok.

● **TOGO**
Librairie du Bon Pasteur
B.P. 1164, Lomé.

● **TUNISIE**
Société tunisienne de diffusion
5, avenue de Carthage, Tunis.

● **TURKEY**
Kultur Yayiniari is - Turk Ltd Sti.
Ataturk Bulvari No. 191,
Kat. 21, Ankara
Bookshops in Istanbul and Izmir.

● **UNITED KINGDOM**
Her Majesty's Stationery Office
49 High Holborn,
London WC1V 6HB (callers only).
HMSO Publications Centre,
Agency Section
51 Nine Elms Lane,
London SW8 5DR (trade and London area mail orders);
13a Castle Street,
Edinburgh EH2 3AR
80 Chichester Street,
Belfast BT1 4JY;
Brazennose Street,
Manchester M60 8AS;
258 Broad Street,
Birmingham B1 2HE;
Southey House, Wine Street,
Bristol BS1 2BQ.

● **URUGUAY**
Librería Agropecuaria S.R.L.
Alzaibar 1328, Casilla Correo 1755,
Montevideo.

● **VENEZUELA**
Tecni-Ciencia Libros, S.A.
Torre Phelps-Mezzanina,
Plaza Venezuela
Caracas
Tamanaco Libros Técnicos
Centro Comercial Ciudad Tamanaco
Nivel C-2
Caracas
Tecni-Ciencia Libros, S.A.
Centro Comercial Shopping Center
Av. Andrés Eloy, Urb. El Prebo
Valencia, Edo. Carabobo.

● **YUGOSLAVIA**
Jugoslovenska Knjiga, Trg.
Republike 5/8,
PO Box 36, 11001 Belgrade
Cankarjeva Zalozba
PO Box 201-IV, 61001 Ljubljana.
Prosveta
Terazije 16, Belgrade.

● **ZAMBIA**
Kingstons (Zambia) Ltd
Kingstons Building,
President Avenue, PO Box 139, Ndola.

● **Other Countries**
Autres Pays
Otros Países

Distribution and Sales Section, FAO
Via delle Terme
di Caracalla,
00100 Rome, Italy.

Tipo-lito SAGRAF - Napoli